Challenges and Advances in Computational Chemistry and Physics

Volume 19

Series Editor
Jerzy Leszczynski
Department of Chemistry and Biochemistry
Jackson State University Chemistry, Jackson, Mississippi, USA

This book series provides reviews on the most recent developments in computational chemistry and physics. It covers both the method developments and their applications. Each volume consists of chapters devoted to the one research area. The series highlights the most notable advances in applications of the computational methods. The volumes include nanotechnology, material sciences, molecular biology, structures and bonding in molecular complexes, and atmospheric chemistry. The authors are recruited from among the most prominent researchers in their research areas. As computational chemistry and physics is one of the most rapidly advancing scientific areas such timely overviews are desired by chemists, physicists, molecular biologists and material scientists. The books are intended for graduate students and researchers.

More information about this series at http://www.springer.com/series/6918

Steve Scheiner
Editor

Noncovalent Forces

Springer

Editor
Steve Scheiner
Department of Chemistry & Biochemistry
Utah State University
Logan
Utah
USA

Challenges and Advances in Computational Chemistry and Physics
ISBN 978-3-319-14162-6 ISBN 978-3-319-14163-3 (eBook)
DOI 10.1007/978-3-319-14163-3

Library of Congress Control Number: 2015931919

Springer Cham Heidelberg New York Dordrecht London

Printed on acid-free paper

Springer is part of Springer Science+Business Media (www.springer.com)

Contents

Contributors

Ibon Alkorta Instituto de Química Médica (IQM-CSIC), Madrid, Spain

E. Arunan Department of Inorganic and Physical Chemistry, Indian Institute of Science, Bangalore, India

Isaac K. Attah Department of Chemistry, Virginia Commonwealth University, Richmond, VA, USA

Antonio Bauzá Departament de Química, Universitat de les Illes Balears, Palma de Mallorca, Baleares, Spain

Himansu S. Biswal School of Chemical Sciences, National Institute of Science Education and Research, Institute of Physics Campus, Bhubaneswar, Odisha, India

Janet E. Del Bene Department of Chemistry, Youngstown State University, Youngstown, OH, USA

Pere M. Deyà Departament de Química, Universitat de les Illes Balears, Palma de Mallorca, Baleares, Spain

José Elguero Instituto de Química Médica (IQM-CSIC), Madrid, Spain

M. Samy El-Shall Department of Chemistry, Virginia Commonwealth University, Richmond, VA, USA

Department of Chemistry, Faculty of Science, King Abdulaziz University, Jeddah, Saudi Arabia

Antonio Frontera Departament de Química, Universitat de les Illes Balears, Palma de Mallorca, Baleares, Spain

Shridhar R. Gadre Department of Chemistry, Indian Institute of Technology Kanpur, Kanpur, India

Sławomir J. Grabowski Kimika Fakultatea, Euskal Herriko Unibertsitatea UPV/EHU, Donostia International Physics Center (DIPC), Donostia, Euskadi, Spain

IKERBASQUE, Basque Foundation for Science, Bilbao, Spain

Michio Iwaoka Department of Chemistry, School of Science, Tokai University, Hiratsuka-shi, Kanagawa, Japan

Jerelle A. Joseph Department of Biological and Chemical Sciences, The University of the West Indies, Wanstead, Barbados

Jiří Kozelka Laboratoire de Chimie et Biochimie Pharmacologiques et Toxicologiques, UMR 8601 CNRS, Université Paris Descartes, Paris, France

Department of Condensed Matter Physics, Faculty of Science, Masaryk University, Brno, Czech Republic

Anmol Kumar Department of Chemistry, Indian Institute of Technology Kanpur, Kanpur, India

Hai-Bei Li School of Ocean, Shandong University, Weihai, People's Republic of China

Qing-Zhong Li The Laboratory of Theoretical and Computational Chemistry, School of Chemistry and Chemical Engineering, Yantai University, Yantai, People's Republic of China

Devendra Mani Department of Inorganic and Physical Chemistry, Indian Institute of Science, Bangalore, India

Sean A. C. McDowell Department of Biological and Chemical Sciences, The University of the West Indies, Wanstead, Barbados

Jane S. Murray Department of Chemistry, University of New Orleans, New Orleans, LA, USA

Sean P. Platt Department of Chemistry, Virginia Commonwealth University, Richmond, VA, USA

Peter Politzer Department of Chemistry, University of New Orleans, New Orleans, LA, USA

Steve Scheiner Department of Chemistry & Biochemistry, Utah State University, Logan, UT, USA

Osamu Takahashi Institute for sustainable sciences and development, Hiroshima University, Higahshi-Hiroshima, Japan

Stacey D. Wetmore Department of Chemistry and Biochemistry, University of Lethbridge, Lethbridge, Alberta, Canada

Steven E. Wheeler Department of Chemistry, Texas A&M University, College Station, TX, USA

Katie A. Wilson Department of Chemistry and Biochemistry, University of Lethbridge, Lethbridge, Alberta, Canada

Chapter 1
Introduction to the Volume

Steve Scheiner

Abstract The reader is introduced to the definition and diversity of noncovalent forces. The division of these interactions into various subtopics is explained by way of introducing each of the chapters. A brief exposition is provided of the typical means by which computational chemists study these forces, and the language that is commonly used.

Perhaps the first question a reader may have concerns the title of this volume: what is a noncovalent force? By its very name, it is clear that these forces exclude the strong bonds that hold atoms together within a molecule. What is usually meant is the collection of phenomena that attract molecules or ions toward one another. And this collection is indeed diverse. One can think first as one extreme of the strong Coulombic attractions between ions of opposite charge in a crystal of NaCl. On the other end of the spectrum lie the very weak forces between noble gas atoms as for example in an Ar matrix. These two extremes bracket a wide array of forces of intermediate strength. And it is this diverse set of phenomena which this volume is intended to address. Some will be quite familiar, as for example the H-bond, which has been part of the chemistry lexicon for over a century. But this familiar concept has undergone an expansion over the years, and now covers a range of different proton donor and acceptor groups not anticipated earlier. Other forces may be less familiar, as in the case of attractions between electronegative atoms as occurs in halogen, chalcogen, and even pnicogen bonds.

This volume is intended to provide the reader with an exposition of current thinking about the forces that bind molecules together. It is aimed toward a diverse audience. Computational chemists will be familiar with some of the concepts and tools, but may not be fully up to date in terms of the rapidly evolving field of noncovalent forces, and its various subfields. Experimentalists may not be fully cognizant of the ways that modern computational chemists think about these forces, the tools that are used to analyze them, and the power of computations to understand the nature of these forces. The various chapters are thus targeted toward both audiences. The reader will perhaps be impressed by the diversity of the types of noncovalent bonds

S. Scheiner (✉)
Department of Chemistry & Biochemistry,
Utah State University, 84322-0300 Logan, UT, USA
e-mail: steve.scheiner@usu.edu

S. Scheiner (ed.), *Noncovalent Forces*, Challenges and Advances in
Computational Chemistry and Physics 19, DOI 10.1007/978-3-319-14163-3_1

that attract molecules toward one another. But at the same time, there may be surprise at some of the common themes that emerge in this seemingly disparate set of forces.

The various chapters comprise a tour of this diverse array of noncovalent bonds. Some of the interactions will be well known, such as hydrogen bonds (HBs). But our understanding of these venerable bonds has evolved over the past years, and encompasses a range of new types which provide the focus here. The involvement of sulfur as both proton donor and acceptor atom has been developed in recent years, as described in Chap. 2. The next two chapters shift attention to the nominally weak proton donor atom C, dividing into π system acceptors in Chap. 3, and lone pairs of electronegative atoms in Chap. 4. The concept of HBs is expanded to open shell, radical systems in Chap. 5, and the ability of metal centers to serve as proton acceptor is explored in the following chapter. The normal expectation of a partially positively charged bridging H atom is reversed in Chap. 7, which describes not only dihydrogen bonds, but others where the bridge is better thought of as a hydride.

The next section expands further on the idea of HBs, in that there are a range of electronegative atoms which can replace H as the bridge. Atoms from Group V offer one such example, and the idea of the correspondingly named pnicogen bonds is developed in Chap. 8. Chalcogen atoms (O, S, etc) represent the topic of Chap. 9, and the underpinnings of halogen bonds are described in Chap. 10. Even the C atom gets into the act, as is explored in Chap. 11. Just like the old school HBs, these sorts of pnicogen, halogen, etc bonds can occur in aggregates of molecules, and strengthen one another through the phenomenon of cooperativity, the subject of Chap. 12. The next chapter brings to our consciousness the idea that lone pairs of one molecule can interact attractively with the π systems of another, and the mechanism for how this might be.

The extended π-clouds of aromatic systems represent an entire subject of their own. Chap. 14 explores the way in which substituents on one π-system affect its interaction with another. The presence of a charge on the system can be expected to have a major effect on interactions involving π-systems, and these perturbations are described in Chap. 15. The systems are enlarged in Chap. 16, with the focus remaining on ion-π interactions. Finally Chap. 17 brings us closer to the macroscopic world of biology, with its analysis of the interactions between DNA and proteins.

There are a number of computational tools and procedures that are particularly common in the computational study of molecular interactions. These are briefly introduced below, so that they might be familiar to the reader when encountered in the ensuing chapters.

1.1 Energetics

Probably the most often asked question about any particular interaction has to do with its "strength". But that is a multifaceted issue, that can have several different answers. Most commonly, the questioner is implicitly wondering how much energy is required to break the noncovalent bond in question. Taking the hydrogen bond between proton donor AH and B as an example, reaction 1 describes the association or binding, so is typically exothermic, with a negative ΔE, which is computed as the difference between the energy of the HB complex and the sum of the energies of the two monomers AH and B, Eq. (1.2)

$$AH + B \rightarrow AH \cdots B \tag{1.1}$$

$$\Delta E = E(AH \cdots B) - \{E(AH) + E(B)\} \tag{1.2}$$

But there are some devils in the details of the computation of ΔE. In the first place, it is assumed that the geometry of AH $\cdot\cdot$ B will be fully optimized, as will those of the isolated monomers. But an alternate prescription would leave AH and B in the geometries which they adopt within the AH $\cdot\cdot$ B complex. There is a generally, but by no means universally accepted, convention that the former procedure is called the binding energy. The latter is commonly termed the interaction energy because it more directly accesses the interaction of the two monomers within the complex. In the majority of cases, the geometrical rearrangement of AH and B within the complex is minor, and has only a small effect on their energies, so that the binding and interaction energies are usually fairly similar. But the reader should be aware that this is not always the case, so some attention to computational detail may be called for. It is also important to recognize that different authors may use different nomenclatures; for example ΔE is the most common symbol but sometimes refers to binding and sometimes to interaction energy. As written, ΔE will be negative for reaction 1, so will typically appear as such in the literature. But this is not always the case, particularly within the written text where it is not unusual for authors to say, for example, that the binding energy of the water dimer is 5 kcal/mol, but a value of − 5 kcal/mol may appear for ΔE in the pertinent table.

A second complication arises in that most calculations utilize a finite basis set when computing ΔE. Specifically, a given set of orbitals will be used to describe the electronic structure of AH, and another set for B. The AH $\cdots$ B complex, however, will utilize the union of the two latter sets of orbitals. The larger basis set of the complex gives it an unfair advantage in the sense that the variation principle tells us that the energy of a system becomes more negative as the basis is enlarged. A more negative energy for AH $\cdot\cdot$ B will leads to an overly negative ΔE for reaction 1. This spurious inflation of the binding energy, called basis set superposition error, is usually corrected by a counterpoise procedure [1] that dates back to 1970, wherein the energy of monomer AH is computed in a basis set that includes not only its own orbitals, but also those of B, the so-called "ghost orbitals" since neither the nuclei nor the electrons of B are present (and vice versa for monomer B).

The interaction energy up to this point refers only to the electronic contribution to the thermodynamic ΔE. The full ΔE includes also vibrational, rotational, and translational terms. Of these, the term that introduces the largest contribution is the zero-point vibrational energy (ZPE), which takes account of the various vibrational modes of reactants and product. Following this correction, it is common to denote the new quantity as ΔE + ZPE. The quantum calculation of the various entities also permits the ready computation of ΔH, ΔS, and ΔG. So even in energetic terms alone, the strength of the bond can be measured by any of the above quantities, depending upon one's inclination.

The reader may have noticed that AH and B in reaction 1 are separate molecules. Suppose one is interested in the strength of an *intra* molecular interaction. Reaction 1 is no longer relevant as there is no AH and B; that is the two constituents of the noncovalent bond cannot be fully separated from one another. How then can the system with this bond broken be defined? Unfortunately, there is no simple answer to this thorny question. This situation has motivated a number of different definitions over the years, but all provide somewhat different answers, so the reader should be cautious in such cases.

1.2 Energy Dissection

There is often some curiosity expressed as to what exactly is the nature of the binding. What is meant by this can be vague, and there is a long list of different terms in common usage: electrostatic, covalent, van der Waals, London forces, charge transfer, donor/acceptor, and on and on. Some of these terms have clearer definition than others and it is not unusual to see these terms applied differently by different authors. There have been numerous attempts to dissect the total interaction energy into its constituent parts. One of the earliest was due to Kitaura and Morokuma [2–4] who defined the electrostatic component as the Coulombic attraction (or repulsion) between the charge distributions of the two monomers in their pristine state, i.e. before the two molecules are allowed to perturb one another's charge clouds. This definition, which seems sensible and along the lines of what most would suggest, implicitly contains within it dipole-dipole, dipole-quadrupole, and higher terms in the multipole expansion. Using the same frozen wave functions, these authors defined an exchange, or steric, repulsion that results when the charge clouds penetrate one another.

Additional attractive forces are connected with the modification of each molecular charge distribution, and it is here that the different energy decomposition schemes differ most. Kitaura and Morokuma differentiated between charge that crossed a boundary from one molecule to another, which they termed charge transfer, and charge redistributions that remained on a single molecule, referred to as polarization. It must be understood however, that such a distinction is artificial, dependent upon where the boundary is drawn, as well as other factors. This fact motivated others to provide a more rigorous means to separate the two phenomena [5–10]. Other schemes avoid this distinction, leaving the charge redistribution as a single term which goes by several names, including induction, orbital interaction, or simply polarization.

The last major contributor to the attractive force between molecules originates in the instantaneous fluctuations of charge of one molecule, and its effect upon its partner. This phenomenon is commonly dubbed dispersion, but also goes by London or even van der Waals force. With respect to calculations, dispersion does not appear at the SCF level, but only when electron correlation is added. For that reason, some refer to any attraction found at the correlated level, over and above SCF attraction, as dispersion, but this is not correct. Correlation causes other sorts of perturbations as well, as for example, modifications of multipole moments within each monomer which in turn change the electrostatic energy. So while dispersion is indeed contained within the correlation energy, it represents only part of the latter. While dispersion energy is by definition attractive, there is no such restriction on the full correlation contribution.

Due in part to its limitation to the SCF level, and therefore inability to address dispersion, the KM scheme faded in usage over the years. It was replaced by other schemes, perhaps the most widespread of which is symmetry-adapted perturbation theory [11–14] or SAPT. This formalism provides electrostatic (ES), induction (IND), and dispersion (DISP) energies, but the effects of exchange reside not only in a first-order exchange energy, but also its effects upon other terms in the form of exchange-induction and exchange-dispersion. SAPT is particularly flexible in that it can be applied at progressively higher levels of perturbation theory, leading to additional terms, and to wave functions computed at either the Hartree-Fock or higher levels. Although SAPT does not normally attempt to divide induction into charge transfer and polarization, there have been some attempts [15–17] to do so, although not frequently applied in the literature.

Among other energy decomposition schemes in common usage, there is LMO-EDA which is based [18] on localized orbitals. In addition to electrostatic, polarization, and dispersion terms, LMO-EDA provides separate attractive exchange and repulsion terms. Another method that has found wide application is based on Natural Bond Orbital (NBO) treatment, and is termed NEDA [19–22]. While the preceding may be the methods in most widespread use, it is by no means an all-encompassing list; there are also other procedures that have been proposed [23–31].

It is stressed that the reader should exercise caution in comparing the components derived from one method to those from another, as different schemes can result in discrepant data. One may wonder which procedure is best, which most accurately reflects reality. It must be understood that just as a frog can be dissected in various ways, there is no single correct way to partition the total interaction energy; each method contains a certain degree of arbitrariness. Nonetheless, the use of these methods has provided key insights into the physical nature of the interactions, as is described in the following chapters.

1.3 Cooperativity

Most chemical systems contain more than just two molecules. For example, liquid water is a rapidly fluctuating system of H-bonded molecules, where most are engaged in H-bonds with more than one other molecule. It has long been recognized that the whole is greater than the sum of its parts. That is, the total binding energy of a AH···AH···AH trimer is greater than twice the binding energy of the simple AH···AH dimer. This amplification is frequently denoted as cooperativity. To be more exact, this enhancement through multiple noncovalent bonds can be termed positive cooperativity, or sometimes as synergistic. But the formation of a trimer does not always result in a stronger binding. If the central molecule, for example, acts as double proton donor, as in $H_2O \cdots HOH \cdots OH_2$, the total binding energy is less than that of a pair of simple dimers. This negative cooperativity is also sometimes called antagonistic or diminutive.

Another way in which computational chemists address the specifics of cooperativity is via multibody terms [32, 33]. Taking the $H_2O \cdots HF \cdots HCl$ system as an example, the trimer is first fully optimized. Holding the geometry fixed, the total interaction energy ΔE is computed as the difference between the energy of the trimer, and the sum of the energies of the separate monomers. Pairwise energies are also computed for each pair. $\Delta E^2(H_2O \cdots HF)$ is defined as the binding energy of the indicated pair, at the geometry of the trimer, but with the third molecule excluded. Likewise, $\Delta E^2(HF \cdots HCl)$ and $\Delta E^2(H_2O \cdots\cdots HCl)$ make up the other two pairwise energies. (Note that the last term will be small due to the distance between H_2O and HCl.) If there were no cooperativity at all, then ΔE would be equal to the sum of the three ΔE^2 quantities. But there is usually some degree of cooperativity, and three-body interaction ΔE^3 is defined as the difference between ΔE and $\sum \Delta E^2$. This same logic can be used to define four and higher body terms in complexes larger than a trimer.

1.4 Electrostatic Potentials

As stated earlier, the electrostatic energy is usually defined as the Coulombic interaction between the charge distributions of the two monomers. There are various means to visualize this interaction, the most common of which presents the electrostatic potential around each molecule in a color format. Red usually indicates the most negative regions, and blue the most positive (but this pattern is not always followed, so the reader should carefully read the figure caption). But in addition to the color scheme, there are other matters that may differ from one work in the literature to the next. For example, the potential is sometimes presented on a surface of constant electron density, which is frequently 0.001 au, but can just as easily be another isocontour. Alternately, the surface may be that which corresponds to the van der Waals surface surrounding each atom, or in other cases a longer distance from the nuclei may be deemed more useful, so can be twice the vdW radius, for example. On any given surface, the potential will have one or more minima and maxima, whose values can provide a quantitative means of comparing one molecule with another. It is common for these extrema to be denoted $V_{s,min}$ and $V_{s,max}$. Still another measure dispenses with the idea of a fixed density or atomic radius and instead presents an isopotential surface. In other words, the figure may illustrate a surface on which the potential is equal to a preselected constant, for example 0.01 au.

Other measures of the electrostatic potential do not require a visual presentation. For example, a charge can be assigned to each atom. But of course a table of atomic charges provides a much cruder picture of the full potential. This approach can be improved by adding atomic dipoles, or even quadrupoles. There remains, however, the question of how to make these assignments. There are numerous prescriptions for this purpose, some of the most common of which are Mulliken, NBO, and AIM schemes. Comparison shows that different schemes often provide very different charges, so some caution is necessary in the interpretation of this data.

1.5 Atoms in Molecules (AIM)

The electron density contains a great deal of information about the character of the bonding within any system. One means of extracting the character of chemical bonding rests on the topology of the density ρ, and most particularly its Laplacian $\nabla^2\rho$. This "atoms-in-molecules" (AIM) analysis [34–37] leads to the concept of basins that surround each atom and separate it from the others, and the total density within this basin can then lead to the assignment of AIM atomic charges. Also of importance, there are zero-gradient curves, termed bond paths, between atoms that are viewed as noncovalent bonds in the context of intermolecular interactions. Along each path there is a bond critical point, more or less midway between the two atoms. Bond paths are not restricted to pairs of nuclei, but can also connect other areas, such as π bonds [38–40]. The numerical value of the density ρ and Laplacian $\nabla^2\rho$ at the bond critical point, has been shown in many cases to correlate nicely with other measures of the strength of the bond, e.g. distance or energetics. However, the reader should be aware that the identification of a bond between two atoms via AIM is not necessarily entirely consistent with other means of analysis, particularly in the case of weak noncovalent bonds. There are examples in the literature where AIM data are at odds with other bond indicators [41–53].

1.6 NBO

The natural bond orbital (NBO) method transforms the fully delocalized wave function into one more in line with chemists' conventional ideas about individual chemical bonds and lone pairs. In terms of noncovalent forces, NBO analysis of a dimeric complex provides a series of second-order perturbation energies E(2) that correspond to the energetic consequence of the transfer of charge from one orbital of the first molecule, to another orbital of the second. In the case of an H-bonded complex $AH \cdots B$, for example, there is typically a sizable E(2) that corresponds to the transfer of charge from the lone pair of B to the σ^* antibonding orbital of AH: $n_B \rightarrow \sigma^*(AH)$. It is this buildup of density in the latter antibond which has been implicated as the source of the weakening and stretching of the A–H covalent bond in the H-bonded complex. Other orbitals can be involved as well, for example π and π^*, in other systems, some of which are described in the ensuing chapters. The NBO scheme offers a very useful picture by which to understand the nature of binding in various sorts of noncovalent interactions.

The intermolecular charge transfers of the NBO prescription would best fall under the rubric of induction energy. But it must be understood that induction is by definition a broader quantity, including all motions of electron density, in all orbitals, and encompasses both inter and intramolecular transfers. As such, it would be a mistake to equate a single value of E(2) with the full measure of induction or charge transfer energy. This scheme also offers a means of evaluating atomic charges, which are often used in the literature.

1.7 Electron Density Redistributions

The mutual perturbations of each molecule upon its partner can be explicitly visualized as a density shift map. One starts with the total electron density of the complex and then subtracts from it the sum of densities of the monomers, prior to the interaction occurring, analogous to Eq. 1.1 above.

$$\Delta\rho = \rho(\mathrm{AH}\cdots\mathrm{B}) - \{\rho(\mathrm{AH}) + \rho(\mathrm{B})\} \qquad (1.3)$$

The resulting map offers a three-dimensional perspective on shifts within each monomer, as well as density transfers from one molecule to the other. Maps such as these have clearly displayed, for example, the loss of density that occurs around the bridging proton in HBs, and the increase within the region occupied by the lone pair of the electron donor atom.

Quantitative encapsulations of the charge shifts are often provided by comparisons of atomic charges before and after the interaction occurs. But of course a single number assigned to an atom can miss important details that are clearly visible in the full map. Another measure of density shifts is associated with the dipole moment. Parallel with Eq. (1.3), the enhancement of the dipole moment is assessed by comparison with the vector sum of the dipoles of the unperturbed monomers.

1.8 Perturbations of the Monomers

When two molecules engage in an interaction with one another, even relatively weak ones, they exert a force which perturbs the internal properties of one another. One manifestation is the change in internal geometries. Perhaps the most famous example of this effect is the stretch undergone by the A–H covalent bond when it participates in a AH· · B H-bond. But other geometrical changes can occur as well, including stretches and contractions of other bonds, as well as modifications of internal bond angles.

The rearrangement of electron density that accompanies formation of a noncovalent bond has repercussions on the spectral features of each monomer as well. Taking HBs as an example once again the reduction of electron density in the vicinity of the bridging proton leads to a diminution of the NMR chemical shielding σ and consequently to a downfield shift of the signal of this proton. Another example from the H-bonding interaction is the red shift of the A–H stretching frequency in the vibrational spectrum of AH $\cdots$ B, along with an intensification of this band. This reduced frequency is connected with a weakening of the covalent A–H bond, which has been attributed to the increased population of the σ^*(AH) antibonding orbital. The latter is consistent with the idea that charge transfer into this antibonding orbital takes place from the lone pair of the electron donor atom B. The reader is forewarned, however, that most calculations of vibrational frequencies employ a fully harmonic approximation, so should make direct comparison to experimental quantities with caution. The magnitudes of the downfield NMR shift as well as the red shift of ν(AH) have been shown in numerous studies to correlate with energetic and geometric measures of the strength of the HB. These spectroscopic perturbations offer a particularly useful bridge with experimental observations, to assess the accuracy of the calculations.

The reader is now hopefully prepared for the chapters that follow with some understanding of the lexicon and computational tools that are being applied by the various contributing authors. You will probably note that there is some disagreement from one set of authors to the next as to the precise mechanisms underlying the molecular attractions. This disagreement is a healthy sign that the topic of this volume represents a vibrant field that is still encompassing new ideas. It is hoped that you will leave your reading desk with a well-rounded comprehension of modern views of the forces that hold molecules together: May the Noncovalent Force be with you.

References

1. Boys SF, Bernardi F (1970) The calculation of small molecular interactions by the differences of separate total energies. Some procedures with reduced errors. Mol Phys 19:553–566
2. Morokuma K (1971) Molecular orbital studies of hydrogen bonds. III. C=O$\cdots$ H–O hydrogen bond in $H_2CO\cdots H_2O$ and $H_2CO\cdots 2H_2O$. J Chem Phys 55:1236–1244
3. Kitaura K, Morokuma K (1976) A new energy decomposition scheme for molecular interactions within the Hartree-Fock approximation. Int J Quantum Chem 10:325–340
4. Morokuma K (1977) Why do molecules interact? The origin of electron donor-acceptor complexes, hydrogen bonding, and proton affinity. Acc Chem Res 10:294–300
5. Stevens WJ, Fink WH (1987) Frozen fragment reduced variational space analysis of hydrogen bonding interactions. Application to the water dimer. Chem Phys Lett 139:15–22
6. Chen W, Gordon MS (1996) Energy decomposition analyses for many-body interaction and applications to water complexes. J Phys Chem 100:14316–14328
7. van der Vaar A, Merz KM (1999) Divide and conquer interaction energy decomposition. J Phys Chem A 103:3321–3329
8. Salvador P, Duran M, Mayer I (2001) One- and two-center energy components in the atoms in molecules theory. J Chem Phys 115:1153–1157
9. Fedorov DG, Kitaura K (2007) Pair interaction energy decomposition analysis. J Comput Chem 28:222–237
10. Khaliullin RZ, Bell AT, Head-Gordon M (2008) Analysis of charge transfer effects in molecular complexes based on absolutely localized molecular orbitals. J Chem Phys 128:184112
11. Szalewicz K, Jeziorski B, Rybak S (1991) Perturbation theory calculations of intermolecular interaction energies. Int J Quantum Chem QBS 18:23–36

12. Jeziorski B, Moszynski R, Szalewicz K (1994) Perturbation theory approach to intermolecular potential energy surfaces of van der waals complexes. Chem Rev 94:1887–1930
13. Szalewicz K, Jeziorski B (1997) In: S. Scheiner (ed), Molecular Interactions. From Van der Waals to strongly bound complexes. Wiley, New York, p 3–43
14. Parrish RM, Sherrill CD (2014) Spatial assignment of symmetry adapted perturbation theory interaction energy components: the atomic SAPT partition. J Chem Phys 141:044115
15. Stone AJ (1993) Computation of charge-transfer energies by perturbation theory. Chem Phys Lett 211:101–109
16. Stone AJ, Misquitta AJ (2009) Charge-transfer in Symmetry-Adapted Perturbation Theory. Chem Phys Lett 473:201–205
17. Misquitta AJ (2013) Charge Transfer from Regularized Symmetry-Adapted Perturbation Theory. J Chem Theory Comput 9:5313–5326
18. Su P, Li H (2009) Energy decomposition analysis of covalent bonds and intermolecular interactions. J Chem Phys 131:014102
19. Glendening ED, Streitwieser A (1994) Natural energy decomposition analysis: An energy partitioning procedure for molecular interactions with application to weak hydrogen bonding, strong ionic, and moderate donor–acceptor interactions. J Chem Phys 100:2900–2909
20. Glendening ED (1996) Natural energy decomposition analysis: explicit evaluation of electrostatic and Polarization effects with application to aqueous clusters of alkali metal cations and neutrals. J Am Chem Soc 118:2473–2482
21. Schenter GK, Glendening ED (1996) Natural energy decomposition analysis: the linear response electrical self energy. J Phys Chem 100:17152–17156
22. Glendening ED (2005) Natural energy decomposition analysis: extension to density functional methods and analysis of cooperative effects in water clusters. J Phys Chem A 109:11936–11940
23. Ziegler T, Rauk A (1977) On the calculation of bonding energies by the Hartree Fock Slater method. Theor Chim Acta 46:1–10
24. Bagus PS, Hermann K, Bauschlicher CWJ (1984) A new analysis of charge transfer and polarization for ligand–metal bonding: Model studies of Al_4CO and Al_4NH_3. J Chem Phys 80:4378–4386
25. van der Vaart A, Merz KM (1999) Divide and conquer interaction energy decomposition. J Phys Chem A 103:3321–3329
26. Mo Y, Gao J, Peyerimhoff SD (2000) Energy decomposition analysis of intermolecular interactions using a block-localized wave function approach. J Chem Phys 112:5530–5538
27. Khaliullin RZ, Head-Gordon M, Bell AT (2006) An efficient self-consistent field method for large systems of weakly interacting components. J Chem Phys 124:204105
28. Mitoraj M, Michalak A (2007) Natural orbitals for chemical valence as descriptors of chemical bonding in transition metal complexes. J Mol Model 13:347–355
29. Reinhardt P, Piquemal J-P, Savin A (2008) Fragment-Localized Kohn–Sham orbitals via a singles configuration-interaction procedure and application to local properties and intermolecular energy decomposition analysis†. J Chem Theory Comput 4:2020–2029
30. Wu Q, Ayers PW, Zhang Y (2009) Density-based energy decomposition analysis for intermolecular interactions with variationally determined intermediate state energies. J Chem Phys 131:164112
31. Horn PR, Sundstrom EJ, Baker TA, Head-Gordon M (2013) Unrestricted absolutely localized molecular orbitals for energy decomposition analysis: Theory and applications to intermolecular interactions involving radicals. J Chem Phys 138:134119
32. Xantheas SS, Dunning THJ (1993) Ab initio studies of cyclic water clusters $(H_2O)_n$, $n = 1$–6. I. Optimal structures and vibrational spectra. J Chem Phys 99:8774–8792
33. Xantheas SS (1994) Ab initio studies of cyclic water clusters $(H_2O)_n$, $n = 1$–6. II. Analysis of many-body interactions. J Chem Phys 100:7523–7534
34. Bader RFW (1990) Atoms in molecules, a quantum theory. Clarendon Press, Oxford
35. Bader RFW, Cheeseman JR, Laidig KE, Wiberg KB, Breneman C (1990) Origin of rotation and inversion barriers. J Am Chem Soc 112:6530–6536

36. Popelier PLA (2000) Atoms in molecules. An introduction. Prentice Hall, Harlow
37. Popelier PLA, Bader RFW (1992) The existence of an intramolecular C–H–O hydrogen bond in creatine and carbamoyl sarcosine. Chem Phys Lett 189:542–548
38. Domagala M, Grabowski SJ (2009) XH··· π and X–H··· N hydrogen bonds—acetylene and hydrogen cyanide as proton acceptors. Chem Phys 363:42–48
39. Grabowski SJ, Ugalde JM (2010) Bond Paths Show Preferable Interactions: Ab Initio and QTAIM Studies on the X–H··· π Hydrogen Bond. J Phys Chem A 114:7223–7229
40. Grabowski SJ (2013) Dihydrogen bond and X–H··· σ interaction as sub-classes of hydrogen bond. J Phys Org Chem 26:452–459
41. Lane JR, Contreras-García J, Piquemal J-P, Miller BJ, Kjaergaard HG (2013) Are bond critical points really critical for hydrogen bonding? J Chem Theory Comput 9:3263–3266
42. Jablonski M (2012) Energetic and geometrical evidence of nonbonding character of some intramolecular halogen ··· oxygen and other Y ··· Y interactions. J Phys Chem A 116:3753–3764
43. Jablonski M, Palusiak M (2013) The halogen···oxygen interaction in 3-halogenopropenal revisited—the dimer model vs. QTAIM indications. Chem Phys 415:207–213
44. Varadwaj PR, Varadwaj A, Jin B-Y (2014) Significant evidence of CO and CC long-range contacts in several heterodimeric complexes of CO with CH_3–X, should one refer to them as carbon and dicarbon bonds! Phys Chem Chem Phys 16:17238–17252
45. Cormanich RA, Moreira MA, Freitas MP, Ramalho TC, Anconi CPA, Rittner R, Contreras RH, Tormena CF (2011) $^{1h}J_{FH}$ coupling in 2-fluorophenol revisited: Is intramolecular hydrogen bond responsible for this long-range coupling? Magn Reson Chem 49:763–767
46. Johnson ER, Keinan S, Mori-Sanchez P, Contreras-Garcia J, Cohen AJ, Yang W (2010) Revealing noncovalent interactions. J Am Chem Soc 132:6498–6506
47. Azofra LM, Scheiner S (2014) Complexation of n SO_2 Molecules ($n = 1,2,3$) with formaldehyde and thioformaldehyde. J Chem Phys 140:034302
48. Azofra LM, Scheiner S (2014) Complexes containing CO_2 and SO_2. mixed dimers, trimers and tetramers. Phys Chem Chem Phys 16:5142–5149
49. Weinhold F, Schleyer PvR, McKee WC (2014) Bay-type H ··· H "bonding" in cis-2-butene and related species: QTAIM versus NBO description. J Comput Chem 35:1499–1508
50. Alkorta I, Sanchez-Sanz G, Elguero J (2014) Pnicogen bonds between $X{=}PH_3$ ($X{=}O$, S, NH, CH_2) and phosphorus and nitrogen bases. J Phys Chem A 118:1527–1537
51. Foroutan-Nejad C, Shahbazian S, Marek R (2014) Toward a consistent interpretation of the QTAIM: tortuous link between chemical bonds, interactions, and bond/line paths. Chem Eur J 20:10140–10152
52. Ma F, Li A (2014) A computational study of pnicogen−hydride interaction in complexes XH_2P···HBeY. Comput Theor Chem 1045:78–85
53. Cormanich RA, Rittner R, O'Hagan D, Bühl M (2014) Analysis of CF ··· FC interactions on cyclohexane and naphthalene frameworks. J Phys Chem A 118:7901–7910

Part I
Hydrogen Bonds

Chapter 2
Hydrogen Bonds Involving Sulfur: New Insights from ab Initio Calculations and Gas Phase Laser Spectroscopy

Himansu S. Biswal

Abstract The hydrogen bonds involving sulfur (sulfur center hydrogen bonds; SCHBs) are generally regarded as weak H-bonds in comparison with the conventional N–H$\cdots$O, O–H$\cdots$O, N–H$\cdots$N and O–H$\cdots$N H-bonds. One of the reasons being considered for this is the smaller electronegativity of S than O or N. However, recent high resolution laser spectroscopy in combination with quantum chemical calculations reveals that SCHBs can be as strong as conventional H-bonds. Surprisingly, in the case of methionine containing dipeptides the amide-N–H$\cdots$S H-bonds are even stronger than amide-N–H$\cdots$O$=$C H-bonds. Sulfur is not only a potential H-bond acceptor, but the S–H group is also a very good H-bond donor and capable of forming a variety of H-bonds. For example, the S–H$\cdots\pi$ H-bond between H_2S and indole/benzene is found to be the strongest H-bond among O–H$\cdots\pi$, O–H$\cdots\pi$, and C–H$\cdots\pi$ H-bonds. In general the SCHBs are dispersive in nature. This chapter details about few SCHB systems, many more systems need to be studied extensively and carefully to unravel many facts and facets about SCHBs. The major challenge for the experimentalists is to accurately determine the intra- and intermolecular H-bond energies and for the theoreticians to propose a universal H-bond descriptor.

2.1 Introduction

The IUPAC's new definition of hydrogen bonding (H-bonding) [1] is stated as "The hydrogen bond is an attractive interaction between a hydrogen atom from a molecule or a molecular fragment X–H in which X is more electronegative than H, and an atom or a group of atoms in the same or a different molecule, in which there is evidence of bond formation." As is mentioned in the definition electro negativities of X and Y in X–H$\cdots$Y regulates the H-bond strength, i.e. the H-bond strength increases with increasing the electronegativity values of donor and acceptor atoms [2, 3]. This is exactly followed by the H-bonds involving second row elements in general and

H. S. Biswal (✉)
School of Chemical Sciences, National Institute of Science Education and Research, Institute of Physics Campus, PO: Sainik School, 751 005 Bhubaneswar, Odisha, India
e-mail: himansu@niser.ac.in

S. Scheiner (ed.), *Noncovalent Forces,* Challenges and Advances in Computational Chemistry and Physics 19, DOI 10.1007/978-3-319-14163-3_2

oxygen and nitrogen in specific. In fact, N–H⋯O, O–H⋯O, N–H⋯N and O–H⋯N H-bonds are very essential in supramolecular chemistry, crystal engineering, designing new materials and structure and function of biomolecules [4–10]. Apart from nitrogen and oxygen, halogens (F, Cl, Br and I) are found to be potential H-bond acceptors and form weaker H-bonds with NH, OH and CH protons [11–16]. Similarly the possibility of H-bond formation by the higher group members of chalcogens (sulfur and selenium) have also been explored and documented in the literature [17–19]. As the atomic number in the chalcogen family increases, the electronegativity of the group members decreases and their metallic character increases. The electro negativities of sulfur (2.58) and selenium (2.55) in the Pauling scale are much smaller than that of oxygen atom (3.44) and comparable with the electronegativity of carbon (2.55). The CH group is regarded as a weak hydrogen bond donor [20–25] and very recently it is reported that sp^3-C can be involved in non-covalent interactions very similar to halogen and hydrogen bond termed as carbon bond [26–28]. Comparable electronegativity of sulfur with that of carbon perhaps leads to the general consensus that the hydrogen bonds involving sulfur (Sulfur Center Hydrogen Bond, SCHB) are very weak. Surprisingly nature has chosen two amino acids such as cysteine and methionine bearing sulfur atom in their side chains. It is in fact observed in the protein structure data bank that sulfur can form many non-covalent interactions including H-bonds that influence structure and function of proteins [17, 29–39]. However the strength (weak or strong), nature (electrostatic or dispersive), directionality (linear or non-linear) of SCHBs are still debatable and need to be investigated at the molecular level.

This chapter summarizes recent progress in the assessment of hydrogen bonds involving sulfur. Hydrogen sulfide dimer (H_2S–H_2S) is the simplest model system for intermolecular SCHBs. In this case S–H⋯S hydrogen bond is formed between the two monomer units. The computed S–H⋯S H-bond energy at semi-empirical level is 0.71 kcal/mol [40], which is much smaller than the conventional hydrogen bond energies. In this example sulfur acts as a H-bond acceptor as well as a donor. There are few matrix isolation IR spectroscopy studies on the SCHBs [41–44]. The experimental results demonstrate the sulfur atom of Methanethiol (MeSH) and dimethyl sulfide (DMS) can be a potential hydrogen bond acceptor, although a weaker acceptor than oxygen [41]. However, the work by M. Wierzejewska [42] demonstrated that the sulfur atom of DMS is better hydrogen bond acceptor than the sulfur atom of dimethyldisulfide (DMDS) and hydrogen sulfide (H_2S) and comparable to the oxygen atom of dimethylether (DME) as observed in the DMS–HNO_3 and DME–HNO_3 binary complexes [43]. Apart from the intermolecular S–H⋯S and O–H⋯S hydrogen bonds, intramolecular SCHBs are frequently observed in the crystals [45–48]. The spectroscopic evidence for intramolecular SCHB is the microwave study of the conformers of jet-cooled thiodiglycol (TDG) [49]. It is observed that intramolecular O–H⋯S hydrogen bond provides a highly compact and folded structure to the most stable conformer of TDG. Sulfur is also capable of forming S–H⋯π π-type H-bonds along with S–H⋯S, N–H⋯S and O–H⋯S σ-type H-bonds as cited above. For instance, H_2S forms strong S–H⋯π H-bonded complex with benzene π-electrons. The computed H-bond energy in H_2S and benzene dimer at CCSD(T)/CBS level is − 2.85 kcal/mol. This binding energy is comparable to that of O–H··· π and N–H··· π H-bond energy and almost two times of C–H··· π H-bond energy [50].

Many theoretical approaches have been adopted to shed light on the experimental observations on SCHBs. The main focus is to understand the directionality, nature and hydrogen bond energies of SCHBs [18, 44, 51, 52, 53–56]. One of the detailed theoretical study by Platts et al. [18] on SCHBs suggest that sulfur atoms of H_2S and H_2CS are weaker H-bond acceptor than oxygen atoms of H_2O and H_2CO. Sulfur prefers to form perpendicular H-bonds where as oxygen forms linear hydrogen bonds, the reason being (X) S··· H–F H-bond is dominated by charge (H)-quadrupole (S) interaction while charge-charge attraction is the main contributor for the (X)O··· H–F H-bonded complexes. In other words SCHBs are dispersive in nature. One of the important finding by them is that the Laplacian and charge density derived from atoms in molecules (AIM) theory is unable to predict the directionality of SCHBs in these complexes. The H-bond energy in (X)S··· H–F is almost half of that in (X)O··· H–F complexes. On contrary in the case of methanol–dimethylsulfide (OH··· S H-bond) and methanol–dimethylether (OH··· O H-bond), the H-bond energies are very similar. The estimated H-bond energy of OH··· S H-bond at the coupled cluster level is − 5.46 kcal/mol which is slightly smaller than the H-bond energy of − 5.97 kcal/mol for OH··· O H-bond. Similarly in a recent work on the nature of the SCHBs, Singh and co-workers [56] from the symmetry adopted perturbation theory analysis (SAPT) of variety of intermolecular H-bond complexes with DMS suggest that electrostatic component of the stabilization energy is the major contributor in the SCHBs rather than the dispersive energy component as mentioned by Barcke and co-workers [18]. One thing emerges out from these few examples is that the hydrogen bond acceptor strength of sulfur, directionality and the electrostatic/dispersive nature of SCHB are not very conclusive and still debatable.

Apart from spectroscopic and computational evidences, the existences SCHBs have been realized in many organic crystals, peptides and proteins [48, 57–73, 74–76]. The intrachain S–H···O = C H-bonds involving sulfur atom of cysteine side chain and backbone carbonyl oxygen have been observed for globular proteins [17, 77]. The strength and occurrence of SCHBs in proteins have recently been investigated by Zhou et al. [78]. By analyzing the geometrical parameters of 500 high resolution protein structures, they concluded that "(i) SCHBs regulate secondary structure of peptides, (ii) SCHBs have longer hydrogen bond length (d) and smaller hydrogen bond angle (θ), (iii) sulfur atoms in the side chain of cysteine, half-cysteine and methionine forms weaker hydrogen bond, (iv) the disulfide bonds are capable of forming SCHBs, (v) the SH group of cysteine acts as a hydrogen bond donor and forms weak S–H··· π hydrogen bond, and (vi) methionine sulfur is a poor hydrogen bond acceptor." Some of these findings are very pertinent and useful while some need to be studied at the molecular level. For example, looking at the electro negativities of acceptor atoms and distance between X and Y in X–H··· Y H-bond systems, it is very difficult to say whether Y (in this case sulfur) will be weak or strong H-bond acceptor. A comprehensive search of crystal structure data base by Allen et al. [19] and Steiner [79] emphasizes that except for few thioethers sulfur is a poor hydrogen bond acceptor than oxygen and the role of SCHBs in bimolecular structure function are minimal. On the contrary, Francois et al. [61] found very strong N–H$\cdots$S hydrogen bond in the crystals of (Triazocyclononane)$_2$Fe$_2$S$_6$. They also observed that the strength of N–H$\cdots$S and N–H$\cdots$O H-bonds highly depend on the hydrogen bond angle (θ). For the mentioned crystal structure the N–H$\cdots$S hydrogen bond strength is optimal for $\theta \sim 80^0$, where as N–H$\cdots$O hydrogen bond is favorable for $\theta > 115^0$. Hence, one can conclude that the strength of SCHB is dependent on the systems and geometry of hydrogen bond. Similarly the role of SCHB in biomolecular structure and function cannot be neglected. For example, in a recent report [80] methionine is found to be very essential for the catalytic role of phosphite dehydrogenase (PTDH), that catalyses the oxidation of phosphite to phosphate. In fact, the N–H$\cdots$S hydrogen bond interaction between His292 and Met53 stabilizes the transition state of the reaction, thereby lowering the activation energy barrier and increasing the catalytic activity of PTDH.

The above cited few examples from the literature manifest many facts and facets of SCHBs, that are still unexplored and need to be addressed very accurately. It is very clear that sulfur can form hydrogen bond and SCHBs are ubiquitous in proteins and organic crystals. For the last few years we have been trying to unravel SCHBs at molecular level using the arsenal of high resolution gas phase laser spectroscopy and computational methods. Hope our results will mark a point with respect to vexata quaestio of the acceptor strength of sulfur, directionality and nature of SCHBs. The aim is to mimic the SCHBs observed in the organic crystals, peptides, and proteins by choosing relevant simpler model compounds. The model compounds and their clusters are prepared in the supersonic jet cooled condition by exploiting isentropic phenomenon. The UV and IR spectroscopy of these compounds and their clusters are recorded in the jet cooled condition. Supersonic jet spectroscopy has many advantages over the conventional spectroscopy and used by many research groups to study molecular complexes that ranges from weakly bound van der Waal's complexes to very strongly bound ionic-hydrogen bonded complexes [81–87]. One of the major benefits of this technique is that it reduces thermal broadening of the spectra and helps the molecular complexes to form at a lower temperature that cannot be possible in the ambient temperature and pressure. Combining this with TOF-Mass spectrometer allows us to detect the complexes of our interest and studying them in isolated conditions. Many double resonance spectroscopic methods are used to record the high resolution conformer specific UV and IR spectra. The computational methods are very handy in assigning the spectra, estimating the H-bond energy and predicting the nature of the SCHBs. The experimental data are also very useful in benchmarking the computational methods and provide plenty opportunities to develop new theoretical methods to understand and correctly predict non-covalent interactions such as SCHBs. In the next few sections we will discuss how the marriage between the computational methods and high resolution laser spectroscopic methods help us to shed light on the H-bonds sulfur atoms; sulfur being a H-bond acceptor or S–H as a H-bond donor.

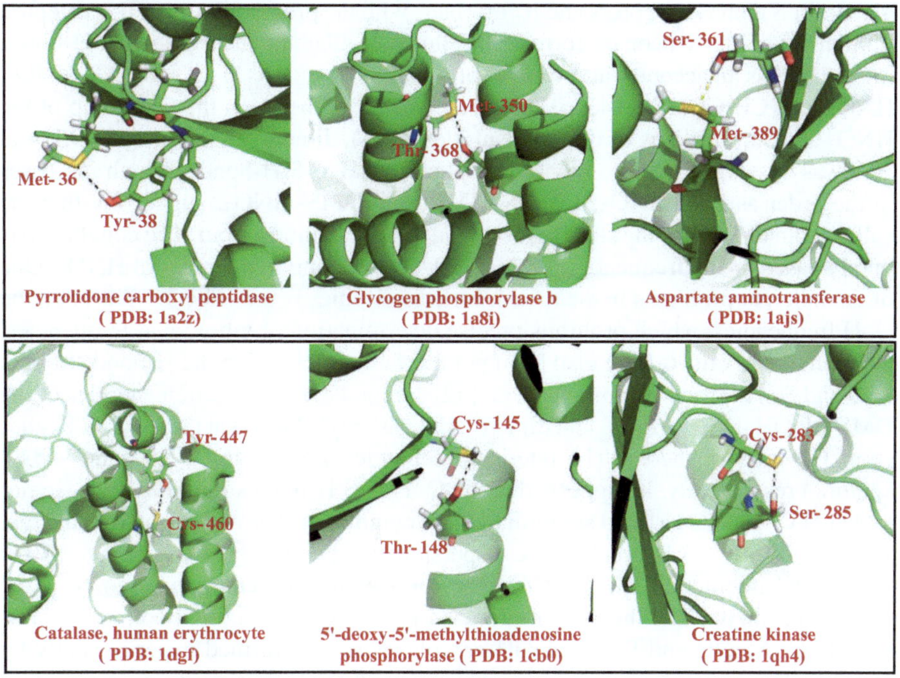

Fig. 2.1 O–H$\cdots$S Hydrogen bonds involving sulfur atoms of methionine and cysteine in proteins

2.2 "Sulfur" as a H-Bond Acceptor

Sulfur can be a potential H-bond acceptor. The mostly studied H-bonds with sulfur as a H-bond acceptor are O–H$\cdots$S and N–H$\cdots$S H-bonds. More detail of these H-bonds will be discussed in the subsequent sections.

2.2.1 O–H$\cdots$S Hydrogen Bonding

It is evident from the protein structure data bank that sulfur atoms of methionine and cysteine are capable of forming O–H$\cdots$S H-bonds with the side chain OH groups of tyrosine, threonine and serine. Few selected examples are shown in the Fig. 2.1.

Several computational and spectroscopic studies have been devoted to mimic and understand such interactions at the molecular level [18, 41–43, 88–99]. The matrix isolated FTIR spectroscopy in combination with ab initio calculation provides valuable information to categories and characterize SCHBs. Maes and coworkers used the extent of the red shift of H–Cl frequency in 1:1 complexes of HCl and YR_2 ($Y = O$, S, Se and $R = CH_3$, C_2H_5) as the measure of the strength of H-bond, more is the red shift stronger is the H-bond [90]. The red shift in H-Cl stretching frequency is higher for S than O acceptors and it correlates with the proton affinity of the acceptor. Similar studies have been performed for intermolecular and intramolecular H-bonds involving S as an acceptor and various proton donors such as H_2O, HF, HCl, HNO_2, HNO_3, and CF_3CCH [42, 43]. The O–H$\cdots$S H-bond between the OH group of the HNO_3 and sulfur atoms of dimethylsulfide (DMS), dimethyldisulfide (DMDS) and H_2S was observed. In these binary complexes O–H$\cdots$S H-bond strength increases in the order as H_2S–HNO_3, DMDS–HNO_3 and DMS–HNO_3, suggesting that the sulfur atom of DMS is the strongest H-bond acceptor among them. The conclusion is purely based on the frequency shift in the O–H stretching frequencies of HNO_3. One of the inherent problems in the matrix isolation studies is the effect of matrix on the O–H frequencies, which prohibits precise determination of vibrational frequencies. Computational efforts have also been put forward to determine the H-bond strength of O and S acceptors. Wennmohs et al. [52] computed the interaction energies of the DMS-MeOH, DME-MeOH and dimethylthiocarbonyl–MeOH at the CCSD (T)/aug-cc-pVQZ level of theory. The interaction energies are -5.46, -5.97 and -5.33 kcal/mol respectively. It suggests that the O–H$\cdots$S H-bond strength is very similar to that of O–H$\cdots$O H-bond strength in these complexes. The authors also found that in case of O–H$\cdots$S H-bonding the dispersion energy contribution is about $\sim 70\,\%$ to the total interaction energy. However, the electrostatic component controls the H-bond geometry. The theoretical predication of S as potential H-bond acceptor as O in DMS-MeOH and DME-MeOH complexes was confirmed experimentally by Daryl L. Howard and Henrik G. Kjaergaard [99]. From the vapor phase infrared spectroscopic study of the complexes of MeOH it was claimed that S is nearly equivalent to O as H-bond acceptor. The problem with the vapor phase IR spectra is that they are very broad and no control over mass selectivity. Hence it is very difficult to obtain precise information about the shift in O−H vibrational frequency in DMS-MeOH and DME-MeOH complexes.

The importance of SCHBs and some debatable conceptions about SCHBs prompted us to study them systematically. A combined experimental and computational approach has been deployed to extract molecular level information about SCHBs. The computational methods and experimental techniques are complementary to each other. For example computed vibrational frequencies and structure of molecules are very helpful in assigning the experimental spectral features while the experimental data can be used for bench marking and developing computational methods. To mimic the O–H$\cdots$S interaction between the side chain of tyrosine and S of methionine, para-cresol (p-CR) and thioethers are taken as the respective model compounds. As shown in Fig. 2.2, the alkyl chain length of the thioethers is systematically varied to see its effect on the nature and strength of the SCHBs.

Laser spectroscopy of jet-cooled molecules and molecular clusters is very handy to extract precise experimental data on non-covalent bonded clusters, in this case the H-bonded clusters. In most of the cases double resonance spectroscopy such as fluorescence dips infrared (FDIR) spectroscopy and resonant ion dip infrared spectroscopy (RIDIR) coupled with mass spectrometry are used to get the X–H stretching frequencies (here, O–H, N–H, S–H stretching frequencies) of the monomers and their H-bonded complexes. The relative changes in the X-H stretch in the complexes are

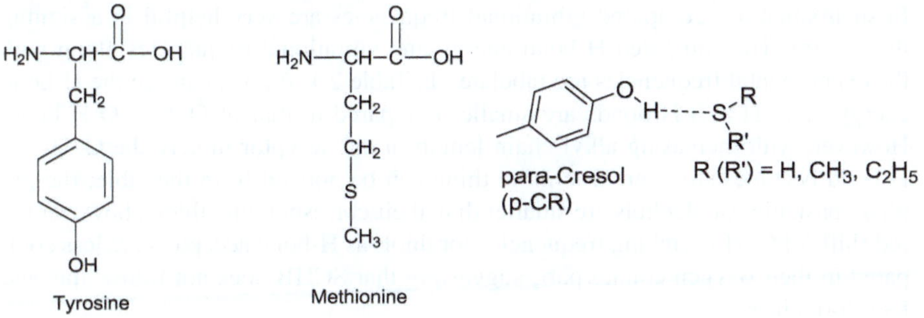

Fig. 2.2 Model compounds used to study O–H···S H-bond between OH group of tyrosine and sulfur atoms of methionine and cysteine

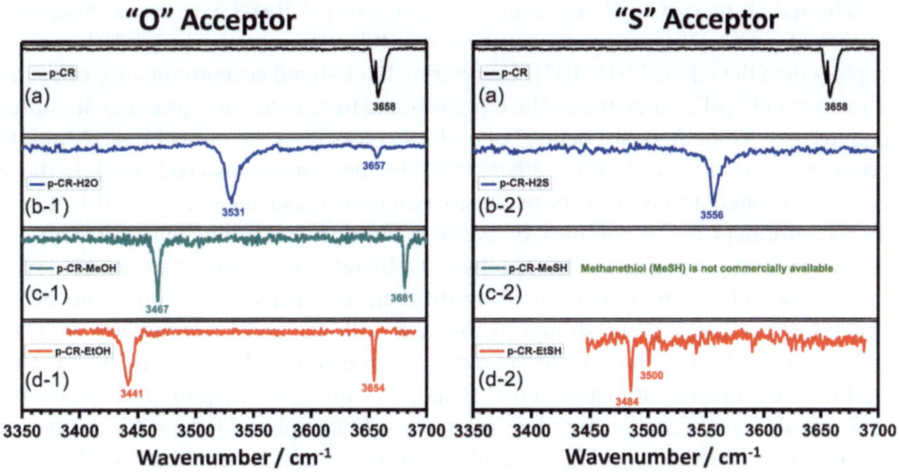

Fig. 2.3 FDIR spectra of 1:1 complexes of p-CR and YHR (Y = O, S and R = H, CH_3, and C_2H_5). H_2O, MeOH and EtOH are used as "O" H-bond acceptors and H_2S and EtSH are used as "S" H-bond acceptors

generally taken as a measure of the H-bond strength. It is a general practice for the conventional H-bond that more red shift in the X–H stretch, stronger is the X–H···Y H-bond. The X–H···Y H-bonds becomes stronger if Y becomes more basic.

Figure 2.3 presents the FDIR spectra of 1:1 complexes of p-CR and YHR (Y = O, S and R = H, CH_3, and C_2H_5). Here the OH group of p-CR acts as a H-bond donor and alcohols and thiols as H-bond acceptor.

The OH stretching frequencies of p-CR are red shifted in the complexes irrespective of the acceptor atoms, suggesting formation of O–H···O(S) H-bonds. The formations of O–H···O(S) H-bonds are also confirmed by DFT calculations [100, 101]. If one goes by the red shift of OH stretch, it is very clear from the figure that O of H_2O and alcohols are better H-bond acceptors than the S of H_2S and corresponding thiols. There are multiple conformers observed experimentally.

In such situations computed vibrational frequencies are very helpful in assigning the spectra. The computed H-bond energy and vibrational frequencies along with the experimental frequencies are tabulated in Table 2.1. As one can see the H-bond energy of O–H· · · S H-bonds are smaller compared to that of O–H· · · O H-bond. However, with increasing alkyl chain length in the acceptor moiety the O–H· · · S H-bond becomes stronger. One more thing can be noticed from the table; the gas phase basicities of alcohols are smaller than their corresponding thiols; however the red shifts of O–H stretching frequencies for thiols as H-bond acceptors are less compared to their oxygen counterpart, suggesting that SCHB does not follow the acid base formalism.

The situation is little different in case of ethers and thioethers as H-bond acceptor. The FDIR spectra of 1:1 complexes of p-CR with ethers and thioethers are shown in Fig. 2.4.

The red shifts of O– H stretching frequency of p-CR•DES is larger compared to that of p-CR•DEE, suggesting the acyclic thioethers can be better H-bond acceptors than the ethers [102, 103]. Comparing the H-bond acceptor quality of cyclic thioether and cyclic ether, the cyclic ether is found to be better acceptor than its sulfur counterpart. Thioethers are better H-bond acceptors than the thiols [104]. Multiple conformers are observed for the ethers and thioether complexes [102–104]. In those cases DFT calculations are very useful to assign the experimental spectral features. Two conformers in the complex of cyclic ether, tetrahydrothiophene (THT), with p-CR are found experimentally. The two O–H stretching frequencies are observed for p-CR•THT complex while probing different electronic transitions. The FDIR spectra of p-CR•THT are shown in Fig. 2.5 with C2 and Cs symmetry in THT These two conformers have different electronic transitions, but same IR transitions. In this case it is difficult to assign the IR spectra without the help of quantum chemical calculations. The non-covalent interaction (NCI) plots [105, 106] and atoms in molecules (AIM)[107–111] molecular graph assist the assignment. As shown in the figure conformer A of p-CR•THT does not show any vibronic coupled band in the FDIR spectra, while conformer B shows a strong vibronic coupled band. The vibronic couple band arises because of the coupling between the O– H oscillator of p-CR and C–H oscillator of THT. This is in fact happens for the p-CRTHT (Cs) conformer as predicted by NCI as well as AIM topology.

The binding and the red shift in O—H in the H-bonded complexes are the net result of several fundamental interactions, such as charge-dipole, dipole-dipole, dipole-induced dipole and the higher order multipole interactions. The magnitudes of these interactions are very difficult to determine experimentally. However theoretical insights are useful to account for the individual energy contribution to the total interaction energy. Various energy decomposition analyses such as. Natural energy decomposition analysis (NEDA) [112–114], Kitaura and Morokuma (KM) [115, 116], and reduced variational space self-consistent field (RVS) [117] decomposition analyses are used to solve the purpose. It is noticed that dispersion energy component is very important in SCHBs [100–104]. Table 2.2 summarizes dispersion energy contribution to the total interaction energy in the O–H· · · O and O–H· · · S H-bonded intermolecular complexes.

Table 2.1 Experimental red shifts of O–H stretching frequencies, red shifts of band origin (*BO*) positions, gas phase basicity of acceptors (*GB*), computed O–H shifts, changes in the O–H bond lengths (Δr) and bond dissociation energies (D_0) of p-CR•YRR' (Y = O and S, R(R') = H, CH_3 and C_2H_5) complexes

	Species	O–H Shift (Expt) in cm^{-1}	BO shift (Expt) in cm^{-1}	GB in kJ/mol	O–H shift (B3LYP/aDZ) in cm^{-1}	Δr (B3LYP/aDZ) in Å	BE[a] (MP2/aDZ) in kcal/mol
"O" Acceptors	p-CR•H_2O	– 127	– 357.0	660.0	– 192	0.0095	– 4.29
	p-CR•MeOH	– 191	– 425.0	724.5	– 246	0.0120	– 5.76
	p-CR•EtOH	– 217	– 422.4	746.0	– 256	0.0125	–6.33
	p-CR•DEE	– 280	– 451.3	801.0	– 292	0.0140	–7.67
	p-CR•THF	– 315	– 482.0	794.7	– 335	0.0160	–7.42
"S" Acceptors	p-CR•H_2S	– 102	– 239.0	673.0	– 138	0.0064	–2.49
	p-CR•MeSH			742.0	– 210	0.0096	–4.08
	p-CR•EtSH	– 174	– 316.3	758.4	– 219	0.0102	–4.70
	p-CR•DES	– 292	– 384.5	827.6	– 299	0.0146	–6.23
	p-CR•THT	– 285	– 373.0	819.3	– 296	0.0142	–6.15

[a] Bond dissociation energies (D_0) are computed at MP2/aug-cc-pVDZ//B3LYP-aug-cc-pVDZ level of theory, the basis set superposition error (*BSSE*) and zero point energy (*ZPE*) corrections are also applied. aDZ: aug-cc-pVDZ, *DME* dimethylether, *DEE* diethylether, *THF* tetrahydrofuran, *DMS* dimethylsulfide, *DES* diethylsulfide, and *THT* tetrahydrothiophene

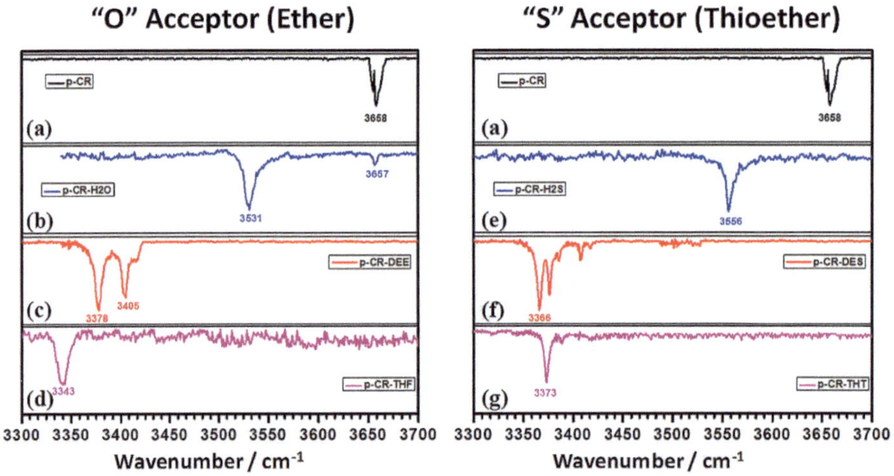

Fig. 2.4 FDIR spectra of 1:1 complexes of *p-CR* and YR ($Y = O$, S and $R = H$, CH_3, and C_2H_5). H_2O, DEE and THT are used as "O" H-bond acceptors and H_2S DES, and THT are used as "S" H-bond acceptors

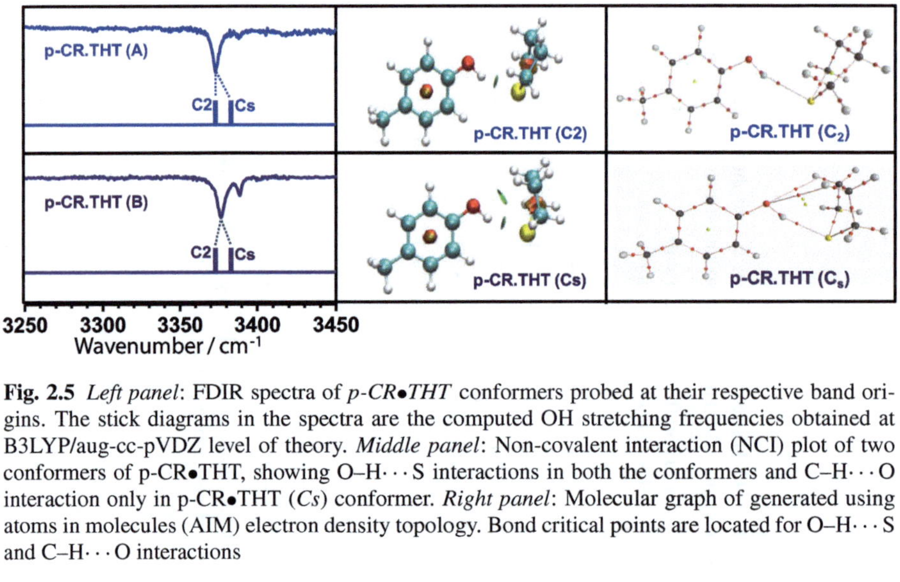

Fig. 2.5 *Left panel*: FDIR spectra of *p-CR•THT* conformers probed at their respective band origins. The stick diagrams in the spectra are the computed OH stretching frequencies obtained at B3LYP/aug-cc-pVDZ level of theory. *Middle panel*: Non-covalent interaction (NCI) plot of two conformers of p-CR•THT, showing O–H· · · S interactions in both the conformers and C–H· · · O interaction only in p-CR•THT (*Cs*) conformer. *Right panel*: Molecular graph of generated using atoms in molecules (AIM) electron density topology. Bond critical points are located for O–H· · · S and C–H· · · O interactions

In general the O–H· · · S HBs are dispersive in nature while O–H· · · O HBs are electrostatic. However, the dispersion energy contribution in the O–H· · · O HB case increases with the increase of alkyl chain length of the hydrogen bond acceptors. In the case of the O–H· · · S HB, the dispersion energy contribution decreases from H_2S

Table 2.2 Total interaction energy without the addition of ΔZPE (kcal/mol) at the MP2 level (ΔE_{int}^{MP2}), according to the NEDA, KM, and RVS energy decomposition analyses (ΔE_{int}), and dispersion energy (ΔE_{disp}) for the X = O vs S complexes. Numbers in parenthesis denote the percentage of contribution of dispersion interaction to the total interaction energy. In all the cases single point energy calculation was done at the MP2/aug-cc-pvDZ for the respective B3LYP/aug-cc-pvDZ optimized structures

	Species	ΔE_{int}^{MP2}	ΔE_{int}			ΔE_{disp}		
			KM	RVS	NEDA	KM	RVS	NEDA
"O" Acceptors	p-CR•H_2O	−6.19	−4.77	−4.87	−4.77	−1.42 (23 %)	−1.32 (21 %)	−1.42 (23 %)s
	p-CR•MeOH	−7.25	−5.20	−5.34	−5.20	−2.05 (28 %)	−1.91 (26 %)	−2.05 (28 %)
	p-CR•EtOH (anti)	−7.79	−5.27	−5.44	−5.27	−2.52 (32 %)	−2.35 (30 %)	−2.52 (32 %)
	p-CR•DME	−7.91	−5.21		−5.21	−2.70 (34 %)		−2.70 (34 %)
	p-CR•DEE (TT)	−9.26	−4.99	−5.28	−4.99	−4.27 (46 %)	−3.98 (43 %)	−4.27 (46 %)
	p-CR•THF	−9.02	−6.65	−6.27	−6.04	−2.37 (26 %)	−2.75 (30 %)	−2.98 (33 %)
"S" Acceptors	p-CR•H_2S	−3.76	−1.19	−1.23	−1.19	−2.57 (68 %)	−2.53 (67 %)	−2.57 (68 %)
	p-CR•MeSH	−5.16	−2.40	−2.61	−2.40	−2.76 (53 %)	−2.55 (49 %)	−2.76 (53 %)
	p-CR•EtSH (anti)	−5.74	−2.64	−2.84	−2.64	−3.10 (54 %)	−2.90 (51 %)	−3.10 (54 %)
	p-CR•DMS	−6.40	−3.05		−3.05	−3.35 (52 %)		−3.35 (52 %)
	p-CR•DES (TT)	−7.30	−3.30	−3.57	−3.30	−4.00 (55 %)	−3.73 (51 %)	−4.00 (55 %)
	p-CR•THT	−7.20	−3.98	−3.60	−3.37	−3.22 (45 %)	−3.60 (50 %)	−3.83 (53 %)

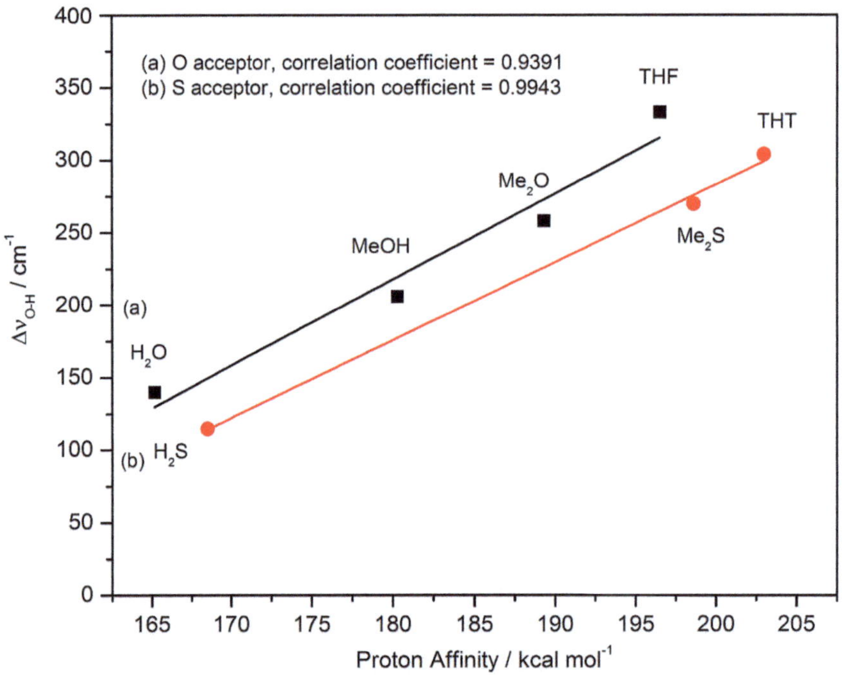

Fig. 2.6 The correlation plots of red shift in O−H frequency ($\Delta\nu_{O-H}$) vs proton affinity (*PA*) for the O−H· · · O and O−H· · · S bound complexes. (Reprinted with permission from ref 121. Copyright 2013 American Chemical Society)

to MeSH (from 68 to 53 %), but remains unchanged with further increase of alkyl chain length of the hydrogen bond acceptors.

Another aspect about SCHBs, is to look for H-bond descriptor. The H-bond descriptors are the computational and experimental molecular properties that are used to predict the H-bond strength. One such parameter is the gas phase proton affinity (PA) and/or gas phase basicity of the H-bond donor. It is expected that for a particular H-bond donor, H-bond acceptors with higher PA will form stronger H-bonds. PA has been a well established H-bond descriptor for the conventional O–H· · · O H-bonded systems [118–120] where electrostatic interaction is dominant. For example a good correlation between the red shift in O–H frequency of phenol with the PA of acceptor bases has been observed [118, 119]. Very recently Bhattacharyya et al. [121] explored the possibility of extending the acid-bas formalism to the O–H· · · S H-bonds. para-Fluorophenol (FP) was considered as H-bond donor for several S containing solvents of varying PA such as, H_2S, Me_2S, MeSH, and THT. The correlation plots between the experimentally determined red shifts of O–H stretching frequencies and PA of O and S acceptors are shown in the Fig. 2.6.

Excellent linear correlations are observed between red shift of O−H stretching frequency ($\Delta\nu_{O-H}$) and PA for the individual O and S acceptor groups. However,

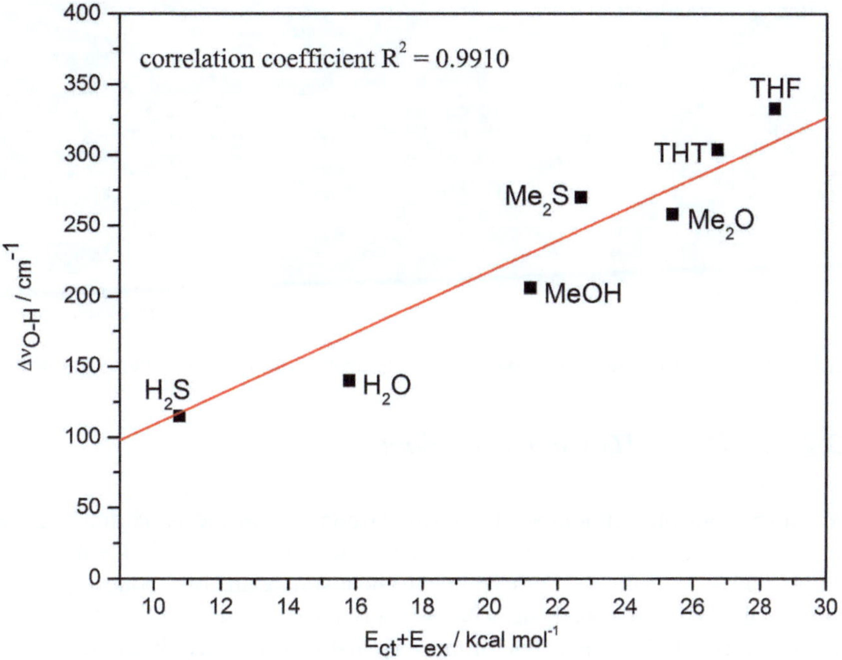

Fig. 2.7 Correlation plot of $(E_{ct} + E_{ex})$ (kcal/mol) vs red shift in O–H frequency $(\Delta\nu_{O-H})$ for the O–H···O and O–H···S bound complexes. (Reprinted with permission from ref 121. Copyright 2013 American Chemical Society)

there is no linear correlation exist between the red shift of O–H stretching frequency $(\Delta\nu_{O-H})$ and PA if one combines both the acceptor group members. Nevertheless these correlations are useful in predicting $\Delta\nu_{O-H}$ from the PA of the H-bond acceptor, e. g. $\Delta\nu_{O-H}$ of FP-MeSH can be predicted to be 202 cm^{-1} from the PA of MeSH 184.8 kcal mol^{-1}. They also observed very good correlation between the dissociation energies (D_0) and $\Delta\nu_{O-H}$ for the individual O and S acceptor groups. The energy decomposition analysis showed that irrespective of O and S acceptor groups there is a unified correlation exist between the sum of the charge transfer (E_{ct}) and the exchange (E_{ex}) energy component of the total binding energy and red shift of O–H stretching frequency. The correlation graph is depicted in Fig. 2.7.

The major challenge is to obtain the interaction energy of O–H···S H-bond experimentally. Mass analyzed threshold ionization (MATI) spectroscopy [122–129] and zero kinetic energy photoelectron spectroscopy (ZEKE-PES) [130–134] are the two popular laser spectroscopy techniques used to obtain the bond dissociation energies (D_0) for H-bonded complexes, but the success rates are limited. Few attempts have been made by our group to get the experimental D_0 values of O–H···S H-bonded complexes. These experimental D_0 will be used for benchmarking the quantum chemical methods.

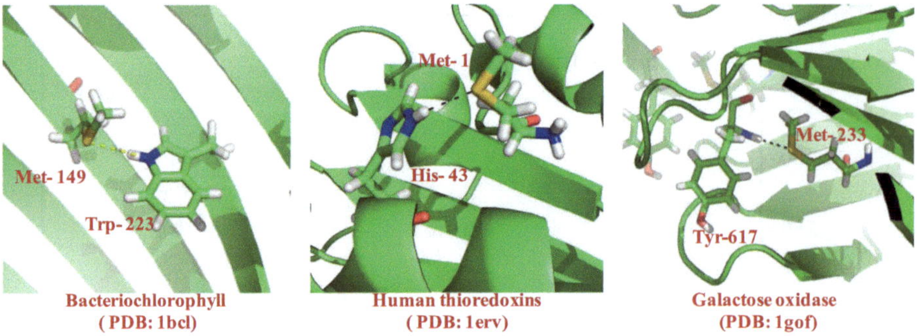

Fig. 2.8 N–H⋯S H-bonds involving sulfur atoms of methionine and N–H groups in proteins

2.2.2 N–H⋯S Hydrogen Bonding

The sulfur atom of methionine also forms H-bonds with the N–H donor groups. Figure 2.8 displays some of the cases where methionine sulfur is involved in H-bonding with side chain N–H of tryptophan and histidine and backbone amide N–H.

One of the exhaustive analysis of N–H⋯S H-bond was reported by Wategaonkar and co workers [135]. This is one of the very first reports where both the computational and experimental techniques are used to explain the nature of N–H⋯S H-bond between NH of indole and S of DMS. In fact, indole is used as the model compound of tryptophan and to probe local environment and dynamics of proteins in solution [136–138]. The IR spectra of H-bonded complexes of indole (IND) with various H-bond acceptors are shown in Fig. 2.9. The highest red shift of N–H stretching frequency is observed for IND-DMS complex. The red shift of N–H for this complex is the highest among other complexes studied. The red shift for N–H⋯S H-bond is very similar to N–H⋯O H-bond in IND-DME complex, but almost two times larger than that for N–H⋯O H-bond in IND-H_2O complex and ~3.5 times larger than N–H⋯π H-bond in IND-Benzene complex [139]. If one goes by the red shift of N–H stretching frequency, then N–H⋯S H-bond should be considered the strongest H-bond among all other H-bonds shown in the figure.

The bond dissociation energies (D_0) of indole-DMS and 3-methylindole-DMS complexes are determined at the MP2/CBS and compared with the experimental determined D_0 of IND-H_2O and IND-Bz [135]. The D_0 of IND-DMS at MP2/CBS is − 5.59 kcal/mol which is little higher than that of IND-H_2O (− 4.67 kcal/mol) [122] and IND-Bz (− 5.21 kcal/mol) [140]. Similar D_0 values and very different N–H red shift in N–H⋯S, N–H⋯O and N–H⋯π H-bond complexes urge us to investigate the nature of these three different type H-bonds. It has already been established that X–H⋯π H-bonds are dispersive in nature [141–145]. Energy decomposition analysises of IND-DMS and IND-H_2O complexes suggest that without dispersion interaction N–H⋯S H-bonded complexes are not stable complexes. The dispersion energy is about 115 % of the total interaction energy [135]. It can be inferred that the nature of N–H⋯S is very different from N–H⋯O H-bond and somewhat similar to

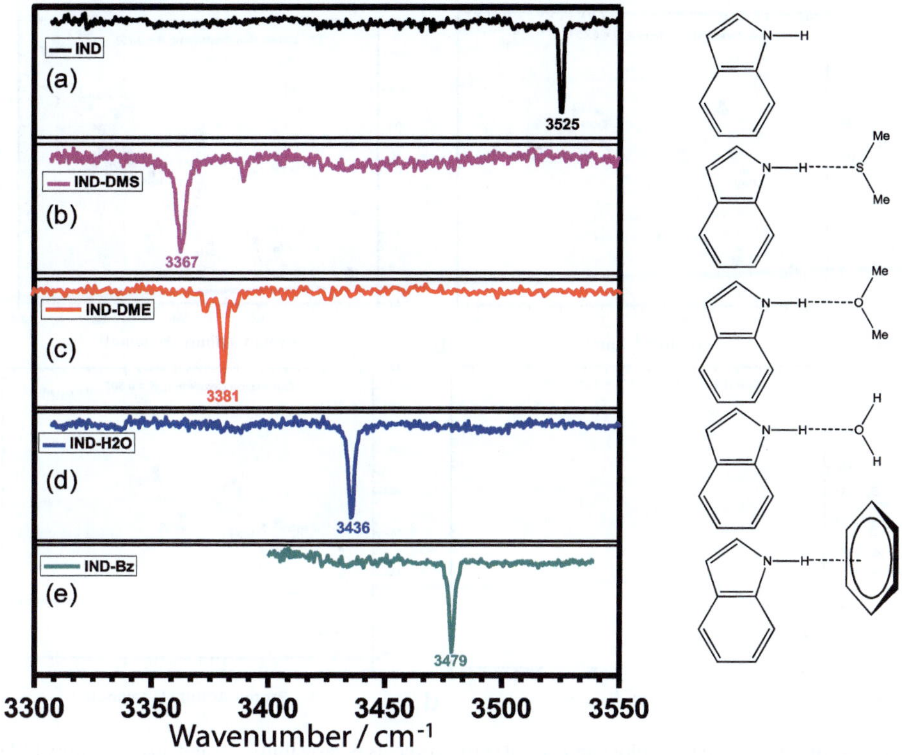

Fig. 2.9 FDIR spectra of H-bonded complexes of indole with various H-bond acceptors. *IND* Indole, *DMS* Dimethylsulfide, *DME* Dimethylether, *Bz* Benzene

N–H$\cdots\pi$ H-bond. The correlation plots of band origin shift (δ) versus polarizabilty (α) and proton affinity (PA) for different types of N–H$\cdots$Y H-bonded and van der Waals complexes of indole as shown in Fig. 2.10 also confirms that IND-DMS can be categorized with the IND-Bz and van der Waals complexes.

Going by the correlation presented above, one can put N–H$\cdots$S H-bond and N–H$\cdots\pi$ H-bond in the same class, but the 3.5 times larger red shift of N–H stretching frequency in IND-DMS than in IND-Bz is completely unexplained by this correlations. One can safely say that N–H$\cdots$S H-bond (in IND-DMS) is very different from the conventional N–H$\cdots$O (in IND-H_2O), N–H$\cdots$N (in IND-NH_3) H-bonds and unconventional N–H$\cdots\pi$ (in IND-Bz) H-bond, but how N–H$\cdots$S H-bond is different from others is a challenging topic of further explorations both by theories and experiments.

The intra and intermolecular amide N–H$\cdots$S H-bonds are frequently observed in the crystals of proteins [57, 146, 147] and simple organic molecules [58, 65, 148, 149]. The H-bond length of N–H$\cdots$S H-bond varies from system to system and it depends on the spatial arrangement of the crystals and the orientation of amide N–H towards sulfur atom. By considering H-bond distances in most of these cases the

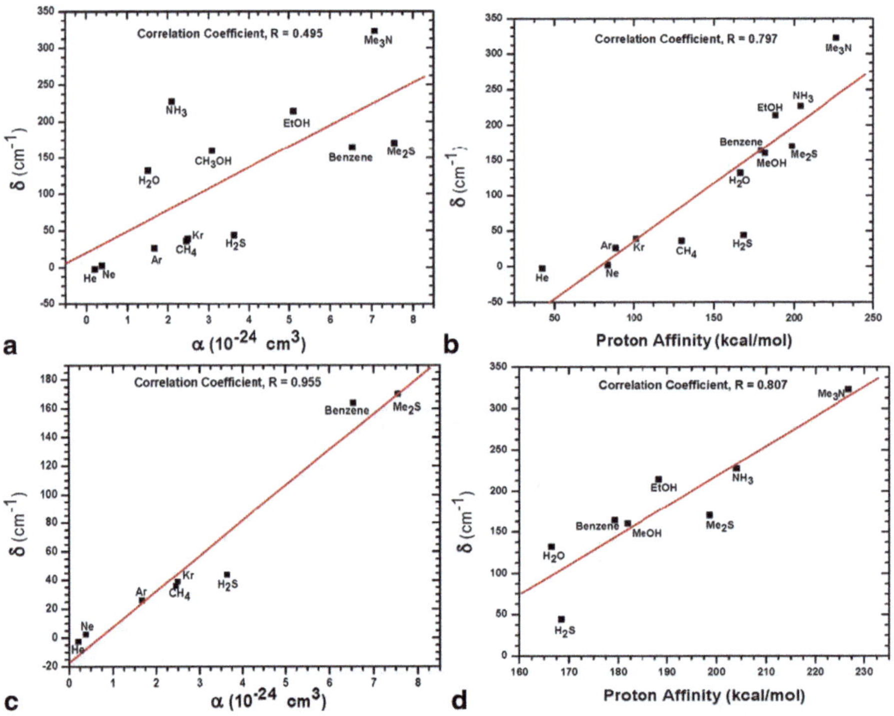

Fig. 2.10 The correlations plot band origin shift (δ) versus polarizabilty (α) and proton affinity (*PA*) for different types of N–H··· Y hydrogen-bonded and van der Waals complexes of indole. **a** The band origin shift (δ) versus polarizabilty (α). **b** the band origin shift (δ) versus proton affinity (*PA*) plots for all of the complexes. **c** the band origin shift (δ) versus polarizabilty (α) for the van der Waals complexes and M·Me_2S, M·H_2S, and M·benzene. **d** the band origin shift (δ) versus proton affinity (*PA*) for the N–H··· O and N–H··· N type HB complexes and M·Me_2S, M·H_2S, and M·benzene. (Reprinted with permission from ref 135. Copyright 2009 American Chemical Society)

amide N–H··· S H-bond is regarded as a weak H-bond like N–H··· π H-bond, even if the statistical analysis of methionine and cysteine containing proteins supports N–H··· S H-bond as a weak H-bond [78]. There are hardly any comprehensive study of amide N–H··· S H-bond using laser spectroscopy and high level ab intio calculations. Recently Mons and co-workers studied amide N–H··· S H-bonds in a systematic manner using methionine containing model peptides [150]. The model peptides are capped dipeptides of phenylalanine and methionine residues. They are N-acetyl-Lphenylalaninyl-L-methionine-amide (FM) and N-acetyl-L-methioninyl-L-phenylalanine-amide (MF) as shown in Fig. 2.10. In these peptides a variety of H-bonds can be possible with the amide N–H. The amide N–H forms N–H··· O=C H-bond with the back bone carbonyls and it provides different folding patterns and structures to the peptides. The N–H··· S and N–H··· π H-bonds are also formed with the side chain sulfur atom of methionine and π electron cloud of phenylalanine, respectively. Hence, these peptides are ideal peptides to study the N–H··· S H-bonds

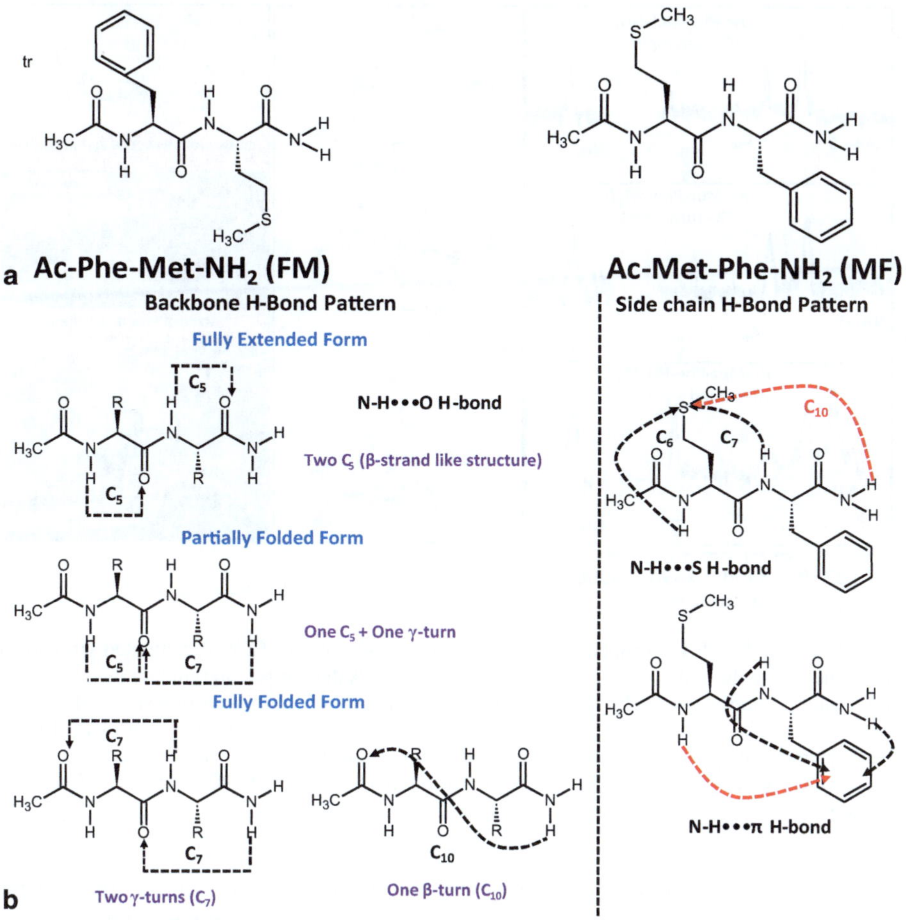

Fig. 2.11 **a** Model peptides, N-acetyl-Lphenylalaninyl-L-methionine-amide (*FM*) and N-acetyl-L-methioninyl-L-phenylalanine-amide (*MF*). **b** *Left Panel*: Selected sterically allowed backbone H-bond patterns. *Right panel*: Local H-bonding patterns of the side-chain of methionine (N–H$\cdots$S H-bond), or phenylalanine (N–H$\cdots\pi$ H-bond). The H-bond patterns are labelled C_n, according to the "n" number of atoms present in the ring formed by the H-bond

and compare them with other type H-bonds existing in the same system. As shown in Fig. 2.11b, the sulfur atom of methionine can either form an intra residue five membered (C_5) N–H$\cdots$S H-bond with its own residue or inter six membered (C_6) and ten membered (C_{10}) N–H$\cdots$S H-bonds with "i" and "i + 1" residues, respectively. The IR-UV double resonance spectroscopy was employed to get the IR spectral signature of these peptides in the N–H stretch region. Since, these peptides are studied in isolated gas phase conditions, the effect of solvent environment and crystal packing pattern on N–H$\cdots$S, N–H$\cdots$O = C, and N–H$\cdots\pi$ H-bonds are avoided. In most of the proteins the methionine residues reside in the hydrophobic core of

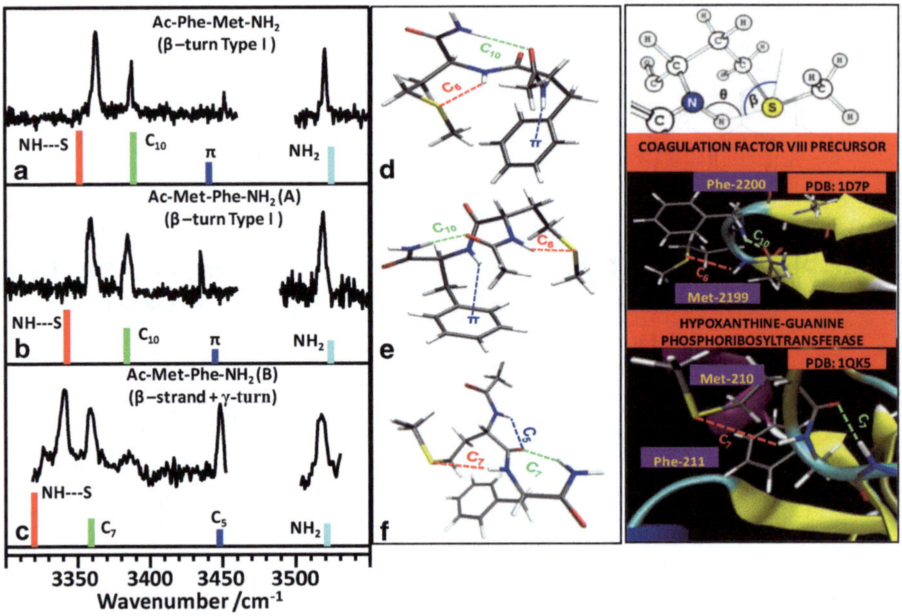

Fig. 2.12 *Left Panel*: Experimental infrared spectra versus B97-D/TZVPP computed vibrational frequencies of **a** the main conformer of FM and **b, c**, the stick diagrams represent the computed N–H frequencies. *Middle panel*: The H-bond pattern and structures of conformers of FM and MF whose computed N–H stretching frequencies best fit the experimental ones. *Right panel*: The H-bond parameters and examples of locally folded methionine residues in proteins as obtained from X-ray diffraction crystal. (Reprinted with permission from ref 150. Copyright 2012 American Chemical Society)

the protein. Hence, studying them by gas phase spectroscopy without any solvent (very low dielectric medium) will allow us to mimic the environment of methionine in proteins.

The experimental IR spectra and the computed N–H stretching frequencies of assigned conformers of FM and MF at B97-D/TZVPP level are shown in Fig. 2.12.

The most red shifted N–H stretch is observed for N–H··· S H-bond. The shift is even stronger than that observed for N–H··· O = C H-bonds. It is really surprising that sulfur can form very strong H-bonds as opposed to earlier observation that it is as weak as N–H··· π H-bonds. The B97-D methods are very helpful in computing the vibrational frequencies of a large number of conformers of the model peptides and those frequencies are used to assign the correct conformer to experimental IR spectra. The folding pattern as observed in these small model peptides are also found in proteins. Two of the examples are shown in the Fig. 2.11. The major challenge is to determine the intramolecular N–H··· S, N–H··· O = C, and N–H··· π H-bond energies. The donor-acceptor overlap energies and electron density at the bond critical point as obtained from NBO and AIM calculations do not correlate with the experimental observed red shifts of N–H stretching frequencies. The donor-acceptor overlap energies and electron density at the bond critical point for N–H··· O = C

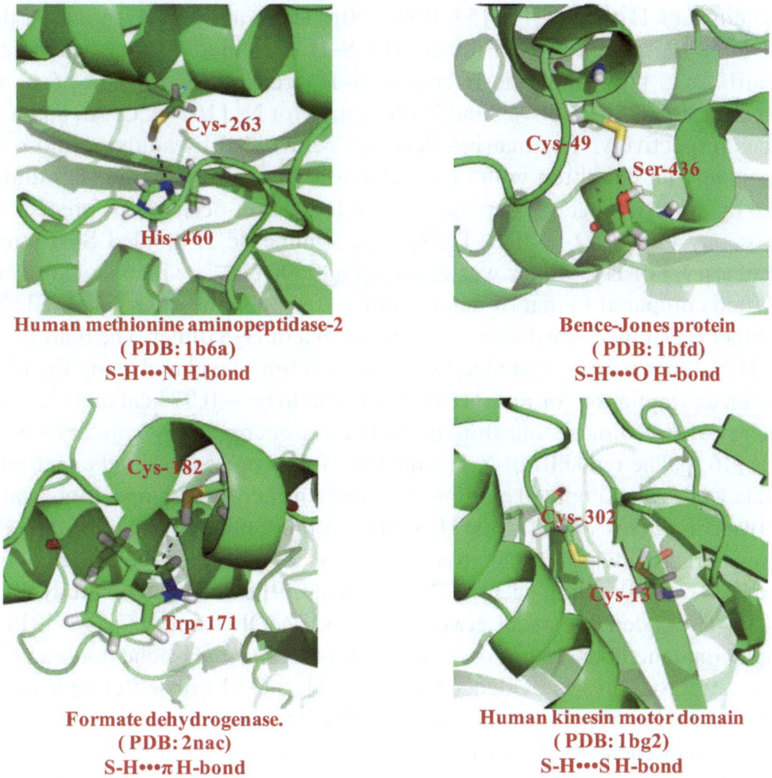

Fig. 2.13 *S–H· · · N, S–H· · · O, S–H· · · S, and S–H· · · π* H-bonds involving S–H group of cysteine

H-bonds are always higher than for N–H· · · S H-bonds. Molecular tailoring [151] and fragmented molecular orbital [152] methods may be useful in getting the exact intramolecular N–H· · · S H-bond energies.

2.3 S–H as a H-Bond Donor

Sulfur is not only a potential H-bond acceptor forming N–H· · · S and O–H· · · S H-bonds as strong as N–H· · · O and O–H· · · O H-bonds, S–H also participates as H-bond donor with many different types of acceptors. From a detailed survey of the protein structure data bank it is observed that S–H group of cysteine can make S–H· · · N, S–H· · · O, S–H· · · S, and S–H· · · π H-bonds. Few examples of SCHBs involving cysteine S–H group are shown in Fig. 2.13.

The simplest example where S–H acts as an H-bond acceptor is the H_2S dimer. The structural similarity between H_2O and H_2S has attracted many researchers to compare their molecular properties. One of them is the H-bond acceptor and donor potentialities of H_2O and H_2S [53, 153–160]. There are many reports on the matrix isolated IR spectra of H_2S dimer. The S–H stretching frequencies varies with the matrix; e.g. the symmetric H-bonded S–H stretching frequencies (ν_1) are observed at 2580.3, 2567, 2569.5, and 2566.4 cm^{-1} for N_2 [161], CO, Ar, and Kr [162] matrices respectively. The matrix effect can be avoided by studying those dimers in supersonic-jet condition with mass selection. Very recently, the H_2S dimer was studied by Wategaonkar's group using VUV ionization-detected IR predissociation spectroscopy (VUV-ID-IRPDS) [159]. The symmetric H-bonded S–H stretching frequencies(ν_1) of H_2S dimer was observed at 2590 cm^{-1}, which is about 24 cm^{-1} red shifted compared to that of the monomer [163]. On the other hand red shift for H_2O dimer is 56 cm^{-1} which is $\sim$2.5 times of that of H_2S dimer. The binding energy of the H_2S dimer was estimated at the MP2/CBS limit level and using the Helgaker two-point extrapolation formula [164]. It is found to be − 0.97 kcal mol^{-1}. The binding energy of H_2S dimer is one third of the binding energy of H_2O dimer (− 3.16 kcal mol^{-1}) [165]. The red shift of X-H stretching frequencies observed experimentally correlate well with computed binding energies. An energy decomposition analysis of these two complexes suggests that H_2S dimer is predominantly stabilized by the dispersion interaction where as electrostatic energy component is the major contributor for the H_2O dimer binding energy. The authors [159] extended their study further to S–H···O H-bonded complex between H_2S and MeOH using VUV-ID-IRPDS and computational methods. In this case H_2S prefers to act as a H-bond donor rather than H-bond acceptor. This is evidenced as the S–H···O H-bonded complex between H_2S and MeOH is not observed experimentally. The symmetric H-bonded S–H (ν_1) stretching frequencies of H_2S-MeOH complex is observed at 2577 cm^{-1} which is 37 cm^{-1} red shifted compared to the monomer stretching frequencies. The red shift is larger compared to the H_2S–H_2S case. The binding energy of H_2S-MeOH (S–H···O H-bonded dimer) complex at MP2/CBS level is − 2.65 kcal mol^{-1} which is 0.21 kcal mol^{-1} higher than that of the O–H···S H-bonded dimer. The difference in the stability comes from the electrostatic component. The electrostatic interaction in S–H···O H-bonded dimer is more than in the O–H···O H-bonded dimer. The computed binding energy of S–H···O=C H-bond ($\sim$-4 kcal mol^{-1}) [166] is comparable with that of N–H···S H-bonds and stronger than C–H···O and C–H···S H-bonds.

The S–H···N H-bond between H_2S and different amines has been characterized by Scheiner and coworkers [167] using ab intio methods. The bond dissociation energies (D_0) of H_2S-NH_3 and H_2S-N $(CH_3)_3$ are − 1.76 and − 3.55 kcal mol^{-1}, respectively. These values are quite larger compared to the bond dissociation energy of H_2S dimer. The complexes are linear in structure with H-bond angle (θ) very close to 180°. The S–H bond length of H_2S increases in all the complexes and red shifts of the S–H stretching frequencies are observed. In another report Scheiner and coworkers mentioned blue-shifted S–H···N and S–H···P H-bonds [168]. Here the S–H of SHN is the H-bond donor and amines and phosphines are the H-bond acceptors. The maximum blue shift (110 cm^{-1}) was observed NSH···NH_3 complex. Such a large blue shift is really surprising and needs further rather experimental evidences.

The S–H$\cdots\pi$ H-bonds involving the S–H group of cysteine and π electron density of aromatic amino acids are very important as they are frequently observed in proteins [34, 76, 138, 166, 169, 170]. Most commonly observed X–H$\cdots\pi$ H-bonds in nature are O–H$\cdots\pi$, N–H$\cdots\pi$, and C–H$\cdots\pi$ interactions and they are also vital in predicting and dictating structural and conformational preferences of biomolecules [138, 141, 142, 144, 171–173]. Many efforts have been made to characterize S–H$\cdots\pi$ H-bond and compare it with other X–H$\cdots\pi$ H-bonds. In most of the cases benzene and H_2S have been considered as the model compounds to study such H-bonds [50, 145, 155, 157, 174, 175].

We reported first ever IR spectroscopic studies of the S–H$\cdots\pi$ H-bonds in the gas phase in complexes of H_2S with indole (IND) and 3-methylindole (3-MI) [176]. The red shift in the band origin of the IND–H_2S (44 cm^{-1}) and 3-MI–H_2S (71 cm^{-1}) complexes were found to be much smaller than those of their respective H_2O complexes (132 and 239 cm^{-1}) which are N–H$\cdots$O H-bonded complexes. On the other hand band origin shift of IND–H_2S is very similar to that of the IND–CH_4 complex (76 cm^{-1}) which is a C–H$\cdots\pi$ H-bonded complex [177]. This provides an indirect evidence of H_2S forming S–H$\cdots\pi$ H-bond with indole and 3-MI. The direct evidences come from the IR spectra of the complexes (Fig. 2.14). The N–H stretching frequency is almost unchanged upon the complex formation, thereby indicating that the N–H stretch remains free in the complex. At the same time both complexes furnished an H-bonded symmetric S–H stretching frequency at 2593 and 2589 cm^{-1}, respectively. The H-bonded S–H stretch for IND–H_2S was red shifted by 21 cm^{-1}, while that for the 3-MI–H_2S was red shifted by 25 cm^{-1}. This gives a confirmatory inference that S–H group of H_2S prefers to act as an H-bond donor to the π electron density of IND and 3-MI.

The bond dissociation energies of IND-H_2S and 3-MI-H_2S (π-type H-bonded complexes) were calculated at the MP2/CBS and compared with those of IND-H_2O and 3-MI-H_2O complexes which form σ-type H-bonds. Table 2.3 summarizes the bond dissociation energies, band origin shift and S(O)-H stretching frequency shifts for the above mentioned complexes.

The S–H$\cdots\pi$ H-bond energies of both the complexes are larger than the N–H$\cdots$O H-bond energies of IND.H_2O and 3-MI-H_2O complexes. Moreover, comparison of the computed binding energies of the π-type H-bonded complexes of IND and benzene with H_2O, H_2S, NH_3, and CH_4 showed that the X–H$\cdots\pi$ H-bond energies lie in the order as S–H$\cdots\pi$ > O–H$\cdots\pi$ > N–H$\cdots\pi$ > C–H$\cdots\pi$. H-bonds [176]. Energy decomposition analysis of the H-bonded complexes with IND revealed the net electrostatic components to lie in the order $H_2O \approx H_2S > NH_3 > CH_4$ whereas the dispersion contribution was found to follow the order $H_2S > CH_4 > NH_3 > H_2O$. Very recently attempts have been made by Boxer and coworkers to quantify the electrostatic contribution in X–H$\cdots\pi$ H-bonds using vibrational Stark spectroscopy in combination with DFT calculations [178, 179]. It is observed that the red shifts of the H-bonded O H stretching frequencies in O–H(phenol)$\cdots\pi$ H-bonds correlate linearly with the strength of the applied electric fields indicating that they are primarily governed by electrostatic interactions. However in the case of N–H(indole)$\cdots\pi$ and S–H(thiophenol)$\cdots\pi$ H-bonds deviations from the linear

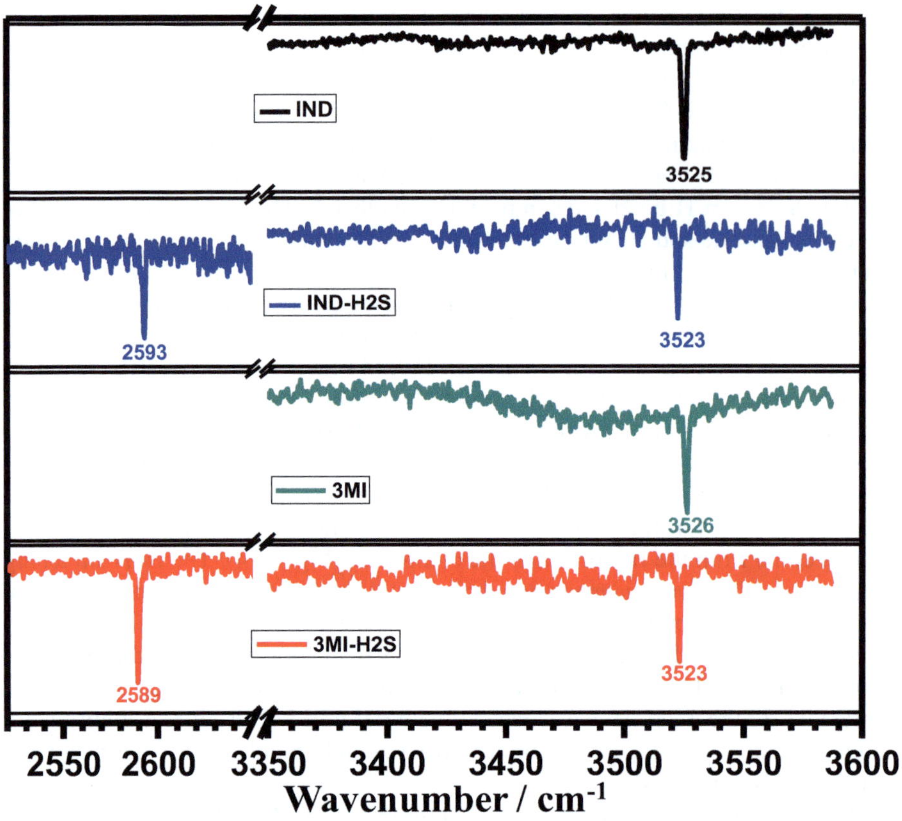

Fig. 2.14 FDIR spectra of indole (*IND*) and 3-methylindole (*3-MI*) and their H-bonded complexes with H_2S probing at their respective band origins. (Reprinted with permission from ref 176. Copyright 2009 American Chemical Society)

Table 2.3 Experimentally observed parameters such as XH stretching frequency shift, the band origin shift, The S_1 state lifetime, and binding energy of indole·L and 3-MI·L (L = H_2O and H_2S) σ- and π-type hydrogen-bonded complexes. (Reprinted with permission from ref 176. Copyright 2009 American Chemical Society)

Complex	HB type	$\Delta\nu_{NH}$ (cm^{-1})	$\Delta\nu_{XH}$ $(cm^{-1})_{X=O,S}$	ΔE_{BO} (cm^{-1})	τ (ns)	ΔE_0 (kcal/mol)
IND·H_2O	N–H···O σ HB	– 89	– 5[a]	– 132	22.8	– 4.67[a]
IND·H_2S	S–H··· π π HB	– 2	– 21	– 44	11.6	– 4.89[b]
3-MI·H_2O	N–H···O σ HB	– 84	– 5[a]	– 239	14.8	– 4.49[a]
3-MI·H_2S	S–H··· π π HB	– 3	– 25	– 71	7.9	– 5.17[b]

[a] Experimental value from ref 124

[b] Binding energy computed at the CBS limit

correlations were observed. These observations were found to be consistent with the atomic polarizabilities of the associated X–H groups and hence can be attributed to the dispersion dominance in N–H (indole)$\cdots\pi$ and S–H(thiophenol)$\cdots\pi$ H-bonds.

Acknowledgment I am very much thankful to Prof. Sanjay Wategaonkar and Dr. Michel Mons, who introduced me to the exciting field of laser spectroscopy. My special thanks to Dr. Rudresh Acharya for stimulating discussions and for his help in rendering figures for SCHBs in protein structures. I am also grateful to my coworkers, students and authors of the cited references who have contributed to this work in many ways. This work is financially supported by DST-Inspire faculty fellowship, Department of Science and Technology (DST, Govt. of India) and National Institute of Science Education and Research, (Department of Atomic Energy, DAE, Govt. of India).

References

1. Arunan E, Desiraju GR, Klein RA, Sadlej J, Scheiner S, Alkorta I, Clary DC, Crabtree RH, Dannenberg JJ, Hobza P, Kjaergaard HG, Legon AC, Mennucci B, Nesbitt DJ (2011) Defining the hydrogen bond: an account (IUPAC Technical Report). Pure Appl Chem 83(8):1619–1636
2. Pauling L (1960) The nature of the chemical bond, 3rd edn. Cornell University Press, Ithaca
3. Pimentel GC, McClellan AL (1960) The hydrogen bond. W. H. Freeman and Company, New York
4. Jeffrey GA, Saenger W (1991) Hydrogen bonding in biological structures. Springer-Verlag, Berlin
5. Jeffrey G A (1997) An introduction to hydrogen bonding. Oxford University Press, New York
6. Scheiner S (1997) Hydrogen bonding: a theoretical perspective. Oxford University Press, New York
7. Desiraju GR, Steiner T (1999) The weak hydrogen bond in structural chemistry and biology. Oxford University Press, New York
8. Grabowski SJ (2006) Hydrogen bonding—new insights challenges and advances in computational chemistry and physics, vol 3, 1st edn. Springer, New York
9. Gilli G, Gilli P (2009) The nature of the hydrogen bond: outline of a comprehensive hydrogen bond theory. Oxford University Press, New York
10. Hobza P, Müller-Dethlefs K (2010) Non-covalent interactions: theory and experiment. Royal Society of Chemistry, Cambridge
11. Dunitz JD, Taylor R (1997) Organic fluorine hardly ever accepts hydrogen bonds. Chem Eur J 3:89–98
12. Dunitz JD (2004) Organic fluorine: odd man out. ChemBioChem 5:614–621
13. Frey JA, Leist R, Leutwyler S (2006) Hydrogen Bonding of the Nucleobase Mimic 2-Pyridone to Fluorobenzenes: an ab Initio Investigation. J Phys Chem A 110:4188–4195
14. Toth G, Bowers SG, Truong AP, Probst G (2007) The role and significance of unconventional hydrogen bonds in small molecule recognition by biological receptors of pharmaceutical relevance. Curr Pharm Des 13:3476–3493
15. Zhu Y-Y, Yi H-P, Li C, Jiang X-K, Li Z-T (2008) The N–H$\cdots$X (X = Cl, Br, and I) hydrogen-bonding pattern in aromatic amides: a crystallographic and 1H NMR study. Cryst Growth Des 8:1294–1300
16. Koller AN, Bozilovic J, Engels JW, Gohlke H (2010) Aromatic N versus aromatic F: bioisosterism discovered in RNA base pairing interactions leads to a novel class of universal base analogs. Nucleic Acids Res 38:3133–3146
17. Gregoret LM, Rader SD, Fletterick RJ, Cohen FE (1991) Hydrogen bonds involving sulfur atoms in proteins. Proteins 9(2):99–107

18. Platts JA, Howard ST, Bracke BRF (1996) Directionality of hydrogen bonds to sulfur and oxygen. J Am Chem Soc 118(11):2726–2733
19. Allen FH, Bird CM, Rowland RS, Raithby PR (1997) Hydrogen-bond acceptor and donor properties of divalent sulfur (Y–S–Z and R–S–H). Acta Crystallogr, Sect B: Struct Sci 53:696–701
20. Sutor DJ (1963) Evidence for existence of C–H··· O hydrogen bonds in crystals. J Chem Soc (FEB) 1105–1110
21. Taylor R, Kennard O (1982) Crystallographic evidence for the existence of CH··· O, CH··· N and CH··· Cl hydrogen bonds. J Am Chem Soc 104(19):5063–5070
22. Brandl M, Weiss MS, Jabs A, Suhnel J, Hilgenfeld R (2001) C–H··· π -interactions in proteins. J Mol Biol 307(1):357–377
23. Scheiner S, Kar T, Pattanayak J (2002) Comparison of various types of hydrogen bonds involving aromatic amino acids. J Am Chem Soc 124(44):13257–13264
24. Manikandan K, Ramakumar S (2004) The occurrence of C–H··· O hydrogen bonds in alpha-helices and helix termini in globular proteins. Proteins 56(4):768–781
25. Sarkhel S, Desiraju GR (2004) N–H··· O, O–H··· O, and C–H··· O hydrogen bonds in protein-ligand complexes: strong and weak interactions in molecular recognition. Proteins 54(2):247–259
26. Mani D, Arunan E (2013) The X-C··· Y (X = O/F, Y = O/S/F/Cl/Br/N/P) 'carbon bond' and hydrophobic interactions. Phys Chem Chem Phys 15:14377–14383
27. Bauza A, Mooibroek TJ, Frontera A (2014) Influence of ring size on the strength of carbon bonding complexes between anions and perfluorocycloalkanes. Phys Chem Chem Phys. Ahead of Print. doi:10.1039/c4cp01983k
28. Thomas SP, Pavan MS, Guru Row TN (2014) Experimental evidence for 'carbon bonding' in the solid state from charge density analysis. Chem Commun 50:49–51
29. Adman E, Watenpaugh KD, Jensen LH (1975) NH··· S Hydrogen bonds in Peptococcus aerogenes ferredoxin, clostridium pasteurianum rubredoxin, and chromatium high potential iron protein. Proc Nat Acad Sci U S A 72(12):4854–4858
30. Morgan RS, Tatsch CE, Gushard RH, McAdon JM, Warme PK (1978) Chains of alternating sulfur and p-bonded atoms in 8 small proteins. Int J Pept Protein Res 11(3):209–217
31. Reid KSC, Lindley PF, Thornton JM (1985) Sulfur-aromatic interactions in proteins. FEBS Lett 190(2):209–213
32. Li HM, Thomas GJ (1991) Cysteine conformation and sulfhydryl interactions in proteins and viruses.1. Correlation of the Raman S–H B and with hydrogen-bonding and intramolecular geometry in model compounds. J Am Chem Soc 113(2):456–462
33. Renault L, Nassar N, Vetter I, Becker J, Klebe C, Roth M, Wittinghofer A (1998) The 1.7 angstrom crystal structure of the regulator of chromosome condensation (RCC1) reveals a seven-bladed propeller. Nature 392(6671):97–101
34. Iwaoka M, Takemoto S, Okada M, Tomoda S (2002) Weak nonbonded S–X (X = O, N, and S) interactions in proteins. Statistical and theoretical studies. Bull Chem Soc Jpn 75(7):1611–1625
35. Pantoja-Uceda D, Shewry PR, Bruix M, Tatham AS, Santoro J, Rico M (2004) Solution structure of a methionine-rich 2S albumin from sunflower seeds: Relationship to its allergenic and emulsifying properties. Biochem 43(22):6976–6986
36. Brosnan JT, Brosnan ME (2006) The sulfur-containing amino acids: An overview. J Nutr 136(6):1636S–1640S
37. Nicoletti FP, Comandini A, Bonamore A, Boechi L, Boubeta FM, Feis A, Smulevich G, Boffi A (2010) Sulfide binding properties of truncated hemoglobins. Biochem 49(10):2269–2278
38. Iwaoka M, Isozumi N (2012) Hypervalent nonbonded interactions of a divalent sulfur atom. Implications in protein architecture and the functions. Molecules 17(6):7266–7283
39. Kolluru GK, Shen XG, Bir SC, Kevil CG (2013) Hydrogen sulfide chemical biology: pathophysiological roles and detection. Nitric Oxide-Biol Ch 35:5–20
40. Sabin JR (1971) Hydrogen bonds involving sulfur. I. Hydrogen sulfide dimer. J Am Chem Soc 93(15):3613–3620

41. Li S, Li YS (1991) FTIR spectra of matrix isolated complexes between sulfur-compounds. Spectrochim Acta Part A 47(2):201–209
42. Wierzejewska M (2000) FTIR matrix isolation studies of complexes of dimethylsulfide, dimethyldisulfide and hydrogen sulfide with nitric acid. Vib Spectrosc 23(2):253–262
43. Wierzejewska M (2000) Infrared matrix isolation studies of complexes formed between dimethylsulfide, dimethyldisulfide and nitrous acid. J Mol Struct 520:199–214
44. Wierzejewska M, Saldyka M (2004) Are hydrogen bonds to sulfur and oxygen different? Theoretical study of dimethylsulfide and dimethylether complexes with nitric acid. Chem Phys Lett 391(1–3):143–147
45. Galardon E, Roger T, Deschamps P, Roussel P, Tomas A, Artaud I (2012) Synthesis of a (FeSH)-S-II complex stabilized by an intramolecular N–H···S hydrogen bond, which acts as a H_2S donor. Inorg Chem 51(19):10068–10070
46. Okamura T-a, Kunisue K, Omi Y, Onitsuka K (2013) Strong NH···S hydrogen bonds in molybdoenzyme models containing anilide moieties. Dalton Trans 42:7569–7578
47. Okamura T-a, Ushijima Y, Omi Y, Onitsuka K (2013) Systematic investigation of relationship between strength of NH···S hydrogen bond and reactivity of molybdoenzyme models. Inorg Chem 52:381–394
48. Zhou YJ, Zhang MM, Wang XS (2013) The N–H···X hydrogen bonds in the crystal structures of (Thio)Isochromene Derivatives. J Chem Crystallogr 43(1):26–30
49. Xu LH, Liu Q, Suenram RD, Lovas FJ, Walker ARH, Jensen JO, Samuels AC (2004) Rotational spectra, conformational structures, and dipole moments of thiodiglycol by jet-cooled FTMW and ab initio calculations. J Mol Spectrosc 228(2):243–250
50. Crittenden DL (2009) A systematic CCSD(T) study of long-range and noncovalent interactions between benzene and a series of first- and second-row hydrides and rare gas atoms. J Phys Chem A 113(8):1663–1669
51. Cabaleiro-Lago EM, Rodriguez-Otero J (2002) Methanethiol dimer and trimer. An ab initio and DFT study of the interaction. J Phys Chem A 106(32):7440–7447
52. Wennmohs F, Staemmler V, Schindler M (2003) Theoretical investigation of weak hydrogen bonds to sulfur. J Chem Phys 119(6):3208–3218
53. Vila A, Mosquera RA (2006) Are the hydrogen bonds involving sulfur bases inverse or anomalous? Int J Quantum Chem 106(4):928–934
54. Jezierska A, Panek JJ, Mazzarello R (2009) Structural and electronic structure differences due to the O–H···O and O–H···bond formation in selected benzamide derivatives: a first-principles molecular dynamics study. Theor Chem Acc 124(5–6):319–330
55. Steudel R, Steudel Y (2009) Microsolvation of thiosulfuric acid and its tautomeric anions $[HSSO_3]^-$ and $[SSO_2(OH)]^-$ Studied by B3LYP-PCM and G3X(MP2) calculations. J Phys Chem A 113(36):9920–9933
56. Kaur D, Aulakh D, Khanna S, Singh H (2014) Theoretical study on the nature of S···H and O···H hydrogen bonds. J Sulfur Chem 35(3):290–303
57. Ueyama N, Nishikawa N, Yamada Y, Okamura T, Oka S, Sakurai H, Nakamura A (1998) Synthesis and properties of octaethylporphinato(arenethiolato)iron(III) complexes with intramolecular NH–S hydrogen bond: chemical function of the hydrogen bond. Inorg Chem 37(10):2415–2421
58. Nangia A, Desiraju GR (1999) Axial and equatorial conformations of penicillins, their sulphoxides and sulphones: the role of N–H···S and C–H···O hydrogen bonds. J Mol Struct 65–79
59. Word JM, Lovell SC, LaBean TH, Taylor HC, Zalis ME, Presley BK, Richardson JS, Richardson DC (1999) Visualizing and quantifying molecular goodness-of-fit: small-probe contact dots with explicit hydrogen atoms. J Mol Biol 285(4):1711–1733
60. Word JM, Lovell SC, Richardson JS, Richardson DC (1999) Asparagine and glutamine: using hydrogen atom contacts in the choice of side-chain amide orientation. J Mol Biol 285(4):1735–1747
61. Francois S, Rohmer MM, Benard M, Moreland AC, Rauchfuss TB (2000) The N–H–S hydrogen bond in $(TACN)_2Fe_2S_6$ (TACN = triazacyclononane) and in model systems involving the persulfido moiety: an ab initio and DFT study. J Am Chem Soc 122(51):12743–12750

62. Steiner T (2000) S–H··· S hydrogen-bond chain in thiosalicylic acid. Acta Crystallogr Sect C-Cryst Struct Commun 56:876–877
63. Krepps MK, Parkin S, Atwood DA (2001) Hydrogen bonding with sulfur. Cryst Growth Des 1(4):291–297
64. Hambley TW, Hibbs DE, Turner P, Howard ST, Hursthouse MB (2002) Insights into bonding and hydrogen bond directionality in thioacetamide from the experimental charge distribution. J Chem Soc-Perkin Trans 2(2):235–239
65. Lynch DE, McClenaghan I, Light ME, Coles SJ (2002) The solid-state packing of sulfur substituted 2-aminopyrimidines and the occurrence of N–H-S hydrogen-bonding associations. Cryst Eng 5(1):79–94
66. Usman A, Fun HK, Ganguly NC, Datta M, Ghosh K (2003) 2-(2-hydroxyphenyl)-1,3-dithiane. Acta Crystallogr Sect E-Struct Rep Online 59:O773–O775
67. Sundaresan CN, Dixit S, Venugopalan P (2004) A supramolecular assembly dominated by N–H··· S hydrogen bonds: structure of 2-thioureidobenzoxazole by single crystal X-ray diffraction. J Mol Struct 693(1–3):205–209
68. Valdes-Martinez J, Hernandez-Ortega S, Rubio M, Li DT, Swearingen JK, Kaminsky W, Kelman DR, West DX (2004) Study of the sulfur atom as hydrogen bond acceptor in N(2)-pyridylmethyl-N-arylthioureas. J Chem Crystallogr 34(8):533–540
69. Moroder L (2005) Isosteric replacement of sulfur with other chalcogens in peptides and proteins. J Pept Sci 11(4):187–214
70. Tutughamiarso M, Egert E (2011) Cocrystals of 6-propyl-2-thiouracil: N–H··· O versus N–H··· S hydrogen bonds. Acta Crystallogr, Sect C: Cryst Struct Commun 67:O439–O445
71. Naktode K, Kottalanka RK, Panda TK (2012) N-(2,6-Dimethylphenyl)diphenylphosphinamine chalcogenides (S, Se) and a zirconium complex possessing phosphanylamide in the coordination sphere. New J Chem 36(11):2280–2285
72. Minkov VS, Boldyreva EV (2013) Weak hydrogen bonds formed by thiol groups in N-Acetyl-L-Cysteine and their response to the crystal structure distortion on increasing pressure. J Phys Chem B 117(46):14247–14260
73. Olivella M, Caltabiano G, Cordomi A (2013) The role of Cysteine 6.47 in class A GPCRs. BMC Struct Biol 13:3
74. AlDamen MA, Sinnokrot M (2014) Crystallographic and theoretical studies of 1-(1-naphthyl)-2-thiourea with intermolecular N–H··· S heteroatom interaction and N–H··· π interaction. J Struct Chem 55(1):53–60
75. Beck BW, Xie Q, Ichiye T (2001) Sequence determination of reduction potentials by cysteinyl hydrogen bonds and peptide dipoles in 4Fe-4S ferredoxins. Biophys J 81(2):601–613
76. Duan GL, Smith VH, Weaver DF (2001) Characterization of aromatic-thiol π-type hydrogen bonding and phenylalanine-cysteine side chain interactions through ab initio calculations and protein database analyses. Mol Phys 99(19):1689–1699
77. Pal D, Chakrabarti P (1998) Different types of interactions involving cysteine sulfhydryl group in proteins. J Biomol Struct Dyn 15(6):1059–1072
78. Zhou P, Tian FF, Lv FL, Shang ZC (2009) Geometric characteristics of hydrogen bonds involving sulfur atoms in proteins. Proteins Struct Funct Bioinf 76(1):151–163
79. Steiner T (2002) The hydrogen bond in the solid state. Angew Chem Int Ed 41(1):48–76
80. Ranaghan KE, Hung JE, Bartlett GJ, Mooibroek TJ, Harvey JN, Woolfson DN, van der Donk WA, Mulholland AJ (2014) A catalytic role for methionine revealed by a combination of computation and experiments on phosphite dehydrogenase. Chem Sci 5(6):2191–2199
81. Duncan MA (2003) Infrared spectroscopy to probe structure and dynamics in metal ion-molecule complexes. Int Rev Phys Chem 22(2):407–435
82. Ebata T, Fujii A, Mikami N (1998) Vibrational spectroscopy of small-sized hydrogen-bonded clusters and their ions. Int Rev Phys Chem 17(3):331–361
83. Fujii A, Mizuse K (2013) Infrared spectroscopic studies on hydrogen-bonded water networks in gas phase clusters. Int Rev Phys Chem 32(2):266–307
84. Fujii M, Dopfer O (2012) Ionisation-induced site switching dynamics in solvated aromatic clusters: phenol-(rare gas)(n) clusters as prototypical example. Int Rev Phys Chem 31(1):131–173

85. Kang C, Pratt DW (2005) Structures, charge distributions, and dynamical properties of weakly bound complexes of aromatic molecules in their ground and electronically excited states. Int Rev Phys Chem 24(1):1–36
86. Simons JP, Jockusch RA, Carcabal P, Hung I, Kroemer RT, Macleod NA, Snoek LC (2005) Sugars in the gas phase. Spectroscopy, conformation, hydration, co-operativity and selectivity. Int Rev Phys Chem 24(3–4):489–531
87. Wild DA, Bieske EJ (2003) Infrared investigations of negatively charged complexes and clusters. Int Rev Phys Chem 22(1):129–151
88. Buckingham AD, Fowler PW (1985) A model for the geometries of van der Waals complexes. Can J Chem 63:2018–2025
89. Graindourze M, Maes G (1985) Matrix isolation vibrational spectra of alkyl chalcogenides complexed with hydrogen chloride. Matrix isolation IR spectra as a guide for the analysis of solution IR spectra of alkyl sulfide and alkyl selenide complexes with hydrogen chloride. J Mol Spectrosc 114:97–104
90. Maes G, Graindourze M (1985) Matrix isolation vibrational spectra of alkyl chalcogenides complexed with HCl: structure of alkyl sulfide and alkyl selenide complexes with hydrochloric acid in ar matrices from infrared spectra. J Mol Spectrosc 113(2):410–425
91. Maes G, Graindourze M (1985) Conformational isomerism of ethyl chalcogenides in inert matrixes: influence on the νs band structure of hydrogen-bonded complexes with hydrogen chloride. J Mol Spectrosc 114:280–288
92. Andrews L, Arlinghaus RT, Hunt RD (1986) FTIR spectra of dialkyl sulfide and alkanethiol complexes with hydrogen fluoride in solid argon. Inorg Chem 25(18):3205–3209
93. Barnes AJ, Wright MP (1986) Molecular complexes of hydrogen halides with ethers and sulfides studied by matrix isolation vibrational spectroscopy. J Mol Struct: Theochem 28:21–30
94. Jeng MLH, Ault BS (1990) Infrared matrix isolation study of hydrogen bonds involving carbon–hydrogen bonds: alkynes with bases containing second- and third-row donor atoms. J Phys Chem 94(4):1323–1327
95. Jeng MLH, Ault BS (1990) Infrared matrix isolation studies of hydrogen bonds involving carbon–hydrogen bonds: alkenes with selected bases. J Phys Chem 94:4851–4855
96. Li S, Li YS (1991) FT-IR spectra of matrix isolated complexes between some alkanethiols and sulfur dioxide. J Mol Struct 248:79–88
97. Kim KS, Tarakeshwar P, Lee JY (2000) Molecular clusters of π-systems: theoretical studies of structures, spectra, and origin of interaction energies. Chem Rev 100(11):4145–4185
98. Li S, Kurtz H, Korambath P, Li YS (2000) Infrared spectra, photochemistry, and ab initio calculations of matrix isolated methanethiol/sulfur dioxide complex. J Mol Struct 550–551:235–244
99. Howard DL, Kjaergaard HG (2008) Hydrogen bonding to divalent sulfur. Phys Chem Chem Phys 10(28):4113–4118
100. Biswal HS, Shirhatti PR, Wategaonkar S (2009) O–H⋯O versus O–H⋯S hydrogen bonding I: experimental and computational studies on the p-Cresol·H_2O and p-Cresol·H_2S complexes. J Phys Chem A 113(19):5633–5643
101. Biswal HS, Shirhatti PR, Wategaonkar S (2010) O–H⋯O versus O–H⋯S hydrogen bonding 2: alcohols and thiols as hydrogen bond acceptors. J Phys Chem A 114:6944–6955
102. Biswal HS, Chakraborty S, Wategaonkar S (2008) Experimental evidence of O–H⋯S hydrogen bonding in supersonic jet. J Chem Phys 129(18):184311–184317
103. Biswal HS, Wategaonkar S (2010) O–H⋯O versus O–H⋯S Hydrogen Bonding. 3. IR–UV double resonance study of hydrogen bonded complexes of p-cresol with diethyl ether and its sulfur analog. J Phys Chem A 114(19):5947–5957
104. Biswal HS, Wategaonkar S (2011) OH⋯X (X = O, S) hydrogen bonding in tetrahydrofurane and tetrahydrothiophene. J Chem Phys 135(13):134306
105. Contreras-García J, Johnson ER, Keinan S, Chaudret R, Piquemal J-P, Beratan DN, Yang W (2011) NCIPLOT: a program for plotting noncovalent interaction regions. J Chem Theory Comput 7(3):625–632

106. Otero-de-la-Roza A, Johnson ER, Contreras-Garcia J (2012) Revealing non-covalent interactions in solids: NCI plots revisited. Phys Chem Chem Phys 14(35):12165–12172
107. Bader RFW, Nguyen-Dang TT (1981) Quantum theory of atoms in molecules—Dalton revisited. Adv Quantum Chem 14:63–124
108. Bader RFW (1990) Atoms in molecules: a quantum theory. Clarendon Press, Oxford
109. Bader RFW (1991) A quantum theory of molecular structure and its applications. Chem Rev 91(5):893–928
110. Koch U, Popelier PLA (1995) Characterization of C–H–O hydrogen bonds on the basis of the charge density. J Phys Chem 99(24):9747–9754
111. Popelier P (2000) Atoms in molecules, an introduction. Prentice Hall, Englewood Cliffs
112. Glendening ED, Streitwieser A (1994) Natural energy decomposition analysis: an energy partitioning procedure for molecular-Interactions with application to weak hydrogen bonding, strong ionic, and moderate donor-dcce tor interactions. J Chem Phys 100(4):2900–2909
113. Glendening ED (1996) Natural energy decomposition analysis: explicit evaluation of electrostatic and polarization effects with application to aqueous clusters of alkali metal cations and neutrals. J Am Chem Soc 118(10):2473–2482
114. Schenter GK, Glendening ED (1996) Natural energy decomposition analysis: the linear response electrical self energy. J Phys Chem 100(43):17152–17156
115. Kitaura K, Morokuma K (1976) A new energy decomposition scheme for molecular interactions within the Hartree-Fock approximation. Int J Quantum Chem 10(2):325–340
116. Umeyama H, Morokuma K (1977) The origin of hydrogen bonding. An energy decomposition study. J Am Chem Soc 99(5):1316–1332
117. Stevens WJ, Fink WH (1987) Frozen fragment reduced variational space analysis of hydrogen bond in interactions-application to the water dimer. Chem Phys Lett 139(1):15–22
118. Jouvet C, Lardeux-Dedonder C, Richard-Viard M, Solgadi D, Tramer A (1990) Reactivity of molecular clusters in the gas phase: proton-transfer reaction in neutral phenol-(ammonia)n and phenol-(ethanamine)n. J Phys Chem 94:5041–5048
119. Iwasaki A, Fujii A, Watanabe T, Ebata T, Mikami N (1996) Infrared spectroscopy of hydrogen-bonded phenol-amine clusters in supersonic jets. J Phys Chem 100(40):16053–16057
120. Biswas N, Wategaonkar S, Watanabe T, Ebata T, Mikami N (2004) Fluorescence, REMPI, hole-burning, and FDIR spectroscopy of para-cyanophenol-water1 complex. Chem Phys Lett 394:61–67
121. Bhattacharyya S, Bhattacherjee A, Shirhatti PR, Wategaonkar S (2013) O–H$\cdots$S hydrogen bonds conform to the acid-base formalism. J Phys Chem A 117(34):8238–8250
122. Braun JE, Mehnert T, Neusser HJ (2000) Binding energy of van der Waals- and hydrogen-bonded clusters by threshold ionization techniques. Int J Mass Spectrom 203(1–3):1–18
123. Braun J, Neusser HJ, Hobza P (2003) N–H$\cdots$p interactions in Indole-Benzene-h_6,d_6 and Indole-Benzene-h_6,d_6 radical cation complexes. Mass Analyzed threshold ionization experiments and correlated ab initio quantum chemical calculations. J Phys Chem A 107(19):3918–3924
124. Georgiev S, Neusser HJ (2004) Investigation of hydrogen bonding in 3-methylindole–H_2O cluster by mass analyzed threshold ionization. Chem Phys Lett 389(1–3):24–28
125. Zhang B, Li C, Su H, Lin JL, Tzeng WB (2004) Mass analyzed threshold ionization spectroscopy of p-fluorophenol cation and the p-fluoro substitution effect. Chem Phys Lett 390:65–70
126. Georgiev S, Neusser HJ (2005) Mass analyzed threshold ionization of hydrogen bonded clusters of biological molecules: the 3-methylindole center.C_6H_6 complex. J Electron Spectrosc Relat Phenom 142(3):207–213
127. Shibasaki K, Fujii A, Mikami N, Tsuzuki S (2006) Magnitude of the CH/π interaction in the gas phase: Experimental and theoretical determination of the accurate interaction energy in benzene-methane. J Phys Chem A 110(13):4397–4404
128. Huang JH, Huang K, Liu SQ, Luo Q, Tzeng WB (2007) Molecular structures and vibrations of cis and trans m-cresol in the electronically excited S_1 and cationic D_0 states. J Photochem Photobiol A 188(2–3):252–259

129. Shirhatti PR, Wategaonkar S (2012) Mass analyzed threshold ionization (MATI) spectroscopy of p-cresol. Indian J Phys 86(3):159–164
130. Cordes E, Dopfer O, Wright TG, Mullerdethlefs K (1993) Vibrational spectroscopy of the phenol-ethanol cation. J Phys Chem 97(29):7471–7479
131. Vondrak T, Sato S, Kimura K (1997) Cation vibrational spectra of indole and indole-argon van der Waals complex. A zero kinetic energy photoelectron study. J Phys Chem A 101(13):2384–2389
132. Dessent CEH, Muller-Dethlefs K (2000) Hydrogen-bonding and van der Waals complexes studied by ZEKE and REMPI spectroscopy. Chem Rev 100(11):3999–4021
133. Zierhut M, Dummler S, Roth W, Fischer I (2003) Multiphoton ionization and zero kinetic energy photoelectron spectroscopy of the 1-naphthol(H_2O) cluster. Chem Phys Lett 381:346–353
134. Cockett MCR (2005) Photoelectron spectroscopy without photoelectrons: twenty years of ZEKE spectroscopy. Chem Soc Rev 34(11):935–948
135. Biswal HS, Wategaonkar S (2009) Nature of the N–H···S hydrogen bond. J Phys Chem A 113(46):12763–12773
136. Ballew RM, Sabelko J, Gruebele M (1996) Direct observation of fast protein folding: the initial collapse of apomyoglobin. Proc Nat Acad Sci U S A 93(12):5759–5764
137. Desfrancois C, Carles S, Schermann JP (2000) Weakly bound clusters of biological interest. Chem Rev 100(11):3943–3962
138. Meyer EA, Castellano RK, Diederich F (2003) Interactions with aromatic rings in chemical and biological recognition. Angew Chem Int Ed 42(11):1210–1250
139. Biswal HS, Gloaguen E, Mons M, Bhattacharyya S, Shirhatti PR, Wategaonkar S (2011) Structure of the indole-benzene Dimer Revisited. J Phys Chem A 115(34):9485–9492
140. Braun JE, Grebner TL, Neusser HJ (1998) Van der Waals versus hydrogen-bonding in complexes of indole with argon, water, and benzene by mass-analyzed pulsed field threshold ionization. J Phys Chem A 102(19):3273–3278
141. Tsuzuki S, Honda K, Uchimaru T, Mikami M, Tanabe K (2000) Origin of the attraction and directionality of the NH/π interaction: Comparison with OH/π and CH/π interactions. J Am Chem Soc 122(46):11450–11458
142. Tsuzuki S, Honda K, Uchimaru T, Mikami M, Tanabe K (2000) The magnitude of the CH/p interaction between benzene and some model hydrocarbons. J Am Chem Soc 122(15):3746–3753
143. Morita S-i, Fujii A, Mikami N, Tsuzuki S (2006) Origin of the attraction in aliphatic C–H/π interactions: Infrared spectroscopic and theoretical characterization of gas-phase clusters of aromatics with methane. J Phys Chem A 110(36):10583–10590
144. Tsuzuki S, Honda K, Uchimaru T, Mikami M, Fujii A (2006) Magnitude and directionality of the interaction energy of the aliphatic CH/π interaction: Significant difference from hydrogen bond. J Phys Chem A 110(33):10163–10168
145. Alberti M, Aguilar A, Huarte-Larranaga F, Lucas JM, Pirani F (2014) Benzene-hydrogen bond (C_6H_6-HX) Interactions: the influence of the X nature on their strength and anisotropy. J Phys Chem A 118(9):1651–1662
146. Ueyama N, Nishikawa N, Yamada Y, Okamura T, Nakamura A (1996) Cytochrome P-450 model (porphinato)(thiolato)iron(III) complexes with single and double NH···S hydrogen bonds at the thiolate site. J Am Chem Soc 118(50):12826–12827
147. Suzuki N, Higuchi T, Urano Y, Kikuchi K, Uekusa H, Ohashi Y, Uchida T, Kitagawa T, Nagano T (1999) Novel iron porphyrin-alkanethiolate complex with intramolecular NH···S hydrogen bond: Synthesis, spectroscopy, and reactivity. J Am Chem Soc 121(49):11571–11572
148. Liu S-G, Li Y-Z, Zuo J-L, You X-Z (2004) N, N'-Bis[2-(methylsulfanyl)phenyl]pyridine-2,6-dicarboxamide. Acta Cryst E 60(9):o1527–o1529
149. Du P, Jiang XK, Li Z-T (2009) Five- and six-membered N–H···S hydrogen bonding in aromatic amides. Tetrahedron Lett 50(3):320–324
150. Biswal HS, Gloaguen E, Loquais Y, Tardivel B, Mons M (2012) Strength of (NHS)-S-···hydrogen bonds in methionine residues revealed by gas-phase ir/uv spectroscopy. J Phys Chem Lett 3(6):755–759

151. Khedkar JK, Deshmukh MM, Gadre SR, Gejji SP (2012) Hydrogen bond energies and cooperativity in substituted calix[n]arenes ($n = 4, 5$). J Phys Chem A 116(14):3739–3744
152. Kitaura K, Ikeo E, Asada T, Nakano T, Uebayasi M (1999) Fragment molecular orbital method: an approximate computational method for large molecules. Chem Phys Lett 313(3–4):701–706
153. Amos RD (1986) Structures, harmonic frequencies and infrared intensities of the dimers of H_2O and H_2S. Chem Phys 104(1):145–151
154. Gutowsky HS, Emilsson T, Arunan E (1997) Rotational spectra, structure, and internal dynamics of Ar–H_2S isotopomers. J Chem Phys 106(13):5309–5315
155. Arunan E, Emilsson T, Gutowsky HS, Fraser GT, de Oliveira G, Dykstra CE (2002) Rotational spectrum of the weakly bonded C_6H_6–H_2S dimer and comparisons to C_6H_6–H_2O dimer. J Chem Phys 117(21):9766–9776
156. Goswami M, Mandal PK, Ramdass DJ, Arunan E (2004) Rotational spectra and structure of the floppy C_2H_4–H_2S complex: bridging hydrogen bonding and van der Waals interactions. Chem Phys Lett 393(1–3):22–27
157. Hermida-Ramon JM, Cabaleiro-Lago EM, Rodriguez-Otero J (2005) Theoretical characterization of structures and energies of benzene–$(H_2S)_n$ and $(H_2S)_n$ ($n = 1$–4) clusters. J Chem Phys 122(20):204315
158. Goswami M, Arunan E (2011) Microwave spectrum and structure of $C_6H_5CCH\cdots H_2S$ complex. J Mol Spectrosc 268(1–2):147–156
159. Bhattacherjee A, Matsuda Y, Fujii A, Wategaonkar S (2013) The intermolecular S–H$\cdots$Y (Y = S, O) hydrogen bond in the H_2S Dimer and the H_2S-MeOH complex. Chemphyschem 14(5):905–914
160. Goswami M, Neill JL, Muckle M, Pate BH, Arunan E (2013) Microwave, infrared-microwave double resonance, and theoretical studies of $C_2H_4\cdots H_2S$ complex. J Chem Phys 139(10):104303
161. Tursi AJ, Nixon ER (1970) Infrared spectra of matrix-isolated hydrogen sulfide in solid nitrogen. J Chem Phys 53(2):518–521
162. Barnes AJ, Howells JDR (1972) Infra-red cryogenic studies. Part 7 hydrogen sulphide in matrices. J Chem Soc Faraday Trans 68(4):729–736
163. Lechugafossat L, Flaud JM, Camypeyret C, Johns JWC (1984) The spectrum of natural hydrogen sulfide between 2150 and 2950 cm^{-1}. Can J Phys 62(12):1889–1923
164. Helgaker T, Klopper W, Koch H, Noga J (1997) Basis-set convergence of correlated calculations on water. J Chem Phys 106:9639–9646
165. Rocher-Casterline BE, Ch'ng LC, Mollner AK, Reisler H (2011) Communication: determination of the bond dissociation energy (D_0) of the water dimer, $(H_2O)_2$, by velocity map imaging. J Chem Phys 134:211101/211101–211101/211104
166. Adhikari U, Scheiner S (2012) Contributions of various noncovalent bonds to the interaction between an amide and S-containing molecules. Chemphyschem 13:3535–3541
167. Solimannejad M, Scheiner S (2011) Unconventional H-bonds: SH$\cdots$N interaction. Int J Quantum Chem 111:3196–3200
168. Solimannejad M, Gharabaghi M, Scheiner S (2011) SH$\cdots$N and SH$\cdots\pi$ blue-shifting H-bonds and N$\cdots$P interactions in complexes pairing HSN with amines and phosphines. J Chem Phys 134:024312/024311–024312/024316
169. Steiner T, Koellner G (2001) Hydrogen bonds with π-acceptors in proteins: frequencies and role in stabilizing local 3D structures. J Mol Bio 305(3):535–557
170. Ringer AL, Senenko A, Sherrill CD (2007) Models of S/π interactions in protein structures: comparison of the H_2S-benzene complex with PDB data. Protein Sci 16(10):2216–2223
171. Tsuzuki S, Uchimaru T (2006) Magnitude and physical origin of intermolecular interactions of aromatic molecules: recent progress of computational studies. Curr Org Chem 10(7):745–762
172. Vaupel S, Brutschy B, Tarakeshwar P, Kim KS (2006) Characterization of weak NH-π intermolecular interactions of ammonia with various substituted π-systems. J Am Chem Soc 128(16):5416–5426

173. Salonen LM, Ellermann M, Diederich F (2011) Aromatic rings in chemical and biological recognition: energetics and structures. Angew Chem Int Ed 50(21):4808–4842
174. Tauer TP, Derrick ME, Sherrill CD (2005) Estimates of the ab initio limit for sulfur-π interactions: the H_2S-benzene dimer. J Phys Chem A 109(1):191–196
175. Wang YX, Paulus B (2007) A comparative electron correlation treatment in H_2S-benzene dimer with DFT and wavefunction-based ab initio methods. Chem Phys Lett 441(4–6):187–193
176. Biswal HS, Wategaonkar S (2009) Sulfur, not too far behind O, N, and C: S–H$\cdots$π hydrogen bond. J Phys Chem A 113(46):12774–12782
177. Hager J, Wallace SC (1983) Laser spectroscopy and photodynamics of indole and indole-vanderwaals molecules in a supersonic beam. J Phys Chem 87(12):2121–2127
178. Saggu M, Levinson NM, Boxer SG (2011) Direct measurements of electric fields in weak OH$\cdots$π hydrogen bonds. J Am Chem Soc 133(43):17414–17419
179. Saggu M, Levinson NM, Boxer SG (2012) Experimental quantification of electrostatics in X-H$\cdots$π hydrogen bonds. J Am Chem Soc 134(46):18986–18997

Chapter 3
CH··· π Interaction in Organic Molecules

Osamu Takahashi

Abstract CH··· π interaction, which is one of weak non-covalent hydrogen bonds (H-bonds), plays an important role in physics, chemistry, and biology. First, definition of the H-bond is introduced, and a position of the CH··· π H-bond is confirmed. Next, several experimental evidence of the CH··· π H-bond are described. Then, consequence of the CH··· π H-bond in organic chemistry such as conformation, reaction selectivity, and molecular recognition is discussed especially from recent results.

3.1 Introduction

A molecule consists of atoms linked with various kinds of chemical bonds. A covalent bond, the strongest chemical bond, connects atoms in the molecule with shared two or more electrons, while non-covalent bonds are weaker but play a dominant role in determining molecular shape. The non-covalent bond also plays important roles in various chemical phenomena such as structure of supramolecules, chiroptical property, enantiomeric selectivity, etc. Importance of weak hydrogen bonds has been dealt with in monographs written by Scheiner [1] and Nishio et al. [2].

Importance of the non-covalent bond depends not only on its strength but also the directionality. A hydrogen bond (H-bond) is one of the non-covalent bonds, which plays important roles in physics, chemistry, and biology.

In the present chapter, definition of the H-bond is given first and the position of the CH··· π interaction or CH··· π H-bond is confirmed. Next, several experimental evidences for the CH··· π H-bond in organic chemistry and spectroscopy with theoretical calculations are summarized. As evidences in organic chemistry, some topics related to conformation, reaction selectivity, molecular recognition, and chiroptical properties are discussed, especially from recent reports.

O. Takahashi (✉)
Institute for sustainable sciences and development, Hiroshima University, 739-8526 Higahshi-Hiroshima, Japan
e-mail: shu@hiroshima-u.ac.jp

S. Scheiner (ed.), *Noncovalent Forces*, Challenges and Advances in Computational Chemistry and Physics 19, DOI 10.1007/978-3-319-14163-3_3

3.2 What is the CH$\cdots\pi$ Interaction?

Definition of the H-bond in early days has been changed and extended [3, 4]. According to Pauling, the H-bond originates from the Coulombic or dipole-dipole interaction between two polar atoms (X, Y): XH$\cdots$Y. The XH (hydrogen donor) is a hydrogen donor and positively polarized, while Y (hydrogen acceptor) bears lone-pair electrons and negatively polarized.

3.2.1 Definition of the CH$\cdots\pi$ Interaction

A modern and more comprehensive definition of the H-bond has been presented, in 1960, by Pimentel and McClellan [5]. According to this definition, there is no restriction to the nature of X and Y, energy, and directionality of the H-bond. Recently, the IUPAC commission has published a technical report to define the H-bond: *This is an attractive interaction between a hydrogen atom from a molecule or a molecular fragment X–H in which X is more electronegative than H, and an atom or a group of atoms in the same or a different molecule, in which there is evidence of bond formation* [6]. In their report, six criteria and characteristics of H-bonds are listed. A target system in the present chapter, CH$\cdots\pi$ H-bond, is also included in this definition, as the weakest limit of H-bonds.

There are several types of H-bonds. The OH$\cdots$O bond is a typical one and relatively strong. Not only the OH and NH groups but also CH groups are hydrogen donors for XH groups in the second row elements of periodic table of the elements. Further, not only a lone pair provided by O and N atoms but also π electrons are good candidates as hydrogen acceptor. For example, CH$\cdots$O, OH$\cdots\pi$, and NH$\cdots\pi$ H-bonds are weaker than the conventional H-bond; the interaction energy is from ca. 2 to 4 kcal mol^{-1}. The CH$\cdots\pi$ interaction is the weakest H-bond and its interaction energy ranges from ca. 0.5 to 2 kcal mol^{-1}; this is comparable with the thermal energy, however, plays a very important role in organic chemistry and biology. A number of review articles were written and its importance is widely recognized [4, 7–11].

3.2.2 Energy Decomposition Scheme

The interaction energy between groups in H-bonds can be decomposed into several physical components. Several energy decomposition schemes have been reported. The first one is the so-called Kitaura-Morokuma partitioning [12], which was developed within the framework of Hartree-Fock approximation. Following their pioneering work, some alternative schemes were proposed. Reed et al. [13] presented another energy decomposition scheme based on natural bond orbitals (NBOs). Methods based on electron density in a molecule have also been developed. For instance, Bader's "atoms in molecules" (AIM) analysis is applied to partition the electron

Table 3.1 Energy decomposition for various types of hydrogen bonds using SAPT method (in kcal mol^{-1}).

Type of H-bond	Example	E^{a}_{total}	E^{a}_{es}	E^{a}_{er}	E^{a}_{ind}	E^{a}_{disp}	E^{a}_{other}	E_{es}/E_{total}	E_{disp}/E_{total}
OH··· O[b]	$H_2O \cdots H_2O$	− 4.60	− 8.22	8.30	− 1.38	− 2.36	− 0.94	1.79	0.51
OH··· N[b]	$H_2O \cdots NH_3$	− 6.10	− 11.15	11.76	− 2.10	− 3.05	− 1.55	1.83	0.5
OH··· π[b]	$H_2O \cdots C_2H_4$	− 2.57	− 3.82	4.96	− 0.78	− 2.24	− 0.68	1.49	0.87
NH··· O[b]	$NH_3 \cdots H_2O$	− 2.15	− 3.81	4.14	− 0.50	− 1.61	− 0.37	1.77	0.75
NH··· N[b]	$NH_3 \cdots NH_3$	− 2.75	− 5.14	5.84	− 0.79	− 2.05	− 0.60	1.87	0.75
NH··· π[b]	$NH_3 \cdots C_2H_4$	− 1.34	− 1.79	2.55	− 0.23	− 1.59	− 0.28	1.33	1.19
CH··· O[b]	$CH_4 \cdots H_2O$	− 0.45	− 0.89	1.61	− 0.10	− 0.98	− 0.09	2.00	2.18
CH··· N[b]	$CH_4 \cdots NH_3$	− 0.65	− 1.32	2.32	− 0.30	− 1.17	− 0.18	2.03	1.8
CH··· π[b]	$CH_4 \cdots C_2H_4$	− 0.81	− 0.57	1.45	− 0.03	− 1.58	− 0.07	0.71	1.95
OH··· π[c]	$H_2O \cdots C_6H_6$	− 2.86	− 2.94	3.78	− 1.05	− 3.18		1.03	1.11
NH··· π[c]	$NH_3 \cdots C_6H_6$	− 2.08	− 2.05	3.40	− 0.60	− 3.32		0.98	1.60
CH··· π[c]	$CH_4 \cdots C_6H_6$	− 1.27	− 0.86	2.23	− 0.10	− 2.4	− 0.16	0.68	1.89
CH··· π[d]	$C_2H_2 \cdots C_6H_6$	− 2.54	− 2.89	5.46	− 0.55	− 3.75	− 0.72	1.14	1.48

[a] The total interaction (E_{total}), electrostatic (E_{es}), exchange repulsion (E_{er}), induction (E_{ind}), dispersion (E_{disp}), and other effects (E_{other}) energies, respectively
[b] This work
[c] Reference [18]
[d] Reference [19]

density of a many-electron system into basins uniquely [14]. In these schemes, symmetry adapted perturbation theory (SAPT) [15–17] is widely applied to study the intermolecular interaction within the framework of density functional theory (DFT). Some typical H-bonds and their energy components are listed in Table 3.1. The electrostatic energy (E_{es}) is dominant for relatively strong H-bonds such as OH··· O and the dispersion energy (E_{disp}) is minor. On the other hand, E_{disp} is dominant for weak H-bonds such as CH··· π.

3.2.3 AIM Analysis

The AIM analysis is a valuable tool in hands of both experimental and theoretical chemists. Using this analysis, topological properties of electron density in the system can be evaluated [14, 20]. This is an elegant and sophisticated theory of chemical bonding based on the topological analysis of the electron density. The electron density can be measured by either X-ray and neutron diffraction crystallography or computation with the aid of high-level ab initio and DFT calculation. It is possible to decompose and analyze quantitatively the intra- and inter-molecular interactions that

Table 3.2 Topological parameters of some types of H-bonds[a]

Type of H-bonds	Example	ρ_{XH}	$\nabla^2\rho_{XH}$	$\rho_{H...Y}$	$\nabla^2\rho H...Y$
OH···O	$H_2O\cdots H_2O$	0.356	– 2.512	0.023	0.091
OH···N	$H_2O\cdots NH_3$	0.348	– 2.450	0.028	0.085
OH···π	$H_2O\cdots C_2H_2$	0.283	– 1.033	0.014	0.052
CH···π	$C_2H_2\cdots C_2H_2$	0.284	– 1.029	0.007	0.019

[a] Reference [22]

characterize any molecular systems. The aspect of AIM theory is that it redefines the concept of the chemical bond in terms of the topological properties of $\rho(r)$, namely its gradient field, $\nabla\rho$(r), and its curvature or Laplacian, $\nabla^2\rho$(r) [3]. A bond critical point (BCP) is a point along the trajectory of the gradient path connecting two local electron density maxima with $\nabla\rho(\boldsymbol{r})=0$ (nuclei) and lying at the borderline of the two atomic basins involved. The theory shows that the properties of $\nabla^2\rho(\boldsymbol{r})=0$ at this BCP can discriminate among the different types of chemical bonds. Koch and Popelier proposed empirical criteria for XH···Y interactions to establish true H-bonds, based on the AIM analysis [21]. In Table 3.2, typical topological parameters for H-bonds are compiled. All cases are satisfied with the Popelier's criteria.

3.2.4 Cambridge Structural Database (CSD) Analysis

In studying the weak H-bond, a definite evidence is provided by database analysis using the Cambridge Structural Database (CSD) [23]. Various weak H-bonds were studied such as CH···O [24], OH···π, NH···π [25], and CH···π H-bonds [26, 27].

Umezawa et al. [27] searched for short intramolecular and intermolecular CH···π H-bonds contacts. They found that ca. 29 % of compounds in their entries are contacting with short CH···π (aromatic) distance. The corresponding values for OH···π and NH···π interactions are 1.4 and 2.7 %, respectively. Note that these values were estimated with CSD version 515, 181309 entries. If CSD data of other version were used, the results would be changed. However the number of the compound entries is so large, the final conclusions would not be affected even if the recent CSD was used. The same trends were obtained for intermolecular XH···π H-bonds. It is obvious that the CH···π H-bond is quite important to describe the intra- and intermolecular structure in crystal.

As described in the previous section, the H-bond shows directionality, in which the three atoms XH···Y usually tend toward linearity. For the CH···π H-bond, the same trends have been found, computationally and spectroscopically. A minimum of potential energy surface as a function of bending angle of XH···Y showed linear from ab initio calculations although its depth was shallow as already discussed in the section of energy decomposition. However, according to an analysis using the CSD, the directionality and the CH···π-plane distance (D_{pln}) have been found to

Table 3.3 Structural database analysis for CH· · · π interaction[a]

Contact type	D_{pln}/Å[b]	α/degree[c]	α(corrected)/degree[d]
$Cl_3CH\cdots\pi$	2.53 ± 0.17	157 ± 12	169 ± 11
$Cl_2CH\cdots\pi$[e]	2.62 ± 0.15	151 ± 13	159 ± 14
sp-CH· · · π	2.62 ± 0.13	152 ± 13	159 ± 13
sp^2-CH· · · π[f]	2.73 ± 0.11	148 ± 11	154 ± 13
sp^2-CH· · · π[g]	2.70 ± 0.11	146 ± 9	149 ± 11
$CCH_3\cdots\pi$	2.75 ± 0.10	148 ± 13	157 ± 15

[a] Reference [29]
[b] The mean values and the standard deviations of CH· · · π distance
[c] The mean values and the standard deviations of CH· · · π access angle
[d] The mean values and the standard deviations of CH· · · π access angle α corrected by a factor of 1/sinα
[e] Duplicate hits by two (Cl_2CH_2) or three (CCH_3) hydrogen atoms to a single aromatic rung were not counted
[f] Organic crystals with no disorder and $R \leq 5\,\%$
[g] Neutron data including organometallic compounds

correlate, depending on the strength (or acidity) of the proton donor [27, 28]. In Table 3.3, results of a CSD analysis for CH· · · π interactions are shown. Histograms in Fig. 3.1 show the distribution of CH hydrogens as a function of C–H· · · π access angle (α) [29]. Correlation diagrams between D_{pln} and α are given in Fig. 3.2. In crystals, the directionality may not be linear by various limitations. Although energy components of CH· · · π interaction is mostly dispersive, (no directionality), it is obvious that the CH· · · π H-bond has directionality.

3.3 Evidence of CH· · · π H-bond

A number of experimental evidences of CH· · · π H-bond have accumulated by various spectroscopic techniques. In the present section, especially recent results of spectroscopic evidence are reviewed.

3.3.1 Infra-red (IR) Spectroscopy

IR spectra of hydrogen-bonded systems have been studied extensively. It is popular to examine a peak shift of XH stretching mode and to determine the thermodynamic quantities of CH· · · π interaction [30]. Recently, direct observation of intermolecular CH· · · π H-bonds of benzene with acetylene [31], benzene with methane [32], and benzene with ethane [33] was reported by Fujii et al., using infrared-ultraviolet double resonance spectroscopic technique in the gas phase. In Fig. 3.3, observed

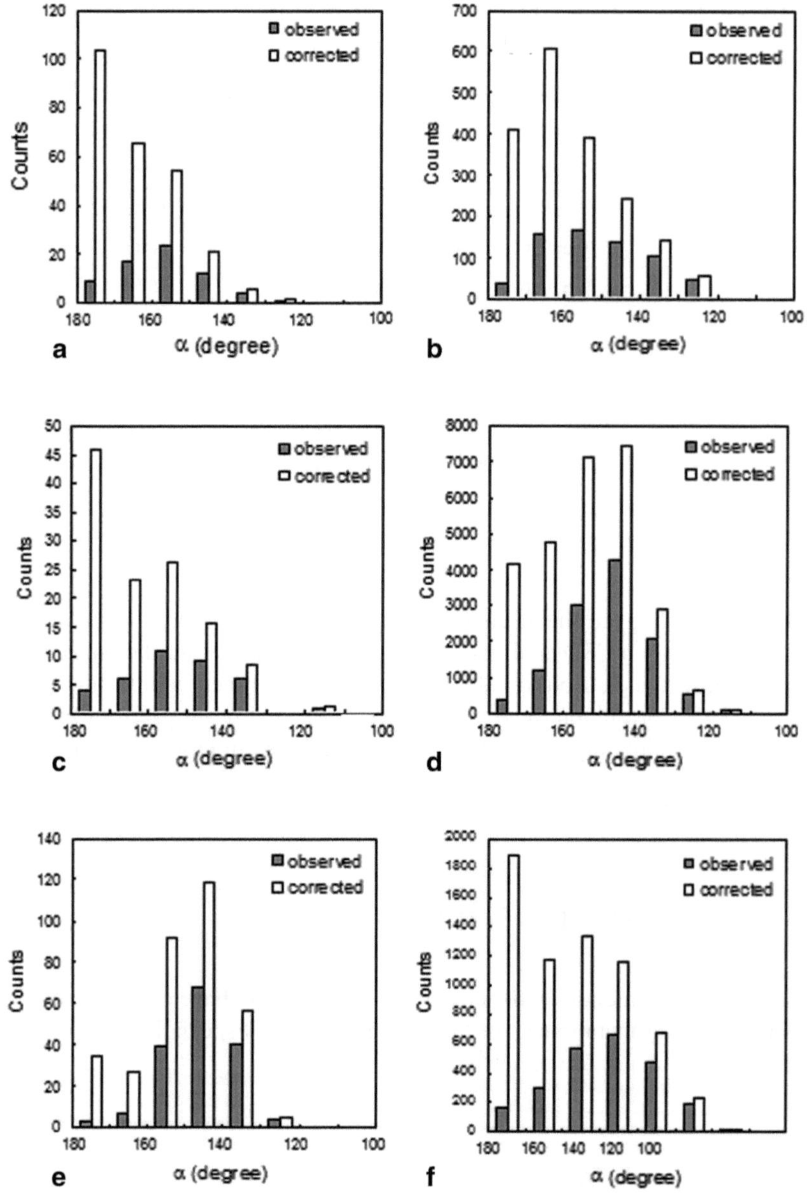

Fig. 3.1 Histograms showing the distribution of CH hydrogen atoms ($D_{pln} < 2.9$ Å) against the distance from the center of the ring (D_{px1}). **a** Cl_3CH, **b** Cl_2CH_2, **c** *sp*-CH, **d** sp^2-CH (aromatic CH), **e** sp^2-CH (aromatic CH, neutron data), **f** CCH_3. The data of only organic crystals with no disorder and $R \leq 5\,\%$ were used for (d) and (f). *Open bars*: observed. *Shaded bars*: corrected by a factor of $1/D_{px1}$. D_{px1}: Horizontal distance of H from the center of the ring. (Reprinted (adapted) with permission from the Chemical Society of Japan. Copyright 2001 the Chemical Society of Japan)

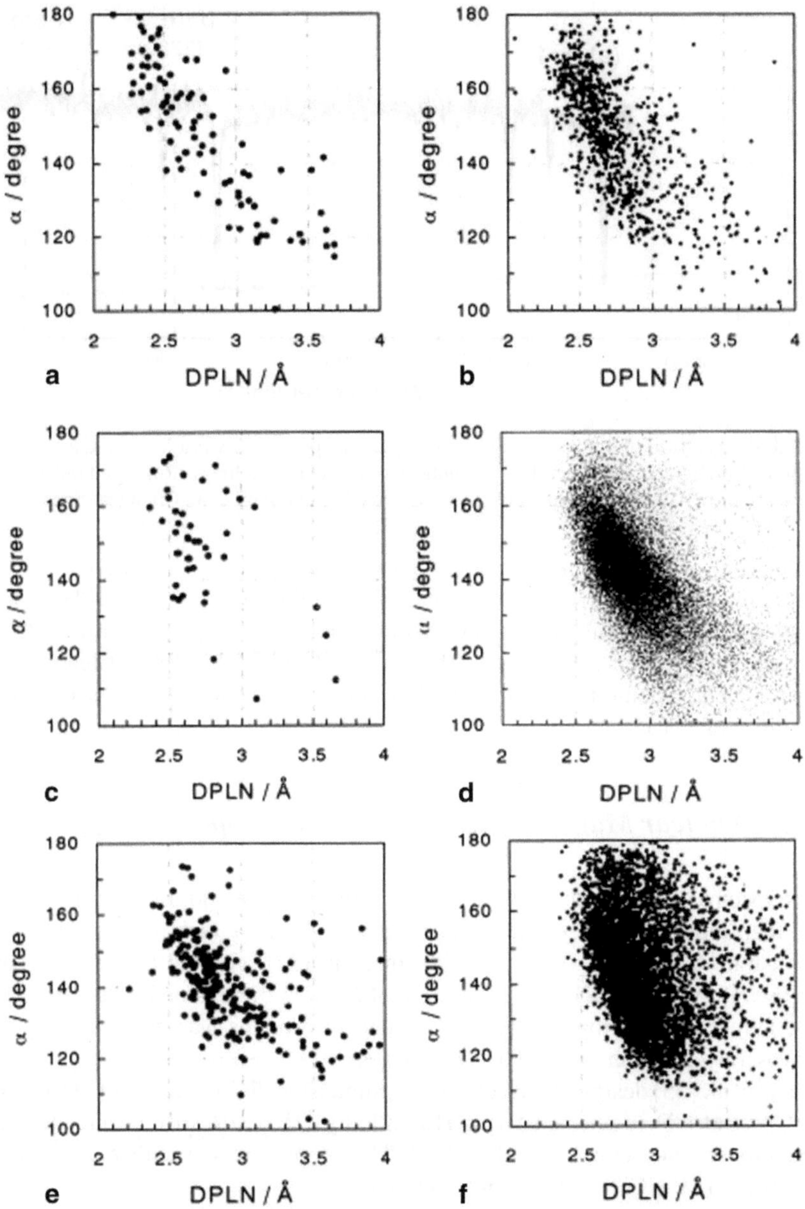

Fig. 3.2 Histograms showing the distribution of CH hydrogen atoms ($D_{px1} < 1.4$ Å, $D_{pln} < 2.9$ Å) against the C–H· · · π access angle (α). **a** Cl_3CH, **b** Cl_2CH_2, **c** *sp*-CH, **d** sp^2-CH (aromatic CH), **e** sp^2-CH (aromatic CH, neutron data), **f** CCH_3. The data of only organic crystals with no disorder and $R \leq 5$ % were used for (d) and (f). *Open bars*: observed. Shaded bars: corrected by a factor of 1/sinα. (Reprinted (adapted) with permission from the Chemical Society of Japan. Copyright 2001 the Chemical Society of Japan)

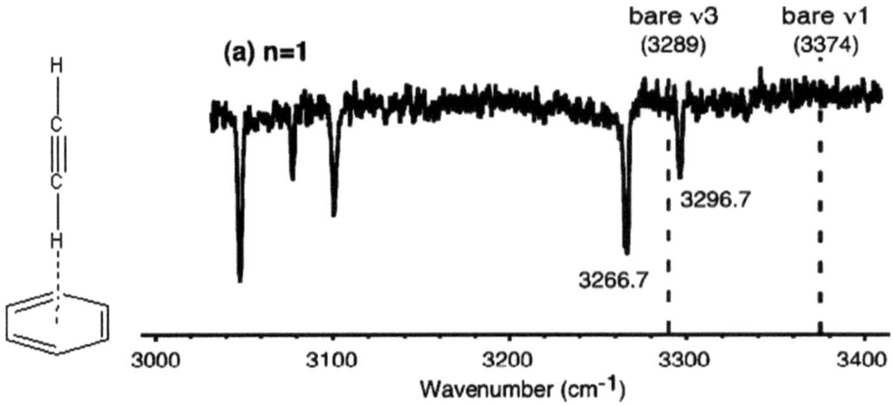

Fig. 3.3 IR spectrum of benzene-acetylene complex. Unperturbed vibrational frequencies of CH stretching vibrations in bare acetylene are indicated by the dashed lines. (Reprinted (adapted) with permission from American Chemical Society. Copyright 2004 American Chemical Society)

IR spectrum of benzene-acetylene complex is shown. The observed large peak of $3266.7\,cm^{-1}$ was assigned to the ν_3 band, which is the anti-symmetric CH stretching vibration mode, and this band shows a remarkable low-frequency shift of $22\,cm^{-1}$ from the unperturbed frequency of the bare molecule. This special feature of the complex clearly demonstrates that the intermolecular bond between acetylene and benzene is of a π H-bond type.

3.3.2 Nuclear Magnetic Resonance (NMR) Spectroscopy

NMR spectroscopy is one of the most powerful tools in studying chemistry; structure, thermodynamic properties, and reactivity. Various techniques were used to demonstrate CH··· π H-bond [30]. Plevin et al. [34] reported that weak H-bonds such as OH, NH, and CH··· π are important in three-dimensional structures of proteins. Direct observation of CH··· π H-bond in proteins was performed by NMR spectroscopy, together with DFT calculations.

Zhao et al. [35] designed a system to investigate the difference of strength between CH··· π and CD (deuterium)··· π H-bonds by ^{1}H NMR spectra and theoretical calculations. They concluded that the D/H isotope effect is either very small or nonexistent in the CH··· π H-bond (Fig. 3.4).

3.3.3 Microwave Spectroscopy

Microwave spectroscopy is used to examine the rotational motion of a molecule, and accurate molecular structures can be determined by analyzing spectral data. Direct microwave spectra of complexes including CH··· π H-bonds have been recorded.

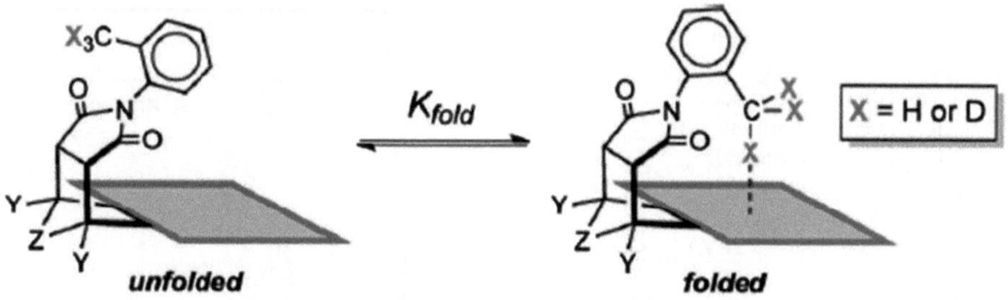

Fig. 3.4 Scheme showing the unfolded, folded conformational equilibrium of the molecular balances, which can be used to measure changes in the strength of the intramolecular CH$\cdots\pi$ H-bond in the folded conformer. (Reprinted (adapted) with permission from American Chemical Society. Copyright 2014 American Chemical Society)

Table 3.4 Structual parameters for $C_6H_6 \ldots C_2H_2^a$

	$r_0(R_{cm})$	r_s	ab initio
R_{cm}/ Å[b]	4.1546	4.1430	4.0387
$R_{CH\ldots\pi}$/Å[c]	2.4921	2.4717	2.3694

[a] Reference [38]
[b] The center of mass distance
[c] Perpendicular distance from H atom to the benzene plane

Tubergen et al. [36] observed rotational spectra of 1-phenyl-2-propanol, methamphetamine, and 1-phenyl-2-propanone. The lowest energy conformations of these species were found to be stabilized by weak OH$\cdots\pi$, NH$\cdots\pi$, and CH$\cdots\pi$ intramolecular H-bonds. The folded structure for 1-phenyl-2-propanol was predicted by ab initio calculations previously [37]. Very recently, Ulrich et al. [38] studied the rotational spectra of benzene-acetylene complex; their results are consistent with intermolecular CH$\cdots\pi$ H-bonds observed by other spectroscopies and theoretical calculations. In Table 3.4, experimentally determined structural parameters are compiled together with theoretical ones.

3.3.4 X-Ray Electron Spectroscopy

Recent progress in synchrotron radiation facility enabled to detect weak H-bonds in molecule using a soft X-ray. Energy of soft X-ray is in the range of 100 eV to a few keV, and 1s electron for most of the second and third row elements can be excited directly. Binding energy of 1s electron calls core-electron binding energy (CEBE), and its value is atom-specific, i.e., 290 eV for carbon and 410 eV for nitrogen, etc. And even if the same atomic spices are included in a molecule, the CEBE is varied with chemical environment of an excited atom. So such properties can be useful for various chemical analysis, and are widely used as electron spectroscopy for chemical analysis (ESCA) [39].

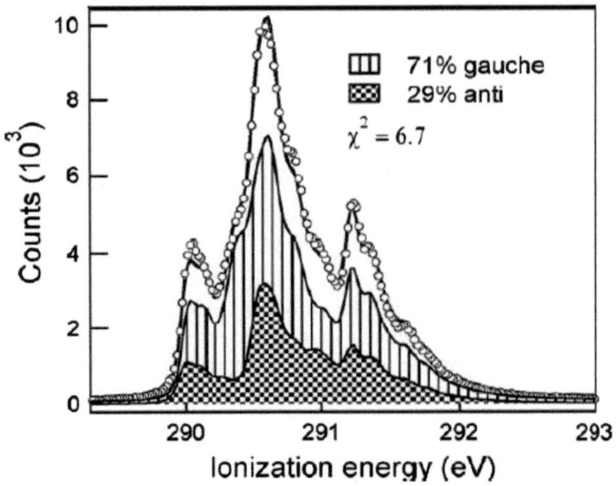

Fig. 3.5 The experimental C1s photoelectron spectrum of 1-pentyne (circles) is shown together with a model that takes into account both anti and gauche confomers. (Reprinted (adapted) with permission from Elsevier B.V. Copyright 2009 Elsevier B.V.)

Holme et al. [40] studied the conformations of 1-pentyne using X-ray photoelectron spectroscopy (XPS) at the C K-edge. They decomposed the observed spectra for anti and gauche conformations together with the theoretical calculations, and their analysis yields 29 ± 3 % anti and 71 ± 3 % gauche conformers. In Fig. 3.5, the experimental XPS spectrum of 1-pentyne and decomposed ones are depicted.

3.4 Consequent Phenomena Under CH$\cdots\pi$ H-bond

As described in the previous sections, although the CH$\cdots\pi$ H-bond is so weak and is comparable to the thermal energy, it plays a critical effect in various phenomena in organic chemistry.

3.4.1 Conformation

Stabilities of conformation are one of key factors to determine the molecular structure. A lot of reviews have appeared relating to this issue [10, 11].

According to textbooks of organic chemistry, discussions regarding to the stability of conformation start from anti-gauche form of *n*-butane, as shown in Fig. 3.6. The C–C bond axis in the center of the molecule can rotate freely, and the anti conformer, in which two methyl groups are located at the opposite sites is known to be more stable than the gauche. This is interpreted generally as the steric repulsion between two methyl groups, i.e., electrostatic repulsion between two bulkier moieties than hydrogen atoms. However, for molecules by substituting one methyl group to another, the anti is not necessarily more stable than the gauche due to the internal H-bond or

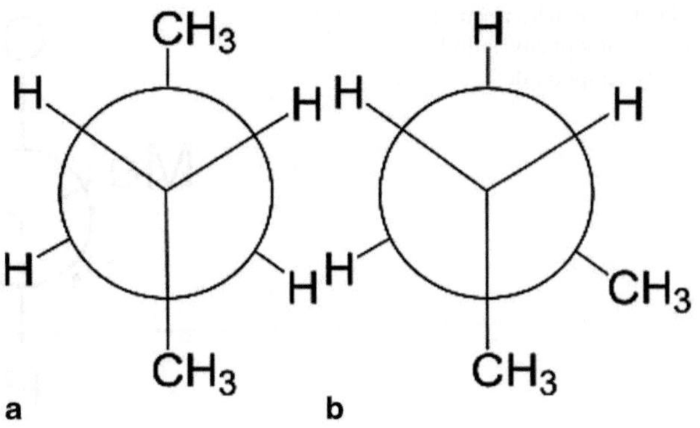

Fig. 3.6 Conformation of *n*-butane. **a** anti and **b** gauche forms

Table 3.5 Difference in the Gibbs energy between the anti- and gauche conformers of CH_3-Y-CH_2-X (X; OH, OCH_3, F, Cl, CN, CCH, and C_6H_5: Y = O, CH_2) at MP2/6-311++G(3df,3pd) with thermal corrections and G3 theory[a]

	Y = 0		Y = CH_2^b	
X	MP2	G3	MP2	G3
OH	2.35	2.20	0.04	0.06
OCH_3	2.86	2.45	1.30	0.11
F	4.02	3.76	0.12	0.12
Cl	4.00	3.71	0.17	0.10
CN	1.68	1.26	0.24	0.14
CCH	0.98	0.94	0.09	0.05
C_6H_5	1.09[c]		0.07[c]	

[a] Reference [43]
[b] This work
[c] MP2/6-311G(d, p) level of approximation

the gauche effect [41]. It is well-known that the gauche for ethylenediamine is more stable than the anti [42]. Takahashi et al. studied the stability of the gauche form for CH_3-Y-CH_2-X system (X = OH, OCH_3, F, Cl, CN, CCH, and C_6H_5; Y = O) using high-level ab initio calculation [43]. Together with the result for Y = CH_2, difference in the Gibbs energy between the anti- and gauche conformers at MP2/6-311++G(3df,3pd) and G3 level of theory are listed in Table 3.5. Note that these combinations include weak H-bonds such as CH···O, CH···N, and also CH···π. The energy difference between the stabilization energy for X = O is larger than that for X = CH_2 because the distances between functional groups become shorter. For these molecules, it is interpreted that an attractive force is operating between functional groups.

Next intramolecular interactions related to the CH···π H-bond will be discussed. Iitaka et al. reported the X-ray structure of 1-(*p*-bromophenyl)ethyl *t*-butyl sulphoxide; there, its *t*-butyl group was found to orient itself gauche to the phenyl group [44]. Later, their group studied the conformation of related compounds, and found that folded conformers are the most favored [[7], Fig. 3.7].

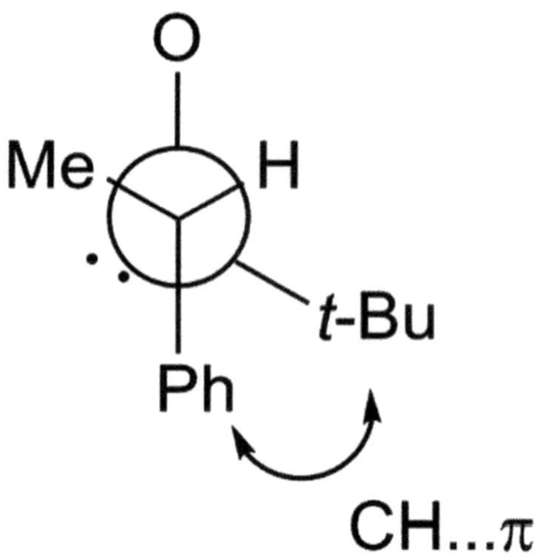

Fig. 3.7 Gauche form of 1-(p-bromophenyl)ethyl t-butyl sulphoxide

They also carried out computational studies using molecular mechanics (MM) techniques for 1-substituted 2-phenylpropanols to reproduce the order of the stable rotamers from Lanthanide-induced shift technique [7]. However, MM results sometimes failed to reproduce the order of the rotamers because of the simplicity of computational model. In the 2000s, Takahashi et al. tried high-level ab initio calculations at MP2 level of approximation for a series of compounds such as Ph-CH_2-X-R and Ph-CHMe-X-R (X = CO, CH_2, O, S, SO, SO_2; R = H, Me, Et, *i*-Pr, *t*-Bu) to find stable rotamers and reproduce the order of them [45–49]. The rotamers with R$\cdots$Ph in gauche relationship are generally more stable than the R$\cdots$Ph anti rotamers. Contrary to the MM calculations, the energy order of rotamers can be predicted more accurately by ab initio calculations with electron correlation. The attractive CH$\cdots\pi$ H-bond has been suggested to be a dominant factor in determining the conformation of this series of compounds. Success of the use of ab initio calculations suggests that the importance of the dispersion force is important, which can be described with the above computational level (Fig. 3.8).

The CH$\cdots\pi$ H-bond operate not only in acyclic molecules but also in cyclic molecules [50–53]. It is well-known that an isopropyl group in unsaturated cyclohexane derivatives prefers to be axial for several terpenic compounds including isomenthone **1** and isocarvomenthone **2**. This unusual preference can be understood by CH$\cdots\pi$ H-bonds operating between CHs in this groups and C = O π-system. Takahashi et al. [50] calculated the conformational energy of these compounds. They found that the most stable conformer has been found to have the axial isopropyl group (Fig. 3.9).

Furthermore, the genesis of stabilization of the axial conformers in 2- and 3-alkyl cyclohexanones, as compared to the structurally corresponding cyclohexane derivatives, was sought in the context of the attractive CH... π(C = O) H-bond.

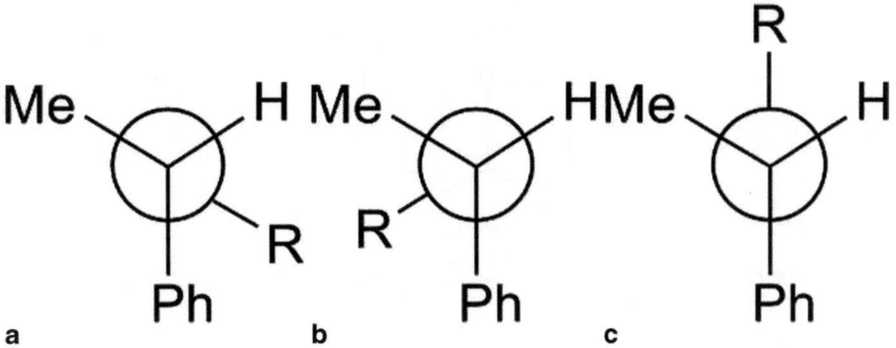

Fig. 3.8 Stable rotamers of Ph-CHMe-X-R (X = CO, CH_2, O, S, SO, SO_2; R = H, Me, Et, *i*-Pr, *t*-Bu)

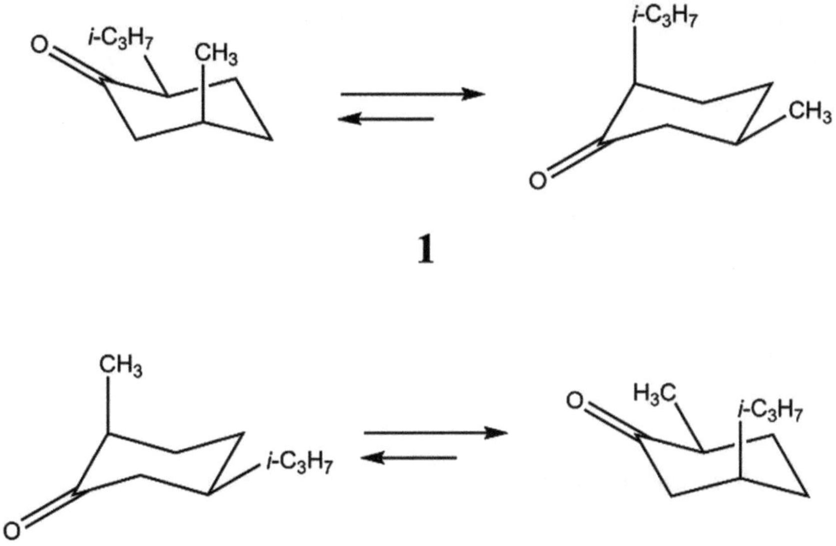

1

2

Fig. 3.9 Conformation of isomenthone *1* and isocarvomenthone *2*

Occasionally the folded conformations are observed in steroidal compounds. Burgstahler et al. [54] reported that levopimaric acid **3** exists in the folded conformation. The reason may be explained by intramolecular CH. . . π H-bond between CH in the 10β angular methyl group and sp^2 carbons of the conjugated diene ring. Takahashi et al. [53] performed ab initio calculations using model compounds of **3** and concluded that the folded conformers were more stable than the extended ones (Fig. 3.10).

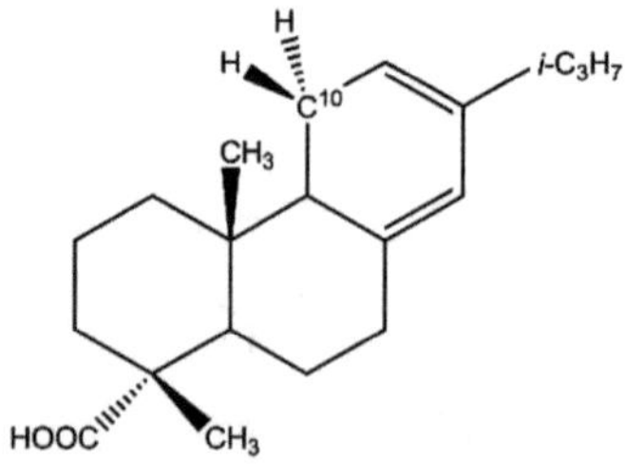

3

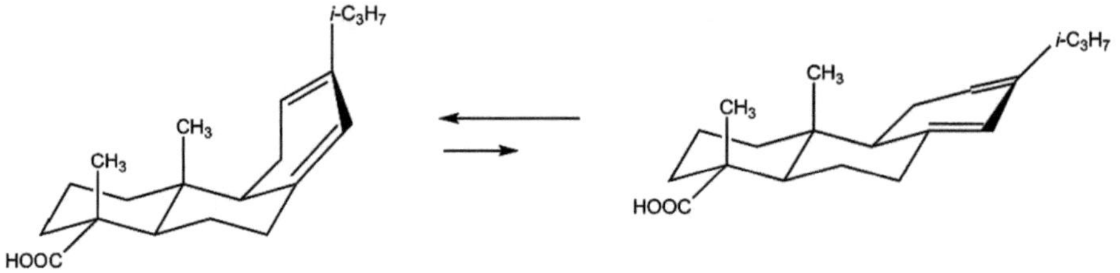

Fig. 3.10 Folded and extended conformations of levopimaric acid *3*

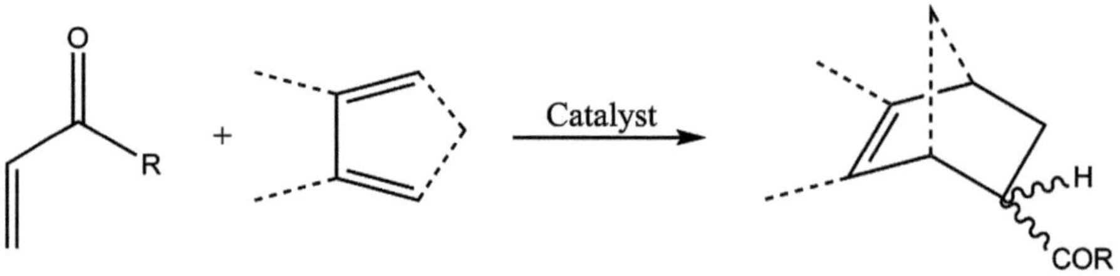

Fig. 3.11 Diels-Alder reactions of vinyl ketones and dienes

3.4.2 Reaction Selectivity

Folded conformer by weak intramolecular H-bond sometimes plays a critical role in chemical reaction, i.e., stereoselective reactivity and chiral recognition. A great number of examples have reported already and some of them were compiled in review articles [9]. In the present, recent results are mainly introduced.

The Diels-Alder reaction is one of the most versatile synthetic transformations for the construction of the cyclohexane framework. Carmona et al. [55] prepared the aqua complexes $[(\eta^5\text{-}C_5Me_5)M(PROPHOS)\text{-}(H_2O)][SbF_6]_2$ (M = Rh, Ir; PROPHOS = diphenylphosphane), and found that these complexes efficiently catalyze the asymmetric Diels-Alder reaction between ketones and dienes. The structure of the intermediate complexes of these reactions indicates that the CH$\cdots\pi$ H-bond is operating between a phenyl group of the PROPHOS ligand and the α-vinyl proton of the ketones (Fig. 3.11).

Allen et al. [56] examined NHC-catalyzed [4+2] cycloaddition between an enolate derived from α,β-unsaturated aldehydes or α-functionalized aldehydes and an

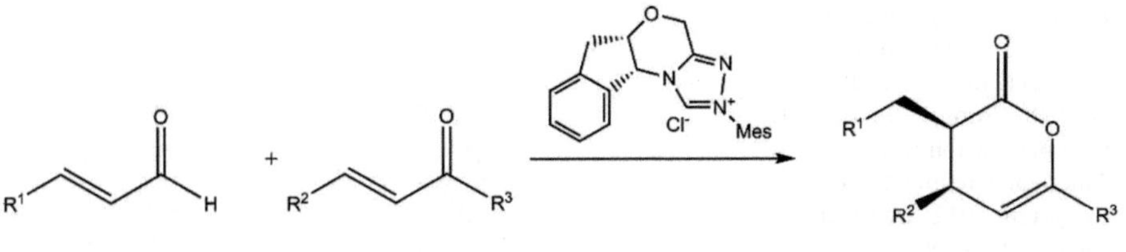

Fig. 3.12 Diels-Alder reactions of α,β-unsaturated aldehydes and an enone as the diene catalized by N-heterocyclic carbenes

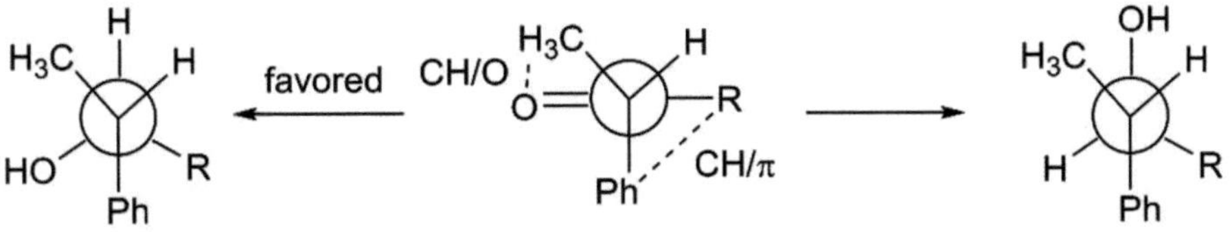

Fig. 3.13 Diastereo-differentiating reactions with $LiAlH_4$

enone as the diene, and these reactions displayed remarkable diastereo- and enantioselectivity, producing γ,δ-unsaturated δ-lactones in up to 99 % ee and greater than 20:1 de, as well as near-quantitative yields. They also examined mechanism using ab initio calculations and concluded that an oxyanion-steering and a CH$\cdots\pi$ H-bond play important roles to determine the high stereoselectivity (Fig. 3.12).

As have been done by Allen et al., computational chemistry can now explain this kind of reaction mechanisms. To investigate the Diels-Alder reaction theoretically, multireference procedure such as CASSCF may be used to reproduce a correct behavior of the reaction, because more than two electrons are contributed to the reaction mechanism [57, 58]. However, target systems including the CH$\cdots\pi$ H-bond are still too large for recent computer resources. Thus most of recent theoretical studies for Diels-Alder reaction often used single reference methodology such as DFT.

As another reaction mechanism, a diastereo-differentiating reactions was investigated. Takahashi et al. [59] calculated the transition states of a model reaction of alkyl 1-phenylethyl ketones with $LiAlH_4$. The transition-state geometries leading to the predominant product are similar to those of the ground-state conformation. In geometries leading to the minor product, the relevant torsion angles are twisted to avoid unfavorable steric interaction. Short CH$\cdots\pi$ and CH$\cdots$O distances suggest that these weak hydrogen bonds are operating in stabilizing the structure of the transition state (Fig. 3.13).

3.4.3 Molecular Recognition

During more than 40 years, a lot of studies related to molecular recognition operating by CH$\cdots\pi$ interaction in liquid and solid phases have been accumulated, and the knowledge contributed to the development of host-guest chemistry. Several inclusion host compounds such as cyclodextrin, calixarene, cyclophane, pseudorotaxane, and

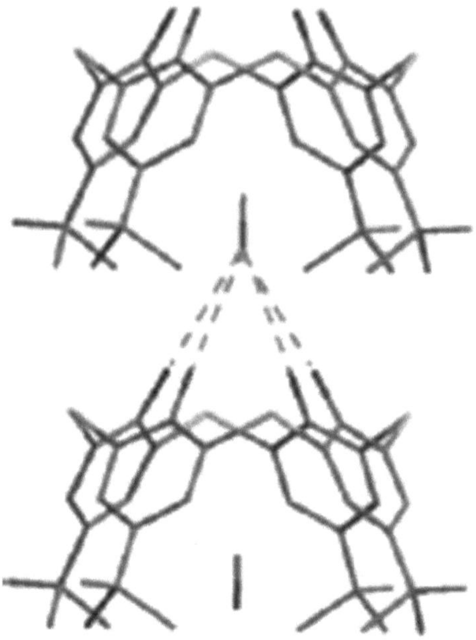

Fig. 3.14 X-ray structures of **4**·MeOH. Dotted lines represents intermolecular H-bond. (Reprinted (adapted) with permission from American Chemical Society. Copyright 2012 American Chemical Society)

catenane have been synthesized. To interpret the driving force of the complex formation and the directionality, weak interactions such as π-π, CH···π, CH···O H-bond are important [8].

In the solid phase, molecular orientation can be easily determined by X-ray and neutron diffraction analysis. The molecular orientation is discussed with a help of computational chemistry.

Morohashi et al. [60] synthesized powdery crystals of *p*-*tert*-butylthiacalix[4] arene **4** which selectively include EtOH from 1:1 mixtures of MeOH-EtOH and Et-PrOH, and EtCOOH from HCOOH-EtCOOH. Through CH···π and other H-bonds, EtOH entered into two neighboring host molecules in a head-to-tail manner to construct an infinite columnar structure along the *c*-axis (Fig. 3.14).

Goel et al. [61] reported CH···π H-bond-driven self-assembly in π-conjugated skeletons based on oligophenylenevinylenes (OPVs) and trace the origin of interactions at the molecular level by using single-crystal structures. OPVs were designed with appropriate pendants in the aromatic core and varied by hydrocarbon or fluorocarbon tails along the molecular axis. Single-crystal structures of hydrocarbon OPVs provided direct evidence for the existence of CH···π interaction (Fig. 3.15).

3.4.4 Chiral Recognition and Chiroptical Properties

Molecules bearing an asymmetric atom are optically active and rotate the plane of polarized light. As a consequence, the absorption or emission intensity by left- or right-handed light becomes different. Now circular dichroism (CD) which observe

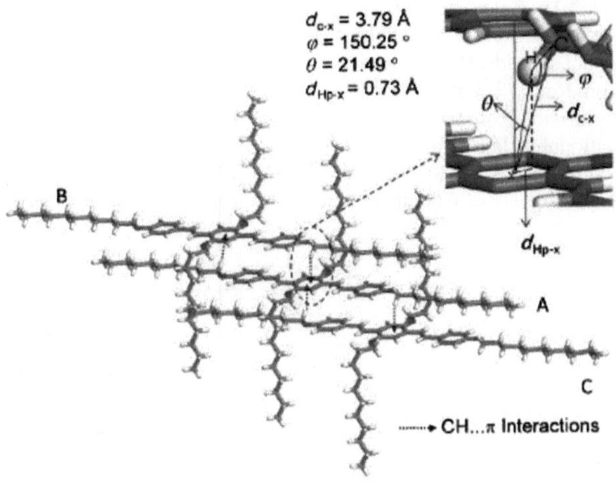

Fig. 3.15 CH· · · π H-bond in an OPV single crystal. (Reprinted (adapted) with permission from American Chemical Society. Copyright 2004 American Chemical Society)

difference of intensity between left- or right-handed light is popular. There has been a long history in the investigation for the relationship between molecular structure and chiroptical property. The theoretical background of chiroptical spectroscopy is described in Ref. [62–64]. Various spectroscopy has been developed, such as electronic CD (ECD) and vibrational CD (VCD), respectively, for the visible/ultraviolet and infrared region [65, 66].

CD spectral properties are useful in determining the absolute configuration of optically active molecules. Several empirical rules were proposed. Moffitt et al. [67] presented an empirical rule that the sign of the rotatory intensity of a $n \rightarrow \pi^*$ transition of a carbonyl chromophore in saturated chiral molecules: the so-called octant rule. Later, Scott et al. [68] proposed another octant rule for the $\pi \rightarrow \pi^*$ transition of olefins. Recently, theoretical calculations by ab initio calculation have been developed, and calculations for relatively large molecules can be done to determine the absolute configuration.

Intramolecular interaction between groups affects optical properties of molecule; this is reflected on the CD spectra. In 1976, Burgstahler et al. [69, 70] presented evidence that the effect of an axial alkyl group allylic to a C = C double-bond was more important than that of the skewness of the diene chromophore (so-called diene helicity rule). Introduction of an axial methyl group to a bridgehead carbon next to the π-group has shown to enhance the CD amplitude of an exomethylene steroids, such as 4- and 6-methylene-5a-estrane at ca. 200 nm. Takahashi et al. [71] investigated the effect of a methyl group on the rotational strength of these steroids by time-dependent density functional theory at the M06-2X/6-311 + + G(d, p)//MP2/6-31G(d, p) level of approximation. Replacement of the bridgehead hydrogen atom of these steroids by a methyl group influenced the CD amplitude at the $\pi \rightarrow \pi^*$ transition. They [72] also demonstrated that the theoretical CD amplitude of 1,3-cyclohexadiene compounds increased by introduction of a methyl group at the bridgehead carbon. In Fig. 3.16, calculated CD amplitudes of three model compounds of tricyclic 1,3-dienes are shown. The result suggests that introduction of a methyl group increased the CD amplitude at around 270 nm; this is fully consistent with the experimental data.

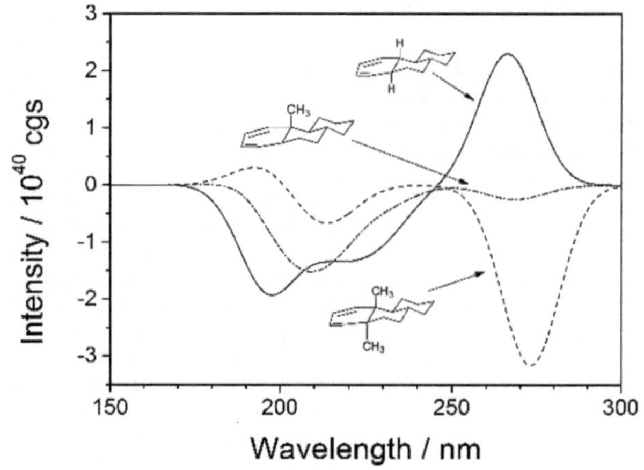

Fig. 3.16 Calculated CD spectra of tricyclic 1,3-diene model steroids. Reproduced from Ref. [72] with permission from the Centre National de la Recherche Scientifique (CNRS) and The Royal Society of Chemistry

3.5 Summary and Perspective

More than 40 years have been passed since the weak H-bond such as CH$\cdots\pi$ interaction was proposed for the first time. Now, this interaction is widely recognized in physics, chemistry, and biology. Weak H-bonds play important roles in various fields of organic chemistry. In the present chapter, several experimental and computational results of the CH$\cdots\pi$ H-bond are introduced. The theoretical results obtained by high-level ab initio and DFT calculations can support the experimental data. Joint studies experiments and MO calculations will increase more and more to realize reaction mechanism and biomolecules, which did not described in the present chapter.

Acknowledgment The author thanks Dr. Motohiro Nishio for his continuous encouragements and reading the manuscript. This work was supported by the Cooperative Research Program of the Network Joint Research Center for Materials and Devices of MEXT, the Asahi Glass Foundation, and a Grant-in-Aid for Scientific Research from JSPS. The author thanks the Information Media Center at Hiroshima University for the use of a grid of high-performance PCs and the Research Center for Computational Science in Okazaki, Japan.

References

1. Scheiner S (1997) Hydrogen bonding. Oxford Universty Press, New York
2. Nishio M, Hirota M, Umezawa Y (1998) The CH/π Interaction. Wiley-VCH, New York
3. Gilli G, Gilli P (2009) The nature of the hydrogen bond: outline of a comprehensive hydrogen bond theory. Oxford University Press, Oxford
4. Nishio M, Umezawa Y, Fantini J, Weiss MS, Chakrabarti P (2014) CH-π hydrogen bonds in biological macromolecules. Phys Chem Chem Phys 16(25):12648–12683. doi:10.1039/c4cp00099d
5. Pimentel GC, McClellan AL (1960) The Hydrogen Bond. Freeman, San Francisco

6. Arunan E, Desiraju GR, Klein RA, Sadlej J, Scheiner S, Alkorta I, Clary DC, Crabtree RH, Dannenberg JJ, Hobza P, Kjaergaard HG, Legon AC, Mennucci B, Nesbitt DJ (2011) Defining the hydrogen bond: An account (IUPAC Technical Report). Pure Appl Chem 83:1619. doi:10.1351/pac-rep-10-01-01
7. Nishio M, Hirota M (1989) CH/π interaction: implications in organic chemistry. Tetrahedron 45(23):7201–7245. doi:10.1016/S0040-4020(01)89185-7
8. Nishio M (2004) CH/π hydrogen bonds in crystals. CrystEngComm 6(27):130–158. doi:10.1039/b313104a
9. Nishio M (2005) CH/π hydrogen bonds in organic reactions. Tetrahedron 61(29):6923–6950. doi:10.1016/j.tet.2005.04.041
10. Takahashi O, Kohno Y, Nishio M (2010) Relevance of weak hydrogen bonds in the conformation of organic compounds and bioconjugates: evidence from recent experimental data and high-level ab Initio MO calculations. Chem Rev 110(10):6049–6076. doi:10.1021/cr100072x
11. Nishio M (2011) The CH/π hydrogen bond in chemistry. Conformation, supramolecules, optical resolution and interactions involving carbohydrates. Phys Chem Chem Phys 13(31):13873–13900. doi:10.1039/C1CP20404A
12. Kitaura K, Morokuma K (1976) A new energy decomposition scheme for molecular interactions within the Hartree-Fock approximation. Int J Quantum Chem 10(2):325–340. doi:10.1002/qua.560100211
13. Reed AE, Curtiss LA, Weinhold F (1988) Intermolecular interactions from a natural bond orbital, donor-acceptor viewpoint. Chem Rev 88(6):899–926. doi:10.1021/cr00088a005
14. Bader RFW (1990) Atoms in molecules: A quantum theory. Oxford University Press, Oxford
15. Jeziorski B, Moszynski R, Szalewicz K (1994) Perturbation theory approach to intermolecular potential energy surfaces of van der waals complexes. Chem Rev 94(7):1887–1930. doi:10.1021/cr00031a008
16. Chen W, Gordon MS (1996) Energy decomposition analyses for many-body interaction and applications to water complexes. J Phys Chem 100(34):14316–14328. doi:10.1021/jp960694r
17. Williams HL, Chabalowski CF (2000) Using Kohn–Sham orbitals in symmetry-adapted perturbation theory to investigate intermolecular interactions. J Phys Chem A 105(3):646–659. doi:10.1021/jp003883p
18. Bloom JWG, Raju RK, Wheeler SE (2012) Physical nature of substituent effects in XH/π interactions. J Chem Theory Comput 8(9):3167–3174. doi:10.1021/ct300520n
19. Mishra BK, Karthikeyan S, Ramanathan V (2012) Tuning the C–H···π interaction by different substitutions in benzene-acetylene complexes. J Chem Theory Comput. doi:10.1021/ct300100h
20. Popelier P (2000) Atoms in molecules: an introduction. Prentice Hall
21. Koch U, Popelier PLA (1995) Characterization of C–H–O hydrogen bonds on the basis of the charge density. J Phys Chem 99(24):9747–9754. doi:10.1021/j100024a016
22. Grabowski SJ (2001) A new measure of hydrogen bonding strength—ab initio and atoms in molecules studies. Chem Phys Lett 338(4–6):361–366. doi:10.1016/S0009-2614(01)00265-2
23. Allen F (2002) The Cambridge structural database: A quarter of a million crystal structures and rising. Acta Crystallographica Sect B 58(3 Part 1):380–388. doi:10.1107/S0108768102003890
24. Taylor R, Kennard O (1982) Crystallographic evidence for the existence of CH···O, CH···N and CH···Cl hydrogen bonds. J Am Chem Soc 104(19):5063–5070. doi:10.1021/ja00383a012
25. Malone JF, Murray CM, Charlton MH, Docherty R, Lavery AJ (1997) X-H···π (phenyl) interactions theoretical and crystallographic observations. J Chem Soc, Faraday Trans 93(19):3429–3436. doi:10.1039/a700669a
26. Umezawa Y, Tsuboyama S, Takahashi H, Uzawa J, Motohiro (1999) CH/π interaction in the conformation of peptides. A database study. Biorg Med Chem 7(9):2021–2026. doi:10.1016/S0968-0896(99)00123-6
27. Umezawa Y, Tsuboyama S, Takahashi H, Uzawa J, Nishio M (1999) CH/π interaction in the conformation of organic compounds. A database study. Tetrahedron 55(33):10047–10056. doi:10.1016/S0040-4020(99)00539-6

28. Takahashi H, Tsuboyama S, Umezawa Y, Honda K, Nishio M (2000) CH/π Interactions as demonstrated in the crystal structure of host/guest compounds. A database study. Tetrahedron 56(34):6185–6191. doi:10.1016/S0040-4020(00)00575-5
29. Takahashi O, Kohno Y, Iwasaki S, Saito K, Iwaoka M, Tomoda S, Umezawa Y, Tsuboyama S, Nishio M (2001) Hydrogen-bond-like nature of the CH/π interaction as evidenced by crystallographic database analyses and ab initio molecular orbital calculations. Bull Chem Soc Jpn 74(12):2421–2430. doi:10.1246/bcsj.74.2421
30. Nishio M, Hirota M, Umezawa Y (1998) The CH/π interaction. Evidence, nature, and consequences. Wiley-VCH, New York
31. Fujii A, Morita S, Miyazaki M, Ebata T, Mikami N (2004) A molecular cluster study on activated ch/π interactions: infrared spectroscopy of aromatic molecule–acetylene clusters. J Phys Chem A 108(14):2652–2658. doi:10.1021/jp049946b
32. Morita S, Fujii A, Mikami N, Tsuzuki S (2006) Origin of the attraction in aliphatic C-H/π interactions: infrared spectroscopic and theoretical characterization of gas-phase clusters of aromatics with methane. J Phys Chem A 110(36):10583–10590. doi:10.1021/jp064297k
33. Fujii A, Hayashi H, Tsuzuki S (2012) Preference of the monodentate contact in the CH/π interaction between an alkyl group and a single phenyl ring: stable structures of benzene–ethane clusters. Chem Phys Lett 537(0):11–15. doi:10.1016/j.cplett.2012.04.005
34. Plevin MJ, Bryce DL, Boisbouvier J (2010) Direct detection of CH/π interactions in proteins. Nat Chem 2(6):466–471. doi:10.1038/nchem.650
35. Zhao C, Parrish RM, Smith MD, Pellechia PJ, Sherrill CD, Shimizu KD (2012) Do deuteriums form stronger CH–π interactions? J Am Chem Soc 134(35):14306–14309. doi:10.1021/ja305788p
36. Tubergen MJ, Lavrich RJ, Plusquellic DF, Suenram RD (2006) Rotational spectra and conformational structures of 1-Phenyl-2-propanol, methamphetamine, and 1-phenyl-2-propanone. J Phys Chem A 110(49):13188–13194. doi:10.1021/jp064810u
37. Takahashi O, Saito K, Kohno Y, Suezawa H, Ishihara S, Nishio M (2004) The conformation of alkyl benzyl alcohols studied by ab initio MO calculations—A comparison with IR and NMR spectroscopic data. Eur J Org Chem (11):2398–2403. doi:10.1002/ejoc.200300801
38. Ulrich NW, Seifert NA, Dorris RE, Peebles RA, Pate BH, Peebles SA (2014) Benzeneacetylene: a structural investigation of the prototypical CH…π interaction. Phys Chem Chem Phys 16(19):8886–8894. doi:10.1039/c4cp00845f
39. Siegbahn K, Nordling C, Fahlman A, Nordberg R, Hamrin K, Hedman J, Johansson G, Bergmark T, Karlsonn S, Lindgren I, Lindberg B (1966) ESCA-atomic, molecular and solid state structure studied by means of electron spectroscopy. North-Holland Publishing Company
40. Holme A, Sæthre LJ, Børve KJ, Thomas TD (2009) Carbon 1s photoelectron spectroscopy of 1-pentyne conformers. J Mol Struct 920(1–3):387–392. doi:10.1016/j.molstruc.2008.11.035
41. Wolfe S (1972) Gauche effect. Stereochemical consequences of adjacent electron pairs and polar bonds. Acc Chem Res 5(3):102–111. doi:10.1021/ar50051a003
42. Kazerouni MR, Hedberg L, Hedberg K (1994) Conformational analysis. 16. Ethylenediamine. An electron-diffraction and ab initio investigation of the molecular structure, conformational composition, anti-gauche energy and entropy differences, and implications for internal hydrogen bonding. J Am Chem Soc 116(12):5279–5284. doi:10.1021/ja00091a036
43. Takahashi O, Yamasaki K, Kohno Y, Ueda K, Suezawa H, Nishio M (2009) The origin of the generalized anomeric effect: possibility of CH/n and CH/π hydrogen bonds. Carbohydr Res 344(10):1225–1229. doi:10.1016/j.carres.2009.04.011
44. Iitaka Y, Kodama Y, Nishihata K, Nishio M (1974) X-Ray structure determination of 1-(p-bromophenyl)ethyl t-butyl sulphoxide. Configuration and conformation. J Chem Soc Chem Commun (10):389–390. doi:10.1039/c39740000389
45. Takahashi O, Yasunaga K, Gondoh Y, Kohno Y, Saito K, Nishio M (2002) The conformation of 2-phenylpropionaldehyde and alkyl 1-phenylethyl ketones as evidenced by ab initio calculations. Relevance of the CH/π and CH/O interactions in stereochemistry. Bull Chem Soc Jpn 75(8):1777–1783. doi:10.1246/bcsj.75.1777

46. Takahashi O, Gondoh Y, Saito K, Kohno Y, Suezawa H, Yoshida T, Ishihara S, Nishio M (2003) The alkyl/phenyl-folded conformation of alkyl 1-phenylethyl sulfides and sulfones as evidenced by ab initio MO calculations. Implication for the 1,2-asymmetric induction. New J Chem 27(11):1639–1643. doi:10.1039/B303268J
47. Takahashi O, Kohno Y, Gondoh Y, Saito K, Nishio M (2003) General preference for alkyl/phenyl folded conformations. Relevance of the CH/π and CH/O interactions to stereochemistry as evidenced by ab initio MO calculations. Bull Chem Soc Jpn 76(2):369–374. doi:10.1246/bcsj.76.369
48. Takahashi O, Kohno Y, Saito K, Nishio M (2003) Prevalence of the alkyl/phenyl-folded conformation in benzylic compounds $C_6H_5CH_2$-X-R (X = O, CH_2, CO, S, SO, SO_2): significance of the CH/π interaction as evidenced by high-level ab initio MO calculations. Chem Eur J 9(3):756–762. doi:10.1002/chem.200390084
49. Takahashi O, Saito K, Kohno Y, Suezawa H, Ishihara S, Nishio M (2003) The conformation of 1-alkyl-2-phenylpropan-1-ols studied by ab initio MO calculations. Relevance of the CH/π and OH/π hydrogen bonds. Bull Chem Soc Jpn 76(11):2167–2173. doi:10.1246/bcsj.76.2167
50. Takahashi O, Yamasaki K, Kohno Y, Kurihara Y, Ueda K, Umezawa Y, Suezawa H, Nishio M (2008) The conformation of alkyl cyclohexanones and terpenic ketones. Interpretation for the 'alkylketone effect' based on the CH/π(CO) hydrogen bond. Tetrahedron 64(10):2433–2440. doi:10.1016/j.tet.2007.12.052
51. Takahashi O, Yamasaki K, Kohno Y, Ueda K, Suezawa H, Nishio M (2008) Origin of the axial-alkyl preference of (R)α-phellandrene and related compounds investigated by high-level ab initio MO calculations. Importance of the CH/π hydrogen bond. Tetrahedron 64(24):5773–5778. doi:10.1016/j.tet.2008.04.007
52. Takahashi O, Yamasaki K, Kohno Y, Ueda K, Suezawa H, Nishio M (2009) The origin of the relative stability of axial conformers of cyclohexane and cyclohexanone derivatives: importance of the CH/n and CH/π hydrogen bonds. Bull Chem Soc Jpn 82(2):272–276. doi:10.1246/bcsj.82.272
53. Takahashi O, Yamasaki K, Kohno Y, Ueda K, Suezawa H, Nishio M (2009) The conformation of levopimaric acid investigated by high-level ab initio MO calculations. Possibility of the CH/π hydrogen bond. Tetrahedron 65(17):3525–3528. doi: 10.1016/j.tet.2009.02.014
54. Burgstahler AW, Ziffer H, Weiss U (1961) The configurations of levopimaric acid and α-phellandrene; interpretation of their rotatory dispersions. J Am Chem Soc 83(22):4660–4661. doi:10.1021/ja01483a039
55. Carmona D, Viguri F, Asenjo A, Lahoz FJ, García-Ordua P, Oro LA (2012) Enantioselective catalytic Diels–Alder reactions with enones as dienophiles. Organometallics 31(12):4551–4557. doi:10.1021/om300346s
56. Allen SE, Mahatthananchai J, Bode JW, Kozlowski MC (2012) Oxyanion steering and CH–π interactions as key elements in an n-heterocyclic carbene-catalyzed [4 + 2] cycloaddition. J Am Chem Soc 134(29):12098–12103. doi:10.1021/ja302761d
57. Li Y, Houk KN (1993) Diels-Alder dimerization of 1,3-butadiene: An ab initio CASSCF study of the concerted and stepwise mechanisms and butadiene-ethylene revisited. J Am Chem Soc 115(16):7478–7485. doi:10.1021/ja00069a055
58. Sakai S (2000) Theoretical analysis of concerted and stepwise mechanisms of diels–alder reaction between butadiene and ethylene. J Phys Chem A 104(5):922–927. doi:10.1021/jp9926894
59. Takahashi O, Saito K, Kohno Y, Suezawa H, Ishihara S, Nishio M (2004) Origin of the diastereofacial selectivity in the nucleophilic addition to chiral acyclic ketones. An ab initio MO study. New J Chem 28(3):355–360. doi:10.1039/B310173H
60. Morohashi N, Noji S, Nakayama H, Kudo Y, Tanaka S, Kabuto C, Hattori T (2011) Unique Inclusion Properties of Crystalline Powder p-tert-Butylthiacalix[4]arene toward Alcohols and Carboxylic Acids. Org Lett 13(13):3292–3295. doi:10.1021/ol200506p
61. Goel M, Jayakannan M (2012) C–H/π-interaction-guided self-assembly in π-conjugated oligomers. Chem Eur J 18(10):2867–2874. doi:10.1002/chem.201102670
62. Schellman JA (1968) Symmetry rules for optical rotation. Acc Chem Res 1(5):144–151. doi:10.1021/ar50005a003

63. Buckingham AD, Stiles PJ (1974) Theory of natural optical activity. Acc Chem Res 7(8):258–264. doi:10.1021/ar50080a004
64. Hansen AE, Bouman TD (1980) Natural chiroptical spectroscopy: theory and computations. In: Adv Chem Phys 44 Wiley, pp 545–644. doi:10.1002/9780470142639.ch5
65. Nicu VP, Autschbach J, Baerends EJ (2009) Enhancement of IR and VCD intensities due to charge transfer. Phys Chem Chem Phys 11(10):1526–1538. doi:10.1039/B816151H
66. Nicu VP, Heshmat M, Baerends EJ (2011) Signatures of counter-ion association and hydrogen bonding in vibrational circular dichroism spectra. Phys Chem Chem Phys 13(19):8811–8825. doi:10.1039/C0CP02701D
67. Moffitt W, Woodward RB, Moscowitz A, Klyne W, Djerassi C (1961) Structure and the optical rotatory dispersion of saturated ketones. J Am Chem Soc 83(19):4013–4018. doi:10.1021/ja01480a015
68. Scott AI, Wrixon AD (1970) A symmetry rule for chiral olefins. Tetrahedron 26(15):3695–3715. doi:10.1016/S0040-4020(01)92948-5
69. Burgstahler AW, Weigel LO, Gawronski JK (1976) Cotton effect behavior of skewed 1,3-cyclohexadienes. Evidence for dominance of homoannular allylic chirality contributions. J Am Chem Soc 98(10):3015–3016. doi:10.1021/ja00426a057
70. Burgstahler AW, Jahansouz H, Véliz EA, Takasugawa F (2002) Diene chirality and cotton effects of nonhomoannular cisoid conjugated dienes: circular dichroism and crystal structure of a steroidal 19-nor-1(10),9(11)-diene derived from westphalen's diol diacetate. Chirality 14(2–3):180–186. doi:10.1002/chir.10067
71. Takahashi O, Yamasaki K, Kohno Y, Ueda K, Nishio M (2012) Importance of the CH/π hydrogen bond in the enhancement of CD amplitude of exomethylene steroids. RSC Advances 2(29):10891–10898. doi:10.1039/C2RA21446F
72. Takahashi O, Kohno Y, Ueda K, Nishio M (2013) Influence of the CH/π hydrogen bond on the enhancement of circular dichroism (CD) amplitude of 1,3-cyclohexadiene compounds. New J Chem 37(3):843–849. doi:10.1039/C2NJ40955K

Chapter 4
The CH··O H-Bond as a Determining Factor in Molecular Structure

Steve Scheiner

Abstract The factors that affect the strength of CH··O hydrogen bonds (HBs) are enumerated, along with the source of their stability, and the structural and spectroscopic features that signal their presence. Their influence upon a number of chemical and biological processes is discussed. Within the context of proteins, the $C^{\alpha}H$ group of protein residues can engage in CH··O HBs, as can the CH groups of a number of amino acid side chains including aromatic residues. CH··O HBs have the potential to make a major contribution to protein folding, particularly in an aqueous environment. These unconventional HBs may play as important a role in the formation of protein β-sheet structures as do NH··O HBs. The strength of CH··O HBs is magnified several-fold by the presence of positive charge on the proton donor. The implications of these principles are discussed for a number of specific biological and chemical problems, which include the catalytic mechanisms of methyltransferases and serine proteases, and the structural properties of fluoroamides.

4.1 Introduction

Although it is widely thought that the first ideas of H-bonds (HBs) were limited to O, N, and F as proton donor or acceptor atoms, the idea that the much less electronegative C atom could also serve as donor appeared in the literature very soon after the concept of a HB was originally proposed. Perhaps the first suggestion [1] was derived from boiling point measurements. A related work [2] was published in 1937, arising from studies of the liquid phase for systems like $CHCl_3$ and ethers, followed 2 years later by a study [3] that used this concept to explain the high dissociation constant of o-toluic and n-butyric acids. Heat of mixing data for a range of molecules by Marvel et al led this group to suggest [4] CH··X bonds in 1940, and to propose that neighboring groups such as Cl, Br, and phenyl could activate the CH donor. The idea was extended [5] to biological systems such as proteins like collagen shortly thereafter. The 1950s yielded multiple confirmatory reports [6–12] that included not

S. Scheiner (✉)
Department of Chemistry & Biochemistry,
Utah State University, 84322-0300 Logan, UT, USA
e-mail: steve.scheiner@usu.edu

S. Scheiner (ed.), *Noncovalent Forces,* Challenges and Advances in Computational Chemistry and Physics 19, DOI 10.1007/978-3-319-14163-3_4

only additional heat of mixing results, but also NMR, IR, and X-ray data. More accurate data of various sorts [13–22] continued to buttress the idea in the 1960s, and greatly extended the range of systems in which they were observed to include collagen and polyglycine among others.

The data emanating from the experimental work was highly suggestive of the formation of CH··O HBs. X-ray diffraction data, for example, had documented the fairly close approach of the C and O atoms. Unexpectedly high heats of mixing pointed to stronger intermolecular attractions than might have been anticipated without invoking CH··O HBs. Shifts of the bridging proton's NMR chemical shift were consistent with what was seen in conventional HBs, as were the changes in the CH stretching frequency. But this evidence was inconclusive in that alternate explanations could be offered. The proximity of the C and O atoms could be the result of crystal packing forces, and not the product of a genuine attractive force, particularly as the position of the bridging H was left unresolved by the refinement of the diffraction data. The change in the CH stretching frequency might be dictated instead by intermolecular geometry [11] or perhaps steric crowding [23].

Quantum chemical methods are well suited to fill this gap, and to definitively answer the question as to whether a CH··O interaction is an attractive one, and to quantitatively assess its strength. Prototype systems can be designed in which a pair of molecules is held together solely by the interaction of a CH donor and O acceptor, eliminating any ambiguities associated with larger systems with multiple points of contact. The interaction energy can be dissected into its constituent components, which can then be compared to those of conventional HBs of the OH··O variety. The effects of the interaction upon the NMR chemical shielding or the C–H stretching frequency can be analyzed so as to determine whether they are due to a HB or to some other phenomenon. Even small changes in molecular geometry that are signatures of HBs can be determined by these calculations, in many cases too small to be observed experimentally. And the wave function derived from these calculations is amenable to analysis in a number of ways, which can directly answer the question as to whether or not a HB is present.

4.2 Early Calculations

Due to the generally weak nature of CH··O HBs, it was not until the 1970s that quantum chemical techniques had matured to the point where it was reasonable to take up this question. Even so, many of the earliest calculations [24–30] were limited by the constraints at the time to semiempirical methods, small basis sets, or failure to include electron correlation. Despite these limitations, the calculations supported the idea of CH··O as an attractive force, even if the computational level was inadequate for quantitative purposes. The coming of age of some of the newer quantum chemistry codes, along with improved computers, catalyzed a blossoming of work in this area in the 1980s. Relevant publications since that time are far too numerous to describe in any detail here, so only a representative sample will be summarized.

As a strong acid, it was no surprise that HCN could serve as a potent proton donor [31–36], as could the triply bonded C in HC≡CH [37–40]. The sp^2-hybridized C of the vinyl [41] or phenyl group [42–45] could also donate a proton, as could other aromatic systems such as imidazole [46], pyrimidine [47] or nucleic acid bases [48–50]. Other work concluded that the very simple CH_4 could engage in a HB [51–61], albeit a weak one. An early study of chlorosubstituted methanes [62] indicated that the CH··O HB with water would be progressively strengthened by each exchange of H with Cl. Other electron-withdrawing substituents like F, NO_2, etc have a similar strengthening effect [63–70].

4.3 Recent Work

It was at this point in time that this group became involved in this issue. Some of our earlier work up through 2006 has been summarized already [71–73] so will only be briefly outlined here, with emphasis on major findings. Beginning with the simplest system which might contain a CH··O HB, the binding energy between CH_4 and OH_2, is very weak [74], on the order of 0.5 kcal/mol or less. But there is a strong substituent effect in that electron withdrawing agents on the proton donor molecule polarize the CH bond, making it a more potent donor. In numerical terms with each replacement of a H atom on CH_4 by F, the HB energy climbs by about 1 kcal/mol, to the point where the $F_3CH{\cdot\cdot}OH_2$ HB is very nearly as strong as that in the water dimer. As typical of HBs, the progressive strengthening is also accompanied by a shortening; in this case R(H··O) contracts by about 0.14 Å with each F substitution. Comparison of a number of computational protocols showed [74] that the trends are largely independent of the level of theory, and even the quantitative data show an insensitivity to basis set. These findings were later confirmed by others [75, 76] for very similar systems. This idea of inductive effects strengthening a CH··O HB is not limited to F, but has been demonstrated for a range of other electron-withdrawing substituents like Cl [70, 77–79].

Unlike conventional HBs such as OH··N where the OH bond is stretched upon formation of the complex, and its stretching frequency shifted to the red, there were some indications that CH··O and CH··N HBs commonly behave in a contrary fashion [80–86]. The work [74, 87] addressed the matter of C–H bond contraction and blue shift, specifically to determine whether such behavior might exclude this interaction from its HB classification. It was determined first that in all other respects, the CH··O interaction fits into this rubric. Electron density shifts are characteristic of typical HBs, and the NMR signal of the bridging proton shifts downfield [88] by an amount proportional to the HB strength. Also in line with HB expectations are the components of the interaction energy arising from an energy decomposition. In fact, this sort of dissection provided some insight into the nature of this contrary behavior. All HBs, whether conventional or CH··O, exhibit a large electrostatic attraction, which is complemented by smaller polarization, charge transfer, and dispersion terms, all balanced by a repulsive exchange energy. And in all cases, a small stretch of the

CH/OH covalent bond magnifies all of these factors. The difference resides only in the *amount* of this magnification. The combined magnification of the attractive components outweighs the greater repulsion energy when the OH bond is stretched, albeit by only a small amount, so this bond is stretched when the OH··O bond is formed. The opposite is true for CH··O where the enhanced repulsion overwhelms the magnified attractive term, forcing the CH bond to contract. This central idea, that the direction of shift of the CH stretching frequency resides in a fine balance between two sets of opposing forces, has been confirmed by other research groups, each using a different definition of these forces [78, 89–114].

As one might expect, O is not the only atom that can accept a CH proton. Calculations [80, 87, 115–122] have shown that N serves an equivalent purpose. In fact, due its greater basicity, CH··N HBs tend to be stronger than CH··O [123]. Like their CH··O counterparts, the NMR signal of the bridging proton is also deshielded in CH··N HB [41]. Despite the generic weakness of CH as a proton donor, computations have shown that it is strong enough to engage F as an acceptor [79, 103, 124–131]. Other proton-accepting atoms are S [100, 132–136], even as strong as 7 kcal/mol [137, 138], Cl [139–142] or I [143]. Even an atom as weakly basic as P [144, 145] can serve this function as can carbenes [146]. And there is an extensive literature concerning CH··π HBs, where the electron donor is an aromatic π system [147–152]. Such CH··π HBs can serve to guide structural changes as in N-heterocyclic carbene palladium complexes [153].

Another major factor which influences the direction of C–H stretching frequency shift is the hybridization of the C atom. Calculations [87] show that whereas the sp^3 CH of alkanes commonly (but not universally) leads to a blue shift, the opposite result of a red shift occurs for the sp-hybridized alkynes, which generally form stronger CH··O HBs. Given their intermediate sp^2 hybridization, and HB strengths intermediate between alkanes and alkynes, it was not surprising to learn that alkenes manifest only very small shifts, sometimes to the red and other times to the blue. But again, whether alkane, alkene, or alkyne, all properties of these CH··O HBs, e.g. NMR chemical shifts or electron density shifts, mimic those of conventional HBs such as OH··O. Comparable trends were noted later with other systems [111, 132, 154], some of which contained the analogous CH··N HBs [155].

HBs are well known for their ability to reinforce one another. That is, a string of n H-bonded molecules is commonly bound together by a force which exceeds that occurring within $n - 1$ simple dimers, in what is frequently referred to as cooperativity. A detailed examination [156] revealed that this same phenomenon is common to CH··O HBs as well. This cooperativity is observed not only in the energetics, but also in the HB lengths and in the electron density shifts resulting from the HB formation. On the other hand, there is little evidence of cooperativity within the context of CH covalent bond contraction or CH stretching frequency shifts. Whether conventional or CH··O, the cooperativity in either sort of HB is reduced when the system is immersed in a model solvent. The cooperative aspects of CH··O HBs have been the subject of some inquiry by others as well [157–162]. The overriding conclusion is that these HBs act much like any others in the sense that synergistic, positive cooperativity will occur if a central molecule acts simultaneously as both electron

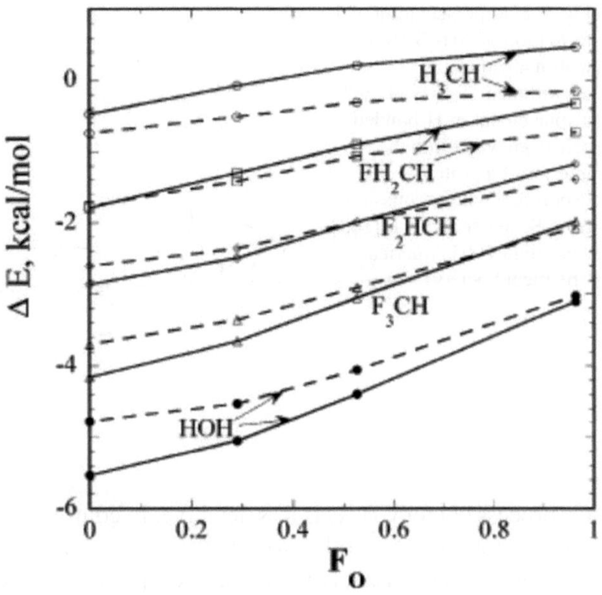

Fig. 4.1 Interaction energies of each indicated proton donor molecule with H_2O as proton acceptor. Horizontal axis $F_o = (\varepsilon - 1)/(\varepsilon + 2)$ measures polarizability of surrounding medium, where ε is dielectric constant. B3LYP/6-31 + G** and MP2/aug-cc-pVTZ results indicated by solid and broken lines, respectively. (Reprinted with permission from Scheiner and Kar [163]. Copyright 2005 American Chemical Society)

donor and acceptor, while double donor or acceptor activity will result in an overall weakening or negative cooperativity.

4.4 Biological Systems

Turning once more to the simple fluoromethanes as model proton donor, consideration of how its CH··O HBs are affected by solvent provides some insight into the situation within a large biomolecule such as a protein. Figure 4.1 shows how the binding energies of any of the F_nCH_{4-n} molecules with water are progressively weakened as the polarizability of the surrounding medium is increased [163], and the same is true of the conventional OH··O HB in the water dimer.

What are the implications of this observation for protein folding? The contribution of any HB to folding may be thought of as the process that begins with the two subunits disengaged in an unfolded protein, and thus both exposed to aqueous solvent. They then approach one another and form a connecting HB, but now within the confines of a protein, i.e. in a less polarizable medium than within water. In other words, the impact of the HB is a combination of two conceptually separate processes. As illustrated schematically in Fig. 4.2, the two subunits (1) form a HB within water, and then (2) the H-bonded pair is removed from water and placed instead in a protein interior. The exothermicity of the former process (1) is countered in part by the endothermicity of the latter (2), sometimes referred to as a desolvation penalty.

How does this scenario play out for CH··O vs OH··O HBs? Whether *in vacuo* or in solution, the CH··O HB is weaker [163] than is the conventional OH··O, so ΔE

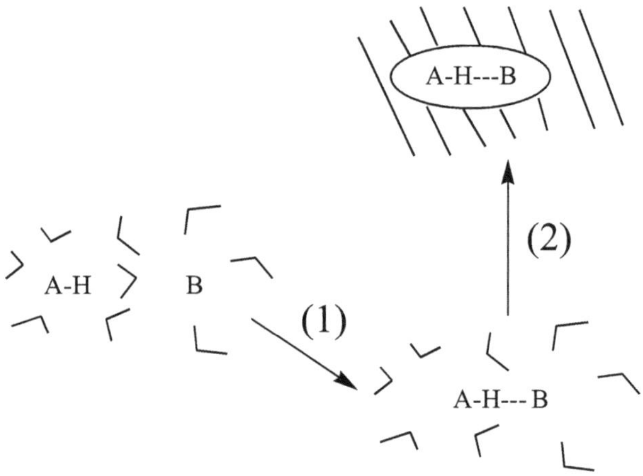

Fig. 4.2 Representation of *1* formation of AH··B HB within an aqueous environment, followed by *2* displacement of H-bonded pair from water into the interior of a protein. (Reprinted with permission from Scheiner and Kar [163]. Copyright 2005 American Chemical Society)

for process (1) for CH··O is less negative. In contrast, however, ΔE of step (2) is less positive for CH··O. In other words, it takes less energy to move a CH··O bond from water to a less polarizable medium, like a protein interior, than it does to desolvate a OH··O HB. The net result is that it is more favorable for a CH··O HB to participate in the protein folding process than for OH··O.

From a quantitative perspective, the binding energy of $F_3CH + OH_2$ is 3.7 kcal/mol in vacuum, and diminished to 2.1 kcal in water; an intermediate value of 2.9 kcal/mol occurs in a protein interior, modeled with dielectric constant of 4. The corresponding values for the OH··O HB in the water dimer are all larger: 5.5, 3.1, and 4.1 kcal/mol. The desolvation penalty of the CH··O HB, viz. the energy required to take this complex from water to protein is 3.3 kcal/mol, but this quantity is higher for the OH··O HB, 5.4 kcal/mol. When these pieces are all assembled, the contribution of the CH··O HB is 1.0 kcal/mol greater than that for OH··O.

Of course, a continuum homogeneous polarizable medium is only a rough approximation of the interior of a protein. In order to bring the model one step closer to the real situation, a number of discrete water molecules were placed [163] around the H-bonded systems, forming a first solvation sphere. HB energies in this primitive solvated system were rather close to the same quantities computed with the dielectric continuum model. A second issue with the latter solvation model is the choice as to what value of dielectric constant most correctly models a protein interior. It is not uncommon in the literature for a value of $\varepsilon = 4$ to be considered to simulate a generic protein interior but of course each protein is different, and even within a single protein, some domains will be more polarizable than others. But in any case, the conclusion that a CH··O HB can be as much of a contributor to protein folding as OH··O has a certain degree of experimental support [164].

4.4.1 Amino Acids

$F_3CH + OH_2$ is only the roughest of models of a CH··O HB within a protein, even if it does provide some fundamental insights. A first effort to study a more realistic model donor started [165] with a set of amino acids, $NH_2C^{\alpha}HRCOOH$, all of which contain a $C^{\alpha}H$ group. Gly, Ala, Val, Ser, and Cys were all paired with a water molecule as they engaged in a $C^{\alpha}H{\cdot\cdot}OH_2$ HB. There was very little sensitivity to the nature of the sidechain R, with binding energies all in the 1.9–2.5 kcal/mol range. Also rather constant was the $R(C^{\alpha}{\cdot\cdot}O)$ distance, 3.31–3.35 Å, and the NMR downfield chemical shift of the $C^{\alpha}H$ proton which varied between − 1.35 and − 1.71 ppm. There was a bit more sensitivity in the contraction of the C–H bond, from 0.3 to 3.1 mÅ; all of the C–H stretching frequencies were to the blue, in the 14–56 cm^{-1} range, so this pattern fits the idea of blue stretches accompanying sp^3 hybridization

In order to expand the scope to charged amino acids, Lys^+ was modeled [165] by the R sidechain of $(CH_2)_4NH_3{}^+$ and Asp^- by CH_2COO^-. As one might expect, the presence of a positive charge, even one that is removed from the $C^{\alpha}H$ by a hydrocarbon chain, enhanced the binding energy with OH_2 up to 4.9 kcal/mol. But perhaps surprisingly, this stronger bond did not result in much change in any of the other parameters: $R(C^{\alpha}{\cdot\cdot}O)$, $\Delta r(C^{\alpha}H)$, and $\Delta\sigma_H$ were in line with the values obtained for the neutral amino acids. Likewise for the anionic Asp^-, even though its binding to OH_2 is much weaker.

4.4.2 Dipeptides

Within the context of the amino acid model, the $C^{\alpha}H$ is surrounded by a $-NH_2$ on one side and –COOH on the other. An expansion of each to a full peptide group leads to a glycyl dipeptide $CHONHC^{\alpha}H_2CONH_2$ that better represents the setting of this central group within a protein. The ability of this $C^{\alpha}H$ group to participate in a HB was tested [166], this time using the carbonyl O of formamide H_2NCHO as a more representative proton acceptor within a protein.

The dipeptide model introduces a good deal of flexibility into the donor molecule. This flexibility is represented primarily by the dihedral angles φ and ψ, that are commonly used to denote the rotation of the two peptide groups around the C–C^{α} and C^{α}–N bonds. This work [166] centered around the two conformations of dipeptides that represent minima on their potential energy surface. The C7 minimum derives its name from the presence of a seven-membered ring that contains an intramolecular NH··O HB, while a five-membered ring occurs in the C5 structure. Note that neither internal HB directly involves the $C^{\alpha}H_2$ group which is available to participate in a CH··O HB with the neighboring formamide.

The interaction energy of a conventional NH··O HB with the formamide carbonyl O atom of the C7 dipeptide is 7.5 kcal/mol, considerably larger than the 2.3 kcal/mol of the $C^{\alpha}H{\cdot\cdot}O$ HB. In the case of the C5 structure, however, the NH··O HB is cut by a factor of three to only 2.5 kcal/mol, whereas the $C^{\alpha}H{\cdot\cdot}O$ HB is slightly larger, at 3.8 kcal/mol. In other words, the CH··O HB is stronger than the NH··O HB for the C5 geometry of the dipeptide. This dipeptide structure with (φ,ψ) = (180°,180°) represents an extended structure of a polypeptide backbone, so is certainly an important segment of the Ramachandran (φ,ψ) region, not far from the β-sheet geometry (more about the β-sheet below).

What can account for this surprising and remarkable sensitivity of the NH··O HB to the conformation of the dipeptide? The threefold reduction of this HB strength is particularly puzzling as the geometry of this bond, including R(NH··O), is nearly the same in the C5 and C7 conformers. A detailed inquiry [167] expanded the (φ,ψ) conformational space of the dipeptide to cover the entire Ramachandran map, not just the C5 and C7 areas, as shown in Fig. 4.3. It was found that the NH··O HB energy is nearly uniform over the majority of the (φ,ψ) map, as indicated by the red and orange sections of Fig. 4.3, but becomes progressively weaker as one moves toward the fully extended (−180°,180°) structure in the upper left corner, at which point the HB energy nearly vanishes entirely. The region of weakened NH··O HB extends over a fairly wide area, which may be categorized as $-180° < \varphi < -100°$, and $100° < \psi < 180°$.

Careful scrutiny of the data [167] pointed to one particular feature as the prime culprit. In the extended structure of the dipeptide, Fig. 4.4a shows that the proton donor NH is close to the carbonyl O of the neighboring peptide unit. (It is in fact this proximity which leads to the common characterization of this structure as C5.) A negative region of electrostatic potential emanates from this carbonyl O atom, which acts as a shield of sorts, pushing an approaching proton acceptor away from the NH group as indicated by the red double arrow in Fig. 4.4a, and thereby weakening the incipient NH··O HB. As the dipeptide curls away from the (−180°,− 180°) extended structure, the neighboring carbonyl O moves away from the NH, leaving this group exposed to the approaching proton acceptor group, and allowing the NH··O HB to achieve its normal potential, as for example in Fig. 4.4b where (φ,ψ) = (− 80°, + 80°).

An important conclusion arising from this work is that it is incorrect to consider the strength of any particular HB, NH··O or otherwise, as a given or constant. The actual binding energy can be heavily influenced by the conformation adopted by the protein, particularly if the structure places the proton donor group in the vicinity of another group with a strong electrostatic potential.

Of course, the peptide units that interact with one another within proteins are not necessarily adjacent along the polypeptide backbone. The folding of the protein brings peptide groups from quite different segments of the backbone into close coincidence, so it is important to consider a fuller range of interpeptide geometries than the restricted low-energy sections of the (φ,ψ) space. A full search of the potential energy surface of a pair of $CH_3NHCOCH_3$ (NMA) molecules was thus undertaken [168] so as to identify any minimum-energy structures, free of the restrictions imparted by a connecting unit.

The primary minimum pictured in Fig. 4.5a unsurprisingly contained a strong NH··O HB [168]. This bond was able to achieve full strength since there was no peptide adjacent to the NH group whose carbonyl O could impede the approach of the carbonyl from the other NMA molecule. The less expected result concerned a second

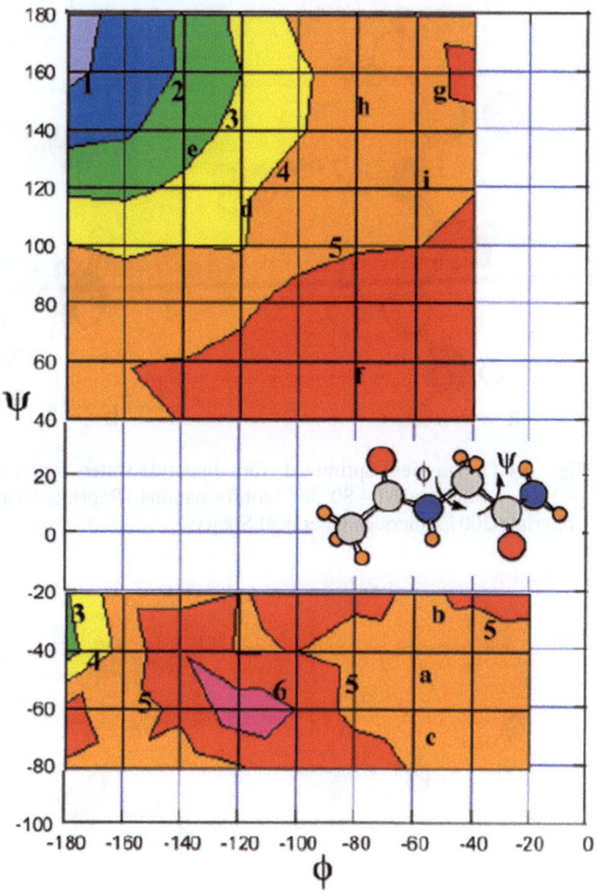

Fig. 4.3 Binding energy of water to $CH_3CONHCH_2CONH_2$, as a function of internal dihedral angles φ and ψ of glycyl dipeptide. Numerical labels indicate binding energy (kcal/mol); Letters indicate standard locations of **a** α-helix, **b** 3_{10}-helix, **c** π-helix, **d** parallel β-sheet, **e** antiparallel β-sheet, **f** 2.27 ribbon, **g** collagen triple helix, **h** PPII, and **i** type II β-bend. (Reprinted with permission from Scheiner [167]. Copyright 2007 American Chemical Society)

pair of minima, almost as stable as the first. In the first of these structures, Fig. 4.5b, the planes of the two NMA molecules lie parallel to one another. Quite similar in energy is Fig. 4.5c which contains an anti-parallel, but still stacked, geometry. Analysis of the binding forces led to two primary components. First, there is a π → π* transfer of charge from the π(CO) orbital of one molecule to the π*(CO) antibonding orbital of the other amide, and vice versa. This attraction is augmented by CH··O HBs between the methyl H atoms of one molecule and the carbonyl O of the other. It is noted that this attractive force would be strengthened in a protein where the $–CH_3$ group would be replaced by a $–CH_2$-methylene, surrounded on both sides by electron-withdrawing amide units. And further, that the directions of the two NMA molecules are immaterial: both parallel and antiparallel geometries are equally stable.

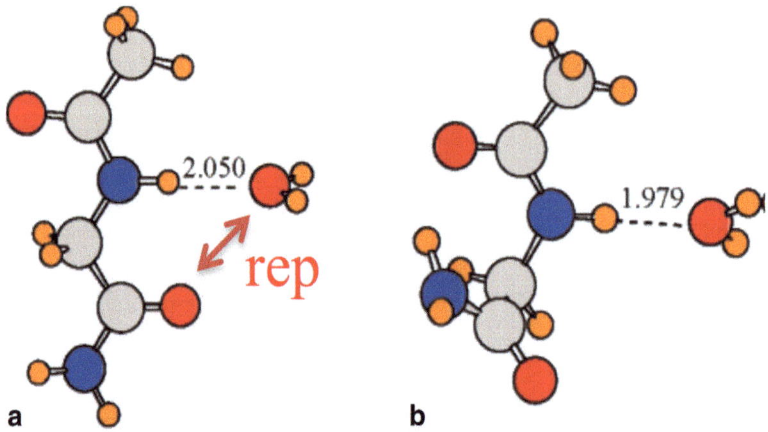

Fig. 4.4 Geometries optimized for dipeptide-water system, with dipeptide in **a** (φ,ψ)) ($-180°,-180°$) and **b** ($-80°,80°$) conformations. (Reprinted with permission from Scheiner [167]. Copyright 2007 American Chemical Society)

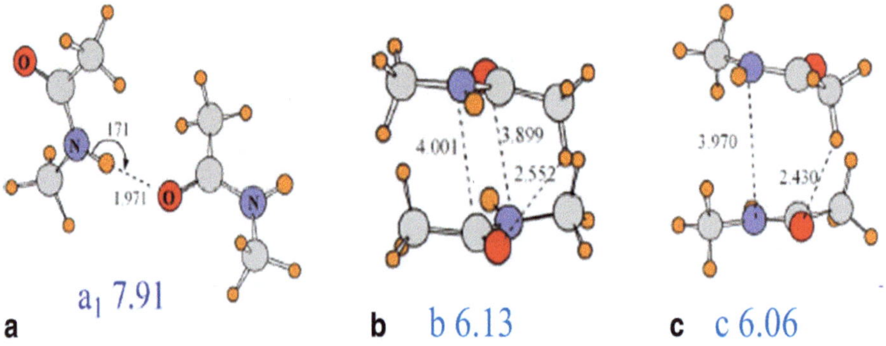

Fig. 4.5 Optimized geometries of N-methylacetamide dimers, with binding energies (kcal/mol) shown in blue. Distances and angles in Å and degs. (Reprinted with permission from Adhikari and Scheiner [168]. Copyright 2013 American Chemical Society)

4.4.3 Longer Chains

Given the variability of NH··O HB strengths in polypeptides, and its weakness in particular in extended conformations, one is naturally led to consider β-sheets, wherein each strand adopts a fairly extended geometry. The conventional wisdom holds that the strands are held together by interstrand NH··O HBs, but even an idealized visualization of the structure of these sheets shows that CH groups of one strand lie in close proximity to the carbonyl O atoms of the next strand. Could not the ensuing CH··O HBs contribute to the stability of the interchain linkages, just as the NH··O HBs do?

This question was specifically addressed [169] in both parallel and anti-parallel configurations of β-sheets. Beginning with the geometry of a full double-strand extent of anti-parallel polyglycine, a piece was excised such that each strand contains both CH and NH donor groups, lying opposite the carbonyl O of the other strand. The $HCONHCH_2CONH_2$ dimer was then optimized to yield both NH··O and CH··O HBs. The former were shorter than the latter, with R(NH··O) = 1.97 Å and R(CH··O) = 2.57 Å, as displayed in Fig. 4.6. But as shown above, the length can be deceiving, as a short HB is not necessarily a strong one. It was necessary to extract the energetic contribution of each sort of HB, separate from the others. The question of how to disentangle individual interactions, when both are present simultaneously, has been a thorny problem for some time.

The issue was addressed [169] by removing one proton donor group at a time, and then computing the interaction energy of the remaining dimer. As indicated in Fig. 4.6a, the NH··O HB was deleted by replacing the terminal NH_2 groups on each strand by a H atom, leaving a CH stump that is both a weak proton donor, and too far away from the carbonyl O to engage in a HB in any case. The interaction between the two $HCONHC^{\alpha}H_2COH$ molecules consists only of $C^{\alpha}H$··O HBs. The CH··O HBs were deleted by simply removing the central CH_2 group of each strand, which leaves only a set of $HCONH_2$ molecules, bound only by NH··O HBs, as indicated in Fig. 4.6b. It should be stressed that the geometry of the dimer was left unchanged by each deletion, to avoid contamination of the results by changes in HB geometry. In this manner, it was determined that the pair of NH··O HBs contribute 14 kcal/mol, as compared to 10 kcal/mol for the CH··O HBs.

The location where the strands are cut is an arbitrary one. If this location is shifted down a bit, the energetic contributions of the NH··O HBs drop from 14 to only 10 kcal/mol, placing it precisely on a par with the CH··O HBs. And it should be stressed that this result is not an artifact of the use of a pair of dipeptides; extension to tripeptides has no substantive effect on the results. The situation is somewhat different in the case of a parallel β-sheet. An equivalent partitioning into separate NH··O and CH··O HB energetic contributions shows that the latter makes a *larger* contribution than does the former. Specifically, the 8.3 kcal/mol contribution from the pair of NH··O HBs is superseded by a 9.5 kcal/mol contribution from the CH··O HBs.

In summary, whichever model is adopted, whether parallel or anti-parallel, whether dipeptide or tripeptide, and wherever the cut is made to excise the system for study, it can hardly be said that the interstrand binding is solely due to NH··O HBs. The CH··O HBs clearly make contributions which are comparable to, and perhaps even larger than, those of NH··O HBs.

Later calculations verified some of the primary conclusions. For example, the importance of the $C^{\alpha}H\cdots O=C$ HB to the interaction energy in a β-sheet was supported by examinations [170, 171] of β-sheet models of $(Gly)_n$ and $(Ala)_n$ which found CH··O HB to be almost as strong as NH··O in the anti-parallel structure, but stronger in parallel. Additional confirmation came from Guo et al. in 2009 [172]. In that same year, the $C^{\alpha}H$··O HB was computed to be stronger than NH··O in dipeptide and tripeptide models of the parallel β-sheet geometry of Ala [173]. In fact, CH··O HBs have shown some propensity to stabilize the α-helix as well [173–175]. There

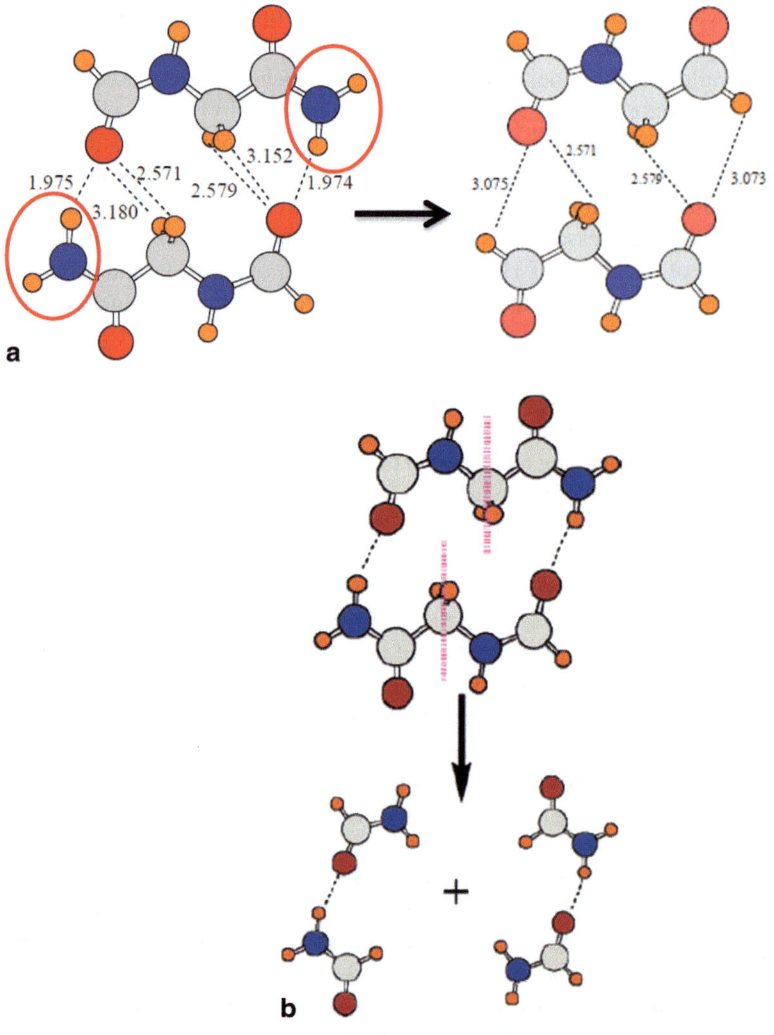

Fig. 4.6 Atom excisions made in $HCONHCH_2CONH_2$ dimer. **a** Replacement of terminal NH_2 groups by H, to leave only CH··O HBs. **b** Removal of central CH_2 groups, leaving behind only NH··O. (Reprinted with permission from Scheiner [169]. Copyright 2006 American Chemical Society)

is also experimental support for the importance of CH··O HBs is β-sheets [176]; the β-pleated sheet structure for β-sulfidocarbonyls [177] contains not only CH··O, but also CH··S HBs.

With the realization that CH··O HBs have the potential to make significant contributions to the stability of fully formed β-sheets, the next question might relate to their ability to actually influence the formation of such sheets. This subject was addressed [178] by the design of a novel molecule in Fig. 4.7a which contains first a dipeptide segment $-NHCOCH_2NHCOCH_3$ which has the full (φ,ψ) Ramachandran space

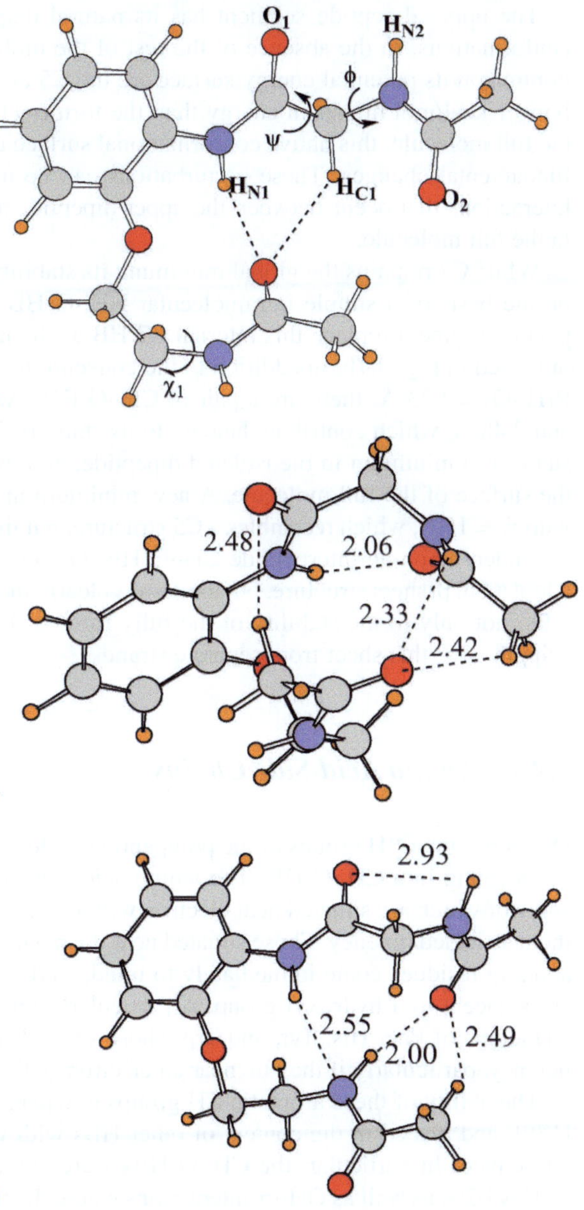

Fig. 4.7 System which contains both upper dipeptide and lower peptide unit. **a** Schematic diagram showing atomic labeling. **b** Global minimum with interatomic distances in Å, **c** secondary minimum with $\psi = 150°$. (Reprinted with permission from Adhikari and Scheiner [178]. Copyright 2013 American Chemical Society)

available to it. Another segment (the lower one in Fig. 4.7a) of the same molecule contains a –$CH_2NHCOCH_3$ peptide unit with which the first dipeptide might engage in NH··O and/or CH··O H-bonding. These two segments are both attached to a phenyl/ether connector unit which holds them together, but whose nature and rigidity prevents its participation in any HBs of its own.

The upper dipeptide segment has its natural disposition in terms of preferred conformations, in the absence of the rest of the molecule. As a dipeptide, the two minima on its potential energy surface are the C5 and C7 structures, with the latter lying 1 kcal/mol higher in energy than the former [178]. Within the framework of the full molecule, this native conformational surface undergoes some important and fundamental changes. These perturbations can be understood on the basis of the interactions that occur between the upper dipeptide and the lower peptide segment in the full molecule.

While C7 remains the global minimum, its stability can no longer be understood on the basis of a simple intramolecular NH··O HB. Instead, HBs with the lower peptide segment replace this internal C7 HB as the most important component. As indicated in Fig. 4.7b, in addition to the conventional interpeptide NH··O HB, with R(H··O) = 2.33 Å, there are a pair of CH··O HBs, with R(H··O) distances of 2.42 and 2.48 Å, which contribute heavily to its stability. The C5 configuration loses its status as a minimum in the isolated dipeptide, and in fact becomes a maximum on the surface of the full molecule. A new minimum in Fig. 4.7c appears in its place, with $\psi = 150°$, which resembles a C5 structure, but like the global minimum, is also dependent upon an interpeptide CH··O HB. Of course, this system is too small to adopt a full β-sheet structure, but the results clearly indicate the importance of CH··O HBs, not only to the stability of the fully formed sheet, but also to the process of "zipping up" this sheet from separate strands.

4.4.4 Amino Acid Side Chains

Of course, the $C^{\alpha}H$ groups of the polypeptide skeleton are not the only ones capable of engaging in a CH··O HB. The amino acid sidechains also contain CH protons, some of which are situated near electron withdrawing groups that ought to impart to them an added potency. Those situated near the positively charged termini of the Lys and Arg residues come immediately to mind, as do the $C^{\beta}H$ hydrogens of Ser that are adjacent to a hydroxyl group. A particularly interesting class are the aromatic CH atoms of Phe, His, Tyr, and Trp whose sp^2 hybridization should enhance their potency, particularly if they lie near an electron-withdrawing atom.

The ability of these aromatic CH groups to engage in a CH··O HB was assessed [179], and placed in the context of other HBs with which these side chains might participate. In particular, the CH··O HBs were compared to conventional NH··O, OH··N HBs, as well as OH··π interactions where the π system of the aromatic group serves as electron donor; water was used as the partner molecule.

As anticipated the conventional HBs were the strongest, with binding energies varying between 4 and 7 kcal/mol. OH··π HBs were weaker, between 2 and 4 kcal/mol, and CH··O slightly weaker still lying in the 1–2 kcal/mol range. The HB lengths correlated with these binding energies, with R(H··O) varying between 2.9 and 3.0 Å for conventional HBs, up to 3.3–3.4 Å for CH··O. Consistent with the sp^2 hybridization of the proton donor C atom, both contractions and stretches were observed for the CH··O HBs, and stretching frequency changes were small, and both red and blue. As in the case of other HBs, the isotropic NMR signal of the bridging CH··O proton shifted downfield. The OH··π proton's NMR shift went in the opposite direction, due to the usual magnetic field currents of the aromatic ring above which it lies. The CH protons lying in closer proximity to electron-withdrawing atoms, such as the N atoms of His, displayed a somewhat greater propensity to engage in CH··O interactions.

Of particular interest was the effect of placing a charge on the aromatic proton donor. The protonated imidazole model of His formed a very strong HB, amounting to 10–11 kcal/mol. Consonant with this greater strength was a HB length that was reduced by 0.3 Å, and a strong blue shift of its CH stretching frequency, of 78–118 cm^{-1}. The NMR shift of this CH proton was also enhanced, by a factor of 2. Other work has supported these ideas. A CH^{+}··O HB of protonated imidazole was in part responsible [180] for the self-assembly of a triple helical structure.

An early calculation of the interaction involving a methylpyridinium cation with dimethyl ether [181] also found a strong CH^{+}··O HB with a binding energy of as much as 13 kcal/mol. Regarding other amino acid side chains, the $C^{\delta}H$ group of proline engages in CH··O HBs, within the context of real protein geometries [182].

4.4.5 Effect of Charge

Indeed placing a charge on either subunit has been known for some time [183, 184] to amplify the binding in HBs, so it is natural to expect that the same ought to be true for CH··O HBs as well. And in fact, there was some evidence this might be true in a few cases [185, 186]. This idea was probed systematically [187] in a series of systems that all employed the carbonyl O of N-methylacetamide (NMA) as the proton acceptor. Beginning with neutral proton donors $S(CH_3)_2$ and $N(CH_3)_3$ as a point of reference, both formed optimized complexes with NMA in which a CH of each methyl group present was engaged in a CH··O HB. The total binding energies amounted to 4.9 and 2.1 kcal/mol, respectively for the S and N systems. The situation changed dramatically when an extra methyl group was added to each, so that the new donors were the $S(CH_3)_3^+$ and $N(CH_3)_4^+$ cations. The various CH··O HBs all contracted by 0.2–0.4 Å, and the binding energies rose to 20.6 and 18.8 kcal/mol, for the S and N systems, respectively, an amplification by a factor of 4–9. (The optimized geometries for the amines are illustrated in Fig. 4.8a, 4.8b) These HBs are very strong, stronger than any neutral conventional HBs, and in fact on a par with ionic HBs of the OH^{+}··O or NH^{+}··O sort.

It is understood that displacing any HB from a gas-phase situation to a solvated environment will weaken it. And such was found [187] to be the case with all HBs studied, ionic as well as neutral. Indeed, the binding strengths of the ionic systems were more dramatically reduced with increasing dielectric constant ε of the surrounding polarizable continuum model of solvation. But nevertheless, the ionic systems remained more tightly bound than the neutral complexes, even for high ε of 78 that simulates water.

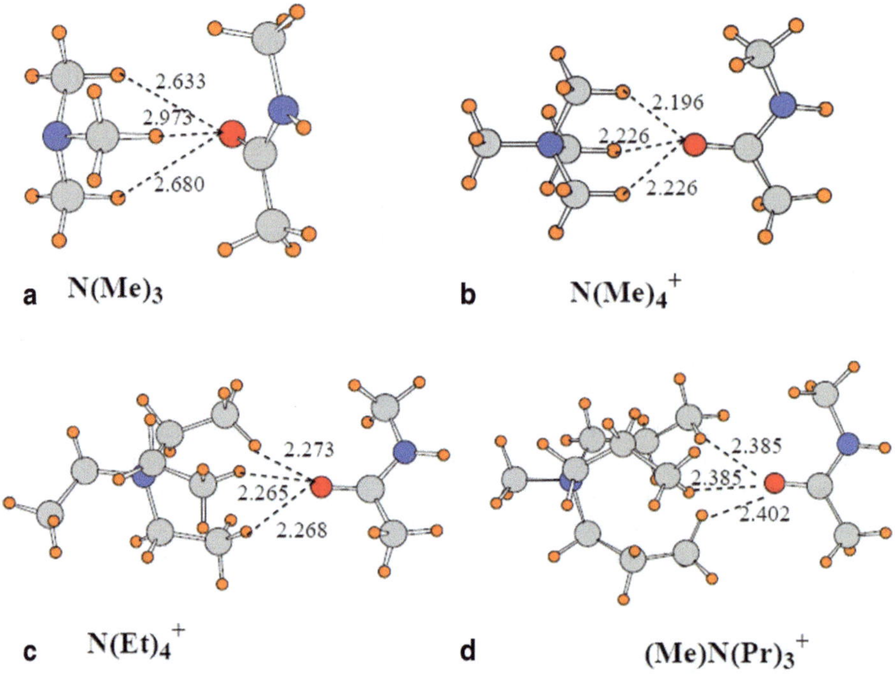

Fig. 4.8 Optimized geometries of indicated amines with N-methylacetamide (NMA). Distances in Å. (Reprinted with permission from Adhikari and Scheiner [187]. Copyright 2013 American Chemical Society)

As the methyl groups on these cations are lengthened to ethyl, propyl, etc, the alkyl H atoms occur at various distances from the heteroatomic center of charge. For example, the methylene CH_2 hydrogens lie adjacent to N/S, while the terminal methyl protons are removed by one additional C–C linkage. This distinction was found [187] to make an important difference. If the CH··O HBs involved the terminal methyl groups, then each chain lengthening, i.e. methyl → ethyl → propyl, very substantially reduced the binding energy with the NMA acceptor. Taking the N series as an example, the binding energy of 20.3 kcal/mol of $N(CH_3)_4^+$ was reduced to 14.1 kcal/mol for $N(Et)_4^+$, and then to 10.5 kcal/mol for $MeN(Pr)_3^+$. As displayed in Fig. 4.8, this reduced binding energy is accompanied by elongations of the relevant $R(CH^+ \cdot\cdot O)$ HBs. This decrease is much more gradual if instead of terminal methyl groups, the HBs are rather formed with the CH_2 protons lying adjacent to S/N. The elongation to ethyl and then to propyl results in binding energies of 18.2 and 17.5 kcal/mol, greatly diminished reductions. One might anticipate that the positive charge on CH protons drops as one moves along the alkyl chain further from the heteroatomic center of charge, which would help to explain this distinction. And indeed, careful scrutiny of the electrostatic potential confirmed this suspicion.

Since these are charged systems, it is tempting to presume that the two entities are bound together primarily by Coulombic attraction. And indeed a SAPT decomposition of the interaction energy reveals a strong electrostatic component, exceeding 20 kcal/mol. However, one cannot ignore other attractive forces: induction and dispersion together contribute between 13 and 16 kcal/mol. The induction signals its presence in a number of ways, that also confirm that there are indeed bona fide HBs present in these ionic complexes. First of all, there are large NBO values of E(2) for the charge transfer from the proton-accepting O atom to the σ^*(CH) antibonding orbitals of the donors, as much as 18 kcal/mol, characteristic of HBs. Second, maps of electron density redistribution that accompany complexation contain the characteristic trademarks of HBs: losses of density around the bridging proton and gains in the area of the lone pair of the proton acceptor atom. The identity of these interactions as true HBs is further confirmed by downfield shifts of the NMR signal of the bridging protons. These shifts are as large as 2 ppm for the ionic systems, more than twice the magnitude of the corresponding quantities in the neutral counterparts.

As indicated above, the optimal arrangements of these ionic systems contain a trifurcated HB wherein one H atom from each of three different methyl groups interacts directly with the proton acceptor O. The binding is weakened if this trifurcation involves three protons from the *same* methyl group, by as much as 35 %. This reduction likely has a geometric cause in that there is a good deal of deviation of each θ(CH··O) angle from linearity when all H atoms come from the same methyl group. This sort of nonlinearity can be avoided if there is but a single proton involved in the CH··O bond. While this single linear HB is slightly superior to a trifurcated interaction with a single methyl group, it remains weaker than the optimal arrangement of three CH··O HBs arising from three separate methyls.

4.5 Specific Examples of Biological Implications

4.5.1 Methyltransferases

The strength of ionic CH··O HBs has some direct applications to biochemistry. As one example, consider methylation of proteins which is essential to the metabolism of amino acids, cofactors, hormones, lipids, nucleic acids, and polysaccharides as well as covalent modification of DNA and proteins. Many of the enzymes that carry out this function are dependent upon S-adenosylmethionine (AdoMet) which transfers its methyl group to the acceptor substrate via an S_N2 reaction, wherein the C of the transferring methyl adopts a planar transition state. There has been some evidence that the process is aided by CH··O HBs in the active site of SET domain class of protein lysine methyltransferases. These HBs help to enhance the binding of AdoMet, to hold the transferring methyl group in its proper orientation, and to stabilize the transition states partial positive charge.

This issue was addressed [188] by combining quantum calculations with experimental measurements. Human SET7/9 lysine methyltransferase, was chosen as representative of this class of proteins. This system displays two highly conserved AdoMet methyl CH··O HBs that are conserved in the SET domain methyltransferase class. One of the proton acceptors is the backbone carbonyl oxygen atom of His-293, whereas the second acceptor is the hydroxyl group of Tyr-335, the invariant tyrosine in the active site of SET domain enzymes. The tyrosine OH accepts a number of CH··O HBs from the AdoMet methyl group, methylene groups, and adenine C8 atom, in addition to donating a conventional HB to a carbonyl group of Ala-295 in the enzyme active site.

The calculations were carried out by modeling the AdoMet donor by $S(Et)_2Me^+$; a N-methylacetamide (NMA) molecule was used in place of the His-293 peptide linkage that engages in a HB with the AdoMet, and phenol replaced the full Tyr-335 residue. In order to first derive a sense of how strongly the AdoMet might bind to each proton acceptor, each was allowed to separately interact directly with $S(Et)_2Me^+$. An optimization of the AdoMet··NMA pair yielded a geometry much like that in the earlier work [187] with the very similar $S(Et)_3^+$, involving three CH··O HBs to the NMA oxygen, with R(H··O) of 2.2 Å, and a binding energy of 20.5 kcal/mol. The O of phenol is a weaker proton acceptor, which binds to the AdoMet model by 7.4 kcal/mol. The three HB lengths vary between 2.4 and 2.9 Å.

Previous studies of SET domain methyltransferases had demonstrated that the invariant tyrosine is critical to enzyme function but its roles in catalysis and AdoMet recognition remained unresolved. For example, mutation of this residue to a phenylalanine (Y335F) severely impaired AdoMet binding affinity, demonstrating the importance of the CH··O hydrogen bonds to substrate recognition. In order to examine this question, the Tyr was mutated to Phe in the experiments; this mutation was modeled in the calculations by replacing the phenol model by benzene, and also by the isosteric aniline. Experimental analysis demonstrated that these mutations retain the structure and protein substrate binding properties of the wild-type enzyme.

The effects of charge were first modeled [188] by comparing the binding of the charged AdoMet model $S(Et)_2Me^+$ with its neutral $S(Et)_2$ analog. The active site model also contained phenol in place of Tyr335, NMA modeled Ala-295, and the adenine group of the cofactors, as displayed in Fig. 4.9a, b (adenine has been deleted for purposes of clarity). Consistent with experimental data, the binding energy of the ionic sulfonium was stronger than that of its neutral analogue by 11 kcal/mol. Breakdown of the total into pairwise interaction energies allowed this difference to be traced to the attraction of the sulfonium to the NMA (Ala) and phenol groups. The H-bond distances in Fig. 4.9a, b support this idea: there are two CH··O distances in Fig. 4.9a that are shorter than 2.4 Å, both marked by broken red lines. This selectivity, i.e. the difference in binding energy between sulfonium and thioether, was considerably reduced when phenol was replaced by benzene. As indicated in Fig. 4.9c, d, the HB lengths for the charged and uncharged AdoMet model differ very little, consistent with the much smaller selectivity. The primary cause of this change is the loss of HBs involving the OH group of phenol. This loss is particularly striking in the cases of the CH··O HBs involving the sulfonium.

The replacement of the phenol by aniline substitutes –OH with $-NH_2$, which in principle can also serve as both proton donor and acceptor. So it was initially puzzling that the aniline mutant, like benzene, also lost selectivity for the sulfonium.

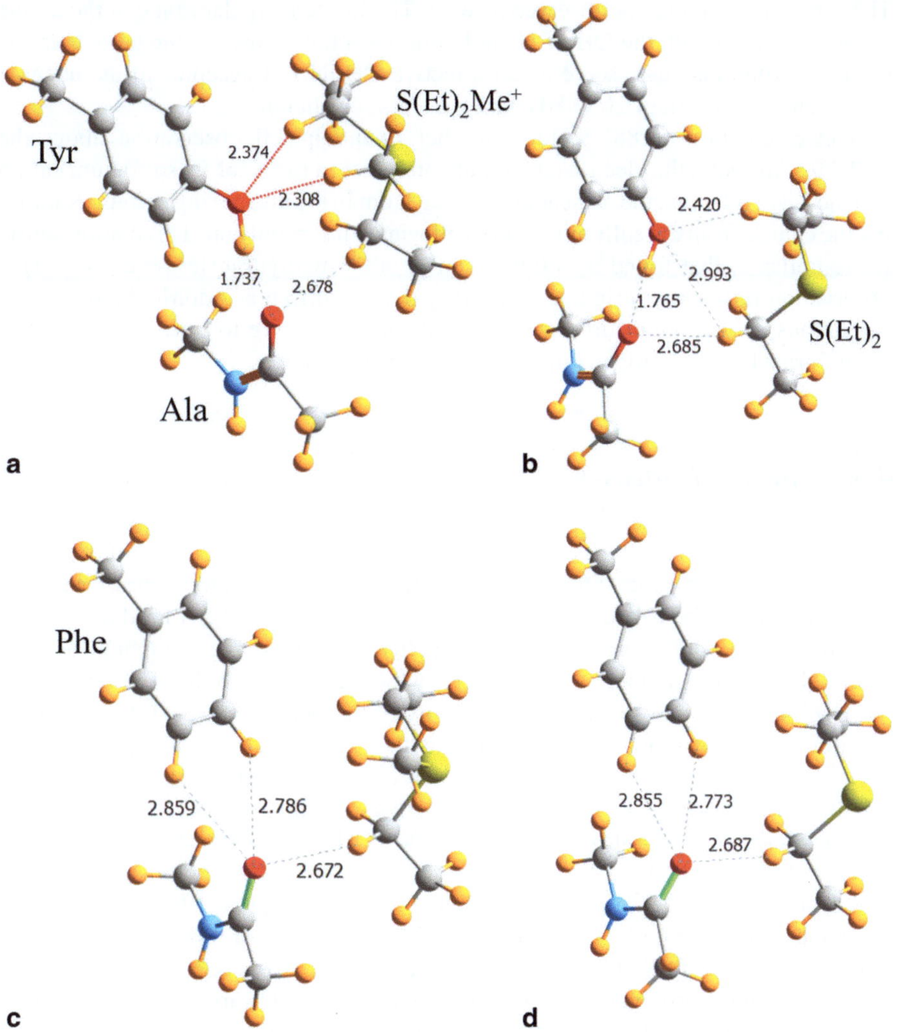

Fig. 4.9 Optimized geometries of model active site of SET7/9, using phenol model of Tyr in (**a** and **b**), and benzene model of Phe in (**c** and **d**). $S(Et)_2Me^+$ models AdoMet in (**a** and **c**); neutral $S(Et)_2$ in (**b** and **d**). HB lengths in Å. (Reprinted with permission from Horowitz et al. [188]. Copyright 2013 American Chemical Society)

And indeed, the N forms a strong CH··N HB with the sulfonium, with R(CH··N) only 2.30 Å in length. A more thorough analysis revealed, however, that this HB is not without cost. In order to accommodate this HB, the amine group rotates 58° about the plane of the aniline ring (compared to the optimized monomer) into a less energetically favorable position. Accordingly, there is an energetic penalty incurred by HB formation in the active site, thus sacrificing the lowest-energy aniline amine conformation in order to better accommodate CH··N hydrogen bonding to AdoMet.

There was a second factor observed as well. The limited angular range of the amine group compared with the OH of phenol eliminates the ability for the aniline amine group to simultaneously serve as an effective CH··N HB acceptor to the adenine group, and negates the CH··O HB which occurs for phenol.

The calculations [188] yielded another unanticipated observation about the SET7/9 active site: the electrostatic repulsion between the AdoMet sulfonium cation and the C8 of adenine, the most acidic carbon atom in the purine ring system. Removing the charge from the sulfonium cation alleviated this repulsion. This conformation is dissimilar to that found in solution and in other methyltransferase classes. This observation raises the intriguing possibility that the different AdoMet binding conformations in various methyltransferase classes may serve to tune the substrate's methyl transfer reactivity and merits further investigation.

4.5.2 Serine Proteases

Another application arises in connection with the catalytic mechanism of the serine protease class of enzymes. One common feature of these enzymes is the presence of what has come to be called a charge relay system. An Asp-His-Ser triad of residues are situated adjacent to one another in the active site. The Asp is thought [189, 190] to initially be in its anionic –COO^- state, which acts through the intermediacy of the His imidazole ring to facilitate the ability of the Ser O^{γ} nucleophile to attack a C atom of the substrate. The His residue is thought to act in part by picking up a proton in one stage of the catalytic cycle and delivering it to another site later. The orientation of this His ring is thus very important for this process.

There had been some strong indications that a CH··O HB between the $C^{\varepsilon1}$H of His and the O of a neighboring Ser residue could be an important functional component of the mechanism [191–193]. This idea led to the proposal [194, 195] of a "ring-flip" mechanism involving a 180° rotation of the His. A rotation of this sort is disfavored in the initial state of the enzyme, as it would delete one of the four HBs in which the His is thought to participate, two of which are of the CH··O variety. Upon formation of the tetrahedral intermediate, on the other hand, the His becomes protonated which now permits retention of all 4 of these HBs, thus facilitating the rotation of the imidazole ring of His, and the entire enzymatic process, by better positioning it for the next step in this reaction.

These ideas were tested in a set of calculations that modeled each of the relevant residues by a smaller, and computationally tractable, molecule. The business end of His-57 was modeled by methylimidazole, Ser-195 by ethanol, the aspartate residue by CH_3COO^-, and the peptide group of Ser-214 by formamide [196]. In order to permit the necessary geometry optimizations that allow the groups a range of motion, while maintaining the positions of each of these groups within the active site structure, a couple of atoms of each residue were frozen in their X-ray coordinate positions.

The initial optimized structure is illustrated in Fig. 4.10a, which does indeed suggest a $C^{\varepsilon1}$H··O HB to be present, along with the expected conventional HBs. A flip of the ring, to the rotated configuration of His-57 in Fig. 4.10b, does indeed result in a minimum, but this process is rather endothermic, on the order of 15 kcal/mol. While the $C^{\varepsilon1}H$ does not have a proton acceptor partner in 10b, the same was true of $C^{\delta2}H$ in 10a. More importantly, the potent $N^{\delta1}H$ donor has lost its partner in b. Consistent with the hypothesis, the ring flip is more favorable energetically after the tetrahedral intermediate has been formed, going from Fig. 4.10c–4.10d. In connection with Ser-214 and its purported CH··O HB with His-57, although this HB appears to be quite weak in the initial structure, it does appear to help stabilize the tetrahedral intermediate. Further, the CH··O HB acts to hold the His ring in a position appropriate for the reaction to proceed. Arguing against the validity of the ring-flip are the small populations of the relevant configurations, based on Boltzmann factors. All told, while providing some structural data, the calculations were unable to establish unequivocally the validity of the ring-flip mechanism.

One unanticipated finding of the calculations relates to the second O atom of the Asp-102, the one that is not itself involved in a HB to the $N^{\delta1}$ of His-57. This atom serves as proton acceptor in a CH··O HB with the $C^{\beta}H$ of His-57, with a length of 2.47 Å in Fig. 4.10a. This HB persists throughout the catalytic cycle, whether or not the His-57 ring has flipped around by 180°. This HB is fairly strong with R(H··O) in the 2.4–2.6 Å range.

In another biomolecular context, CH··O HBs appear to be quite prominent in oligosaccharides and carbohydrates [197] where they can amount to some 40 % of the total interaction energy, and can lock particular configurations [198]. The interactions between RNA bases and phosphate include CH··O HBs [199] as do 2-deoxyribonucleosides [200, 201]. Studies of a model protein backbone combined with nucleic acid bases showed evidence of CH··O HBs as well [202].

4.5.3 *Other Systems*

The importance of CH··O to biomolecular structure and function has become more clear with each passing year. A statistical analysis [203] of protein-ligand complexes provided geometrical evidence of their widespread occurrence. There is evidence of their role also in interhelical packing forces [204]. The $C^{\alpha}H\cdots O{=}C$ HB has been shown to be a major driving factor in interhelical interactions [205], where Gly residues are beneficial due to the lesser steric repulsions arising from the lack of an R group. This finding reinforces the earlier idea [206] that fibrils of poly-Glu are stabilized by bifurcated H-bonds and the frequency [207] with which Gly participates in CH··O HBs at protein-ligand interfaces. Quadrupole couplings and ^{2}H solution NMR data suggest $C^{\alpha}H\cdots O{=}C$ HBs in ubiquitin [208]. They have been proposed [209] to act as the principal driving force for folding of β,γ-hybrid model peptide and of the structure of amicyanin [210]. Ultrahigh resolution neutron diffraction data of proteins [211] have noted CH··O HBs with an average HB length of 2.0 Å. These bonds are essential ingredients of oligosaccharides and carbohydrates where they have been estimated [198] to account for as much as 40 % of the total interaction energy. In another related case, there appear to be as many as 24 CH··O HBs in cellotetraose [212].

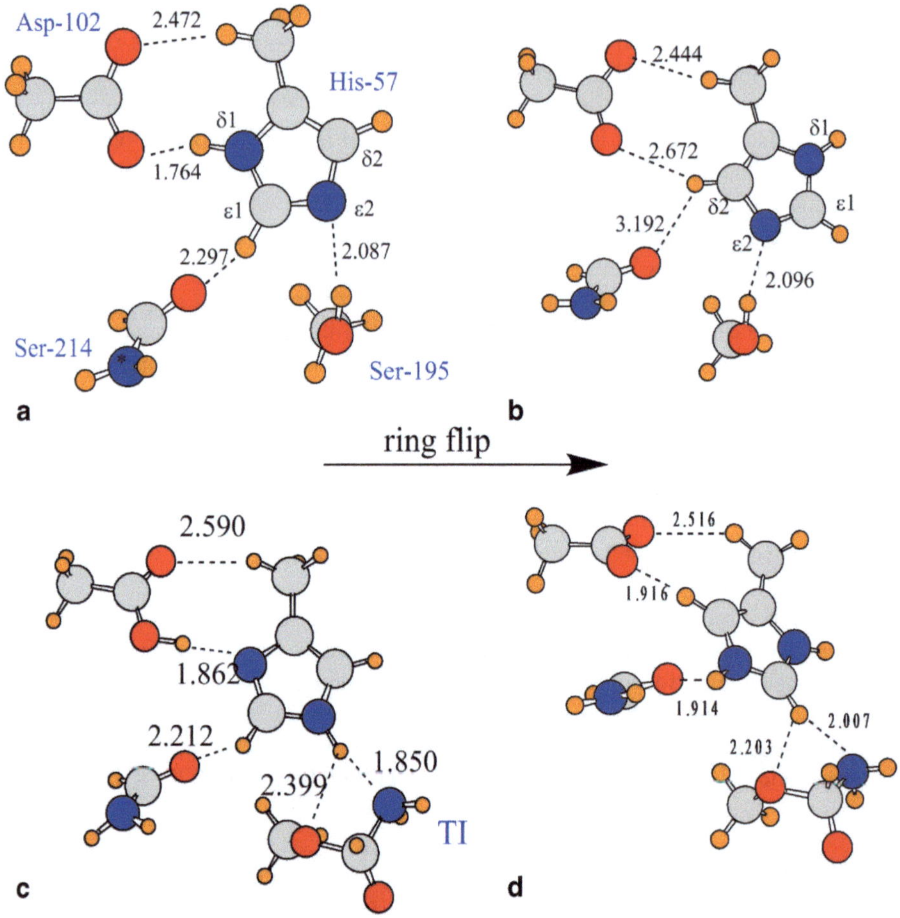

Fig. 4.10 Optimized structures relevant to ring-flip hypothesis of serine proteases. Residue numbers are displayed in (**a**) for each model unit. **a** and **b** represent the initial binding, before and after His-57 flip, respectively. **c** and **d** following formation of tetrahedral intermediate. Distances in Å. (Reprinted with permission from Scheiner [196]. Copyright 2008 American Chemical Society)

4.6 Implications for Organic Chemistry

4.6.1 Fluoroamides

There is a strong tendency of the F atom in an α-fluoroamide to adopt a position anti to the carbonyl O. This trend is likely due at least in part to an electrostatic repulsion between the F and O atoms, both of which carry a partial negative charge. Another contributing factor may be a weak HB between the NH of the amide and the neighboring F atom [213, 214]. But as with any other general tendency, one can envision over-riding factors that might push the system in another direction. Suppose, for instance, that the molecule containing the α-fluoroamide, had appended to it another group, one which contained a proton-acceptor atom. Consider the conformation wherein this proton acceptor sits opposite the carbonyl O of the fluoroamide. In this position, it would of course be unable to interact in an attractive way with the F atom. But if the CH_2F group of the fluoroamide were to rotate around its C–C bond such that one of the H atoms might approach the acceptor, one has the possibility of a CH··O HB. The strength of this HB might be sufficient to hold the F atom up near the carbonyl O, in violation of the general tendency for an anti configuration. Further, if a second electron-withdrawing F atom were added to the terminal group, the resulting CHF_2 group would likely be an even stronger proton donor, enhancing the intergroup HB, and magnifying the push away from the usual anti structure.

These ideas were tested in a combined quantum chemical and experimental approach [215]. The first order of business was to assess the native trend toward the anti configuration in the absence of any intergroup HB. The potential energy surface for a rotation around the C–C bond in $CH_3NHCOCH_2F$ was calculated and the anti structure was indeed found to be its minimum, as displayed in Fig. 4.11a. Internal rotation to place one of the CH_2F protons opposite the carbonyl O costs 6 kcal/mol. An analogous calculation of the difluorosubstituted $CH_3NHCOCHF_2$ shows this system (Fig. 4.11b) also prefers to keep a F atom anti to O; the cost to rotate a H atom into this anti position is 4 kcal/mol.

Can an intergroup CH··O be strong enough to counter this native pull? In order to answer this question, the fluoroamide was placed onto a larger molecule, to which was also added a carbamate group whose carbonyl O could serve as proton acceptor, as displayed in Fig. 4.12a. Focusing first on the monofluoro CH_2F group, full geometry optimization led to a global minimum with the F anti to the carbonyl O, as would be expected were there no proton accepting group. However, in contrast to the small $CH_3NHCOCH_2F$ model, there is a second minimum as well, one which contains a CH··O HB to the carbamate, with $R(CH\cdot\cdot O) = 2.49$ Å, as shown in Fig. 4.12a. And whereas such a structure is 6 kcal/mol higher in energy than the anti configuration in the small model, it lies only 0.9 kcal/mol higher in this larger system. It might be concluded then that the presence of the CH··O HB in the secondary minimum stabilizes this structure by the difference, some 5 kcal/mol.

As indicated above, the addition of a second F atom ought to strengthen any intergroup CH··O HB. The same system was therefore again studied, but with the CH_2F replaced by CHF_2. The CH··O HB has now been empowered to reverse the native tendency for an anti conformation, and the structure shown in Fig. 4.12b with the intergroup CH··O HB now represents the global minimum. The anti structure is now relegated to secondary minimum status, even if only higher in energy by 0.1 kcal/mol.

To be sure that it is the intergroup HB that causes this reversal, and not some artifact of the large connecting group, the carbamate was deleted [215] and the geometry optimized. Confirming the influence of the CH··O HB, in its absence the geometry reverts to the anti conformation displayed in Fig. 4.12c, with no hint of a minimum when the H atom lies opposite the carbonyl O. Another verification of the

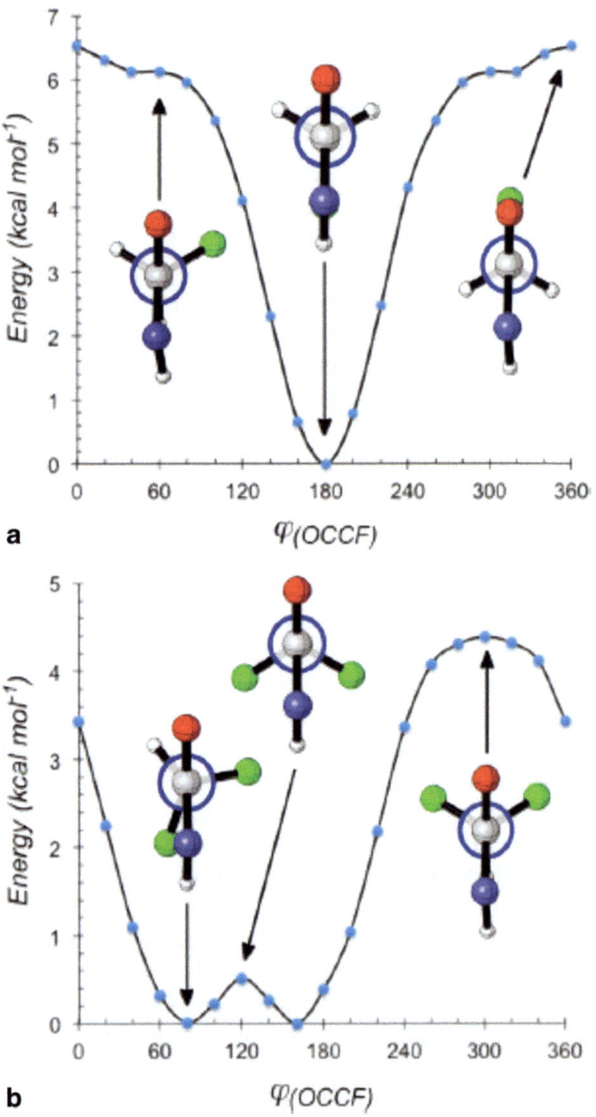

Fig. 4.11 Rotational profiles computed for **a** monofluoro and **b** difluoroamides. (Reprinted with permission from Jones et al. [215]. Copyright 2012 American Chemical Society)

influence of the CH··O HB arises when the terminal CHF_2 is replaced by a simple methyl group, as in Fig. 4.12d. The absence of an electron-withdrawing substituent would severely weaken any CH··O HB. And indeed, following this substitution, the methyl group can rotate freely, with a nearly flat rotational profile.

The energy difference between the conformations with the CH syn and anti to the carbonyl O offers only one estimate of the CH··O HB energy. There are of course other factors that influence this energy difference, e.g. differing interaction between the neighboring NH and either the CH or CF, depending upon the conformation chosen, or interactions between the CF_2H and the terminal methyl group on the

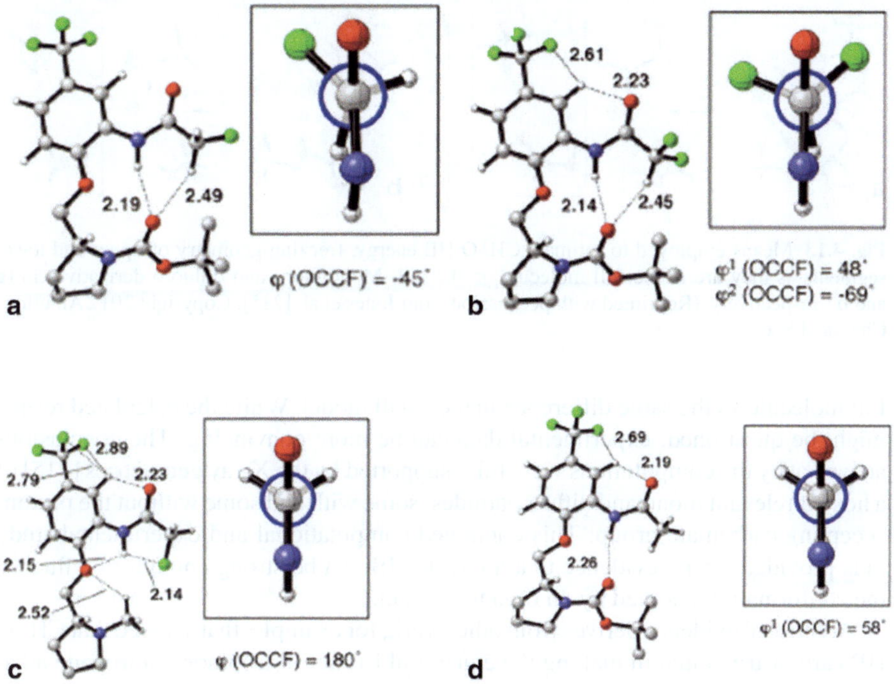

Fig. 4.12 Optimized geometries of molecules containing fluoroamide units; insets focus on the rotation around the C–C bond of the fluoroamide. (Reprinted with permission from Jones et al. [215]. Copyright 2012 American Chemical Society)

carbamate. A second opinion regarding this quantity of interest, as it were, was obtained by focusing only on the groups participating in the CH··O HB, and deleting the remaining groups and their complicating effects.

More specifically, employing a scheme that was also used to analyze protein β-sheet energetics [169], the upper CH donor was reduced to only a $CHOCFH_2$ molecule and the lower carbamate to HCOOH. Importantly, both of these groups were forced to retain the precise alignment which they had in the full molecular system. So for example, the R(CH··O) distance of 2.49 Å in the fully optimized molecule was held constant in the smaller $CHOCFH_2$/HCOOH dimer, as illustrated in Fig. 4.13a. The same is true of other geometrical aspects, e.g. the θ(CH··O) angle and even internal bond lengths. This procedure not only eliminates spurious and non-relevant interactions, but also provides a complex containing two separate molecules, whose interaction energy can be computed unambiguously, something which is not possible when the two groups are parts of the same molecule.

The binding energies of these complexes, which are directly attributable solely to the CH··O HB, were found in this manner to be 3.0 kcal/mol for the monofluorosubstituted molecule, and 3.5 kcal/mol for the analogous disubstituted system. It was gratifying to note that the latter value is quite close to the 4.0 kcal/mol obtained in the less direct way in the full molecule, by comparing syn-anti energy differences in the

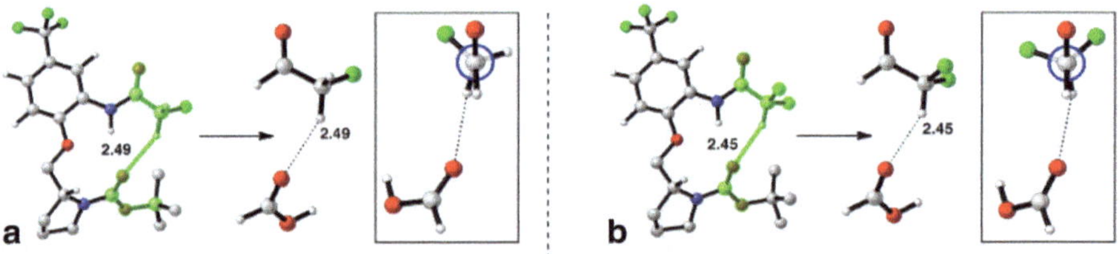

Fig. 4.13 Means employed to estimate CH··O HB energy, freezing geometry of upper and lower segments as they are in the full molecule on the left. Monofluoro and difluoro derivatives in (**a** and **b**), respectively. (Reprinted with permission from Jones et al. [215]. Copyright 2012 American Chemical Society)

full molecule vs the same difference in the small model. While the calculated results might be questioned, experimental data can be more convincing. The conclusions arrived at by the computations were fully supported by the X-ray geometries [215] of a host of relevant mono and difluoroamides, some with and some without the proton-accepting carbamate group. This combined computational and experimental study thus provided strong evidence that a CH··O HB can be strong enough to influence the conformation adopted by an organic system.

Additional evidence derives from other work, for example, that showed that CH··O HBs are instrumental in making the equatorial isomer of tropinone, a tropane alkaloid, dominant in its complex with water [216]. There is a CH··O/S intramolecular HB that influences the conformation of Meldrum's acid derivatives, a bond that persists in the gas phase [137]. CH··O HBs can control the stereochemistry during palladium-catalyzed arylation and vinylation of lactones [217, 218]. They are part of the mechanism by which aminophosphonate diesters adopt their equilibrium structures [219]. Intramolecular bonds of this type can be utilized to induce aromatic 1,2,3-triazole oligomers to form [220] folded and helical secondary structures, containing a 18 Å diameter cavity. In terms of catalytic function and design, CH··O HBs are intimately involved in binding a substrate in thiourea organocatalysis [221]. Bonds of this sort also lower the barrier to H transfer of keto-enol tautomerization in β-cyclohexanedione [222] and differentially dictate the conformation in solid and solution of substituted cyclohexanes, driving them toward the boat structure [223], or [224] enhance chiral recognition.

4.7 Perspective

It is clear that the CH group can be a potent donor within the context of HBs. Motivated by early experimental indications of the presence of CH··O HBs, computations over recent years have provided solid support for their presence, as well as important details. These interactions behave in most respects like other more conventional HBs: electron-withdrawing substituents strengthen their proton-donating power, as does the change in hybridization from sp^3 to sp^2 and then to sp. Their strength derives primarily from a combination of Coulombic attraction and charge transfer from the

O lone pairs into the σ*(CH) antibonding orbital. Also like standard HBs, CH··O interactions display cooperativity when aligned properly.

Just like any HB, CH··O bonds can have profound influence on molecular structure. Calculations suggest that they actively participate in the folding of proteins, particularly in the assembly of β-sheets. In addition to the ubiquitous $C^{\alpha}H$ group on the protein backbone, the sidechains of numerous protein residues also engage in CH··O HBs. Regardless of the HB considered, its strength is very substantially magnified when the proton donor carries a positive charge. This effect is attenuated as the CH donor occurs further from the center of charge, but is still present even when several atoms removed.

CH··O HBs are directly involved in the catalytic mechanism of several enzymes, including methyltransferases and serine proteases as specific examples. This interaction is also capable of influencing the conformation of organic molecules, with mono and di-fluoroamides taken as simple examples.

It seems clear that the study of CH··O HBs will continue to accelerate, from both experimental and computational perspectives. As in the past, numerous cases will emerge where these interactions have been present all along but unrecognized and unappreciated. Future work will likely identify new aspects of these noncovalent bonds that will guide the synthesis of molecular systems with novel and useful properties.

Acknowledgments I am indebted to my coworkers whose names are listed in the references of the individual papers, and who made very substantial contributions. Collaborators Professors Ray Trievel and Martin Smith inspired the work on the methyltransferases and fluoroamides. Some of this work was supported financially by the NIH GM57936 and NSF CHE-1026826.

References

1. Kumler WD (1935) The effect of the hydrogen bond on the dielectric constants and boiling points of organic liquids. J Am Chem Soc 57:600–605
2. Glasstone S (1937) The structure of some molecular complexes in the liquid phase. Trans Faraday Soc 33:200–214
3. Dippy JFJ (1939) The dissociation constants of monocarboxylic acids; their measurement and their significance in theoretical organic chemistry. Chem Rev 25:151–211
4. Marvel CS, Copley MJ, Ginsberg E (1940) Hydrogen bonds involving the C–H link. XI. Effect of structure on bonding of donor and acceptor molecules. J Am Chem Soc 62:3109–3112
5. Huggins ML (1943) The structure of fibrous proteins. Chem Rev 32:195–218
6. Kosolapoff GM, McCullough JF (1951) Comparison of hydrogen bonding abilities of some organic compounds of phosphorus. J Am Chem Soc 73:5392–5393
7. Dougill MW, Jeffrey GA (1953) The structure of dimethyl oxalate. Acta Cryst 6:831–837
8. Parry GS (1954) The crystal structure of uracil. Acta Cryst 7:313–320
9. Huggins CM, Pimentel GC, Shoolery JN (1955) Proton magnetic resonance studies of chloroform in solution: evidence for hydrogen bonding. J Chem Phys 23:1244–1247
10. Pinchas S (1955) Infrared absorption of the aldehydic C–H group. Anal Chem 27:2–6
11. Schneider WG, Bernstein HJ (1956) Molecular association and infra-red spectrum of solid formaldehyde and acetaldehyde. Trans Faraday Soc 52:13–18

12. Pinchas S (1957) Infrared absorption of aldehydic C–H group. Ortho-substituted benzaldehydes. Anal Chem 29:334–339
13. Brown I, Smith F (1960) Liquid-vapour equilibria. X. The systems acetone + nitromethane and acetone + acetonitrile at 45 °C. Austr. J Chem 13:30–37
14. Sutor DJ (1962) The C–H··· O hydrogen bond in crystals. Nature 195:68–69
15. Sutor DJ (1963) Evidence for the existence of C–H··· O hydrogen bonds in crystals. J Chem Soc 1105–1110
16. Allerhand A, Schleyer PvR (1963) A survey of C–H groups as proton donors in hydrogen bonding. J Am Chem Soc 85:1715–1723
17. Pinchas S (1963) Intramolecular hydrogen bonding in o-nitrobenzaldehyde and related compounds. J Phys Chem 67:1862–1865
18. Ferguson G, Tyrrell J (1965) C–H··· O hydrogen bonding. Chem Commun 195–196
19. Bednowitz AL, Post B (1966) Direct determination of the crystal structure of β-fumaric acid. Acta Cryst 21:566–571
20. Krimm S (1967) Hydrogen bonding of C–H··· O = C in proteins. Science 158:530–531
21. Krimm S, Kuroiwa K (1968) Low temperature infrared spectra of polyglycines and C–H. . . O = C hydrogen bonding in polyglycine II. Biopolymers 6:401–407
22. Ramachandran GN, Chandrasekharan R (1968) Interchain hydrogen bonds via bound water molecules in the collagen triple helix. Biopolymers 6:1649–1658
23. Forbes WF (1962) The study of hydrogen bonding and related phenomena by spectroscopic methods. Part VII. Intramolecular hydrogen bonding and steric interactions in o-nitrobenzaldehyde and related compounds. Can J Chem 40:1891–1898
24. Morokuma K (1971) Molecular orbital studies of hydrogen bonds. III. C = O··· H–O hydrogen bond in H_2CO··· H_2O and H_2CO··· $2H_2O$. J Chem Phys 55:1236–1244
25. van Duijneveldt-van de Rijdt JGCM van Duijneveldt FB (1971) Perturbation calculations on the hydrogen bonds between some first-row atoms. J Am Chem Soc 93:5644–5653
26. Johansson A, Kollman P, Rothenberg S (1972) Hydrogen bonding and the structure of the HF-HCN dimer. Chem Phys Lett 16:123–127
27. Johansson A, Kollman P, Rothenberg S (1972) The electronic structure of the HCN dimer and trimer. Theor Chim Acta 26:97–100
28. Bonchev D, Cremaschi P (1974) C–H group as proton donor by formation of a weak hydrogen bond. Theor Chim Acta 35:69–80
29. Kollman P, McKelvey J, Johansson A, Rothenberg S (1975) Theoretical studies of hydrogen-bonded dimers. Complexes involving HF, H_2O, NH_3, HCl, H_2S, PH_3, HCN, HNC, HCP, CH_2NH, H_2CS, H_2CO, CH_4, CF_3H, C_2H_2, C_4H_4, C_6H_6, F^-, and H_3O^+. J Am Chem Soc 97:955–965
30. Vishveshwara S (1978) Ab-initio molecular orbital studies on CH. . . X hydrogen bonded systems. Chem Phys Lett 59:26–29
31. Frisch MJ, Pople JA, Del Bene JE (1983) Hydrogen bonds between first-row hydrides and acetylene. J Chem Phys 78:4063–4065
32. Somasundram K, Amos RD, Handy NC (1986) *Ab initio* calculation for properties of hydrogen bonded complexes H_3N··· HCN, HCN··· HCN, HCN··· HF, H_2O··· HF. Theor Chim Acta 69:491–503
33. Hinchliffe A (1986) Ab initio study of the hydrogen-bonded complexes NH_3··· HCN, PH_3··· HCN, AsH_3··· HCN, H_2O··· HCN, H_2S··· HCN, H_2Se··· HCN. J Mol Struct (Theochem) 136:193–199
34. Chattopadhyay S, Plummer PLM (1990) Ab initio studies on the dimer of sulfur dioxide and hydrogen cyanide. J Chem Phys 93:4187–4191
35. Peralta JE, Ruiz de Azua MC, Contreras RH (1999) Natural bond orbitals analysis of C–H··· O interactions in NCH/H_2O and NCH/OCH_2, and their effect on nuclear magnetic shielding constants. J Mol Struct (Theochem) 491:23–31
36. Heikkilä A, Pettersson M, Lundell J, Khriachtchev L, Räsänen M (1999) Matix isolation and ab initio studies of 1:1 hydrogen-bonded complexes HCN–H_2O and HNC–H_2O produced by photolysis of formaldoxime. J Phys Chem A 103:2945–2951

37. Turi L, Dannenberg JJ (1993) Molecular orbital studies of C–H··· O H-bonded complexes. J Phys Chem 97:7899–7909
38. Schulz B, Botschwina P (1996) Ab initio calculations for propyne and the hydrogen-bonded complex NH_3··· H–CaC–CH_3. Mol Phys 89:1553–1565
39. Simon S, Duran M, Dannenberg JJ (1996) How does basis set superposition error change the potential surfaces for hydrogen-bonded dimers? J Chem Phys 105:11024–11031
40. Philp D, Robinson JMA (1998) A computational investigation of cooperativity in weakly hydrogen-bonded assemblies. J Chem Soc Perkin Trans II 1643–1650
41. Afonin AV, Vashchenko AV, Takagi T, Kimura A, Fujiwara H (1999) Specific intramolecular interactions C–H··· N in heteroaryl vinyl ethers and heteroaryl vinyl sulfides studied by ^{1}H, ^{13}C, and ^{15}N NMR spectroscopies and by ab initio calculations on molecular structures as well as on nuclear shieldings. Can J Chem 77:416–424
42. Cense JM, Agafonov V, Ceolin R, Ladure P, Rodier N (1994) Crystal and molecular structure analysis of flutamide. Bifurcated helicoidal C–H··· O hydrogen bonds. Struct Chem 5:79–84
43. Kim K, Friesner RA (1997) Hydrogen bonding between amino acid backbone and side chain analogues: a high-level ab initio study. J Am Chem Soc 119:12952–12961
44. Hobza P, Spirko V, Selzle HL, Schlag EW (1998) Anti-hydrogen bond in the benzene dimer and other carbon proton donor complexes. J Phys Chem A 102:2501–2504
45. Samanta U, Chakrabarti P, Chandrasekhar J (1998) Ab initio study of energetics of X-H··· π (X = N, O, and C) interactions involving a heteroaromatic ring. J Phys Chem A 102:8964–8969
46. Ornstein RL, Zheng Y (1997) Ab initio quantum mechanics analysis of imidazole C–H. . . O water hydrogen bonding and a molecular mechanics forcefield correction. J Biomol Struct Dyn 14:657–665
47. McCarthy W, Smets J, Adamowicz L, Plokhotnichenko AM, Radchenko ED, Sheina GG, Stepanian SG (1997) Competition between H-bonded and stacked dimers of pyrimidine: IR and theoretical ab-initio study. Mol Phys 91:513–525
48. Kratochvil M, Engkvist O, Sponer J, Jungwirth P, Hobza P (1998) Uracil dimer: potential energy and free energy surfaces. Ab initio beyond Hartree-Fock and empirical potential surfaces. J Phys Chem A 102:6921–6926
49. Shishkin OV, Sponer J, Hobza P (1999) Intramolecular flexibility of DNA bases in adenine-thymine and guanine-cytosine Watson-Crick base pairs. J Mol Struct 477:15–21
50. Hobza P, Sponer J (1999) Structure, energetics, and dynamics of nucleic acid base pairs: nonempirical *ab initio* calculations. Chem Rev 99:3247–3276
51. Woon DE, Zeng P, Beck DR (1990) Ab initio potentials and pressure second virial coefficients for CH_4–OH_2 and CH_4–SH_2. J Chem Phys 93:7808–7812
52. Novoa JJ, Whangbo M-H, Williams JM (1990) Ab initio computational study of the C–H··· donor and C–H··· anion contact interactions in organic donor salts. Mol Cryst Liquid Cryst 181:25–42
53. Legon AC, Roberts BP, Wallwork AL (1990) Rotational spectra and geometries of the gas-phase dimers (CH_4,HF) and (CH_4,HCl). Chem Phys Lett 173:107–114
54. Novoa JJ, Tarron B, Whangbo M-H, Williams JM (1991) Interaction energies associated with short intermolecular contacts of C–H bonds. Ab initio computational study of the C–H..O contact interaction in CH_4..OH_2. J Chem Phys 95:5179–5186
55. Wiberg KB, Waldron RF, Schulte G, Saunders M (1991) Lactones. 1. X-ray crystallographic studies of nonanolactone and tridecanoloactone: nature of CH··· O nonbonded interactions. J Am Chem Soc 113:971–977
56. Novoa JJ, Whangbo M-H, Williams JM (1991) Interaction energies associated with short intermolecular contacts of C–H bonds. Structure and energetics of the interaction between CH_4 and CN^-. Chem Phys Lett 177:483–490
57. Sennikov PG, Sharibdjanov RI, Khoudoinazarov K (1992) The molecular orbital study of the structures and energies of bimolecular complexes of CH_4 and SiH_4 with water. J Mol Struct 270:87–97
58. Szczesniak MM, Chalasinski G, Cybulski SM, Cieplak P (1993) Ab initio study of the potential energy surface of CH_4–H_2O. J Chem Phys 98:3078–3089

59. Vizioli C, de Azúa MCR, Giribet CG, Contreras RH, Turi L, Dannenberg JJ, Rae ID, Weigold JA, Malagoli M, Zanasi R, Lazzeretti P (1994) Proximity effects on nuclear spin-spin coupling constants. 1.^{1}J(CH) couplings in the vicinity of an atom bearing lone pairs. J Phys Chem 98:8858–8861
60. Novoa JJ, Planas M, Rovira MC (1996) On the usefulness of the counterpoise method on hydrogen-bonded complexes: a numerical test using near complete basis sets on H_2O... HF, $(H_2O)_2$, $(HF)_2$, and CH_4... H_2O. Chem Phys Lett 251:33–46
61. Rovira C, Novoa JJ (1997) Strength and directionality of the C(sp^3)–H··· S(sp^3) interaction. An ab initio study using the H_2S··· CH_4 model complex. Chem Phys Lett 279:140–150
62. Hobza P, Sandorfy C (1984) Quantum chemical and statistical thermodynamic investigations of anesthetic activity. 3. The interaction between CH_4, CH_3Cl, CH_2Cl_2, $CHCl_3$, CCl_4, and an O–H... O hydrogen bond. Can J Chem 62:606–609
63. Kumpf RA, Damewood JR (1988) Do nitromethane and malonitrile form C–H··· O hydrogen bonds? Implications for molecular recognition by crown ethers. J Chem Soc Chem Commun 621–622
64. Neuheuser T, Hess BA, Reutel C, Weber E (1994) Ab initio calculations of supramolecular recognition modes. Cyclic versus noncyclic hydrogen bonding in the formic acid/formamide system. J Phys Chem 98:6459–6467
65. Erickson JA, McLoughlin JI (1995) Hydrogen bond donor properties of the difluoromethyl group. J Org Chem 60:1626–1631
66. Turi L, Dannenberg JJ (1995) Molecular orbital studies of the nitromethane-ammonia complex. An unusually strong C–H··· N hydrogen bond. J Phys Chem 99:639–641
67. Alkorta I, Maluendes S (1995) Theoretical study of CH··· O hydrogen bonds in H_2O–CH_3F, H_2O–CH_2F_2, and H_2O–CHF_3. J Phys Chem 99:6457–6460
68. Novoa JJ, Mota F (1997) Substituent effects in intermolecular C(sp^3)–H··· O(sp^3) contacts: how strong can a C(sp^3)–H··· O(sp^3) hydrogen bond be? Chem Phys Lett 266:23–30
69. Boldeskul IE, Tsymbal IF, Ryltsev EV, Latajka Z, Barnes AJ (1997) Reversal of the usual ν(C–H/D) spectral shift of haloforms in some hydrogen-bonded complexes. J Mol Struct 436:167–171
70. Cubero E, Orozco M, Luque FJ (1999) Electron density topological analysis of the C–H··· O anti-hydrogen bond in the fluoroform-oxirane complex. Chem Phys Lett 310:445–450
71. Scheiner S (2000) CH· · · O hydrogen bonding. In: Hargittai M, Hargittai I (eds) Advances in molecular structure research, JAI Press, Stamford, pp 159–207
72. Scheiner S (2005) The CH--O hydrogen bond. A historical account. In: Dykstra C, Frenking G, Kim K, Scuseria G (eds) Theory and applications of computational chemistry: the First 40 years, Elsevier, Amsterdam, p. 831–857
73. Scheiner S (2006) Contribution of CH··X hydrogen bonds to biomolecular structure. In: Grabowski S. J. (ed) Hydrogen bonding—new insights. Springer, pp 263–292
74. Gu Y, Kar T, Scheiner S (1999) Fundamental properties of the CH··· O interaction: is it a true hydrogen bond? J Am Chem Soc 121:9411–9422
75. Alkorta I, Rozas I, Elguero J (2000) Effects of fluorine substitution on hydrogen bond interactions. J Fluor Chem 101:233–238
76. Martins JBL, Politi JRS, Braga AD, Gargano R (2006) Complexes of water with the fluoromethanes. Chem Phys Lett 431:51–55
77. Kryachko ES, Zeegers-Huyskens T (2001) Theoretical study of the CH··· O interaction in fluromethanes·H_2O and chloromethanes·H_2O complexes. J Phys Chem A 105:7118–7125
78. Delanoye SN, Herrebout WA, van der Veken BJ (2002) Improper or classical hydrogen bonding? A comparative cryosolutions infrared study of the complexes of $HCClF_2$, $HCCl_2F$, and $HCCl_3$ with dimethyl ether. J Am Chem Soc 124:7490–7498
79. van der Veken BJ Delanoye SN Michielsen B Herrebout WA (2010) A cryospectroscopic study of the blue-shifting C–H··· O bonded complexes of pentafluoroethane with dimethyl ether-d_6, acetone-d_6 and oxirane-d_4. J Mol Struct 976:97–104
80. Bedell BL, Goldfarb L, Mysak ER, Samet C, Maynard A (1999) Matrix isolation infrared and ab initio study of the 1:1 complexes of bromocyclopropane with NH_3 and $(CH_3)_3N$: evidence for a novel C–H··· N hydrogen bond. J Phys Chem A 103:4572–4579

81. Karger N, Amorim da Costa AM, Ribeiro-Claro JA (1999) C–H··· O bonded dimers in liquid 4-methoxybenzaldehyde: a study by NMR, vibrational spectroscopy, and ab initio calculations. J Phys Chem A 103:8672–8677
82. Hobza P, Havlas Z (2000) Blue-shifting hydrogen bonds. Chem Rev 100:4253–4264
83. Marques MPM, da Costa AMA, Ribeiro-Claro PJA (2001) Evidence of C–H··· O hydrogen bonds in liquid 4-ethoxybenzaldehyde by NMR and vibrational spectroscopies. J Phys Chem A 105:5292–5297
84. Reimann B, Buchhold K, Vaupel S, Brutschy B, Havlas Z, Spirko V, Hobza P (2001) Improper, blue-shifting hydrogen bond betwen fluorobenzene and fluoroform. J Phys Chem A 105:5560–5566
85. Hobza P, Havlas Z (2002) Improper, blue-shifting hydrogen bond. Theor Chem Acc 108:325–334
86. Gopi R, Ramanathan N, Sundararajan K (2014) Experimental evidence for blue-shifted hydrogen bonding in the fluoroform–hydrogen chloride complex: a matrix-isolation infrared and ab initio study. J Phys Chem A 118:5529–5539
87. Gu Y, Kar T, Scheiner S (2000) Comparison of the CH··· N and CH··· O interactions involving substituted alkanes. J Mol Struct 552:17–31
88. Gu Y, Kar T, Scheiner S (2000) Evaluation of the H-bonding properties of CH··· O interactions based upon NMR spectra. J Mol Struct (Theochem) 500:441–452
89. Masunov A, Dannenberg JJ, Contreras RH (2001) C–H bond-shortening upon hydrogen bond formation: influence of an electric field. J Phys Chem A 105:4737–4740
90. Pejov L, Hermansson K (2003) On the nature of blueshifting hydrogen bonds: ab initio and density functional studies of several fluoroform complexes. J Chem Phys 119:313–324
91. Qian W, Krimm S (2002) Vibrational spectroscopy of hydrogen bonding: origin of the different behavior of the C–H··· O hydrogen bond. J Phys Chem A 106:6628–6636
92. Qian W, Krimm S (2002) C–H··· O and O–H··· O hydrogen bonding in formic acid dimer structures: a QM/MM study confirms the common origin of their different spectroscopic behavior. J Phys Chem A 106:11663–11671
93. Hermansson K (2002) Blue-shifting hydrogen bonds. J Phys Chem A 106:4695–4702
94. Delanoye SN, Herrebout WA, van der Veken BJ (2002) Blue shifting hydrogen bonding in the complexes of chlorofluoro haloforms with acetone-d_6 and oxirane-d_6. J Am Chem Soc 124:11854–11855
95. Li X, Liu L, Schlegel HB (2002) On the physical origin of blue-shifted hydrogen bonds. J Am Chem Soc 124:9639–9647
96. Alabugin IV, Manoharan M, Peabody S, Weinhold F (2003) Electronic basis of improper hydrogen bonding: a subtle balance of hyperconjugation and rehybridization. J Am Chem Soc 125:5973–5987
97. Alabugin IV, Manoharan M, Weinhold FA (2004) Blue-shifted and red-shifted hydrogen bonds in hypervalent rare-gas FRg–H···Y. J Phys Chem A 108:4720–4730
98. Rutkowski KS, Rodziewicz P, Melikova SM, Herrebout WA, van der Veken BJ, Koll A (2005) Blue shifted $F_3CH\cdots FCD_3$ and $Cl_3CH\cdots FCD_3$ weakly H-bound complexes. Cryospectroscopic and ab initio study. Chem Phys 313:225–243
99. Yang Y, Zhang W-J, Gao X-M (2006) Theoretical study on N–H···O blue-shifted H-bond for $HNO\cdots H_2O_2$ complex. Chin. J Chem 24:887–893
100. Jablonski M (2007) Blue-shifting intramolecular C–H··· O(S) contacts in sterically strained systems. J Mol Struct (Theochem) 820:118–127
101. Li AY (2007) Chemical origin of blue- and redshifted hydrogen bonds: intramolecular hyperconjugation and its coupling with intermolecular hyperconjugation. J Chem Phys 126:154102
102. Joseph J, Jemmis ED (2007) Red-, blue-, or no-shift hydrogen bonds: a unified explanation. J Am Chem Soc 129:4620–4632
103. Michielsen B, Herrebout WA, van der Veken BJ (2008) C–H bonds with a positive dipole gradient can form blue-shifting hydrogen bonds: the complex of halothane with methyl fluoride. ChemPhysChem 9:1693–1701

104. Li AY (2008) Theoretical study of linear and bifurcated H-bonds in the systems Y··· H_2CZ_n (n = 1,2; Z = O, S, Se, F, Cl, Br; Y = Cl^-,Br^-). J Mol Struct (Theochem) 862:21–27
105. Karpfen A, Kryachko ES (2009) On the intramolecular origin of the blue shift of A–H stretching frequencies: triatomic hydrides HAX. J Phys Chem A 113:5217–5223
106. Grabowski SJ (2011) Red- and blue-shifted hydrogen bonds: the Bent rule from quantum theory of atoms in molecules perspective. J Phys Chem A 115:12789–12799
107. Karpfen A, Kryachko ES (2005) Blue-shifted hydrogen-bonded complexes. II. CH_3F··· $(HF)_{1\leq n\leq 3}$ and CH_2F_2··· $(HF)_{1\leq n\leq 3}$. Chem Phys 310:77–84
108. Kryachko ES, Karpfen A (2006) Theoretical force-field model for blue-shifted hydrogen bonds with fluoromethanes. Chem Phys 329:313–328
109. Karpfen A (2011) Blue-shifted A–H stretching frequencies in complexes with methanol: the decisive role of intramolecular coupling. Phys Chem Chem Phys 13:14194–14201
110. Chandra AK, Zeegers-Huyskens T (2013) Theoretical study of the cooperativity in substituted dimethyl ethers complexed with two water molecules. Red or blue shifts of the ν(CH) vibrations? Chem Phys 410:66–70
111. Donoso-Tauda O, Jaque P, Santos JC (2011) Theoretical analysis based on X–H bonding strength and electronic properties in red- and blue-shifting hydrogen-bonded X–H··· π complexes. Phys Chem Chem Phys 13:1552–1559
112. Mo Y, Wang C, Guan L, Braïda B, Hiberty PC, Wu W (2014) On the nature of blueshifting hydrogen bonds. Chemistry. Eur J 20:8444–8452
113. Jabłoński M (2014) Red and blue shifted hydridic bonds. J Comput Chem 35:1739–1747
114. Struble MD, Kelly C, Siegler MA, Lectka T (2014) Search for a strong, virtually "No-Shift" hydrogen bond: a cage molecule with an exceptional OH–F interaction. Angew Chem Int Ed 53:8924–8928
115. Gou Q, Feng G, Evangelisti L, Caminati W (2014) Interaction between Freons and Amines: the C–H···N weak hydrogen bond in quinuclidine–trifluoromethane. J Phys Chem A 118:737–740
116. Dey A, Mondal SI, Patwari GN (2013) Binary complexes of ammonia with phenylacetylenes: a combined experimental and computational approach to explore multiple minima on intermolecular potentials. Chem Phys Chem 14:746–753
117. Kharat B, Deshmukh V, Chaudhari A (2012) Hydrogen-bonding interactions in acetonitrile oligomers using density functional theory method. Struct Chem 23:637–644
118. Maity S, Dey A, Patwar GN, Karthikeyan S, Kim KS (2010) A combined spectroscopic and abinitio investigation of phenylacetylene-methylamine complex. Observation of σ and π type hydrogen-bonded configurations and fluorescence quenching by weak C–H···N hydrogen bonding. J Phys Chem A 114:11347–11352
119. Howard AA, Tschumper GS, Hammer NI (2010) Effects of hydrogen bonding on vibrational normal modes of pyrimidine. J Phys Chem A 114:6803–6810
120. Jablonski M, Palusiak M (2010) Basis set and method dependence in atoms in molecules calculations. J Phys Chem A 114:2240–2244
121. Liu G, Wang H, Li W (2006) Solvent effect on the type (red-shifted or blue-shifted) of hydrogen bond. J Mol Struct (Theochem) 772:103–108
122. Mons M, Dimicoli I, Tardivel B, Piuzzi F, Brenner V, Millié P (2002) Energetics of a model NH-π interaction: the gas phase benzene-NH_3 complex. Phys Chem Chem Phys 4:571–576
123. Shirhatti PR, Maity DK, Wategaonkar S (2013) C–H···Y hydrogen bonds in the complexes of p-cresol and p-cyanophenol with fluoroform and chloroform. J Phys Chem A 117:2307–2316
124. Zhang G, He W, Chen D (2014) On difference of properties between organic fluorine hydrogen bond C–H···F–C and conventional hydrogen bond. Mol Phys 112:1736–1744
125. Christenholz CL, Obenchain DA, Peebles RA, Peebles SA (2014) Rotational spectroscopic studies of C–H···F interactions in the vinyl fluoride···difluoromethane complex. J Phys Chem A 118:1610–1616
126. Lu N, Ley RM, Cotton CE, Chung W-C, Francisco JS, Negishi E-I (2013) Molecular tuning of the closed shell C–H···F–C hydrogen bond. J Phys Chem A 117:8256–8262
127. Biller MJ, Mecozzi S (2012) A high level computational study of the CH_4/CF_4 dimer: how does it compare with the CH_4/CH_4 and CF_4/CF_4 dimers? Mol Phys 110:377–387

128. Vincent MA, Hillier IH (2011) The structure and interaction energies of the weak complexes of $CHClF_2$ and CHF_3 with HCCH: a test of density functional theory methods. Phys Chem Chem Phys 13:4388–4392
129. Saar BG, O'Donoghue GP, Steeves AH, John W., Thoman J (2006) Evidence for a blue-shifting intramolecular hydrogen bond in the vibrational overtone spectrum of 1H-nonafluorobutane. Chem Phys Lett 417:159–163
130. Favero LB, Giuliano BM, Melandri S, Maris A, Ottaviani P, Velino B, Caminati W (2005) CH... O and CHF links form the cage structure of dioxane-trifluoromethane. J Phys Chem A 109:7402–7404
131. Lu P, Liu G-Q, Li J-C (2005) Existing problems in theoretical determination of red-shifted or blue-shifted hydrogen bond. J. Mol. Struct (Theochem) 723:95–100
132. Domagala M, Grabowski SJ (2010) Hydrocarbons as proton donors in C–H··· N and C–H··· S hydrogen bonds. Chem Phys 367:1–6
133. Wolstenholme DJ, Weigand JJ, Cameron EM, Cameron TS (2008) The progression of strong and weak hydrogen bonds in a series of ethylenediammonium dithiocyanate derivatives—A new bonding protocol for macromolecules? Phys Chem Chem Phys 10:3569–3577
134. Domagala M, Grabowski SJ (2005) C–H··· N and C–H··· S hydrogen bonds—influence of hybridization on their strength. J Phys Chem A 109:5683–5688
135. Afonin AV, Toryashinova DD, Schmidt EY (2004) Investigation of C–H··· X (X = N, O, S) intramolecular hydrogen bond in 1-vinyl-2-(2'-heteroaryl)pyrroles by ab initio calculations. J Mol Struct (Theochem) 680:127–135
136. McDowell SAC (2006) Blue and red shifts in F_3C–H··· B (B = FH, ClH, OH_2, SH_2 and Cl^-) complexes predicted by a perturbative model. Chem Phys Lett 424:239–242
137. Hopkins WS, Hasan M, Burt M, Marta RA, Fillion E, McMahon TB (2014) Persistent intramolecular C–H···X (X = O or S) hydrogen-bonding in benzyl meldrum's acid derivatives. J Phys Chem A 118:3795–3803
138. Tamasi G, Botta F, Cini R (2006) DFT-molecular modeling analysis of C–H··· N and C–H··· S hydrogen bond type interactions in selected platinum–purine/pyrimidine complexes. J Mol Struct (Theochem) 766:61–72
139. Feng G, Gou Q, Evangelisti L, Vallejo-López M, Lesarri A, Cocinero EJ, Caminati W (2014) Competition between weak hydrogen bonds: C–HCl is preferred to C–HF in CH_2ClF–H_2CO, as revealed by rotational spectroscopy. Phys Chem Chem Phys 16:12261–12265
140. Ito F (2012) Observation of CHF_2Cl trimer in Xe matrix and calculations of stable isomers by density functional theory. J Mol Struct 1012:43–49
141. Castro M, Nicolas-Vazquez I, Zavala JI, Sanchez-Viesca F, Berros M (2007) Theoretical study of intramolecular CH–X (X = N, O, Cl) hydrogen bonds in thiazole derivatives. J Chem Theory Comput 3:681–688
142. Balamurugan V, Mukherjee J, Hundal MS, Mukherjee R (2007) Supramolecular architectures with ladder and lamellar topologies using metal-ligand coordination units via C–H···Cl and O–H···Cl hydrogen bonding. Struct Chem 18:133–144
143. Pazout R, Houskova J, Dusek M (2011) Platinum precursor of anticancer drug: a structure fixed by long intermolecular N–H··· I and C–H··· I hydrogen bonds. Struct Chem 22:1325–1330
144. Michielsen B, Verlackt C, van der Veken BJ, Herrebout WA (2012) C–H··· X (X = S, P) hydrogen bonding: the complexes of halothane with dimethyl sulfide and trimethylphosphine. J Mol Struct 1023:90–95
145. Ramasami P, Ford TA (2012) Ab initio studies of some hydrogen-bonded complexes of fluoroform—evidence for blue-shifted behaviour. J Mol Struct 1023:163–169
146. Howard ST, Abernethy CD (2004) Intramolecular CH... $C_{carbene}$ hydrogen bonds and competing interactions in monoprotonated tripodal carbenes. J Comput Chem 25:649–659
147. Yu F (2013) Intermolecular interactions of formic acid with benzene: energy decomposition analyses with ab initio MP2 and double-hybrid density functional computations. Int J Quantum Chem 113:2355–2360
148. Karthikeyan S, Ramanathan V, Mishra BK (2013) Influence of the substituents on the CH··· π interaction: benzene–methane complex. J Phys Chem A 117:6687–6694

149. Majumder M, Mishra BK, Sathyamurthy N (2013) CH··· π and π··· π interaction in benzene-acetylene clusters. Chem Phys 557:59–65
150. Michielsen B, Dom JJJ, van der Veken BJ, Hesse S, Xue Z, Suhm MA, Herrebout WA (2010) The complexes of halothane with benzene: the temperature dependent direction of the complexation shift of the aliphatic C–H stretching. Phys Chem Chem Phys 12:14034–14044
151. Shirhatti PR, Wategaonkar S (2010) Blue shifted hydrogen bond in 3-methylindole·CHX_3 complexes (X = Cl, F). Phys Chem Chem Phys 12:6650–6659
152. Takahashi O, Kohno Y, Gondoh Y, Saito K, Nishio M (2003) General preference for alkyl/phenyl folded conformations. Relevance of the CH/π and CH/O interactions to stereochemistry as evidenced by ab initio MO calculations. Bull Chem Soc Jpn 76:369–374
153. Xu X, Pooi B, Hirao H, Hong SH (2014) CH–π and CF–π interactions lead to structural changes of N-heterocyclic carbene palladium complexes. Angew Chem Int Ed 53:1283–1287
154. Wang Y, Balbuena PB (2001) Associations of alkyl carbonates: intermolecular C–H··· O interactions. J Phys Chem A 105:9972–9982
155. Wetmore SD, Schofield R, Smith DM, Radom L (2001) A theoretical investigation of the effects of electronegative substitution on the strength of C–H··· N hydrogen bonds. J Phys Chem A 105:8718–8726
156. Kar T, Scheiner S (2004) Comparison of cooperativity in CH··O and OH··O hydrogen bonds. J Phys Chem A 108:9161–9168
157. Malenov DP, Janjic GV, Veljkovic DŽ, Zaric SD (2013) Mutual influence of parallel, CH/O, OH/π and lone pair/π interactions in water/benzene/water system. Comput. Theor. Chem 1018:59–65
158. Gao X, Liu Y, Li H, Bian J, Zhao Y, Cao Y, Mao Y, Li X, Xu Y, Ozaki Y, Wu J (2013) A cooperative hydrogen bonding system with a C–H··· O hydrogen bond in ofloxacin. J Mol Struct 1040:122–128
159. Lee K-M, Chen JCC, Chen H-Y, Lin IJB (2012) A triple helical structure supported solely by C–H··· O hydrogen bonding. Chem Commun 48:1242–1244
160. Solimannejad M, Malekani M, Alkorta I (2011) Cooperativity between the hydrogen bonding and halogen bonding in F_3CX ··· NCH(CNH) ··· NCH(CNH) complexes (X = Cl, Br). Mol Phys 109:1641–1648
161. Samanta AK, Pandey P, Bandyopadhyay B, Chakraborty T (2010) Cooperative strengthening of an intramolecular O–H···O hydrogen bond by a weak C–H···O counterpart: matrix-isolation infrared spectroscopy and quantum chemical studies on 3-methyl-1,2-cyclohexanedione. J Phys Chem A 114:1650–1656
162. Li Q, An X, Gong B, Cheng J (2007) Cooperativity between OH··· O and CH··· O hydrogen bonds involving dimethyl sulfoxide-H_2O–H_2O complex. J Phys Chem A 111:10166–10169
163. Scheiner S, Kar T (2005) Effect of solvent upon CH··O hydrogen bonds with implications for protein folding. J Phys Chem B 109:3681–3689
164. Shi Z, Olson CA, Bell AJ, Kallenbach NR (2002) Non-classical helix-stabilizing interactions: C–H··· O H-bonding between Phe and Glu side chains in α-helical peptides. Biophys Chem 101–102:267–279
165. Scheiner S, Kar T, Gu Y (2001) Strength of the $C^{\alpha}H$··O hydrogen bond of amino acid residues. J Biol Chem 276:9832–9837
166. Scheiner S (2005) Relative strengths of NH··O and CH··O hydrogen bonds between polypeptide chain segments. J Phys Chem B 109:16132–16141
167. Scheiner S (2007) The strength with which a peptide group can form a hydrogen bond varies with the internal conformation of the polypeptide chain. J Phys Chem B 111:11312–11317
168. Adhikari U, Scheiner S (2013) Preferred configurations of peptide-peptide interactions. J Phys Chem A 117:489–496
169. Scheiner S (2006) Contributions of NH··O and CH··O H-bonds to the stability of β-sheets in proteins. J Phys Chem B 110:18670–18679
170. Vener MV, Egorova AN, Fomin DP, Tsirelson VG (2007) QTAIM study of the closed-shell interactions in peptide secondary structures: a cluster treatment of oligo- and polyalanines. Chem Phys Lett 440:279–285

171. Parthasarathi R, Raman SS, Subramanian V, Ramasami T (2007) Bader's electron density analysis of hydrogen bonding in secondary structural elements of proteins. J Phys Chem A 111:7141–7148
172. Guo H, Gorin A, Guo H (2009) A peptide-linkage deletion procedure for estimate of energetic contributions of individual peptide groups in a complex environment: application to parallel β-sheets. Interdiscip Sci Comput Life Sci 1:12–20
173. Vener MV, Egorova AN, Fomin DP, Tsirel'son VG (2009) A quantum-topological analysis of noncovalent interactions in secondary polyalanine structures. Russ. J Phys Chem B 3:541–547
174. LaPointe SM, Farrag S, Bohórquez HJ, Boyd RJ (2009) QTAIM study of an α-helix hydrogen bond network. J Phys Chem B 113:10957–10964
175. Park H, Yoon J, Seok C (2008) Strength of C_{α}–H··· O = C hydrogen bonds in transmembrane protein. J Phys Chem B 112:1041–1048
176. Pohl G, Plumley JA, Dannenberg JJ (2013) The interactions of phenylalanines in β-sheet-like structures from molecular orbital calculations using density functional theory (DFT), MP2, and CCSD(T) methods. J Chem Phys 138:245102
177. Hussain S, Das G, Chaudhuri MK (2007) Intermolecular hydrogen bonded and self-assembled β-pleated sheet structures of β-sulfidocarbonyls. J Mol Struct 837:190–196
178. Adhikari U, Scheiner S (2013) First Steps in growth of a polypeptide toward β-sheet structure. J Phys Chem B 117:11575–11583
179. Scheiner S, Kar T, Pattanayak J (2002) Comparison of various types of hydrogen bonds involving aromatic amino acids. J Am Chem Soc 124:13257–13264
180. Chang H-C, Lee KM, Jiang J-C, Lin M-S, Chen J-S, Lin IJB, Lin SH (2002) Charge-enhanced C–H··· O interactions of a self-assembled triple helical spine probed by high-pressure. J Chem Phys 117:1723–1728
181. Raymo FM, Bartberger MD, Houk KN, Stoddart JF (2001) The magnitude of [C–H··· O] hydrogen bonding in molecular and supramolecular assemblies. J Am Chem Soc 123:9264–9267
182. Guo H, Beahm RF, Guo H (2004) Stabilization and destabilization of the C^{δ}–H··· O = C hydrogen bonds involving proline residues in helices. J Phys Chem B 108:18065–18072
183. Deakyne CA, Meot-Ner M (1999) Ionic hydrogen bonds in bioenergetics. 4. Interaction energies of acetylcholine with aromatic and polar molecules. J Am Chem Soc 121:1546–1557
184. Cannizzaro CE, Houk KN (2002) Magnitude and chemical consequences of R_3N^+ –C–H··· O = C hydrogen bonding. J Am Chem Soc 124:7163–7169
185. Liu T, Gu J, Tan X-J, Zhu W-L, Luo X-M, Jiang H-L, Ji R-Y, Chen K-X, Silman I, Sussman JL (2002) The relationship between binding models of TMA with furan and imidazole and the molecular electrostatic potentials: DFT and MP2 computational studies. J Phys Chem A 106:157–164
186. Shishkin OV, Palamarchuk GV, Gorb L, Leszczynski J (2008) Opposite charges assisted extra strong C–H··· O hydrogen bond in protonated 2'-deoxyadenosine monophosphate. Chem Phys Lett 452:198–205
187. Adhikari U, Scheiner S (2013) The magnitude and mechanism of charge enhancement of CH··O H-bonds. J Phys Chem A 117:10551–10562
188. Horowitz S, Dirk LMA, Yesselman JD, Nimtz JS, Adhikari U, Mehl RA, Scheiner S, Houtz RL, Al-Hashimi HM, Trievel RC (2013) Conservation and functional importance of carbon-oxygen hydrogen bonding in AdoMet-dependent methyltransferases. J Am Chem Soc 135:15536–15548
189. Kraut J (1977) Serine proteases: structure and mechanism of catalysis. Ann Rev Biochem 46:331–358
190. Wang JH (1970) Directional character of proton transfer in enzyme catalysis. Proc Natl Acad Sci U S A 66:874–881
191. Topf M, Várnai P, Richards WG (2002) Ab initio QM/MM dynamics simulation of the tetrahedral intermediate of serine proteases: insights into the active site hydrogen-bonding network. J Am Chem Soc 124:14780–14788

192. Derewenda ZS, Derewenda U, Kobos PM (1994) (His)C^{ε}–H··· O = C < hydrogen bond in the active sites of serine hydrolases. J Mol Biol 241:83–93
193. Ishida T, Kato S (2003) Theoretical perspectives on the reaction mechanism of serine proteases: the reaction free energy profiles of the acylation process. J Am Chem Soc 125:12035–12048
194. Ash EL, Sudmeier JL, Day RM, Vincent M, Torchilin EV, Haddad KC, Bradshaw EM, Sanford DG, Bachovchin WW (2000) Unusual ^{1}H NMR chemical shifts support (His) $C^{\varepsilon 1}$–H··· O = C H-bond: proposal for reaction-driven ring flip mechanism in serine protease catalysis. Proc Natl Acad Sci U S A 97:10371–10376
195. Bachovchin WW (2001) Contributions of NMR spectroscopy to the study of hydrogen bonds in serine protease active sites. Magn Reson Chem 39:39S199–S213
196. Scheiner S (2008) Analysis of catalytic mechanism of serine proteases. Viability of ring-flip hypothesis. J Phys Chem B 112:6837–6846
197. Chen F, Selvam L, Wang F (2010) Blue shifted intramolecular C–H···O improper hydrogen bonds in conformers of zidovudine. Chem Phys Lett 493:358–363
198. Zierke M, Smieško M, Rabbani S, Aeschbacher T, Cutting B, Allain FH-T, Schubert M, Ernst B (2013) Stabilization of branched oligosaccharides: Lewis benefits from a nonconventional C–H···O hydrogen bond. J Am Chem Soc 135:13464–13472
199. Zgarbová M, Jurecka P, Banás P, Otyepka M, Sponer JE, Leontis NB, Zirbel CL, Sponer J (2011) Noncanonical hydrogen bonding in nucleic acids. benchmark evaluation of key base-phosphate interactions in folded RNA molecules using quantum-chemical calculations and molecular dynamics simulations. J Phys Chem A 115:11277–11292
200. Shishkin OV, Palamarchuk GV, Gorb L, Leszczynski J (2006) Intramolecular hydrogen bonds in canonical 2'-deoxyribonucleotides: an atoms in molecules study. J Phys Chem B 110:4413–4422
201. Shishkin OV, Gorb L, Zhikol OA, Leszczynski J (2004) Conformational analysis of canonical 2-deoxyribonucleotides. 1. Pyrimidine nucleotides. J Biomol Struct Dyn 21:537–554
202. Alkorta I, Elguero J (2003) Interaction of protein backbone with nucleic acid bases. J Phys Chem B 107:5306–5310
203. Liu Z, Wang G, Li Z, Wang R (2008) Geometrical preferences of the hydrogen bonds on protein-ligand binding interface derived from statistical surveys and quantum mechanics. J Chem Theory Comput 4:1959–1973
204. Hildebrand PW, Günther S, Goede A, Forrest L, Frömmel C, Preissner R (2008) Hydrogen-bonding and packing features of membrane proteins: functional implications. Biophys J 94:1945–1953
205. Mueller BK, Subramanian S, Senes A (2014) A frequent, GxxxG-mediated, transmembrane association motif is optimized for the formation of interhelical Cα–H hydrogen bonds. Proc Natl Acad Sci U S A 111:E888–E895
206. Fulara A, Dzwolak W (2010) Bifurcated hydrogen bonds stabilize fibrils of poyl(L-glutamic) acid. J Phys Chem B 114:8278–8283
207. Panigrahi SK, Desiraju GR (2007) Strong and weak hydrogen bonds in the protein-ligand interface. Proteins 67:128–141
208. Sheppard D, Li D-W, Godoy-Ruiz R, Brschweiler R, Tugarinov V (2010) Variation in quadrupole couplings of a deuterons in ubiquitin suggests the presence of C^{α}–H^{α}···O = C hydrogen bonds. J Am Chem Soc 132:7709–7719
209. Venugopalan P, Kishore R (2013) Unusual folding propensity of an unsubstituted β, γ-hybrid model peptide: importance of the C–H. . . O intramolecular hydrogen bond. Chem Eur J 19:9908–9915
210. Sukumar N, Mathews FS, Langan P, Davidson VL (2010) A joint x-ray and neutron study on amicyanin reveals the role of protein dynamics in electron transfer. Proc Natl Acad Sci U S A 107:6817–6822
211. Chen JC-H, Hanson BL, Fisher SZ, Langan P, Kovalevsky AY (2012) Direct observation of hydrogen atom dynamics and interactions by ultrahigh resolution neutron protein crystallography. Proc Natl Acad Sci U S A 109:15301–15306

212. Yoneda Y, Mereiter K, Jaeger C, Brecker L, Kosma P, Rosenau T, French A (2008) van der Waals versus hydrogen-bonding forces in a crystalline analog of cellotetraose: cyclohexyl 4'-O-cyclohexyl β-D-cellobioside cylohexane solvate. J Am Chem Soc 130:16678–16690
213. Banks JW, Batsanov AS, Howard JAK, O'Hagan D, Rzepa HS, Martin-Santamaria S (1999) The preferred conformation of α-fluoroamides. J Chem Soc Perkin Trans II 2409–2411
214. Tormena CF, Amadeu NS, Rittner R, Abraham RJ (2002) Conformational analysis in N-methylfluoroamides. A theoretical, NMR and IR investigation. J Chem Soc Perkin Trans II 773–778
215. Jones CR, Baruah PK, Thompson AL, Scheiner S, Smith MD (2012) Can a C–H···O interaction be a determinant of conformation. J Am Chem Soc 134:12064–12071
216. Écija P, Vallejo-López M, Evangelisti L, Fernández JA, Lesarri A, Caminati W, Cocinero EJ (2014) O–H··· N and C–H··· O hydrogen bonds control hydration of pivotal tropane alkaloids: Tropinone··· H_2O complex. Chem Phys Chem 15:918–923
217. Huang Z, Chen Z, Lim LH, Quang GCP, Hirao H, Zhou JS (2013) Weak arene C–H··· O hydrogen bonding in palladium-catalyzed arylation and vinylation of lactones. Angew Chem Int Ed 52:5807–5812
218. Huang Z, Lim LH, Chen Z, Li Y, Zhou F, Su H, Zhou JS (2013) Arene CH–O hydrogen bonding: a stereocontrolling tool in palladium-catalyzed arylation and vinylation of ketones. Angew Chem Int Ed 52:4906–4911
219. Juribasic M, Bellotto L, Tusek-Bozic L (2012) N–H···O=P hydrogen-bonded dimers as the main structural motif of aminophosphonate diesters. Struct Chem 23:257–266
220. You L-Y, Chen S-G, Zhao X, Liu Y, Lan W-X, Zhang Y, Lu H-J, Cao C-Y, Li Z-T (2012) C–H··· O hydrogen bonding induced triazole foldamers: efficient halogen bonding receptors for organohalogens. Angew Chem Int Ed 51:1657–1661
221. Lippert KM, Hof K, Gerbig D, Ley D, Hausmann H, Guenther S, Schreiner PR (2012) Hydrogen-bonding thiourea organocatalysts: The privileged 3,5-bis(trifluoromethyl)phenyl group. Eur J Org Chem 2012:5919–5927
222. Bandyopadhyay B, Pandey P, Banerjee P, Samanta AK, Chakraborty T (2012) CH···O interaction lowers hydrogen transfer barrier to keto-enol tautomerization of β-cyclohexanedione: combined infrared spectroscopic and electronic structure calculation study. J Phys Chem A 116:3836–3845
223. Vibhute AM, Gonnade RG, Swathi RS, Sureshan KM (2012) Strength from weakness: opportunistic CH··O hydrogen bonds differentially dictate the conformational fate in solid and solution states. Chem Commun 48:717–719
224. Solà J, Riera A, Verdaguer X, Maestro MA (2005) Phosphine-substrate recognition through the C–H··· O hydrogen bond: application to the asymmetric Pauson-Khand reaction. J Am Chem Soc 127:13629–13633

Chapter 5
Hydrogen Bonds Involving Radical Species

Qing-Zhong Li and Hai-Bei Li

Abstract In this chapter, we focused on the structures, patterns, energies, and nature of hydrogen bonds involving radicals, such as H_3C, OH, BH_2, and BeH, based on the fact that hydrogen-bonded complexes involving radicals may be formed in the related reactions and processes, and are useful for understanding their mechanisms. Theses radicals as the proton donor and acceptor may participate in the formation of different types of hydrogen bonds, including single-electron hydrogen bonds with the single electron of radicals as the proton acceptor, dihydrogen bonds with the hydridic hydrogen of radicals as the proton acceptor, conventional hydrogen bonds with the lone-pair electron of radicals as the proton acceptor or with the proton of radicals as the proton donor. In addition, a covalent interaction is also formed between radicals and the other molecule. The formation of these interactions was understood from the view of HOMO and LUMO of radicals, and their nature was analyzed by the energy decomposition scheme, showing similar nature in most cases with conventional hydrogen bonds. We paid a particular attention to the cooperative effect of single-electron hydrogen bond with other types of interactions as well as the competition among different types of interactions involving radical species.

5.1 Introduction

The radicals, such as hydroxyl radical (HO), hydroperoxyl radical (HOO), organic peroxy radical (ROO), alkyl radical (R), particularly methyl radical (H_3C), have been considered to be extremely important in many chemical reactions and the related processes, including combustion chemistry and celestial chemistry. For instance, ROO radicals, usually from the atmospheric oxidation of hydrocarbons, are responsible for

Q.-Z. Li (✉)
The Laboratory of Theoretical and Computational Chemistry,
School of Chemistry and Chemical Engineering, Yantai University,
264005 Yantai, People's Republic of China
e-mail: liqingzhong1990@sina.com

H.-B. Li (✉)
School of Ocean, Shandong University, 264209 Weihai, People's Republic of China
e-mail: lihaibei@sdu.edu.cn

S. Scheiner (ed.), *Noncovalent Forces,* Challenges and Advances in
Computational Chemistry and Physics 19, DOI 10.1007/978-3-319-14163-3_5

the formation of tropospheric ozone, and also the generation of HO_x (HO and HOO) radicals [1]. Reactive free radical, HOO, plays significant roles in the stratospheric chemistry and oxidation processes in the troposphere, such as the generation of sulphuric acid through the oxidation of SO_2 [2]. In these reactions or processes, radicals probably form the complexes through hydrogen bonds or other types of interactions.

Hydrogen bond is one of the most interesting topics of intermolecular interactions because it plays an important role in molecular recognition, crystal engineering, and chemical reactions [3]. Based on the electronic character of the monomers, two types of hydrogen bond interactions are defined, that is, closed-shelled and open-shelled ones, in which the latter has attracted a growing attention in recent years. The open-shelled hydrogen bonds are able to regulate the electron transfer processes in many enzymatic systems [4] and affect the chemical properties of many radical species, such as bond dissociation energies and reduction/oxidation potentials [5]. It has been demonstrated that radical species plays the roles of proton acceptor/donor in hydrogen-bonded complexes. A single-electron hydrogen bond could form if the single electron of a radical acts as the proton acceptor [6].

A great number of theoretical and experimental investigations of the complexes involving radicals [7] have been performed to understand the mechanisms of the related reactions and processes, such as the proton transfer between neutral molecules and radical species in organic reactions. In the following sections, we introduced different types of hydrogen bonds involving radicals, and paid more attention to the effects of competition and cooperativity between them and other types of interactions. In order to compare the complexes at the same level, we recalculated all the complexes using Gaussian 09 program [8] at the UMP2/aug-cc-pVTZ(PP) level, even some of them have been studied before. The binding energy of these complexes is obtained using supermolecular method, that is, the difference between the sum of the energy of the monomers and the total energy of the complex. In general, the energy of the monomer is obtained from the optimized geometries of the isolated molecules with the exception if the significant variations in geometry occur between the isolated monomers and the ones in the complex, the geometry of the monomer in the complex will be used.

5.2 Single-Electron Hydrogen Bonds

5.2.1 Alkyl Radicals as Proton Acceptors

Methyl radical (H_3C) is a simple prototype for a wide class of organic radicals. It usually acts as the proton acceptor in single-electron hydrogen bonds and plays important roles as an intermediate in the field of chemistry and biochemistry involving methyl radical. There are two types of the highest occupied molecular orbital (HOMO) and the lowest unoccupied molecular orbital (LUMO) for H_3C radical (Fig. 5.1): alpha and beta. The shape of alpha-LUMO is similar to that of beta-LUMO, and both orbitals are near degenerate due to their slight difference in energy. However, alpha- and beta-HOMOs of H_3C radical are different. The alpha-HOMO is perpendicular to

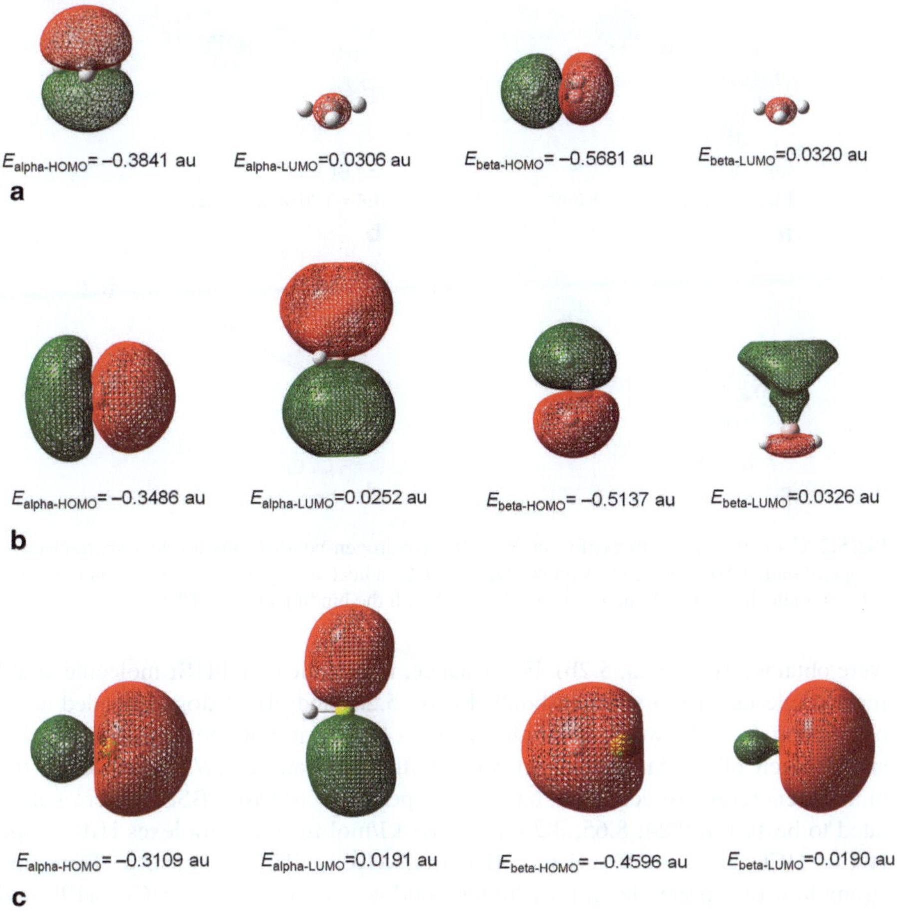

Fig. 5.1 HOMOs and LUMOs of CH_3 (**a**), H_2B (**b**), and HBe (**c**) radicals with the orbital energy (E)

the plane of methyl radical, describing the distribution of the single free electron on the C atom, while the beta-HOMO is in the plane of methyl radical, which describes the C–H σ-bonds. In combination with the higher energy of the former than that of the latter, it is predictable that the alpha-HOMO will provide the unpaired electron when H_3C radical forms a single-electron hydrogen bond with the proton donors.

$H_3C\cdot\cdot\cdot HF$ complex was detected first in the reaction of F atoms with methane by matrix isolation infrared spectroscopy [9], characterized by a red shift of $-198\ \text{cm}^{-1}$ for the H–F stretching vibration [10], and later was further confirmed by the hyperfine coupling constant using electron paramagnetic resonance [11]. Theoretical calculations showed that this complex forms a single-electron hydrogen bond with C_{3v} symmetry (Fig. 5.2a). The bind energy of the complex is around $8 \sim 13$ kJ/mol, depending on the computational levels. [6, 11, 12]. The complexes of H_3C radical with other hydrogen halides were also studied, and different types of interactions

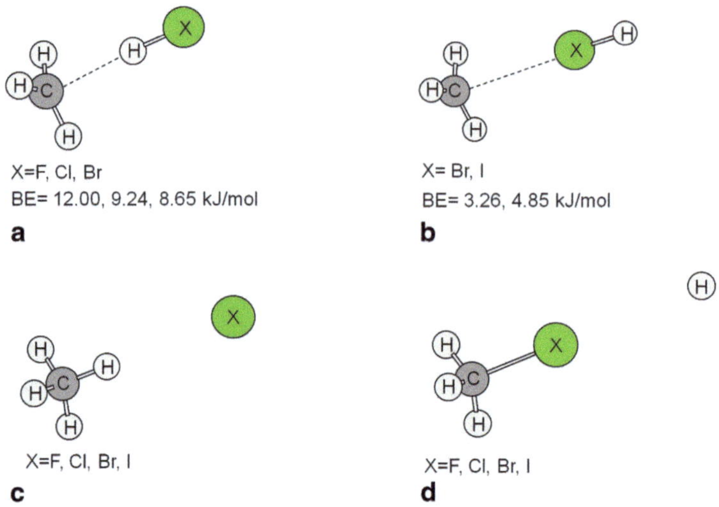

Fig. 5.2 Geometrical structures of **a** single-electron hydrogen-bonded complexes, **b** single-electron halogen-bonded complexes, reaction products of H_3C radical and hydrogen halide, **c** methane and halogen radical, and **d** halomethane and H radical with the binding energy (BE)

were obtained (Fig. 5.2a, 5.2b). For instance, H_3C radical and HBr molecule could form single electron hydrogen-bonded (Fig. 5.2a) and also halogen-bonded complexes (Fig. 5.2b) between the single electron of the former monomer and the proton and halogen of the latter, respectively. At the MP2/aug-cc-pVTZ(PP) level, the binding energies corrected for basis set superposition error (BSSE) were calculated to be 12.00, 9.24, 8.65, 3.26, and 4.85 kJ/mol for the complexes $H_3C\cdots HF$, $H_3C\cdots HCl$, $H_3C\cdots HBr$, $H_3C\cdots BrH$, and $H_3C\cdots IH$, respectively. Thus, the strength of the single electron hydrogen bond is in order of HF > HCl > HBr, and it is in order of HBr < HI for the single electron halogen bond, where the former is related with the electronegativity of the halogen and the latter is with the magnitude of σ-hole, a region of positive electrostatic potential on the outer of the covalently bonded halogen's surface [13]. In nature, this character is consistent with that of the conventional hydrogen bonds and halogen bonds. It has been suggested that the complexes $H_3C\cdots HX$ and $H_3C\cdots XH$ (X = F, Cl, Br, and I) are the intermediates of the reactions: $CH_3 + HX \rightarrow CH_4 + X$ (Fig. 5.2c) and $CH_3 + HX \rightarrow CH_3X + H$ (Fig. 5.2d) [14]. With the development of the high-resolution detective technique, these preliminary precursors of the reactions are expected to be observed in the near future. Wang et al. performed a detailed theoretical study on the single electron hydrogen-bonded complexes of $H_3C\cdots HF$ and $H_3C\cdots HCCH$, and suggested that the single-electron hydrogen bond has many similar features with the conventional one, such as the lengthening of the X–H (X = F and C) bond of the proton donor and the red shift of the stretching vibrational mode [6]. Raghavendra and Arunan [15] compared the single-electron hydrogen, lithium, and chlorine bonds with H_3C as the electron donor and HF, LiF, and ClF as the electron acceptor, respectively, that is, $H_3C\cdots HF$, $H_3C\cdots LiF$, and $H_3C\cdots ClF$. Interestingly, they found the similar features of the interactions in all these complexes using Atoms in Molecules (AIM) theoretical analysis.

Besides the complexes $H_3C\cdots HX$ (X = F, Cl, Br, and I), many other single-electron hydrogen-bonded complexes involving H_3C radical have been studied, such as $H_3C\cdots HCN$, $H_3C\cdots HNC$, $H_3C\cdots HCCH$, and $H_3C\cdots H_2O$. Figures 5.3, and 5.4 depict the optimized structures of complexes $H_3C\cdots Y$ (Y = HCN, HNC, H_2O, H_2S, and NH_3), where H_3C radical also plays the role of the proton acceptor in the complexes. In most cases, H_3C radical forms the weak single-electron hydrogen bonds with the other neutral proton donors with the binding energies less than 12 kJ/mol. On the other hand, the strength of this weak hydrogen bonding interaction is able to be regulated by the substitution effects. Generally, both electron-donating substituents in H_3C radical and electron-withdrawing ones in the proton donor could strengthen the single-electron hydrogen bonds. Furthermore, the single-electron hydrogen bonds become much strong if the proton donor is protonated. For example, the binding energy is about 62 kJ/mol in the complex $H_3C\cdots H_3O^+$ at the MP2/aug-cc-pVTZ level.

Complex $H_3C\cdots HCN$ (Fig. 5.3a) was suggested to be a post-reactive intermediate of the reaction $CN + CH_4 \rightarrow HCN + CH_3$ [16], and it has been studied using many experimental and theoretical methods. This complex was characterized by a red shift of $-45.51\ cm^{-1}$ for the C–H stretching vibration in helium nanodroplets using infrared laser spectroscopy [16]. Theoretically, the red shift of C–H bond was estimated to be -48, -38, and $-23\ cm^{-1}$ at the UCCSD(T)/6-311++G(2d,2p) [17], UMP2/6-311++G(2d,2p) [17], and UMP2/6-311++G(d, p) levels [16], respectively. It is obvious that this red shift is closely correlated with the theoretical methods applied and is well reproduced at the UCCSD(T)/6-311++G(2d,2p) level, but underestimated at the UMP2/6-311++G(d, p) level. This indicates that it is necessary to apply the high level methods considering the electron correlation in combination with the large basis sets to quantitatively reproduce the properties of open-shelled systems [18]. On the other hand, complex $H_3C\cdots HNC$ (Fig. 5.3b) forms a stronger single-electron hydrogen bond than that of $H_3C\cdots HCN$ complex, due to the higher acidity of HNC monomer where the proton connects to atom N with the higher electronegativity than that of C in HCN. This is consistent with the variation trend of the strength of hydrogen bond in the complexes $H_3C\cdots HX$ (X = F, Cl, Br, and I). The binding energies were estimated to be 6.40 and 11.41 kJ/mol for $H_3C\cdots HCN$ and $H_3C\cdots HNC$ complexes, respectively.

The reaction of OH radical with methane is important in combustion and the atmosphere, which produces one H_3C radical and one water molecule via direct abstraction of one H atom from methane by OH [19], and its channel complex $H_3C\cdots H_2O$ (Fig. 5.4a) has been evidenced in helium nanodroplets using infrared laser spectroscopy [20]. The $H_3C\cdots H_2O$ complex was characterized with a red shift of the H–O–H symmetric stretching vibration compared with that of H_2O monomer. Further *ab initio* calculations [21] predicted that it is a weak single electron hydrogen-bonded complex with C_s symmetry, which is in good agreement with the experiment results. Complexes $H_3C\cdots H_2S$ (Fig. 5.4d) and $H_3C\cdots H_3N$ (Fig. 5.4g) were also studied in order to compare with $H_3C\cdots H_2O$, and the results show that the stability

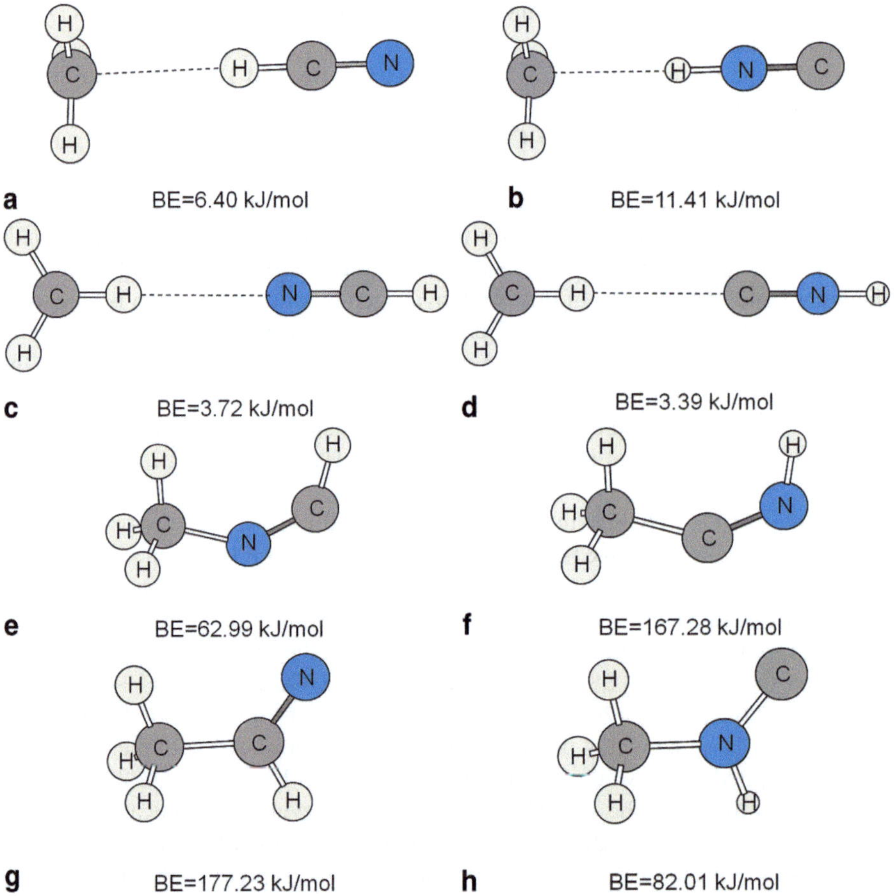

Fig. 5.3 Complexes with a single-electron hydrogen bond (**a** and **b**), **a** hydrogen bond with H_3C as a proton donor (**c** and **d**), and a covalent interaction (**e**–**h**) formed between the H_3C radical and HCN or HNC with binding energy (BE)

of these complexes is in order of $H_3C\cdots H_3N < H_3C\cdots H_2S < H_3C\cdots H_2O$, which is consistent with the acidity of molecules NH_3, H_2S and H_2O.

The effects of substitution, hybridization, and solvent on the properties of C$\cdots$H–O single-electron hydrogen bond in the $H_3C\cdots H_2O$ complex have been investigated by means of quantum chemical calculations [22]. This hydrogen bond is able to be strengthened by the halogenations of the proton donor, such as $H_3C\cdots$HOCl. The methyl group in the proton donor and acceptor plays different roles in the formation of the C$\cdots$H–O hydrogen bond, where the former plays the role of electron-withdrawing and the latter is electron-donating. Both cases make positive contributions to the enhancement of the hydrogen bonding interaction, but the latter has larger one than the former. When the proton acceptor varies from methyl radical to vinyl one, the binding energy increases by 1 kJ/mol. On the other hand, the binding energy also increases due to the enhancement of the acidity of the proton

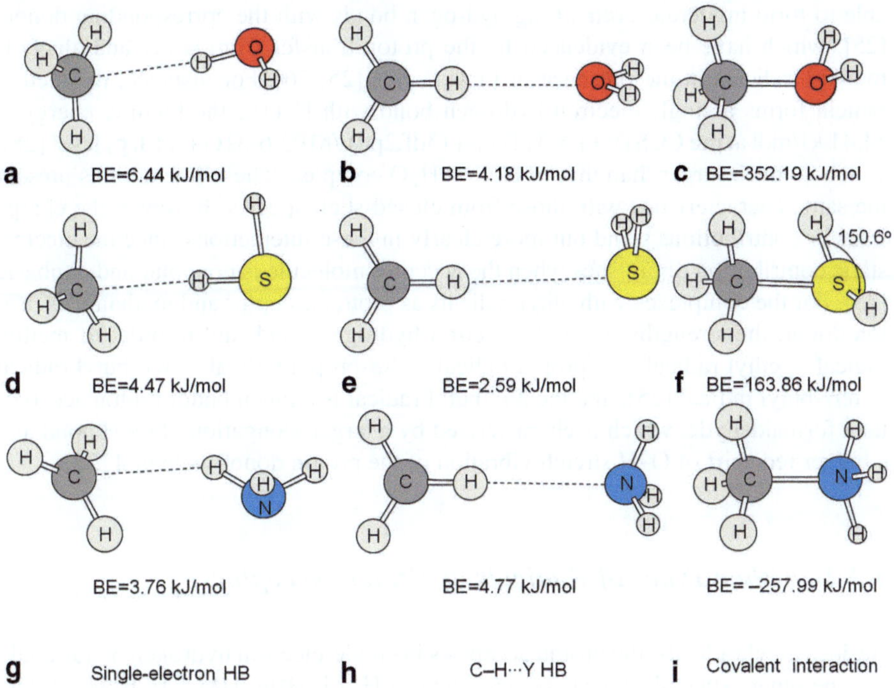

Fig. 5.4 Complexes of a single-electron hydrogen bond (**a**, **d**, **g**), **a** hydrogen bond with H_3C as a proton donor (**b**, **e**, **h**), and a covalent interaction (**c**, **f**, **i**) formed between the H_3C radical and H_2O, H_2S, or NH_3 with binding energy (BE)

donor, such as the proton donor changing from water to vinyl alcohol [22]. Furthermore, it has been confirmed that the solvents also have an enhancing effect on the strength of the single-electron hydrogen bond, as evidenced by the increase of the binding energy and the shortening of the binding distance [22].

On the basis of the measurement of the binding distance and binding energy, the nonadditivity effect of methyl group has been studied in the single electron hydrogen-bonded complexes with H_3C, $(H_3C)H_2C$, $(CH_3)_2HC$, and $(CH_3)_3C$ as the proton acceptors and H_2O as the proton donor at the UMP2/6-311++G(2df,2p) level [23]. It is found that the increase of the number of the methyl substituents in the proton acceptor results in a shorter binding distance and a larger binding energy, indicating that there is nonadditivity effect of the methyl groups in the single electron hydrogen-bonded complexes. Furthermore, this nonadditivity is negative, that is, the contribution of each methyl group decreases with the increase of the number of the methyl group in the proton donor. This is different from the positive nonadditivity of methyl groups in $H_3C\cdots BrH$ complex [24].

As mentioned above, alkyl radicals form the weak single-electron hydrogen bonds with the neutral proton donors, such as HF, HCl, HBr, H_2O, NH_3, H_2S, HCN, HNC, and HCCH. When ionic proton donors are present, however, alkyl radicals are able to form moderate even strong hydrogen bonds with the corresponding donors [25], which have been evidenced by the proton transfer from water and alcohols to alkyl radicals in the presence of Lewis acids [25, 26]. For instance, the methyl radical forms a single-electron hydrogen bond with H_3O^+, the binding energy is 51.41 kJ/mol at the CCSD(T)/6-311G++(3df,2pd)//MP2/6-31G++(d, p) level [28], which is much larger than that in $H_3C\cdots H_2O$ complex. These interactions present the same characteristics with those from closed-shell species; however, the charge transfer contributions stand out more clearly in these interactions since the electrostatic contribution diminishes when the acceptor molecule is nonpolar and nonbasic [25]. For the complexes with alkyl radicals as proton acceptor and methanol as proton donor, the strengths of single-electron hydrogen bonds are in order of methyl radical < ethyl radical < *n*-propyl radical < *iso*-propyl radical < *sec*-butyl radical < *tert*-butyl radical [25], and the *tert*-butyl radical is a much better proton acceptor than formaldehyde, which is characterized by a larger elongation of O–H bond and a bigger red shift of O–H stretch vibration of the proton donor methanol [25].

5.2.2 Other Types of Radicals as Proton Acceptors

Besides alkyl radicals, the proton acceptors in single-electron hydrogen bonds could also be other types of radical species, such as H, Li, HBe, HMg, H_2B, H_2Al, and H_2Ga. Similar with those of H_3C radical, the alpha- and beta-HOMOs of HBe and H_2B are different in energy and shape for both radicals (Fig. 5.1). The alpha-HOMOs of H_2B and HBe also direct to another hybridization orbital occupied by the single free electron, while the beta-HOMOs describe the Y–H σ-bonds (Y = B and Be). Both the higher energy and the shape of alpha-HOMO indicate that H_2B and HBe radicals are favorable to act as the proton acceptor with the single electron participating in the formation of single-electron hydrogen bonds. On the basis of the similarity between H_2B (HBe) and H_3C radicals, it is predictable that the complexes with H_2B (HBe) and H_3C as the proton acceptor might have similar features. Figures 5.5a–5.5c show the optimized structures of the single electron hydrogen-bonded complexes of $H_2B\cdots HCN$, $H_2B\cdots HNC$, and $H_2B\cdots HF$, where the proton is introduced to the single electron along the direction of the alpha-HOMO of H_2B radical. Solimannejad and Alkorta [29] performed *ab initio* study of the single-electron hydrogen bonds with AH_2 radicals (A = B, Al, and Ga) as the proton acceptors and HX (X = F, Cl, Br, CN, and CCH) as the proton donors to investigate their binding energies, frequency shifts, and geometrical properties. The stability of complexes is in order of $H_2B\cdots HX >$ $H_2Al\cdots HX > H_2Ga\cdots HX$, and with the same proton acceptor, the binding energy is $H_2A\cdots HF > H_2A\cdots HCl > H_2A\cdots HBr > H_2A\cdots HCN > H_2A\cdots HCCH$ [29]. For instance, the complex $H_2B\cdots HNC$, with the binding energy of 13.71 kJ/mol and the binding distance of 2.390 Å, is more stable than $H_2B\cdots HCN$, which has the binding energy and distance of 7.19 kJ/mol and 2.684 Å, respectively [30]. This is similar to the complexes of H_3C radical with HCN and HNC.

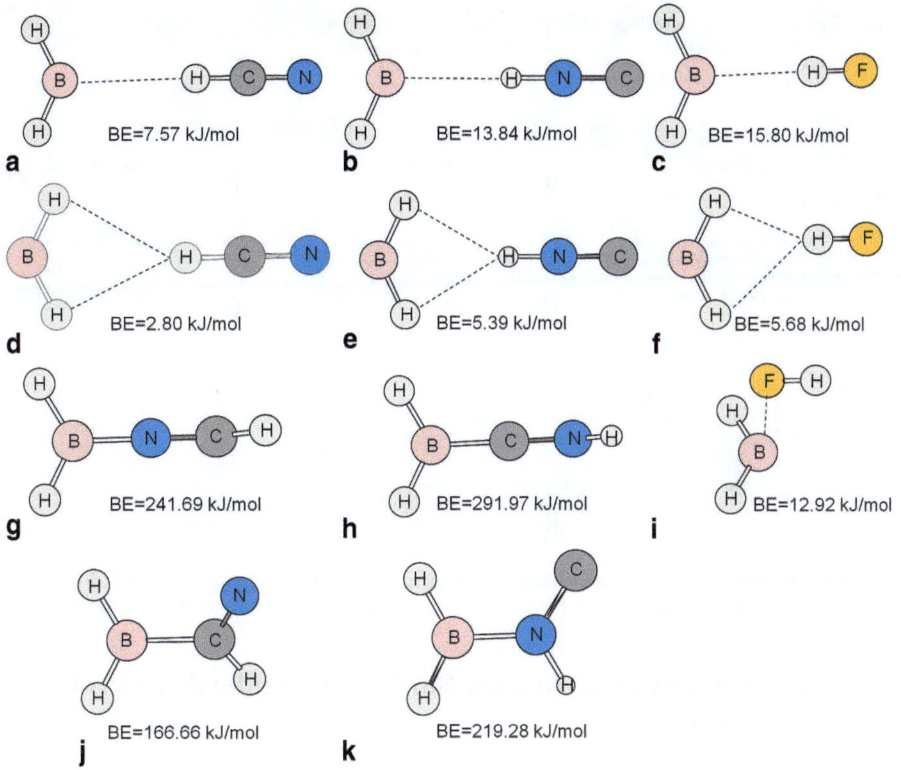

Fig. 5.5 Complexes of a single-electron hydrogen bond (**a–c**), a dihydrogen bond (**d–f**), and a covalent interaction (**g–k**) formed between BH_2 and HCN, HNC, and HF with binding energy (BE)

Figure 5.6 presents the optimized structures of the single electron hydrogen-bonded complexes of HBe··· HCN, HBe··· HNC, and HBe··· HF with the binding energies of 5.81, 11.12, and 13.46 kJ/mol, respectively. Thus, the stability of the complexes is in the same order as that of H_2B as the proton acceptor. In order to compare the capability of accepting protons of HBe, H_2B, and H_3C in the single electron hydrogen-bonded complexes with HF as the proton donor, we found that the binding energy increases in order of $H_3C < HBe < H_2B$ [31]. Additionally, theses complexes were studied at the levels of QCISD/aug-cc-pVTZ and CCSD(T)/aug-cc-pVTZ, and the results showed that the properties of open-shelled systems are significantly affected by the computational levels, which is consistent with the conclusion proposed by Qi et al. [18].

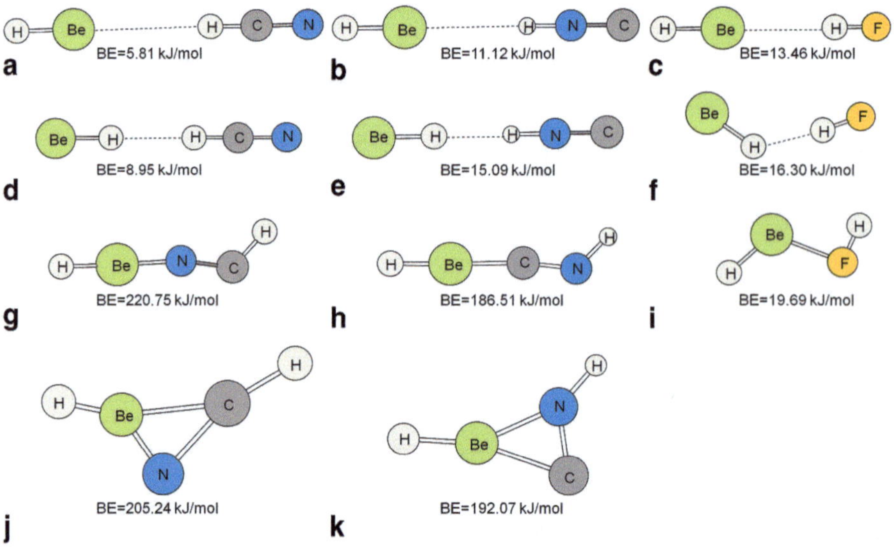

Fig. 5.6 Complexes with a single-electron hydrogen bond (**a–c**). a dihydrogen bond (**d–f**), and a covalent interaction (**g–k**) formed of BeH with HCN, HNC, and HF with binding energy (BE)

5.3 Other Types of Hydrogen Bonds with Radicals as Proton Acceptors

For a radical, it is probable to have one lone pair electron besides a single electron. In this case, the lone pair electron will be much more favorable to interact with the proton of the other monomer compared with the single electron. Thus, an interesting point is raised about the capability of accepting protons for such radicals and the neutral molecules.

The complex $OH\cdots H_2O$ has attracted much attention [32] due to its significance in chemical processes in the Earth's atmosphere [33], crystalline and amorphous ices [34], and aqueous environments including biological systems [35]. Both OH radical and H_2O molecule are able to be taken as the proton donor and acceptor in the formation of hydrogen bond, therefore, they could form two different types of hydrogen-bonded complexes: OH radical as a proton donor and H_2O molecule as the proton acceptor giving rise to the complex with C_s symmetry (Fig. 5.7a), and on the contrary, the complex (Fig. 5.7c) with OH as a proton acceptor and H_2O as the proton donor. The former is much more stable than the latter (the binding energy 22.61 *vs.* 13.38 kJ/mol), in which the former stable complex has been identified in the gas phase via microwave spectroscopy [32]. Thus, molecule H_2O is a better proton acceptor than OH in the complexes composed of H_2O and OH. Lai and Chou [36] performed a detailed study of the above two complexes composed of OH and H_2O at the UCCSD/6-31++G(d, p) level, and figured out the nature of OH and H_2O as proton acceptor in the complexes using NBO analysis. The $LP_O \rightarrow BD^*_{O-H}$ orbital

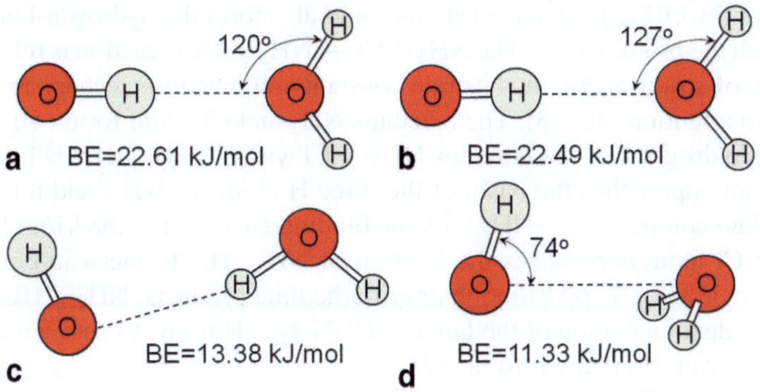

Fig. 5.7 Complexes formed between OH radical and H_2O with binding energy (BE)

interaction plays an important role in stabilizing the complexes (LP and BD* denote the lone pair and anti-bonding orbital, respectively), where OH radical possesses a lower LP_O and BD^*_{O-H} in energy than H_2O. They also performed the calculations to compare the complexes between $H_2O\cdots HO$ and $H_2S\cdots HO$ [36]. The binding energies were estimated to be 14.96 and 4.01 kJ/mol for the proton acceptor H_2O and H_2S, respectively, which are well correlated with the basicity of H_2O and H_2S as a proton acceptor.

In order to a get deeper understanding of such interactions, we studied the complexes composed of radicals OH/SH and molecules HCN/HNC [36], which are important species in interstellar space. The results further confirmed the conclusion that neutral molecules are a better proton acceptor than the radicals, and the center atom O of H_2O has greater ability of donating electron than the S atom of H_2S. However, in the complexes, $O(S)\cdots HCN(HNC)$, S atom becomes a stronger proton acceptor than O [36]. The binding energy was estimated to be 4.93 and 6.35 kJ/mol for the complexes $O\cdots HCN$ and $S\cdots HCN$, respectively, at the QCISD/6-311++G(2df,2p) level. Comparing them with that of the complexes $HO\cdots HCN$ (11.79 kJ/mol) and $HS\cdots HCN$ (7.65 kJ/mol), we come to the conclusion that the H atom in OH and SH plays an electron-donating role in the proton acceptor, which is favorable for the formation of hydrogen bond. This is the first report about the role of H in the monomer as the proton acceptor.

The interaction mode between OH and H_2O in Fig. 5.7b is the same as that in Fig. 5.7a, but the structure in the former has C_{2v} symmetry with one imaginary frequency of 102 cm^{-1} at the MP2/aug-cc-pVTZ level. The structure in Fig. 5.7d is stabilized by a two-center three-electron bonding (hemibonding) between the singly occupied local-π orbital on OH and the doubly occupied local-π lone pair on H_2O [36]. The hemibonding interaction between OH and H_2O was considered to have some contribution to the major ultraviolet absorption band of hydroxyl radical in water [40], even there is some controversy over its existence and the hemibonded water structure was considered to be an artifact of the exchange-correlation functional applied [41].

Besides the OH radical, naked atoms could also form the hydrogen-bonded complexes with the proton donor. The $NH_3 + Cl \rightarrow NH_2 + HCl$ reaction is relevant to the chemistry of stratosphere, thus its rate constant and reaction mechanism have been paid much attention [42, 43]. The reactants NH_3 and Cl could form two intermediates. One hydrogen-bonded complex $NH_3 \cdots Cl$ with C_{3v} symmetry is formed when the Cl atom approaches the basin of the three H atoms of NH_3, and a more stable intermediate complex $Cl \cdots NH_3$ with the binding energy of 29.68 kJ/mol is formed when the Cl atom approaches the N atom of NH_3. The former can eliminate the HCl molecule via a 32.60 kJ/mol barrier to the final products, $NH_2 + HCl$, whereas the further decomposition of the latter to $Cl\text{–}NH_2 + H$ needs to pass a much higher potential barrier of 203.57 kJ/mol [43].

It has been well established that halogen oxide radicals (ClO) play an important role in the ozone depletion events during the artic polar springtime [44], thus much attention has been paid to the hydration of ClO. Experimental investigations have confirmed the formation of $ClO \cdots H_2O$ complex [45], and its structures have also been unveiled by the theoretical calculations. Two isomers were found: a halogen-bonded structure with the O atom of water as the electron donor and with the Cl atom as the electron acceptor, and a hydrogen-bonded structure with the O atom of ClO as the proton acceptor and water as the proton donor [46]. It has been demonstrated that the halogen-bonded complex is more stable than the hydrogen-bonded one. The hydrated structures of $ClO(H_2O)_n$ with $n = 2\text{–}6$, were also studied and found that ClO radical is bonded to water clusters by means of the formation of the $OCl \cdots OH$ and $ClO \cdots HO$ network structures and give rise to the global minimum cyclic complex.

For HBe, HMg, H_2B, H_2Al, and H_2Ga radicals, they are able to form hydrogen bonds via not only the single electron on the center atom but also the hydridic hydrogen atom. These radicals could form dihydrogen bonds, which is an attractive interaction between the protonic hydrogen and the hydridic hydrogen [47]. The structures of dihydrogen-bonded complexes of H_2B and HBe with HCN, HNC, and HF are presented in Figs. 5.5 and 5.6, respectively. The $BeH \cdots HCN$ and $BeH \cdots HNC$ complexes are linear, and the $BeH \cdots HF$ complex is bent with the largest binding energy 16.30 kJ/mol. For H_2B radical, instead of the formation of the conventional dihydrogen bonds, the bifurcate dihydrogen bonds are formed with two hydrogen atoms participating in the $H \cdots H$ contacts (Fig. 5.5) [29]. The bifurcate dihydrogen bond is symmetric in $BH_2 \cdots HCN$ and $BH_2 \cdots HNC$ complexes but is asymmetric in $BH_2 \cdots HF$.

5.4 Hydrogen Bonds with Radicals as Proton Donors

For radical species, such as CH_3 and OH, they possess not only the single electron but also the acidic protons. Such species, generally, could also provide the protons to form hydrogen bonds. For instance, the methyl radical, it is able to form the single-electron hydrogen bonds as a proton acceptor (Fig. 5.4b, 5.4e, 5.4h), and to supply the proton forming the hydrogen bonds as a proton donor.

We compared the ability of donating protons for different types of radicals and the corresponding neutral molecules in the complexes H_mX–H$\cdots NH_3$ and H_{m-1} X–H$\cdots NH_3$ (X = O, S, Se, m = 1; X = N, m = 2; and X = C, m = 3) using the MP2, QCISD, and CCSD(T) methods [48]. The following conclusions were obtained: (1) the methyl radical could act as the proton donor besides the proton acceptor in the formation of hydrogen bonds, (2) the methyl radical is more facile to provide the proton than methane because the binding energy is much higher for $H_2CH\cdots NH_3$ complex (4.39 kJ/mol) than that for $H_3CH\cdots NH_3$ complex (2.34 kJ/mol) at the QCISD/aug-cc-pVTZ level, (3) Other radicals are also a better proton donor than the corresponding neutral molecules, (4) the ability of the radicals to provide the proton is in order of C–H < N–H < O–H and Se–H < S–H < O–H, which is similar with that of the neutral molecules, (5) the difference of donating protons between the radicals and the corresponding neutral molecules is related to both the acidity of proton donor and the basicity of proton acceptor. The binding energy is larger for the stronger acidic species, and on the other hand, the stronger proton acceptor leads to a bigger difference in the interaction strength between the complexes composed of $H_{m-1}X$–H and H_mX–H, respectively.

As discussed above, the CH_3 radical is able to act as the proton acceptor and donor in the formation of hydrogen bonds, thus, there is a competition between both types of interaction modes. CH_3 radical is facile to act as the proton acceptor when it interacts with water because the binding energy is 6.44 kJ/mol for $H_3C\cdots HOH$ complex and 4.18 kJ/mol for $H_2CH\cdots OH_2$ (Fig. 5.4b). The similar trend is also found in the systems of H_3C with HCN, HNC, and H_2S. On the contrary, CH_3 radical is more favorable to act as the proton donor when it interacts with NH_3 with the binding energy of 3.76 kJ/mol for $H_3C\cdots HNH_2$ complex and 4.77 kJ/mol in $H_2CH\cdots NH_3$ (Fig. 5.4h). Thus the ability of donating the proton and electron for CH_3 radical is also related to the acidity and basicity of the other monomer in the complexes.

The spectroscopic properties of the SH radical and its hydrated complex are crucial in understanding sulfur transformation in the atmosphere. Du and Francisco [49] performed the high level quantum chemical calculations on the equilibrium geometries, binding energies, and spectroscopic properties of the complex $SH\cdots H_2O$. Two stable isomers were found with the SH radical as the proton donor forming the global minimum complex and as the proton acceptor forming the local minimum one. The binding energies of both complexes are 12.16 and 10.95 kJ/mol, respectively. The global minimum of $SH\cdots H_2O$ complex is less stable than that of $OH\cdots H_2O$ due to less overlap of the lone pair orbital in water LP_O with the BD^*_{S-H} antibonding orbital.

Solimannejad and Ghafari [50] analyzed the intermolecular interactions in ternary radical–molecule complexes between molecules HCN(HNC) and radicals HO(HS) in gas phase and in water media at the MP2/cc-pVTZ level. There are three structures for each ternary system with the radical as the proton donor, the proton acceptor, and a dual role of both donor and acceptor, respectively. The geometries of the first two are chained with a favorable cooperative effect, whereas the last one is cyclic with a diminutive effect. Many-body interaction analyses indicate that hydrogen bonding between two HCN (HNC) molecules gives more stability to the triads than hydrogen bonding between HCN (HNC) and OH (SH) species. The water media has an enhancing effect on the stabilities of the complexes. Solimanejad and Scheiner [51] compared the ability of donating proton of SH and H_2S in the SH$\cdots$N hydrogen bond with a series of small molecules containing nitrogen atom as the proton acceptor. They found that the SH$\cdots$N interaction involving with SH radical is slightly stronger than the $H_2S\cdots N$ hydrogen bond in which the closed-shell H_2S serves as the proton donor.

5.5 Competition Between Single-Electron Hydrogen Bonds and Other Interactions

It is possible for some radicals that there are several sites of Lewis acid/base, thus, when they form complexes with other monomers, several types of interactions probably coexist. Unavoidably, there will be competition among the formation of different types of interactions. We selected some examples to address this phenomenon.

Besides the formation of the single electron hydrogen-bonded complex, CH_3 radical is able to form a covalent-bonded complex with HCN and HNC (Fig. 5.3e, 5.3f). The binding energies are 62.99 and 167.37 kJ/mol, respectively, which are much higher than that of single-electron hydrogen bonds [17]. H_3C radical essentially plays different roles in both interactions. It acts as a Lewis acid in the covalent-bonded complex (Fig. 5.3e, 5.3f), and a Lewis base in the single electron hydrogen-bonded one (Fig. 5.3a, 5.3b). In the covalent interaction, the CH_3 radical interacts with the Lewis base through the alpha-LUMO. The stronger covalent interaction leads to a more remarkable change of H_3C geometry, varying from the plane structure in the isolated monomer to the umbrella-like configuration in the complex. Simultaneously, the geometries of HCN and HNC also have a big change, varying from a linear structure in the monomer to a bent one in the complex. Another covalent-bonded structure is also formed as shown in Fig. 5.3g, 5.3h, where the middle C atom of HCN and the middle N atom of HNC are introduced into the alpha-LUMO of the CH_3 radical. Similarly, the geometries of this type covalent-bonded complex also changed significantly. Compared the binding energies of four complexes (Fig. 5.3e, 5.3f, 5.3g, 5.3h), the strength of the covalent-bonded C–N bond is much weaker than that of the C–C bond, which is in contrast with the bonding energy of the conventional C–C and C–N bonds.

In a similar way, H_2B radical also forms a covalent-bonded complex (Fig. 5.5g, 5.5h) with HCN and HNC, besides the single electron hydrogen-bonded complex [30]. The binding energy in the covalent-bonded complex is 241.69 kJ/mol and 291.97 kJ/mol at the MP2/aug-cc-pVTZ level for $H_2B\cdots$HCN and $H_2B\cdots$HNC, respectively, which are much higher than that in the single electron hydrogen-bonded complexes (Fig. 5.5a, 5.5b). The covalent-bonded complex of $H_2B\cdots$HNC is more stable than that of $H_3C\cdots$HNC with the binding energy of 167.28 kJ/mol at the same level. As a consequence, the H_2B radical is not only a stronger Lewis base in the hydrogen bond but also a stronger Lewis acid in the covalent interaction than the H_3C radical. Similar with H_3C, H_2B radical also interacts with the middle C atom of HCN and the middle N atom of HNC giving rise to the other patterns of the covalent-bonded structures presented in Fig. 5.5j, 5.5k. Both types of covalent-bonded complexes, resulting in a change of geometry of HCN and HNC, are more stable for the HNC complex than for the HCN complex. During the formation of both types of covalent-bonded complexes, H_2B provides the beta-LUMO to interact with HCN and HNC, even the energy of the beta-LUMO is higher than that of the alpha-LUMO. This is different from that in the H_3C counterpart. However, the F atom of HF attacks the alpha-LUMO of H_2B, giving rise to a complex (Fig. 5.5i) with the binding energy of 12.92 kJ/mol, which can compete with the single electron hydrogen-bonded complex with the binding energy of 15.80 kJ/mol.

HBe also forms two types of covalent-bonded complexes with HCN and HNC by its beta-LUMO, and its alpha-LUMO is used as the Lewis acid to interact with HF, as shown in Fig. 5.6g–5.6k. However, different from the complexes composed of H_2B and CH_3 radicals, instead of singly covalent-bonded to middle atoms of HCN and HNC, HBe radical could be bicovalent-bonded to both middle and side N/C atoms and forms tricyclic structures (Fig. 5.6j, 5.6k). Both covalent interactions between HBe and HCN are stronger than those in the HNC counterpart, which are in contrast with the complexes of H_2B with HCN(HNC). On the other hand, the conformation of the HBe covalent-bonded complex with HF is different from that with HCN and HNC. The covalent interaction between HBe and HF is roughly equal to that of the single-electron hydrogen bond and the dihydrogen bond (discussed in the following paragraph), indicating that there is a competition during the formation of the three-type complexes between HBe radical and HF molecule.

Besides the formation of the single-electron hydrogen bond, for radicals HBe, HMg, H_2B, H_2Al, and H_2Ga, they could also form dihydrogen bonds with the basic H atom participating in the hydrogen bond formation. Radicals H_2B, H_2Al, and H_2Ga have two types of dihydrogen bond structures: one has C_{2v} symmetry with the two basic H atoms interacting with the proton of the other monomer, and the other is of C_s symmetry with one basic H atom to interact with the proton. The stability of the dihydrogen-bonded complex is in order of $H_2B < H_2Ga < H_2Al$. For the BH_2 radical, the single-electron hydrogen bond is stronger than the dihydrogen bond, while the opposite phenomenon is found for the AlH_2 and GaH_2 systems [29].

In $H_3C\cdots H_2O$ complex, a local minimum (Fig. 5.4b) is also found with one of the hydrogen atoms of H_3C pointing towards the oxygen atom [20], which is lower in binding energy than the most stable $H_3C\cdots H_2O$ complex with a single-electron hydrogen bond. Similar with the covalent bonded complexes of H_3C radical with molecules HCN and HNC, the covalent-bonded complex of H_3C and H_2O is also obtained (Fig. 5.4c). This covalent-bonded complex is weaker than the H_2S counterpart (Fig. 5.4f), which is reverse to the electron-donating ability of O and S atoms. In the covalent-bonded complex of H_3C and H_2S, the strong covalent interaction leads to a big change of H_3C and H_2S geometries, and the H–S–H angle has a large increase from the monomer (92.2°) to the complex (150.6°). Similarly, the methyl radical also forms a covalent bonded complex with NH_3 (Fig. 5.44i) with the C$\cdots$N distance of 1.497 Å, which is much smaller than the sum of van der Waals of the C and N atoms (3.2 Å). However, its binding energy is negative, indicating that it is not favorable to form this complex.

When the C atom in the CH_3 radical is replaced with other atom in the Group IV, the corresponding radicals H_3A (A = Si, Ge, Sn, and Pb) could also form the single-electron hydrogen bond with H_3O^+ although it is weaker than that of H_3C. For the heavier Group IV atom, the hydrogen atoms of H_3A have a partial negative charge, thus they are able to form a dihydrogen bond with the proton in H_3O^+ [28]. For the complexes composed of radicals H_3Si and H_3Ge, the single-electron hydrogen bond is about 8 $\sim$ 17 kJ/mol more stable than the dihydrogen bond, whereas the dihydrogen bond is 33.44 kJ/mol more stable than the single electron hydrogen bond for the H_3Sn complex. The corresponding dihydrogen bond involving H_3Pb was not obtained because it changed to be other products such as H_2 in the optimization process, due to the strong dihydrogen bond. These results suggested that oxidation can induce formation of molecular hydrogen through the formation of unusual hydrogen-bonded systems [28].

5.6 Nature of Hydrogen Bonds Involving Radicals

Many studies have been performed for hydrogen bonds involving radicals; however, their nature has not been investigated systematically. Here, we applied the localized molecular orbital energy decomposition analysis (LMOEDA) [52] implemented in the GAMESS program [53] to decompose the binding energy of hydrogen bonds involving radical species into five physical components: electrostatic (E^{ele}), exchange (E^{ex}), repulsion (E^{rep}), polarization (E^{pol}), and dispersion (E^{disp}) energies, and the results are listed in Table 5.1. It is obvious from Table 5.1 that for each type of interaction the E^{ex} contribution is the largest among the four attractive terms. The exchange interaction is correlated with the overlap of the molecular orbitals, thus, the larger E^{ex} means the larger orbital interactions between two monomers. On the other hand, the large overlap of orbitals indicates that two interacting monomers are close in spatial distance, which results in the high repulsive energy E^{rep}. This has been confirmed by the positive correlation between E^{ex} and E^{rep} in their absolute value (Table 5.1), where the absolute value of the latter is nearly two times higher than that of the former. Considering the dependency of E^{rep} on E^{ex} and the fact that both terms are often not discussed separately in most energy analyses, we paid our main attention to comparing the contribution of E^{ele}, E^{pol}, and E^{disp} attractive terms to the stability of the complexes.

For the single electron hydrogen-bonded complexes $H_3C\cdots HX$ (X = F, Cl, and Br), the E^{ele} term is the most negative, indicating a dominant electrostatic contribution in stabilizing these complexes. With the increase of X atomic number, the E^{disp} term becomes more negative, and the most negative E^{ele} and E^{pol} occur in $H_3C\cdots HBr$, which is inconsistent with the smallest binding energy in this complex. This abnormal result is different from most conventional hydrogen bonds, and is rationalized by

Table 5.1 Electrostatic (E^{ele}), exchange (E^{ex}), repulsion (E^{rep}), polarization (E^{pol}), and dispersion (E^{disp}) energies in the complexes at the MP2/aug-cc-pVTZ level. All are in kJ/mol

	E^{ele}	E^{ex}	E^{rep}	E^{pol}	E^{disp}
Fig. 5.2a-F	– 16.93	– 25.67	47.23	– 11.12	– 16.93
Fig. 5.2a-Cl	– 16.47	– 37.16	65.12	– 11.04	– 16.47
Fig. 5.2a-Br	– 20.65	– 53.75	94.72	– 15.63	– 20.65
Fig. 5.2b-Br	– 6.69	– 16.72	28.51	– 1.80	– 6.69
Fig. 5.2b-I	– 12.50	– 31.48	53.71	– 4.60	– 12.50
Fig. 5.3a	– 9.45	– 13.21	23.78	– 4.14	– 9.45
Fig. 5.3b	– 14.88	– 22.82	42.22	– 8.86	– 14.88
Fig. 5.3c	– 4.31	– 7.48	12.37	– 1.21	– 4.31
Fig. 5.3d	– 4.35	– 8.61	14.09	– 1.17	– 4.35
Fig. 5.4a	– 10.07	– 17.56	30.68	– 4.22	– 10.07
Fig. 5.4b	– 6.02	– 9.45	15.80	– 1.63	– 6.02
Fig. 5.4d	– 7.44	– 19.40	32.52	– 3.34	– 7.44
Fig. 5.4e	– 3.64	– 8.90	14.46	– 0.92	– 3.64
Fig. 5.4g	– 5.35	– 10.37	17.39	– 1.30	– 5.35
Fig. 5.4h	– 7.69	– 13.50	22.11	– 2.30	– 7.69
Fig. 5.5a	– 12.50	– 13.75	24.91	– 4.10	– 12.50
Fig. 5.5b	– 21.65	– 28.26	52.50	– 10.07	– 21.65
Fig. 5.5c	– 27.42	– 37.12	–68.30	– 14.50	– 27.42
Fig. 5.5d	– 2.68	– 7.06	12.54	– 2.38	– 2.68
Fig. 5.5e	– 4.77	– 12.16	22.07	– 4.81	– 4.77
Fig. 5.5f	– 5.85	– 13.75	24.83	– 6.14	– 5.85
Fig. 5.6a	– 9.78	– 11.08	20.15	– 3.34	– 9.78
Fig. 5.6b	– 17.18	– 23.07	42.76	– 8.07	– 17.18
Fig. 5.6c	– 22.78	– 32.23	59.23	–12.46	– 22.78
Fig. 5.6d	– 12.54	– 13.79	24.66	– 4.72	– 12.54
Fig. 5.6e	– 17.85	– 23.24	42.80	– 10.12	– 17.85
Fig. 5.6f	– 24.08	– 34.65	63.66	– 15.76	– 24.08

the largest E^{rep} in $H_3C\cdots HBr$. However, the single electron halogen-bonded complexes $H_3C\cdots XH$ (X = Br and I) are jointly stabilized by electrostatic and dispersion energies, which respectively accord with the positive electrostatic potential on the halogen atom [54] and the atomic radius.

For the single electron hydrogen-bonded complexes $H_3C\cdots HCN$ and $H_3C\cdots HNC$, the values of E^{pol} and E^{disp} are almost equal, but both of them are smaller than E^{ele}, indicative of the electrostatic nature of the single-electron hydrogen bond. For the C–H$\cdots$N/C hydrogen bonds in $H_2CH\cdots NCH$ and $H_2CH\cdots CNH$ complexes, the contribution of E^{disp} is comparable to that of E^{ele}, and the relatively large E^{disp} is consistent with the weak interaction.

The dominant role of electrostatic interaction is the same case for the single electron hydrogen-bonded complexes of H_2B and BeH with HCN, HNC, and HF, and this conclusion is similar to that in hydrogen bond of water dimer [52]. With the increase of binding energy, the ratio of E^{pol} to E^{ele} is also increased from 0.33 in $H_2B\cdots HCN$ to 0.53 in $H_2B\cdots HF$, and from 0.34 in $HBe\cdots HCN$ to 0.55 in $HBe\cdots HF$. The larger E^{pol} means that the orbitals undergo more observable changes in their shapes. The E^{disp} contribution is the smallest in these complexes, although it is greater for the stronger interaction.

For the weak single-electron hydrogen bonds in the complexes of $H_3C\cdots HOH$, $H_3C\cdots HSH$, and $H_3C\cdots HNH_2$, the contribution of E^{disp} is close to that of E^{ele}. A similar result is also found for the weak C–H$\cdots$Y (Y = O, S, and N) hydrogen bonds.

The dihydrogen bonds in the complexes of BeH with HCN, HNC, and HF are comparable in strength to the corresponding single-electron hydrogen bonds, thus the three attractive terms (E^{ele}, E^{pol}, and E^{disp}) show similar roles in both types of hydrogen bonds. However, for the weak dihydrogen bonds in the complexes of H_2B with HCN, HNC, and HF, the three attractive terms make almost equal contribution to stabilize these complexes.

5.7 Cooperative Effects

Cooperative effect is an important property of hydrogen bonds, which plays an important role for the applications of hydrogen bonds in molecular recognition, crystal engineering, and chemical reactions. Thus much attention has been paid to the cooperativity among hydrogen bonds or with other types of non-covalent interactions.

An *ab initio* calculation at the MP2/aug-cc-pVTZ level has been performed for the complexes $H_3C\cdots HCN\cdots HCN$ and $H_3C\cdots HNC\cdots HNC$ [55], where a single-electron hydrogen bond coexists with another type of hydrogen bond. The equilibrium structures, frequency shifts, NMR chemical shifts, binding energies, atom charges, charge transfers, and electron densities in both trimers were studied in order to investigate the cooperative effect between the single-electron hydrogen bond and the N$\cdots$H–C and C$\cdots$H–N hydrogen bonds in $H_3C\cdots HCN\cdots HCN$ and $H_3C\cdots HNC\cdots HNC$, respectively. In $H_3C\cdots HCN\cdots HCN$ trimer, the binding energy of single-electron hydrogen bond increases by 22 % compared to that of dimer $H_3C\cdots HCN$, whereas that of N$\cdots$H–C hydrogen bond increases by 7 %. These results show significant cooperativity between both types of hydrogen bonds, as well as the changes in the binding distances and bond lengths. A similar result is also found for the $H_3C\cdots HNC\cdots HNC$ trimer.

In $H_2B\cdots HCN\cdots HCN$ and $H_2B\cdots HNC\cdots HNC$ trimers, the single-electron hydrogen bond exhibits a similar cooperative effect with another type of hydrogen bond [30]. The binding energies are 7.19 and 19.40 kJ/mol in $H_2B\cdots HCN$ and $HCN\cdots HCN$ dimers, respectively, that is, the single-electron hydrogen bond in $H_2B\cdots HCN$ dimer is weaker than the hydrogen bond N$\cdots$H–C in $HCN\cdots HCN$. The formation of the trimer enhances the binding energy of single-electron hydrogen bond by 170 % in $H_2B\cdots HCN\cdots HCN$ trimer and 110 % in $H_2B\cdots HNC\cdots HNC$, and by 34 and 37 % for the hydrogen bond in $H_2B\cdots HCN\cdots HCN$ and $H_2B\cdots HNC\cdots HNC$ trimers, respectively. These results support the conclusion that the stronger interaction has a greater effect on the weaker one [56]. Compared the binding energy of different system, the enhancement of single-electron hydrogen bond in $H_2B\cdots HCN\cdots HCN$ and $H_2B\cdots HNC\cdots HNC$ trimers is more prominent than that in $H_3C\cdots HCN\cdots HCN$ and $H_3C\cdots HNC\cdots HNC$ ones.

In $H_2O\cdots H_3O^+\cdots C(CH_3)_3$ complex, where H_3O^+ as a double proton donor interacts with H_2O and $C(CH_3)_3$, respectively, both O$\cdots$H–O and C$\cdots$H–O hydrogen bonds in the ternary system are less strong than those in the corresponding binary system, giving rise to smaller bond length changes and red-shifts, thus a negative cooperative effect is present [25].

5.8 Conclusions

The hydrogen bonds involving radical species have been discussed in this chapter. In many chemical reactions and processes, it is possible that the radicals form the complexes with other molecules via different types of interactions, especially the hydrogen bonds. Great progress has been achieved in the understanding of the hydrogen bonds involving radical species, both experimentally and theoretically. Many radical species as the proton acceptor are able to form the single electron hydrogen-bonded complexes with the neutral molecules. The free single electron in radical species participates in the formation of the open-shelled hydrogen bond with its direction towards the proton of the other molecule. The binding energies of the complexes are generally smaller than 20 kJ/mol. The magnitude of the binding energy is proportional to the acidity of the proton donor and will be enhanced significantly with the protonation of the proton donor or the substitutions of the radical species and neutral molecules with electron-withdrawing and electron-denoting groups, respectively. Theoretically, the high level computational method applied is important to reproduce the properties of the open-shelled hydrogen-bonded complexes, for instance, UCCSD(T)/6-311++G(2d,2p) method, including the large basis set and electron strong correlation effect, could give rise to a good result [18]. On the other hand, radicals could also form other types of hydrogen bonds including dihydrogen bonds, with binding energy usually smaller than that of the single-electron hydrogen bonds. Therefore, there will be competitions to form different types of interactions between radical species and other molecules. Essentially, the nature of the hydrogen bonds involving radical species is similar with the conventional ones where the attractive contribution depends on the properties of both the radical itself and the other monomer. Furthermore, there also exists the cooperativity in the trimers or larger clusters involving radicals, with the larger effect on the weaker interaction, which is coincident with that of the conventional hydrogen bonding interactions.

Acknowledgements This work was supported by the Outstanding Youth Natural Science Foundation of Shandong Province (JQ201006) and the Program for New Century Excellent Talents in University (NCET-2010-0923).

References

1. Wallington TJ, Dagaut P, Kurylo MJ (1992) Chem Rev 92:667–710
2. Calvert JG, Lazrus A, Kok GL, Heikes BG, Walega JG, Lind J, Cantrell CA (1985) Nature 317:27–35
3. Jeffrey GA (1997) An introduction to hydrogen bonding; Oxford University Press: New York
4. de Visser SP Shaik S (2003) J Am Chem Soc 125:7413–7424
5. Lucarini M, Mugnaini V, Pedulli GF, Guerra M (2003) J Am Chem Soc 125:8318–8329
6. Wang BQ, Li ZR, Wu D, Hao XY, Li RJ, Sun CC (2003) Chem Phys Lett 375:91–95
7. Sander W, Roy S, Polyak I, Ramirez-Anguita JM, Sanchez-Garcia E (2012) J Am Chem Soc 134:8222–8230
8. Frisch MJ, Trucks GW, Schlegel HB, Scuseria GE, Robb MA, Cheeseman JR, Montgomery JA Jr, Vreven T, Kudin KN, Burant JC, Millam JM, Iyengar SS, Tomasi J, Barone V, Mennucci B, Scalmani G, Cossi M, Rega N, Petersson GA, Nakatsuji H, Hada M, Ehara M, Toyota K, Fukuda R, Hasegawa J, Ishida M, Nakajima T, Honda Y, Kitao O, Nakai H, KleneMLX, Knox JE, Hratchian HP, Cross JB, Adamo C, Jaramillo J, Gomperts R, Stratmann RE, Yazyev O, Austin AJ, Cammi R, Pomelli C, Ochterski JW, Ayala PY, Morokuma K, Voth GA, Salvador P, Dannenberg JJ, Zakrzewski VG, Dapprich S, Daniels AD, Strain MC, Farkas O, Malick DK, Rabuck AD, Raghavachari K, Foresman JB, Ortiz JV, Cui Q, Baboul AG, Clifford S, Cioslowski J, Stefanov BB, Liu G, Liashenko A, Piskorz P, Komaromi I, Martin RL, Fox DJ, Keith T, Al-Laham MA, Peng CY, Nanayakkara A, Challacombe M, Gill PMW, Johnson B, Chen W, Gonzalez C, Wong MW, Pittsburgh PA, Pople JA (2009) Gaussian 09, revision A02. Gaussian Inc, Wallingford
9. Jacox ME (1979) Chem Phys 42:133–148
10. Johnson GL, Andrews L (1980) J Am Chem Soc 102:5736–5741
11. Misochko EY, Benderskii VA, GoldschlegerAU, Akimov AV, Shestakov AF (1995) J Am Chem Soc 117:11997–11998
12. Tachikawa H (1998) J Phys Chem A 102:7065–7069
13. Clark T, Hennemann M, Murray JS, Politzer P (2007) J Mol Model 13:291–296
14. Chen Y, Tschuikow-Roux E, Rauk A (1991) J Phys Chem 95:9832–9836
15. Raghavendra B, Arunan E (2007) J Phys Chem A 111:9699–9706
16. Rudić S, Merritt JM, Miller RE (2006) J Chem Phys 124:104305
17. Solimannejad M, Alikhani ME (2005) Chem Phys Lett 406:351–354
18. Qi XJ, Liu L, Fu Y, Guo QX (2005) Struct Chem 16:347–353
19. Zhang B, Shiu W, Liu K (2005) J Phys Chem A 109:8983–8988
20. Rudic S, Merritt JM, Miller RE (2009) Phys Chem Chem Phys 11:5345–5352
21. Hashimoto T, Iwata S (2002) J Phys Chem A 106:2652–2658
22. An XL, Liu HP, Li QZ, Gong BA, Cheng JB (2008) J Phys Chem A 112:5258–5263
23. Li QZ, Zhu HJ, An XL, Gong BA, Cheng JB (2009) Int J Quantum Chem 109:605–611
24. Li QZ, An XL, Gong BA, Cheng JB (2008) J Mol Struct Theochem 866:11–14

25. Hammerum S (2009) J Am Chem Soc 131:8627–8635
26. Spiegel DA, Wiberg KB, Schacherer LN, Medeiros MR, Wood JL (2005) J Am Chem Soc 127:12513–12515
27. Pozzi D, Scanlan EM, Renaud P (2005) J Am Chem Soc 127:14204–14205
28. Gil A, Sodupe M, Bertran (2004) Chem Phys Lett 395:27–32
29. Solimannejad M, Alkorta I (2006) J Phys Chem A 110:10817–10821
30. Wang WJ, Li QZ (2011) Comput Theor Chem 966:128–133
31. Li QZ, Li R, Yi SC, Li WZ, Cheng JB (2012) Struct Chem 23:411–416
32. Ohshima Y, Sato K, Sumiyoshi Y, Endo Y (2005) J Am Chem Soc 127:1108–1109
33. Aloisio S, Francisco JS (2000) Acc Chem Res 33:825–830
34. Langford VS, McKinley AJ, Quickenden TI (2000) Acc Chem Res 33:665–671
35. Hamad S, Lago S, Mejias JA (2002) J Phys Chem A 106:9104–9113
36. Lai CH, Chou PT (2007) J Comput Chem 28:1357–1363
37. Jing B, Li QZ, Gong BA, Cheng JB, Li WZ, Liu ZB (2010) Mol Phys 108:1655–1664
38. Jing B, Li QZ, Gong BA, Liu ZB, Li WZ, Cheng JB, Sun JZ (2011) Mol Phys 109:831–838
39. Yamaguchi M (2011) J Phys Chem A 115:14620–14628
40. Chipman DM (2011) J Phys Chem A 115:1161–1171
41. Vande J, Sprik M (2005) Phys Chem Chem Phys 7:1363–1367
42. Gao Y, Alecu IM, Hsieh PC, Morgan BP, Marshall P, Krasnoperov LN (2006) J Phys Chem A 110:6844–6850
43. Xu ZF, Lin MC (2007) J Phys Chem A 111:584–590
44. Sander SP, Friedl RR, Yung YL (1989) Science 245:1095–1098
45. McKeachie JR, Appel MF, Kirchner U, Schindler RN, Benter T (2004) J Phys Chem B 108:16786–16797
46. Galvez O, Gomez PC (2007) Chem Phys Lett 448:16–23
47. Custelcean R, Jackson JE (2001) Chem Rev 101:1963–1980
48. Li QZ, Kou H, Li R, Li WZ, Cheng JB (2011) Comput Theor Chem 976:83–87
49. Du S, Francisco JS (2009) J Chem Phys 130:124304
50. Solimannejad M, Ghafari S (2013) Struct Chem 24:1493–1498
51. Solimanejad M, Scheiner S (2011) Int J Quantum Chem 111:3196–3200
52. Su PF, Li H (2009) J Chem Phys 131:014102
53. Schmidt MW, Baldridge KK, Boalz JA, Elbert ST, Gorden MS, Jensen JH, Koseki S, Matsunaga N, Nguyen KA, Su SJ, Windus TL, Dupuis M, Montgomery JA Jr (1993) J Comput Chem 14:1347–1363
54. Politzer P, Murray JS, Clark T (2010) Phys Chem Chem Phys 12:7748–7757
55. Li QZ, An XL, Luan F, Li WZ, Gong BA, Cheng JB, Sun JZ (2008) J Chem Phys 128:154102
56. Li QZ, Lin QQ, Li WZ, Cheng JB, Gong BA, Sun JZ (2008) ChemPhysChem 9:2265–2269

Chapter 6
Agostic and Hydrogen-Bonding X–H$\cdots$M Interactions Involving a d^8 Metal Center: Recent Advances Towards Their Understanding

Jiří Kozelka

Abstract The binding of d^8 transition metal ions to X–H bonds (X = non-metal) has been subject of intense research in the last two decades. Two different types of orbital interactions can stabilize X–H$\cdots$M bonds: (1) charge transfer from a filled orbital of the metal into the empty σ^*-antibonding orbital of the X–H bond; (2) charge transfer from the filled σ-bonding orbital of the X–H bond into an empty orbital of the metal. The first type corresponds to a hydrogen bond, whereas the second is commonly designated as an agostic bond. The present article analyses experimental and theoretical approaches to the characterization of these two interaction types in d^8 transition metal complexes, points out some assignment errors that occurred in the past, and summarizes recent advances towards the understanding of the structure, dynamics, and physical origin of these weak interactions.

6.1 Introduction

Metal centers can interact with a hydrogen atom which is bound to another atom, thus forming a weak M$\cdots$H–X bond. Such bonds can have two different origins. In the first case, they result from the donation of electron density of the X–H σ bond to an empty orbital of the metal. Such bonds, named agostic bonds, usually fill a vacant coordination site, i.e., complete the coordination sphere of the metal ion. In the second type, the σ^*-antibonding X–H orbital interacts with a lone-pair of the metal, thus forming a (non-conventional) hydrogen bond. Here, a metal complex *extends* its usual coordination sphere by the hydrogen-bonding interaction. Whereas the agostic bonds are typical of electron-deficient metal centers, the non-conventional hydrogen bonds occur in electron-rich centers. According to the number of electrons involved

J. Kozelka (✉)
Laboratoire de Chimie et Biochimie Pharmacologiques et Toxicologiques, UMR 8601 CNRS, Université Paris Descartes, 45, rue des Saints-Pères, 75270 Paris, France
e-mail: kozelka.jiri@gmail.com

Department of Condensed Matter Physics, Faculty of Science, Masaryk University, Kotlářská 2, 611 37 Brno, Czech Republic

S. Scheiner (ed.), *Noncovalent Forces*, Challenges and Advances in Computational Chemistry and Physics 19, DOI 10.1007/978-3-319-14163-3_6

in the bonding of the system, agostic bonds are formally 3-center-2-electron (3c–2e) bonds, whereas hydrogen bonds correspond to 3-center-4-electron (3c–4e) bonds.

A particularly interesting class is constituted by square-planar d^8 complexes. These possess sterically accessible filled nd_{xz}, nd_{yz}, and nd_{z2} orbitals (z axis assumed to be perpendicular to the coordination plane) which can serve as hydrogen bond acceptors, but they can also form agostic bonds with their vacant nd_{x2-y2} and $(n+1)s$ orbitals, when no ligands are available to form more stable bonds. Figure 6.1 shows representative examples of C–H···Pt interactions, allowing the two interaction types to be distinguished at first glance: whereas the hydrogen-bonding interaction (**1**) involves a coordinatively saturated square-planar metal ion and a C–H bond contacting Pt(II) along the z-axis, in the agostic interactions (**2–4**), the C–H bonds assume the function of the fourth ligand in an otherwise coordinatively unsaturated Pt(II) center. Both interaction types involve charge transfer (CT), but in opposite directions: in the hydrogen-bonding case, from a filled orbital of platinum to the σ^*-antibonding orbital of the C–H bond, and in the agostic interactions, from the C–H σ-bond to the nd_{x2-y2} orbital.

Several review articles on agostic [1–4] and hydrogen-bonding [5–7] X–H···M interactions have appeared. In the present review, we focus on contributions to the understanding of the physical basis of X–H···M interactions involving a d^8 metal center that were published during the last two decades.

6.2 Structural Studies Revealing Agostic or Hydrogen-Bonding X–H···M(d^8) Interactions

Although it is apparent from Fig. 6.1 that a straightforward distinction between agostic and hydrogen bonding X–H···M(d^8) interactions can be made simply by considering whether the interaction is part of the square-planar coordination sphere (yes: agostic, no: hydrogen-bonding), some confusion occurred in the past. Already the seminal paper by Brookhart and Green in which the designation "agostic bond" was introduced [12] contains two erroneous assignments of C–H···M(d^8) hydrogen bonds observed in the structures of the palladium(II) complexes *trans*-$[PdBr\{C_4(CO_2Me)_4H\}(PPh_3)_2]$ (Fig. 6.2, **5**) [13] and *trans*-$[PdI_2(PMe_2Ph)_2]$ [14] to agostic bonds. It is interesting to note that Maitlis et al., who reported the structure of **5**, already suggested that the C–H···Pd interaction may be considered as a hydrogen bond.

In the following decade, Albinati, Pregosin et al. characterized a number of Pt(II), Pd(II), and Rh(I) complexes featuring a saturated square-planar coordination sphere and showing an extra intramolecular axial contact with a C–H or an N–H bond [15–20]. The authors observed that these interactions show spectroscopic features different from those of typical agostic interactions but did not identify them as hydrogen bonds, rather, they designated them as "weak" or "pregostic". Nevertheless, they noted that the interaction is 3c–4e in nature [20], and thus similar to hydrogen bonding. Similar intramolecular C–H···M contacts were identified in 8-methylquinoline

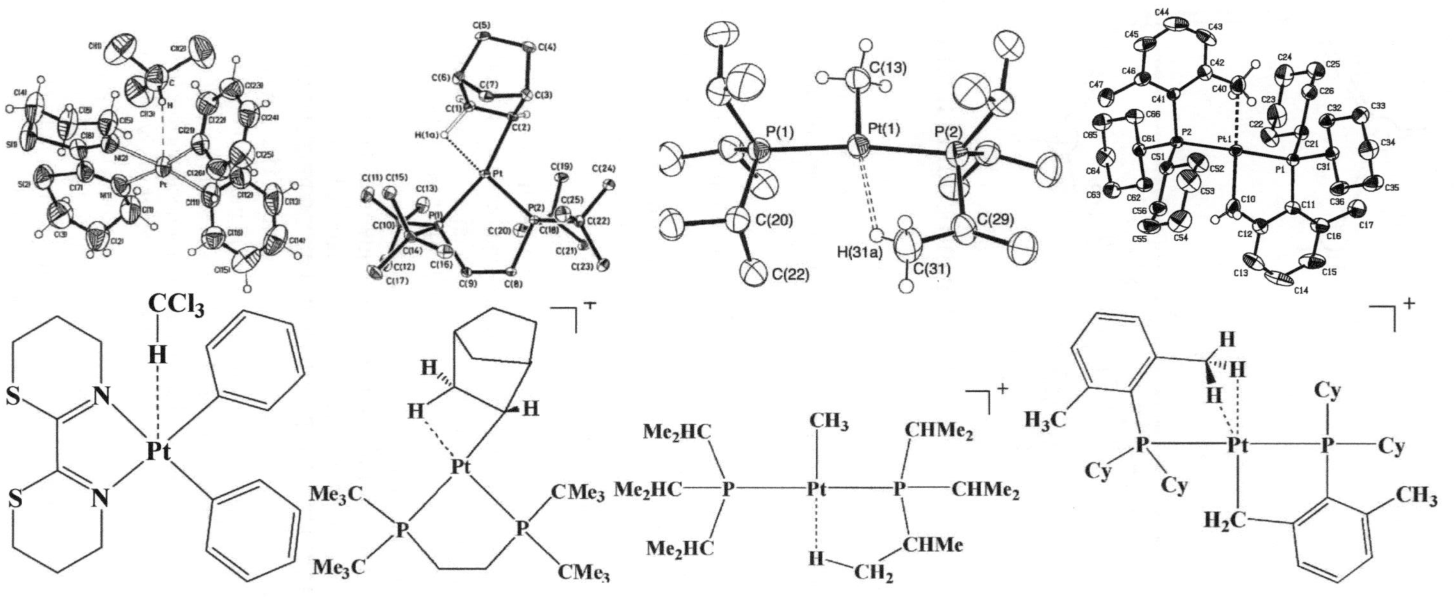

1 hydrogen-bonding

2 β-agostic

3 γ-agostic

4 bifurcated δ-agostic

Fig. 6.1 Crystal structures and structural formulas of [Pt(C_6H_5)$_2$(2,2'-Bi-5,6-dihydro-4*H*-1,3-thiazine-*N, N'*)].$CHCl_3$ (**1**), [Pt(norbornyl) (*P, P'*–$Bu^t{}_2PCH_2$–$CH_2PBu^t{}_2$)]$^+$(**2**), [Pt(Me)(P^iPr_3)$_2$]$^+$(**3**) and [Pt{P(2,6-$Me_2C_6H_3$)Cy_2}{κ^2-*P, C*–P(2-Me-6-CH_2–C_6H_3)Cy_2}]$^+$(**4**), exemplifying hydrogen-bonding (**1**) and agostic (**2–4**) C–H··· Pt interactions, respectively. The ORTEP figures were adapted with permission from refs. [8] (http://journals.iucr.org/), [9–11], respectively

Fig. 6.2 Crystal structure details of *trans*-$[PdBr\{C_4(CO_2Me)_4H\}(PPh_3)_2]$ (**5**) showing the hydrogen-bonding interaction which was later assigned erroneously as agostic [12]. Phenyl substituents on the phosphorus atoms and ester groups on the butadienyl carbons were omitted for clarity. Reproduced with permission from [13]

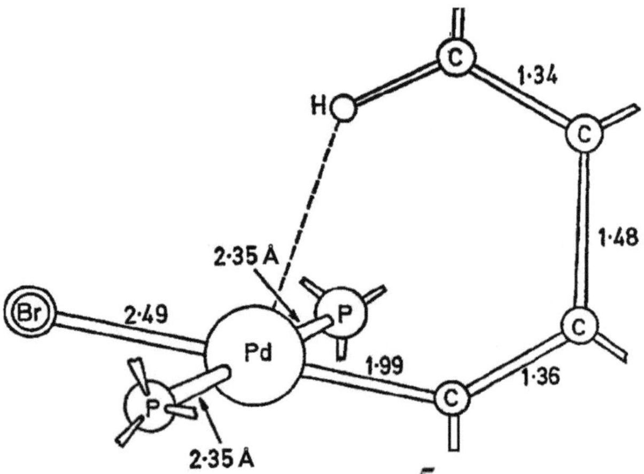

5
hydrogen-bonding interaction

complexes of Rh(I) and Ir(I) by Neve et al. [21, 22]; these authors called the interactions "weak" or "remote agostic", and incorrectly interpreted them as interactions with "the unsaturated rhodium (or iridium) center", although these d^8 metal centers were coordinatively saturated. Obviously, the principal distinction between the situation where a C–H bond donates electron density into a vacant orbital of an unsaturated metal center (such as in **2–4**), and the opposite case where a lone-pair of a saturated metal center donates electrons to a vacant, antibonding orbital of a C–H bond (such as in **1**), was not yet made that time. Adding to the confusion, the designations "pregostic" and "remote agostic" have misled many subsequent authors [2,23–27] to assume that there exists a class of C–H$\cdots$M(d^8) interactions intermediate between agostic and hydrogen-bonding, or constituting a "prestage" to agostic bonding [28]. The assumption of such an intermediate class gained support from a study by Crabtree et al. in which the authors examined the structural characteristics of 50 selected X–H$\cdots$M(d^8) interactions (X = N, C) [29]. In their selection of compounds, most of the N–H$\cdots$M angles were over 160°, whereas the C–H$\cdots$M angles were generally lower. The authors concluded that N–H$\cdots$M(d^8) systems were "best regarded as hydrogen bonds", but the C–H$\cdots$M(d^8) systems were "intermediate in geometry between the hydrogen bonded N–H$\cdots$M systems and true agostic bonds". The problem of the structural analysis by Crabtree et al. [29] was that the compounds examined included almost exclusively *intramolecular* X–H$\cdots$M contacts. Thus, the X–H$\cdots$M angles reflected primarily geometrical constraints of the ligands and only in the second place the electronic requirements of the M$\cdots$H bond. In this context, it is worth mentioning that *intramolecular* X–H$\cdots$M contacts are not necessarily attractive but can also be repulsive. An example of such a repulsive contact is given by the complex *cis*-$[PtCl(NH_3)_2$(*N*9-9-aminoacridine)$]^+$(**6**, Fig. 6.3), where it is manifest by the widened Pt–N9-C9 bond angle of 136° [30]. Although in **6**, the N–H$\cdots$M contact has an attractive hydrogen-bonding component and the compound shows NMR-spectroscopic characteristics of hydrogen bonds, the geometrical constraints of the

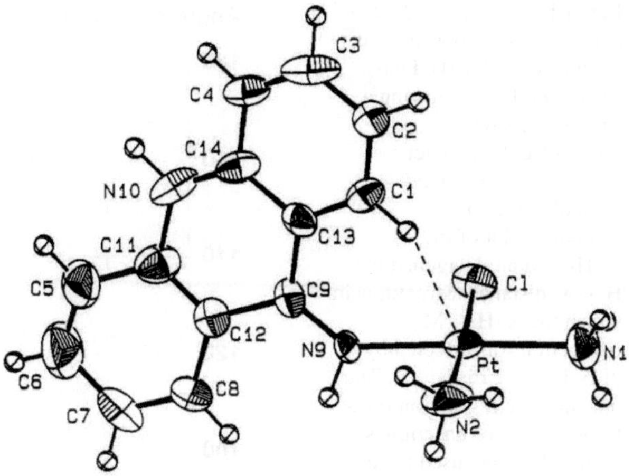

Fig. 6.3 Crystal structure of *cis*–$[PtCl(NH_3)_2(N9\text{-}9\text{-}aminoacridine)]^+$(**6**) showing a repulsive H1··· Pt contact (*dashed line*). That this contact is repulsive in spite of a hydrogen-bonding component can be inferred from the widened C9-N9-Pt angle (136°). Adapted with permission from [30]

6
hydrogen-bonding component, but overall repulsive interaction because of ligand-binding constraints

ligand obviously force the hydrogen atom to a repulsive, (i.e., too short) distance, and the repulsion has to be relieved by widening the Pt–N–C bond angle. Similar distortions of bond angles, apparently caused by repulsive C–H··· Pt contacts, have been seen in [Pd(benzo[*h*]quinoline)(H_2O){2-(dimethylaminomethy1)phenyl-*N*}]$^+$[31], as well as in complexes reported by Albinati, Pregosin et al. [16–18], as it has been noted by Hambley [32]. It appears clear from these examples that extracting geometrical preferences of X–H··· M bonding contacts from crystal structures of *intramolecular* H··· M associations is problematic.

More pertinent was therefore a structural study by Desiraju et al. [33] in which the authors examined *intermolecular* bonding X–H··· M (X = C, N, O) interactions, searching the Cambridge Structural Database (CSD) for H··· M contacts shorter than 3.2 Å. In this case, the examined sample was statistically relevant and the scattering of X–H··· M bond angles did reflect intrinsic structural preferences of the individual (X = C, N, or O) groups. Since agostic interactions observed in crystals are almost exclusively intramolecular (with the exception of the so-called "intermolecular pseudo-agostic bonds"[1] [33]), the search yielded predominantly non-agostic interactions. The conclusion of Desiraju et al. [33] was exactly opposite to that of Crabtree et al. [29]: C–H··· M interactions follow the same trend as N–H··· M and O–H··· M interactions. Since it is of interest to deplete from each other the structural preferences of early and late transition metals (which the study of Desiraju et al. [33] did not separate), and since in the present review, we are interested principally in d^8 metal complexes, we have carried out our own search of the 2014 version of the CSD database for *intermolecular* X–H··· M (M = Pd(II), Pt(II), Rh(I), Ir(I); X = C, N, O)

[1] In contrast to "pregostic" or "remote agostic" interactions, pseudoagostic interactions share all characteristics with agostic interactions; their only particularity is being intermolecular.

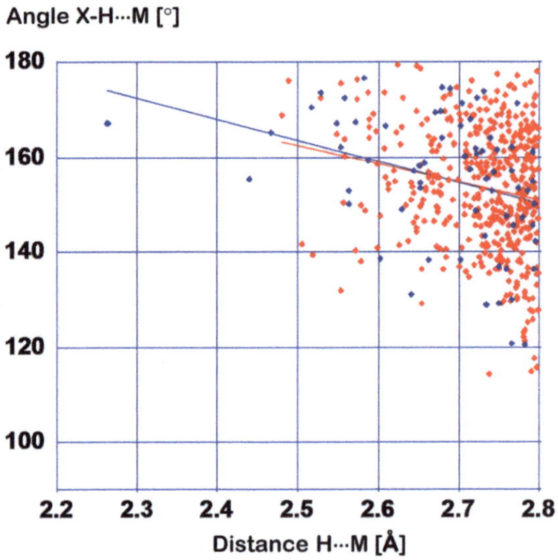

Fig. 6.4 Analysis of CSD entries for tetracoordinate d^8 complexes of Pd(II), Pt(II), Rh(I), and Ir(I), featuring a short (≤ 2.8 Å) intermolecular contact X–H···M; X = C (*red* symbols) or N (*blue* symbols). Plot of the X–H···M angle against the H···M distance. Structures in which the X–H···M interaction was affected by disorder were discarded. The *full* lines are regression lines through the *two* ensembles of points. The data used for the plot are available as supporting material

contacts (M···H ≤ 2.8 Å) within square-planar, coordinatively saturated complexes (such as **1**). This search yielded 317 C–H···M, 15 N–H···M, and 9 O–H···M contacts for Pd(II), 132 C–H···M, 52 N–H···M, and 5 O–H···M contacts for Pt(II), 45 C–H···M, 4 N–H···M, and 0 O–H···M contacts for Rh(I), and 8 C–H···M, 2 N–H···M, and 0 O–H···M contacts for Ir(I). The positions of hydrogen atoms in the O–H···M interactions proved generally ill-defined, therefore, we analyzed only the C–H···M and N–H···M interactions. The CSD codes of the selected structures and the corresponding H···M distances and X–H···M angles are listed in the Supporting Material. Figure 6.4 shows the plot of the X–H···M angle as a function of the M···H distance. It can be seen that both subclasses have a preference for a linear X–H···M angle, and this preference becomes more stringent with decreasing M···H length. The scattering of the values for C–H···M and N–H···M interactions is similar and the regression lines through the two subsets of points virtually coincide. This result confirms that the conclusion of Crabtree et al. [29], according to which C–H···M(d^8) interactions involving a coordinatively saturated d^8 metal atom are structurally different from N–H···M(d^8) interactions and tend to be bent, is invalid. Nevertheless, the myth of intrinsically bent non-agostic C–H···M(d^8) bonds has propagated until the present time.

Although the suggestion that the C–H···M(d^8) interaction seen in the crystal structure of compound **5** might be considered as a hydrogen bond appeared as early as in 1972 [13], the insight that coordinatively saturated M(d^8) centers can function as acceptors of hydrogen bonding met growing recognition only after the publication of the neutron diffraction structure of $[NPr^n_4]_2\{[PtC1_41 \cdot$ *cis*$-[PtC1_2(NH_2Me)_2]\}$ by Brammer et al. [34] which showed a textbook example of a nearly linear N–H···Pt hydrogen bond between the Pt-coordinated methylamine and the Pt center of $[PtC1_41^{2-}$. This intermolecular hydrogen bond is, of course, strongly aided by electrostatic attraction, since both the (formally cationic) methylamine and the

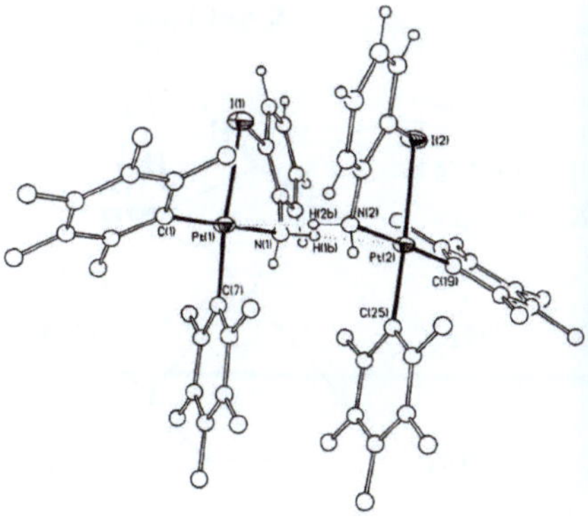

7

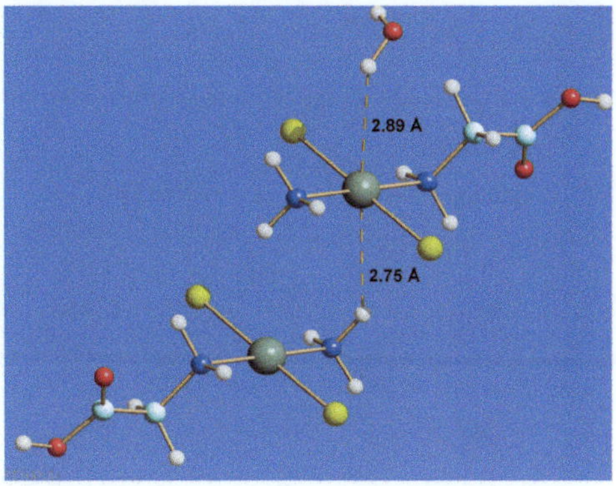

8

Fig. 6.5 X-ray structure of *cis*–[Pt(C_6F_5)$_2$(*I, N*-2-iodoaniline)] (**7**) and neutron diffraction structure of *trans*–[$PtCl_2$(NH_3)(*N*–glycine)]·H_2O (**8**), showing intermolecular N–H··· Pt and O–H··· Pt hydrogen bonds, respectively, between electrically neutral moieties. Reproduced with permission from [35] and [36], respectively

tetrachloroplatinate dianion bear net charges. However, subsequent crystal structures revealed cases of intermolecular hydrogen bonds between electrically neutral partners: **1** [8], **7** [35] and **8** [36] (Fig. 6.5) are such examples showing intermolecular C–H··· Pt, N–H··· Pt, and O–H··· Pt hydrogen bonds, respectively. These structures are particularly convincing demonstrations that d^8 centers can be hydrogen-bonding acceptors from O–H, N–H, and C–H bonds.

6.3 Spectral Characteristics of Agostic Versus Hydrogen-Bonding X–H··· M(d^8) Interactions

In both agostic and hydrogen-bonding X–H··· M interactions the X–H bond is weakened which can be detected by measuring a reduced J_{XH} coupling constant and/or a red-shifted X–H stretching frequency [12, 17, 20]. For both types J_{MH}-coupling can be frequently detected (if the M nuclide has a nonzero nuclear spin), reflecting the metal-hydrogen orbital interaction. X–H··· M interactions also affect the chemical shift of the metal-bound proton, as outlined in detail below.

The strong anisotropy of square-planar complexes of d^8 metal ions, and in particular, of their electronic d^8 system, becomes manifest in the magnetic field surrounding the complex [37, 38]. Thus, nuclei, and most sensitively those of protons, experience a deshielding effect if they occupy the region above or below the coordination plane, close to the z-axis (if the z-axis is defined as the normal to the coordination plane through the metal atom). In contrast, nuclei located close to the coordination plane experience a shielding effect.

The shielding effect on in-plane nuclei has been found to be particularly strong in platinum(II) hydride complexes whose hydridic protons show chemical shifts

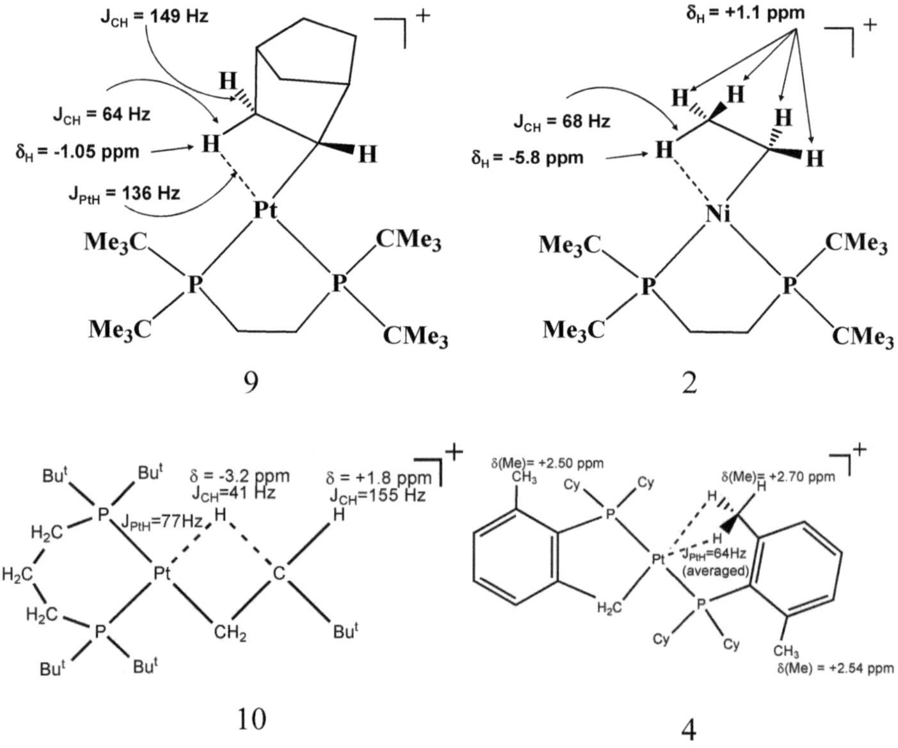

Fig. 6.6 NMR parameters of agostic complexes of d^8 metal ions. Note that in the crystal structures of **4** and in **9**, the methyl protons were localized from the difference Fourrier maps. In the NMR spectra of **9**, the agostic proton could be resolved from the non-agostic ones at 175 K. For **4**, the individual H-resonances of the interacting methyl group could not be resolved

between − 5 and − 24 ppm [39–41]. Buckingham and Stevens [37] used ligand field theory calculations to show that the observed chemical shifts originate mainly from paramagnetic shielding from the incompletely filled d-shell of platinum, and not from high electron density at the hydrogen nucleus as suggested by Chatt et al. [39]. Concurring with this result, d^0 and d^{10} metal hydrides do not show such shielding and the hydridic protons exhibit positive δ-values [42–44]. In a brilliant paper, Ziegler et al. reported state-of-the-art DFT-GIAO calculations for a number of octahedral d^6 hydride complexes to show that their negative hydride chemical shifts originated exclusively from paramagnetic shielding by the metal d^6 system [45]. The calculations which perfectly reproduced the experimental chemical shifts showed that if only diamagnetic shielding operated, the chemical shifts would be positive (with respect to TMS) in all cases.

Hydrogen atoms in typical agostic d^8 metal complexes, such as those displayed in Fig. 6.6, show negative chemical shifts. Referring to the negative chemical shifts of metal hydrides, these negative shifts were sometimes called “hydridic shifts” [46]. As in the case of the hydride complexes, these negative shifts arise from paramagnetic shielding, and not from enhanced electron density at the hydrogen nucleus. This has been shown explicitly by Scherer et al. [46] for several agostic diphosphine-stabilized Ni(II) alkyl complexes of the type $[Ni(alkyl)(P, P'–Bu^t{}_2PCH_2–CH_2PBu^t{}_2)]^+$(alkyl = ethyl (**9**) [47], norbornyl, dicyclopentadienyl). For these compounds, DFT/AIM-derived atomic charges carried by the agostic hydrogen were close to zero, and DFT-calculated shielding constants, which reproduced well the experimental values, demonstrated that the experimentally found negative shifts arise from paramagnetic shielding, as in the case of the hydride shifts. For the dicyclopentadienyl derivative, these results could be also confirmed by an experimental charge-density analysis [46]. Another experimental proof that the negative chemical shifts observed for β-agostic complexes (e.g., **1**, **9**, **10**) arise from the d^8 anisotropy and not from enhanced electron density is provided by the δ-agostic complex **4**, whose NMR parameters are included in Fig. 6.6. We can see that in the three β-agostic complexes, the agostic proton, which lies close to the coordination plane, is shielded ($\delta < 0$). The difference between the chemical shifts of the agostic and non-agostic hydrogen atoms attached to the same carbon is – 6.9 ppm in **9** and – 5.0 ppm in **10**. In contrast, in the δ-agostic complex **4** where the agostic interaction is shared between two methyl protons both of which lie outside the coordination plane (as shown by the X-ray structure where the methyl protons were localized from the difference Fourrier maps [11], Fig. 6.1), the methyl group containing the agostic protons shows a positive chemical shift (2.70 ppm), comparable to that of the other, non-agostic *ortho* methyl group of the same dimethylphenyl ligand (2.54 ppm). The fact that only those agostic protons which lie in the coordination plane show a negative chemical shift supports the above conclusion that negative shifts of agostic protons arise from paramagnetic shielding and not from increased electron density on the proton.

Figure 6.6 displays also salient spin-spin coupling constants. The agostic interaction is manifest in a decrease of the $^1J_{CH}$ coupling constant from ~ 150 to 40–70 Hz, consistent with a weakening of the agostic C–H bond. In the platinum complexes, we observe J-coupling of the agostic hydrogen with the ^{195}Pt nucleus (136 Hz in **2**, 77 Hz in **10**, 64 Hz in **4** (average value for the three methyl Hs)), clearly indicating orbital interaction.

Hydrogen bonds to d^8 metal centers occur exclusively at the apical site of the metal ion, and the hydrogen atoms experience therefore always paramagnetic deshielding from the metal atom [37, 38]. This deshielding could, in principle, contain a contribution from a decrease of electron density at the hydrogen nucleus, as observed for classical hydrogen bonds; however, it has not yet been shown whether this contribution is significant in X–H··· M(d^8) hydrogen bonds. So far, NMR manifestations of X–H··· M(d^8) hydrogen bonding have been reported only for intramolecular bonds, where these hydrogen bonds have sufficient lifetimes to be observed by NMR. The first report of low-field shifts of hydrogen atoms interacting with a d^8 center by hydrogen bonding was that by Miller et al. [38] who reported that nickel(II) complexes of styrene derivatives such as **11** (Fig. 6.7) show deshielding of protons that are likely to approach the nickel center from the apical site. Miller et al. brought convincing arguments, referring to the calculations of Buckingham and Stevens [37], to show that the

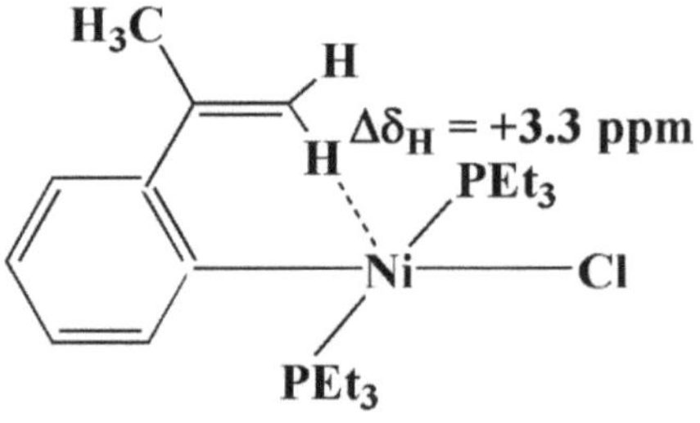

11

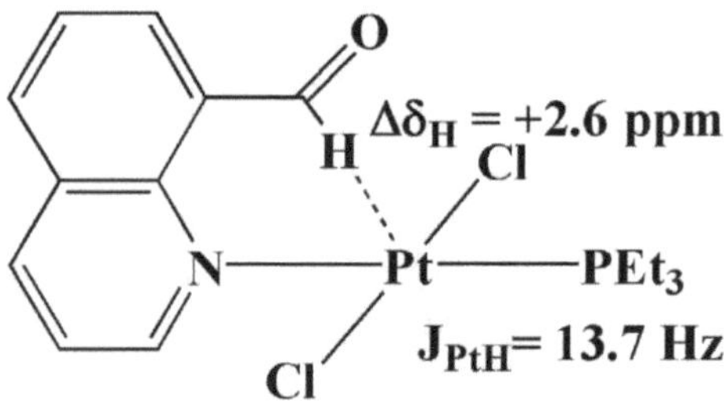

12

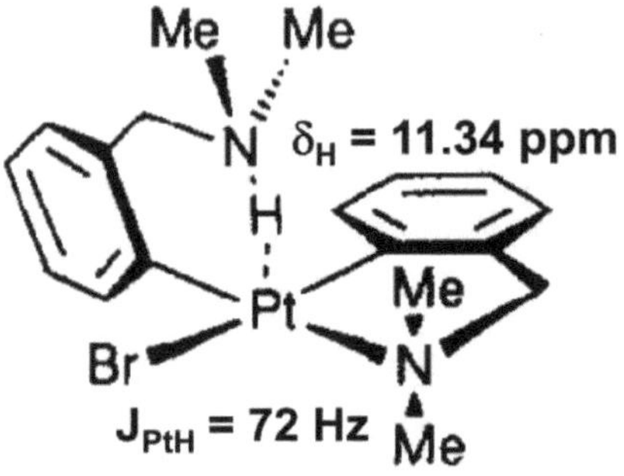

13

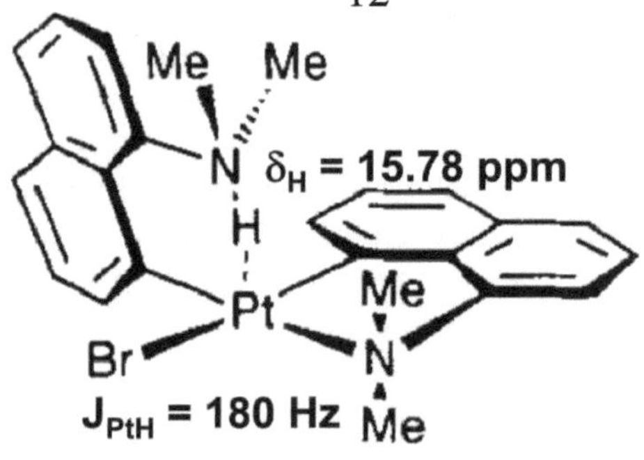

14

Fig. 6.7 Salient NMR parameters of d^8 metal complexes displaying C–H$\cdots$Ni (**11**) [38], C–H$\cdots$Pt (**12**) [17], and N–H$\cdots$Pt (**13, 14**) [48] hydrogen bonding. Note that for **11** and **12**, the chemical shift *difference* between the complexed and uncomplexed ligand is given, whereas for **13** and **14,** the chemical shift in the complex is indicated (the chemical shift in the free ligand has not been reported). The high δ values in **13** and **14** show nevertheless clearly that the N–H hydrogen experiences a deshielding effect

apical proton experiences paramagnetic deshielding from the metal. However, the authors did not recognize the interaction as hydrogen-bonding. Albinati, Pregosin et al. subsequently examined d^8 complexes such as **12**, where C-H bonds undergo similar interactions with the central atom, and identified the direct Pt$\cdots$H interaction by measuring the $^1J_{PtH}$ spin-spin coupling constants. Although X–H$\cdots$M(d^8) hydrogen bonding is expected to weaken the X–H bond, the Albinati-Pregosin complexes with C–H$\cdots$M(d^8) hydrogen bonds showed $^1J_{CH}$ coupling constants which were virtually unaffected (within experimental error) by the coordination to the metal. On the other hand, in the complexes **13** and **14** (Fig. 6.7) featuring N–H$\cdots$Pt hydrogen-bonding interactions, a reduction of the $^1J_{NH}$ coupling constants by $\sim$20 % with respect to the free ligands was observed, and the J_{PtH} coupling constants were larger, reaching the order of magnitude of J_{PtH} observed in agostic complexes (Fig. 6.6). It has to be emphasized that in **13** and **14**, the interaction is aided by the negative charge on the Pt center and the formally positive charge of the ammonium hydrogen-bond donor, which adds a large stabilizing electrostatic component to the interaction.

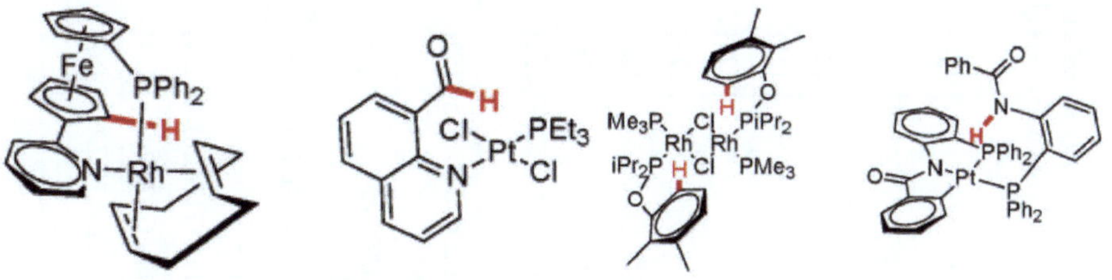

15 16 17 18

Fig. 6.8 Four d^8 complexes containing H··· M interactions studied by DFT and AIM theory by Oldfield et al. Reproduced with permission from [25]

6.4 Characterization of Agostic and Hydrogen-Bonding X–H··· M(d^8) Interactions Using AB Initio and DFT Methods

6.4.1 AIM Analyses of X–H··· M(d^8) Interactions

The assumption that "weak" or "pregostic" C–H··· M(d^8) interactions constitute a separate interaction type between agostic bonds and hydrogen bonds (Sect. 6.2) has prompted several groups to carry out computational studies of this presumed intermediate class. For instance, Oldfield et al. [25] studied three d^8 complexes suspected to show such intermediate C–H··· M(d^8) interactions using DFT calculations and the AIM theory (Fig. 6.8): complex **15**, that was reported to show a "weak" C–H··· Rh(I) interaction ascribed tentatively (and incorrectly) to a donation of C–H σ electrons to the (coordinatively saturated) Rh(I) center, [49], complexes **16** [17] and **17** [24] designated as "pregostic", and complex **18** which was initially assigned as agostic [50] but later identified as hydrogen-bonding by Brammer et al. [34]. Oldfield et al. showed that all four complexes had similar electron densities at the bond critical points (BCP) of the H··· M bonds; the Laplacians of the electron density were also similar (Table 6.1), and conform to expected ranges for hydrogen bonds (see the discussion below). Calculations of the chemical shifts indicated deshielding of the protons contacting M, in agreement with reported NMR data. Calculation of the total energy density at the BCP yielded slightly negative values for **15** and **18**, for which partial covalent character of the H··· M bond was concluded, but slightly positive values for **16** and **17**, for which a purely electrostatic interaction was concluded. However, this conclusion was in contradiction with a significant overlap found between the metal $d_{xz/yz}$ orbitals and the C–H σ^* orbital in these complexes. In fact, as pointed out in the recent review by Weinhold and Klein [51], a hydrogen bond *always* has a covalent n→ σ^* component, thus, considering a hydrogen bond a purely electrostatic interaction seems problematic anyway. Interestingly, the orbitals overlapping with the C–H σ^* orbital in **16** and **17** were $d_{xz/yz}$, and not d_{z2}. This might be related to the fact that in all four complexes studied the X–H··· M interaction was *intramolecular*,

Table 6.1 Electron density $\rho(r_c)$ and the Laplacian of electron density $\nabla^2(\rho(r_c))$ at the H··· M bond critical point, determined from AIM analyses for selected hydrogen-bonding and agostic X–H··· M(d^8) complexes. For hydrogen-bonding complexes, the stabilization energies SE calculated by the supermolecule method (where available), and their estimates from $\rho(r_c)$ (Fig. 6.9) are also given.

Complex	interaction	el. density [e/a_0^3]	Laplacian of el. density [e/a_0^5]	SE[a] [$kcal.mol^{-1}$]	SE(estimate)[b] [$kcal.mol^{-1}$]	Level of theory	Ref.
a) Hydrogen bonding complexes							
15	C–H··· Rh	0.024	0.067		4.8	[c]	[25]
16	C–H··· Pt	0.023	0.072		4.6	[c]	[25]
17	C–H··· Rh	0.012	0.031		2.4	[c]	[25]
18	N–H··· Pt	0.025	0.059		5.0	[c]	[25]
[Ni(bhac)(phim)][d]	O–H··· Ni	0.011	0.039		2.2	B3LYP/6-311G(d,p)	[26]
[Ni(ahac)(phim)][d]	O–H··· Ni	0.012	0.041		2.4	B3LYP/6-311G(d,p)	[26]
7	O–H··· Pt	0.009	0.025		1.8	B3LYP/LANL2DZ/6-311++G (d,p)[e]	[52]
7	N–H··· Pt	0.012	0.034		2.4	B3LYP/LANL2DZ/6-311++G (d,p)[e]	[52]
cis-$[PtCl_2(NH_3)_2].H_2O$	O–H··· Pt	0.024	0.061	4.7	4.8	MP2/LANL2DZ(f)/6-311++G (2d,f,p)	[53]
cis-$[PtCl_2(NH_3)_2].H_2O$	O–H··· Pt	0.020	0.047	6.0	4.0	MP2/DP/6-311+G (2d,2p)[f]	[54]
trans-$[PtCl_2(NH_3)_2].H_2O$	O–H··· Pt	0.020	0.047	4.1	4.0	MP2/DP/6-311+G (2d,2p)[f]	[54]
cis-$[Pt(OH)_2(NH_3)_2].H_2O$	O–H··· Pt	0.020	0.045	7.6	4.0	MP2/DP/6-311+G (2d,2p)[f]	[54]
trans-$[Pt(OH)_2(NH_3)_2].H_2O$	O–H··· Pt	0.020	0.045	5.5	4.0	MP2/DP/6-311+G (2d,2p)[f]	[54]
$[Pt(C_6F_6)_3(8\text{-OH-quinaldine})]^-$	O–H··· Pt	0.021	0/089		4.2	B3LYP/ LANL2DZ/6-311++G(d,p)[g]	[55]

Table 6.1 (continued)

Complex	interaction	el. density $[e/a_0^3]$	Laplacian of el. density $[e/a_0^5]$	SE[a] [kcal.mol^{-1}]	SE(estimate)[b] [kcal.mol^{-1}]	Level of theory	Ref.
$[Pt(C_6F_6)(bzq)(8\text{-}OH\text{-}qn)]^h$	C–H···Pt	0.034	0.070		6.8	M06/SDD/6-31G*/6-31G**	[56]
$[Pt(C_6F_6)(bzq)(2\text{-}Me\text{-}8\text{-}OH\text{-}qn)]^i$	C–H···Pt	0.036	0.066		7.2	M06/SDD/6-31G*/6-31G**	[56]
$[Pt(C_6F_6)_3(8\text{-}OH\text{-}qn)]^{-h}$	C–H···Pt	0.035	0.070		7.0	M06/SDD/6-31G*/6-31G**	[56]
$[Pt(C_6F_6)_3(bzq)(2\text{-}Me\text{-}8\text{-}OH\text{-}qn)]^{-i}$	C–H···Pt	0.039	0.079		7.8	M06/SDD/6-31G*/6-31G**	[56]
$[Pt(C_6F_6)_3(8\text{-}Me\text{-}quinoline)]^-$	C–H···Pt	0.018	0.064		3.6	B3LYP/LANL2DZ/6-31G(d,p)	[57]
1	C–H···Pt	0.018	0.048		3.6	B3LYP/LANL2DZ/6-31G(d,p)	[57]
b) Agostic complexes							
2	C–H···Pt	0.059	0.183			B3LYP/LANL2DZ/6-31G(d,p)	[57]
3	C–H···Pt	0.013	0.049			B3LYP/LANL2DZ/6-31G(d,p)	[57]
4	C–H···Pt	0.041	0.134			B3LYP/LANL2DZ/6-31G(d,p)	[57]
$[Pt(Et)P,P'\text{-}Bu^t{}_2P(CH_2)_3PBu^t{}_2)]^+$	C–H···Pt	0.053	0.176			B3LYP/LANL2DZ/6-31G(d,p)	[57]
$[PdBr(Ph)(1\text{-}adamantylP^tBu_2)]$	C–H···Pd	0.023	0.075			BP86/ECP1/6-31G*	[58]

Table 6.1 (continued)

Complex	interaction	el. density [e/a_0^3]	Laplacian of el. density [e/a_0^5]	SE[a] [kcal.mol^{-1}]	SE(estimate)[b] [kcal.mol^{-1}]	Level of theory	Ref.
[PdBr(Ph)(2-adamantylP^{t}Bu$_2$)]	C–H···Pd	0.029	0.083			BP86/ECP1/6-31G*	[58]
[PdBr(Ph)(P^{t}Bu$_3$)]	C–H···Pd	0.018	0.058			BP86/ECP1/6-31G*	[58]
[PdI(Ph)(P^{t}Bu$_3$)]	C–H···Pd	0.017	0.053			BP86/ECP1/6-31G*	[58]
19	C–H···Rh	0.026	− 0.021			B3LYP/DZVP/6-31G*	[59]
	C–H···Rh	0.021	− 0.016				

[a] Calculated using the supramolecule approach, BSSE-corrected
[b] According to the relationship SE = 200*$\rho(r_c)$ determined from Fig. 6.10
[c] mPW1PW91/SDD for M, 6-311++G(2d,2p) for the interacting H and its neighboring atoms, 6-31G* and 3-21G* for the other non-metal atoms
[d] H_2bhac = acetylacetone-benzoylhydrazone; H_2ahac = acetylacetone-acetylhydrazone; phim = 2-phenylimidazole
[e] One unit cell from the experimental neutron-diffraction crystal structure [36] was used without geometry optimization
[f] DP = Dolg-Pélissier pseudopotential/pseudoorbital basis including two polarization f orbitals [60, 61]
[g] Experimental X-ray coordinates, only the position of the interacting H-atom was optimized
[h] bzq = 7,8-benzoquinolinate, 8-OH–qn = 8-hydroxyquinoline
[i] bzq = 7,8-benzoquinolinate, 2-Me-8-OH–qn = 2-methyl-8-hydroxyquinoline

and the sterical constraints did not allow the X–H bond to get oriented exactly along the z axis. Overall, Oldfield's study has confirmed that C–H··· M and N–H··· M contacts *involving a coordinatively saturated* d^8 *center* are of the same type, i.e., they correspond to hydrogen-bonding interactions.

Another study inspired by the presumed existence of an intermediate class of square-planar d^8 complexes showing an apical C–H··· M interaction whose nature was "not very clear" was that of Mukhopadhyay and Pal [26]. The authors synthesized two square-planar Ni(II) complexes bearing a 2-phenylimidazole ligand which contacts the Ni(II) center with one of its *ortho* hydrogen atoms. A natural population analysis based on wavefunctions obtained from DFT calculations showed that the *ortho* hydrogen atom involved in the interaction with the Ni(II) atom bears a higher positive charge than the other *ortho* hydrogen atom of the same phenyl ring, in line with a hydrogen bond-like interaction. The topological parameters of the C–H··· Ni interaction calculated using the AIM theory (see Table 6.1) were also in agreement with hydrogen bonding.

Thakur and Desiraju [57] examined twenty crystal structures of metal complexes from the Cambridge Structural Database (CSD) presenting C–H··· M interactions, and investigated the ability of the AIM and NBO theories to distinguish between agostic and hydrogen-bonding interactions. In particular, using AIM analyses, they checked whether the criteria of Popelier et al. [62, 63] for the electron density at the bond critical point and for its Laplacian were applicable. Popelier et al. had reported ranges $0.002\ e/a_0^3 \leq \rho(r_c) \leq 0.035\ e/a_0^3$ and $0.024\ e/a_0^5 \leq \nabla^2\rho(r_c) \leq 0.139\ e/a_0^5$ for hydrogen bonds, whereas for agostic interactions the values were reported to lie outside these ranges, namely $0.04\ e/a_0^3 \leq \rho(r_c)$ and $0.15\ e/a_0^5 \leq \nabla^2\rho(r_c) \leq 0.25\ e/a_0^5$ [63]. Among the twenty complexes examined by Thakur and Desiraju figured also six X–H··· M(d^8) interactions (M = Pt(II)), two hydrogen-bonding and four agostic. The data for these interactions constitute the two bottom hydrogen-bonding entries and the four top agostic entries of Table 6.1. It can be seen that compound **3**, a textbook example of a γ-agostic interaction, shows $\rho(r_c)$ and $\nabla^2\rho(r_c)$ values rather typical of hydrogen bonds. The agostic interaction in **3** is particularly weak, as apparent from the rapid exchange between the methyl groups of all isopropyl substituents observed in NMR solution spectra even at 190 K [10]; however, the crystal structure (Fig. 6.1) leaves no doubt that it is an agostic, and not a hydrogen-bonding interaction. Thus, it is apparent that the BCP analysis only allows for the distinction between weaker and stronger X–H··· M interactions, but is incapable of indicating whether the metal is electron donor or acceptor. The Popelier criteria simply exploit the fact that hydrogen bonding is usually weaker than agostic interactions, however, they cannot be used to distinguish between a hydrogen bond and a particularly weak agostic interaction. Also, Thakur and Desiraju concluded [57] that "the BCP analysis, based upon the Popelier criteria, showed poor prediction of the interaction type, when compared to the experimental reports". They suggested that the bond ellipticity at the H··· M BCP might provide a supplementary parameter, being higher for agostic than for hydrogen-bonding interactions. Nevertheless, even this is not a general rule, since, for instance, for the agostic complex **4**, the ellipticity at BCP is 0.273 a.u., a value more typical for hydrogen bonds than for agostic interactions (Table 9 of [57]).

Fig. 6.9 Crystal structure of [RhCl(CO)(CAAC)] (CAAC = 2-(2,6-di-isopropylphenyl)-6-isopropyl-3,3,9-trimethyl-2-azaspiro(4.5)decan-1-ylidene) (**19**) displaying the *two* agostic H· · · Rh bonds as *dashed* lines. Reproduced with permission from [64]

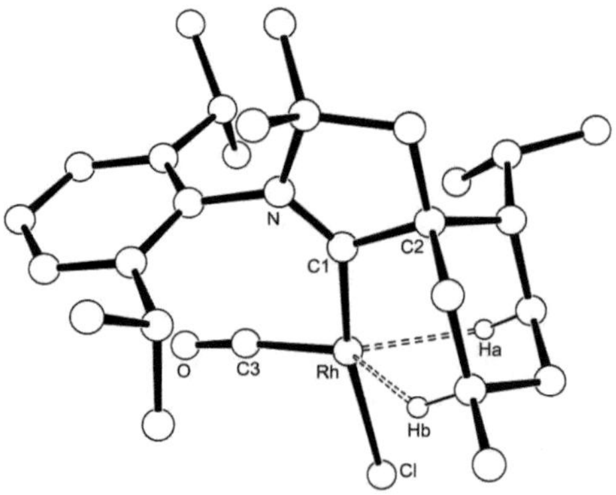

19

The last two entries in Table 6.1 show BCP data for the rhodium complex **19** [59] featuring two particularly strong agostic C–H· · · Rh interactions that fill jointly the fourth coordination site (Fig. 6.9) [64]. Here the Laplacians of the BCP electron densities are negative, indicating rather strong covalent interactions. It is interesting to recall that very strong hydrogen bonds also show negative $\nabla^2\rho(r_c)$ values [51]. However, those strong hydrogen bonds show also high $\rho(r_c)$ values which is not the case here.

Zhang et al. performed DFT (B3LYP/6-311++G(d,p)) calculations and an AIM analysis on *trans*–[PtCl$_2$(NH$_3$)(*N*-glycine)]·H$_2$O (**8**), and determined $\rho(r_c)$ and $\nabla^2\rho(r_c)$ values for the two hydrogen-bonding X–H· · · Pt (X = O, N) interactions occurring in the crystal lattice (Fig. 6.5) [52]. The electron densities and their Laplacians (Table 6.1) are rather low, compared to other hydrogen-bonding interactions shown in Table 6.1. However, this may be partly caused by the fact that the calculations were performed on the neutron diffraction coordinates without optimization. The same team carried out a similar study on the [Pt(C$_6$F$_5$)$_3$(8-hydroxyquinaldine)]$^-$ complex showing an intramolecular O–H· · · Pt hydrogen bond. For this complex, $\rho(r_c)$ and $\nabla^2\rho(r_c)$ values were comparable to those of other X–H· · · Pt hydrogen bonds (Table 6.1). For an intramolecular hydrogen bond, a calculation of the interaction energy using the supermolecule approach is problematic; therefore, the authors estimated the interaction energy using the observation that the hydrogen-bond energy is proportional to the electron density at the BCP, as reported by different authors. In fact, Parthasarathi et al. [65] and Weinhold and Klein [51] have determined, for various hydrogen-bonded pairs ranging from very weak to very strong, the stabilization energy (SE) using the supramolecule approach, and found that it is proportional to the electron density at the hydrogen bond BCP. Although the level of theory was not the same in the two studies (MP2/aug-cc-pVDZ versus B3LYP/aug-cc-pVTZ), the correlation diagrams are remarkably similar, as shown in Fig. 6.10. In particular, for weak hydrogen bonds (SE < 10 kcal.mol^{-1}), the regression lines virtually coincide, with a proportionality constant of $\sim$200 kcal.mol^{-1}/ea$_0^3$. As the investigated hydrogen-bonded pairs were constituted exclusively from main-group atoms, we decided to test whether the proportionality would hold also for X–H· · · M(d^8)

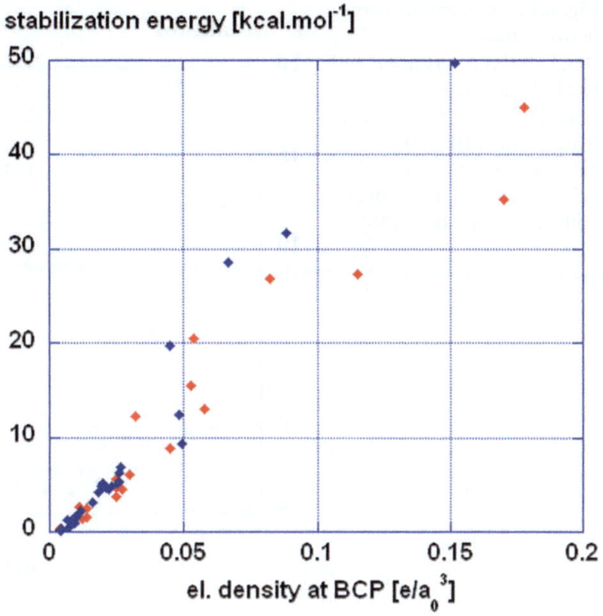

Fig. 6.10 Hydrogen-bonding energy (determined by the supermolecule approach) as a function of the electron density at the hydrogen bond BCP, for various hydrogen-bonded pairs. *Blue* symbols: Data from MP2/aug-cc-pVDZ calculations by Parthasarathi et al. [65]. *Red* symbols: Data from B3LYP/aug-cc-pVTZ calculations by Weinhold and Klein [51]

hydrogen bonds, where the lone-pairs at the acceptor atom are very different, and have included, in Table 6.1, the estimates for hydrogen-bonding energy from $\rho(r_c)$ with the above proportionality constant. It can be seen that the agreement with the calculated interaction energy (where available) is relatively reasonable, the two values never diferring more than by a factor of 2.

6.4.2 *Quantification of Dispersion and Charge-Transfer Components of X–H··· M(d^8) Interactions*

Our own interest in hydrogen bonding involving d^8 metal ions as acceptors resulted from our Hartree-Fock and MP2 calculations on platinum(II) complexes interacting with an axial water molecule [66], which indicated that an uncharged Pt(II) complex such as *trans*–[Pt(OH)$_2$(NH$_3$)$_2$] undergoes associations with water molecules preferentially with the H-atom oriented towards the filled 5d$_{z2}$ orbital of platinum, thus forming an O–H···Pt hydrogen bond. The energy well of ~4 kcal.mol^{-1} observed in the MP2 binding curve was only very shallow in the HF curve, indicating a major dispersion contribution to the interaction. A similar result was obtained for *trans*–[PtCl$_2$(NH$_3$)(*N*-glycine)] interacting with one water molecule (**8**) (Figs. 6.5, 6.11); in this case, the hydrogen-bonding orientation of the water molecule was confirmed by a neutron diffraction analysis [36]. A hydrogen-bonding orientation with a major dispersion component resulted also as the most favorable geometry from high level *ab initio* calculations on axial water associations of

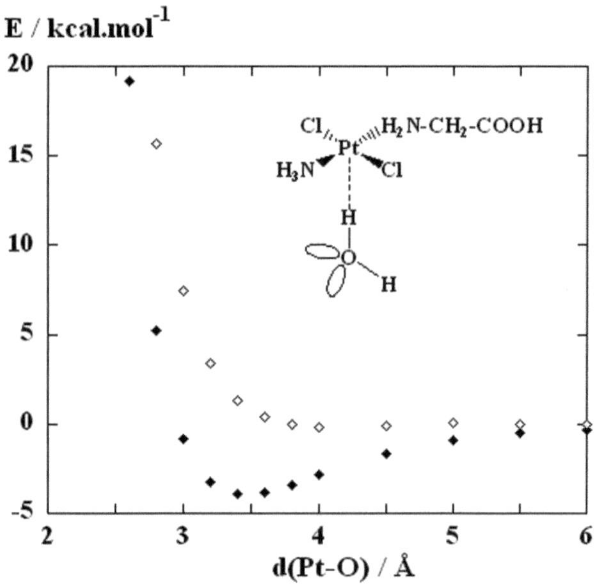

Fig. 6.11 Interaction energy between *trans*–[$PtCl_2(NH_3)$(*N*-glycine)] and H_2O (**8**) calculated as a function of Pt· · · O distance. MP2 (◇) and HF (◇) calculations were BSSE-corrected. Reproduced with permission from [36]

cis–[$PtCl_2(NH_3)_2$] by de Almeida et al. [53, 67], and from MP2 calculations on axial water interactions with other electrically neutral Pt(II) complexes [54]. Comparing the axial water association of *cis*–[$PtCl_2(NH_3)_2$] with that of the aquated derivatives *cis*–[$PtCl(NH_3)_2(H_2O)]^+$and *cis*–[$Pt(NH_3)_2(H_2O_2)]^{2+}$, de Almeida et al. found that whereas for *cis*–[$PtCl_2(NH_3)_2$], the hydrogen-bonding (H-ahead) orientation of the axial water molecule is favored both by electrostatic and dispersion forces, for the aquated species, the O-ahead orientation is strongly favored due to electrostatics [67]. This result is in agreement with our calculations on water interaction with *trans*–[$Pt(OH)_2(NH_3)_2$] and $[Pt(NH_3)_4)]^{2+}$where electrostatic forces favor the H-ahead orientation in the former case but the O-ahead orientation in the latter, and significant dispersion interaction is found only in the former case [66].

De Almeida et al. also used perturbation methods to calculate, for the axial hydrogen-bonding water interaction with *cis*–[$PtCl_2(NH_3)_2$], the electrostatic, induction, and dispersion components of the interaction energy [53]. They showed that at the MP2 energy minimum, the electrostatic and induction components (the latter including the charge-transfer energy) are not negligible. Our conclusion that the polarization and charge-transfer components are not important in axial water associations of *trans*-[$Pt(OH)_2(NH_3)_2$] [66] and *trans*–[$PtCl_2(NH_3)$(*N*-glycine)] [36] was therefore probably premature. That at the MP2 energy minimum the HF energy is close to zero for these interactions (see Fig. 6.11) could be due to the fact that the sum of electrostatic, polarization, and charge-transfer components is counterbalanced by exchange-repulsion. Supporting this interpretation, our later AIM analysis of these interactions indicated significant charge transfer for the H-ahead water orientation [54].

Zhang et al. performed DFT (B3LYP/6-311++G(d, p)) calculations and an NBO analysis of *trans*–[$PtCl_2(NH_3)$(*N*-glycine)]·H_2O (**8**) to quantify the charge transfer

Table 6.2 NBO analysis of donor-acceptor interactions for hydrogen-bonding and agostic C–H··· Pt(II) complexes[a]

Complex	Interaction type[b]	$\Delta E^{(2)}_{M\to\sigma^*(C-H)}$ [kcal.mol^{-1}]	$\Delta E^{(2)}_{\sigma(C-H)\to M}$ [kcal.mol^{-1}]	CSD Refcode
$[Pt(C_6F_{6)3}(8\text{-mehq})]^{-c}$	HB	4.38	2.75	TOXJEV
1	HB	8.57	1.52	KEKZAB
2	agostic	16.66	39.06	GELKAJ
3	agostic	1.43	3.96	BIMHAH
4	agostic	8.81	20.62	WURJAU
[Pt(Et)*P, P'*–$Bu^t{}_2P(CH_2)_3PBu^t{}_2)]^+$	agostic	3.78	28.28	KILKOF10

[a] From [57]
[b] Considering that $\Delta E^{(2)}_{M\to\sigma^*(C-H)} > \Delta E^{(2)}_{\sigma(C-H)\to M}$ for hydrogen bonds, $\Delta E^{(2)}_{M\to\sigma^*(C-H)} < \Delta E^{(2)}_{\sigma(C-H)\to M}$ for agostic interactions
[c] 8-mehq = 8-methylquinoline

(CT) from platinum lone-pairs to the σ* antibond of the interacting water molecule, and to the σ* antibond of the NH_3^+group of a second complex molecule interacting from the other apical site (Fig. 6.5). The CT to the O–H and N–H σ* antibonds was 0.0051 and 0.020 e, respectively, and the corresponding second-order perturbative energy $\Delta E^{(2)}_{n\to\sigma^*}$, summed over the three platinum lone-pairs (presumably $5d_{z2}$, $5d_{xz}$ and $5d_{yz}$) was calculated as 0.97 and 1.42 kcal.mol^{-1}, respectively [52]. In a subsequent NBO analysis of the $[Pt(C_6F_5)_3(8\text{-hydroxyquinaldine})]^-$ complex showing an O–H··· Pt hydrogen bond from the 8-hydroxy group, the same authors determined the CT from the platinum lone pairs to the O–H antibond of 0.0054 e, and the sum of the $\Delta E^{(2)}_{n\to\sigma^*}$values over the platinum lone-pairs of 2.06 kcal.mol^{-1} [55]. These values are similar to those calculated for the water-to-platinum hydrogen bond in **8**, which may appear surprising, as in the hydroxyquinaldine complex the hydrogen bond is aided by the net − 1 charge on the platinum center, and thus, considerably stronger charge transfer would be expected.

Thakur and Desiraju [57] used B3LYP/6-31G(d, p) calculations and an NBO analysis to examine the power of the NBO method to distinguish between agostic and hydrogen-bonded C–H··· M systems. They found that of all examined methods, NBO analysis "was in best agreement with experimental characterizations by NMR spectroscopy and X-ray crystallography". As shown in Table 6.2 for the six investigated C–H··· Pt(II) interactions, the agostic bonds are distinguished from hydrogen bonds by the relative CT energies corresponding to C–H to metal CT versus metal to C–H CT: for agostic complexes, the former is larger whereas for hydrogen-bonded complexes, the latter is larger. This criterion allows 19 of the 20 complexes investigated by Thakur and Desiraju to be correctly assigned. The assignment fails for the agostic complex *cis, trans*-$[RuH_2$(1,3-di-*tert*-butylimidazol-2-ylidene)$(PPh_3)_2]$ where the coordination geometry is a distorted octahedron and the calculated σ →M and M→ σ* CT energies (20 and 29 kcal.mol^{-1}) are similar.

6.4.2.1 Relationship Between Transferred Charge and Charge-Transfer Energy

In the 2013 paper, Zhang et al. made the attractive suggestion that one could make use of the proportionality between transferred charge and charge-transfer energy, and calculate one value from the other on the basis of a known proportionality constant [55]. This interesting but somewhat problematic issue deserves a comment here. In the zero-overlap approximation, the charge transfer q can be expressed as $q = \Delta E^{(2)}_{n\rightarrow\sigma*}/(\varepsilon_{\sigma*} - \varepsilon_n)$, where $\Delta E^{(2)}_{n\rightarrow\sigma*}$ is the second-order perturbative energy and $\varepsilon_{\sigma*}$ and $\varepsilon_n$are the unperturbed NBO orbital energies [68]. If variations in $(\varepsilon_{\sigma*} - \varepsilon_n)^{-1}$can be neglected with respect to $\Delta E^{(2)}_{n\rightarrow\sigma*}$, then a proportionality between q and $\Delta E^{(2)}_{n\rightarrow\sigma*}$ should be observed. In fact, B3LYP/aug-cc-pVTZ calculations and an NBO analysis of a series of hydrogen-bonding systems ranging from very weak ($\Delta E^{(2)}_{n\rightarrow\sigma*} \approx 0.4$ kcal.mol^{-1}) to very strong ($\Delta E^{(2)}_{n\rightarrow\sigma*} \approx 180$ kcal.mol^{-1}) by Weinhold and Klein has shown a very good linear correlation (R = 0.994) passing through 0, with a proportionality constant of 417 kcal.mol^{-1}/electron transferred, when the very strong hydrogen bonds ($\Delta E^{(2)}_{n\rightarrow\sigma*} > 100$ kcal.mol^{-1}) were excluded (Table 6.2 of [51]). An earlier, similar study done at the HF/6–31G* level of theory also yielded a good proportionality dependence (R = 0.987 with a proportionality constant of 683 kcal.mol^{-1}/electron transferred [69]). This indicates that, except for very strong hydrogen bonds, for which the zero overlap approximation obviously does not hold, $\Delta E^{(2)}_{n\rightarrow\sigma*}$is in fact proportional to q (i.e. the variations in $(\varepsilon_{\sigma*} - \varepsilon_n)^{-1}$are not strong enough to perturb the proportionality); however, the proportionality constant depends substantially on the level of theory. A still different proportionality constant was derived by Cappelletti et al. [70]. These authors used high-resolution molecular-beam scattering experiments to determine *experimentally* the intermolecular potential for various gas-phase binary complexes, including water complexes with rare gases, H_2, N_2 and O_2. They evaluated the CT energy to be proportional to the CT (calculated either using NBO theory or the "charge displacement analysis" [71]) with a rough proportionality constant of 2–3 eV (46–69 kcal.mol^{-1}) per electron transferred. The authors also performed NBO analyses of the hydrogen-bonded systems at a very high level of theory (CCSD/aug-cc-pVQZ), but did not report the $\Delta E^{(2)}_{n\rightarrow\sigma*}$ CT energies; therefore, it is not clear whether the discrepancy between their proportionality constant and that obtained by Weinhold and Klein [51] is due to a discrepancy between the experimental and theoretical (NBO) CT energies, or from the fact that the relationship is different for the extremely weak hydrogen bonds ($\Delta E_{CT} \leq 0.1$ kcal.mol^{-1}) for which the experimental CT energies were determined [70].

Cappelletti et al. have also compared the CT in water-noble gas complexes evaluated by the "charge displacement analysis" [71], NBO and AIM methods, and found that the NBO charges agreed with those from the charge displacement analysis reasonably well [70, 72], whereas the AIM charges were considerably larger [70]. This finding goes in line with considerably larger CT from *trans*-[PtCl_2(NH_3)(*N*-glycine)] to water determined using AIM calculations (0.04 e) [54] as compared to that calculated using an NBO analysis (0.005 e) [52], although the latter value was obtained on neutron diffraction coordinates [36] without optimization, which could reduce the calculated CT on its own. Clearly, care has to be taken when comparing CT values obtained with different methods and from different geometries.

6.4.3 *Ab-Initio Molecular Dynamics (AIMD) Simulations of X–H··· M(d^8) Interactions*

6.4.3.1 Water Solvation of Pt(II) Complexes Probed by AIMD Simulations

Vidossich et al. reported an *ab initio* molecular dynamics (AIMD) study of the complex *trans*–[$PtCl_2(NH_3)$(*N*-glycine)] (**8**) with ~100 explicit water molecules under periodic boundary conditions [73], in order to test whether the axial hydrogen-bonding interaction with a water molecule seen in the neutron diffraction structure [36] would hold in aqueous solution. As can be seen from the radial distribution functions (rdf) of the axial region (Fig. 6.12, top), the AIMD calculations predict water molecules to assume preferentially the H-ahead (i.e., hydrogen-bonding) orientation with respect to the platinum atom. This result is in agreement with static *in vacuo* calculations on **8** [36] and on other uncharged Pt(II) complexes [53, 54, 66], and suggests that O–H··· Pt hydrogen bonds between water and electrically neutral platinum complexes may hold also in solution.

Similar rdf diagrams indicating hydrogen-bonding orientation of solvating water molecules towards platinum were derived by Autschbach et al. from AIMD simulations of *cis*–[$PtCl_2(NH_3)_2$] in a water bath consisting of 64 water molecules [74] and of solvated dianionic complexes $[PtX_4]^{2-}$ (X = Cl, Br, CN) [75].

Lau and Ensing [76] carried out AIMD simulations on *cis*–[$PtCl_2(NH_3)_2$] and its two hydrolyzed derivatives, *cis*–$[PtCl(NH_3)_2(H_2O)]^+$ and *cis*–$[Pt(NH_3)_2(H_2O)_2]^{2+}$, each solvated with ~50 explicit water molecules. For the electroneutral complex *cis*–[$PtCl_2(NH_3)_2$], they observed a somewhat smaller population of hydrogen-bonding axial water molecules than the previous authors. For *cis*–$[PtCl(NH_3)_2(H_2O)]^+$, the corresponding rdf peak near Pt–H distance of 2.5 Å had an even smaller amplitude, as expected for the increased positive complex charge, and for *cis*–$[Pt(NH_3)_2(H_2O)_2]^{2+}$, it was completely absent (Fig. 6.12, middle). These results are in perfect agreement with the *in vacuo* calculations of de Almeida et al. [67], and also with the observation that rate constants for the hydrolysis of chlorido Pt(II) complexes depend only very slightly on the complex charge ([77], see also discussion in [36]).

In contradiction to the above results are AIMD simulations of $[Pt(H_2O)_4]^{2+}$ solvated in a periodic box of 70 water molecules by Marcos et al. [78, 79]. In these simulations of a *dicationic* complex, axial water molecules significantly populated the hydrogen-bonding orientation (Fig. 6.12, bottom).

There is a neutron diffraction structure of a dicationic Pt(II) complex having axial interactions available: that of $[Pt(py)_4]Cl_2.3H_2O$ [80]. In this structure, Pt(II) interacts axially with one Cl^- counter-ion from one side and with a water molecule

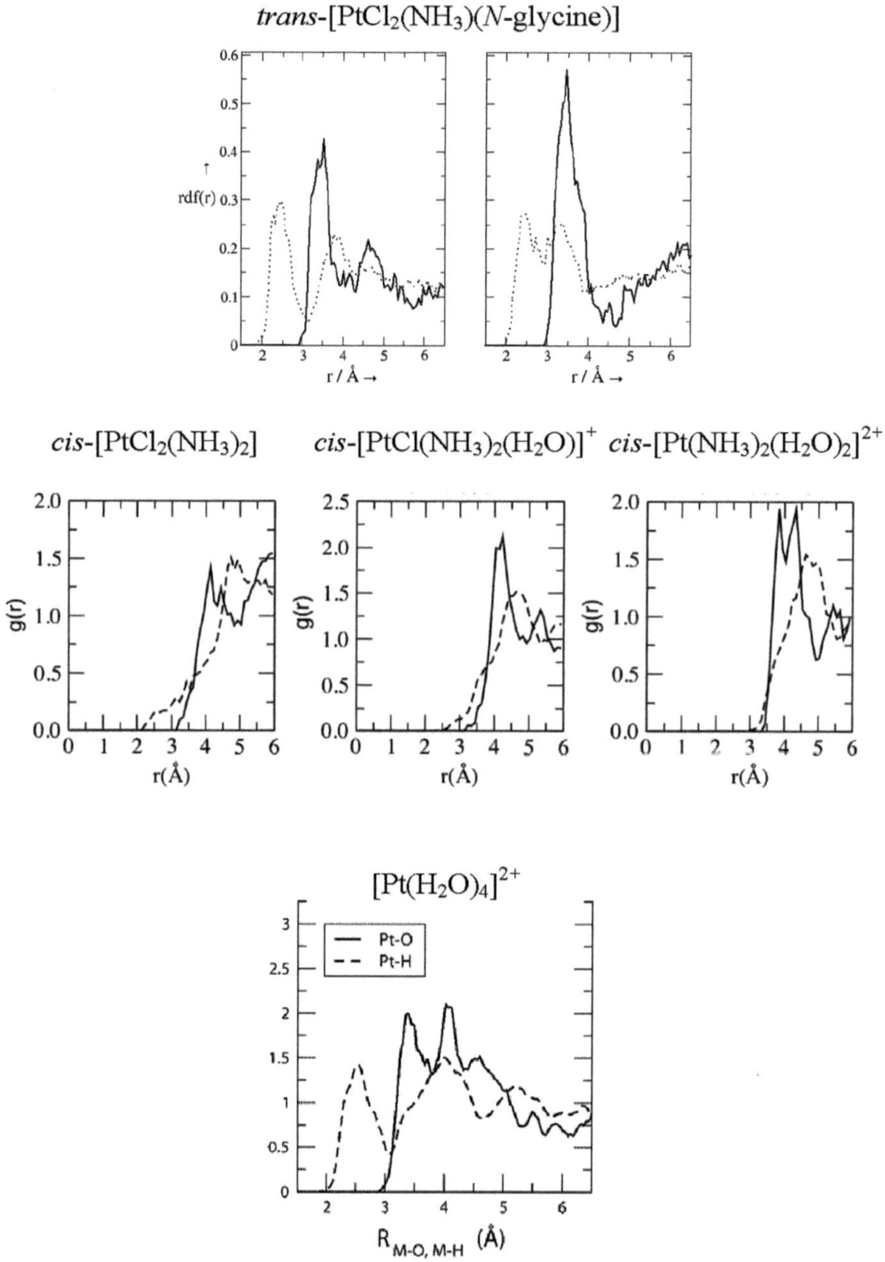

Fig. 6.12 Radial distribution functions (rdf) in the axial regions of different Pt(II) complexes solvated in water, obtained from AIMD simulations. Pt-O distance distributions are shown as *solid line*, Pt-H distance distributions as *dashed* or *pointed line*. *Top*: *trans*–[$PtCl_2(NH_3)$(*N*-glycine)] from AIMD simulations based on BLYP (*left*) and BLYP–D3 (*right*) functionals. Reproduced with permission from [73]. *Middle*: *cis*–[$PtCl_2(NH_3)_2$], *cis*-[$PtCl(NH_3)_2(H_2O)]^+$, and *cis*–[$Pt(NH_3)_2(H_2O)_2]^{2+}$ from AIMD simulations based on the B3LYP functional. Reproduced with permission from [76]. *Bottom*: [$Pt(H_2O)_4]^{2+}$ from AIMD simulations based on the PBE functional. Reproduced with permission from [79]

oriented towards platinum with the oxygen lone-pairs from the other side. Obviously, in a dicationic complex, the axial sites above and below the Pt atom are sites of highly positive electrostatic potential and favor a lone-pair-ahead orientation of axial water molecules, as also indicated by *in vacuo* calculations of interaction energy [66, 67]. Although Marcos et al. [78, 79] argued that the bulk solvent favors the hydrogen-bonding orientation, it is difficult to imagine that it would overcompensate the effect of the electrostatic potential. In any case, the AIMD simulation of Lau and Ensing of *cis*–$[Pt(NH_3)_2(H_2O)_2]^{2+}$does not indicate any tendency of this similar dicationic complex to assume the hydrogen-bonding orientation. The difference observed by these authors between the behavior of *cis*–$[PtCl_2(NH_3)_2]$ *versus* *cis*–$[Pt(NH_3)_2(H_2O)_2]^{2+}$in solution perfectly matches the switch from the hydrogen-bonding orientation seen in the neutron diffraction structure of *trans*–$[PtCl_2(NH_3)($*N*-glycine$)].H_2O$ (Fig. 6.5, [36]) to the lone-pair-to-platinum orientation seen for $[Pt(py)_4]Cl_2.3H_2O$ [80].

6.4.3.2 Probing the Solution Dynamics of Agostic Interactions in Pt(II) Complexes with AIMD Simulations

Tricoordinated complexes of d^8 metal ions are intermediates in many organometallic transformations ([58], [81], and references therein). Agostic interactions are frequently involved in the stabilization of these unsaturated complexes. In solution, and even in crystal structures, these interactions can be fluxional, as exemplified by the crystal structure of $[Pd(4\text{-}OMe\text{–}Ph)\{N(3,5\text{-}(CF_3)_2C_6H_3)_2\}(P^tBu_3)]$ where two molecules are present in the asymmetric unit, one without and one with an agostic interaction involving a *t*-butyl substituent of P^tBu_3 [82]. The solution dynamics of agostic interactions is related to their chemical transformations. Occupation of the vacancy in unsaturated complexes is an important issue since it can be directly related to their reactivity. Vidossich, Lledós et al. have therefore performed QM/MM molecular dynamics simulations of three representative Pt(II) complexes displaying β-, δ-, and remote, ξ-agostic interactions, respectively, in explicit dichloromethane solvent [83]. The metal complexes were treated quantum mechanically (PBE/DZVP), whereas ~ 1000 dichloromethane solvent molecules and the counterions were described using molecular mechanics. The simulations of 15 ps showed a stable agostic bond for the β-agostic interaction, and exchange between C–H bonds from different methyl groups of a *t*-butyl group for the δ-agostic interaction. For the ξ-agostic interaction, two types of movement were observed, exchange between the three C–H bonds of the agostic methyl group, and movements of this methyl group in and out of the coordination sphere. Although more studies and longer simulations are needed to make more specific statements and to correlate the simulations with NMR experiments, this simulation certainly represents a promising initial step towards detailed dynamic description of agostic bonds at an atomistic level.

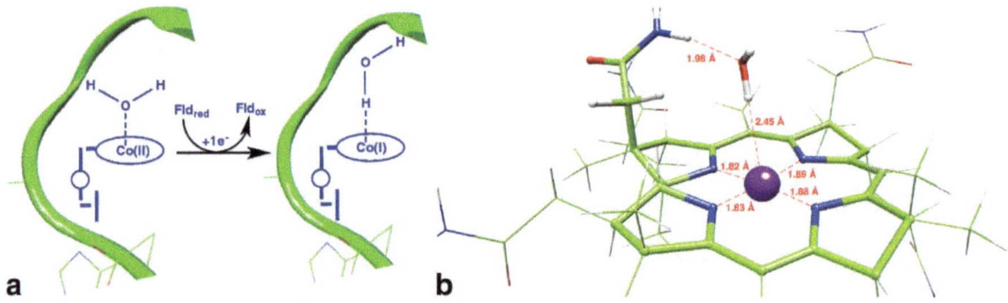

Fig. 6.13 **a** Alternate mechanistic pathway for MetH-bound Co(II)/Co(I) reduction. **b** BP86-optimized Co^{+}Cbi with an axial water molecule. Reproduced with permission from [90] and [87], respectively

6.5 Co(I) In Cobalamin-Dependent Transferases: A Biologically Relevant d^8 Acceptor of Hydrogen-Bonding?

Cobalamin-dependent methyltransferases and adenosyltransferases share a common mechanistic feature: reduction of cob(II)alamin (Co^{2+}Cbi) to cob(I)alamin (Co^{+}Cbi). For example, in the reactivation cycle of *E. coli* cobalamin-dependent methionine synthase (MetH), Co^{2+}Cbi has to be reduced to Co^{+}Cbi when Co^{+}Cbi is accidentally oxidized to Co^{2+}Cbi [84]. Co^{+}Cbi is an extremely powerful reducing agent (the midpoint potential for the Co^{2+}/Co^{+}couple at pH 7 is -490 mV against SHE for the wild-type *E. Coli* MetH [85]) and it is not clear how the Co^{2+}Cbi→Co^{+}Cbi reduction is achieved with common biological reducing agents whose midpoint potentials are higher [86]. A generally accepted hypothesis, based on spectroscopic and structural studies, invokes the stabilization of the Co^{+}state by creating a tetracoordinate environment favoring the d^8 configuration ([84] and references therein). Although this hypothesis plausibly fits the known principles of the ligand field theory, Kozlowski et al. challenged it asking why the enzyme should put up with the energetic disadvantage of creating an empty space on both sides of the Co-corrin plane and leave the Co^{+}ion vulnerable to unwanted nucleophilic attack, when it has the possibility to stabilize the d^8 state by an axial O-H$\cdots$Co(I) hydrogen bond, either from an inversely coordinated water ligand (Fig. 6.13a) [87–89], or from a nearby tyrosine ligand [90]. As a test, the authors have used various DFT functionals to optimize Co^{2+}Cbi and Co^{+}Cbi with an axial water molecule. The calculations have shown that whereas Co^{2+}Cbi prefers classical coordination with oxygen lone-pairs, Co^{+}Cbi attracts water with one of its hydrogen atoms, forming an O–H$\cdots$Co(I) hydrogen bond (Fig. 6.13b) [87]. The hydrogen bond was found to be thermodynamically stable and to consist of charge-transfer, correlation, dispersion, and electrostatic components.

6.6 Conclusion

A considerable amount of experimental and theoretical data has accumulated on d^8 metal complexes presenting X–H··· M agostic or hydrogen-bonding interactions. Some initially unclear aspects could be clarified, and general characteristics distinguishing these two different interaction types can be now delineated:

i) In agostic interactions, one (or two in some cases) C-H bond interacts with a tricoordinate d^8 center, completing thus the square-planar tetracoordination typical of the d^8 complexes. Only intramolecular agostic complexes of d^8 metals have been characterized to date. In contrast, hydrogen bonding contacts occur exclusively between a coordinatively saturated d^8 center and an intra- or intermolecular X–H hydrogen bond donor (X = C, N, O) along the z-axis of the coordination square. The two cases are exemplified by the structures shown in Fig. 6.1.

ii) Both interaction types involve a covalent component. In agostic bonds, the latter arises from the charge transfer from the filled C–H σ-orbital to antibonding orbitals of the metal. In hydrogen bonds, the transfer goes from the filled d_{z2}, d_{xz}, and/or d_{yz} metal orbitals to the X–H σ∗-antibonding orbital. Counting the valence electrons involved, agostic bonding corresponds to a 3-center-2-electron bond and hydrogen-bonding to a 3-center-4-electron bond. The donor-acceptor orbital overlap is the principal determinant of the directionality of X–H··· M(d^8) bonding: hydrogen bonds tend to be linear, whereas agostic bonds tend to be bent. There is no intermediate class of C–H··· M(d^8) bonds involving a coordinatively saturated d^8 center; all interactions previously interpreted as pregostic, preagostic etc., were shown to be of the hydrogen-bonding type, and as all hydrogen bonds, tend to linearity. In intramolecular hydrogen bonds, however, a non-linear C–H··· M angle can be imposed by the structural constraints of the ligand.

iii) The magnetic anisotropy of the d^8 electronic system causes shielding of nuclei in the coordination plane and deshielding along the z-axis. H-atoms involved in hydrogen bonding with d^8 centers are therefore always deshielded. The effect can be enhanced by the induction effect typical of all hydrogen bonds; however, its relative contribution to the overall deshielding in X–H··· M(d^8) hydrogen bonds has not been evaluated so far. Agostic H-atoms are usually located in the coordination plane and are accordingly shielded. However, in particular cases such as that of complex **4**, the agostic hydrogens lie outside the coordination plane and are not deshielded.

iv) The Bond Critical Point (BCP) analysis using the AIM theory does not allow for an unambiguous distinction between agostic bonding and hydrogen bonding. Agostic bonds do usually show $\rho(r_c)$ and $\nabla^2\rho(r_c)$ values outside the range observed for hydrogen bonds, however, weak agostic bonds can show values in the hydrogen bonding range. Nevertheless, $\rho(r_c)$ has been shown to correlate with hydrogen-bonding energy, and can be therefore used as an estimate of strength of hydrogen bonding.

v) Natural Bond Orbital (NBO) analysis has been instrumental in evaluating the σ →M and M→ σ* charge transfer, and the concomitant charge-transfer energies. Generally, the former is larger for agostic bonds, and the latter for hydrogen bonds.

Recent computational advances towards the understanding of agostic & hydrogen bonding X–H$\cdots$M(d^8) interactions include *ab initio* molecular dynamics simulations, which allow insights into the dynamic behavior of these systems in solution. A particularly interesting result from computational studies is the prediction of models for the Co^+"supernucleophile" occurring in the active site of cobalamin-dependent methyl and adenosyl transferases, where the d^8 state is stabilized by tetracoordination complemented with axial hydrogen bonding [87, 90]. It will be a challenging task to test these models experimentally against currently accepted models.

Supporting Material Table of CSD Refcodes and geometrical parameters of X–H$\cdots$M contacts used in Fig. 6.4

Acknowledgment I would like to thank all students and colleagues whom I was lucky to work with. Their names figure in the correponding references. I acknowledge support with platinum chemicals from W. C. Heraeus GmbH and financial support from the Hubert-Curien program "Galileo" and from COST (Action D39/0004/06).

References

1. Clot E, Eisenstein O (2004) Agostic interactions from a computational perspective: one name, many interpretations. Struct Bond 113:1–36
2. Brookhart M, Green MLH, Parkin G (2007) Agostic interactions in transition metal compounds. Proc Nat Acad Sci USA 104:6908–6914
3. Lein M (2009) Characterization of agostic interactions in theory and computation. Coord Chem Rev 253:625–634
4. Scherer W, Herz V, Hauf C (2012) On the nature of β-agostic interactions: a comparison between the molecular orbital and charge density picture. Struct Bond 146:159–208
5. Martín A (1999) Hydrogen bonds involving transition metal centers acting as proton acceptors. J Chem Ed 76:578–583
6. Brammer L (2003) Metals and hydrogen bonds. Dalton Trans 3145–3157
7. Calhorda MJ (2006) In: Grabowski SJ (ed) Hydrogen bonding- new insights, vol 3. Springer, Dordrecht, pp 245–262
8. Bruno G, Lanza S, Nicolò F (1990) Structure of $[Pt(C_6H_5)_2(btz\text{-}N, N')].CHCl_3$, btz=2,2'-Bi-5,6-dihydro-4*H*-1,3-thiazine. Acta Cryst C 46:765–767
9. Carr N, Dunne BJ, Orpen AG, Spencer JL (1988) Co-ordinatively unsaturated diphosphine platinum(II) alkyl cations: a new class of β-agostic complexes. J Chem Soc Chem Commun 926–928
10. Ingleson MJ, Mahon MF, Weller AS (2004) $[PtMe(^iPr_3P)_2]^+$: a Pt(II) complex with an agostic interaction that undergoes C–H activation. Chem Commun 2398–2399
11. Baratta W, Stoccoro S, Doppiu A, Herdtweck E, Zucca A, Rigo P (2003) Novel T-shaped 14-electron platinum(II) complexes stabilized by one agostic interaction. Angew Chem Int Ed 42:105–108
12. Brookhart M, Green MLH (1983) Carbon-hydrogen-transition metal bonds. J Organomet Chem 250:395–408
13. Roe DM, Bailey PM, Moseley K, Maitlis PM (1972) Structure of bromobis(triphenylphosphine)-(1,2,3,4-tetrakisrnethoxycarbonylbuta-1,3-dienyl)palladium and evidence for a C–H$\cdots$Pd interaction. J Chem Soc Chem Commun 1273–1274
14. Bailey NA, Jenkins JM, Mason R, Shaw BL (1965) Unusual co-ordination of the palladous ion in the structure of *trans*-di-iodobis(dimethylphenylphosphine)palladium(II). J Chem Soc Chem Commun 237

15. Albinati A, Anklin CG, Pregosin PS (1984) Platinum induced C–H activation in aromatic aldehydes. Unusual J(Pt, H) coupling constants and structure of *trans*-dichloroquinoline-8-carboxaldehyde triethylphosphine platinum(II). Inorg Chim Acta 90:L37–L38
16. Albinati A, Arz C, Pregosin PS (1987) ^nJ(Pt, H) and Pt–H–C interactions in Schiff base complexes of 2-(benzylideneamino)-3-methylpyridine. Molecular structures of dichloro(2-((2,4,6-trimethylbenzylidene)amino)-3-methylpyridine)(triethylarsine)platinum(II) and dichloro-(2-amino-3-methylpyridine)(triethylphosphine)palladium(II). Inorg Chem 26:508–513
17. Albinati A, Anklin CG, Ganazzoli F, Rüegg H, Pregosin PS (1987) Preparative and ^{1}H NMR spectroscopic studies on palladium(II) and platinum(II) quinoline-8-carbaldehyde (1) complexes. X-ray Structures of the cyclometalated acyl complex $PdCl(C(O)C_9H_6N)(PPh_3).PPh_3$ and *trans*-$PtCl_2(1)(PEt_3)$. Inorg Chem 26:503–508
18. Albinati A, Pregosin PS, Wombacher F (1990) Weak Pt–H–C interactions. Extensions to 8-methylquinoline, benzoquinoline, and a tetralone schiff base. X-ray crystal structure of *trans*-$PtCl_2$(benzoquinoline)(PEt_3). Inorg Chem 29:1812–1817
19. Bortolin B, Bucher U, Ruegger H, Venanzi LM, Albinati A, Lianza F (1992) Synthesis and 2D NMR studies of {bis(pyrazolyl)borato}rhodium(I) complexes with weak Rh··· H–C interactions and the x-ray crystal structure of {(cyclooctane-1,5-diyl)bis(pyrazol-1-yl)borato}(1,5-cyclooctadiene)rhodium. Organometallics 11:2513–2521
20. Albinati A, Lianza F, Pregosin PS, Müller B (1994) New N–H··· Pt Interactions. The nature of the bond. Inorg Chem 33:2522–2526
21. Neve F, Ghedini M, Crispini A (1992) Weak Rh–H–C interactions. Molecular structure of [*trans*–Rh(CO)(8-methylquinoline)(PPh_3]BF_4. Organometallics 11:3324–3327
22. Neve F, Ghedini M, De Munno G, Crispini A (1991) Aromatic and benzylic C–H activation. Synthesis and structural characterization of iridium 2-phenylpyridine and 8-methylquinoline complexes. Organometallics 11:1143–1148
23. Calhorda MJ (2000) Weak hydrogen bonds: theoretical studies. Chem Comun 801–809
24. Lewis JC, Wu J, Bergman RG, Ellman JA (2005) Preagostic Rh–H interactions and C–H bond functionalization: a combined experimental and theoretical investigation of rhodium(I) phosphinite complexes. Organometallics 24:5737–5746
25. Zhang Y, Lewis JC, Bergman RG, Ellman JA, Oldfield E (2006) NMR shifts, orbitals, and M··· H–X bonding in d^8 square planar metal complexes. Organometallics 25:3515–3519
26. Mukhopadhyay A, Pal S (2006) Intramolecular apical C–H··· M interactions in square-planar nickel(II) complexes with dianionic tridentate ligands and 2-phenylimidazole. Eur J Inorg Chem 4879–4887
27. Taubmann C, Öfele K, Herdweck E, Hermann WA (2008) Complexation of (*5H*)-dibenzo[*a, d*]cyclohepten-5-ylidene to palladium(II) via the diazo route and evidence of C–H··· Pd Interactions. Organometallics 28:4254–4257
28. Schöler S, Wahl MH, Wurster NIC, Puls A, Hättig C, Dyker G (2014) Bidentate cycloimidate palladium complexes with aliphatic and aromatic anagostic bonds. Chem Comun 50:5909–5911
29. Yao W, Eisenstein O, Crabtree RH (1997) Interactions between C–H and N–H bonds and d^8 square planar metal complexes: hydrogen-bonded or agostic? Inorg Chim Acta 254:105–111
30. Sundquist WI, Bancroft DP, Lippard SJ (1990) Synthesis, characterization, and biological activity of *cis*-diammineplatinum(II) complexes of the DNA intercalators 9-aminoacridine and chloroquine. J Am Chem Soc 112:1590–1596
31. Deeming AJ, Rothwell IP, Hursthouse MB, New L (1970) Comparison of 8-methylquinoline and benzo[*h*]quinoline complexes of palladium(II) with those of related ligands. Crystal and molecular structure of aqua(benzo[*h*]quinoline)[2-(dimethylaminomethyl)phenyl-*N*]palladium(II) perchlorate. J Chem Soc Dalton Trans 1489–1496
32. Hambley TW (1998) Van der Waals radii of Pt(II) and Pd(II) in molecular mechanics models and an analysis of their relevance to the description of axial M··· H(–C), M··· H(–N), M··· S and M··· M (M=Pd(II) or Pt(II)) interactions. Inorg Chem 37:3767–3774
33. Braga D, Grepioni F, Tedesco E, Biradha K, Desiraju GR (1997) Hydrogen bonding in organometallic crystals. 6. X--H---M hydrogen bonds and M---(H--X) pseudo-agostic bonds. Organometallics 16:1846–1856

34. Brammer L, Charnock JM, Goggin PL, Goodfellow RJ, Koetzle TF, Orpen AG (1987) Hydrogen bonding by cisplatin derivatives: evidence for the formation of N–H··· Cl and N–H··· Pt bonds in $[NPr^n_4]\{[PtCl_4]\cdot$ *cis*–$[PtCl_2(NH_2Me)_2]\}$. J Chem Soc Chem Commun 443–445
35. Casas JM, Falvello LR, Forniés J, Martín A (1996) Syntheses and structures of the complexes *cis*–$[M(C_6F_5)_2(N–X)]$ (M=Pd, Pt; N–X=2-iodoaniline, 2-benzoylpyridine) containing N–X acting as a didentate chelating ligand and displaying I–M or O–M interactions. Inorg Chem 35:56–62
36. Rizzato S, Bergès J, Mason SA, Albinati A, Kozelka J (2010) Dispersion-driven hydrogen bonding: predicted hydrogen bond between water and platinum(II) identified by neutron diffraction. Angew Chem Int Ed 49:7440–7443
37. Buckingham AD, Stephens JP (1964) Proton chemical shifts in the nuclear magnetic resonance spectra of transition-metal hydrides: square-planar platinum(II) complexes. J Chem Soc 4583–4587
38. Miller RG, Stauffer RD, Fahey DR, Parnell DR (1970) Alkenaryl compounds of nickel(II) and palladium(II). Influence of the transition metal on ligand proton chemical shifts. J Am Chem Soc 92:1511–1521
39. Chatt J, Duncanson LA, Shaw BL (1957) A volatile chlorohydride of platinum. Proc Chem Soc 343
40. Chatt J, Shaw BL (1962) Hydrido-complexes of platinum(II). J Chem Soc 5075–5084
41. Church MJ, Mays MJ (1968) Spectroscopic studies on some new cationic complexes of platinum(II). J Chem Soc (A):3074–3078
42. Bercaw JE, Marvich RH, Bell LG, Brintzinger HH (1972) Titanocene as an intermediate in reactions involving molecular hydrogen and nitrogen. J Am Chem Soc 94:1219–1238
43. Manriquez JM, McAlister DR, Sanner RD, Bercaw JE (1978) Reduction of carbon monoxide promoted by alkyl and hydride derivatives of permethylzirconocene. J Am Chem Soc 100:2716–2724
44. Caulton KG, Goeden GV (1981) Soluble copper hydrides: solution behavior and reactions related to CO hydrogenation. J Am Chem Soc 103:7354–7355
45. Ruiz-Morales Y, Schreckenbach G, Ziegler T (1996) Origin of the hydridic ^1H NMR chemical shift in low-valent transition-metal hydrides. Organometallics 15:3920–3923
46. Scherer W, Herz V, Brück A, Hauf C, Reiner F, Altmannshofer S, Leusser D, Stalke D (2011) The nature of β-agostic bonding in late-transition-metal alkyl complexes. Angew Chem Int Ed 50:2845–2849
47. Conroy-Lewis FM, Mole L, Redhouse AD, Lister SA, Spencer JL (1991) Synthesis of coordinatively unsaturated diphosphine nickel(II) and palladium(II) β-agostic ethyl cations: X-ray crystal structure of $[Ni\{Bu^t_2P(CH_2)_2PBu^t_2\}(C_2H_5)][BF_4]$. J Chem Soc Chem Commun 1601–1603
48. Pregosin PS, Rüegger H, Wombacher F, van Koten G, Grove DM, Wehman-Ooyevaar ICM (1992) New Pt··· H–N bonds characterized by ^{15}N-filtered and 2D NOESY ^1H NMR spectroscopy. Magn Reson Chem 30:548–551
49. Yoshida T, Tani K, Yamagata T, Tatsuno Y, Saito T (1990) Preparation and structure of $[Rh\{(h^5–C_5H_4(2-C_5H_4N))(h^5–C_5H_4PPh_2)\}(cod)]PF_6$ and $[Ir(H)\{Fe[h^5–C_5H_3(2-C_5H_4N)](h^5–C_5H_4PPh_2)\}(cod)]PF_6$; a Rh^I complex having a C–H··· Rh^I interaction and a hydrido Ir^{III} complex (where cod = cyclo-octa-1,5-diene). Chem Commun 292–294
50. Hedden D, Roundhill DM, Fultz WC, Rheingold AL (1986) Reaction chemistry of some new hybrid phosphine amide complexes of platinum(II) and palladium(II). Isolation and X-ray structure determination of an ortho-metalated platinum(II) complex derived from a chelated phosphine amide complex of platinum(II). Organometallics 5:336–343
51. Weinhold F, Klein RA (2012) What is a hydrogen bond? Mutually consistent theoretical and experimental criteria for characterizing H-bonding interactions. Mol Phys 110:565–579
52. Li Y, Zhang G, Chen D (2012) Theoretical investigation of hydrogen bonding between water and platinum(II): an atom in molecule (AIM) study. Mol Phys 110:179–184
53. Fedoce Lopes J Da Silva JCS Rocha WR De Almeida WB Dos Santos HF (2011) Quantum chemical study of cisplatin-water complexes: an investigation of electron correlation effects. J Chem Theory Comput 10:371–391

54. Bergès J, Fourré I, Pilmé J, Kozelka J (2013) A quantum chemical topology study of the water-platinum(II) interaction. Inorg Chem 52:1217–1227
55. Zhang G, Li X, Li Y, Chen D (2013) Electron density characteristics and charge transfer effect of hydrogen bond O–H··· Pt(II): atoms in molecules study and natural bond orbital analysis. Mol Phys 111:3276–3282
56. Baya M, Belio U, Martin A (2014) Synthesis, characterization, and computational study of complexes containing Pt··· H hydrogen bonding interactions. Inorg Chem 53:189–200
57. Thakur TS, Desiraju GR (2007) Theoretical investigation of C–H··· M interactions in organometallic complexes: a natural bond orbital (NBO) study. J Mol Struct: THEOCHEM 810:143–154
58. Stambuli JP, Incarvito CD, Bühl M, Hartwig JF (2004) Synthesis, structure, theoretical studies, and ligand exchange reactions of monomeric, T-shaped arylpalladium(II) halide complexes with an aditional, weak agostic interaction. J Am Chem Soc 126:1184–1194
59. Sassmanshausen J (2011) Agostic or not? Detailed density functional theory studies of the compounds [LRh(CO)Cl], [LRh(COD)Cl] and [LRhCl] (L = cyclic (alkyl)(amino)carbene, COD = cyclooctadiene). Dalton Trans 40:136–141
60. Andrae D, Häussermann U, Dolg M, Stoll H, Preuss H (1990) Energy-adjusted *ab initio* pseudopotentials for the second and third row transition elements. Theor Chim Acta 77:123–141
61. Kozelka J, Bergès J (1998) *Ab Initio* calculations on *cis*–[$PtCl_2(PMe_3)_2$]: search for a model chemistry for platinum(II) complexes. J Chim Phys 95:2226–2240
62. Koch U, Popelier PLA (1995) Characterization of C–H–O hydrogen bonds on the basis of the charge density. J Phys Chem 99:9747–9754
63. Popelier PLA, Logotheties G (1998) Characterization of an agostic bond on the basis of the electron density. J Organomet Chem 555:101–111
64. Lavallo V, Canac Y, DeHope A, Donnadieu B, Bertrand G (2005) A rigid cyclic (alkyl)(amino)carbene ligand leads to isolation of low-coordinate transition-metal complexes. Angew Chem Int Ed 44:7236–7239
65. Parthasarathi R, Subramanian V, Sathyamurthy N (2006) Hydrogen bonding without borders: an atoms-in-molecules perspective. J Phys Chem A 110:3349–3351
66. Kozelka J, Bergès J, Attias R, Fraitag J (2000) O–H··· Pt(II): hydrogen bond with a strong dispersion component. Angew Chem Int Ed Engl 39:198–201
67. Fedoce Lopes J, Rocha WR, Dos Santos HF, De Almeida WB (2008) Theoretical study of the potential energy surface for the interaction of cisplatin and their aquated species with water. J Chem Phys 128:165103
68. Reed AE, Curtiss LA, Weinhold F (1988) Intermolecular interactions from a natural bond orbital, donor-acceptor viewpoint. Chem Rev 88:899–926
69. Reed AE, Weinhold F, Curtiss LA, Pochatko DJ (1986) Natural bond orbital analysis of molecular interactions: theoretical studies of binary complexes of HF, H_2O, NH_3, N_2, O_2, F_2, CO, and CO_2 with HF, H_2O and NH_3. J Chem Phys 84:5687–5705
70. Cappelletti D, Ronca E, Belpassi L, Tarantelli F, Pirani F (2012) Revealing charge-transfer effects in gas-phase water chemistry. Acc Chem Res 45:1571–1580
71. Belpassi L, Infante I, Tarantelli F, Visscher L (2008) The chemical bond between Au(I) and the noble gases. Comparative study of NgAuF and $NgAu^+$ (Ng = Ar, Kr, Xe) by density functional and coupled cluster methods. J Am Chem Soc 130:1048–1060
72. Belpassi L, Tarantelli F, Pirani F, Candori P, Cappelletti D (2009) Experimental and theoretical evidence of charge transfer in weakly bound complexes of water. Phys Chem Chem Phys 11:9970–9975
73. Vidossich P, Ortuno MA, Ujaque G, Lledos A (2011) Do metal··· water hydrogen bonds hold in solution? Insight from *ab initio* molecular dynamics simulations. Chem Phys Chem 12:1666–1668
74. Truflandier LA, Sutter K, Autschbach J (2011) Solvent effects and dynamic averaging of ^{195}Pt NMR shielding in cisplatin derivatives. Inorg Chem 50:1723–1732
75. Truflandier LA, Autschbach J (2010) Probing the solvent shell with ^{195}Pt chemical shifts: density functional theory molecular dynamics study of Pt^{II} and Pt^{IV} anionic complexes in aqueous solution J Am Chem Soc 132:3472–3483

76. Lau JK-C, Ensing B (2010) Hydrolysis of cisplatin—a first-principles metadynamics study. Phys Chem Chem Phys 12:10348–10355
77. Martin DS Jr (1967) Anomalies in ligand exchange reactions for platinum(II) complexes. Inorg Chim Acta Reviews 87–97
78. Beret EC, Pappalardo RR, Doltsinis NL, Marx D, Sánchez Marcos E (2008) Aqueous Pd^{II} and Pt^{II}: anionic hydration revealed by Car-Parrinello simulations. ChemPhysChem 9:237–240
79. Beret EC, Martínez JM, Pappalardo RR, Sánchez Marcos E, Doltsinis NL, Marx D (2008) Explaining asymmetric solvation of Pt(II) versus Pd(II) in aqueous solution revealed by *ab initio* molecular dynamics simulations. J Chem Theory Comput 4:2108–2121
80. Wei CH, Hingerty BE, Busing WR (1989) Structure of tetrakis(pyridine)platinum(II) chloride trihydrate: unconstrained anisotropic least-squares refinement of hydrogen and non-hydrogen atoms from combined X-ray-neutron diffraction data. Acta Cryst C 45:26–30
81. Urtel H, Meier C, Eisenträger F, Rominger F, Joschek JP, Hofmann P (2001) A neutral three-coordinate alkylrhodium(I) complex: Stabilization of a 14-electron species by γ-C–H agostic interactions with a saturated hydrocarbon group. Angew Chem Int Ed 40:781–784
82. Yamashita M, Hartwig JF (2004) Synthesis, structure, and reductive elimination chemistry of three-coordinate arylpalladium amido complexes. J Am Chem Soc 126:5344
83. Ortuno MA, Vidossich P, Ujaque G, Conejero S, Lledos A (2013) Solution dynamics of agostic interactions in T-shaped Pt(II) complexes from *ab initio* molecular dynamics simulations. Dalton Trans 42:12165–12172
84. Koutmos M, Datta, Pattridge KA, Smith JL, Matthews RG (2009) Insights into the reactivation of cobalamin-dependent methionine synthase. Proc Nat Acad Sci USA 106:18527–18532
85. Jarrett JT, Choi CY, Matthews RG (1997) Changes in protonation associated with substrate binding and cob(I)alamin formation in cobalamin-dependent methionine synthase. Biochemistry 36:15739–15748
86. Jarrett JT, Hoover DM, Ludwig ML, Matthews RG (1998) The mechanism of adenosylmethionine-dependent activation of methionine synthase: a rapid kinetic analysis of intermediates in reductive methylation of cob(II)alamin enzyme. Biochemistry 37:12649–12658
87. Kumar M, Kozlowski PM (2011) A biologically relevant $Co^{1+}\cdots$ H bond: possible implications in the protein-induced redox tuning of Co^{2+}/Co^{1+} reduction. Angew Chem Int Ed 123:8861–8864
88. Kumar M, Kumar N, Hirao H, Kozlowski PM (2012) Co^{2+}/Co^{+} redox tuning in methyltransferases induced by a conformational change at the axial ligand. Inorg Chem 51:5533–5538
89. Kumar M, Hirao H, Kozlowski PM (2012) Co^{+}–H interaction inspired alternate coordination geometries of biologically important cob(I)alamin: possible structural and mechanistic consequences for methyltransferases. J Biol Inorg Chem 17:1107–1121
90. Kumar M, Kozlowski PM (2013) Can the local enzyme scaffold act as an H-donor for a Co(I)–H bond formation? The curious case of methionine synthase-bound cob(I)alamin. J Inorg Biochem 126:26–34

Chapter 7
What is Common for Dihydrogen Bond and H· · ·σ Interaction—Theoretical Analysis and Experimental Evidences

Sławomir J. Grabowski

Abstract Two types of the hydrogen bond are described and compared here; the A–H· · ·H–B dihydrogen bond and the A–H· · ·σ interaction. In a case of the dihydrogen bond the H· · ·H contact between the hydrogen atoms characterized by the opposite charges is observed; i.e. between the protonic (A)H and hydridic (B)H hydrogens. For the A–H· · ·σ hydrogen bond the A–H proton donating bond interacts with the σ-electrons of the molecular hydrogen. These interactions are topologically different since for DHB the bond path linking the attractors of H-atoms with the corresponding bond critical point is observed. For the A–H· · ·σ interaction the bond path between the (A)H-atom attractor and the bond critical point of the H–H bond of the molecular hydrogen is observed. Both types of the hydrogen bond are characterized by the significant $\sigma \rightarrow \sigma^*$ orbital-orbital interaction, $\sigma_{BH} \rightarrow \sigma_{AH}{}^*$ in a case of DHB and $\sigma_{HH} \rightarrow \sigma_{AH}{}^*$ for A–H· · ·σ. There are also evidences that A–H· · ·H–B and A–H· · ·σ may be classified as the hydrogen bonds. The examples of complexes characterized by the mentioned above types of the hydrogen bond are analyzed in this chapter, the theoretical as well as experimental examples are presented.

7.1 Introduction

It was 20 years ago when R. H. Crabtree and coworkers noted that "Typical hydrogen bond acceptors possess an O or N lone pair, but more recently, the σ-electron pair of a transition metal hydride has been shown to give intramolecular N–H· · ·H–M and O–H· · ·H–M hydrogen bonds of an unconventional type, for which we suggest the term dihydrogen bonds. These interactions have bond strengths of 4–6 kcal/mol and H· · ·H distances of ca. 1.7–1.9 Å and seem to play a role in proton transfer, fluxional processes and other reactions [1]." The authors also wrote in the footnote of the article that "a referee points out that these unconventional hydrogen bonds require a

S. J. Grabowski (✉)
Kimika Fakultatea, Euskal Herriko Unibertsitatea UPV/EHU, Donostia International Physics Center (DIPC), P.K. 1072, 20080 Donostia, Euskadi, Spain
e-mail: s.grabowski@ikerbasque.org

IKERBASQUE, Basque Foundation for Science, Maria Diaz de Haro 3, 48013 Bilbao, Spain

S. Scheiner (ed.), *Noncovalent Forces,* Challenges and Advances in Computational Chemistry and Physics 19, DOI 10.1007/978-3-319-14163-3_7

Scheme 7.1 The hydrogen bond and dihydrogen bond interactions

hydrogen bond	**A-H $^{+\delta}\ldots^{-\delta}$ B**
dihydrogen bond	**A-H $^{+\delta}\ldots^{-\delta}$ H-B**

new name, for which we suggest the term dihydrogen bond [1]." The authors refer to the intramolecular N–H···H–M and O–H···H–M interactions but later articles concern also intermolecular dihydrogen bonds. Early studies on dihydrogen bond are briefly described in the next section of this chapter. Additionally the dihydrogen bonds are classified as "hydrogen bonds of an unconventional type" with typical proton donating bond, O–H or N–H, and an unconventional proton acceptor, another hydrogen atom of the hydride bond. Even more, the authors claim that the σ-electron pair of the hydride bond plays the role of the Lewis base center. The H···H distances are in the range of 1.7–1.9 Å, less than the sum of the van der Waals radii of H-atoms (∼ 2.0–2.4 Å). The range of the interaction energy for dihydrogen bonds, 4–6 kcal/mol [2], is situated around the hydrogen bond energy for the water dimer, ca. 5 kcal/mol [3], where typical, conventional O–H···O hydrogen bond exists.

In the article mentioned above [1] and other early studies of Crabtree and co-workers [4, 5] the dihydrogen bond (DHB) term was reserved for O–H···H–M and N–H···H–M interactions (where M designates the transition metal or boron) but the authors have claimed that possible elements connected with the hydridic hydrogen are "currently known as boron and the transition metals; other cases will no doubt be discovered in the future [6]." And really further studies have shown that numerous other interactions can be classified as DHBs [7], even C–H···H–C contacts were considered as the interactions of this type [8, 9]. It was also found that for numerous H···H links the energy of interaction exceeds 10 or even 20 kcal/mol [10–16]. Scheme 7.1 presents the main difference between hydrogen and dihydrogen bond. For the first interaction the Lewis base center (B) is usually electronegative atom possessing at least one lone electron pair, for the dihydrogen bond the hydridic negatively charged H-atom plays the role of the Lewis base center. The same types of A–H proton donating bonds with the excess of the positive charge on H-atom are observed for both interactions (Scheme 7.1).

One should note that the dihydrogen bond plays a substantial role in numerous processes; controlling reactivity and selectivity in solution, σ-bond metathesis, hydride reduction or the ligand attachment to transition metal clusters [7, 17]. Its role in the hydrogen storage design should be also mentioned [18–21]. The dihydrogen bond may be also treated as the preliminary stage of the reaction leading to the separation of the gaseous hydrogen [17, 22–25]. Reversely, the activation of the molecular hydrogen may lead to further processes through the dihydrogen bond formation [26–28]. The dihydrogen bond seems to be important phenomenon since there are numerous studies concerning this kind of interaction, one can mention review articles [7, 29–32], book chapters [33–35] and even the book monograph [36]. This chapter refers to older concepts and to recent studies to present the current state of knowledge on the dihydrogen bond interaction.

It is interesting to note that Crabtree and coworkers claimed in their early studies that the σ-bond is the electron donor for dihydrogen bonded systems [37]. They have presented the concept of the A–H···σ hydrogen bond where A–H is the proton donating bond (Lewis acid) and σ-electrons (Lewis base) are the proton acceptor. This is why it is also discussed in this chapter if the A–H···σ and A–H···H–B refer to the same kind of interactions, i.e. if they may be classified as the dihydrogen bonds, or if these are different interactions.

7.2 Early Studies and First Concepts on Dihydrogen Bond Interaction

In 1934 Zachariasen and Mooney reported results on the crystal structure of ammonium hypophosphite and they have written that "the hydrogen atoms of the hypophosphite group behave toward ammonium as if they were H^- ions" [38]. This is probably the first study where the existence of the interaction between hydrogen atoms characterized by the opposite charges was detected. In the later study Burg analyzed the IR spectrum of the liquid $(CH_3)_2NH–BH_3$ and suggested the existence of $N–H···H_3B$ interactions possessing the characteristics similar to hydrogen bonds [39]. Titov and coworkers analyzed the reaction of aminoboranes which leads to the loss of molecular hydrogen. They have claimed that such a reaction is an effect of "close spatial arrangement of the oppositely charged hydrogen atoms" [40]. This is a very important study since it was pointed out there that the interaction between the opposite charged H-atoms, named in later studies as the dihydrogen bond, is the preliminary stage of the reaction of the separation of the molecular hydrogen. This was analyzed in the later studies that the dihydrogen bond may be considered as the initiation of the separation of the gaseous hydrogen [17, 22–25]. Similarly the hydrogen bond interaction is often treated as the preliminary stage of the proton transfer process [41, 42] or the σ-hole bond as the initiation of the S_N2 reaction [43].

In one of early studies Brown and Heseltine have reported the infrared evidence for a moderately strong and specific interaction between the coordinated BH_3 group in amine-boranes and a protonic H atom [44]. The authors have concluded that the appearance of a strong, broad, lower frequency stretching absorption in the infrared spectrum is the proof of the existence of the hydrogen bond. They have also noted that neither the BH_3 group nor the amine-borane molecule possesses a lone pair of electrons or any electron rich system thus this is the untypical kind of the hydrogen bond. In another study, Brown and coworkers [45] have reported the appearance of low-frequency O–H stretching absorptions upon addition of Me_3N and BH_3, to dilute solutions of methanol and phenol in carbon tetrachloride. The authors concluded that these absorptions are due to the formation of hydrogen bonded complexes in which the BH_3 group acts as a proton acceptor. They have suggested possible explanations of the complex formation: the formation of weak B···H–O bonds or the formation of weak B–H···H–O bonds between the hydridic H atoms attached to boron and the protonic H atom of OH group or this kind of interaction is a property of the whole

Fig. 7.1 The structure of $ReH_2(CO)(NO)(PMe_3)_2$ with indole; the H· · ·H contact corresponding to the N–H· · ·H–Re dihydrogen bond is presented (*broken line*), this structure was taken from the Cambridge Structural database [54] followed ref. [53]

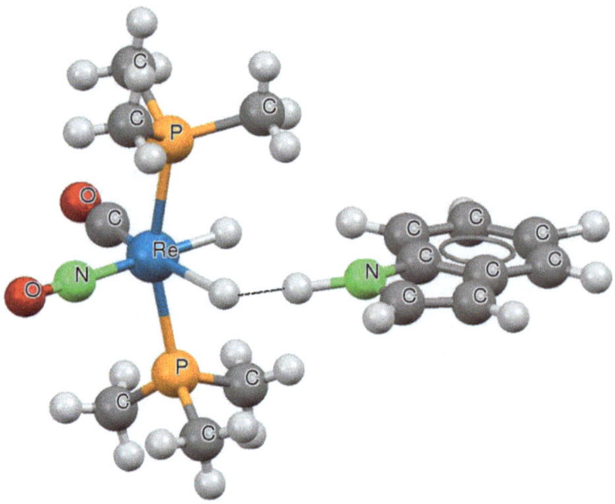

BH_3 group. One can see that one of those explanations suggest the existence of the B–H· · ·H–O connections, i.e. the dihydrogen bonds (Scheme 7.1).

The important studies justifying definitively the existence of the attractive interaction between hydrogen atoms of the opposite charges concern crystal structures. The first studies on structures containing such interactions were performed in nineties of the last century. In 1990 R. C. Stevens and coworkers reported the hydridohydroxy complex *cis*-$[IrH(OH)(PMe_3)_4]PF_6$ crystal structure analyzed at 20 K by single-crystal neutron diffraction [46]. The neutron diffraction results confirm earlier X-ray study on this crystal structure [47]. The authors have postulated, on the basis of geometrical results, the existence of an attractive $H^{-\delta}\ldots{}^{+\delta}H$ interaction between the hydridic Ir–H bond and the electron-deficient hydrogen of the OH group. In 1994 R.H.Morris and coworkers have found the Ir–$H^{-\delta}\ldots{}^{+\delta}$H–N intramolecular link between an iridium hydride and the protonated nitrogen of a sulfur-bonded thiopyridine ligand in the crystal structures analyzed and they have claimed that this link is a new type of interaction [48]. It seems that in the middle of nineties of the last century two groups of researchers have performed pioneering crystal structure analyses where inter- and intramolecular attractive H· · ·H interactions were detected; the group of Morris [48–50] and that one of Crabtree and coworkers [1, 4–6, 37, 51, 52]. One should also mention the early experimental studies on dihydrogen bonds of Epstein and Shubina and coworkers [25, 29, 30, 33, 53] and of others [7]. Figure 7.1 presents fragment of the crystal structure where the dihydrogen bond interaction is presented.

The dihydrogen bond was usually defined as an attractive interaction between the H-atoms of the opposite charge or as an interaction between a conventional hydrogen bond donor such as an NH or OH bond as the weak acid component and an element-hydride bond as the weak base component, where the element in question can be a transition metal or boron [5]. One can see that two concepts are presented here, the center-center interaction between the acidic and basic hydrogen atoms and the concept of the unconventional hydrogen bond where the σ-electrons of the hydridic bond play the role of the Lewis base. The latter kind of the hydrogen bond may be designated as A–H···σ, where A–H is the proton donating bond; such understanding of the dihydrogen bond interaction was proposed by Crabtree and coworkers [37]. The authors have also pointed out that there is a possibility that in a case of M–H σ-bonds, where M designates the transition metal, the metals possess nonbonding d_π electrons which can act as the alternative proton acceptors. Hence they have analyzed the N–H···H–B interactions (B designates boron here) to exclude the possibility of the interaction with d_π electrons [37]. They have performed the search through Cambridge Structural Database [54] which has shown twenty-six intermolecular interactions of that kind in crystal structures with the corresponding H···H distances in the range 1.7–2.2 Å. The N–H···H angle for those contacts tends to be close to linearity with the majority of angles in the range 150–170°, B–H···H angle tends to be bent with the majority of angles in the range 95–115°. This is why the authors have pointed out that for such interactions the N–H proton donating bond is directed towards the σ-B–H hydridic bond but not to the negatively charged H-atom. This is worth mentioning that the neutron diffraction results on the crystal structure of ammonia-borane are in line with these findings since N–H···H–B dihydrogen bonds were found there with the N–H···H and B–H···H angles equal to 156(3)° and 106(1)°, respectively, and with the H···H intermolecular distance amounting 2.02(3) Å [37]. However the mentioned above concept of the A–H···σ hydrogen bond was based only on the geometry of the structures considered. It was described and discussed later [32] that there are other energetic and topological results which do not support the idea of the B–H and M–H σ-bonds acting as a whole as proton acceptors in hydrogen bonded systems.

The BH_3NH_3 ammonia-borane may be treated as the model example for the explaining of the significance of the dihydrogen bond interactions [37]. One can see that ethane and ammonia-borane are isoelectronic but there are large differences in their properties; the melting point is −181 C and +104 °C for ethane and ammonia-borane, respectively. Such a difference between ethane and ammonia-borane is partly related to the polarity of BH_3NH_3 (5.2 D). However the melting point for also polar CH_3F (1.8D) and also isoelectronic with C_2H_6 and BH_3NH_3, is equal to −140°C. This means that for ammonia-borane the other factors influence additionally on its properties, not only polarity, and that these factors are related to the existence of the dihydrogen bond interactions. This situation is similar to that one of water where the existence of hydrogen bonds influences its physical properties; water is often compared to the isoelectronic species of methane.

The next studies have shown that the range of dihydrogen bonding interactions may be extended; in one of the first reviews on DHBs, and especially on crystal structures containing such interactions, Custelcean and Jackson have pointed out that DHB is an interaction between hydridic hydrogens of M–H bonds, where M = Al, B, Ga, Ir, Mo, Mn, Os, Re, Ru, W) and traditional X–H proton donors where X = F, O, N, C [7]. Bakhmutov, in the monograph on dihydrogen bonded systems, describes more types of the dihydrogen bond [36]. Even the C–H···H–C dihydrogen bonds were analyzed by Wolstenholme and Cameron in the crystal structures of tetraphenylphosphonium squarate, bianthrone, and bis(benzophenone)azine [9].

7.3 First Theoretical Calculations on the Dihydrogen Bond

Liu and Hoffmann analyzed theoretically the $[Ir\{H(\eta^1\text{-}SC_5H_4NH)(PCy_3)\}_2]^+BF_4^-$ structure [55], synthesized earlier by Morris and coworkers [48], where Cy designates cyclohexyl. The BF_4^- ion was neglected in the theoretical study to simplify the calculations; also the PCy_3 group was replaced with the PH_3 group in calculations [55]. The calculations were performed with the use of the extended Hückel method. The authors have found for the cation considered the weakly attractive intramolecular Ir–H···H–N interaction, with the H···H distance amounting 1.75 Å. Liu and Hoffmann analyzed also the linear F–H···H–Li system performing calculations at the RHF/6-31G* level and finding the H···H interaction energy of 9.3 kcal/mol [55]. These are ones of the first calculations carried out for the intra- and intermolecular dihydrogen bonded systems.

The BH_3NH_3 crystal structure containing dihydrogen bonds was mentioned in the previous section; also the unusual physical properties of BH_3NH_3 which reflect the existence of the DHB interactions were presented. The ammonia-borane and its derivatives were the subject of numerous analyses. For example, in one of the first theoretical studies on DHBs [1, 5], the BH_3NH_3 dimer in the gas phase was considered theoretically since the PCI-80/B3LYP [56, 57] calculations were performed for it. The authors have found the structure being in energetic minimum where molecules are linked through two identical B–H···H–N interactions. The energy corresponding to each H···H interaction in this structure was evaluated to amount 6.1 kcal/mol.

More extended and systematic calculations, up to the MP2/6-31G** level, were performed on a series of A–H···H–B dihydrogen bonds (A = B, Li, Be; B = N, C) [10]. The interaction energies for the systems analyzed are in the range typical for the hydrogen bonds. Even in cases of the $BH_4^-\cdots HCN$ and $BeH_2\cdots NH_4^+$ complexes strong charge assisted interactions are observed since the binding energies for them are equal to 18.0 and 9.3 kcal/mol, respectively (BSSE correction included).

Ab initio calculations on intermolecular and intramolecular DHBs for Mo, W, Ru, Re and Ir complexes have been carried out showing the H···H interaction energy greater than 5 kcal/mol and sometimes greater than 10 kcal/mol [12–14], similarly as for strong conventional O–H···O and N–H···O hydrogen bonds. There are other early theoretical studies on dihydrogen bonded systems, for example the calculations (MP2, MP4, QCISD(T) and B3LYP methods) were performed on the dihydrogen bonded complexes between the hydrides LiH, NaH, BeH_2, MgH_2, CH_4, SiH_4, GeH_4, SnH_4 and hydrofluoric acid and the correlations between H···H distance and the binding energy were presented [58, 59].

7.4 Dihydrogen Bonds as a Sub-Class of Hydrogen Bonds

It was pointed out in few early studies that the dihydrogen bond being the special type of the hydrogen bond is mainly electrostatic in nature interaction. However such findings generally do not concern the dihydrogen bonded systems but only the samples of complexes analyzed. For example, in one of early studies, the decomposition of the energy of interaction within the Kitaura-Morokuma scheme [60] was performed for two dihydrogen bonded systems, LiH···HF and LiH···HOH [12]. The same type of the decomposition was also performed for two complexes linked through conventional F–H···O and O–H···O hydrogen bonds in HNO···HF and HNO···HOH complexes for comparison [12]. It was found that in both cases, of DHBs and of the conventional hydrogen bonds, the electrostatic interaction energy (ES) is the most important attractive term followed by the charge transfer term (CT) and next by the polarization (PL). If one considers the absolute values of the mentioned above energies thus these results show that |CT| is less than 30 % of |ES| if all four hydrogen and dihydrogen bonded systems are taken into account. If only dihydrogen bonded systems are taken into account thus |CT| is less than 25 % of |ES|. This is probably the first study on DHBs where the decomposition of the energy of interaction was performed to deepen the understanding of those interactions [12].

The IMPPT (intermolecular Møller–Plesset perturbation theory) decomposition scheme [61] was applied for the H···H interactions in LiH···H_2, LiH···CH_4, LiH···C_2H_6 and LiH···C_2H_2 complexes calculated up to the MP2/aug-cc-pVTZ level [62]. The energy partitioning was carried out to the second order. This type of decomposition was also performed for the water dimer linked through the O–H···O hydrogen bond for comparison [62]. The authors have pointed out that a very similar picture is presented for the LiH···C_2H_2 complex and the water dimer because for both complexes the main binding contribution comes from electrostatic interaction. For these complexes the weights of the attractive electrostatic (negative) and the repulsive exchange (positive) energies are such that their sum, i.e. the Heitler-London interaction energy term, is negative. However for the remaining complexes, characterized by weak interactions, LiH···H_2, LiH···CH_4 and LiH···C_2H_6, the picture is different since the repulsive exchange contribution outweighs the electrostatic term thus the Heitler-London energy is positive and the dispersion energy is the main attractive contribution among the remaining attractive terms. The authors conclude that these results justify the classification of the LiH···C_2H_2 complex to be linked through the dihydrogen bond. However the remaining complexes characterized by the H···H contacts can be classified as weak van der Waals complexes.

The binary complexes of ammonia-borane, aminoborane and ammonia linked through hydrogen and/or dihydrogen bonds were analyzed [63]. The Kitaura-Morokuma decomposition of the energy of interaction [60] performed on the MP2/aug-cc-pVDZ results shows differences between the dihydrogen and hydrogen bond. In the latter case there is much greater significance of "non-electrostatic" attractive terms than in a case of the dihydrogen bond. This means that for the hydrogen bonds the polarization, charge transfer, correlation and the higher order terms of the energy of interaction play more important role than for the dihydrogen bonded systems where the stronger dominance of the electrostatic attraction is observed. However in both cases of dihydrogen and hydrogen bonds the electrostatic interaction is the most important attractive term [63].

One can state that the dihydrogen bond, similarly as the hydrogen bond, is an electrostatic in nature interaction. However the complexes presented so far are linked through weak or at most medium in strength interactions. The strongest dihydrogen bond interaction where the interaction energy partitioning was performed presented here so far occurs for the LiH···HF complex [55, 58, 59]. The situation changes significantly, sometimes drastically, for very strong dihydrogen bonds. Del Bene et al. [64, 65] has carried out the MP2/aug'-cc-pVTZ calculations on the model systems ranging from weak to strong dihydrogen bonds. For example they have calculated the binding energy for the $LiNCH^+\cdots HLi$ complex to be equal to 27.1 kcal/mol. This complex is characterized by the short H···H intermolecular contact of 1.309 Å [64].

One of the studies reveals strong dihydrogen bonds within the $NH_4^+\cdots HBeH$, $NF_3H^+\cdots HBeH$ and $NH_4^+\cdots HBeF$ complexes [66]. For the $NF_3H^+\cdots HBeH$ system optimized at the MP2/aug-cc-pVDZ level the short intermolecular H···H contact of 1.132 Å and the binding energy (corrected for BSSE) of 22.8 kcal/mol were calculated. The binding energy for the latter complex calculated by the MP2 method with aug-cc-pVXZ (X = 2,3) extrapolated to the complete basis set (CBS) amounts 21.6 kcal/mol.

The variation-perturbation approach of the decomposition of the energy of interaction [67, 68] was applied for the dihydrogen bonded complexes characterized by very strong interactions [69]. Since the results of this decomposition are presented later in this chapter and this approach is not commonly applied in other studies on intermolecular interactions thus it is described briefly here. The starting wave functions of the subsystems are obtained in this decomposition in the dimer-centred basis set [70]. In contrast to the Kitaura-Morokuma scheme [60] applied also in the analysis of DHB systems [12, 63] the total interaction energy as well as all of its components in the variation-perturbation approach are free of basis set superposition error (BSSE) due to the full counterpoise correction [67, 68]. The following interaction energy terms are the result of this decomposition.

$$\Delta E = E_{EL}{}^{(1)} + E_{EX}{}^{(1)} + E_{DEL}{}^{(R)} + E_{CORR}. \tag{7.1}$$

$E_{EL}{}^{(1)}$ is the first order electrostatic term describing the Coulomb interaction of static charge distributions of both sub-systems of the complex analyzed; $E_{EX}{}^{(1)}$ is the repulsive first order exchange component resulting from the Pauli exclusion principle; and $E_{DEL}{}^{(R)}$ and E_{CORR} correspond to higher order delocalization and correlation terms. Delocalization term contains all classical induction, exchange-induction, etc. from second order up to infinity. Strongly basis set dependent charge transfer term is included in much less basis set sensitive delocalization contribution [67, 68]. The correlation term includes dispersion interactions as well as intramolecular correlated electrostatic, exchange, induction and dispersion contributions. The presented decomposition scheme has been implemented within GAMESS package [71].

The MP2/aug-cc-pVTZ calculations have been performed for the $H_2OH^+\cdots HBeH$, $H_2OH^+\cdots HBeBeH$, $H_2OH^+\cdots HBeF$, $HClOH^+\cdots HBeH$, $Cl_2OH^+\cdots HBeH$ and $Cl_2OH^+\cdots HBeF$ complexes [69]. For all of them very short H···H intermolecular contacts (1.0–1.3 Å) were observed. These are the shortest intermolecular distances which have been ever reported, with binding energies for the corresponding complexes within the range of 13.7–24.3 kcal/mol. The decomposition of the interaction energy (Eq. 7.1) was performed for these complexes and for all of them the first order Heitler-London energy ($\Delta E_{HL} = \Delta E_{EX}{}^{(1)} + \Delta E_{EL}{}^{(1)}$) is positive. For all those complexes linked by DHBs the delocalization term, $\Delta E^{(R)}_{DEL}$, is the most important attractive interaction contribution and it is mostly responsible for the stabilization of the complexes. The correlation energy term ΔE_{CORR} is much smaller in absolute value, about 3–4 times, in comparison with the delocalization term. This is different than for the weak and medium in strength hydrogen and dihydrogen bonds, usually known as electrostatic in nature, where the electrostatic term outweighs the exchange energy and the ΔE_{HL} term is negative [72, 73]. However the significance of the delocalization energy is also observed for the conventional strong hydrogen bonds [72]. This means that the hydrogen and dihydrogen bonds possess similar characteristics since for both for weaker interactions the electrostatic interaction is dominant while for stronger ones the processes connected with the electron charge redistribution being the result of complexation play the main role. It was pointed out that the hydrogen bond is an interaction without borders [74] what means that there is the continuous passage from covalent bonds to H-bonds and further to van der Waals interactions. And the same is observed for the dihydrogen bonds [75].

The other characteristics of the A–H···B hydrogen bonds are commonly known. For the shortening of the H···B distance the elongation of the A–H proton donating bond is observed what is accompanied by the increase of the strength of the hydrogen bond [76, 77]. The same relationships are observed for dihydrogen bonds [75]. For example, the complexes linked through the H···H interactions were analyzed where the HF molecule plays the role of the Lewis acid and different hydrides are the proton acceptors [58]. The non-linear correlations between the H···H distance and the binding energy were found, i.e. for shorter intermolecular H···H contacts the stronger interactions were observed. The elongation of the H–F bond was also observed for these complexes as a result of dihydrogen bond formation. This elongation correlates with the binding energy, i.e. for the stronger interactions the greater elongation of the H–F bond is observed.

For some of hydrogen bonds the shortening of the proton donating bond is observed as a result of complexation with the corresponding stretch shift to higher frequencies (blue-shift); this is why such interactions are often named as the blue shifted hydrogen bonds [78]. The blue shifted dihydrogen bonds were also predicted for a number of complexes: F_3C–H···H–Be–X, F_3C–H···H–Mg–X, F_3C–H···H_4Si, and the analogues complexes where F_3C–H is replaced by F_3Si–H (X = H, F, Cl and CH_3). The calculations for these systems were performed up to the QCISD/6-31+G(d) level of approximation [79]. One can also mention the study on N–H···H–B blue-shifted dihydrogen bonds [80] and on the coexistence of blue and red-shifted

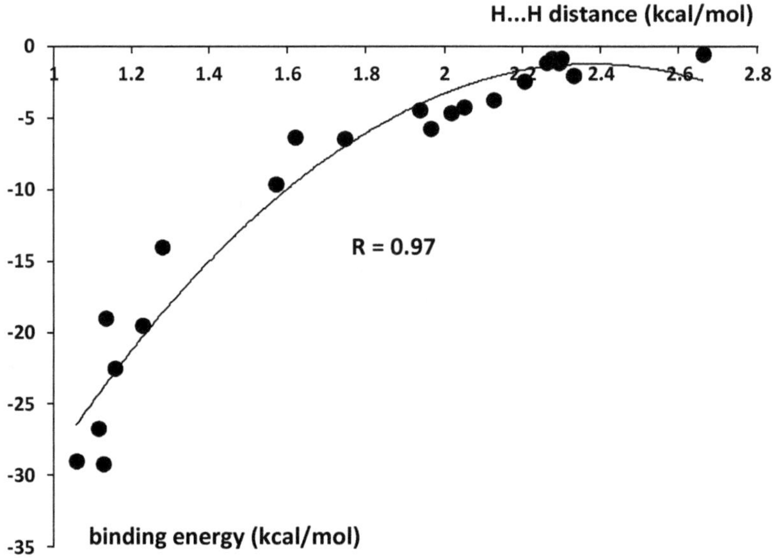

Fig. 7.2 The relationship between H· · ·H distance and the binding energy for different dihydrogen bonded complexes

hydrogen and dihydrogen bonds in valine [81]. However, similarly as for the hydrogen bonds, for the dihydrogen bonds the complexation, for the greater part of interactions, is connected with the elongation of the proton donating bond with the combined shift of the corresponding stretch frequency to lower frequencies. Even the dependence between stretch frequency and the bond length for the proton donating bond was observed for a sample of complexes linked through N–H· · ·H–B and N–H· · ·H–Al contacts [82].

Different types of DHBs were analyzed; from weak H· · ·H interactions classified as van der Waals interactions and with binding energies less than 2 kcal/mol, through DHBs of the medium strength, to strong interactions possessing numerous characteristics of covalent bonds [75]. 22 different complexes were analyzed where C_2H_2, C_2FH, CH_4, CFH_3, CF_2H_2, CF_3H, $CClH_3$, CCl_2H_2, CCl_3H, H_3O^+, $HClOH^+$, Cl_2OH^+, NH_4^+, and F_3NH^+ were chosen as the proton donating systems while different beryllium molecules as well as the LiH molecule as the species delivering the $H^{-\delta}$ negatively charged Lewis base center. The correlation between the H· · ·H distance and the binding energy was found (see Fig. 7.2); it is the non-linear and monotonic relationship (the second order polynomial dependence). This is worth to mention that for numerous hydrogen bonded systems the excellent distance—binding energy linear correlation is observed [83]. However for such studies the homogenous samples are considered, i.e. the samples of complexes formed by related species. For the relation presented in Fig. 7.2 different classes of Lewis acid and Lewis base centers were taken into account. For example, for different dihydrogen bonded complexes analyzed early the distance—binding energy correlation was not

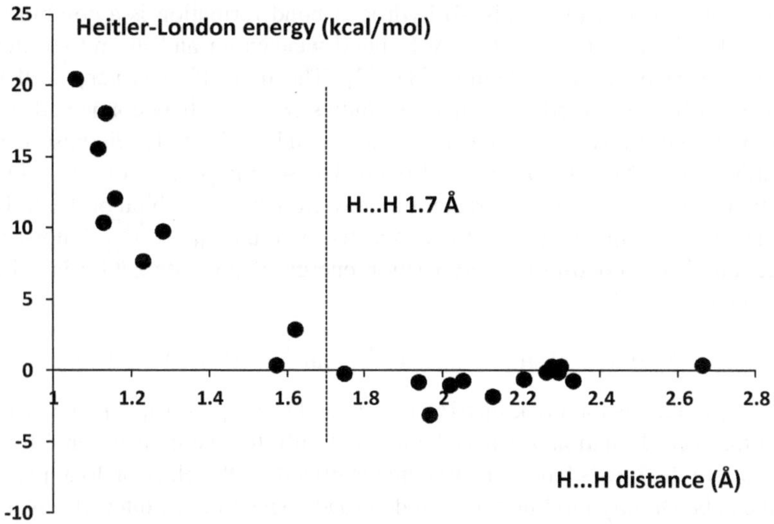

Fig. 7.3 The relationship between H· · ·H distance and the Heitler-London energy for different dihydrogen bonded complexes

found [10]. However in that case various Lewis acid and Lewis base centers were also analyzed.

Figure 7.3 presents the dependence between the H· · ·H distance and the Heitler-London energy ($\Delta E_{HL} = \Delta E_{EX}{}^{(1)} + \Delta E_{EL}{}^{(1)}$) calculated within the variation-perturbation approach (Eq. 7.1). The sample of 22 dihydrogen bonded complexes [75] mentioned earlier here is considered in this relationship. This is interesting that for the H· · ·H contacts shorter than 1.7 Å, i.e. for the charge assisted strong dihydrogen bonds, the ΔE_{HL} term is positive since the electrostatic energy is outweighed here by the exchange repulsion. This means that the delocalization ($\Delta E_{DEL}{}^{(R)}$) is dominant for such strong interactions. And really, for short H· · ·H distances the $\Delta E_{DEL}{}^{(R)}/\Delta E_{EL}{}^{(1)}$ ratio is much greater than for the weaker not charge assisted DHBs [72, 75]. It was also found within the variation-perturbation decomposition scheme that in general for strong and very strong hydrogen and dihydrogen bonds the ΔE_{HL} term is positive and the delocalization interaction is much more important than the electrostatic one [72, 73].

7.5 Could the Dihydrogen Bond Be Classified as the A–H. . . σ Interaction?

It is interesting to analyze processes of the electron charge redistribution being a result of the formation of the dihydrogen bond. Such processes were often analyzed for the hydrogen bonded systems [83]. It was pointed out, within the Natural Bond Orbitals scheme [84, 85], that the A–H···B hydrogen bond formation is a combination of two effects: the hyperconjugative A–H bond weakening and the rehybridization-promoted A–H bond strengthening [84–87]. The first effect is very well known and it was often analyzed in numerous studies [83–85]. It is connected with the electron charge transfer from the lone pair orbital of the B Lewis base center to the antibonding σ* orbital of the A–H bond. The second process of rehybridization leads to the increase in the s-character of the A-atom hybrid orbital of the A–H bond [86]. The hyperconjugative effect is connected with the $n_B \rightarrow \sigma^*_{AH}$ donor-acceptor interaction. The second-order perturbation energy of this interaction is calculated according to Eq. 7.2.

$$\Delta E\,(n_B \rightarrow \sigma_{AH}{}^*) = -2\,\langle n_B\,|F|\,\sigma_{AH}{}^*\rangle^2/(\varepsilon(\sigma_{AH}{}^*) - \varepsilon(n_B)) \quad (7.2)$$

where $\langle n_B|F|\sigma^*_{AH}\rangle$ is the Fock matrix element and $(\varepsilon(\sigma^*_{AH}) - \varepsilon(n_B))$ is the orbital energy difference. Equation 7.2 may be applied only for the conventional hydrogen bonds where the Lewis base center is characterized by the electron lone pair. However it can be slightly modified to include broader spectrum of interactions recently classified as hydrogen bonds as well as to include the other Lewis acid—Lewis base interactions. For example, the $\pi_B \rightarrow \sigma^*_{AH}$ orbital-orbital interaction corresponds to the A–H···π hydrogen bond where π-electrons play the role of the Lewis base center [32].

It is interesting that for the A–H···H–B dihydrogen bonds [32], the $\sigma_{B-H} \rightarrow \sigma^*_{AH}$ donor-acceptor interaction within the Natural Bond Orbitals (NBO) method is the most important term connected with the electron charge transfer from the Lewis base to the Lewis acid. The Eq. (7.2) may be modified to be applied for the dihydrogen bonded systems:

$$\Delta E\,(\sigma_{BH} \rightarrow \sigma_{AH}{}^*) = -2\langle\sigma_{BH}\,|F|\,\sigma_{AH}{}^*\rangle^2/(\varepsilon(\sigma_{AH}{}^*) - \varepsilon(\sigma_{BH})). \quad (7.3)$$

The terms of Eq. 7.3 are defined in the similar way as those of Eq. 7.2. It seems that the $\sigma_{B-H} \rightarrow \sigma^*_{AH}$ interaction energy may be treated as a measure of the strength of dihydrogen bond. For example, the H···H intramolecular contacts possessing characteristics of the dihydrogen bond were analyzed recently theoretically (MP2/6-311++G(d, p) level) [88]. Different classes of dihydrogen bonds were considered, among them intramolecular C–H···H–C interactions. However for two classes of intramolecular interactions, O–H···H–B and O–H···H–Al the most important changes being the result of the formation of dihydrogen bonds were detected. Figure 7.4 presents the relationships between the H···H distance and the energy expressed by Eq. 7.3 for two samples mentioned above and containing BH_2 or AlH_2 hydridic group acting as the Lewis base. The excellent exponential correlations show that the $\sigma_{B-H} \rightarrow \sigma^*_{AH}$ energy of interaction may be treated as the indicator of the strength of dihydrogen bond here.

It was found recently that the hyperconjugative and rehybridization processes may be considered as steering the formation of numerous non-covalent interactions such as dihydrogen bond, halogen bond, dihalogen bond, hydride bond etc, not only the hydrogen bond [89–91]. Let us analyze in detail the rehybridization process in a case

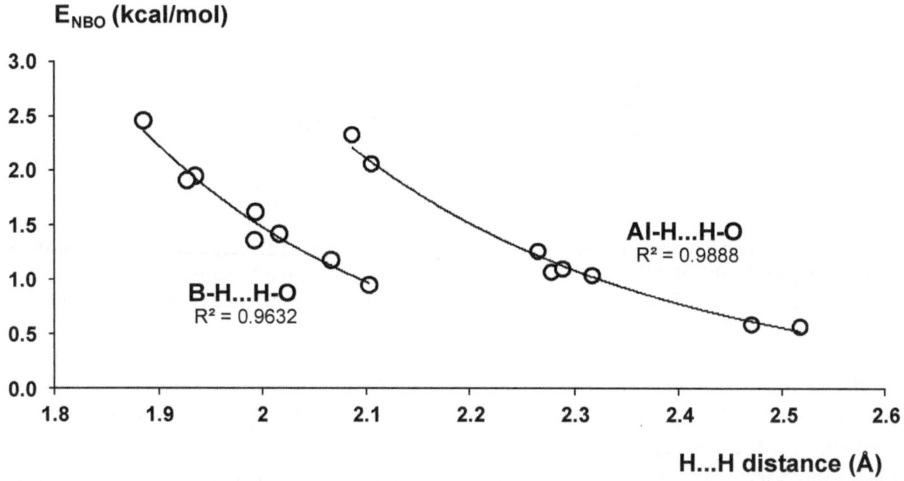

Fig. 7.4 The exponential dependences between the H···H distance and the $\sigma_{B-H} \rightarrow \sigma^*_{AH}$ energy of interaction (designated as E_{NBO} in the figure) for two classes of intramolecular dihydrogen bonds; O–H···H–B and O–H···H–Al; this figure was prepared on the basis of the results of ref. [88]

of the hydrogen bond formation. It leads to the increase in s-character of A hybrid orbital in the A–H bond; this is in line with Bent's rule which states that atoms maximize the s-character in hybrid orbitals aimed toward electropositive substituents and the atoms minimize such character (maximize the p-character) toward more electronegative substituents [92]. The A–H···B hydrogen bond formation usually leads to the increase of the positive charge of the H-atom [83] thus this atom may be treated as more electropositive in the hydrogen bond than for the A–H bond not involved in any interaction. Hence the Bent rule may be applied for the hydrogen bonded systems and really it was found that the hydrogen bond formation is connected with the increase in the mentioned above s-character [93]. The same is observed for the A–H···H–B dihydrogen bonds. The N–H···H–Be and N–H···H–Mg dihydrogen bonds were analyzed recently [91] in the NH_4^+···HBeH and NH_4^+···HMgH complexes, respectively (the MP2(full)/6-311++G(3df,3pd) level calculations were performed). The QTAIM atomic integrated charges were calculated for these complexes and for the Lewis acid and Lewis base sub-units not involved in interactions. It was found that the charge of H-atom in the free NH_4^+ cation is equal to 0.509 au while the charge of the H-atom being in contact with the hydridic hydrogen in the NH_4^+···HBeH and NH_4^+···HMgH complexes is equal to 0.528 au in both cases; thus the increase of the positive charge (electropositivity) of H-atom is observed as a result of the formation of the dihydrogen bond. According to the Bent rule the s-character of the nitrogen hybrid orbital should increase. And this situation is really observed since the s-character amounts 25.0 for free ammonia cation and it is equal to 28.6 and 30.2 in the NH_4^+···HBeH and NH_4^+···HMgH complexes, respectively.

Thus NBO results justify that the dihydrogen bond is the sub-class of the hydrogen bond interactions. However the NBO method shows also the difference between those

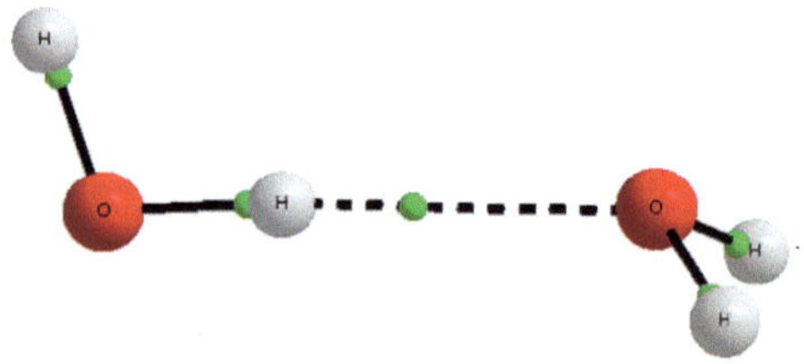

Fig. 7.5 The molecular graph of the water dimer, $(H_2O)_2$, *big circles* correspond to attractors, small ones to the bond critical points, solid and *broken lines* represent the bond paths; the broken one corresponds to the H···O intermolecular contact representing the existence of the hydrogen bond

interactions, the $n_B \rightarrow \sigma^*_{AH}$ and $\sigma_{B-H} \rightarrow \sigma^*{}_{AH}$ are the most important orbital-orbital interactions for the hydrogen and dihydrogen bond, respectively [32]. These findings also support the concept that the dihydrogen bond may be classified as the A–H···σ hydrogen bond since for the $\sigma_{B-H} \rightarrow \sigma^*{}_{AH}$ interaction the BH orbital represents here σ-electrons acting as the Lewis base. However one should note that the mentioned above orbital-orbital interaction is only the part of the charge transfer interaction within the NBO approach. Besides there are the other attractive interactions, as it was discussed earlier here, such as electrostatic, polarization and dispersion. And most often, the electrostatic interaction is the most important attractive one for hydrogen and dihydrogen bonded systems.

Crabtree and coworkers have pointed out that there are conventional A–H···B hydrogen bonds and that the dihydrogen bond may be classified as the A–H···σ unconventional hydrogen bond; similarly as the A–H···π interaction which also possesses characteristics typical for the hydrogen bond [37]. It seems that the QTAIM approach [94, 95] may be useful to clarify the nature of interactions discussed here. The bond path (BP), being the link between attractors, with the corresponding bond critical point (BCP) is often treated as the evidence of the stabilizing interaction [96]. It was pointed out that the molecular graph representing the structure analyzed shows bond paths which are the indicators of the preferable interactions. In a case of the A–H···B hydrogen bond the H···B bond path is usually detected, where H and B designate the attractors of the hydrogen atom and of the B Lewis base center, respectively. Figure 7.5 shows the molecular graph of the water dimer where one can see the H···O bond path corresponding to the O–H···O hydrogen bond. The distinct situation is observed for the A–H···π interactions. For example, for the $C_2H_2 \cdots C_2H_4$ complex the molecules are linked through the C–H···π hydrogen bond (Fig. 7.6) where the H-attractor of the C–H proton donating bond of acetylene is connected with the BCP of the double bond of ethylene. One can say that the C=C bond critical point of C_2H_4 mimics here "the point Lewis base center"; the C_2H_4 molecule possesses two-center (the C=C double bond) electron donating system. However there are numerous studies where benzene or other multicenter aromatic systems play the role of the proton acceptor in the A–H···π hydrogen bonds. Figure 7.7 shows the molecular graph of the $Cl_3CH \cdots C_6H_6$ complex where the C–H proton donating bond is directed to the center of the benzene molecule and

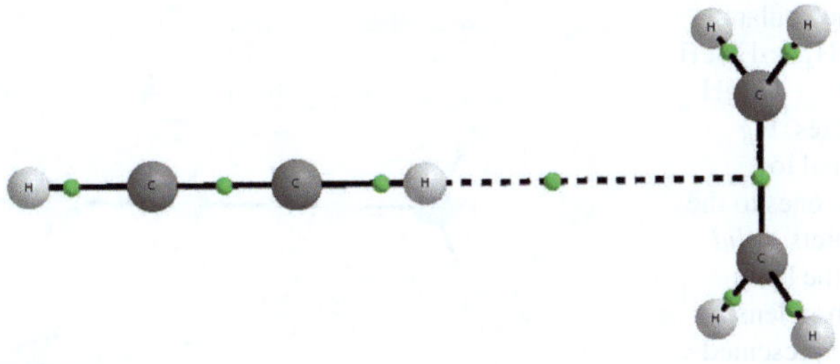

Fig. 7.6 The molecular graph of the $C_2H_2 \cdots C_2H_4$ complex, *big circles* correspond to attractors, small ones to the bond critical points, *solid* and *broken lines* represent the bond paths; the *broken line*—H···BCP bond path, corresponds to the H···π interaction representing the C–H···π hydrogen bond

Fig. 7.7 The molecular graph of the $Cl_3CH \cdots C_6H_6$ complex, *big circles* correspond to attractors, small ones to the bond, ring and cage critical points, solid and *broken lines* represent the bond paths. *Broken lines* represent six H···C bond paths which correspond to the H···π interaction

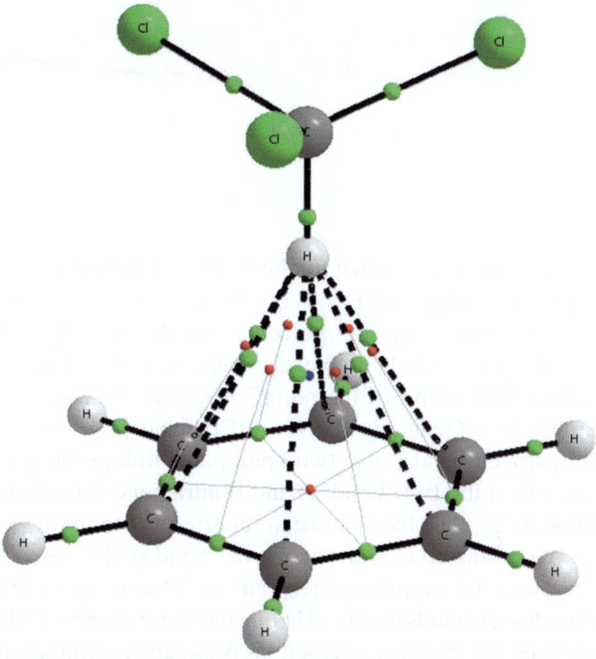

where six H···C bond paths linking the H-atom attractor of the Cl_3CH molecule with the carbon attractors of the benzene ring are observed.

The $\pi \rightarrow \sigma^*_{AH}$ orbital-orbital interaction is the main contribution to the charge transfer interaction for the system linked through A–H... π contact with two-center π-electron system. The situation is more complicated for many-center π-electron systems like benzene and other aromatic species [32, 97]. The QTAIM and NBO approaches were applied recently to discuss different types of the A–H···π hydrogen bonds [97].

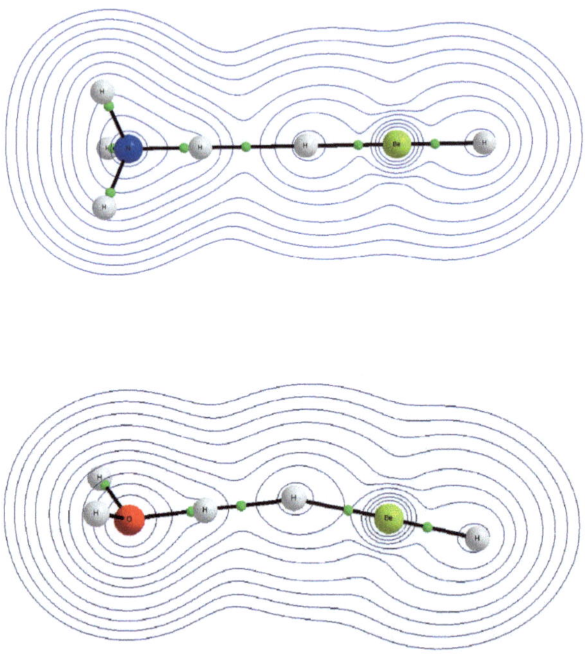

Fig. 7.8 The molecular graphs of the $NH_4^+\cdots HBeH$ (*up*) and $H_2OH^+\cdots HBeH$ (*below*) complexes, *big circles* correspond to attractors, small ones to the bond critical points, *solid lines* represent the bond paths; the electron density isolines are also presented

This is interesting that for the A–H···H–B dihydrogen bonds the molecular graphs show the existence of H···H bond paths with the corresponding bond critical points. Figure 7.8 presents the molecular graphs of $NH_4^+\cdots HBeH$ and $H_2OH^+\cdots HBeH$ dihydrogen bonded complexes. One can see that there are not H···BCP bond paths where BCP mimics "the point" Lewis base center of the Be–H σ-bond for those complexes. Generally the H···BCP bond paths do not exist for the A–H···H–B dihydrogen bonds; also two bond paths linking the protonic hydrogen with the attractors of the B–H bond (H and B attractors) do not exist for DHBs; at least, to my knowledge, they have not been reported in literature so far.

According to earlier studies the bond paths observed for the systems being in the energetic minima correspond to the preferable interactions [96, 97]. For the dihydrogen bonds the H···H bond paths are observed which correspond to the mainly electrostatic in nature interactions between the protonic and hydridic hydrogen atoms. One may conclude that the QTAIM results show that the dihydrogen bonds with contacts between H-atoms of the opposite charges could not be classified as the A–H···σ interactions. However the question arises if the A–H···σ hydrogen bonds exist at all. It is discussed in the next section of this chapter.

The Quantum Theory of Atoms in Molecules (QTAIM) often cited here is one of the most powerful methods to analyze different inter- and intramolecular interactions [94, 95]. Eight QTAIM criteria of the existence of the hydrogen bond were proposed early on by Koch and Popelier [98]. In numerous studies on the A–H···B hydrogen bonds the criterion of the existence of the H···B bond path with the corresponding bond critical point is the most often checked one among the other criteria. The

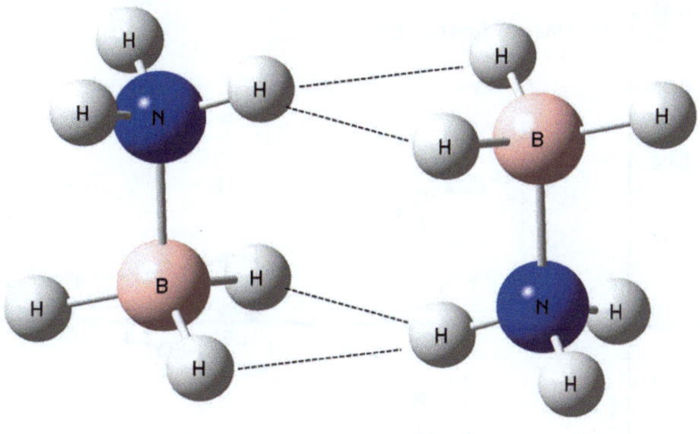

Fig. 7.9 The $(BH_3NH_3)_2$ dimer, $BH_2 \cdots H$–N bifurcated dihydrogen bonds are presented (*broken lines*)

properties of this BCP are also often analyzed in numerous studies. Koch and Popelier have pointed out that the electron density at the H· · ·B BCP should lie within the range (0.002, 0.04) au and also the range for the laplacian was proposed. The decrease of the electron charge of the hydrogen atom as an effect of the A–H· · ·B hydrogen bond formation and the simultaneous decrease of the hydrogen atom volume are the other QTAIM criteria. The latter criteria are in line with the NBO description of the mechanism of the hydrogen bond formation since the same changes of the H-atom charge and volume were found to be the result of the rehybridization process [84, 86]. This is also important that the criteria of ranges of the values of the electron density and its laplacian concern medium in strength and weak hydrogen bonds. Very strong interactions possess numerous features typical for covalent bonds while very weak interactions possess features of van der Waals interactions [72, 74]. This is why in such cases the characteristics of interactions may lie outside of ranges proposed by Koch and Popelier [98].

The QTAIM parameters often applied to characterize hydrogen bonds were also analyzed by Popelier for the $(BH_3NH_3)_2$ dimer and it was found that this dimer is linked through three H· · ·H contacts for which the corresponding bond paths with BCPs exist [99]. The latter contacts may be classified as dihydrogen bonds since numerous criteria proposed earlier for hydrogen bonds are also fulfilled here. However it seems that energetically more stable $(BH_3NH_3)_2$ dimer is characterized by the equivalent BH_3NH_3 molecules related by inversion symmetry where monomers are linked through four H· · ·H contacts corresponding to two bifurcated DHBs (Fig. 7.9). Such a structure was found by Cramer and Gladfelter with the use of the MP2/cc-pVDZ level [100] and later it was analyzed by Scheiner (MP2/aug-cc-pVDZ) [63].

The other characteristics of BCP, not only ρ_{BCP}, and its laplacian, $\nabla^2 \rho_{BCP}$, mentioned earlier here are also often applied to deepen the understanding of the nature of interactions. These are the total electron energy density at BCP (H_{BCP}) and the components of the latter value: the kinetic electron energy density (G_{BCP}—always positive value) and the potential electron energy density (V_{BCP}—always negative value). There are the following relations between the mentioned above values (Eq. 7.4 in atomic units) [95].

$$\nabla^2 \rho_{BCP} = 2G_{BCP} + V_{BCP} \qquad \text{where, } H_{BCP} = G_{BCP} + V_{BCP}. \qquad (7.4)$$

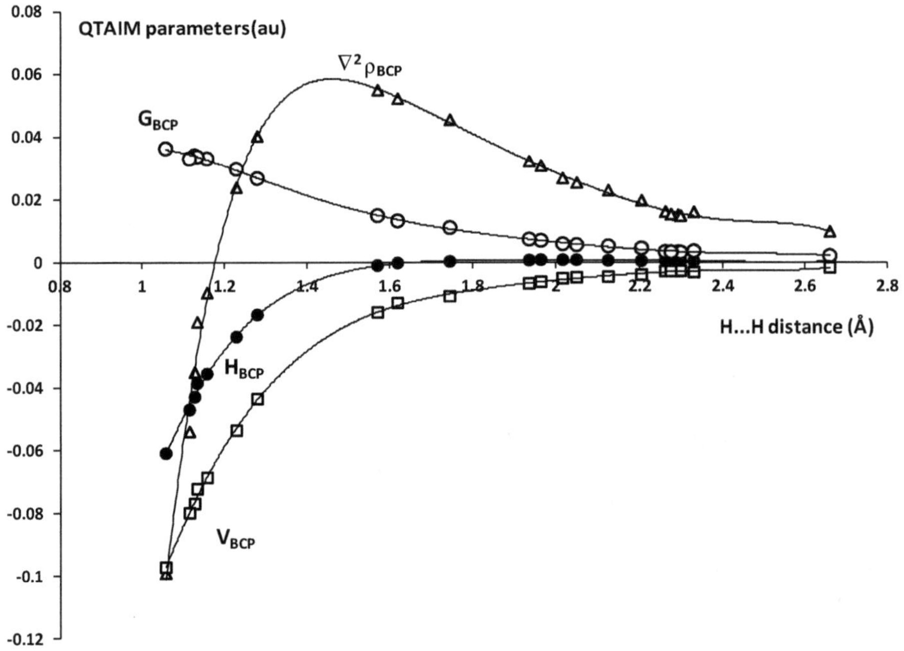

Fig. 7.10 The dependencies between the H· · ·H distance and the characteristics of the corresponding BCP; $\nabla^2 \rho_{BCP}$ (*triangles*), H_{BCP} (*full circles*), V_{BCP} (*squares*) and G_{BCP} (*open circles*)

It is worth mentioning that the negative value of laplacian is attributed to the covalent interactions while its positive value is characteristic for van der Waals interactions, ionic interactions and hydrogen bonds [95]. However for strong hydrogen bonds H_{BCP} is negative and even in extreme cases of very strong hydrogen bonds the laplacian of the electron density at the H· · ·B BCP is negative, similarly as in the case of covalent bonds [72, 73, 101, 102]. The same concerns dihydrogen bonds [75]. Figure 7.10 presents dependencies between the H· · ·H distance and the mentioned above characteristics of the H· · ·H BCP for different dihydrogen bonded systems, the results for the sample containing 22 complexes [75] which was analyzed earlier here are presented in Fig. 7.10. One can see three types of interactions here. Those where $\nabla^2\rho_{BCP}$ is negative may be classified as very strong DHBs, those where $\nabla^2\rho_{BCP}$ is positive and H_{BCP} is negative belong to the class of strong interactions and finally the DHBs where both $\nabla^2\rho_{BCP}$ and H_{BCP} are positive belong to medium in strength or weak interactions. There are the following examples of the dihydrogen bonded complexes belonging to the mentioned above classes; Cl_2OH^+· · ·HBeH, H_3NH^+· · ·HBeH and HCCH· · ·HLi which correspond to very strong ($\nabla^2\rho_{BCP} < 0$ and $H_{BCP} < 0$) strong ($\nabla^2\rho_{BCP} > 0$ and $H_{BCP} < 0$) and weak ($\nabla^2\rho_{BCP} > 0$ and $H_{BCP} > 0$) dihydrogen bonds, respectively. The same classification based on the signs of the $\nabla^2\rho_{BCP}$ and H_{BCP} values was proposed early on for the hydrogen bonds by Rozas et al. [103].

7.6 The A–H···σ Interactions: From Complexes to Clusters

It was described in the previous sections that the A–H···H–B dihydrogen bond is not classified as the A–H···σ interaction since the σ-electrons of the B–H bond do not mimic the one center Lewis base, at least not in the same way as it was observed for π-electrons in the A–H···π hydrogen bonds [104]. For the latter interactions few cases may be considered [97]. One can observe the attractor—bond critical point bond paths, where attractor corresponds to the H-atom of the A–H proton donating bond while BCP mimics the Lewis base center. That situation is observed for the two-center π-electron systems like for the CC bond in acetylene, ethylene and similar molecules. For multicenter π-electron system like in benzene molecule and other aromatic systems the situation is more complicated [97]. The H-attractor of the A–H bond is connected with single atom or with the bond critical point if the unsymmetrical distribution of the electron charge density for the Lewis base sub-unit is observed. The H-attractor may be also connected with few atomic attractors of the Lewis base. ([97], see Fig. 7.7). The case of the existence of attractor—attractor bond path should not be classified as the A–H... π interaction. Such a situation was observed for the $C_5H_5^-$···HF complex where the sub-units are connected by the single H···C bond path and the interaction may be classified as the F–H···C hydrogen bond; the C-atom is the most negatively charged atom of the $C_5H_5^-$ ring and it possesses the properties typical of the Lewis base center [97]. For the A–H···H–B interactions presented so far here the H-atom attractor of the A–H bond is not connected through the bond path with the B–H bond critical point but the H···H bond path is observed.

It seems that the molecular hydrogen is a good candidate as a Lewis base σ-electron system to form links with Lewis acids classified as the A–H···σ hydrogen bonds. Numerous experimental and theoretical studies are known where H_2 molecule interacts as the Lewis base with metals or cations through the σ-electrons [26]. The similar interaction was observed early for the NH_4^+···H_2 complex analyzed theoretically (MP2/aug-cc-pVTZ level) [105]. It was found that the H_2 molecule is approximately perpendicular to the N–H proton donating bond and that the binding energy for this complex is equal to 2.5 kcal/mol. The authors have also analyzed the $NH_4^+ \ldots (H_2)_n$ clusters (*n* up to 8) and they have found that if the number of H_2 molecules in the cluster increases from 1–4 thus the following N–H bonds are saturated by the N–H···H_2 interactions where hydrogen molecules are perpendicular to N–H bonds [105]. Figure 7.11 presents the structure of the NH_4^+···H_2 complex as well as of the $NH_4^+ \ldots (H_2)_2$ and the $NH_4^+ \ldots (H_2)_3$ clusters. If the number of H_2 molecules exceeds 4 thus the additional interactions are observed between the nitrogen center of NH_4^+ and H_2 molecules [105].

The N–H···H_2 T-shaped arrangements were not discussed as possible types of hydrogen bond in the mentioned here study [105]. However such interactions were analyzed later since the more extended ab initio studies have been performed on XH_4^+···H_2 complexes and $XH_4^+ \ldots (H_2)_5$ clusters (X = N, P, As, Sb, Bi) where the MP2/aug-cc-pVTZ calculations were carried out [106]. The N–H···H_2 interactions

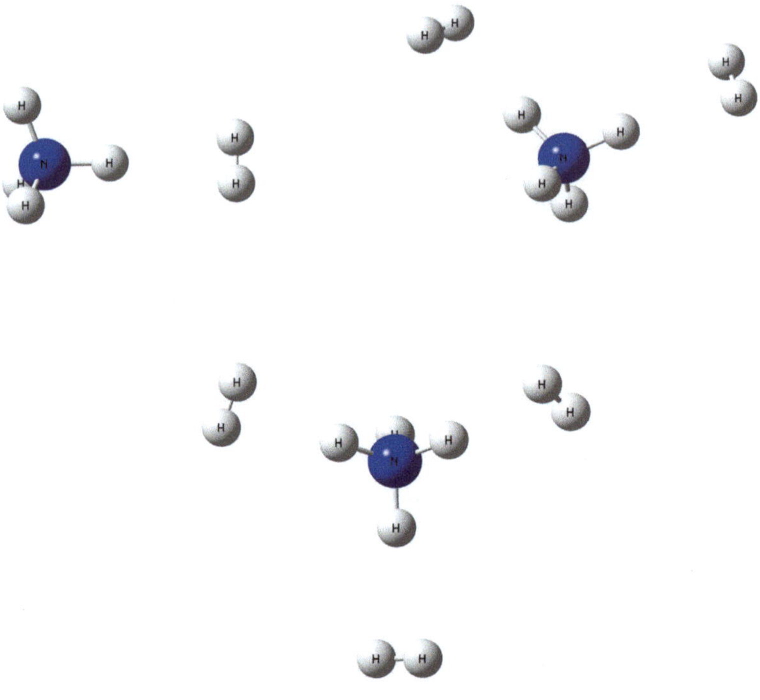

Fig. 7.11 The structure of the $NH_4^+\cdots H_2$ complex and of the $NH_4^+\ldots(H_2)_2$ and $NH_4^+\ldots(H_2)_3$ clusters

were classified as A–H$\cdots\sigma$ hydrogen bonds where σ-electrons of the molecular hydrogen play the role of the proton acceptor. The molecular graphs show here the H$\cdots$BCP bond paths where BCP corresponds to the σ-bond of the hydrogen molecule. Figure 7.12 presents the molecular graphs of the $NH_4^+\cdots H_2$ complex and of the $NH_4^+\ldots(H_2)_4$ cluster.

The decomposition of the energy of interaction performed within the variation-perturbation approach [67, 68] for the $XH_4^+\cdots H_2$ systems shows that the delocalization term is the most important attractive one followed by the electrostatic and dispersion contributions [106]. These findings are similar to those observed for the A–H$\cdots\pi$ hydrogen bonds. It was discussed earlier that for the latter interactions the BCP of the π-electron system mimics the Lewis base center [97, 104]. This is observed here for A–H$\cdots\sigma$ hydrogen bonds where the BCP of H–H bond corresponds to the Lewis base center. There are other similarities between A–H$\cdots\pi$ and A–H$\cdots\sigma$ interactions. For example, the decomposition of the energy of interaction was performed for numerous A–H$\cdots\pi$ hydrogen bonded complexes and it was found that for such interactions the delocalization term is an important attractive contribution [107].

The similar A–H$\cdots\sigma$ interaction was detected for the T-shaped complex of F–H$\cdots H_2$ where the calculations performed at the MP2/6-311++G(3df,3pd) level show the binding energy of 0.8 kcal/mol [108]. Figure 7.13 presents the molecular

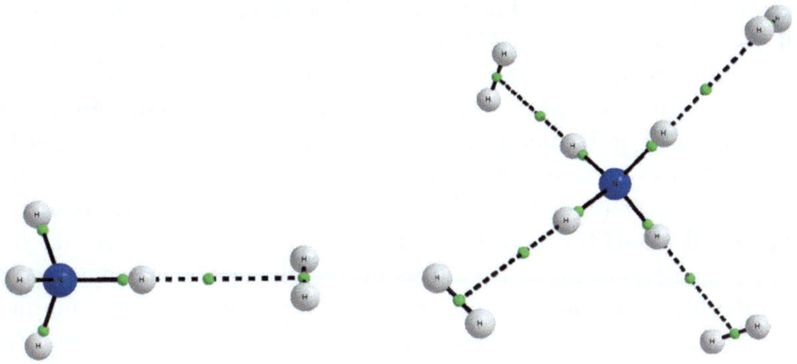

Fig. 7.12 The molecular graph of the $NH_4^+\cdots H_2$ complex (*left*) and of the $NH_4^+\ldots(H_2)_4$ cluster (*right*); *big circles* correspond to attractors, small ones to the bond critical points, *solid and broken lines* represent the bond paths

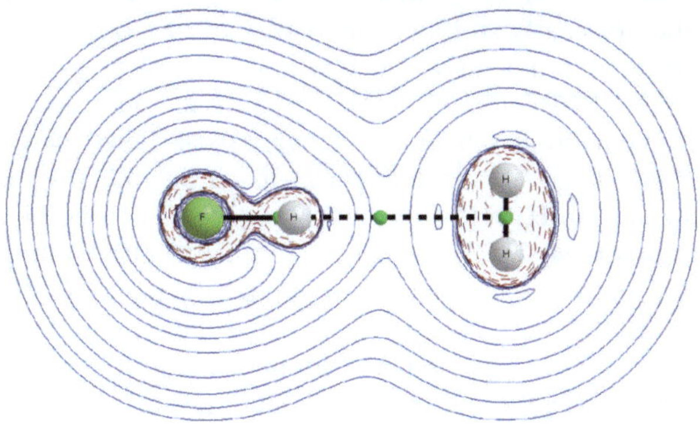

Fig. 7.13 The molecular graph of the $FH\cdots H_2$ T-shaped complex with the isolines of Laplacian of the electron density; positive values are depicted in *solid lines* and negative values in *broken lines* (the laplacian isodensity lines in the plane of the complex)

graph of the complex with the H···BCP bond path corresponding to the H···σ interaction. The laplacian of the electron density isolines in the plane of the complex are presented indicating the regions of the concentration of electron density for the negative laplacian values. The analysis of results of calculations led to the conclusion that the F–H···σ(H_2) interaction possesses properties typical for the hydrogen bond [108]. The following properties were observed there; for the HF and H_2 molecules in the complex there is the elongation of both H–F and H–H σ-bonds by ~0.002 Å as a result of the hydrogen bond formation. Both elongations are connected with the electron charge transfer from the H_2 molecule acting as the Lewis base to the HF Lewis acid unit. It corresponds to the $\sigma \rightarrow \sigma^*$ orbital-orbital interaction. The loss of the electron charge of the H_2 molecule results in the decrease of the occupation of the σ(H–H) orbital and the weakening of the H–H bond while the transfer of the electron charge to the HF molecule results in the increase of the occupation of σ*(H–F) orbital and consequently the weakening and of H–F bond.

The other complexes linked through the A–H···σ hydrogen bonds were also analyzed and they were compared with the analogues complexes with A–H···π interactions; the complexes of acetylene and molecular hydrogen were compared. For all complexes the C_2H_2 and H_2 molecules play the role of the proton acceptor, i.e. of the Lewis base. It was described in this chapter that the characteristics of the A–H···σ interaction are different than those of the A–H···H–B dihydrogen bond. It is also worth to notify that usually there is the strong polarization of the B–H bond in the A–H···H–B system what is not observed for the molecular hydrogen acting as the Lewis base in the A–H···σ interaction. Such a strong polarization was observed for the B–H bond (B designates boron here) in the N–H···H–B contact of the ammonia-borane dimer [37]. It seems that the mentioned above polarization leads to the excess of the negative charge on the H(B)-atom and the directional $H^{-\delta}...^{+\delta}H$ interaction (dihydrogen bond) being mostly electrostatic in nature. In a case of molecular hydrogen acting as the Lewis base the sites of H-atoms are characterized by the positive electrostatic potential while in the direction perpendicular (and nearly so) to the σ-bond the negative electrostatic potential is observed [110]. The atomic charges of the H_2 molecule in the complex are slightly positive because of the electron charge transfer to the Lewis acid and such charges are equal to each other, or nearly so [32]. This is why for the A–H···H_2 interactions, if A–H is the typical proton donating bond with the excess of the positive charge on the H-atom the H···BCP intermolecular bond paths are observed.

The hydrogen molecule is a very weak Lewis base, however sometimes strong or at least medium in strength A–H···σ interactions are observed [109]. For example, for the $H_2OH^+\cdots H_2$ complex (Fig. 7.14) the binding energy is equal to 5.2 kcal/mol (MP2/6-311++G(3df,3pd) level) [109], comparable with the binding energy for the water dimer linked through the typical O–H···O hydrogen bond [3]. For the $H_2OH^+\cdots H_2$ complex there is also the meaningful electron charge transfer from molecular hydrogen to the hydronium cation, 139 milielectrons, while for the water dimer such a transfer from one water molecule to the other one amounts ~20 milielectrons depending on the level of calculation. For the $H_2OH^+\cdots H_2$ complex the total electron energy density at the H···σ(H_2) BCP, H_{BCP}, is negative. Note that the negative H_{BCP} value concerns the BCP situated at the bond path linking the H-atom attractor of H_3O^+ with the BCP of the H–H bond (Fig. 7.14). The latter means that according to QTAIM the O–H···σ interaction may be classified here as a strong one and that such interaction possesses some properties of the covalent bond [83]. The second-order perturbation energy of the $\sigma \rightarrow \sigma^*$ orbital-orbital interaction (see Eq. 7.3) is equal to 16.5 kcal/mol for the $H_2OH^+\cdots H_2$ complex. The analogues n(O) → σ(OH)* orbital-orbital interaction energy for the water dimer, for the interaction typical for the conventional hydrogen bonds (see. Eq. 7.2), is equal to 6.5–7.0 kcal/mol depending on the basis set applied.

This is worth to mention that the A–H···σ interactions were firstly classified as the hydrogen bonds for the $XH_4^+\cdots H_2$ complexes and $XH_4^+\ldots(H_2)_5$ (X = N, P, As, Sb, Bi) clusters [106]; next for the FH···H_2 complex [108] and next such interactions were detected in the other complexes [109]. Even the A–H···σ interactions are classified as hydrogen bonds in new IUPAC definition of hydrogen bond [111].

The A–H···σ interactions were also analyzed experimentally. For example, gas phase measurements of dipole moment and vibrational predissociation lifetimes were

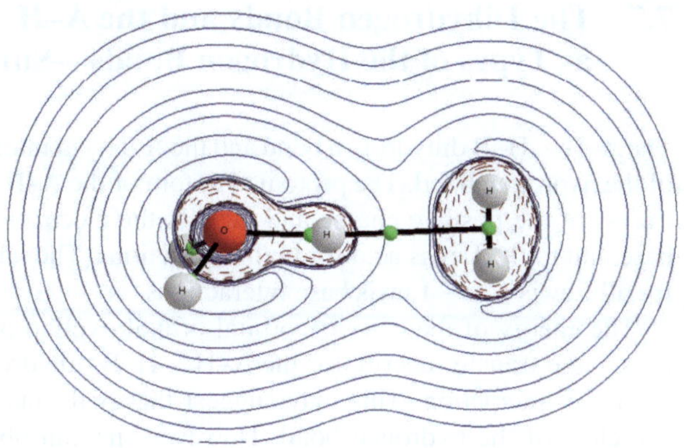

Fig. 7.14 The molecular graph of the $H_2OH^+ \cdots H_2$ complex; the laplacian electron isodensity lines in the plane containing the O–H· · ·σ interaction (the O–H· · ·H_2 fragment) are also presented

performed for the F–H· · ·H_2 complex [112] and the determination of rotational constants confirmed its T-shaped structure. The high-resolution infrared spectra were analyzed for the H_2–HF, D_2–HF and HD–HF systems solvated in helium nanodroplets and also here the existence of the T-shaped structure was confirmed [113–115].

The infrared vibrational predissociation spectra measurements and theoretical calculations were performed for the H_2–HCO^+ complex linked through C–H· · ·σ interaction, where the σ-electrons are those of the molecular hydrogen while C–H proton donating bond is of the HCO^+ ion [116]. The authors explain that the most extensive vibrational bands arise from excitation of the C–H and H_2 stretch vibrations exhibiting rotational structure composed of sub-bands expected for the T-shaped minimum energy structure. The QCISD(T)/6-311G(2df,2pd) calculations show that the energy of the T-shaped conformer of the H_2-HCO^+ complex is lower than the energy of the separated monomers by 4.1 kcal/mol [116]. The authors explain that the experimental spectra of H_2-HCO^+ provide information on two intramolecular vibrations; the H–H stretch localized on H_2 (4060 cm^{-1}) and the C–H stretch localized on HCO^+ (2840 cm^{-1}); the corresponding vibrations of isolated molecules are equal to 4161 and 3089 cm^{-1}, respectively. This is worth to mention that the decrease of the C–H stretching frequency connected with the elongation of the bond is typical for the red-shift hydrogen bond. The authors explain in their study [116] that the H–H stretch decrease is connected with the transfer of the electron charge density from H_2 molecule into HCO^+ ion—this is typical for the hydrogen bond [83] and in general for the Lewis acid—Lewis base interactions [117] where the electron charge transfer from the Lewis base to the Lewis acid is observed. Thus the experimental results are in line with the theoretical findings since they show that the A–H· · ·σ interactions possess numerous characteristics typical for hydrogen bonds. One can also see that the A–H· · ·σ and A–H· · ·H–B (dihydrogen bonds) interactions are different sub-classes of the hydrogen bond.

7.7 The Dihydrogen Bonds and the A–H···σ Interactions as Types of the Hydrogen Bonds—Summary

The A–H···H–B dihydrogen bond and the A–H···σ interaction are classified as types of the hydrogen bond. The protonic H-atom of the A–H bond is characterized by the excess of the positive charge and there is the electron charge shift from the Lewis base unit to the Lewis acid for both interactions. The latter phenomenon is common for all Lewis acid—Lewis base interactions.

The energy of the $\sigma \rightarrow \sigma^*$ orbital-orbital overlap is the most important part of the charge transfer energy for the A–H···H–B dihydrogen bond as well as for the A–H···σ interaction. This may suggest that both interactions belong to the same sub-class of the hydrogen bond. However one can observe numerous differences between them. For the dihydrogen bond the hydridic H-atom is the proton acceptor, i.e. the Lewis base center, while for the A–H···σ interaction the σ-electrons of the molecular hydrogen play the electron donating role. In other words for DHBs there is the "one-atom" Lewis base center ($H^{-\delta}$) while for the A–H···σ interactions there is "two atoms center" of molecular hydrogen. In the latter case the bond critical point corresponding to the H–H bond mimics the one center Lewis base.

For the dihydrogen bonded systems there is the H···H bond path between hydrogen atoms characterized by opposite charges. This is worth to mention that the only A–H···σ interactions known so far are those where the σ-electrons of the molecular hydrogen play the role of electron donors. For such interactions both H-atoms of H_2 molecule are slightly positively charged within the complex as the result of the Lewis acid—Lewis base electron charge shift. For the A–H···H–B dihydrogen bonds the polarization of the B–H bond is observed where H-atom is negatively charged.

The weak and medium in strength dihydrogen bonds are usually electrostatic in nature and the significance of the interaction energy terms related to the electron charge shift increases for stronger interactions. For the A–H···σ interactions the dominance of the electrostatic term is not observed, the delocalization interaction energy term is often the most important attractive contribution.

One can also observe few analogues between A–H···σ and A–H···π interactions on one hand and the analogues between A–H···B conventional hydrogen bond and the A–H···H–B dihydrogen bond on the other hand. For the first pair of interactions there is no one-atom Lewis base center and electrons play the role of the proton acceptor, the Lewis base centers are simulated by the bond critical points. However the situation is more diverse for the A–H···π interactions. For the second pair, the conventional hydrogen bonds and the dihydrogen bonds the one-atom negatively charged Lewis base center is observed.

Acknowledgments Financial support comes from Eusko Jaurlaritza (GIC IT-588-13) and the Spanish Office for Scientific Research (CTQ 2012-38496-C05-04). Technical and human support provided by Informatikako Zerbitzu Orokora—Servicio General de Informatica de la Universidad del Pais Vasco (SGI/IZO-SGIker UPV/EHU), Ministerio de Ciencia e Innovación (MICINN), Gobierno Vasco Eusko Jaurlanitza (GV/EJ), European Social Fund (ESF) is gratefully acknowledged.

References

1. Richardson TB, de Gala S, Crabtree RH (1995) Unconventional hydrogen bonds: intermolecular B–H· · ·H–N interactions. J Am Chem Soc 117:12875–12876
2. Negative values represent binding and interaction energies for stable complexes. For the convenience of discussion the absolute positive values are given in the text
3. Scheiner S (1994) Ab initio studies of hydrogen bonds: the water dimer paradigm. Annu Rev Phys Chem 45:23–56
4. Wessel J, Lee JC Jr, Peris E, Yap GPA, Fortin JB, Ricci JS, Sini G, Albinati A, Koetzle TF, Eisenstein O, Rheingold AL, Crabtree RH (1995) An unconventional intermolecular three-center N–H· · ·H_2Re hydrogen bond in crystalline [$ReH_5(PPh_3)_3$]-indole-C_6H_6. Angew Chem Int Ed Engl 34:2507–2509
5. Crabtree RH, Siegbahn PEM, Eisenstein O, Rheingold AL, Koetzle TFA (1996) A new intermolecular interaction: unconventional hydrogen bonds with element-hydride bonds as proton acceptor. Acc Chem Res 29:348–354
6. Crabtree RH, Eisenstein O, Sini G, Peris E (1998) New types of hydrogen bonds. J Organomet Chem 567:7–11
7. Custelcean R, Jackson JE (2001) Dihydrogen bonding: structures, energetics, and dynamics. Chem Rev 101:1963–1980
8. Robertson KN, Knop O, Cameron TS (2003) C–H· · ·H–C interactions in organoammonium tetraphenylborates: another look at dihydrogen bonds. Can J Chem 81:727–743
9. Wolstenholme DJ, Cameron TS (2006) Comparative study of weak interactions in molecular crystals: H–H bonds vs hydrogen bonds. J Phys Chem A 110:8970–8978
10. Alkorta I, Elguero J, Foces-Foces C (1996) Dihydrogen bonds (A–H· · ·H–B). Chem Commun 1633–1634
11. Remko M (1998) Thermodynamics of dihydrogen bonds (A–H· · ·H–B). Mol Phys 94:839–842
12. Orlova G, Scheiner S (1998) Intermolecular MH· · ·HR bonding in monohydride Mo and W complexes. J Phys Chem A 102:260–269
13. Orlova G, Scheiner S (1998) Intermolecular H· · ·H bonding and proton transfer in semisandwich Re and Ru complexes. J Phys Chem A 102:4813–4818
14. Orlova G, Scheiner S, Kar T (1999) Activation and cleavage of H–R bonds through intermolecular H· · ·H bonding upon reaction of proton donors HR with 18-electron transition metal hydrides. J Phys Chem A 103:514–520
15. Kulkarni SA (1998) Dihydrogen bonding in main group elements: an ab initio study. J Phys Chem A 102:7704–7711
16. Kulkarni SA, Srivastava AK (1999) Dihydrogen bonding in main group elements: a case study of complexes of LiH, BH_3, and AlH_3 with third-row hydrides. J Phys Chem A 103:2836–2842
17. Custelcean R, Jackson JE (1998) Topochemical control of covalent bond formation by dihydrogen bonding. J Am Chem Soc 120:12935–12941
18. Matus MH, Anderson KD, Camaioni DM, Autrey ST, Dixon DA (2007) Reliable predictions of the thermochemistry of boron-nitrogen hydrogen storage compounds: BxNxHy, x = 2, 3. J Phys Chem A 111:4411–4421
19. Miranda CR, Ceder G (2007) Ab initio investigation of ammonia-borane complexes for hydrogen storage. J Chem Phys 126:184703
20. Keaton RJ, Blacquiere JM, Baker RT (2007) Base metal catalyzed dehydrogenation of ammonia–borane for chemical hydrogen storage. J Am Chem Soc 129:1844–1845
21. Staubitz A, Besora M, Harvey JN, Manners I (2008) Computational analysis of amine-borane adducts as potential hydrogen storage materials with reversible hydrogen uptake. Inorg Chem 47:5910–5918
22. Custelcean R, Jackson JE (2000) Topochemical dihydrogen to covalent bonding transformation in $LiBH_4 \cdot$ TEA: a mechanistic study. J Am Chem Soc 122:5251–5257
23. Custelcean R, Vlassa M, Jackson JE (2000) Toward crystalline covalent solids: crystal-to-crystal dihydrogen to covalent bonding transformation in $NaBH_4 \cdot$ THEC. Angew Chem Int Ed Engl 39:3299–3302

24. Kenward AL, Piers WE (2008) Heterolytic H_2 activation by nonmetals. Angew Chem Int Ed Engl 47:38–41
25. Filippov OA, Filin AM, Tsupreva VN, Belkova NV, Lledós A, Ujaque G, Epstein LM, Shubina ES (2006) Proton-transfer and H_2-elimination reactions of main-group hydrides EH_4—(E = B, Al, Ga) with alcohols. Inorg Chem 45:3086–3096.
26. Kubas GJ (2001) Metal Dihydrogen and σ-bond complexes—structure, theory, and reactivity. Kluwer Academic/Plenum Publishers: New York
27. Crabtree RH (2005) The organometallic chemistry of the transition metals. Wiley, Hoboken
28. Alcaraz G, Grellier M, Sabo-Etienne S (2009) Bis σ-bond dihydrogen and borane ruthenium complexes: bonding nature, catalytic applications, and reversible hydrogen release. Acc Chem Res 42:1640–1649
29. Epstein LM, Shubina ES (2002) New types of hydrogen bonding in organometallic chemistry. Coord Chem Rev 231:165–181
30. Belkova NV, Shubina ES, Epstein LM (2005) Diverse world of unconventional hydrogen bonds. Acc Chem Res 38:624–631
31. de Oliveira BG (2013) Structure, energy, vibrational spectrum, and Bader's analysis of π···H hydrogen bonds and $H^{-\delta}\cdots H^{+\delta}$ dihydrogen bonds. Phys Chem Chem Phys 15:37–79
32. Grabowski SJ (2013) Dihydrogen bond and X–H···σ interaction as sub-classes of hydrogen bond. J Phys Org Chem 26:452–459
33. Epstein LM, Belkova NV, Shubina ES (2001) Dihydrogen bonded complexes and proton transfer to hydride ligands by spectral (IR, NMR) studies. In: Peruzzini M, Poli R (eds) Recent advances in hydride chemistry (Chapter). Elsevier, Amsterdam pp 391–418
34. Grabowski SJ, Leszczynski J (2005) Is a dihydrogen bond a unique phenomenon? Chapter in vol 9, a book series: computational chemistry: reviews of current trends. World Scientific Publishing Co: Singapore, pp 195–235
35. Grabowski SJ, Leszczynski J (2009) Dihydrogen bonds: novel feature of hydrogen bond interactions. In: Leszczynski J, Shukla M (eds) Practical aspects of computational chemistry, methods, concepts and applications (Chapter). Springer: Heidelberg Dordrecht, London, New York
36. Bakhmutov VI (2008) Dihydrogen bonds, principles, experiments, and applications. John Wiley & Sons, Inc.: Hoboken, New Jersey
37. Klooster WT, Koetzle TF, Siegbahn PEM, Richardson TB, Crabtree RH (1999) Study of the N–H···H–B Dihydrogen bond including the crystal structure of bh_3nh_3 by neutron diffraction. J Am Chem Soc 121:6337–6343
38. Zachariasen WH, Mooney RCL (1934) The structure of hypophosphite group as determined from the crystal lattice of ammonium hypophosphite. J Chem Phys 2:34–37
39. Burg AB (1964) Enhancement of P–H bonding in a phosphine monoborane. Inorg Chem 3:1325–1327
40. Titov LV, Makarova MD, Rosolovskii VY (1968) Guanidinium borohydride. Dokl Akad Nauk 180:381–382
41. Grabowski SJ, Krygowski TM (1999) The proton transfer path for C=O···H–O systems modelled from crystal structure data. Chem Phys Lett 305:247–250
42. Sobczyk L, Grabowski SJ, Krygowski TM (2005) Interrelation between H-bond and Pi-electron delocalization. Chem Rev 105:3513–3560
43. Grabowski SJ (2014) Tetrel bond—σ-hole bond as a preliminary stage of the S_N reaction. Phys Chem Chem Phys 16:1824–1834
44. Brown MP, Heseltine RW (1968) Co-ordinated BH_3, as a proton acceptor group in hydrogen bonding. Chem Commun 23:1551–1552
45. Brown MP, Heseltine RW, Smith PA, Walker PJ (1970) An infrared study of coordinated BH3 and BH2 groups as proton acceptors in hydrogen bonding. J Chem Soc A 410–414
46. Stevens RC, Bau R, Milstein D, Blum O, Koetzle TF (1990) Concept of the H(δ+)···H(δ−) interaction. A low-temperature neutron diffraction study of cis- [IrH (OH)$(PMe_3)_4$] PF_6. J Chem Soc Dalton Trans 1429–1432

47. Milstein D, Calabrese JC, Williams ID (1986) Formation, structures, and reactivity of cis-hydroxy-, cis-methoxy-, and cis-mercaptoiridium hydrides. Oxidative addition of water to Ir(I). J Am Chem Soc 108:6387–6389
48. Lough AJ, Park S, Ramachandran R, Morris RH (1994) Switching on and off a new intramolecular hydrogen-hydrogen interaction and the heterolytic splitting of dihydrogen. Crystal and molecular structure of [Ir({H(η^1-SC_5H_4NH) } 2(PCy_3)$_2$]BF_4.2.7CH_2Cl_2. J Am Chem Soc 116:8356–8357
49. Gusev DG, Lough AJ, Morris RH (1998) New polyhydride anions and proton-hydride hydrogen bonding in their ion pairs. J Am Chem Soc 120:13138–13147
50. Ramachandran R, Morris RH (1994) A new type of intramolecular H···H···H interaction involving N–H···H(Ir)···H–N atoms. Crystal and molecular structure of [IrH(η^1-SC_5H_4NH)$_2$(η^2-SC_5H_4N) (PCy_3)]BF_4.0.72CH_2Cl_2. J Chem Soc Chem Commun 2201–2202
51. Lee JC, Pens E, Rheingold AL, Crabtree RH (1994) An unusual type of H···H interaction: Ir–H···H–O and Ir–H···H–N hydrogen bonding and its involvement in σ-bond metathesis. J Am Chem Soc 116:11014–11019
52. Lee JC, Rheingold AL, Muller B, Pregosin PS, Crabtree RH (1994) Complexation of an amide to iridium *via* an iminol tautomer and evidence Ir–H···H–O hydrogen bond. J Chem Soc Chem Commun 1021–1022
53. Belkova NV, Shubina ES, Gutsul EI, Epstein LM, Eremenko IL, Nefedov SE (2000) Structural and energetic aspects of hydrogen bonding and proton transfer to $ReH_2(CO)(NO)(PR_3)_2$ and $ReCl(CO)(NO)(PMe_3)_2$ by IR and X-ray studies. J Organomet Chem 610:58–70
54. Allen FH (2002) The Cambridge structural database: a quarter of a million crystal structures and rising. Acta Cryst B58:380–388
55. Liu Q, Hoffmann R (1995) Theoretical aspects of a novel mode of hydrogen-hydrogen bonding. J Am Chem Soc 117:10108–10112
56. Siegbahn PEM, Blomberg MRA, Svensson M (1994) PCI-X, a parametrized correlation method containing a single adjustable parameter X. Chem Phys Lett 223:35–45
57. Siegbahn PEM, Svensson M, Boussard PJE (1995) First row bench mark tests of the PCI-X scheme. J Chem Phys 102:5377–5386
58. Grabowski SJ (1999) Study of correlations for dihydrogen bonds by quantum-chemical calculations. Chem Phys Lett 312:542–547
59. Grabowski SJ (2000) High-level ab initio calculations of dihydrogen-bonded complexes. J Phys Chem A 104:5551–5557
60. Kitaura K, Morokuma K (1976) A new energy decomposition scheme for molecular interactions within the Hartree-Fock approximation. Int J Quantum Chem 10:325–340
61. Cybulski SM, Chałasiński G, Moszyński R (1990) On decomposition of MP2 supermolecular interaction energy and basis set effects. J Chem Phys 92:4357–4363
62. Cybulski H, Pecul M, Sadlej J (2003) Characterization of dihydrogen-bonded D–H···H–A complexes on the basis of infrared and magnetic resonance spectroscopic parameters. J Chem Phys 119:5094–5104
63. Kar T, Scheiner S (2003) Comparison between hydrogen and dihydrogen bonds among H_3BNH_3, H_2BNH_2, and NH_3. J Chem Phys 119:1473–1482
64. Alkorta I, Elguero J, Mó O, Yánez M, Del Bene JE (2002) Ab Initio study of the structural, energetic, bonding, and IR spectroscopic properties of complexes with dihydrogen bonds. J Phys Chem A 106:9325–9330
65. Del Bene JE, Perera SA, Bartlett RJ, Alkorta I, Elguero J, Mó O, Yánez M (2002) One-bond ($^{1d}J_{H-H}$) and three-bond ($^{3d}J_{X-M}$) spin-spin coupling constants across X-H···H–M dihydrogen bonds. J Phys Chem A 106:9331–9337
66. Grabowski SJ, Robinson TL, Leszczynski J (2004) Strong dihydrogen bonds—ab initio and atoms in molecules study. Chem Phys Lett 386:44–48
67. Sokalski WA, Roszak S, Pecul K (1988) An efficient procedure for decomposition of the SCF interaction energy into components with reduced basis set dependence. Chem Phys Lett 153:153–159

68. Sokalski WA, Roszak S (1991) Efficient techniques for the decomposition of intermolecular interaction energy at SCF level and beyond. J Mol Struct (Theochem) 234:387–400
69. Grabowski SJ, Sokalski WA, Leszczynski J (2005) How short can the H···H intermolecular contact be? New findings that reveal the covalent nature of extremely strong interactions. J Phys Chem A 109:4331–4341
70. Boys SF, Bernardi F (1970) The calculation of small molecular interactions by the differences of separate total energies. Some procedures with reduced errors. Mol Phys 19:553–566
71. Schmidt MS, Baldridge KK, Boatz JA, Elbert ST, Gordon MS, Jensen JH, Koseki S, Matsunaga N, Nguyen KA, Su SJ, Windus TL, Dupuis M, Montgomery JA (1993) General atomic and molecular electronic structure system. J Comp Chem 14:1347–1363
72. Grabowski SJ, Sokalski WA, Dyguda E, Leszczynski J (2006) Quantitative classification of covalent and noncovalent H-bonds. J Phys Chem B 110:6444–6446
73. Grabowski SJ (2009) Covalent character of hydrogen bonds enhanced by π-electron delocalization. Croat Chem Acta 82:185–192
74. Desiraju GR (2002) Hydrogen bridges in crystal engineering: interactions without borders. Acc Chem Res 35:565–573
75. Grabowski SJ, Sokalski WA, Leszczynski J (2007) Wide spectrum of H···H interactions: van der Waals contacts, dihydrogen bonds and covalency. Chem Phys 337:68–76
76. Jeffrey GA, Saenger W (1991) Hydrogen bonding in biological structures. Springer-Verlag: Berlin
77. Scheiner S (1997) Hydrogen bonding: a theoretical perspective. Oxford University Press: New York
78. Hobza P, Havlas Z (2000) Blue-shifting hydrogen bonds. Chem Rev 100:4253–4264
79. Feng Y, Zhao S-W, Liu L, Wang J-T, Li X-S, Guo Q-X (2004) Blue-shifted dihydrogen bonds. J Phys Org Chem 17:1099–1106
80. Yang Y, Zhang W (2007) Theoretical study of N–H···H–B blue-shifted dihydrogen bonds. J Mol Struct (Theochem) 814:113–117
81. Yu W, Lin Z, Huang Z (2006) Coexistence of dihydrogen, blue and red-shifting hydrogen bonds in an ultrasmall system: valine. ChemPhysChem 7:828–830
82. Trung NT, Hue TT, Nguyen MT, Zeegers-Huyskens T (2008) Theoretical study of the interaction between HNZ (Z = O, S) and H_2XNH_2 (X = B, Al). Conventional and dihydrogen bonds. Phys Chem Chem Phys 10:5105–5113
83. Grabowski SJ (2011) What is the covalency of hydrogen bonding? Chem Rev 11:2597–2625
84. Weinhold F, Landis C (2005) Valency and bonding, a natural bond orbital donor—acceptor perspective. Cambridge University Press: Cambridge, UK
85. Reed AE, Curtiss LA, Weinhold F (1988) Intermolecular interactions from a natural bond orbital, donor-acceptor viewpoint. Chem Rev 88:899–926
86. Alabugin IV, Manoharan M, Peabody S, Weinhold F (2003) Electronic basis of improper hydrogen bonding: a subtle balance of hyperconjugation and rehybridization. J Am Chem Soc 125:5973–5987
87. Weinhold F, Klein R (2012) What is a hydrogen bond? Mutually consistent theoretical and experimental criteria for characterizing H-bonding interactions. Mol Phys 110:565–579
88. Alkorta I, Elguero J, Grabowski SJ (2008) How to determine whether intramolecular H···H interactions can be classified as dihydrogen bonds. J Phys Chem A 112:2721–2727
89. Grabowski SJ (2011) Halogen bond and its counterparts: Bent's rule explains the formation of nonbonding interactions. J Phys Chem A 115:12340–12347
90. Grabowski SJ (2012) QTAIM characteristics of halogen bond and related interactions. J Phys Chem A 116:1838–1845
91. Grabowski SJ (2013) Non-covalent interactions—QTAIM and NBO analysis. J Mol Model 19:4713–4721
92. Bent HA (1961) An appraisal of valence-bond structures and hybridization in compounds of the first-row elements. Chem Rev 61:275–311
93. Grabowski SJ (2011) Red- and blue-shifted hydrogen bonds: the Bent rule from quantum theory of atoms in molecules perspective. J Phys Chem A 115:12789–12799

94. Bader RFW (1985) Atoms in molecules. Acc Chem Res 18:9–15
95. Bader RFW (1990) Atoms in molecules, a quantum theory. Oxford University Press, Oxford
96. Bader RFW (2009) Bond paths are not chemical bonds. J Phys Chem A 113:10391–10396
97. Grabowski SJ, Ugalde JM (2010) Bond paths show preferable interactions: ab initio and QTAIM studies on the X–H··· π hydrogen bond. J Phys Chem A 114:7223–7229
98. Koch U, Popelier PLA (1995) Characterization of C–H–O hydrogen bonds on the basis of the charge density. J Phys Chem 99:9747–9754
99. Popelier PLA (1998) Characterization of a dihydrogen bond on the basis of the electron density. J Phys Chem A 102:1873–1878
100. Cramer CJ, Gladfelter WL (1997) Ab initio characterization of $[H_3N.BH_3]_2$, $[H_3N.A1H_3]_2$, and $[H_3N.GaH_3]_2$. Inorg Chem 36:5358–5362
101. Cremer D, Kraka E (1984) A description of the chemical-bond in terms of local properties of electrodensity and energy. Croat Chem Acta 57:1259–1281
102. Jenkins S, Morrison I (2000) The chemical character of the intermolecular bonds of seven phases of ice as revealed by ab initio calculation of electron densities. Chem Phys Lett 317:97–102
103. Rozas I, Alkorta I, Elguero J (2000) Behavior of ylides containing N, O, and C atoms as hydrogen bond acceptors. J Am Chem Soc 122:1154–11161
104. Domagała M, Grabowski SJ (2009) X–H···π and X–H···N hydrogen bonds—acetylene and hydrogen cyanide as proton acceptors. Chem Phys 363:42–48
105. Urban J, Roszak S, Leszczynski J (2001) Shellvation of the ammonium cation by molecular hydrogen: a theoretical study. Chem Phys Lett 346:512–518
106. Szymczak JJ, Grabowski SJ, Roszak S, Leszczynski J (2004) H···σ interactions—an ab initio and "atoms in molecules" study. Chem Phys Lett 393:81–86
107. Grabowski SJ, Lipkowski P (2011) Characteristics of X–H···π interactions: ab initio and QTAIM studies. J Phys Chem A 115:4765–4773.
108. Grabowski SJ, Sokalski WA, Leszczynski J (2006) Can H···σ, π···H+···σ and σ···H+···σ interactions be classified as H-bonded? Chem Phys Lett 432:33–39
109. Grabowski SJ (2007) Hydrogen bonds with π and σ electrons as the multicenter proton acceptors: high level ab initio calculations. J Phys Chem A 111:3387–3393
110. Grabowski SJ, Alkorta I, Elguero J (2013) Complexes between dihydrogen and amine, phosphine, and arsine derivatives. Hydrogen bond versus pnictogen interaction. J Phys Chem A 117:3243–3251
111. Arunan E, Desiraju GR, Klein RA, Sadlej J, Scheiner S, Alkorta I, Clary DC, Crabtree RH, Dannenberg JJ, Hobza P, Kjaergaard HG, Legon AC, Mennucci B, Nesbitt DJ (2011) Definition of the hydrogen bond. Pure Appl Chem 83:1637–1641
112. Jucks KW, Miller RE (1987) Infrared Stark spectroscopy on the hydrogen-HF binary complex. J Chem Phys 87:5629–5633
113. Moore DT, Miller RE (2003) Dynamics of hydrogen–HF complexes in helium nanodroplets. J Chem Phys 118:9629–9636
114. Moore DT, Miller RE (2003) Solvation of HF by molecular hydrogen: helium nanodroplet vibrational spectroscopy. J Phys Chem A 107:10805–10812
115. Moore DT, Miller RE (2004) Rotationally resolved infrared laser spectroscopy of $(H_2)_n$–HF and $(D_2)_n$–HF ($n = 2$–6) in helium nanodroplets. J Phys Chem A 108:1930–1937
116. Bieske EJ, Nizkorodov SA, Bennett FR, Maier JP (1995) The infrared spectrum of the H2–HCO1 complex. J Chem Phys 102:5152–5164
117. Lipkowski P, Grabowski SJ, Leszczynski J (2006) Properties of the halogen-hydride interaction: an ab initio and "Atoms in Molecules" analysis. J Phys Chem A 110:10296–10302

Part II
Other Bridging Atoms

Chapter 8
The Pnicogen Bond in Review: Structures, Binding Energies, Bonding Properties, and Spin-Spin Coupling Constants of Complexes Stabilized by Pnicogen Bonds

Janet E. Del Bene, Ibon Alkorta and José Elguero

Abstract Extensive ab initio MP2/aug'-cc-pVTZ studies have been carried out in our laboratories to determine the structures, binding energies, bonding properties, and EOM-CCSD spin-spin coupling constants of various series of complexes stabilized by pnicogen bonds. These systematic studies provide insight into the nature of the pnicogen bond, and how changes in this bond are reflected in the properties of these complexes.

8.1 Introduction

The term "pnicogen" or "pnictogen" was originally proposed by Prof. A. E. van Arkel to designate the group 15 elements nitrogen, phosphorus, arsenic, antimony, and bismuth [1, 2]. The name comes from the Greek word meaning choke or choking gas. The symbol Z is used to designate this family of elements, just as the symbols Y and X refer to the group 16 chalcogens and group 17 halogens, respectively [3]. The pnicogen bond which involves a group 15 element acting as an electron acceptor, joins the list of other weak intermolecular interactions such as the hydrogen bond and the halogen bond. The IUPAC definition of a hydrogen bond (HB) is "an attractive interaction between a hydrogen atom from a molecule or a molecular fragment X–H in which X is more electronegative than H, and an atom or a group of atoms in the same or a different molecule, in which there is evidence of bond formation" [4]. The hydrogen bond is often written as X-H...Y, with Y either a σ electron-pair donor or a π electron donor.

J. E. Del Bene (✉)
Department of Chemistry, Youngstown State University,
44555 Youngstown, OH, USA
e-mail: jedelbene@ysu.edu

I. Alkorta (✉) · J. Elguero
Instituto de Química Médica (IQM-CSIC),
Juan de la Cierva, 3, 28006 Madrid, Spain
e-mail: ibon@iqm.csic.es

S. Scheiner (ed.), *Noncovalent Forces,* Challenges and Advances in
Computational Chemistry and Physics 19, DOI 10.1007/978-3-319-14163-3_8

The IUPAC definition of a halogen bond (XB) is as follows: "A halogen bond occurs when there is evidence of a net attractive interaction between an electrophilic region associated with a halogen atom in a molecular entity and a nucleophilic region in another, or the same, molecular entity" [5]. Politzer and Murray have described the halogen atom using the concept of the σ-hole which corresponds to a positive region on the molecular electrostatic potential (MEP) surface [6]. Our definition of the pnicogen bond (ZB) states that this bond is a Lewis acid-Lewis base attractive interaction in which the pnicogen atom is the Lewis acid. While all three bonds are necessarily attractive interactions, there are similarities and differences among them. In the hydrogen bond, the hydrogen atom is positively charged since it is bonded to a more electronegative atom or group of atoms, and it acts as an electron acceptor from an atom or group of atoms, as for example in the F-H...NH_3 complex. The electron-donating partner may donate a lone pair of electrons or π electrons. Similarly, the halogen bond may have the halogen atom bonded to a more electronegative atom or group of atoms as it acts as an electron acceptor through its σ-hole, as illustrated by the complex F-Cl...NH_3. However, it is not necessary that the halogen be bonded to a more electronegative atom, since complexes between I_2 and amines are also known, as well as complexes in which C-Br is the electron acceptor [7, 8]. In a complex stabilized by a pnicogen bond, the pnicogen atom may be bonded to an electronegative or electropositive atom or group of atoms. The pnicogen atom acts as an electron acceptor either through a σ-hole or a π-hole depending on the nature of the pnicogen-containing molecule. Both the σ-hole and the π-hole correspond to positive regions on the MEP surface. In addition, since the pnicogen atom has a lone pair of electrons and in some cases π electrons, it may simultaneous act as an electron donor. Examples of pnicogen-bonded complexes are $H_2FP{:}PFH_2$ and $H_2(CH_3)P{:}P(CH_3)H_2$ [9].

The timeline for research on pnicogen interactions can be divided in two eras, with the year 2011 being the dividing line between them. In 1995 two relevant papers on pnicogen bonds appeared, one by Pyykko et al. on dipnicogen dimers, $(H_2Z\text{-}ZH_2)_2$ [10], and the other by Corriu et al. on intramolecular coordination of phosphorous [11], as in phosphatranes [12]. In 2007, Murray, Politzer, et al. introduced the concept of the σ-hole which was used to describe the halogen bond, and subsequently the pnicogen bond [13]. Two other papers, one on P...P interactions in halo-phosphines [14], and the other on weak interactions between trivalent pnicogen centers in X_3Z...ZX_3 complexes also appeared during this period [15]. The modern era of studies of the pnicogen bond began in 2011 with the publication of two papers, one a landmark paper by Hey-Hawkins and coworkers, with the title "Pnicogen Bonds: A New Molecular Linker?" [9], and the other by Scheiner with the title "A New Noncovalent Force: Comparison of P...N Interactions with Hydrogen and Halogen Bonds" [16]. The subsequent resurgence of interest in the pnicogen bond led to the publication of about one hundred articles on this subject from 2011 to mid-2014.

8.1.1 Experimental Studies

In their 2011 paper, Hey-Hawkins et al. discussed P...P through-space NMR spin-spin couplings [17], and an X-ray structure which suggested the existence of an attractive interaction between P atoms [18]. In this paper they cited as an antecedent the work of Sundberg et al. on 1,2-dicarba-*cloro*-dodecaboranes [19], and based their discussion on the presence of a P...P bond critical point. In addition to these two papers, the most significant experimental data on complexes involving pnicogen bonds is a study of the microwave spectrum, structure, and dipole moment of the $F_3P{:}OH_2$ complex [20]. Politzer, Murray et al. reported a survey of crystal structure data which supported the concept of σ-hole interactions [21].

8.1.2 Theoretical Studies

Studies of selected series of pnicogen-bonded complexes which have been carried out in our laboratories [22–44] will be discussed in detail in this review. In the present section we survey the works of other authors who have made notable contributions to the literature on the pnicogen bond. We begin with the work of Scheiner and his group, who reported studies of a variety of pnicogen-bonded complexes, emphasizing substituent effects on P...N interactions [45–47], the sensitivity of the pnicogen bond to angular distortion [48], and the possibility of N...N interactions [49]. They also compared P...P and P...N bonds with other non-covalent interactions in trimers and a tetramer of PH_3 [50]; compared hydrogen bonding with P...N interactions [51]; investigated the effect of the carbon chain (R) in $RH_2P...NH_3$ complexes [52, 53]; and compared the sensitivity of noncovalent bonds including hydrogen, halogen, chalcogen, and pnicogen bonds to intermolecular separation [54, 55]. Scheiner also published two reviews on pnicogen bonds [56, 57].

The group of Frontera has published papers on pnicogen–π complexes and noted some biological implications [58]; characterized complexes with YO_2Br with Y=N, P, and As [59]; carried out a CSD survey of halogen, chalcogen and pnicogen bonds [60]; and reported a study of pnicogen bonds using different computational methods [44]. Solimannejad has investigated pnicogen, chalcogen, and halogen bonds involving substituted *s*-triazines [61]; cooperativity with lithium bonds [62]; and complexes involving NO_2X [63]. Grabowski reported a study of pnicogen bonds versus hydrogen bonds [64], including large clusters such as $NH_4^+{:}(NCH)_{1-8}$, and $NH_4^+{:}(N_2)_{1-8}$[65]. Contributions from other groups include studies of pnicogen-hydride interactions [66]; cooperativity [67–69]; competition among hydrogen, halogen, and pnicogen bonds [70]; pnicogen interactions involving anions [71]; electron transfer in pnicogen bonds [72]; and characterization of pnicogen bonds using the Laplacian of the electron density [73].

8.2 Methods

The structures of the monomers and complexes reported in this review were optimized at second-order Møller-Plesset perturbation theory (MP2) [74–77] using the aug'-cc-pVTZ basis set [78]. This basis set is the Dunning aug-cc-pVTZ basis set [79, 80] with diffuse functions removed from H atoms. Transition structures were computed for some systems which were found to have double minima along the pnicogen-bonding coordinate. Frequencies were computed to confirm equilibrium and transition structures. These calculations were performed using the Gaussian 09 program [81].

The electron densities of the complexes were analyzed using the Atoms in Molecules (AIM) methodology [82–85] and the Electron Localization Function (ELF) [86], employing the AIMAll [87] and Topmod programs [88]. The topological analysis of the electron density produces the molecular graph of the complex. This graph identifies the location of electron density features of interest, including the electron density (ρ) maxima associated with various nuclei, saddle points which correspond to bond critical points (BCPs), and ring critical points which indicate a minimum electron density within a ring. The zero gradient line which connects a BCP with two nuclei is the bond path. The electron density at the BCP (ρ_{BCP}), the Laplacian of the electron density at the BCP ($\nabla^2\rho_{BCP}$), and the total energy density (H_{BCP}) are additional useful quantities for characterizing the nature of intermolecular interactions [89]. The ELF function illustrates those regions of space at which the electron density is high. All of these measures are useful for identifying bonds and lone pairs, and characterizing bond types. The Natural Bond Orbital (NBO) method [90] was employed to obtain MP2/aug'-cc-pVTZ atomic charges and to analyze the stabilizing charge-transfer interactions in these complexes, employing the NBO-5 and NBO-6 programs [91, 92]. Since MP2 orbitals are nonexistent, the charge-transfer interactions were computed using the B3LYP functional [93, 94] with the aug'-cc-pVTZ basis set at the MP2/aug'-cc-pVTZ geometries, so that at least some electron correlations effects could be included. NBO orbitals were represented with the Jmol program [95] using the tools developed by Marcel Patek. [96]

Spin-spin coupling constants were evaluated using the equation-of-motion coupled cluster singles and doubles (EOM-CCSD) method in the CI (configuration interaction)-like approximation, [97, 98] with all electrons correlated. For these calculations, the Ahlrichs [99] qzp basis set was placed on ^{13}C, ^{15}N, ^{17}O, and ^{19}F, and the qz2p basis set on ^{31}P and ^{35}Cl. The Dunning cc-pVDZ basis set was placed on all ^{1}H atoms except for a hydrogen-bonded ^{1}H, in which case the qz2p basis set was used. All of the EOM-CCSD calculations were carried out using ACES II [100] on the IBM Cluster 1350 (Glenn) at the Ohio Supercomputer Center.

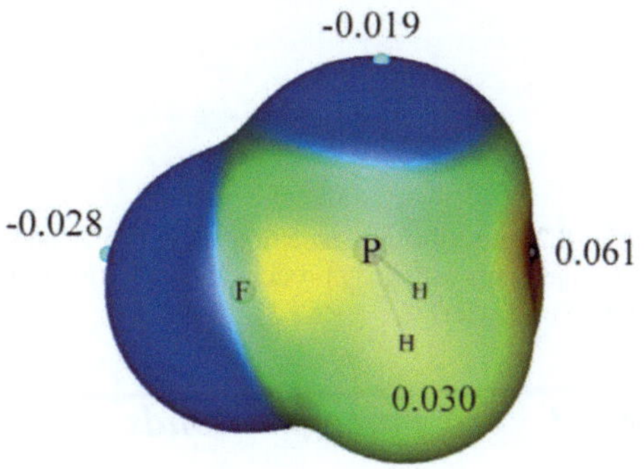

Fig. 8.1 The molecular electrostatic potential of PH_2F showing regions of positive and negative charge. The color code is *Red* > 0.04 > *Yellow* > 0.02 > *Green* > 0.00 > *Blue*. The positions of local minima and maxima are indicated with *light blue* and *black dots*, respectively. (Reprinted with permission from Ref. [22]. Copyright (2012) Elsevier)

8.3 Discussion

The phosphorous atom has been the most studied participant in the pnicogen bond. The molecular electrostatic potentials of molecules containing phosphorus exhibit a region of negative charge associated with the P lone pair, and a region of positive charge at the same P which is the σ-hole. This is illustrated in Fig. 8.1 for the H_2FP molecule. Thus, the P atom can act as both an electron-pair donor and an electron acceptor at P in complexes with pnicogen bonds.

In the following sections, we review some of the research carried out in our laboratories on pnicogen-bonded complexes, with special emphasis on their structures, binding energies, some bonding properties, and spin-spin coupling constants. The first section focuses on complexes with σ-σ pnicogen bonds in dimers $(PH_2X)_2$ [22] and $[H_2C{=}PX]_2$ [34], mixed binary complexes $H_2C{=}(X)P{:}PXH_2$ [35], $H_2XP{:}PCX$ [36], $H_2XP{:}NXH_2$ [23], $X{=}PH_3{:}NY$ [37] and $X{=}PH_3{:}PY$ [37], $H_2FP{:}ClX$ [38], and the molecular anions $H_2YP{:}X^-$[39], with X and Y a variety of substituents. On the $H_2C{=}(X)P{:}PXH_2$ [35] and $H_2XP{:}PCX$ surfaces [36] there are also equilibrium structures stabilized by π-σ pnicogen bonds, and these complexes are examined in the second section. Relationships between the properties of corresponding complexes with σ-σ and π-σ bonds are discussed. Complexes $YN{:}PO_2X$ [33] which are stabilized by π-σ bonds are also included in this section. The final section focuses on ternary and higher-order complexes in which a pnicogen-bonded complex is involved in additional intermolecular interactions [26, 27, 31]. The interest here is to examine how these interactions change the energy of the pnicogen bond, and the nonadditivities of interaction energies. More detailed discussions of the complexes reviewed in this chapter may be found in the original references, in which the geometries, total energies, and molecular graphs are available as Supporting Information. Finally, it should be noted that there may be other equilibrium structures on the potential surfaces in addition to those discussed in this chapter, but these have not been included in this review.

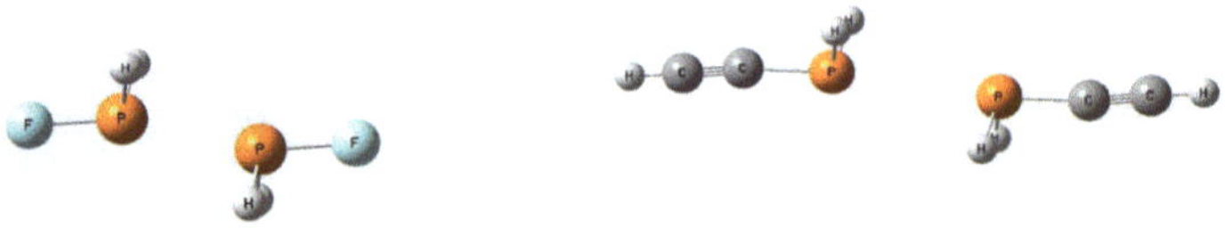

Fig. 8.2 Equilibrium structures of $(PH_2F)_2$ and $[PH_2(CCH)]_2$ illustrating the P-P-F and P-P-C arrangements which approach linearity. (Reprinted with permission from Ref. [22]. Copyright (2012) Elsevier)

8.3.1 σ-σ Pnicogen Bonds

A σ-σ pnicogen bond arises when an atom in one molecule acts as a σ lone-pair electron donor to a group 15 atom which is the electron-pair acceptor through its σ-hole. If both atoms are pnicogen atoms, then it is possible that each will be an electron-pair donor and acceptor. In this section, we discuss some of the properties of complexes stabilized by σ-σ pnicogen bonds.

8.3.1.1 $(PH_2X)_2$

The first pnicogen-bonded complexes that were investigated in our laboratories are those formed between two formally sp^3 hybridized molecules, PH_2X, with the substituents X=F, Cl, OH, NC (bonded at either N or C), CCH, CH_3, and H [22]. Table 8.1 reports selected properties of the $(PH_2X)_2$ complexes, including their binding energies, charge-transfer energies, P-P distances, A-P-P angles with A the atom of X directly bonded to P, and spin-spin coupling constants ${}^{1p}J$(P-P) across the pnicogen bond. The equilibrium structures of these complexes have C_{2h} symmetry, with a P-P-A arrangement that approaches linearity. This arrangement is illustrated in Fig. 8.2 for $(PH_2F)_2$ and $[PH_2(CCH)]_2$.

The intermolecular P-P distances in dimers $(PH_2X)_2$ range from 2.471 to 3.589 Å, which are significantly longer than the P-P covalent bond distance of 2.219 Å in $H_2P\text{-}PH_2$. The binding energies of the dimers lie between –7.1 kJ·mol^{-1} for $(PH_3)_2$ and –34.0 kJ·mol^{-1} for $(PH_2F)_2$, a range which is comparable to that of neutral hydrogen-bonded complexes. As evident from Table 8.1, the more electronegative substituents form the more strongly bound dimers, while the more electropositive substituents form the more weakly bound complexes. The binding energies of these complexes are related to the nature of the stabilizing charge-transfer energies, which arise as electron density is transferred from the lone pair of one P atom to the antibonding σ* P-A orbital of the other. The largest charge-transfer energy is found for the most strongly bound complex $(PH_2F)_2$, and the smallest for $(PH_3)_2$ which has the weakest pnicogen bond. The charge transfer energies and the binding energies are related quadratically, with a correlation coefficient R^2 of 0.982. There are two factors which play a role in determining the magnitude of the charge-transfer energy. The first is the dependence of the charge-transfer energy on the P-P distance. For

Table 8.1 MP2/aug'-cc-pVTZ P-P distances (R, Å), A-P-P angles (∠, °), binding energies (ΔE, kJ·mol^{-1}), charge-transfer energies [$P_{lp}(1) \rightarrow \sigma^*$P-A(2), kJ·mol^{-1}], and spin-spin coupling constants [^{1p}J(P-P), Hz] for complexes $(PH_2X)_2$. (Reprinted with permission from Ref. [22]. Copyright (2012) Elsevier)

Complex	R(P-P)	∠A-P-P[a]	ΔE	$P_{lp}(1) \rightarrow \sigma^*$P-A(2)	^{1p}J(P-P)
$(PH_2F)_2$	2.471	163	− 34.0	131.8	998.6
$(PH_2Cl)_2$	2.771	167	− 22.1	59.9	1119.9
$[PH_2(OH)]_2$	2.851	169	− 20.6	46.6	644.0
$[PH_2(NC)]_2$	3.040	168	− 13.8	31.6	640.3
$[PH_2(CCH)]_2$	3.353	174	− 12.2	11.3	281.9
$[PH_2(CN)]_2$	3.375	171	− 8.4	11.5	300.0
$[PH_2(CH_3)]_2$	3.481	178	− 8.9	11.3	160.9
$(PH_3)_2$	3.589	179	− 7.1	5.6	130.9

[a]A is the atom of group X that is directly bonded to P

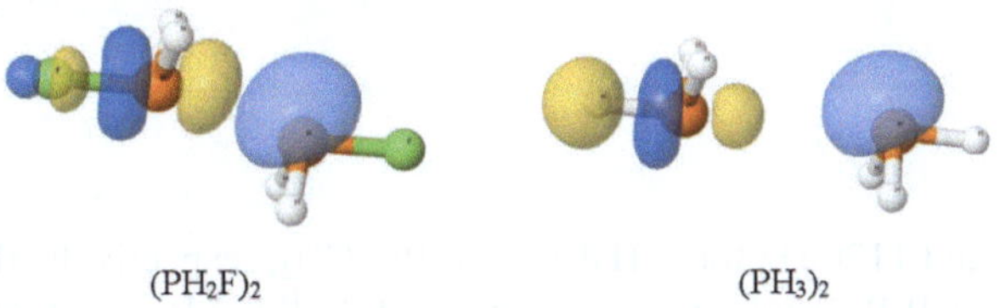

Fig. 8.3 Overlap of the lone-pair orbital on one P atom with the σ* P-A orbital of the other in $(PH_2F)_2$ and $(PH_3)_2$

complexes $(PH_2X)_2$, charge transfer has a quadratic dependence on distance, with a correlation coefficient R^2 of 0.976. The second factor is the degree of overlap between the P lone pair and the σ* P-A orbital, as illustrated in Fig. 8.3 for $(PH_2F)_2$ and $(PH_3)_2$. The importance of the nature of A in determining the properties of pnicogen-bonded complexes can be seen in two papers on $(PHFX)_2$ complexes with F-P...P-F, H-P...P-H, and A-P...P-A approaching linearity [24, 25].

The AIM analysis shows that a bond critical point connects the two P atoms in these complexes. The electron density at the BCP correlates exponentially with the intermolecular P-P distance, with a correlation coefficient R^2 of 0.999. This correlation is found for various types of intermolecular bonds. [101–110] The Laplacian of the electron density and the total energy density at the BCP are positive for these complexes, except for $(PH_2F)_2$, $(PH_2Cl)_2$, and $[PH_2(OH)]_2$, which have negative values of the energy density. This indicates that the P...P bonds in these three complexes have some degree of covalency. There is a significant electron-density buildup in the region between the two phosphorus and at the atoms A, as illustrated in Fig. 8.4 for $(PH_3)_2$. It is the build-up of charge in the region between the two P nuclei that results in the formation of the pnicogen bond.

One-bond spin-spin coupling constants ^{1p}J(P-P) provide additional insight into the pnicogen bond. For complexes $(PH_2X)_2$, ^{1p}J(P-P) values are dominated by the Fermi contact terms which are excellent approximations to total J. Values of ^{1p}J(P-P) are reported in Table 8.1, and can be seen to vary between 131 Hz for $(PH_3)_2$ to 1000

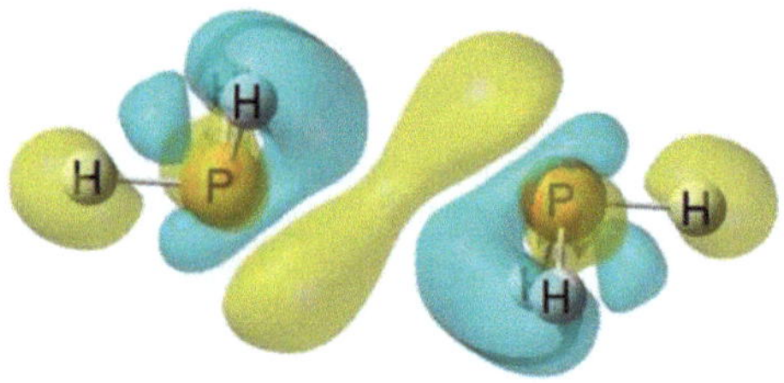

Fig. 8.4 Electron density shifts for $(PH_3)_2$ at the 0.0001 au isosurface. *Blue* and *yellow* indicate regions of decreased and increased electron densities, respectively. (Reprinted with permission from Ref. [22]. Copyright (2012) Elsevier)

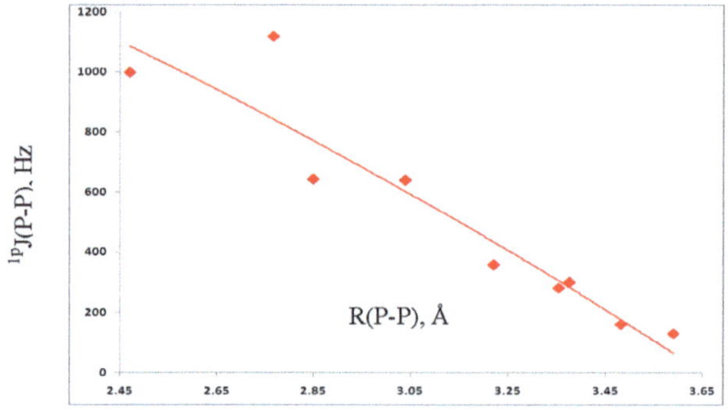

Fig. 8.5 ${}^{1p}J$(P-P) versus the P-P distance for complexes $(PH_2X)_2$. (Reprinted with permission from Ref. [22]. Copyright (2012) Elsevier)

and 1120 Hz for $(PH_2F)_2$ and $(PH_2Cl)_2$, respectively, thus illustrating the sensitivity of P-P coupling to the nature of X. ${}^{1p}J$(P-P) exhibits a quadratic dependence on the P-P distance with a correlation coefficient of 0.887, as illustrated in Fig. 8.5. From this figure is appears that the value of ${}^{1p}J$(P-P) for $(PH_2Cl)_2$ is too large for its intermolecular distance. Using plots such as that of Fig. 8.5, it would be possible to obtain an estimate of the intermolecular distance from the experimental value of the corresponding coupling constant.

8.3.1.2 $(H_2C{=}PX)_2$

Complexes $(H_2C{=}PX)_2$ with formally hybridized sp^2 P atoms may also form pnicogen-bonded complexes of C_{2h} symmetry with P-P-A and/or P-P-C approaching linearity [34]. Complexes $(H_2C{=}PX)_2$ with A-P...P-A linear are designated conformation A complexes. Their intermolecular P-P distances, selected structural parameters, and binding energies are given in Table 8.2, and representative complexes $(H_2C{=}PH)_2$ and $(H_2C{=}PF)_2$ are illustrated in Fig. 8.6. As evident from Table 8.2, the range of binding energies is relatively narrow, from -9.5 to -13.9 kJ·mol^{-1}, much smaller than the range of binding energies of complexes $(PH_2X)_2$. Moreover, the order of decreasing binding energies with respect to the substituent X differs significantly in the two series, since $(PH_2F)_2$ has the highest binding energy among $(PH_2X)_2$ complexes, but $(H_2C{=}PF)_2$ is a relatively weakly bound complex.

Since the binding energies of $(H_2C{=}PX)_2$ complexes do not correlate with the P-P distances, there may be another factor in addition to the pnicogen bond which

Table 8.2 Binding energies (ΔE), charge-transfer energies [$P_{lp}(1)\rightarrow\sigma^*P\text{-}A(2)$, $kJ\cdot mol^{-1}$], P-P and P-H_b distances (R,Å), and coupling constants [$^{1p}J(P\text{-}P)$, Hz] for conformation A complexes. (Reprinted with permission from Ref. [34]. Copyright (2013) American Chemical Society)

$(H_2C{=}PX)_2$	ΔE	R(P-P)	R(P-H_b)[a]	^{1p}J(P-P)	$P_{lp}(1)\rightarrow\sigma^*$P-A(2)
X=CCH	– 13.9	3.510	3.230	140.3	5.4
Cl	– 13.3	3.373	3.305	330.1	8.7
CN	– 11.6	3.456	3.293	187.3	6.9
H	– 11.2	3.618	3.220	74.7	3.6
NC	– 11.1	3.422	3.362	264.1	9.4
CH_3	– 10.8	3.701	3.217	58.5	1.8
F	– 10.4	3.477	3.406	247.9	6.0
OH	– 9.5	3.583	3.370	134.3	4.1

[a] H_b is the H atom of the CH_2 group of one molecule which is closer to the P atom of the other

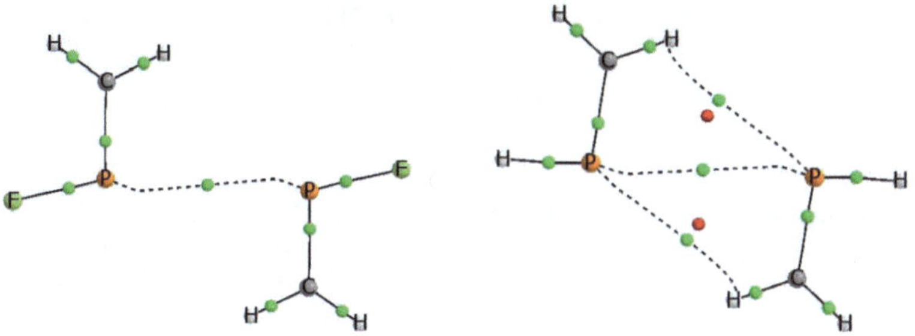

Fig. 8.6 Structures and molecular graphs of $(H_2C{=}PF)_2$ and $(H_2C{=}PH)_2$ illustrating the two types of conformation A dimers. $(H_2C{=}PF)_2$ has only a P...P pnicogen bond, while $(H_2C{=}PH)_2$ has a pnicogen bond and two P...H_b interactions. (Reprinted with permission from Ref. [34]. Copyright (2013) American Chemical Society)

stabilizes these complexes. It is possible to divide the $(H_2C{=}PX)_2$ complexes into two groups: those which are stabilized by the pnicogen bond, and those which are stabilized by the P...P pnicogen bond and in addition by two significant P...H_b interactions, with H_b the H atom of the CH_2 group of one molecule that is closer to the P atom of the other. $(H_2C{=}PF)_2$ in Fig. 8.6 is typical of the first group, and $(H_2C{=}PH)_2$ of the second. A plot of the binding energies of $(H_2C{=}PX)_2$ complexes versus the intermolecular P-P distance is given in Fig. 8.7. The complexes stabilized by pnicogen bonds are those with the more electronegative substituents Cl, CN, F, and OH. These complexes have P-H_b distances of 3.29 Å or greater, and P-P-C angles between 90 and 93°. The remaining complexes with the more electropositive substituents CCH, H, and CH_3 appear in Fig. 8.7 to have binding energies which are too high for their P-P distances. Complexes in this group have shorter P-H_b distances near 3.22 Å and P-P-C angles between 83 and 87°. Since there is a second stabilizing interaction in these three complexes in addition to the P...P pnicogen bond, it should

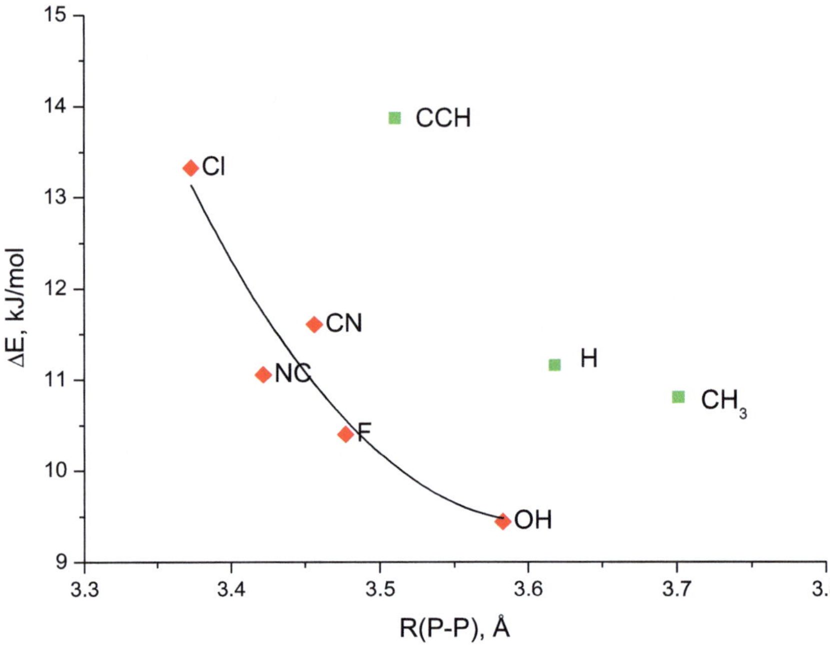

Fig. 8.7 Binding energies of conformation A complexes $(H_2C{=}PH)_2$ versus the P-P distance. ♦ Complexes with P...P pnicogen bonds ■ Complexes involving two P...H_b interactions in addition to a P...P bond. (Reprinted with permission from Ref. [34]. Copyright (2013) American Chemical Society)

not be surprising that the binding energies of $(H_2C{=}PX)_2$ complexes do not correlate with the P-P distances. However, a correlation does exist for the subset of complexes stabilized only by a pnicogen bond, as can be seen in Fig. 8.7.

Table 8.2 also presents $P(1)_{lp}{\rightarrow}\sigma^*$P-A(2) stabilizing charge-transfer energies for complexes $(H_2C{=}PX)_2$, but these energies are relatively small, ranging from 1.8 to 9.4 kJ·mol^{-1}, compared to 100 kJ·mol^{-1} for $(PH_2F)_2$ with F-P...P-F linear. Thus, they reflect the weaker pnicogen bonds in $(H_2C{=}PX)_2$. Charge transfer energies tend to be greater in complexes stabilized only by a pnicogen bond than in those with both a pnicogen bond and two P...H_b interactions.

Although it might have been expected that complexes $(H_2C{=}PX)_2$ with C_{2h} symmetry and C-P...P-C approaching linearity should also be pnicogen-bonded equilibrium structures, such is not the case. The only equilibrium structure of this type is $(H_2C{=}POH)_2$, but it is stabilized primarily by two distorted O-H...P hydrogen bonds. However, pnicogen-bonded complexes do exist which have the carbon of the CH_2 group of one molecule and atom A of the substituent X of the other approaching a nearly linear C-P...P-A arrangement. $(H_2C{=}POH)_2$ and $(H_2C{=}PH)_2$ have C_s symmetry, while the remaining equilibrium complexes with C-P...P-A approaching linearity have C_1 symmetry. These include $(H_2C{=}PCl)_2$, $(H_2C{=}PCH_3)_2$,

Table 8.3 P-P Distances [(R(P-P), Å], binding energies (ΔE, kJ·mol^{-1}), and ^{31}P-^{31}P spin-spin coupling constants [$^{1p}J(P\text{-}P)$, Hz] of equilibrium $(H_2C{=}PX)_2$ structures B and B′. (Reprinted with permission from Ref. [34]. Copyright (2013) American Chemical Society)

Complex	Type[a]	Sym.	R(P-P)	ΔE	$^{1p}J(P\text{-}P)$[b]
$(H_2C{=}POH)_2$	B	C_s	3.738	– 16.4	43.5
$(H_2C{=}PCl)_2$	B′	C_1	3.457	– 11.6	
$(H_2C{=}PCH_3)_2$	B′	C_1	3.696	– 11.4	
	B	C_s	3.712	– 11.4	66.1
$(H_2C{=}PCN)_2$	B′	C_1	3.519	– 11.0	
$(H_2C{=}PNC)_2$	B′	C_1	3.254	– 10.0	
$(H_2C{=}PH)_2$	B	C_s	3.657	– 8.9	99.3
$(H_2C{=}PF)_2$	B′	C_1	3.579	– 7.7	95.9

[a]The arrangement which approaches linearity is C-P(1)...P(2)-A
[b]$^{1p}J(P\text{-}P)$ calculations were not feasible for complexes with C_1 symmetry except $(H_2C{=}PF)_2$

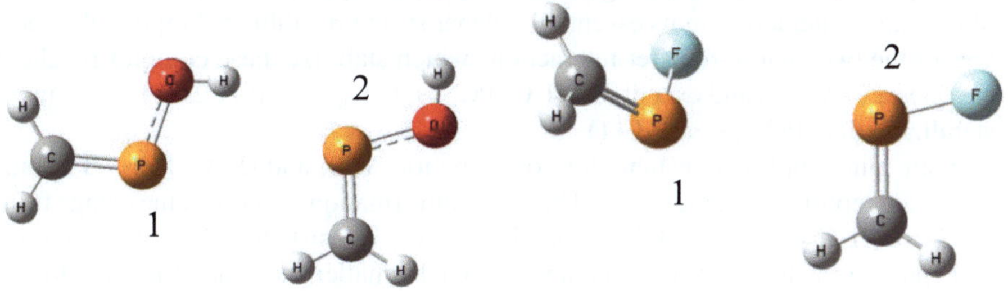

Fig. 8.8 Conformations B $(H_2C{=}POH)_2$ and B′ $(H_2C{=}PF)_2$. Labels identify P(1) and P(2). (Reprinted with permission from Ref. [34]. Copyright (2013) American Chemical Society)

$(H_2C{=}PCN)_2$, $(H_2C{=}PNC)_2$, and $(H_2C{=}PF)_2$, although $(H_2C{=}PCH_3)_2$ with C_1 symmetry is essentially identical structurally and energetically to the C_s dimer. Structural and energetic data for equilibrium complexes with C_s (B) and C_1 (B') symmetry are given in Table 8.3. Figure 8.8 illustrates the structures of $(H_2C{=}POH)_2$ and $(H_2C{=}PF)_2$.

Except for $(H_2C{=}PCH_3)_2$, B and B′ conformations have interestingly different structures. $(H_2C{=}POH)_2$ is stabilized by a P...P pnicogen bond as well as an O-H...P hydrogen bond, with a binding energy of –16.4 kJ·mol^{-1}. The hydrogen bond charge-transfer energy $P(2)_{lp}\rightarrow\sigma^*$O-H(1) is 15.8 kJ·mol^{-1}, significantly more stabilizing than the $P(1)_{lp}\rightarrow\sigma^*$P-O(2) energy of 2.3 kJ·mol^{-1}. $(H_2C{=}PH)_2$ conformation B is stabilized by a pnicogen bond and one P...H_b interaction, and has a binding energy of –8.9 kJ·mol^{-1}. The NBO analysis shows that both molecules are involved as lone pair donors and acceptors for the pnicogen bond with $P(1)_{lp}\rightarrow\sigma^*$P-H(2) and $P(2)_{lp}\rightarrow\sigma^*$P-C(1) energies of 3.6 kJ·mol^{-1}.

Table 8.4 NBO charge-transfer stabilization energies (kJ·mol^{-1}) for $(H_2C{=}PX)_2$ conformation B′ complexes. (Reprinted with permission from Ref. [34]. Copyright (2013) American Chemical Society)

Complex	$P(1)_{lp}\rightarrow\sigma^*P\text{-}A(2)$	$P(2)_{lp}\rightarrow\sigma^*P\text{-}C(1)$
$(H_2C{=}PCl)_2$	5.2	4.9
$(H_2C{=}PCH_3)_2^a$	2.1	1.5
$(H_2C{=}PCN)_2$	3.9	4.2
$(H_2C{=}PNC)_2$	5.6	4.0
$(H_2C{=}PF)_2$	4.0	3.3

[a] $(H_2C{=}PCH_3)_2$ also has a stabilizing P...H_b interaction

In the five conformation B′ complexes with C_1 symmetry, the monomers essentially retain their plane of symmetry, but the plane of one monomer is rotated about the P...P bond relative to the other, as illustrated in Fig. 8.8 for $(H_2C{=}PF)_2$. The binding energies of these complexes decrease in the order $(H_2C{=}PCl)_2 > (H_2C{=}PCH_3)_2 > (H_2C{=}PCN)_2 > (H_2C{=}PNC)_2 > (H_2C{=}PF)_2$. Except for $(H_2C{=}PCH_3)_2$, this is the same order found for the complexes with conformation A. The increased relative stability of $(H_2C{=}PCH_3)_2$ among the B′ complexes may be attributed to the retention of the P...H_b interaction in its essentially planar structure. Table 8.4 reports the energies of the two charge-transfer interactions which stabilize these complexes. These energies are similar, and except for $(H_2C{=}PCN)_2$, $P(1)_{lp}\rightarrow\sigma^*P\text{-}A(2)$ is slightly more stabilizing than $P(2)_{lp}\rightarrow\sigma^*P\text{-}C(1)$.

Spin-spin coupling constants for conformation A, B, and $(H_2C{=}PF)_2$ (B′) complexes are reported in Tables 8.2 and 8.3. For conformation A, these values range from 59 Hz for $(H_2C{=}PCH_3)_2$ to 330 Hz for $(H_2C{=}PCl)_2$. Consistent with binding energies and charge-transfer energies, this range is much smaller than found for the dimers $(PH_2X)_2$. Coupling constants for the conformation B complexes and $(H_2C{=}PF)_2$ lie between 43 and 100 Hz. Coupling constants for $(H_2C{=}PX)_2$ complexes correlate quadratically with the P-P distance, with a correlation coefficient R^2 of 0.933.

8.3.1.3 $H_2C{=}(X)P{:}PXH_2$

Just as the $(H_2C{=}PX)_2$ complexes may exist as conformation A and B dimers, the binary complexes $H_2C{=}(X)P{:}PXH_2$ may also exist as A conformers with C_s symmetry, and B with C_1 symmetry [35]. Intermolecular P-P and P_s-H_b distances, P_d-P_s-A and P_s-P_d-A angles, and binding energies for conformation A complexes are reported in Table 8.5, and a plot of ΔE versus the P-P distance is given in Fig. 8.9. The relatively low correlation coefficient arises from an additional secondary interaction in complexes with the more electropositive substituents, which involves the phosphorus of P_sH_2X and the H atom of the CH_2 group of $H_2C{=}PX$ which is closer to it (H_b). The structural differences between complexes with and without this interaction can be seen in Fig. 8.10 by comparing $H_2C{=}(F)P{:}PFH_2$ and $H_2C{=}(CH_3)P{:}P(CH_3)H_2$, and by comparing P_s-H_b distances and P_s-P_d-A angles in complexes with this interaction compared to complexes with F, Cl, OH, and NC as substituents.

Table 8.5 P-P and P_s-H_b distances (R, Å), binding energies (ΔE, kJ·mol^{-1}), and P_d-P_s-A and P_s-P_d-A angles (∠, deg) of conformation A complexes H_2C=(X)P:PXH_2 with C_s symmetry. (Reprinted with permission from Ref. [35]. Copyright (2013) American Chemical Society)

	R(P-P)	ΔE	R(P_s-H_b)	∠P_d-P_s-A	∠P_s-P_d-A
X=F	3.017	− 17.1	3.548	180	149
Cl	3.099	− 16.5	3.420	179	155
OH	3.204	− 13.9	3.547	177	155
NC	3.228	− 12.2	3.472	178	159
CCH	3.417	− 13.4	3.306	180	168
CN	3.404	− 10.2	3.358	176	168
CH_3	3.572	− 10.2	3.260	175	171
H	3.600	− 9.5	3.254	175	177

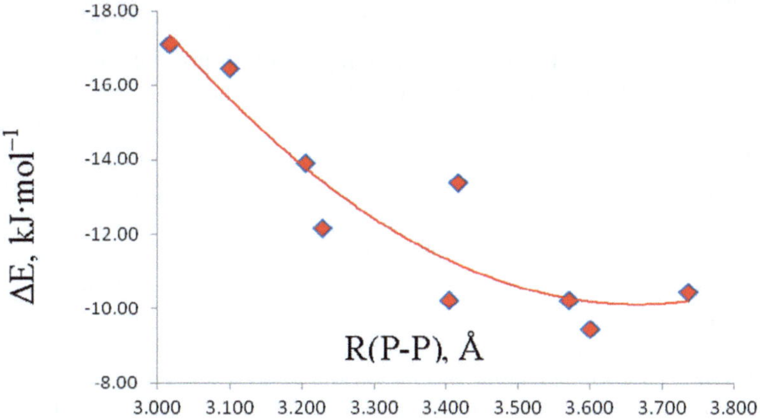

Fig. 8.9 The binding energy (ΔE) versus the P-P distance of H_2C=(X)P:PXH_2 complexes with C_s symmetry. The second-order trendline has a correlation coefficient R^2 of 0.859. (Reprinted with permission from Ref. [35]. Copyright (2013) American Chemical Society)

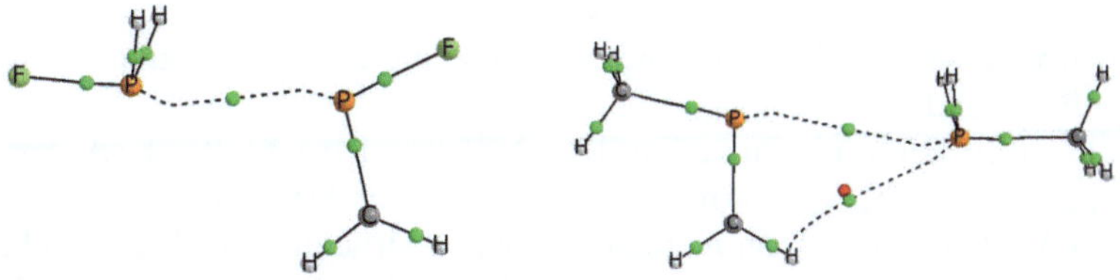

Fig. 8.10 Structures and molecular graphs of conformation A complexes H_2C=(F)P:PFH_2 and H_2C=(CH_3)P:P(CH_3)H_2, indicating a P_s...H_b interaction in the latter but not in the former. (Reprinted with permission from Ref. [35]. Copyright (2013) American Chemical Society)

Table 8.6 NBO charges (au) on $H_2C{=}PX$ and $P_dlp \rightarrow \sigma^*P_s$-A and $P_slp \rightarrow \sigma^*P_d$-A stabilizing charge-transfer energies ($kJ \cdot mol^{-1}$) for conformation A complexes $H_2C{=}(X)P{:}PXH_2$[a]. (Reprinted with permission from Ref. [35]. Copyright (2013) American Chemical Society)

$H_2C{=}(X)P{:}PXH_2$	Charge on $H_2C{=}PX$	$P_dlp \rightarrow \sigma^*P_s$-A	$P_slp \rightarrow \sigma^*P_d$-A
X=F	0.034	36.6	17.1
Cl	0.022	26.2	16.9
OH	0.019	21.7	9.8
NC	0.013	21.9	13.6
CCH	0.005	10.0	6.2
CN	0.005	10.8	7.6
CH_3	0.003	5.6	3.4
H	0.003	2.6	0.2

[a] P_d refers to the phosphorus of $H_2C{=}PX$ and P_s to that of PH_2X. See Fig. 8.11. A is the atom of X which is directly bonded to P

The charges on $H_2C{=}PX$ and the stabilizing charge-transfer energies for A complexes $H_2C{=}(X)P{:}PXH_2$ are reported in Table 8.6. In all complexes, charge transfer from the P_d lone pair to the σ^*P_s-A orbital is always more stabilizing than charge transfer from the lone pair of P_s to the σ^*P_d-A orbital, with P_s and P_d the singly- and doubly-bonded P atoms, respectively. This suggests that the preferred direction of charge transfer is determined by the nature of the electron-pair acceptor σ^* orbital. The dominant direction of charge-transfer is consistent with the positive charges on $H_2C{=}PX$ in these complexes, as indicated in Table 8.6. For complexes not having a $P_s...H_b$ interaction, $P_dlp \rightarrow \sigma^*P_s$-A charge-transfer energies range from 21.7 to 36.6 $kJ \cdot mol^{-1}$, while $P_slp \rightarrow \sigma^*P_d$-A charge-transfer energies range from 9.8 to 17.1 $kJ \cdot mol^{-1}$. For those complexes having this secondary interaction, the charge-transfer energies are much smaller, ranging from 2.6 to 10.8 $kJ \cdot mol^{-1}$ for $P_dlp \rightarrow \sigma^*P_s$-A, and 0.2 and 7.6 $kJ \cdot mol^{-1}$ for $P_slp \rightarrow \sigma^*P_d$-A.

The molecular graphs of these complexes show the presence of bond critical points and associated P...P bond paths. The Laplacians at the BCPs are always positive. However, complexes with the more electronegative substituents F, Cl, OH, and NC have the shortest P-P distances, the largest binding energies, and negative total energy densities at the BCPs, indicating that there is some degree of covalency in these P...P bonds.

$H_2C{=}(X)P{:}P(X)H_2$ conformation B complexes have $C{=}P_d...P_s$-A approaching linearity, with C the carbon of $H_2C{=}PX$. Similar to $(H_2C{=}POH)_2$, only $H_2C{=}(OH)P{:}P(OH)H_2$ has an equilibrium planar structure which is stabilized by a P...P pnicogen bond and an O-H...P_s hydrogen bond. The five remaining σ-σ complexes with X=F, Cl, NC, CCH, and CN are nonplanar with C_1 symmetry, and are stabilized only by pnicogen bonds. Figure 8.11 illustrates the structures of $H_2C{=}(OH)P{:}P(OH)H_2$ and $H_2C{=}(F)P{:}PFH_2$. Stable $H_2C{=}(CH_3)P{:}P(CH_3)H_2$ and $H_2C{=}(H)P{:}PH_3$ B complexes do not exist on the potential surfaces.

The intermolecular P-P distances and binding energies of conformation B complexes are reported in Table 8.7. Not surprisingly, $H_2C{=}(OH)P{:}P(OH)H_2$ which is

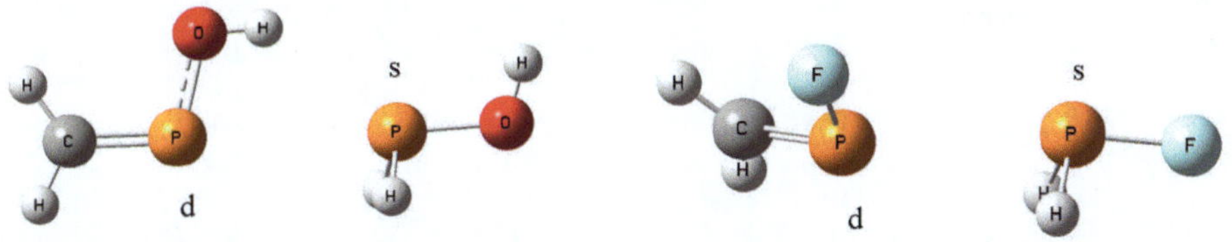

Fig. 8.11 The structures of H_2C=(OH)P:P(OH)H_2 and H_2C=(F)P:PFH_2 with conformation B. (Reprinted with permission from Ref. [35]. Copyright (2013) American Chemical Society)

Table 8.7 Binding energies (ΔE, kJ·mol^{-1}), intermolecular P-P distances (R, Å), NBO charges (au) on H_2C=PX, and $P_d$lp→σ*P_s-A and $P_s$lp→σ*P_d=C stabilizing charge-transfer energies (kJ·mol^{-1}) of conformation B complexes H_2C=(X)P:PXH_2. (Reprinted with permission from Ref. [35]. Copyright (2013) American Chemical Society)

	R	ΔE	H_2C=PX charge	$P_d$lp→σ*P_s-A	$P_s$lp→σ*P_d=C
X=F	3.089	− 13.8	0.036	31.7	8.9
Cl	3.172	− 14.1	0.020	21.0	9.6
OH[a]	3.355	− 21.9	−0.006	13.4	4.3[b]
NC	3.297	− 10.8	0.011	17.2	5.8
CCH	3.423	− 12.2	0.001	8.5	4.5
CN	3.463	− 8.8	0.002	8.9	4.9

[a] Has C_s symmetry. All other complexes have C_1 symmetry
[b] Has an additional stabilizing P_s(lp)→σ*O-H charge-transfer energy of 14.8 kJ·mol^{-1}

stabilized by an O-H...P_s hydrogen bond as well as a P...P pnicogen bond has the highest binding energy of −21.9 kJ·mol^{-1}. Figure 8.12 presents a plot of the binding energies of conformation B complexes versus the intermolecular P-P distances. The point for H_2C=(OH)P:P(OH)H_2 dramatically illustrates that its binding energy is much too high for its P-P distance, a result of the stabilizing effect of the hydrogen bond. Excluding the point for H_2C=(OH)P:P(OH)H_2, the trendline is linear, but the correlation coefficient is only 0.67. With the exception of H_2C=(OH)P:P(OH)H_2, conformation B complexes are less stable than the corresponding conformation A complexes.

For complexes $(H_2C{=}PX)_2$ conformations A and B, the P_d-P_s-A angles tend toward linearity, varying from 175 to 180° in conformation A, and from 168 to 177° in conformation B. However, the P_s-P_d-A and P_s-P_d=C angles exhibit greater deviations, as evident from Tables 8.5 and 8.8. The P_s-P_d-A angles vary between 149 and 177° in conformation A, while the P_s-P_d=C angles vary from 138 to 156° in conformation B. The notable exception is the P_s-P_d=C angle in H_2C=(OH)P:P(OH)H_2 which is 179°, thereby facilitating the formation of the O-H...P_s hydrogen bond. The deviation from linearity particularly of the P_s-P_d=C angles in conformation B may contribute to the lack of correlation between binding energies and P-P distances.

The NBO charges on H_2C=PX and the stabilizing charge-transfer energies for conformation B complexes are reported in Table 8.7. As for conformation A

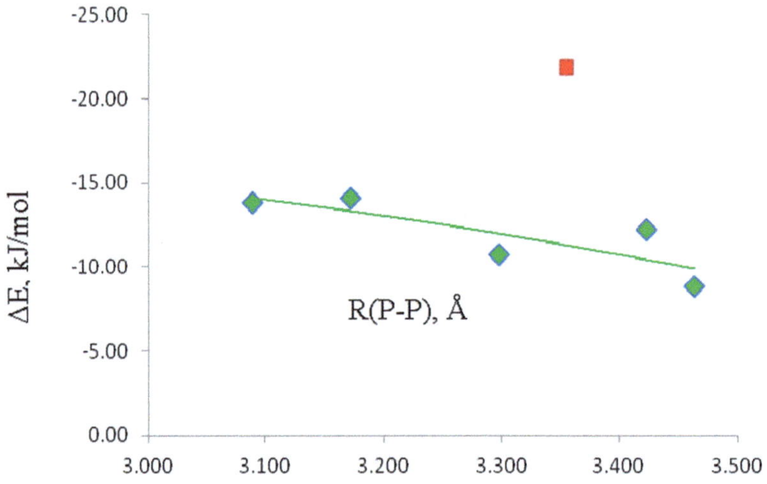

Fig. 8.12 The binding energy (ΔE) versus the P-P distance for conformation B complexes. ■ $H_2C{=}(OH)P{:}P(OH)H_2$. (Reprinted with permission from Ref. [35]. Copyright (2013) American Chemical Society)

Table 8.8 P_d-P_s-A and P_s-P_d=C angles (<, deg) and ^{31}P-^{31}P spin-spin coupling constants [^{1p}J(P-P), Hz] for conformation B complexes $H_2C{=}(X)P{:}PXH_2$. (Reprinted with permission from Ref. [35]. Copyright (2013) American Chemical Society)

$H_2C{=}(X)P{:}PXH_2$	<P_d-P_s-A	<P_s-P_d=C	^{1p}J(P-P)
X=F	173	138	364.8
Cl	172	129	319.5
OH	177	179	174.6
NC	170	150	231.8
CCH	168	144	—[a]
CN	172	156	181.0

[a] Not computed because of computational expense

complexes, charge transfer from the P_d lone pair to the σ^*P_s-A orbital is always more stabilizing than charge transfer from the lone pair of P_s to the σ^*P_d=C orbital. This is consistent with the positive charges on $H_2C{=}PX$ in these complexes, except for $H_2C{=}(OH)P{:}P(OH)H_2$, since the dominant charge-transfer interaction is across the hydrogen bond, $P_s(lp)\rightarrow\sigma^*$O-H.

The P...P BCPs of conformation B complexes have smaller electron densities than those of the corresponding conformation A complexes. This is consistent with the longer P-P distances in conformation B, and with their smaller binding energies, except for $H_2C{=}(OH)P{:}P(OH)H_2$. Only the complexes with X=F and Cl have negative values of H_{BCP}, indicating that the P...P bonds in these two complexes have some degree of covalency. The degree of covalency is less in conformation B than in the corresponding A complex.

Table 8.8 also reports the spin-spin coupling constants ^{1p}J(P-P) for conformation B complexes, and Fig. 8.13 shows the expected correlation between ^{1p}J(P-P) and the P-P distance. It is interesting to note that although $H_2C{=}(OH)P{:}P(OH)H_2$ is structurally

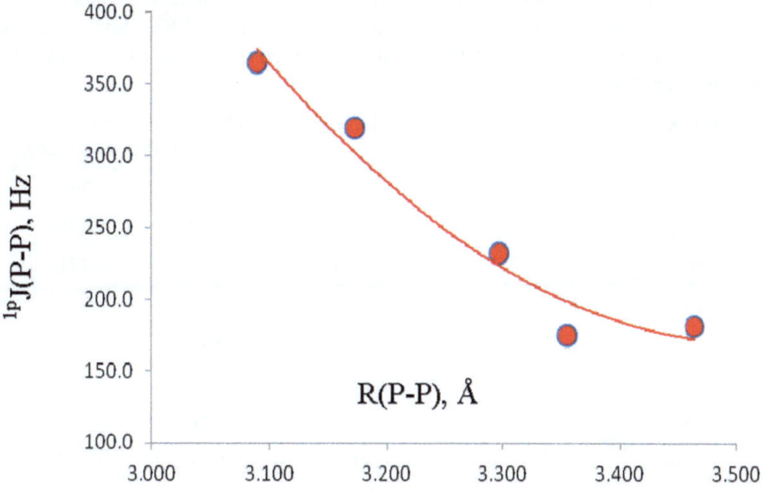

Fig. 8.13 Coupling constants ^{1p}J(P-P) versus the P-P distance for H_2C=(X)P:P(X)H_2 conformation B complexes. (Reprinted with permission from Ref. [35]. Copyright (2013) American Chemical Society)

and energetically quite distinct from the remaining conformation B complexes, these differences do not influence the relationship between ^{1p}J(P-P) and the P-P distance. The second-order trendline shown in Fig. 8.13 has a correlation coefficient R^2 of 0.961.

It is possible to compare the properties of the mixed binary complexes H_2C=(X)P:PXH_2 to those of the dimers $(PH_2X)_2$ and $(H_2C=PX)_2$, all with conformation A. The binding energies of these complexes are reported in Table 8.9. These data indicate that for complexes in which X is one of the more electronegative substituents, the binding energies decrease from left to right, in going from $(PH_2X)_2$ to H_2C=(X)P:PXH_2 to $(H_2C=PX)_2$. When X is an electropositive substituent, binding energies increase from left to right. The ratios of the binding energies $\Delta E(PH_2X)_2/\Delta E(H_2C=PX)_2$ and $\Delta E[H_2C=(X)P:PXH_2]/\Delta E(H_2C=PX)_2$, are also reported in Table 8.9, and plotted in Fig. 8.14. The trendline is a second-order polynomial with a correlation coefficient R^2 of 0.988. This plot indicates that there is a systematic relationship among the relative stabilities of the complexes $(PH_2X)_2$, H_2C=(X)P:PXH_2, and $(H_2C=PX)_2$ as a function of the substituent X, despite the fact that some complexes H_2C=(X)P:PXH_2 and $(H_2C=PX)_2$ have stabilizing interactions in addition to the P...P bond.

The second property of interest in conformation A complexes $(PH_2X)_2$, H_2C=(X)P:PXH_2, and $(H_2C=PX)_2$ with A-P...P-A approaching linearity is the one-bond spin-spin coupling constant ^{1p}J(P-P) as a function of the P-P distance. Table 8.10 reports the P-P distances for these complexes and ^{1p}J(P-P) values, and Fig. 8.15 provides a plot of these variables. Although for a given X, the shorter P-P distance and larger ^{1p}J(P-P) are found in complexes $(PH_2X)_2$ compared to $(H_2C=PX)_2$, there is some overlap, as evident from Fig. 8.15. ^{1p}J(P-P) values cover a large range, from 59 Hz for $(H_2C=PCH_3)_2$ to 1000 Hz for $(PH_2F)_2$ and 1120 Hz for $(PH_2Cl)_2$. The points

Table 8.9 Binding energies (ΔE, $kJ{\cdot}mol^{-1}$) of complexes and the ratios $\Delta E(PH_2X)_2/\Delta E(H_2C{=}PX)_2$ and $\Delta E[H_2C{=}(X)P{:}PXH_2]/\Delta E(H_2C{=}PX)_2$ for complexes with A-P...P-A approaching linearity. (Reprinted with permission from Ref. [35]. Copyright (2013) American Chemical Society)

	$(PH_2X)_2$[a]	$H_2C{=}(X)P{:}PXH_2$[b]	$(H_2C{=}PX)_2$[a]	$(PH_2X)_2/(H_2C{=}PX)_2$	$H_2C{=}(X)P{:}PXH_2/(H_2C{=}PX)_2$
X=F	− 34.0	− 17.1	− 10.4	3.263	1.644
Cl	− 22.1	− 16.5	− 13.3	1.657	1.235
OH	− 20.6	− 13.9	− 9.5	2.175	1.471
NC	− 13.8	− 12.2	− 11.1	1.244	1.100
CCH	− 12.2	− 13.4	− 13.9	0.882	0.965
CH_3	− 8.9	− 10.2	− 10.8	0.821	0.945
CN	− 8.4	− 10.2	− 11.6	0.721	0.880
H	− 7.1	− 9.5	− 11.2	0.634	0.847

[a] C_{2h} symmetry
[b] C_s symmetry

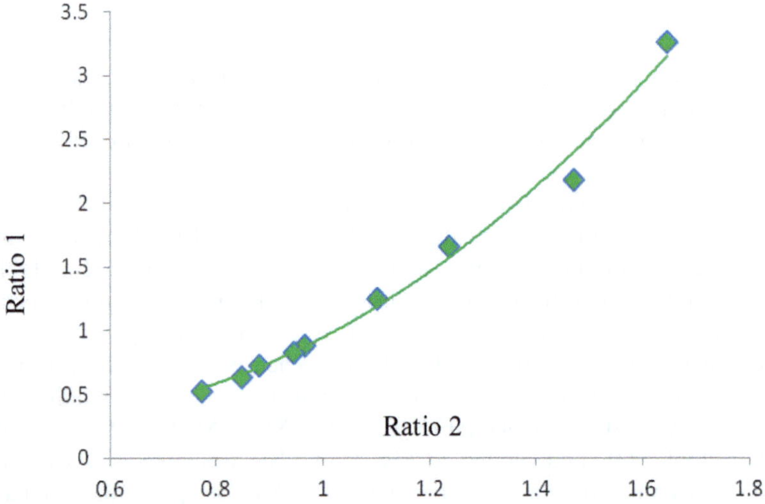

Fig. 8.14 $\Delta E(PH_2X)_2/\Delta E(H_2C{=}PX)_2$ (Ratio 1) versus $\Delta E[H_2C{=}(X)P{:}PH_2X]/\Delta E(H_2C{=}PX)_2$ (Ratio 2) for corresponding complexes with A-P...P-A approaching linearity. (Reprinted with permission from Ref. [35]. Copyright (2013) American Chemical Society)

for the latter two complexes deviate most from the trendline. Nevertheless, Fig. 8.15 indicates that complexes in these three series with similar P-P distances have similar values of ^{1p}J(P-P).

Table 8.10 P-P distances (R, Å) and ^{31}P-^{31}P spin-spin coupling constants [^{1p}J(P-P), Hz] for $(PH_2X)_2$, H_2C=(X)P:PXH_2, and $(H_2C=PX)_2$ conformation A complexes with A-P...P-A approaching linearity. (Reprinted with permission from Ref. [35]. Copyright (2013) American Chemical Society)

	$(PH_2X)_2^a$		$H_2C=(X)P:PXH_2^b$		$(H_2C=PX)_2^a$	
X=	R	^{1p}J(P-P)	R	^{1p}J(P-P)	R	^{1p}J(P-P)
F	2.471	1000	3.017	637	3.477	248
Cl	2.771	1120	3.099	617	3.373	330
OH	2.851	644	3.204	361	3.583	134
NC	3.040	640	3.228	428	3.422	264
CCH	3.353	282	3.417	210	3.510	140
CH_3	3.481	161	3.572	106	3.701	59
CN	3.375	300	3.404	244	3.456	187
H	3.589	131	3.600	101	3.618	75

[a] C_{2h} symmetry
[b] C_s symmetry

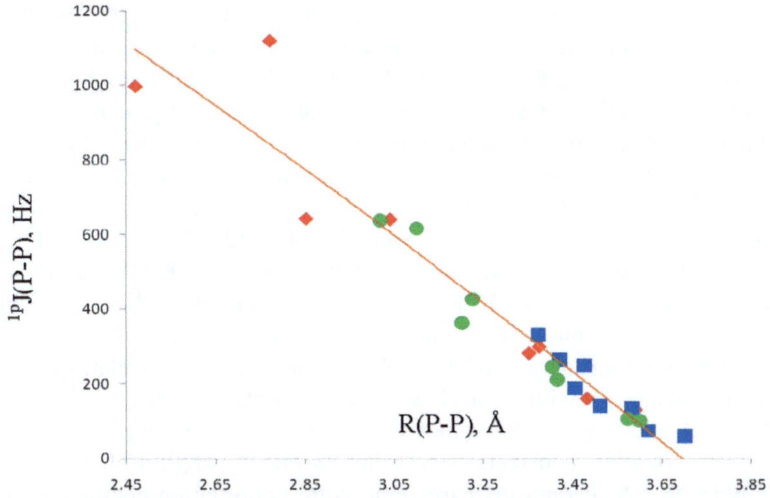

Fig. 8.15 ^{1p}J(P-P) versus the P-P distance for conformation A complexes ♦ $(PH_2X)_2$, ● H_2C=(X)P:PXH_2, and ■ $(H_2C=PX)_2$. The correlation coefficient R^2 is 0.927. (Reprinted with permission from Ref. [35]. Copyright (2013) American Chemical Society)

8.3.1.4 H_2XP:PCX

PH_2X and PCX form pnicogen-bonded complexes only when the substituents are CCH, NC, CN, CH_3, and H [36]. The binding energies, P-P distances, P_t-P_s-A and P_s-P_t-C angles, and charge-transfer energies of the equilibrium H_2XP:PCX complexes are reported in Table 8.11. The binding energies range from -3.1 kJ·mol^{-1}

Table 8.11 Binding energies (ΔE, kJ·mol^{-1}), P-P distances (R, Å), P_t-P_s-A and P_s-P_t-C angles ($\angle$, deg), charge-transfer energies (kJ·mol^{-1}), and spin-spin coupling constants [^{1p}J(P-P), Hz] for H_2XP:PCX conformation A complexes[a]. (Reprinted with permission from Ref. [36]. Copyright (2014) Springer Science and Business Media)

H_2XP:PCX	ΔE	R(P-P)	$\angle P_t$-P_s-A	$\angle P_s$-P_t-C	$P_t(lp)\rightarrow\sigma^*P_s$-A	$P_s(lp)\rightarrow\sigma^*P_t$-C	^{1p}J(P-P)
X=CCH	−7.4	3.594	163	179	6.4	1.4	157.6
NC	−4.2	3.521	166	179	9.4	2.5	209.4
CN	−3.1	3.649	167	176	5.6	1.5	150.4
CH_3	−5.7	3.705	160	176	3.4	1.5	100.8
H	−4.7	3.772	158	175	3.9	1.4	88.3

[a]Complexes with X=F and Cl do not form complexes with σ-σ pnicogen bonds, and $H_2(OH)P$:PCOH is a hydrogen-bonded complex

for $H_2(CN)P$:PCCN to –7.4 kJ·mol^{-1} for $H_2(CCH)P$:PCCCH. These binding energies are smaller than the binding energies of the corresponding complexes $(PH_2X)_2$ and H_2C=(X)P:PXH_2. For a given X, the intermolecular P-P distances in these three sets of complexes decrease in the order H_2XP:PCX > H_2C=(X)P:PXH_2 > $(PH_2X)_2$. The P_s-P_t-C alignment in conformation A complexes closely approaches linearity, while the P_t-P_s-A alignment deviates from linearity to some extent. P_t and P_s are the triply- and singly-bonded P atoms of PCX and PH_2X, respectively.

H_2XP:PCX complexes with pnicogen bonds are stabilized by charge transfer. The more favorable charge-transfer interaction involves donation of the P_t lone pair to the σ* P_s-A orbital. Charge-transfer energies range from 3.4 kJ·mol^{-1} when X=CH_3 to 9.4 kJ·mol^{-1} when X=NC. In contrast, charge-transfer energies from the P_s lone pair to the σ* P=C orbital are 2.5 kJ·mol^{-1} when X=NC, and 1.4 or 1.5 kJ·mol^{-1} for the remaining complexes. The preference for charge transfer to PH_2X further supports the suggestion that the factor which determines the direction of charge transfer is the nature of the σ* P-A orbital. The $P_t(lp)\rightarrow\sigma^*$P-A charge-transfer energies do not correlate with the binding energies of H_2XP:PCX complexes, but correlate quadratically with the P-P distances, with a correlation coefficient R^2 of 0.963.

Table 8.11 also reports the values of ^{1p}J(P-P) for these conformation A complexes. The coupling constants vary from 88 Hz for H_3P:PCH to 209 Hz for $H_2(NC)P$:PCNC. A second-order trendline with a correlation coefficient of 0.961 illustrates the dependence of ^{1p}J(P-P) on the P-P distance.

8.3.1.5 H_2XP:NXH_2, H_2FP:NXH_2, and H_2XP:NFH_2

All of the complexes discussed thus far have P...P pnicogen bonds, but a number of other atoms with lone pairs may act as electron donors to group 15 elements to form pnicogen bonds. Among these is nitrogen, which is also a pnicogen atom that forms a series of complexes H_2XP:NXH_2, for X=F, Cl, OH, CN (bonded at both C and N), CCH, CH_3, and H [23]. The majority of these complexes have C_s symmetry

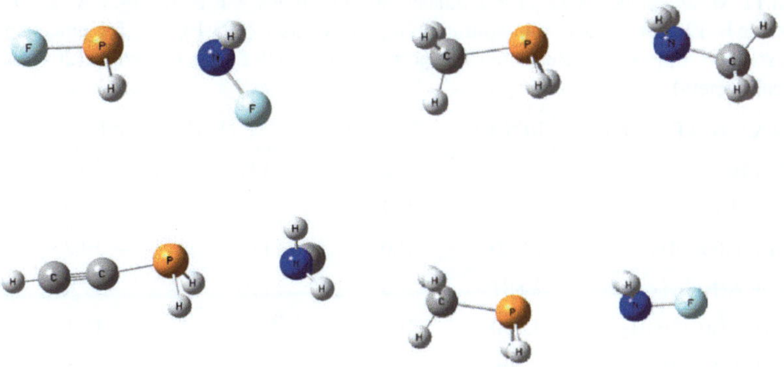

Fig. 8.16 Complexes $H_2FP{:}NFH_2$, $H_2(CH_3)P{:}N(CH_3)H_2$, $H_2(CCH)P{:}N(CCH)H_2$, and $H_2(CH_3)P{:}NFH_2$ with P...N pnicogen bonds. (Reprinted with permission from Ref. [23]. Copyright (2011) American Chemical Society)

with the H atoms of PXH_2 and NXH_2 *trans* with respect to the P-N axis. The complexes with unsaturated substituents $H_2(NC)P{:}N(NC)H_2$, $H_2(CCH)P{:}N(CCH)H_2$, and $H_2(CN)P{:}N(CN)H_2$ have C_1 symmetry and a *gauche* arrangement of these atoms. Figure 8.16 illustrates the structures of selected complexes, and Table 8.12 presents the intermolecular N-P distances, binding energies, and A-P-N and P-N-A angles. Binding energies range from -8 to -27 kJ·mol^{-1}. These binding energies are greater than those of the corresponding complexes $(PH_2X)_2$, except for $(PH_2F)_2$ which is 7 kJ·mol^{-1} more stable than $H_2FP{:}NFH_2$, $[PH_2(OH)]_2$ which is 1 kJ·mol^{-1} more stable than $H_2(OH)P{:}N(OH)H_2$, and $[PH_2(CH_3)]_2$ and $H_2(CH_3)P{:}N(CH_3)H_2$ which have similar stabilities.

The intermolecular N-P distances for the entire set of complexes in Table 8.12 range from 2.448 Å in $H_2FP{:}N(CH_3)H_2$ to 3.292 Å in $H_3P{:}NH_3$, and correlate well with binding energies, as illustrated in Fig. 8.17. The A-P-N angles approach linearity, varying between 160 and 172°. This orientation provides for the overlap of the nitrogen lone pair orbital with the σ* P-A orbital. The P-N-A angles are significantly less than the A-P-N angles, a reflection of the tetrahedral arrangement of the N-A bond and the lone pair at N. However, $H_2(CH_3)P{:}NFH_2$ and $H_3P{:}NFH_2$ have P-N-A angles of 180 and 177°, respectively, and in these two complexes the dominant charge-transfer interaction is from the lone pair on P to the σ* N-A orbital, as can be seen in Table 8.13. $H_2(CH_3)P{:}N(CH_3)H_2$ and $H_3P{:}NH_3$ have P-N-A angles of 163 and 155°, respectively. These four complexes have relatively weak pnicogen bonds. In contrast, complexes $H_2FP{:}NXH_2$ have the largest binding energies of -29 to -39 kJ·mol^{-1}.

Bond critical points and corresponding bond paths link the phosphorous and nitrogen atoms in each complex. The Laplacians at the BCPs are positive, but the energy densities for the complexes of PH_2F, PH_2Cl and $PH_2(OH)$ are negative, indicating that the N...P bonds in complexes involving these molecules have some degree of covalent character [111].

Table 8.12 MP2/aug'-cc-pVTZ N-P distances (R, Å), angles A-P-N and P-N-A (∠, °), binding energies (ΔE, kJ·mol^{-1}), and spin-spin coupling constants [^{1p}J(N-P), Hz] for pnicogen-bonded complexes with N...P bonds. (Reprinted with permission from Ref. [23]. Copyright (2011) American Chemical Society)

Complexes $H_2XP:NXH_2$	R(N-P)	∠A-P-N[a]	∠P-N-A[a]	ΔE	^{1p}J(N-P)
$H_2FP:NFH_2$	2.524	172	131	– 26.7	– 113.6
$H_2ClP:NClH_2$	2.669	167	121	– 24.3	– 90.8
$H_2(OH)P:N(OH)H_2$	2.750	160	112	– 19.2	– 52.0
$H_2(NC)P:N(NC)H_2$	2.887	166	110	– 17.4	– 41.4
$H_2(CCH)P:N(CCH)H_2$	3.140	165	95	– 15.1	– 19.4
$H_2(CN)P:N(CN)H_2$	3.208	166	90	– 12.8	– 16.8
$H_2(CH_3)P:N(CH_3)H_2$	3.257	172	163	– 8.6	– 19.4
$H_3P:NH_3$	3.292	165	155	– 7.8	– 17.5
Complexes $H_2XP:NFH_2$					
$H_2ClP:NFH_2$	2.695	171	130	– 21.0	– 105.0
$H_2(CH_3)P:NFH_2$	3.120	168	180	– 12.8	– 67.9
$H_3P:NFH_2$	3.225	165	177	– 10.7	– 33.1
Complexes $H_2FP:NXH_2$					
$H_2FP:N(CH_3)H_2$	2.448	166	118	– 39.4	– 41.3
$H_2FP:NH_3$	2.609	168	122	– 29.9	– 65.6
$H_2FP:NClH_2$	2.543	169	126	– 29.1	– 95.1

[a]A is the atom of X or X' that is directly bonded to P or N, respectively

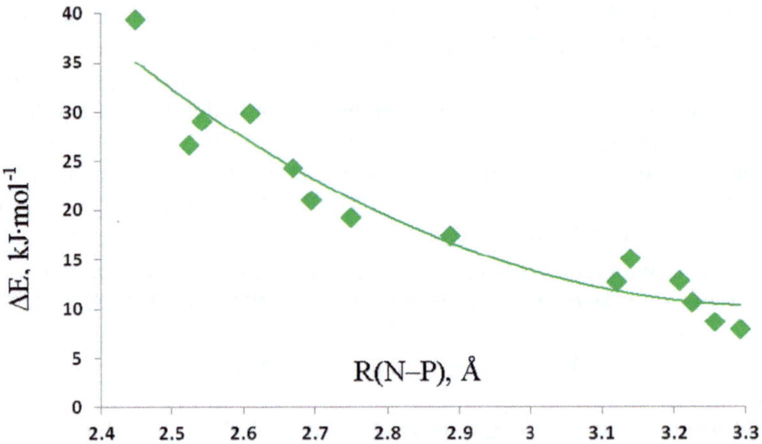

Fig. 8.17 Binding energies of complexes $H_2XP:NXH_2$, $H_2FP:NXH_2$, and $H_2XP:NFH_2$ vs. the N–P distance. The exponential relationship has a correlation coefficient of 0.925. (Reprinted with permission from Ref. [23]. Copyright (2011) American Chemical Society)

Table 8.13 Stabilizing charge-transfer energies ($kJ{\cdot}mol^{-1}$) for complexes $H_2XP{:}NXH_2$, $H_2XP{:}NFH_2$, and $H_2FP{:}NXH_2$. (Reprinted with permission from Ref. [23]. Copyright (2011) American Chemical Society)

Complexes $H_2XP{:}NXH_2$	N(lp)→σ*P-A[a]	P(lp)→σ*N-A[a]
$H_2FP{:}NFH_2$	53.9	12.1
$H_2ClP{:}NClH_2$	42.6	4.0
$H_2(OH)P{:}N(OH)H_2$	33.7	1.6
$H_2(NC)P{:}N(NC)H_2$	23.9	0.7
$H_2(CCH)P{:}N(CCH)H_2$	7.9	0.3
$H_2(CN)P{:}N(CN)H_2$	6.8	0.8
$H_2(CH_3)P{:}N(CH_3)H_2$	4.9	1.5
$H_3P{:}NH_3$	6.5	1.7
Complexes $H_2XP{:}NFH_2$		
$H_2ClP{:}NFH_2$	42.7	5.7
$H_2(CH_3)P{:}NFH_2$	4.7	6.4
$H_3P{:}NFH_2$	0.9	1.1
Complexes $H_2FP{:}NXH_2$		
$H_2FP{:}N(CH_3)H_2$	[b]	[b]
$H_2FP{:}NH_3$	11.6	0.9
$H_2FP{:}NClH_2$	47.4	8.8

[a]A is the atom of X or X′ bonded directly to P or N, respectively
[b]The NBO method considers this complex to be a single molecule

Table 8.12 also reports coupling constants $^{1p}J(N\text{-}P)$ for these complexes. $^{1p}J(N\text{-}P)$ varies from − 17 Hz for $H_2(CN)P{:}N(CN)H_2$ to −114 Hz for $H_2FP{:}NFH_2$. Figure 8.18 presents a plot of $^{1p}J(N\text{-}P)$ versus the N-P distance for complexes with the same

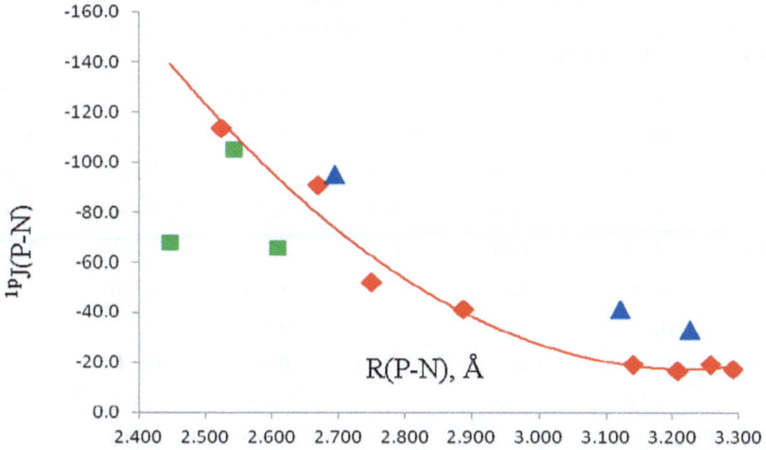

Fig. 8.18 $^{1p}J(P\text{-}N)$ vs. the P-N distance for the pnicogen-bonded complexes ◆ $H_2XP{:}NXH_2$, ▲ $H_2XP{:}NFH_2$, and ■ $H_2FP{:}NXH_2$. (Reprinted with permission from Ref. [23]. Copyright (2011) American Chemical Society)

substituent bonded to both N and P. For these, there is a quadratic increase in ^{1p}J(N-P) as the intermolecular N-P distance decreases, with a correlation coefficient R^2 of 0.973. The points for $H_2FP:NXH_2$ lie below the trendline for the $H_2XP:NXH_2$ complexes, while points for $H_2XP:NFH_2$ lie above the trendline in Fig. 8.18.

8.3.1.6 Complexes X=PH$_3$:NY and X=PH$_3$:PY

The electrostatic potentials of molecules O=PH$_3$, S=PH$_3$, HN=PH$_3$, and H$_2$C=PH$_3$ exhibit a region of positive charge on the side opposite the X=P bond [37]. These regions are analogous to those generated by electronegative atoms, and are σ-holes which are suitable for interaction with a Lewis base. The maximum values of the MEPs decrease with respect to X in the order O > S > NH > CH$_2$. The MEP minima for the nitrogen bases NH_3, NCH, N_2, and the phosphorus base PH_3 indicate a region of negative charge along the principal symmetry axis of each molecule at N for the nitrogen bases and at P of PH_3. However, the MEP of PCH at P along its symmetry axis is positive. This is consistent with the observation that PCH is a poor base, although it does form linear hydrogen-bonded complexes with itself, FH, and ClH.

The P-N distances and binding energies of complexes X=PH$_3$:NY and X=PH$_3$:PY are reported in Table 8.14. For a given X=PH$_3$, the binding energies of the stronger bases decrease in the order NH_3 > NCH > PH_3, while the weaker bases N_2 and PCH have binding energies between − 6 and − 7 kJ·mol^{-1} with all X=PH$_3$. Moreover, for each of the nitrogen bases and PH_3, the binding energies decrease with respect to the acid in the order O=PH$_3$ > S=PH$_3$ > HN=PH$_3$ > H$_2$C=PH$_3$, but there is little variation among binding energies when PCH is the base.

These observations are illustrated in Figs. 8.19 and 8.20. Figure 8.19 provides a plot of the binding energies as a function of the minimum values of the MEPs of

Table 8.14 MP2/aug'-cc-pVTZ binding energies (ΔE, kJ·mol^{-1}) and intermolecular distances (R, Å) of complexes X=PH$_3$:NY and X=PH$_3$:PY. (Reprinted with permission from Ref. [37]. Copyright (2014) American Chemical Society)

	ΔE			R(P-N)		
X=PH$_3$	NH_3	NCH	N_2	NH_3	NCH	N_2
O=PH$_3$	− 20.4	− 17.8	− 7.2	3.145	3.120	3.287
S=PH$_3$	− 18.9	− 17.3	− 6.9	3.241	3.190	3.362
HN=PH$_3$	− 16.2	− 14.3	− 6.3	3.249	3.201	3.357
H$_2$C=PH$_3$	− 13.9	− 12.6	− 5.7	3.331	3.266	3.416
	ΔE			R(P-N)		
X=PH$_3$	PH_3	PCH		PH_3	PCH	
O=PH$_3$	− 12.6	− 6.6		3.629	3.684	
S=PH$_3$	− 11.7	− 6.8		3.732	3.729	
HN=PH$_3$	− 10.5	− 6.4		3.713	3.690	
H$_2$C=PH$_3$	− 9.3	− 6.2		3.767	3.752	

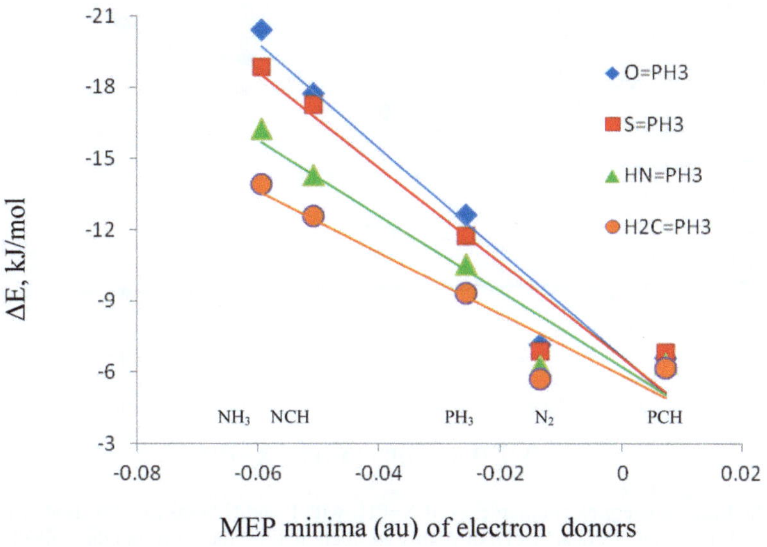

Fig. 8.19 Binding energies of complexes of $X{=}PH_3$ with P and N bases as a function of the MEP minima of the bases. Vertical stacks of points are identified by the corresponding bases. (Reprinted with permission from Ref. [37]. Copyright (2014) American Chemical Society)

the N and P bases, which are identified along the MEP minima axis. The stronger bases NH_3, NCH, and PH_3 have binding energies which differentiate among the acids $X{=}PH_3$, while the binding energies of complexes with the weaker bases N_2 and PCH show little dependence on the nature of the acid. An alternate view of the binding energies can be seen in Fig. 8.20 in a plot of the binding energies against the maximum values of the MEPs for the $X{=}PH_3$ acids, which are identified along the MEP maxima axis. The trendlines indicate that the stronger bases NH_3, NCH, and PH_3 have binding energies which depend on the nature of the acid $X{=}PH_3$. This is also true but to a much lesser extent for the weaker base N_2. However, the trendline for PCH is flat, showing that the binding energy is essentially independent of the acid.

Table 8.15 reports the stabilizing charge-transfer energies for complexes $X{=}PH_3{:}NY$ and $X{=}PH_3{:}PY$. Charge transfer occurs from the lone pair of P or N to the antibonding σ* P=A orbital of $X{=}PH_3$, where A is the atom of X bonded directly to P. The charge-transfer energies for these complexes are within $\pm 3.2\ kJ{\cdot}mol^{-1}$ of the charge-transfer energy of $6.5\ kJ{\cdot}mol^{-1}$ for the N lone pair to the σ* P-H orbital of $H_3P{:}NH_3$. Figure 8.21a illustrates the lone-pair orbital of NH_3 interacting with the σ* P=O orbital of $O{=}PH_3$. For comparison, the N lone pair orbital and the σ* P-H orbital for $H_3P{:}NH_3$ are shown in Fig. 8.21b. In $X{=}PH_3$ complexes with the nitrogen bases, charge transfer energies are largest when $O{=}PH_3$ is the acid, and smallest when $H_2C{=}PH_3$ is the acid. The energies for $S{=}PH_3$ and $HN{=}PH_3$ complexes are intermediate and similar. With all $X{=}PH_3$, NH_3 has the largest charge-transfer energy, and PCH and N_2 have similar, relatively small charge-transfer energies. Charge-transfer energies tend to increase as the binding energies of these complexes increase.

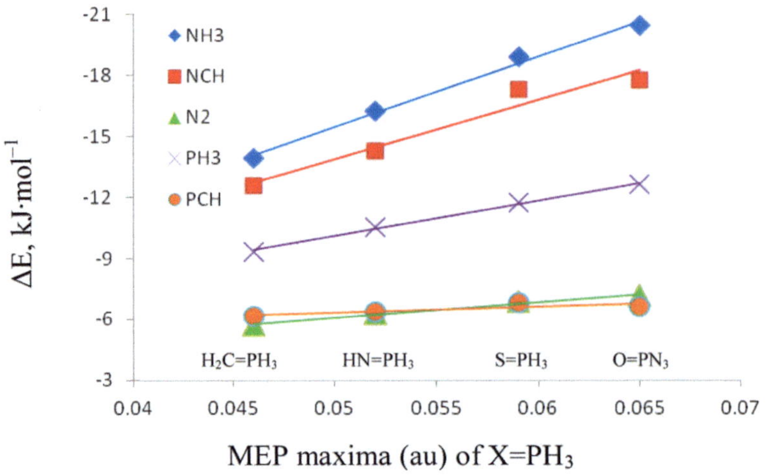

Fig. 8.20 Binding energies of complexes of X=PH_3 with P and N bases as a function of the MEP maxima of the acids. Vertical stacks of points are identified by the corresponding X=PH_3 acids. (Reprinted with permission from Ref. [37]. Copyright (2014) American Chemical Society)

Table 8.15 Stabilizing charge-transfer energies (kJ·mol^{-1}) from the lone pair of the P and N bases to the σ* P=A orbital of X=PH_3[a]. (Reprinted with permission from Ref. [37]. Copyright (2014) American Chemical Society)

Acid/Base	NH_3	NCH	N_2	PH_3	PCH
O=PH_3	9.7	6.2	3.7	6.7	3.4
S=PH_3	7.7	5.1	2.8	4.5	2.6
HN=PH_3	7.8	5.1	3.2	5.4	3.5
H_2C=PH_3	6.2	4.2	2.6	4.1	2.7

[a]A is the atom of X directly bonded to P

In complexes of X=PH_3 with the N and P bases, bond paths connect the N or P of the base to the P and H atoms of the acid, or just to the H atoms or the P-H bonds. This does not mean that the intermolecular bonds are not pnicogen bonds, since the charge-transfer energies from the N or P lone pair to the σ* P-H orbitals of X=PH_3 are negligibly small. The orbital representation of charge transfer for the complex HN=PH_3:NH_3 which has a bond path involving H atoms is illustrated in Fig. 8.21c. The bond paths to H atoms may simply be a consequence of the diffuseness of the σ* P=A orbital of X=PH_3, and should not be interpreted as an indication of the absence of a pnicogen bond.

The total coupling constants ^{1p}J(P-N) and ^{1p}J(P-P) and the corresponding P-N and P-P distances are reported in Table 8.16. Figure 8.22 presents a plot of ^{1p}J(P-N) versus the P-N distance for these complexes, which has a second-order trendline with a correlation coefficient R^2 of 0.868. In this plot there are four sets of two data points with similar P-N distances and coupling constants. How these arise can be seen in Fig. 8.23, which provides a plot of ^{1p}J(P-N) versus the P-N distance as a function

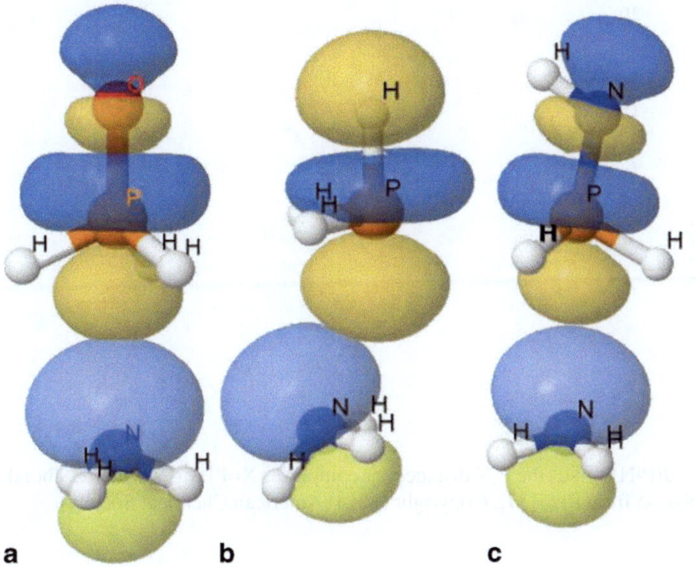

Fig. 8.21 Depiction of the orbitals involved in charge-transfer interactions. NH_3 lone pair with the **a** σ*P=O orbital of $O{=}PH_3$; **b** σ*P-H orbital of PH_3; **c** σ*P=N orbital of $HN{=}PH_3$. (Reprinted with permission from Ref. [37]. Copyright (2014) American Chemical Society)

Table 8.16 Intermolecular distances (R, Å) and ^{1p}J(P-N) and ^{1p}J(P-P) spin-spin coupling constants (Hz) for complexes of $X{=}PH_3$ with nitrogen and phosphorus bases. (Reprinted with permission from Ref. [37]. Copyright (2014) American Chemical Society)

	R(P-N)	^{1p}J(P-N)		R(P-P)	^{1p}J(P-P)
$O{=}PH_3$					
Base=NH_3	3.145	− 19.9	PH_3	3.269	150.8
NCH	3.120	− 17.7	PCH	3.684	92.3
N_2	3.287	− 8.7			
$S{=}PH_3$					
Base=NH_3	3.241	− 14.1	PH_3	3.732	105.9
NCH	3.190	− 12.5	PCH	3.729	68.1
N_2	3.362	− 5.5			
$HN{=}PH_3$					
Base=NH_3	3.249	− 13.9	PH_3	3.713	107.2
NCH	3.201	− 12.6	PCH	3.690	75.4
N_2	3.357	− 6.0			
$H_2C{=}PH_3$					
Base=NH_3	3.331	− 9.6	PH_3	3.767	75.5
NCH	3.266	− 8.9	PCH	3.752	53.7
N_2	3.416	− 4.1			

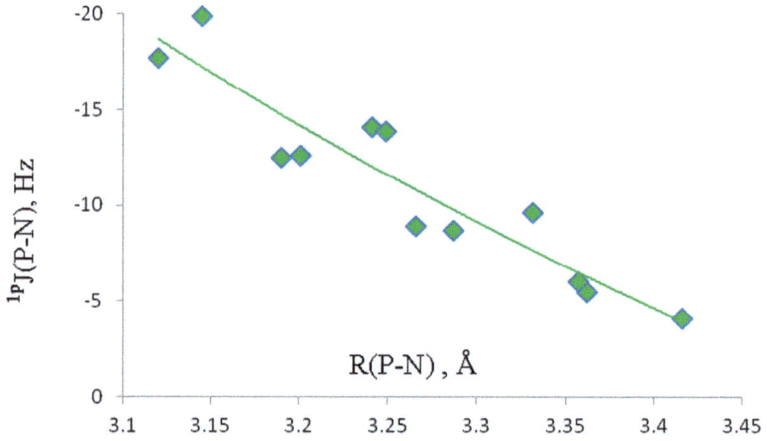

Fig. 8.22 ^{1p}J(P-N) versus the P-N distance for complexes X=PH_3:NY with P...N bonds. (Reprinted with permission from Ref. [37]. Copyright (2014) American Chemical Society)

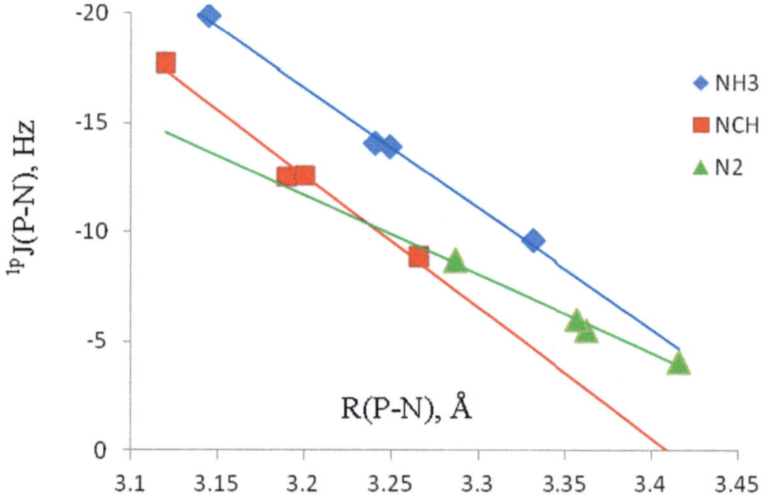

Fig. 8.23 ^{1p}J(P-N) versus the P-N distance for complexes X=PH_3:NY as a function of the nature of the base. (Reprinted with permission from Ref. [37]. Copyright (2014) American Chemical Society)

of the nature of the base. For each base there is one point at a short distance, two at similar intermediate distances, and one at a long distance. Thus, the pairs of points at similar distances in Fig. 8.22 correspond to the same base with two different acids, S=PH_3 and HN=PH_3. The trendlines in Fig. 8.23 have correlation coefficients R^2 between 0.983 and 0.997.

Coupling constants ^{1p}J(P-P) do not correlate with the P-P distance as a function of the nature of the acid, but the correlation is improved as a function of the base.

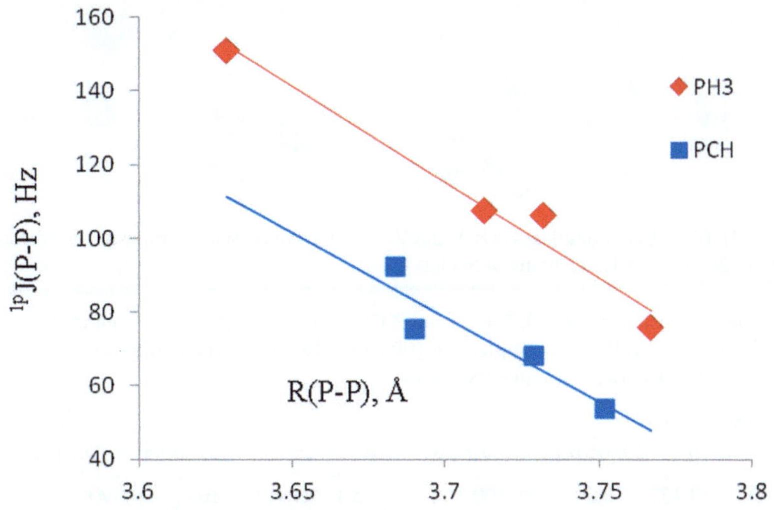

Fig. 8.24 ^{1p}J(P-P) versus the P-P distance for X=PH_3:PY complexes as a function of the base. (Reprinted with permission from Ref. [37]. Copyright (2014) American Chemical Society)

Figure 8.24 shows the variation of ^{1p}J(P-P) with distance for these complexes. The linear trendlines have correlation coefficients of 0.972 for PH_3 and 0.857 for PCH.

8.3.1.7 H_2FP...ClX Pnicogen Bonds

Although both pnicogen bonds and halogen bonds may be formed in complexes H_2FP:ClX, with X=F, Cl, CN (bonded at C and N), CCH, CH_3, and H, here we focus only on pnicogen-bonded complexes [38]. Equilibrium complexes ZB-1 have C_s symmetry with the Cl-A bond *cis* to the bisector of the H-P-H angle, with A the atom of X directly bonded to Cl. Complexes ZB-2 may have either C_s or C_1 symmetry, with the Cl-A bond *trans* or *gauche*, respectively, to the bisector. H_2FP:ClCH_3 ZB-1 and ZB-2 complexes are illustrated in Fig. 8.25. Pnicogen-bonded complexes exist on all H_2FP:ClX surfaces except H_2FP:ClF and H_2FP:ClNC. The very electronegative substituents F and NC withdraw sufficient electron density from Cl so that it can no longer act as an electron-pair donor.

The structures and binding energies of the remaining H_2FP:ClX complexes are reported in Table 8.17. The complexes with the most electropositive substituent, H_2FP:ClCH_3 ZB-1 and ZB-2, have the highest binding energies of – 15.1 and – 18.3 kJ·mol^{-1}, respectively. The binding energies of complexes with substituents H, Cl, and CCH have binding energies between – 10.0 and – 11.6 kJ·mol^{-1}. H_2FP:ClCN complexes have the smallest binding energies of about – 6.4 kJ·mol^{-1}. In each group, the P-Cl distance is shortest in the most strongly bound complex and longest in the most weakly bound. A linear correlation exists between the binding energies and the P-Cl distances in ZB-2 complexes, with a correlation coefficient of 0.958. A

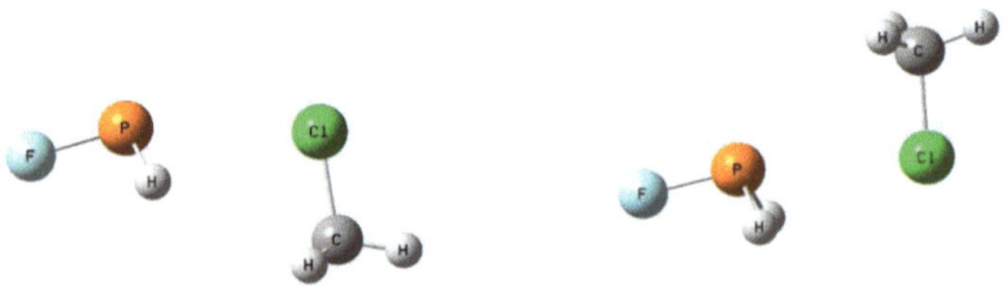

Fig. 8.25 $H_2FP{:}ClCH_3$ complexes ZB-1 and ZB-2. (Reprinted with permission from Ref. [38]. Copyright (2014) American Chemical Society)

Table 8.17 P-Cl distances (R, Å), F-P-Cl and P-Cl-A angles (∠, deg), and binding energies (ΔE, kJ·mol^{-1}) of pnicogen-bonded complexes $H_2FP{:}ClX$. (Reprinted with permission from Ref. [38]. Copyright (2014) American Chemical Society)

$H_2FP{:}ClX$	ZB-1				ZB-2			
	R	∠F-P-Cl	∠P-Cl-A	ΔE	R	∠F-P-Cl	∠P-Cl-A	ΔE
X=CH_3	3.157	163	100	− 15.1	3.121	169	90	− 18.3
H	3.280	167	110	− 10.0	3.277	169	80	− 11.6
Cl	3.169	168	110	− 11.5	3.265	166	89	− 10.8
CCH	3.295	162	93	− 11.0	3.320	165	85	− 10.7
CN	3.365	165	95	− 6.5	3.437	168	81	− 6.3

good correlation between these two variables is not found for the ZB-1 complexes, suggesting that there may be secondary interactions between the substituents and the H atoms of H_2FP in the *cis* orientation.

Values of the F-P-Cl and P-Cl-A angles indicate that $H_2FP{:}ClX$ complexes are stabilized by pnicogen bonds. In ZB-1 and ZB-2 complexes, the F-P-Cl angle characteristically approaches linearity, varying between 162 and 169°. At the same time, values of the P-Cl-A angle are between 93 and 110° in ZB-1 and between 81 and 90° in ZB-2. These angles indicate that no halogen bond exists in these complexes.

In ZB-1 and ZB-2 complexes, the stabilizing charge-transfer interaction occurs from the lone pair of Cl to the σ* P-F orbital of H_2FP. The data of Table 8.18 indicate that the weakest charge transfer interaction in each series occurs in $H_2FP{:}ClCN$, and the strongest in $H_2FP{:}ClCH_3$. However, the charge-transfer energies do not correlate with the binding energies of these complexes, but do correlate linearly with the P-Cl distances, with correlation coefficients R^2 of 0.950 and 0.941 for ZB-1 and ZB-2, respectively. Figure 8.26 provides a pictorial description of the lone-pair orbital on Cl and the σ* P-F orbital which are involved in charge transfer in $H_2FP{:}ClCl$ complexes.

^{1p}J(P-Cl) values for these pnicogen-bonded complexes are reported in Table 8.18. Values range from 25 to 57 Hz for complexes ZB-1, and from 16 to 29 Hz for ZB-2. Values for complexes ZB-1 are always greater than the corresponding ZB-2 values, although the P-Cl distances in ZB-1 complexes are not always shorter than those in ZB-2. The tendency for ^{1p}J(P-Cl) to increase as the P-Cl distance decreases can be seen for all pnicogen bonded complexes, although the data are definitely scattered.

Table 8.18 Cl(lp)→σ*P-F charge-transfer energies (kJ·mol^{-1}) and ^{1p}J(Cl-P) spin-spin coupling constants (Hz) for pnicogen-bonded complexes H_2FP:ClX ZB-1 and ZB-2. (Reprinted with permission from Ref. [38]. Copyright (2014) American Chemical Society)

	Cl(lp)→σ*P-F		^{1p}J(Cl-P)	
H_2FP:ClX	ZB-1	ZB-2	ZB-1	ZB-2
X=CH_3	19.4	20.4	40.6	29.0
H	13.6	13.7	36.8	22.4
Cl	17.7	13.6	56.8	20.2
CCH	9.6	7.9	26.8	18.5
CN	7.4	5.2	24.7	15.6

ZB-1 ZB-2

Fig. 8.26 Representation of the orbitals involved in charge-transfer interactions in H_2FP:ClCl complexes ZB-1 and ZB-2. (Reprinted with permission from Ref. [38]. Copyright (2014) American Chemical Society)

The correlation coefficients for the variation of ^{1p}J(P-Cl) as a function of the P-Cl distance are 0.705 for ZB-1 and 0.928 for ZB-2.

8.3.1.8 Pnicogen-Bonded Anionic Complexes

Anions are strong electron pair donors which can therefore form strong pnicogen bonds [39]. The binding energies of 21 equilibrium structures $H_2YP:X^-$, for X, Y=CH_3, CCH, F, CN (bonded through C and N), and Cl are given in Table 8.19. The binding energies are defined as the negative of the dissociation energies relative to the more stable H_2YP and X^- products. The diagonal elements of Table 8.19 are the energies of the complexes $(H_2XPX)^-$ which have C_{2v} symmetry and symmetric X-P-X bonds. From the arrangement of complexes in Table 8.19, it is possible to determine for any given X, Y pair, which substituent will be covalently bonded to P, and which will form the ion-molecule bond. The anion X^- is determined by the ability of X to accommodate a negative charge, which is

$$Cl > NC > F > CCH > CH_3.$$

The complexes $(H_2FPF)^-$ and $H_2(CH_3)P:F^-$ are illustrated in Fig. 8.27.

Table 8.19 Binding energies (kJ·mol^{-1}) of complexes $H_2YP{:}X^{-}$ [a]. (Reprinted with permission from Ref. [39]. Copyright (2014) American Chemical Society)

$H_2YP{:}X^-$	$H_2(CH_3)P$	$H_2(CCH)P$	H_2FP	$H_2(CN)P$	$H_2(NC)P$	H_2ClP
X=CH_3	− 70.0					
CCH	− 24.0	− 73.1				
F	− 57.4	− 119.8	− 180.5			
CN	− 17.7	− 50.0	− 126.5	− 102.2		
NC	− 19.8	− 50.3	− 98.5	− 93.6	− 131.8	
Cl	− 19.4	− 50.2	− 85.4	− 93.8	− 120.9	− 113.1

[a] Binding energies relative to the more stable monomers $H_2YP + X^-$

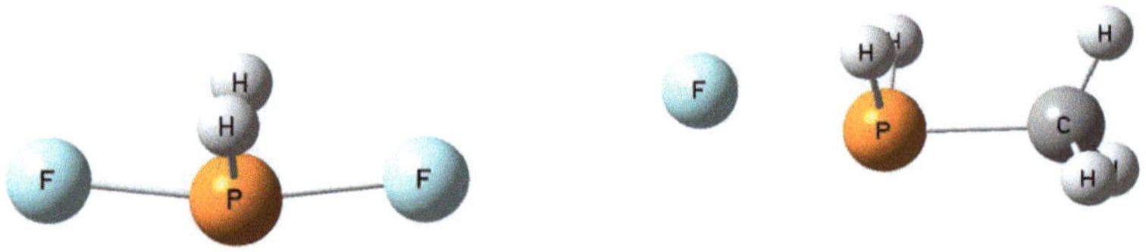

Fig. 8.27 The symmetric molecular anion $(H_2FPF)^-$ and the anionic complex $H_2(CH_3)P{:}F^-$. (Reprinted with permission from Ref. [39]. Copyright (2014) American Chemical Society)

To understand the variation in binding energies, it is advantageous to first examine the structures of complexes $H_2YP{:}X^-$ which have C_s symmetry. These complexes can be described in terms of the P-A and P-A′ distances and the A-P-A′ angle, where A and A′ are the atoms of X and Y, respectively, which are directly bonded to P. In these complexes, the A-P-A′ arrangement tends toward linearity, a recurring feature of complexes with pnicogen bonds. The angle varies between 162 and 169 degrees, except for the complexes of $P(CH_3)H_2$ with CN^-, NC^-, and Cl^-, which have A-P-C angles between 152 and 156 degrees.

Tables 8.20 and 8.21 give the P-A and P-A′ distances, respectively, in complexes $H_2YP{:}X^-$. If the P-A bond is shorter than the P-A bond in the corresponding symmetric complex, or if it is longer but within 0.30 Å of the length of the P-A bond in the symmetric structure, we have assumed that it has some covalent character. Moreover, if the P-A′ bond is within 0.15 Å of the length of the P-A bond in the corresponding symmetric complex, the P-A′ bond has a reduced degree of covalency relative to other corresponding P-A′ bonds, and has some ion-molecule character. The lengths these two bonds suggest that P-A bonds have some covalent character and P-A′ bonds have some ion-molecule character in complexes $H_2(CH_3)P{:}F^-$, $H_2(CCH)P{:}F^-$, $H_2FP{:}CN^-$, $H_2FP{:}NC^-$, $H_2FP{:}Cl^-$, and $H_2(NC)P{:}Cl^-$, as well as in the six symmetric complexes $(H_2XPX)^-$.

That the binding energies of the $H_2YP{:}X^-$ complexes are related to the nature of the P-A and P-A′ bonds can be seen from the data of Table 8.19 by noting that for a fixed H_2YP, as the ability of X to accommodate a negative charge increases

Table 8.20 P-A distances (Å) in complexes $H_2YP:X^{-a,b}$. (Reprinted with permission from Ref. [39]. Copyright (2014) American Chemical Society)

$H_2YP:X^-$	$H_2(CH_3)P$ $P\text{-}CH_3$	$H_2(CCH)P$ P-CCH	H_2FP P-F	$H_2(CN)P$ P-CN	$H_2(NC)P$ P-NC	H_2ClP P-Cl
$X{=}CH_3$	*2.063*					
CCH	2.876	*2.079*				
F	*2.052*	*1.915*	*1.837*			
CN	3.115	2.662	*2.021*	*2.095*		
NC	2.975	2.614	*2.131*	2.357	*2.005*	
Cl	3.302	2.942	*2.595*	2.722	*2.475*	*2.388*

[a] A is the atom of X directly bonded to P
[b] P-A ion-molecule bonds with some covalent character are indicated in italics

Table 8.21 P-A′ distances (Å) in complexes $H_2YP:X^{-a,b}$. (Reprinted with permission from Ref. [39]. Copyright (2014) American Chemical Society)

$H_2YP:X^-$	$H_2(CH_3)P$ $P\text{-}CH_3$	$H_2(CCH)P$ P-CCH	H_2FP P-F	$H_2(CN)P$ P-CN	$H_2(NC)P$ P-NC	H_2ClP P-Cl
$X{=}CH_3$	*2.063*					
CCH	1.903	*2.079*				
F	*1.960*	*1.944*	*1.837*			
CN	1.884	1.848	*1.836*	*2.095*		
NC	1.880	1.829	*1.756*	1.886	*2.005*	
Cl	1.881	1.827	*1.733*	1.874	*1.924*	*2.388*

[a] A′ is the atom of Y directly bonded to P
[b] Pnicogen P-A′ bonds with some ion-molecule character are in italics

down a column, the binding energies decrease or are similar with two exceptions, $H_2(CH_3)P:F^-$ and $H_2(CCH)P:F^-$, which have significantly greater binding energies than the complexes directly above them. In addition, for a given substituent X^-, binding energies tend to increase from left to right across the row to the symmetric complex. There are three exceptions: $H_2FP:CN^-$, $H_2FP:NC^-$, and $H_2(NC)P:Cl^-$. The binding energies of these three complexes are greater than the binding energies of the complexes to their immediate right. These five complexes are five of the six complexes identified as having P-A bonds with partial covalent character and P-A′ bonds with partial ion-molecule character. Thus, the binding energies of these complexes reflect the degree of covalent and ion-molecule character of P-A and P-A′ bonds, respectively.

Charge-transfer in these anionic complexes occurs from a lone pair on A to the σ* P-A′ orbital of H_2YP, and the charge-transfer energies are reported in Table 8.22. No data are reported for the symmetric complexes $(H_2XPX)^-$, or the complexes $H_2FP:NC^-$ and $H_2FP:Cl^-$, since the NBO program considers these as single molecular ions. Neither is a charge-transfer energy reported for $H_2FP:CN^-$ since this

Table 8.22 Charge-transfer stabilization energies (kJ·mol^{-1}) for complexes $H_2YP:X^{-a}$. (Reprinted with permission from Ref. [39]. Copyright (2014) American Chemical Society)

P:X$^-$	$H_2(CH_3)P$	$H_2(CCH)P$	H_2FP	$H_2(CN)P$	$H_2(NC)P$
X=CCH	46.4				
F	150.3	222.3			
CN	23.0	86.1	b		
NC	15.0	103.2	c	107.9	
Cl	16.3	49.4	c	96.2	230.2

[a]The NBO program considers the $(H_2XPX)^-$ complexes as single molecular ions
[b]The NBO program considers this complex as $H_2(CN)P:F^-$
[c]The NBO program considers this complex as a single molecular ion

complex is described as $H_2(CN)P:F^-$. Complexes $H_2(CH_3)P:X^-$ and $H_2(CCH)P:X^-$ have the largest charge-transfer energies when X^- is F^-. For a given X^-, the charge-transfer energies increase across the row from left to right as the P-A bond length decreases, since charge transfer is more effective at shorter distances. The charge-transfer energies are linearly related to the P-Cl distances in $H_2YP:Cl^-$, with a correlation coefficient R^2 of 0.992.

The electron densities at bond critical points are also indicators of the nature of the P-A and P-A′ bonds. Thus, ρ_{BCP} values for the P-A bonds in the six symmetric complexes $(H_2XPX)^-$, and the P-A bonds in $H_2(CH_3)P:F^-$, $H_2(CCH)P:F^-$, $H_2FP:CN^-$, $H_2FP:NC^-$, $H_2FP:Cl^-$, and $H_2(NC)P:Cl^-$ have the largest electron densities of 0.044e or greater. The only other P-A bond in this range is the P-N bond in $H_2(CN)P:NC^-$ which has a BCP density of 0.045e. Moreover, for fixed Y, all P-A′ bonds in these same complexes have electron densities at the BCPs that are smaller than the P-A′ densities for the remaining complexes in the series. The six symmetric complexes $(H_2XPX)^-$ and the complexes $H_2(CH_3)P:F^-$, $H_2(CCH)P:F^-$, $H_2FP:CN^-$, $H_2FP:NC^-$, $H_2FP:Cl^-$, and $H_2(NC)P:Cl^-$ are those that have ion-molecule P-A bonds with increased covalent character, and covalent P-A′ bonds with increased ion-molecule character.

The Laplacians are negative for the P-A bonds in $H_2FP:CN^-$ and in the symmetric $(H_2XPX)^-$ complexes with X=CH_3, CCH, CN, and NC. In addition, these same bonds and the P-A bonds in $(H_2FPF)^-$, $(H_2ClPCl)^-$, $H_2(CH_3)P:F^-$, $H_2(CCH)P:F^-$, $H_2FP:NC^-$, $H_2FP:Cl^-$, and $H_2(NC)P:Cl^-$ have values of the energy density that are also negative and greater in absolute value than 0.01 au, indicating that they have some covalent character. In addition, for fixed Y, P-A′ bonds in these same complexes have less negative energy densities than the other P-A′ bonds in the same series, indicating that the former bonds have lost some covalency. Thus, the properties of the electron densities at bond critical points of P-A and P-A′ bonds are consistent with the characterization of these bonds as having partial covalent character and partial ion-molecule character, respectively, based on binding energies and P-A and P-A′ distances.

Table 8.23 Spin-spin coupling constants ^{1p}J(P-A) and ^{1}J(P-A′) (Hz) for complexes $H_2YP:X^-$. (Reprinted with permission from Ref. [39]. Copyright (2014) American Chemical Society)

Complex	^{1p}J(P-A)	Complex	^{1}J(P-A′)
$H_2YP:Cl^-$	^{1p}J(P-Cl)	$H_2(CH_3)P:X^-$	^{1}J(P-C)
Y=CH_3	21.0	X=CH_3	− 32.9
CCH	51.5	CCH	11.5
F	85.2	F	19.4
CN	70.6	CN	5.3
NC	88.3	NC	5.6
Cl	85.5	Cl	3.6
$H_2YP:NC^-$	^{1p}J(P–N)	$H_2(CCH)P:X^-$	^{1}J(P-C)
Y=CH_3	− 26.0	X=CCH	− 57.8
CCH	− 43.4	F	− 87.2
F	− 3.8	CN	− 46.5
CN	− 36.2	NC	− 38.0
NC	13.7	Cl	− 41.5
$H_2YP:CN^-$	^{1p}J(P-C)	$H_2FP:X^-$	^{1}J(P–F)
Y=CH_3	73.3	X=F	164.1
CCH	99.1	CN	137.2
F	− 81.0	NC	9.4
CN	− 36.9	Cl	− 100.8
$H_2YP:F^-$	^{1p}J(P–F)	$H_2(CN)P:X^-$	^{1}J(P-C)
Y=CH_3	273.7	X=CN	− 36.9
CCH	230.4	NC	− 62.5
F	164.1	Cl	− 67.2
$H_2YP:CCH^-$	^{1p}J(P-C)	$H_2(NC)P:X^-$	^{1}J(P–N)
Y=CH_3	95.8	X=NC	13.7
CCH	− 57.8	Cl	30.9
$H_2YP:CH_3^-$	^{1p}J(P-C)	$H_2ClP:X^-$	^{1}J(P-Cl)
Y=CH_3	− 32.9	X=Cl	85.5

Since the Fermi contact terms are not good approximations to ^{1p}J(P-A) and ^{1}J(P-A′) for all of the molecular anions, all coupling constants reported for these anions are total J values. Coupling constants ^{1p}J(P-Cl) across the pnicogen bonds for complexes $H_2YP:Cl^-$ are reported in Table 8.23, and the P-Cl distances are given in Table 8.20. The symmetric structure $(H_2ClP:Cl)^-$ has the shortest P-Cl distance, followed by the P-Cl distances in $H_2(NC)P:Cl^-$ and $H_2FP:Cl^-$. All of these P...Cl bonds have some covalent character. The values of ^{1p}J(P-Cl) for these three complexes are similar, and are significantly greater than the values for $H_2(CN)P:Cl^-$, $H_2(CCH)P:Cl^-$, and $H_2(CH_3)P:Cl^-$. Figure 8.28 illustrates the very good correlation between ^{1p}J(P-Cl)

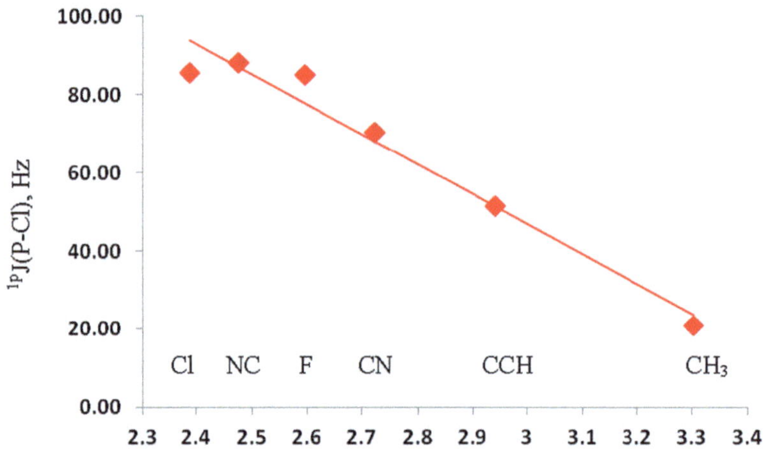

Fig. 8.28 ^{1p}J(P-Cl) versus the P-Cl distance for complexes $H_2YP{:}Cl^-$. Points are identified at the bottom of the graph by the nature of Y. (Reprinted with permission from Ref. [39]. Copyright (2014) American Chemical Society)

and the P-Cl distance.The linear relationship shown has a correlation coefficient R^2 of 0.961, while a second-order curve gives a slightly better fit with a correlation coefficient of 0.979.

Coupling constants ^{1p}J(P-N) for complexes $H_2YP{:}NC^-$ once again differentiate between P...N pnicogen bonds with and without covalent character. Symmetric $[H_2(NC)P{:}NC]^-$ has the shortest P-N distance, followed by the P-N distance in $H_2FP{:}NC^-$. These two complexes have the smallest absolute values of ^{1p}J(P-N), while the remaining three complexes have much larger absolute values. Similarly, ^{1p}J(P-C) values for $H_2FP{:}CN^-$ and $[H_2(CN)P{:}CN]^-$ are negative, while $H_2(CCH)P{:}CN^-$ and $H_2(CH_3)P{:}CN^-$ have positive values of ^{1p}J(P-C) at longer P-C distances. Thus, coupling constants can distinguish between P-A pnicogen bonds with increased covalent character and shorter bond lengths, and P-A bonds that are essentially ion-molecule pnicogen bonds with longer P-A distances.

^{1}J(P-A′) values also differentiate between normal covalent bonds and bonds with reduced covalency and increased ion-molecule character, as evident from Table 8.23. Figure 8.29 presents a plot of ^{1}J(P-C) versus the P-C distance for $H_2(CH_3)P{:}X^-$ complexes, and dramatically illustrates the effect of the loss of covalency and the gain of ion-molecule character by the P-C bond.^{1}J(P-C) for $[H_2(CH_3)P{:}CH_3]^-$ is –33 Hz, but then increases in absolute value as the P-C distance decreases, and exhibits its maximum value for $H_2(CH_3)P{:}F^-$. As this distance further decreases,^{1}J(P-C) decreases in $H_2(CH_3)P{:}CCH^-$, and decreases further as the P-C distance decreases in the remaining three complexes. The correlation coefficient for the second-order curve in Fig. 8.29 is 0.996. The variation of ^{1}J(P-C) would be difficult to understand without some knowledge of the nature of the P-A′ bonds and their variation with distance.^{1}J(P-C) as a function of the P-C distance for $H_2(CCH)P{:}X^-$ is illustrated

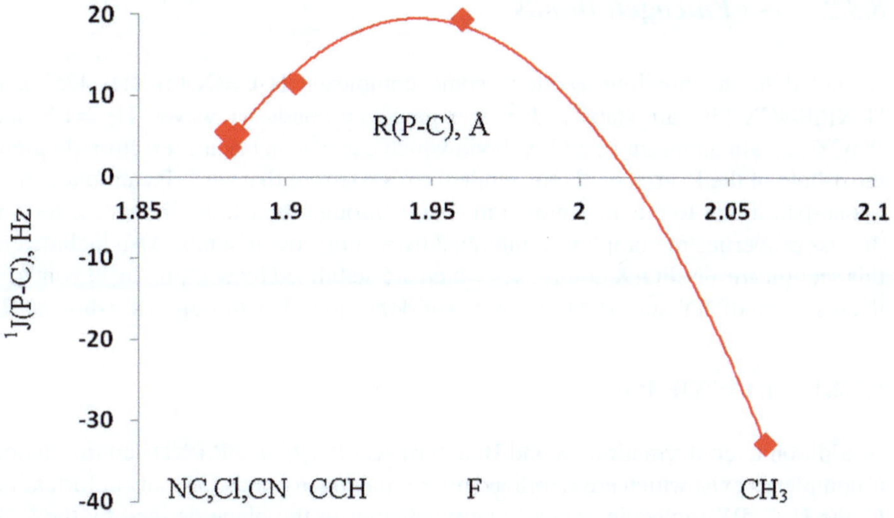

Fig. 8.29 ^{1}J(P-C) versus the P-C distance for complexes $H_2(CH_3)P{:}X^-$. Points are identified at the bottom of the graph by the nature of X. (Reprinted with permission from Ref. [39]. Copyright (2014) American Chemical Society)

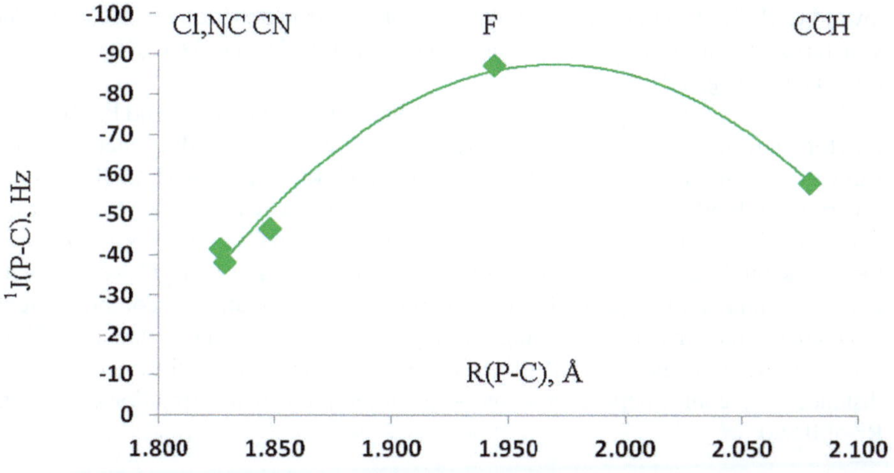

Fig. 8.30 ^{1}J(P-C) versus the P-C distance for complexes $H_2(CCH)P{:}X^-$. Points are identified at the top of the graph by the nature of X. (Reprinted with permission from Ref. [39]. Copyright (2014) American Chemical Society)

in Fig. 8.30, and shows a similar pattern. Thus, the recognition of the partial ion-molecule character of longer P-A′ bonds is essential for understanding the variation of ^{1}J(P-A′).

8.3.2 *π-σ Pnicogen Bonds*

As noted in the previous section, some complexes $H_2C{=}(X)P{:}PXH_2$ [35] and $H_2XP{:}P{\equiv}CX$ [36] are stabilized by σ-σ pnicogen bonds. However, $H_2C{=}PX$ and $P{\equiv}CX$ contain an unsaturated P-C bond which can also act as a π electron donor to the σ-hole of the P atom of H_2XP, while the σ system of that same P can function as a lone-pair donor to the unsaturated molecule through its π-hole. In this section we discuss properties of complexes stabilized by π-σ pnicogen bonds. Also included in this section are YN:PO_2X complexes which are stabilized by π-σ pnicogen bonds as the σ system of NY acts as an electron-pair donor to PO_2X through its π-hole at P.

8.3.2.1 $H_2C{=}(X)P{:}PXH_2$

In addition to conformations A and B, a third set of $H_2C{=}(X)P{:}PXH_2$ conformation C complexes exist which are stabilized by π-σ pnicogen bonds [35]. In conformation C, the $H_2C{=}PX$ molecule is nearly perpendicular to the plane defined by the P...P bond and the bisector of the H-P_s-H angle. P_d-P_s-A angles approach linearity, while P_s-P_d=C angles are less than 90°, dramatically smaller than values in the corresponding conformation A and B complexes. This geometry facilitates interaction of $H_2C{=}PX$ through its π system with PH_2X. Since the C=P π bond is polarized toward C, P_s-P_d=C angles less than 90° promote electron donation by $H_2C{=}PX$. The structures of conformation C complexes $H_2C{=}(F)P{:}PFH_2$ and $H_2C{=}(H)P{:}PH_3$ are illustrated in Fig. 8.31.

The binding energies, intermolecular P-P distances, and P_d-P_s-A and P_s-P_d=C angles for conformation C complexes are reported in Table 8.24. Binding energies range from -8.2 kJ·mol^{-1} for $H_2C{=}(H)P{:}PH_3$ to -20.0 kJ·mol^{-1} for $H_2C{=}(F)P{:}PFH_2$. The order of binding energies of conformation C complexes is similar to that of the corresponding conformation A complexes. Conformation C complexes usually have the largest binding energies, with the order of binding energies being C > A > B with two exceptions. Although the binding energies of conformation C complexes tend to be greater than the corresponding A complexes, intermolecular distances in C are usually longer than those in A. A plot of the binding energy versus the intermolecular distance for C conformers has a second-order trendline with a correlation coefficient R^2 of 0.890.

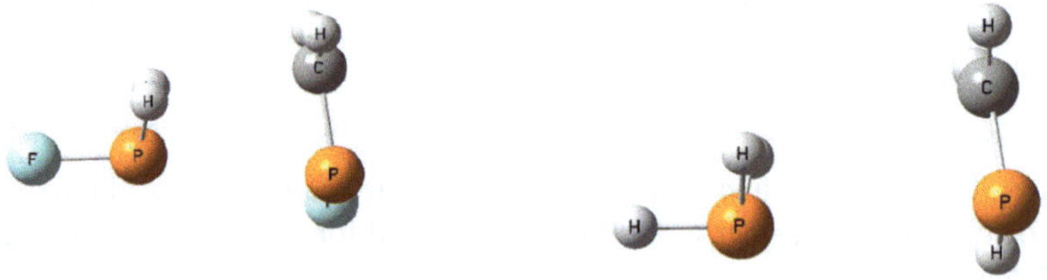

Fig. 8.31 Structures of conformation C complexes $H_2C{=}(F)P{:}PFH_2$ and $H_2C{=}(H)P{:}PH_3$. (Reprinted with permission from Ref. [35]. Copyright (2013) American Chemical Society)

Table 8.24 Binding energies (ΔE, kJ·mol^{-1}), intermolecular P-P distances (R, Å), and P_d-P_s-A and P_s-P_d=C angles (∠, °) for conformation C complexes H_2C=(X)P:PXH_2. (Reprinted with permission from Ref. [35]. Copyright (2013) American Chemical Society)

	R	ΔE	∠P_d-P_s-A	∠P_s-P_d=C
X=F	3.029	– 20.0	173	76
Cl	3.146	– 19.6	174	77
OH	3.236	– 17.6	175	77
NC	3.281	– 15.2	174	77
CCH	3.456	– 15.3	173	78
CN	3.528	– 12.1	175	78
CH_3	3.484	– 12.8	179	83
H	3.595	– 8.2	175	85

Table 8.25 NBO charges (au) on H_2C=PX, πP_d=C→σ^*P_s-A and $P_s$lp→π^*P_d=C stabilizing charge-transfer energies (kJ·mol^{-1}), and spin-spin coupling constants [^{1p}J(P-P), Hz] for conformation C complexes H_2C=(X)P:PXH_2. (Reprinted with permission from Ref. [35]. Copyright (2013) American Chemical Society)

H_2C=(X)P:PXH_2	Charge on H_2C=PX	πP_d=C→σ^*P_s-A	$P_s$lp→π^*P_d=C	^{1p}J(P-P)
X=F	0.010	34.9	16.7	105.1
Cl	0.016	29.3	11.6	125.7
OH	0.006	21.0	9.1	85.3
NC	0.007	18.9	7.8	79.2
CCH	– 0.002	8.9	4.6	—[a]
CN	0.002	8.0	3.3	44.0
CH_3	– 0.003	7.9	5.1	45.0
H	0.001	6.8	4.4	40.2

[a]Value not available because of computational expense

Table 8.25 reports the NBO charges on H_2C=PX and the stabilizing charge-transfer energies πP_d=C→σ^*P_s-A and $P_s$lp→π^*P_d=C. As observed for corresponding conformation A and B complexes, charge transfer from H_2C=PX to PH_2X is again favored. This is consistent with the positive charges on H_2C=PX in these complexes, except for H_2C=(CCH)P:P(CCH)H_2 and H_2C=(CH_3)P:P(CH_3)H_2 which have small negative charges on H_2C=PX. The charge-transfer energy for a conformation C complex is greater than that for the corresponding conformation B, except for H_2C=(CN)P:P(CN)H_2.

The electron density properties at the P...P BCPs have similar characteristics to those observed for conformations A and B. Conformation C complexes with X=F, Cl, OH, and NC have negative values of H_{BCP}, indicating some covalent character of the P...P bond. Moreover, there is an excellent correlation between the electron densities at the P...P BCPs and the P-P distances in A, B, and C complexes. The Laplacian contours and molecular graph, and the ELF of H_2C=(F)P:PFH_2 are illustrated in Fig. 8.32a and 8.32b, respectively. They present a pictorial representation of the

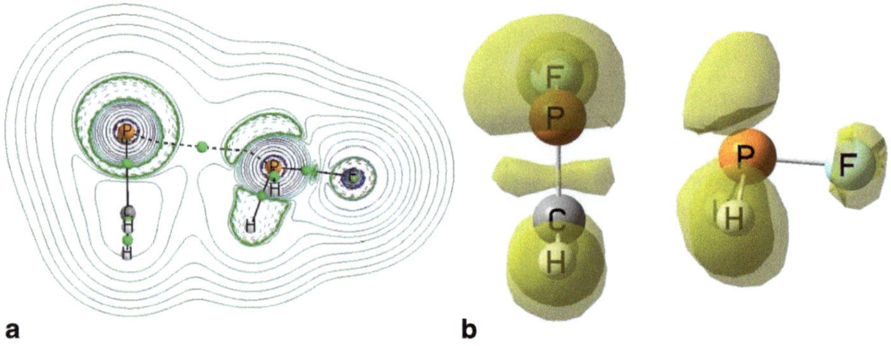

Fig. 8.32 **a** The Laplacian contours and molecular graph of H_2C=(F)P:PFH_2 conformation C complex. The contour plane is defined by the two P atoms and the BCP of the P...P bond. **b** The 0.8 au ELF isosurface of the H_2C=(F)P:PFH_2 conformation C complex. (Reprinted with permission from Ref. [35]. Copyright (2013) American Chemical Society)

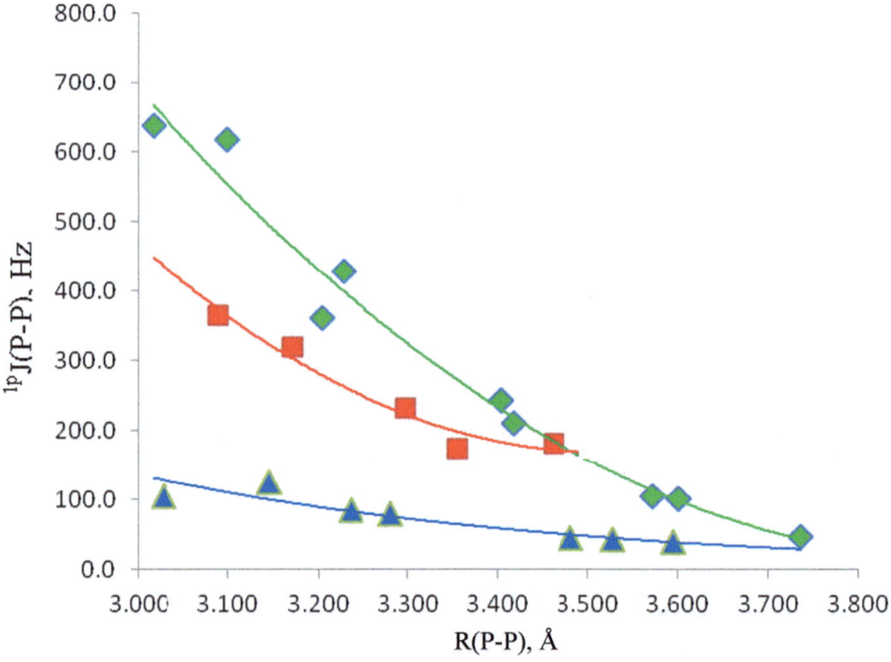

Fig. 8.33 Coupling constants ^{1p}J(P-P) versus the P-P distance for conformations C ▲, A ♦, and B ■ complexes H_2C=(X)P:PXH_2. (Reprinted with permission from Ref. [35]. Copyright (2013) American Chemical Society)

bond path, and of the P_s lone pair and the π electrons which interact with the π- and σ-holes, respectively, consistent with the NBO analysis.

Table 8.25 also reports the ^{31}P-^{31}P spin-spin coupling constants ^{1p}J(P-P) for conformation C complexes, and Fig. 8.33 presents a plot of ^{1p}J(P-P) versus the intermolecular P-P distance for conformers A, B, and C. The usual second-order trendline has a correlation coefficient R^2 of 0.875 for conformation C. However, a decaying exponential, which has a correlation coefficient R^2 of 0.919, provides a better description of the relationship between these two variables. The ordering of the trendlines for the three conformations is A > B > C. This ordering reflects the favorable A-P...P-A arrangement in A, followed by the A-P...P=C arrangement in B. Since the pnicogen bond is formed through the π system of H_2C=PX in C, and ^{1p}J(P-P) is essentially equal to the FC term which depends on s electron densities in ground and excited states, conformation C complexes have the smallest values of ^{1p}J(P-P) at each P-P distance.

8.3.2.2 H_2XP:PCX

In addition to the equilibrium conformation A complexes, H_2XP:PCX complexes may also be stabilized by π-σ bonds in two different conformations B and C [36]. In B complexes the π-σ pnicogen bond has the A-P_s...C alignment approaching linearity, while in conformation C the A-P_s...P_t alignment approaches linearity, as illustrated in Fig. 8.34.

The bond paths of conformation B complexes connect P_s with PCX through the π system at the P≡C C atom. The binding energies, P-P and P_s-C distances, and C-P_s-A angles are reported in Table 8.26. The binding energies of these complexes range from − 8.7 kJ·mol^{-1} for H_3P:PCH to − 16.6 kJ·mol^{-1} for H_2FP:PCF. The P_s-C distances are always shorter than the corresponding P-P distances, but there is no correlation between the binding energies and either the P-P or the P_s-C intermolecular distances.

Conformation B complexes are stabilized by charge-transfer interactions. Since the P≡C bond is polarized toward C, charge transfer occurs from the π bond at C through the σ-hole to P_s, and from the lone pair on P_s to P_t through the π-hole. The dominant charge transfer interaction is πP=C→σ*P-A, as evident from Table 8.27. The single exception occurs when X=H, in which case charge transfer πP=C→σ*P-H is 0.2 kJ·mol^{-1} less stabilizing. The πP=C→σ*P-A charge-transfer energies vary from 3.6 kJ·mol^{-1} for X=CH_3, to 19.8 kJ·mol^{-1} for X=F. Charge-transfer energies from the lone pair on P_s to the π*P=C orbital range from 1.3

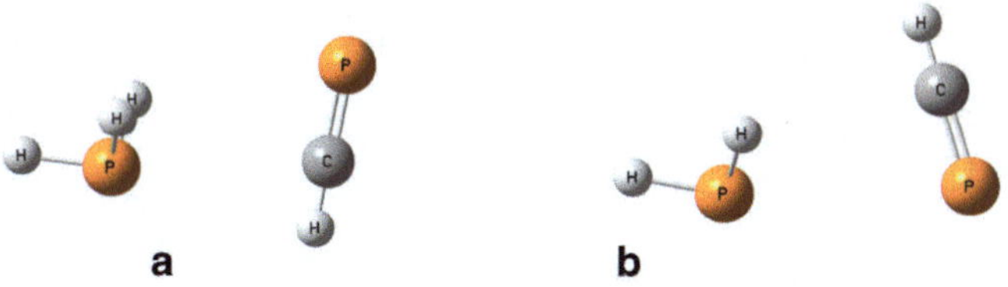

Fig. 8.34 H_3P:PCH complexes with conformations B and C. Complexes have C_s symmetry except for H_2(CN)P:PCCN and H_2(CCH)P:PCCCH B which have C_1 symmetry. (Reprinted with permission from Ref. [36]. Copyright (2014) Springer Science and Business Media)

Table 8.26 Binding energies (ΔE, kJ·mol^{-1}), P-P and P_s-C distances (R, Å), and C-P_s-A angles (∠, deg) for conformation B complexes[a]. (Reprinted with permission from Ref. [36]. Springer Science and Business Media)

H_2XP:PCH	ΔE	R(P-P)	R(P_s-C)	∠C-P_s-A
X=Cl	– 16.4	3.476	3.326	175
F	– 16.6	3.389	3.070	172
CCH	– 13.3	3.827	3.404	161
OH	– 13.2	3.516	3.251	175
NC	– 12.0	3.557	3.241	177
CN	– 9.6	3.941	3.426	165
CH_3	– 12.6	3.850	3.439	177
H	– 8.7	4.071	3.413	171

[a]These complexes have C_s symmetry, except for H_2(CN)P:PCCN and H_2(CCH)P:PCCCH which have C_1 symmetry

kJ·mol^{-1} for X=CCH to 10.0 kJ·mol^{-1} for X=F. The πP=C→σ*P-A charge-transfer energies do not correlate with the binding energies of conformation B complexes or with the P_s-C distances, but do correlate with the intermolecular P-P distances, as illustrated in Fig. 8.35.

Conformation C complexes have bond paths that connect P_s to the π system of PCX, usually but not always at P_t. Table 8.28 reports the P_s-C distances, which are shorter than the P-P distances, although the difference between them is much less than found for conformation B complexes due to the smaller values of the P_s-P_t-C angles. Table 8.28 also reports the binding energies of these complexes, which range from – 7.5 kJ·mol^{-1} for X=H to – 17.6 kJ·mol^{-1} for X=Cl. The binding energies of conformation B and C complexes are similar, differing by 1 to 1.5 kJ·mol^{-1}. The single exception is conformation C of $H_2(CH_3)$P:$PCCH_3$ which is 2.6 kJ·mol^{-1} less stable than B. The binding energies do not correlate with the P-P distances.

Table 8.27 Charge-transfer energies (kJ·mol^{-1}) and spin-spin coupling constants [^{1p}J(P-P), Hz] for conformation B complexes H_2XP_s:P_t≡CX. (Reprinted with permission from Ref. [36]. Copyright (2014) Springer Science and Business Media)

H_2XP:PCX	πP=C→σ*P_s-A	P_s(lp)→π*P=C	^{1p}J(P-P)
X=Cl	14.4	5.1	41.1
F	19.8	10.0	54.3
CCH	4.3	1.3	16.9
OH	11.5	5.1	41.9
NC	11.0	5.6	31.9
CN	3.7	1.5	14.9
CH_3	3.6	3.4	11.9
H	4.2	4.4	7.3

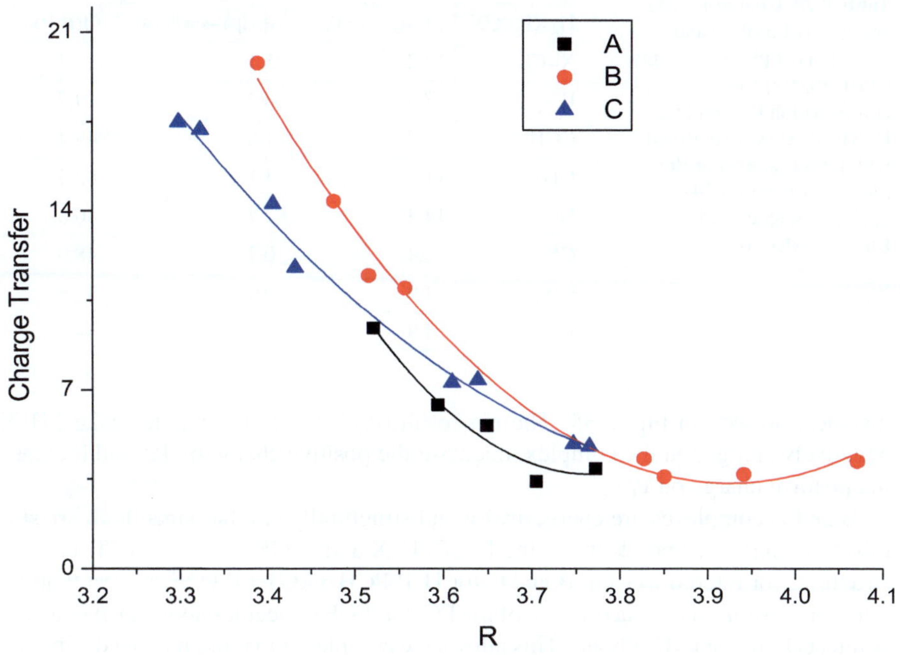

Fig. 8.35 Charge-transfer energies from PCX to PH_2X (kJ·mol^{-1}) versus the P-P distance (R, Å) for complexes with conformations A ■, B ●, and C ▲. Correlation coefficients R^2 are 0.963, 0.992, and 0.990, respectively. (Reprinted with permission from Ref. [36]. Copyright (2014) Springer Science and Business Media)

Table 8.28 Binding energies (ΔE, kJ·mol^{-1}), P-P and P_s-C distances (R, Å), and P_t-P_s-A angles (∠, deg) for conformation C complexes. (Reprinted with permission from Ref. [36]. Springer Science and Business Media)

H_2XP:PCH	ΔE	R(P-P)	R(P_s-C)	∠P_t-P_s-A
X=Cl	− 17.6	3.323	3.313	174
F	− 15.6	3.298	3.270	175
CCH	− 14.7	3.610	3.431	172
OH	− 14.2	3.432	3.360	171
NC	− 13.5	3.407	3.317	175
CN	− 10.6	3.639	3.428	176
CH_3	− 10.0	3.747	3.571	166
H	− 7.5	3.765	3.702	168

The charge-transfer energies reported in Table 8.29 exhibit a pattern similar to that observed for conformation B. The πP=C→σ*P_s-A charge-transfer energies are significantly greater than the P_s(lp)→π*P=C, and are also greater than the corresponding πP=C→σ*P_s-A energies of conformation B complexes, except for H_2FP:PCF which has the largest charge-transfer energy among these complexes. Once again, the πP=C→σ*P_s-A charge-transfer energies correlate with the P-P

Table 8.29 Charge-transfer energies (kJ·mol^{-1}) and spin-spin coupling constants [^{1p}J(P-P), Hz] for conformation C complexes H_2XP_s:$P_t{\equiv}CX$. (Reprinted with permission from Ref. [36]. Copyright (2014) Springer Science and Business Media)

H_2XP:PCX	$\pi P{=}C \rightarrow \sigma^*P_s$-A	$P_s(lp) \rightarrow \pi^*P{=}C$	^{1p}J(P-P)
X=Cl	17.2	5.1	116.7
F	17.5	6.2	119.7
CCH	7.3	1.0	49.4
OH	11.8	4.1	83.3
NC	14.3	2.7	86.8
CN	7.4	0.7	48.0
CH_3	4.9	0.9	32.6
H	4.8	1.5	34.5

distance, as seen in Fig. 8.35. The net result of charge transfer is to make PH_2X negatively charged in the complex, decrease the positive charge on P_s, and increase the positive charge on P_t.

B and C complexes are energetically and structurally similar, since both are stabilized by pnicogen bonds involving P_s of PH_2X and the PCX π system. Transition structures for interconverting B and C for H_3P:PCH and H_2FP:PCF suggest that the interconversion occurs via rotation of the PCH or PCF molecules about an axis which connects P_s to the P=C π bond. This allows the complexes to remain intact during the interconversion. Relative to the less stable conformer C, the barriers to converting C to B are 2.6 and 9.7 kJ·mol^{-1} for H_3P:PCH and H_2FP:PCF, respectively.

Coupling constants for complexes with conformations B and C are given in Tables 8.27 and 8.29, respectively. Interesting relationships can be found by comparing the values of ^{1p}J(P-P) for H_2XP:PCX complexes with conformations A, B, and C. For fixed X, the order of decreasing ^{1p}J(P-P) is A > C > B. The large values of ^{1p}J(P-P) for conformation A complexes may be attributed primarily to the σ-σ nature of the pnicogen bond, and the dependence of the dominant FC term on s electron densities in both ground and excited states. The nature of the FC term is also consistent with the reduced values of ^{1p}J(P-P) for B and C complexes which are stabilized by π-σ pnicogen bonds involving the π electrons of PCX. That ^{1p}J(P-P) for a given X is greater for the conformation C complex compared to B is also consistent with the shorter P-P distances in C, and with the A-P_s-P_t arrangement which approaches linearity. Figure 8.36 presents plots of ^{1p}J(P-P) versus the P-P distance for A, B, and C complexes. The good correlation between these two variables is evident, with the second-order trendlines having correlation coefficients R^2 of 0.961, 0.995, and 0.976, respectively.

8.3.2.3 YN:PO_2X

The molecular electrostatic potentials (MEPs) on the 0.001 au electron density isosurfaces of the monomers PO_2F and PO_2Cl are very negative around the oxygen atoms, slightly negative around the halogen atoms, and positive above and below the

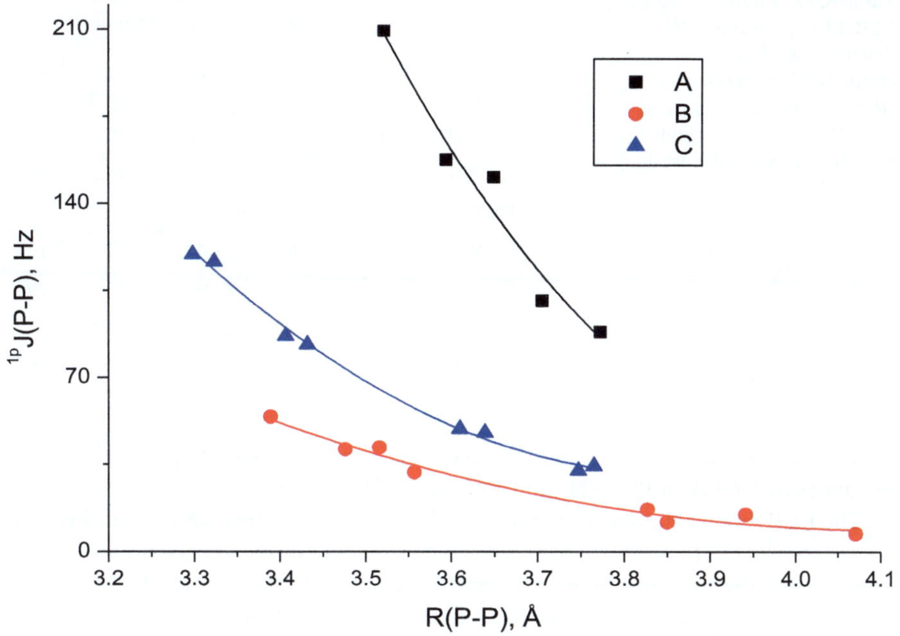

Fig. 8.36 ^{1p}J(P-P) versus the P-P distance for complexes with conformations A, B, and C. (Reprinted with permission from Ref. [36]. Copyright (2014) Springer Science and Business Media)

phosphorous atoms [33]. The MEP values at the π-holes are 0.099 and 0.078 au for PO_2F and PO_2Cl, respectively. It is through the π-holes that N bases can donate a pair of electrons to PO_2F and PO_2Cl to form P...N pnicogen bonds.

The P-N distances and the binding energies of complexes YN:PO_2X for YN=NH_3, H_2C=NH, NH_2F, NP, HCN, FCN, NF_3, and N_2, are reported in Table 8.30. There are large variations among P-N distances, which range from 1.88 to 2.90 Å, and among binding energies, which range from − 13 to − 149 kJ·mol^{-1}. Based on these two properties, the complexes may be subdivided into three groups: the first containing the strongest bases NH_3, H_2C=NH, and NH_2F; the second containing the sp hybridized nitrogen bases PN, HCN, and FCN; and the third group made up of complexes with the weak bases NF_3 and N_2. From Table 8.30 it can also be seen that the binding energy is greater and the P-N distance shorter in YN:PO_2F compared to the corresponding YN:PO_2Cl, consistent with the MEP values for the two acids. Figure 8.37 presents a plot of the binding energies of these complexes versus the P-N distance.

There are two interesting structural features of complexes with the stronger bases. The first is the distortion from planarity of the PO_2X molecules. The degree of distortion can be measured as the sum of the bond angles around P, which is 360° for a planar structure, and 328.4° for a tetrahedral structure. When the base is the strong base H_2C=NH, the sums are 351.7° and 352.2° for complexes with PO_2Cl and PO_2F, respectively. In contrast, this sum is 360.0° and 359.9°, respectively, when the base is N_2. Moreover, in the complexes with the stronger nitrogen bases, the

Table 8.30 Binding energies (ΔE, kJ·mol^{-1}) and P-N distances (R, Å) for complexes YN:PO_2X. (Reprinted with permission from Ref. [33]. Copyright (2013) American Chemical Society)

	PO_2F		PO_2Cl	
	ΔE	R	ΔE	R
H_3N:PO_2X	– 148.5	1.902	– 128.6	1.912
H_2C=(H)N:PO_2X	– 146.2	1.878	– 126.0	1.893
H_2FN:PO_2X	– 105.1	1.953	– 90.2	1.962
PN:PO_2X	– 87.4	1.926	– 70.3	1.949
HCN:PO_2X	– 60.4	2.042	– 42.8	2.161
FCN:PO_2X	– 52.0	2.096	– 36.7	2.295
F_3N:PO_2X	– 21.8	2.381	– 16.5	2.646
N_2:PO_2X	– 15.4	2.725	– 12.9	2.897

P-N distances are very short, and approach the computed MP2/aug'-cc-pVTZ P-N distances of 1.63 Å in PO_2NH_2 and 1.67 Å in PO_2NC.

The ELF representations illustrated in Fig. 8.38 for the most strongly and weakly bound complexes H_3N:PO_2F and N_2:PO_2Cl, respectively, clearly illustrate lone pair donation by the nitrogen base and acceptance through the π-hole of phosphorus. Consistent with this picture, the AIM analysis of the electron density shows the presence of one intermolecular BCP and a corresponding bond path connecting the P and N atoms. The Laplacians at the BCP are always positive, but the total energy densities are negative except for the four most weakly bound complexes with the bases NF_3 and N_2. The relatively large negative values of H_{BCP} are indicative of the covalent character of the stronger P...N bonds.

The NBO stabilizing intermolecular charge-transfer energies are reported in Table 8.31. The evaluation of charge-transfer energies is only possible for complexes with binding energies less than – 71 kJ·mol^{-1}, since the NBO method treats complexes with greater binding energies as single molecules, thereby producing unrealistic charge-transfer energies. In complexes with the weaker bases, charge transfer occurs from the nitrogen lone pair to the antibonding σ* P-O and P-X orbital through the π-hole. Thus, PO_2X gains electron density as the nitrogen base loses electron density.

The spin-spin coupling constants ^{1p}J(P-N) for complexes YN:PO_2F and YN:PO_2Cl are given in Table 8.31, and plots of ^{1p}J(P-N) versus the P-N distance in Fig. 8.39 provide further insight into the nature of the N...P pnicogen bonds in these complexes. The points for each trendline can be grouped in a similar way to the groupings for distances and binding energies. Although the correlation coefficients of the trendlines are not good, the trendlines do provide a clear indication of coupling-constant patterns. At long distances, ^{1p}J(P-N) values increase as the P-N distance decreases, reach a maximum value at shorter distances, and then decrease as the P-N distance further decreases. At the shorter distances, ^{1p}J(P-N) approaches the values of ^{1}J(P-N) for the molecules PO_2NC (-0.3 Hz) and PO_2NH_2 (– 26.6 Hz). ^{1p}J(P-N) values are not useful for estimating P-N distances for these complexes, since two different distances can have the same value of ^{1p}J(P-N).

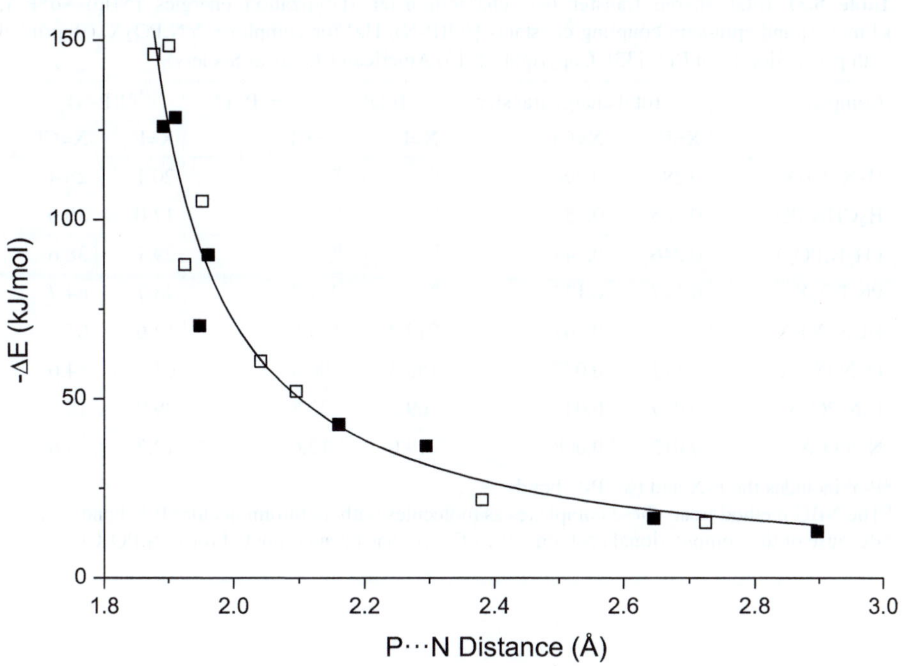

Fig. 8.37 The negative of the binding energy (–ΔE) versus the P-N distance in complexes with PO_2F (□) and PO_2Cl ■. The best-fit trendline has the form $-\Delta E=1/[a+bR(P\text{-}N)]$, with a correlation coefficient R^2=0.94. (Reprinted with permission from Ref. [33]. Copyright (2013) American Chemical Society)

Fig. 8.38 ELF 0.75 au isosurfaces of $H_3N{:}PO_2F$ and $N_2{:}PO_2Cl$ illustrating the interaction between the N lone pair and the π-hole of P. (Reprinted with permission from Ref. [33]. Copyright (2013) American Chemical Society)

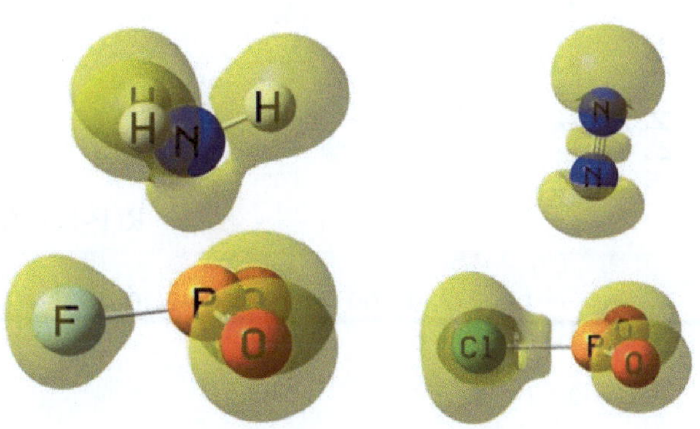

Table 8.31 Total charge transfer (e), charge-transfer stabilization energies [N(lp)→σ*P-Y, kJ·mol^{-1}] and spin-spin coupling constants [^{1p}J(P-N), Hz] for complexes YN:PO_2X. (Reprinted with permission from Ref. [33]. Copyright (2013) American Chemical Society)

Complex	Total charge transfer		Total N(lp)→σ*P-Y[a]		^{1p}J(P-N)	
	X=F	X=Cl	X=F	X=Cl	X=F	X=Cl
$H_3N:PO_2X$	0.286	0.296	[b]	[b]	20.1	28.4
$H_2CHN:PO_2X$	0.245	0.251	[b]	[b]	17.0	30.0
$FH_2N:PO_2X$	0.246	0.244	[b]	[b]	29.1	38.6
$PN:PO_2X$	0.127	0.127	[b]	321.3	48.7	64.7
$HCN:PO_2X$	0.130	0.107	219.2	159.8	57.6	65.3
$FCN:PO_2X$	0.117	0.077	188.3	96.4	64.2	54.6
$F_3N:PO_2X$	0.079	0.043	69.5	32.8	39.9	18.8[c]
$N_2:PO_2X$	0.013	0.008	19.0	12.6	12.2	5.6

[a]P-Y includes the P-X and two P-O bonds
[b]The NBO method treats these complexes as molecules with an intramolecular P-N bond
[c]Because of the computational cost, only the FC term has been computed for $F_3N:PO_2Cl$

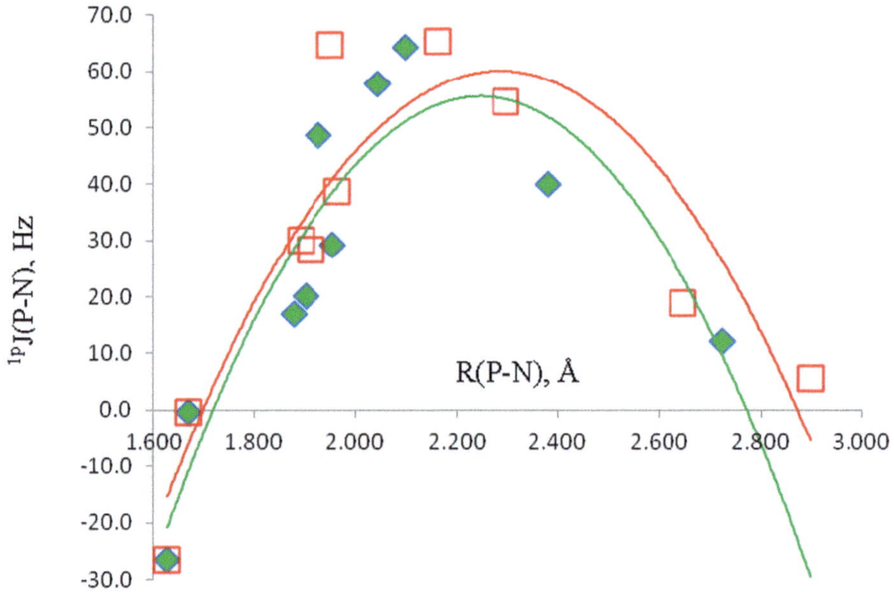

Fig. 8.39 ^{1p}J(P-N) versus R(P-N) for complexes YN:PO_2F ◆ and YN:PO_2Cl □. The points for PO_2NH_2 and PO_2NC at short distances have been included with both series. The correlation coefficients R^2 are 0.84 and 0.83, respectively. (Reprinted with permission from Ref. [33]. Copyright (2013) American Chemical Society)

8.3.3 Ternary and Higher-Order Complexes with σ-σ Pnicogen Bonds

Binary complexes stabilized by σ-σ pnicogen bonds may form ternary and higher-order complexes. In this section we examine the effect of the formation of other intermolecular bonds on the properties of pnicogen bonds, and the cooperativity of intermolecular binding energies. Complexes included in this section are the trimers $(PH_2X)_3$ [31], and pnicogen-bonded complexes that are also hydrogen-bonded, $nFH:(PH_2F)_2$ and $nFH:(H_2FP:NFH_2)$, for n=1 − 3 [26, 27].

8.3.3.1 $(PH_2X)_3$

The equilibrium pnicogen-bonded trimers $(PH_2X)_3$, for X=F, Cl, OH, NC, CCH, CN, CH_3, and H have C_{3h} symmetry, except for $[PH_2(CH_3)]_3$ which has C_1 symmetry, although the C_1 structure is only 0.25 kJ·mol^{-1} more stable than C_{3h}[31]. The structures and molecular graphs of trimers $(PH_2F)_3$, $[PH_2(OH)]_3$ and $(PH_3)_3$ are illustrated in Fig. 8.40. P-P distances, A-P-P angles, and binding energies of these trimers and the corresponding dimers are reported in Table 8.32. The trimer binding energies span a large range, from − 17.1 kJ·mol^{-1} for $(PH_3)_3$ to − 62.9 kJ·mol^{-1} for $(PH_2F)_3$, and are necessarily greater than those of the corresponding dimers. However, the energy per P...P bond is greater in the dimer compared to the corresponding trimer.

The intermolecular P-P distances in the trimers are significantly longer than the corresponding dimer distances, and vary from 2.97 Å in $(PH_2F)_3$ to 3.81 Å in $(PH_3)_3$. Trimer binding energies exhibit a quadratic dependence on the P-P distance, with a correlation coefficient of 0.981. The A-P-P angles are between 159 and 171°, and provide a P-P-A alignment which approaches linearity. This alignment facilitates the stabilizing charge-transfer interactions from the lone pair orbital of one P to the σ* P-A orbital of the P adjacent to the lone pair.

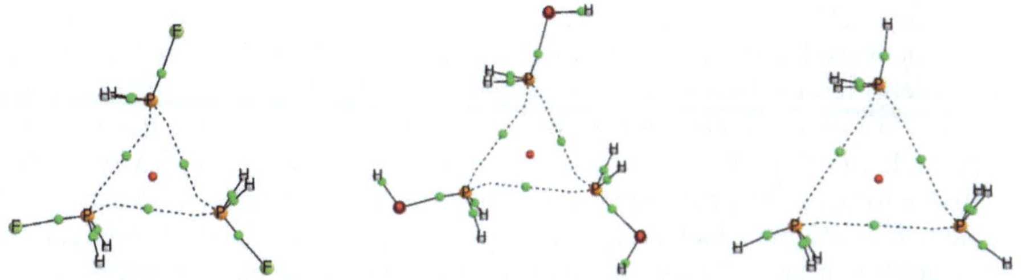

Fig. 8.40 Molecular graphs of trimers $(PH_2F)_3$, $[PH_2(OH)]_3$ and $(PH_3)_3$. Green and red dots indicate the positions of bond and ring critical points, respectively. The dashed lines connecting the P atoms and the BCPs are the bond paths. (Reprinted with permission from Ref. [31]. Copyright (2013) American Chemical Society)

Table 8.32 Binding energies (ΔE, kJ·mol^{-1}), intermolecular P-P distances (R, Å) and A-P-P angles (∠A-P-P,°) for $(PH_2X)_3$ and $(PH_2X)_2$. (Reprinted with permission from Ref. [31]. Copyright (2013) American Chemical Society)

	$(PH_2X)_3$			$(PH_2X)_2$		
Monomer	ΔE	R	∠A-P-P	ΔE	R	∠A-P-P
PH_2F	– 62.9	2.974	171	– 34.0	2.471	163
PH_2Cl	– 49.6	3.144	171	– 22.1	2.771	167
$PH_2(OH)$	– 43.9	3.229	168	– 20.6	2.851	169
$PH_2(NC)$	– 37.4	3.272	170	– 13.8	3.040	168
$PH_2(CCH)$	– 30.3	3.549	166	– 12.2	3.353	174
$PH_2(CN)$	– 23.7	3.541	168	– 8.4	3.375	171
$PH_2(CH_3)$[a]	– 18.8	3.738	161	– 8.9	3.481	178
PH_3	– 17.1	3.809	159	– 7.0	3.589	179

[a] C_{3h} structure

Table 8.33 MBIE components of the interaction energies (kJ·mol^{-1}) for trimers with C_{3h} symmetry. (Reprinted with permission from Ref. [31]. Copyright (2013) American Chemical Society)

	$E_R(1)$[a]	$\sum E_R$	$\Delta^2E(1,2)$[a]	$\sum \Delta^2E$	$\Delta^3E(1,2,3)$
$(PH_3F)_3$	0.90	2.70	– 21.48	– 64.44	– 1.16
$(PH_2Cl)_3$	0.50	1.50	– 16.58	– 49.74	– 1.29
$[PH_2(OH)]_3$	0.24	0.72	– 15.01	– 45.03	+0.45
$[PH_2(NC)]_3$	0.40	1.20	– 12.06	– 36.18	– 2.38
$[PH_2(CCH)]_3$	0.12	0.36	– 10.12	– 30.36	– 0.29
$[PH_2(CN)]_3$	0.19	0.57	– 7.64	– 22.92	– 1.32
$[PH_2(CH_3)]_3$	0.05	0.15	– 6.13	– 18.39	– 0.53
$(PH_3)_3$	0.05	0.15	– 5.61	– 16.83	– 0.36

[a] For these trimers, $E_R(1)=E_R(2)=E_R(3)$, and $\Delta^2E(1,2)=\Delta^2E(1,3)=\Delta^2E(2,3)$

Trimer total binding energies have been decomposed in terms of many-body interaction energies (MBIE)[112, 113] which are reported in Table 8.33. By far, the dominant energy term is the two-body interaction $\Delta^2E(1,2)$. The monomer relaxation energy $E_R(1)$ is destabilizing, while the three-body energy $\Delta^3E(1,2,3)$ is stabilizing, except for $[PH_2(OH)]_3$. An excellent linear correlation exists between $\sum \Delta^2E$ and ΔE.

In the trimers, charge transfer occurs from the lone pair of one P to the σ* P-A orbital of the P atom which is adjacent to the lone pair, with the P…P-A alignment approaching linearity. A linear correlation exists between the binding energies of the trimers and the charge-transfer energies, with a correlation coefficient R^2=0.958. In all trimers, electron population of the σ* P-A orbital leads to a lengthening of the P-A bond relative to the monomer, as evident from the data of Table 8.34.

Table 8.34 P(lp)→σ*P-A[a] charge transfer energies (kJ·mol^{-1}) and changes in P-A bond lengths [δR(P-A), Å] in trimers $(PH_2X)_3$. (Reprinted with permission from Ref. [31]. Copyright (2013) American Chemical Society)

Trimers	P(lp)→σ*P-A	δR(P-A)
$(PH_3F)_3$	50.5	0.006
$(PH_2Cl)_3$	30.6	0.009
$[PH_2(OH)]_3$	26.4	0.004
$[PH_2(NC)]_3$	26.9	0.001
$[PH_2(CCH)]_3$	11.0	0.003
$[PH_2(CN)]_3$	11.8	0.001
$[PH_2(CH_3)]_3$	5.6	0.002
$(PH_3)_3$	5.3	0.002

[a]A is the atom of X which is directly bonded to P, with the P-P-A angle approaching linearity

The ELF isosurface of each trimer shows the location of the P lone pairs, as illustrated in Fig. 8.41a for $(PH_2F)_3$. The MEP on the 0.001 au isosurface of isolated PH_2F in Fig. 8.41b identifies the lone pair and σ-hole for a single molecule, and their interactions in the trimer are depicted schematically in Fig. 8.41c. The molecular graph of $(PH_2F)_3$ in Fig. 8.42 shows the three intermolecular bond critical points that connect the phosphorous atoms, and a ring critical point. The electron densities at P...P BCPs correlate exponentially with the P-P distance with an R^2 value of 0.9996.The P...P bond paths show a significant curvature, especially in the more strongly bound complexes. The simultaneous representation of the Laplacians and bond paths in Fig. 8.42 indicates that each path leads from one P atom where the concentration of lone pair electron density is high and the Laplacian is negative, to

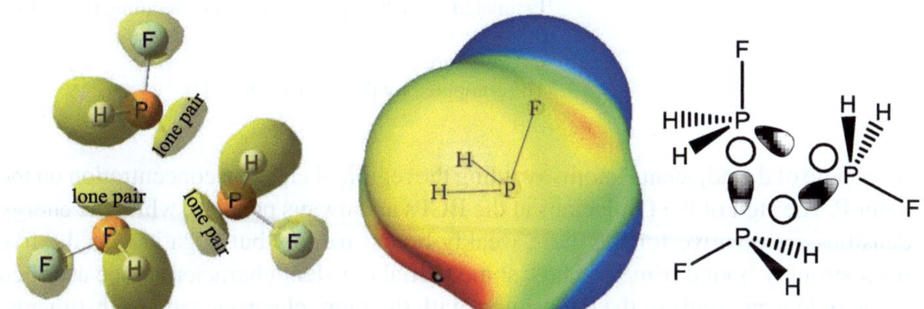

Fig. 8.41 **a** ELF 0.75e isosurface of $(PH_2F)_3$ illustrating the P lone pair basins. **b** The MEP on the 0.001 au electron density isosurface of the isolated PH_2F molecule. The location of the σ-hole maxima (0.060 au) is indicated with a black dot. Color code scale (au): *Red* > 0.03 > *Yellow* > 0.015 > *Green* > 0.0 > *Blue*. **c** Schematic representation of the $(PH_2F)_3$ lone pairs (*shaded lobes*) and σ-holes (*open circles*). (Reprinted with permission from Ref. [31]. Copyright (2013) American Chemical Society)

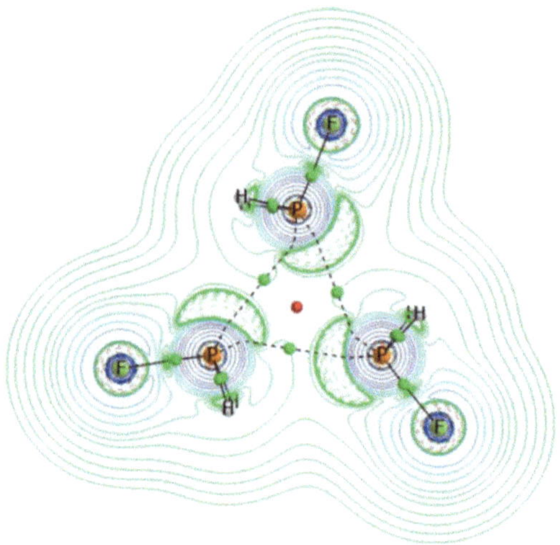

Fig. 8.42 Laplacian contours in the symmetry plane and the molecular graph of $(PH_2F)_3$. Negative values of the Laplacian are indicated with dashed lines. (Reprinted with permission from Ref. [31]. Copyright (2013) American Chemical Society)

Table 8.35 Computed Fermi-contact terms and estimated spin-spin coupling constants [^{1p}J(P-P), Hz] for trimers $(PH_2X)_3$. (Reprinted with permission from Ref. [31]. Copyright (2013) American Chemical Society)

Trimer	Computed FC terms	Estimated ^{1p}J(P-P)^a
$(PH_2F)_3$	502.0	503
$(PH_2Cl)_3$		384
$[PH_2(OH)]_3$	331.6	332
$[PH_2(NC)]_3$		306
$[PH_2(CCH)]_3$		172
$[PH_2(CN)]_3$		175
$[PH_2(CH_3)]_3$	103.6^b	106
$(PH_3)_3$	86.8^c	86

^{a}Estimated from the equation of the FC trendline for the four trimers

^{b}C_{3h} structure

^{c}The computed total ^{1p}J(P-P) is 86.9 Hz

the σ-hole of the adjacent P atom, avoiding the region of electron concentration on the latter P. The sign of the Laplacians at the BCPs are always positive, while the energy densities are positive for the more weakly bound trimers but negative for the five most strongly bound trimers. Thus, some partial covalent character can be ascribed to the pnicogen bonds in the five trimers with the more electronegative substituents.

Because of the computational cost associated with EOM-CCSD calculations, it was not possible to compute coupling data for all trimers. Since Fermi contact terms have been found to be good approximations to P-P coupling constants in pnicogen-bonded complexes, these terms are reported in Table 8.35 for two strongly bound trimers $(PH_2F)_3$ and $[PH_2(OH)]_3$, and the weakly bound trimers $[PH_2(CH_3)]_3$ and $(PH_3)_3$. Total ^{1p}J(P-P) was computed for $(PH_3)_3$. Figure 8.43 provides a plot of the FC terms for the four trimers and ^{1p}J(P-P) for the corresponding dimers $(PH_2X)_2$

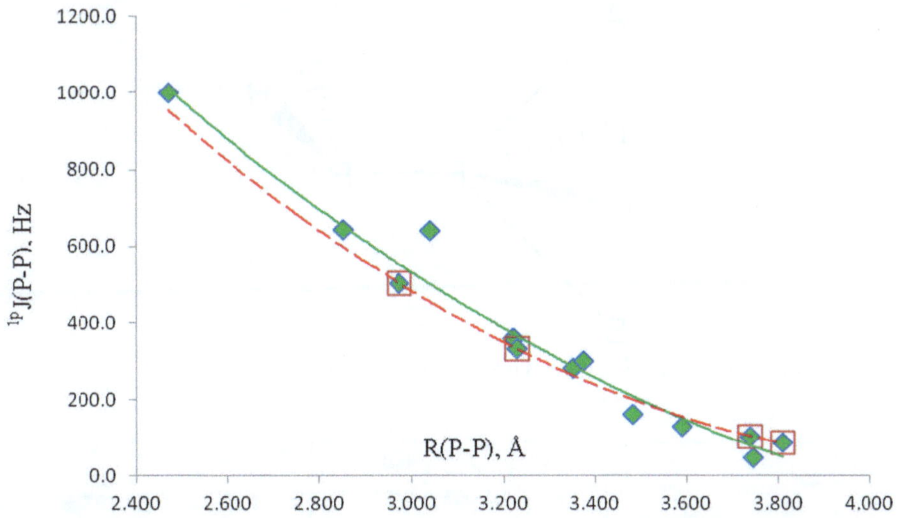

Fig. 8.43 ^{1p}J(P-P) total coupling constants versus the P-P distance for dimers $(PH_2X)_2$, and the FC terms for four trimers $(PH_2X)_3$, versus the P-P distance. The points belonging to the trimers $(PH_3)_3$, $(PH_2F)_3$, $(PH_2OH)_3$,and $(PH_2CH_3)_3$ are enclosed in red boxes. The solid green trendline refers to all complexes, while the dashed red trendline refers only to the trimers. (Reprinted with permission from Ref. [31]. Copyright (2013) American Chemical Society)

against the intermolecular P-P distance. The trendline has a correlation coefficient of 0.967, indicating that the trimer points fit very well with the dimer points. However, the range of ^{1p}J(P-P) for the dimers is very large, and the curvature of the trendline is certainly influenced by the value of ^{1p}J(P-P) for $(PH_2F)_2$ at its short P-P distance. The correlation coefficient of the trendline for the four trimers is 1.000, and this trendline has a reduced curvature. The equation of this trendline has been used to obtain estimates of ^{1p}J(P-P) for the remaining trimers, and these values are also reported in Table 8.35.

8.3.3.2 nFH:$(PH_2F)_2$

The $(PH_2F)_2$ dimer can form higher polymeric structures when the F atoms donate one, two, or three pairs of electrons to FH molecules to form F-H...F hydrogen bonds in ternary (t), quaternary (q), and penternary (p) nFH:$(PH_2F)_2$ complexes, for $n = 1$–3 [26]. To assist in the analysis of the binding energies of these complexes, binding energies of the corresponding nFH:PH_2F binary (b) ternary (t), and quaternary (q) complexes have also been computed. Possible positions of the FH molecule acting as a proton donor to PH_2F have been denoted as A, B, C, and D in Scheme 8.1. These orientations are *cis*, *trans*, and *gauche* at two positions, respectively, to the bisector of the H-P-H angle. In order to distinguish complexes in which FH molecules are bonded to the same or the other PH_2F molecule, a prime has been added to the

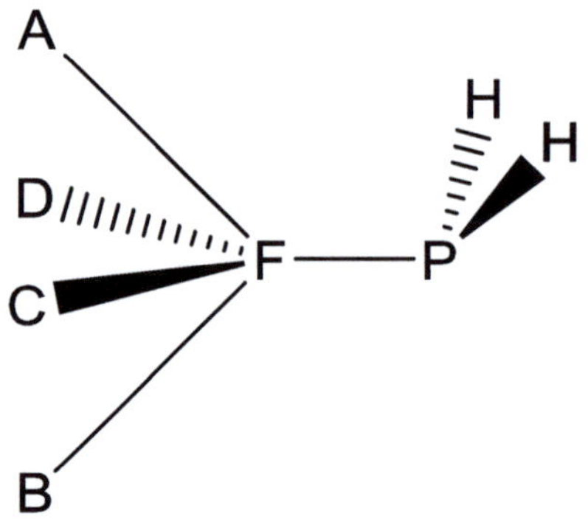

Scheme 8.1 Positions of FH molecules interacting with FPH_2

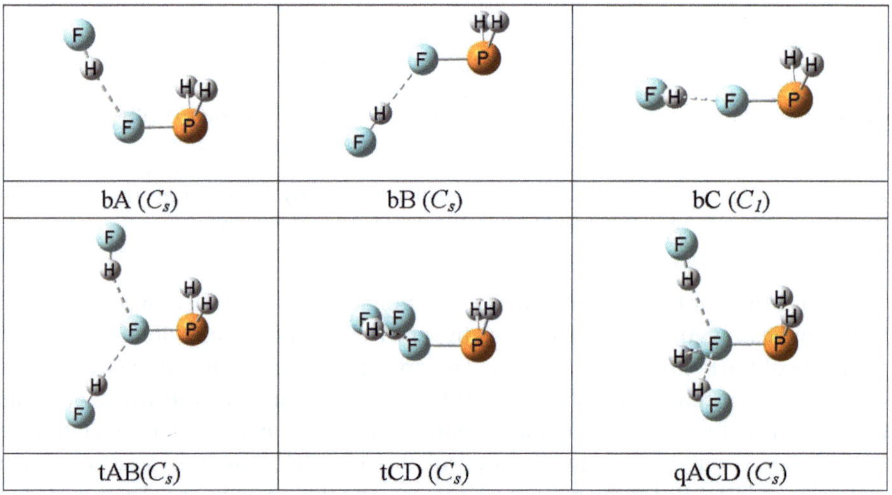

Fig. 8.44 Structures, designations, and symmetries of complexes $nFH:PH_2F$. (Reprinted with permission from Ref. [26]. Copyright (2012) American Chemical Society)

position (A', B', C', D') when the interaction is with the second PH_2F molecule. For complexes having a different number of FH molecules bonded to the two PH_2F molecules, (1) indicates the PH_2F molecule with the greater number of hydrogen bonds, and (2) refers to the molecule with fewer F-H...F bonds. The details of the structures of these complexes and the optimization challenges encountered are not reiterated in this review, but can be found in the original reference. Figs. 8.44 and 8.45 provide the names, symmetries, and pictorial descriptions of the complexes $nFH:(PH_2F)$ and $nFH:(PH_2F)_2$, respectively.

Table 8.36 shows that formation of F-H...F hydrogen bonds in $nFH:(PH_2F)_2$ complexes leads to a decrease in the P-P distance, and that the P-P distance continues to decrease as the number of F-H...F hydrogen bonds increases. Complex qCD with

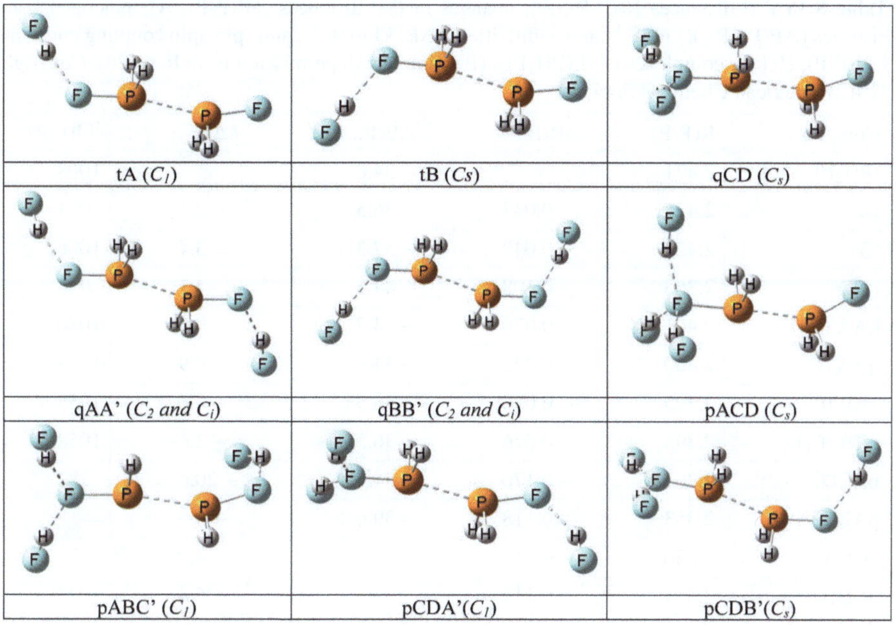

Fig. 8.45 Structures, designations, and symmetries of complexes nFH:$(PH_2F)_2$. (Reprinted with permission from Ref. [26]. Copyright (2012) American Chemical Society)

two FH molecules bonded to F(1) has the shortest P-P distance among the quaternary complexes, while pACD with three FH molecules bonded to F(1) has the shortest P-P distance among the penternary complexes.

The computed P...P binding energies of ternary, quaternary, and penternary complexes are also reported in Table 8.36. The addition of one, two, or three FH molecules hydrogen-bonded to the same or different F atoms of $(PH_2F)_2$ always increases the strength of the pnicogen bond, but the number and positions of the FH molecules determine the extent of that increase. If one FH molecule is bonded to $(PH_2F)_2$, hydrogen bonding at position B is slightly favored. If two FH molecules are bonded to the same F as in qCD, the binding energy further increases, but if the two FH molecules are bonded to different F atoms as in qAA′ and qBB′, the binding energies are similar to but slightly less than the binding energies of tA and tB. If three FH molecules form hydrogen bonds with the same F atom as in pACD, the energy of the P...P bond increases dramatically, and is 15 $kJ{\cdot}mol^{-1}$ greater than the bond energy of other penternary complexes. Bonding two FH molecules to F(1) and one FH to F(2) leads to complexes with binding energies only slightly greater than those of tA and tB.

Table 8.36 reports the nonadditivities of binding energies ($\delta\Delta E$), computed in the standard way given as footnote c. In these complexes the nonadditivities are negative (synergistic), that is, the binding energy of the complex is greater than the sum of the binding energies of the two corresponding components. The penternary complex

Table 8.36 P-P distances [R(P-P), Å], changes in P-P distances [δR(P-P), Å], pnicogen bond energies [ΔE(P...P), kJ·mol^{-1}], nonadditivities (δΔE, kJ·mol^{-1}), and spin-spin coupling constants [^{1p}J(P-P), Hz] for complexes nFH:$(PH_2F)_2$. (Reprinted with permission from Ref. [26]. Copyright (2012) American Chemical Society)

Complex	R(P-P)	δR(P-P)[a]	ΔE(P...P)[b]	δΔE[c]	^{1p}J(P-P)
$(PH_2F)_2$	2.471		– 34.0		1008
tA	2.425	– 0.047	– 36.5	– 2.5	1021
tB	2.422	– 0.049	– 37.3	– 3.4	1006
qCD[d]	2.363	– 0.108	– 44.1	– 4.1	948
qAA′(C_2)	2.401	– 0.070	– 34.7	– 0.8	1080
qAA′(C_i)	2.400	– 0.072	– 34.9	– 0.9	1079
qBB′(C_2)	2.395	– 0.076	– 36.5	– 2.6	1058
qBB′(C_i)	2.395	– 0.076	– 36.5	– 2.6	1058
pACD	2.301	– 0.170	– 54.6	– 20.6	782
pABC′	2.353	– 0.118	– 39.6	– 5.6	—[e]
pCDA′	2.356	– 0.115	– 38.7	– 4.8	—[e]
pCDB′	2.354	– 0.117	– 39.3	– 5.4	1072

[a] δR(P-P)=R(P-P) for nFH:$(PH_2F)_2$-R(P-P) for $(PH_2F)_2$
[b] ΔE(P...P)=E(complex) – {$E_{(1)}$(nFH:PH_2F) + $E_{(2)}$(n′FH:PH_2F)}, where n and n′ indicate the number of FH molecules bonded to molecules (1) and (2), respectively, with $n \geq n'$, and $n + n' \leq 3$
[c] δΔE=ΔE[nFH:$(PH_2F)_2$] – $\Sigma_i \Delta E_i$(binary)
[d] Not an equilibrium structure on the potential surface
[e] EOM–CCSD coupling constants could not be computed for complexes of C_1 symmetry

pACD is unique, having a significantly greater nonadditivity than any other complex. This suggests that the nature of the intermolecular interactions may be changing in this complex.

The topological analysis of the electron density shows the presence of intermolecular bond critical points and corresponding bond paths connecting the pnicogen-bonded P...P atoms and the hydrogen-bonded H...F atoms. Both the Laplacian of the electron density at the P...P BCP and the total energy density at the BCP are negative for all complexes except the parent $(PH_2F)_2$, indicating that these P...P bonds have covalent character. The degree of covalency can be seen in the variation of the Laplacians with the intermolecular distance, which is illustrated in Fig. 8.46. In general, covalency increases as the P-P distance decreases and the number of FH molecules increases. However, complexes qCD and pACD which have two and three FH molecules, respectively, hydrogen bonded to the same F(1) atom, have P...P bonds with significantly greater covalent character than the remaining complexes with the same number of hydrogen bonds. Values of $\nabla^2 \rho_{BCP}$ vary quadratically with the P-P distance, with a correlation coefficient of 0.986.

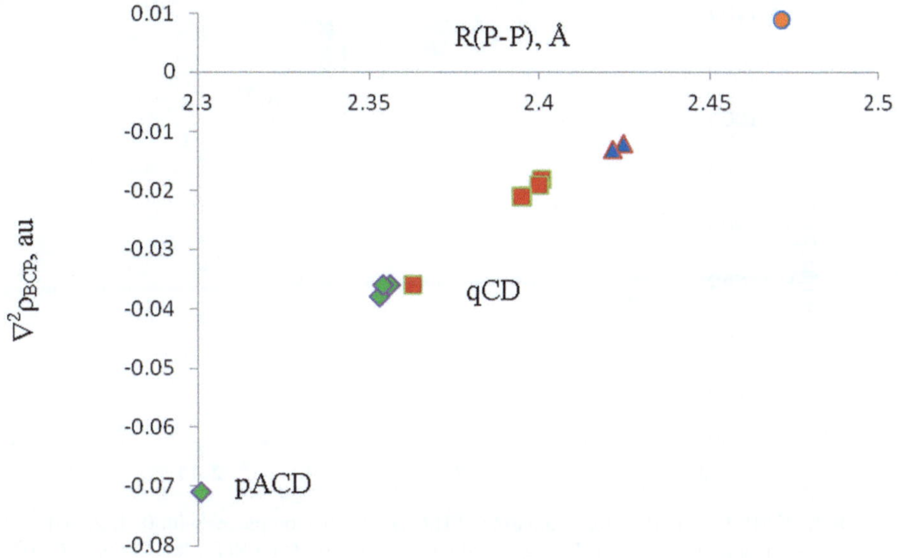

Fig. 8.46 The Laplacian of the electron density versus the P-P distance for complexes ● $(PH_2F)_2$, ▲ $FH:(PH_2F)_2$, ■ $2FH:(PH_2F)_2$, and ◆ $3FH(PH_2F)_2$. (Reprinted with permission from Ref. [26]. Copyright (2012) American Chemical Society)

The default parameters for the NBO analysis usually describe these complexes as single molecules with a formal P...P bond. In order to quantify the orbital interaction energies, it was necessary to impose a Lewis structure with no P...P bond. Unfortunately, the charge-transfer energies obtained are unrealistically high. However, they do show that charge transfer occurs from the PH_2F molecule with fewer F-H...F hydrogen bonds to the molecule with the greater number of hydrogen bonds. Moreover, since charge transfer also occurs from F of PH_2F to the F-H molecule hydrogen-bonded to it, charge-transfer effects will be synergistic in complexes that have a greater number of F-H...F hydrogen bonds at one of the PH_2F molecules.

^{31}P-^{31}P spin-spin coupling constants are also reported in Table 8.36, and are plotted in Fig. 8.47 as a function of the P-P distance. These data exhibit significant scatter. In contrast, the data for a subgroup of complexes including the parent and complexes with one, two or three F-H...F hydrogen bonds at the same F atom, exhibit a very interesting pattern. The trendline for ^{1p}J(P-P) versus the P-P distance for this subset decreases quadratically with decreasing distance, with a correlation coefficient of 0.997.

Some insight into this behavior may be gained by examining the structure of pACD in detail. In this complex, the P-P distance is short, the P(1)-F(1) distance is very long, and the three FH molecules hydrogen bonded to F(1) resemble an anionic cluster $3(FH)F^-$ with approximately local C_{3v} symmetry. This suggests that pACD is approaching the ion-pair complex $H_3F_4^-{:}^+(H_2P{:}PFH_2)$. For comparison, Table 8.37 reports ^{1p}J(P-P) for the cations derived from $(PH_2F)_2$ and pACD, and ^1J(P-P) for the

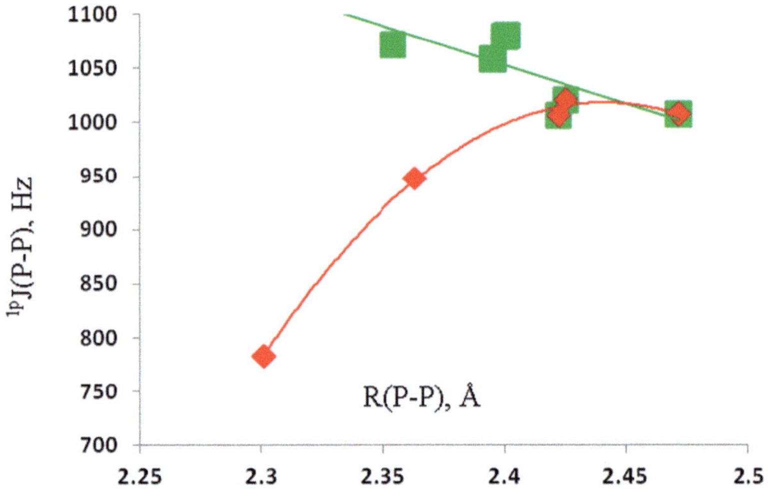

Fig. 8.47 ^{1p}J(P-P) vs R(P-P) for complexes nFH:$(PH_2F)_2$. ♦ Complexes with 0, 1, 2, and 3 FH molecules bonded to the same P-F. (Reprinted with permission from Ref. [26]. Copyright (2012) American Chemical Society)

Table 8.37 P-P distances [R(P-P), Å] and ^{31}P-^{31}P coupling constants [^{1p}J(P-P), Hz] for neutral complexes and corresponding cations $(H_2P{:}PH_2F)^+$, and P-P distances and ^{1}J(P-P) for P_2H_4. (Reprinted with permission from Ref. [26]. Copyright (2012) American Chemical Society)

Neutral	R(P-P)	^{1p}J(P-P)^a	Cation	R(P-P)	^{1p}J(P-P)^a
$(PH_2F)_2$	2.471	1008	$(H_2P{:}PH_2F)^{+b}$	2.471	160
pACD	2.301	783	$(H_2P{:}PH_2F)^{+b}$	2.301	– 147
P_2H_4	R(P-P)	FC	^{1}J(P-P)		
C_2	2.225	– 165	– 115		
C_{2v}	2.266	– 230	– 132		
C_{2h}	2.238	– 45	+ 18		

^{a}Approximated by the FC term
^{b}At the geometry of the corresponding neutral complex

P_2H_4 molecule. These data indicate that the decrease in ^{1p}J(P-P) for qCD and pACD reflects the increasing covalent character of the P...P bonds and the short P-P distances that approach the P-P distance in P_2H_4. However, it should be noted that although the FC term is generally an excellent approximation to ^{1p}J(P-P) in binary complexes, this may not be the case for pACD. At the short P-P distances in P_2H_4, both the PSO and SD terms make non-negligible contributions to the total coupling constant ^{1}J(P-P), which differs significantly from the FC term, as seen from the data of Table 8.37.

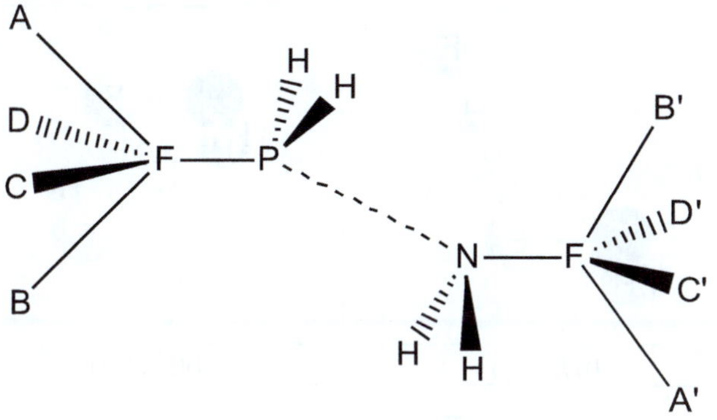

Scheme 8.2 Positions of FH molecules interacting with the binary complex ($H_2FP:NFH_2$)

8.3.3.3 nFH:$H_2FP:NFH_2$

Scheme 8.2 defines the hydrogen-bonding positions for complexes of FH with $H_2FP:NFH_2$ [27]. These are denoted as A, B C, and D when hydrogen bonding occurs at P-F, and A′, B′, C′, and D′ when hydrogen bonding occurs at N-F. b, t, q, and p are defined as for the complexes nFH:PH_2F and nFH:$(PH_2F)_2$. Complexes nFH:NFH_2 and nFH:$H_2FP:NFH_2$ are illustrated in Figs. 8.48 and 8.49, respectively.

Four equilibrium structures FH:($H_2FP:NFH_2$) and four 2FH:($H_2FP:NFH_2$) with FH molecules hydrogen-bonded to P-F and/or N-F have been found on the potential surfaces. In addition, one complex 3FH:($H_2FP:NFH_2$) pACD which is structurally similar to 3FH:$(PH_2F)_2$ pACD, has also been determined. Some complexes with PH_2F or NH_2F needed for energetic comparisons may not be equilibrium structures, but are constrained structures.

Table 8.38 reports the P-N distances in complexes nFH:($H_2FP:NFH_2$). Hydrogen-bond formation at P-F always leads to a decrease of the P-N distance, with the decrease increasing as the number of FH molecules increases. In contrast, hydrogen-bond formation at N-F tends to lengthen the P-N bond, but the changes are smaller than changes in this bond length when hydrogen bonding occurs at P-F. The complex pACD with three FH molecules hydrogen bonded to P-F has the shortest P-N distance of 2.143 Å. However, this distance is significantly longer than the P-N distance of 1.721 Å computed at the same level of theory for the H_2P-NH_2 molecule. P-F bond distances increase relative to the parent complex when hydrogen bonding occurs at P-F, with the degree of lengthening increasing as the number of F-H...F hydrogen bonds increases. The N-F bond lengthens slightly when hydrogen bonding occurs at N-F.

The P...N binding energies of complexes nFH:($H_2FP:NFH_2$) are also reported in Table 8.38. Hydrogen-bond formation only at P-F increases the energy of the P...N bond relative to the parent complex ($H_2FP:NFH_2$), and the binding energy increases as the number of FH molecules increases. In contrast, the energy of the P...N bond decreases slightly when FH molecules hydrogen bond at N-F. This decrease is not very sensitive to the number of FH molecules present, as these binding energies are

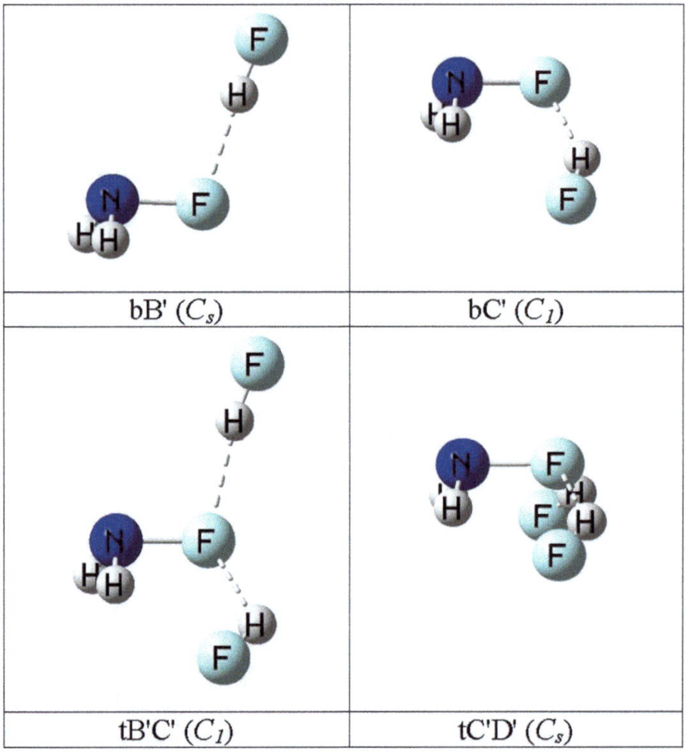

Fig. 8.48 Structures, designations, and symmetries of complexes $FH{:}NFH_2$ and $2FH{:}NFH_2$. (Reprinted with permission from Ref. [27]. Copyright (2012) American Chemical Society)

all approximately -23 kJ·mol^{-1}. The binding energies of the complexes with one FH molecule hydrogen bonded at P-F and one at N-F are intermediate between the ternary complexes with one FH at P-F and those with one FH at N-F.

Cooperative effects in complexes $nFH{:}(H_2FP{:}NFH_2)$ which are stabilized by both pnicogen bonds and hydrogen bonds are reported in Table 8.39. From these data it is apparent that binding energies are synergistic when hydrogen bond formation occurs at the P-F bond. However, the synergistic effect for $3FH{:}(H_2FP{:}NFH_2)$ pACD is less than that for $2FH{:}(H_2FP{:}NFH_2)$ qCD, and significantly less than that for the corresponding complex $3FH{:}(PH_2F)_2$ pACD. In contrast, when hydrogen bonding occurs at N-F, binding energies are diminutive.

Insight into the nonadditivites of binding energies can be gained by examining the charge-transfer energies in these complexes. However, the problem encountered with the NBO method in describing charge-transfer energies in $nFH{:}(PH_2F)_2$ complexes is also encountered for $nFH{:}(H_2FP{:}NFH_2)$, so Lewis structures with no P...N bonds were imposed. The resulting orbital interaction energies are much more realistic than those obtained for the $nFH{:}(PH_2F)_2$ complexes, and are reported in Table 8.40. In the parent complex $(H_2FP{:}NFH_2)$, both N and P have lone pairs of electrons, so

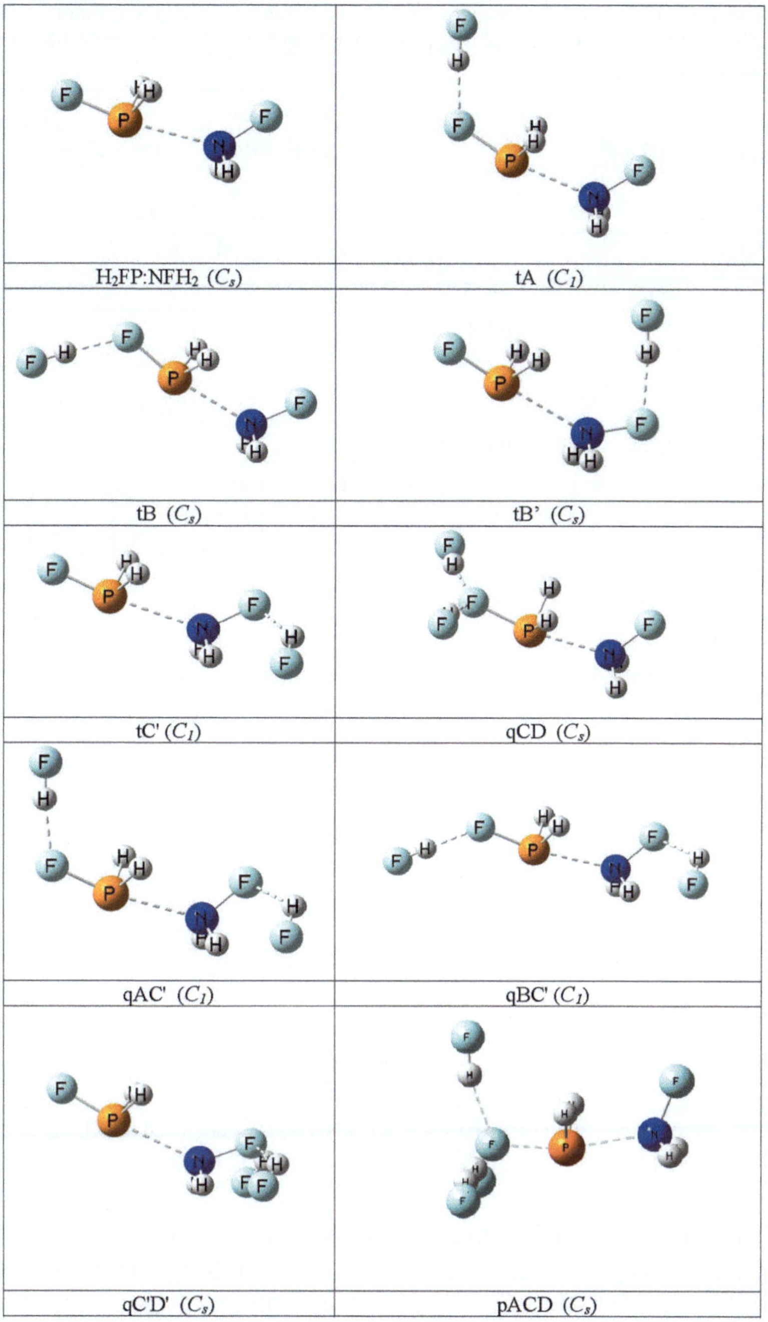

Fig. 8.49 Structures, designations, and symmetries of complexes nFH:(H_2FP:NFH_2). (Reprinted with permission from Ref. [27]. Copyright (2012) American Chemical Society)

Table 8.38 P-N, P-F, and N-F distances (Å) and P...N binding energies ($kJ\cdot mol^{-1}$) for complexes $nFH:(H_2FP:NFH_2)$. (Reprinted with permission from Ref. [27]. Copyright (2012) American Chemical Society)

Structures

Complex	FH at P-F			Same no. of FH at P-F and N-F			FH at N-F		
	P-N	P-F	N-F	P-N	P-F	N-F	P-N	P-F	N-F
$H_2FP:NFH_2$				2.524	1.638	1.420			
tA	2.398	1.678	1.415						
tB	2.397	1.677	1.416						
tB′							2.599	1.629	1.435
tC′							2.546	1.635	1.434
qCD	2.255	1.732	1.411						
qAC′				2.401	1.677	1.425			
qBC′				2.401	1.675	1.427			
qC′D′							2.532	1.634	1.445
pACD	2.143	1.800	1.407						

Binding Energies[a]

	FH at P-F	Same no. of FH at P-F and N-F	FH at N-F
$H_2FP:NFH_2$		– 26.7	
tA	– 34.4		
tB	– 35.3		
tB′			– 22.6
tC′			– 23.7
qCD	– 45.2		
qAC′		– 30.6	
qBC′		– 31.1	
qC′D′			– 23.2
pACD	– 56.8		

[a] $\Delta E(P...N)=E(complex) - \{E(nFH:PH_2F) + E(n'FH:NH_2F)\}$, where n and n′ indicate the number of FH molecules bonded to PH_2F and NH_2F, respectively, with $n + n' \leq 3$

both may act as electron-pair donors. However, since NH_2F is the stronger base, the stabilizing charge-transfer energy for N(lp)→σ*P-F is significantly greater than that for P(lp)→σ*N-F. The N(lp)→σ*P-F charge-transfer energy is very sensitive to the hydrogen-bonding scheme, and in complexes with hydrogen bonding at P-F, this energy increases as the number of hydrogen bonds increases. In contrast, N(lp)→σ*P-F charge-transfer energies decrease when hydrogen bonding occurs at

Table 8.39 Cooperativity of binding energies (δΔE, kJ·mol^{-1})[a] for complexes nFH:(H_2FP:NFH_2). (Reprinted with permission from Ref. [27]. Copyright (2012) American Chemical Society)

Complex	FH at P-F	Same no. of FH at P-F and N-F	FH at N-F
tA	– 7.7		
tB	– 8.6		
tB′			+4.7
tC′			+3.0
qCD	– 12.5		
qAC′		– 4.0	
qBC′		– 4.4	
qC′D′			+9.9
pACD	– 9.2		

[a] δΔE=ΔE[nFH:(H_2FP:NFH_2)] – $\Sigma_i \Delta E_i$(binary)

N-F. If hydrogen bonding occurs at both P-F and N-F, the N(lp)→σ*P-F charge-transfer energy is similar to that for complexes with one FH at P-F. By comparison, N(lp)→σ*P-F charge-transfer energies are not very sensitive to hydrogen bonding.

There are two different charge-transfer interactions in nFH:H_2FP:NFH_2 complexes, one across the P...N pnicogen bond and the other across the F-H...F hydrogen bond. Analysis of the electron densities associated with the molecules PH_2F and NH_2F indicates that when charge-transfer occurs across the pnicogen bond, not all of the charge remains on the PH_2F and NH_2F molecules in nFH:(H_2FP:NFH_2), but

Table 8.40 P(lp)→σ*N-F and N(lp)→σ*P-F charge-transfer energies (kJ·mol^{-1}) for nFH:(H_2FP:NFH_2) complexes. (Reprinted with permission from Ref. [27]. Copyright (2012) American Chemical Society)

Complex	P (lp)→σ*N-F			N (lp)→σ*P-F		
	FH at P-F	Same no. FH at each	FH at N-F	FH at P-F	Same no. FH at each	FH at N-F
H_2FP:NFH_2		12.3			58.7	
tA	9.1			91.3		
tB	9.5			91.4		
tB′			15.4			43.5
tC′			13.9			53.5
qCD	12.0			137.2		
qAC′		9.2			90.3	
qBC′		9.9			89.9	
qC′D′			15.9			54.5
pACD	15.3			195.2		

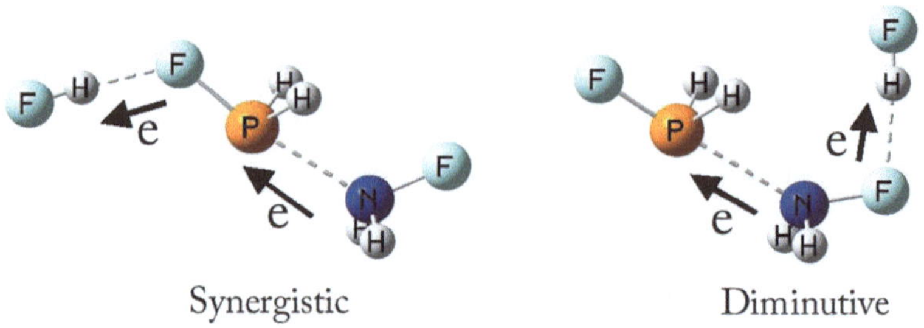

Fig. 8.50 Charge-transfer across pnicogen bonds and hydrogen bonds in complexes FH:(H_2FP:NFH_2) with hydrogen bonding at P-F and N-F. (Reprinted with permission from Ref. [27]. Copyright (2012) American Chemical Society)

some charge is transferred to the FH molecules. Moreover, charge-transfer across hydrogen bonds appears to be the dominant charge-transfer interaction. As a result, when hydrogen bonding occurs at P-F, the direction of charge flow is that preferred in the parent complex across the pnicogen bond from NH_2F to PH_2F, and this is in the same direction as that from PH_2F to FH across hydrogen bonds. These complexes are then more stable than the parent, and the nonadditivities reported in Table 8.39 are synergistic. In contrast, when hydrogen bonding occurs at N-F, the direction of charge flow is from NH_2F to FH, which is opposite that preferred in the parent complex from NH_2F to PH_2F. As a result, the complexes become less stable, and nonadditivities are antagonistic, or diminutive. These two different charge-transfer schemes are illustrated in Fig. 8.50.

Table 8.41 provides intermolecular P-N distances and values of ${}^{1p}J$(P-N), and Fig. 8.51 presents a graphical representation of these data. It is apparent that like ${}^{1p}J$(P-P) for complexes nFH:$(PH_2F)_2$, the usual distance dependence of increasing absolute value of ${}^{1p}J$ with decreasing distance is not found. However, the parent complex (H_2FP:NFH_2) and those with FH molecules hydrogen bonded only at P-F do exhibit a pattern. The trendline which connects ${}^{1p}J$(P-N) for these complexes has a correlation coefficient of 0.981. The changes in the P-N distance and ${}^{1p}J$(P-N) in 3FH:(H_2FP:NFH_2) pACD are reminiscent of changes observed for 3FH:$(PH_2F)_2$ pACD, but are much smaller. While the P-N bond length in 3FH:$(H_2FP$:$NFH)_2$ approaches the P-N bond length in the C_s and C_1 structures of H_2P-NH_2, it is significantly greater than the computed P-N bond lengths of 1.766 and 1.721 Å, respectively, for this molecule. The FC terms for P-N coupling in these structures are 12.9 and 44.5 Hz, respectively. Thus, the FC term does decrease systematically in absolute value in complexes with 1, 2, and 3 FH molecules at P-F. Its value in pACD appears to be approaching the value for the covalent bond in H_2P-NH_2, as can be seen in Fig. 8.51. The decrease in ${}^{1p}J$(P-N) in pACD is a reflection of the changing nature of the P...N pnicogen bond as it acquires increased covalent character.

Table 8.41 R(P-N) (Å) and ${}^{1\text{p}}J$(P-N) (Hz) for complexes nFH:(H_2FP:NFH_2) with n=1–3. (Reprinted with permission from Ref. [27]. Copyright (2012) American Chemical Society)

	FH at PH_2F		FH at PH_2F and NH_2F		FH at NH_2F	
Complex	R(P-N)	${}^{1\text{p}}J$(P-N)	R(P-N)	${}^{1\text{p}}J$(P-N)	R(P-N)	${}^{1\text{p}}J$(P-N)
(H_2FP:NFH_2)			2.524	– 113.5		
tA	2.398	– 114.5				
tB	2.397	– 115.0				
tB′					2.599	– 124.4
tC′					2.546	– 120.7
qCD	2.255	– 111.4				
qAC′			2.401	– 117.3		
qBC′			2.401	– 119.7		
qC′D′					2.532	– 128.2
pACD	2.143	– 99.6				

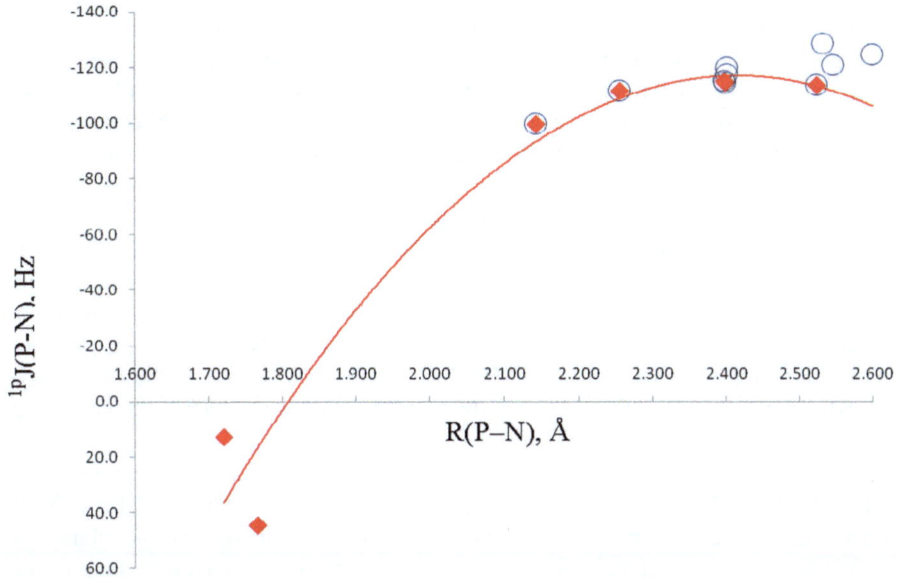

Fig. 8.51 ${}^{1\text{p}}J$(P-N) vs. the P-N distance for complexes (○) nFH:(H_2FP:NFH_2). ♦ Complexes with 0, 1, 2, and 3 FH molecules hydrogen bonded at P-F and the H_2P-NH_2 molecule at two different symmetries. The correlation coefficient R^2=0.951. (Reprinted with permission from Ref. [27]. Copyright (2012) American Chemical Society)

8.4 Summary

The pnicogen bond is a Lewis acid–Lewis base interaction in which a group 15 atom acts as the Lewis acid. Complexes stabilized by pnicogen bonds that have been investigated in our laboratories and included in this review are the dimers $(PH_2X)_2$ and $(H_2C{=}PX)_2$, the binary complexes $H_2C{=}(X)P{:}PXH_2$, $H_2XP{:}PCX$, $H_2XP{:}NXH_2$, $X{=}PH_3{:}NY$, $X{=}PH_3{:}PY$, $H_2FP{:}ClX$, and $YN{:}PO_2X$, and the molecular anions $H_2YP{:}X^-$, for a variety of substituents X and Y. Trimers $(PH_2X)_3$, and ternary and high-order complexes $nFH{:}(PH_2F)_2$ and $nFH{:}(H_2FP{:}NFH_2)$ for $n = 1$–3, have been included as well. Some of the general features of pnicogen-bonded complexes which have emerged from our studies are summarized below.

A σ-σ pnicogen bond arises between two pnicogen atoms when the first atom donates a σ lone pair of electrons to the second through its σ-hole, and that atom in turn donates a σ lone pair of electrons to the first through its σ-hole. If the two pnicogen atoms are both phosphorus, then the structures of the resulting complexes tend to have linear A-P...P-A′ alignments, with A and A′ the atoms of the substituents X and Y that are directly bonded to the two phosphorus atoms. This alignment provides for the overlap of the lone-pair orbital of one P with the σ* P-A orbital of the other, and facilitates P(lp)→σ*P-A charge-transfer across the pnicogen bond. Charge-transfer energies usually correlate with the binding energies of these complexes, provided that there are no other secondary interactions. They may also correlate with the distance across the pnicogen bond, since charge transfer is more efficient at shorter distances. Moreover, since different atoms may be bonded to the two P atoms, the identities of A and A′ can vary, which gives rise to different stable conformers with the same molecular formula, but with different binding energies and charge-transfer energies.

σ-σ pnicogen bonds may also form between P and N in $H_2XP{:}NXH_2$, between P and either P or N in $X{=}PH_3{:}NY$ and $X{=}PH_3{:}PY$, between P and Cl in complexes $H_2FP{:}ClX$, and even between two N atoms in the complex $(NH_2F)_2$. In complexes $H_2XP{:}NXH_2$, the preferred direction of charge transfer is N(lp)→σ*P-A. However, as the binding energies of the $H_2XP{:}NXH_2$ complexes decrease, the two charge-transfer energies approach each other. In the two complexes which have the smallest binding energies, the preferred direction of charge-transfer is P(lp)→σ*N-A. In complexes $X{=}PH_3{:}NY$ and $X{=}PH_3{:}PY$, the only charge-transfer interaction occurs from the N or P base to the $X{=}PH_3$ acid. Similarly, in complexes $H_2FP{:}ClX$, the only stabilizing charge-transfer interaction is Cl(lp)→σ*P-F. In the molecular anions $H_2YP{:}X^-$ charge-transfer always occurs from the lone pair on A to the σ* P-A′ orbital, with A and A′ the atoms of X and Y, respectively, which are directly bonded to P. Molecular anions have relatively large binding energies and charge-transfer energies. Depending on the nature of X and Y, the P-A bonds in these complexes may have reduced ion-molecule character and increased covalent character, and the P-A′ bonds may have reduced covalent character and increased ion-molecule character. These bonding characteristics are reflected in other properties of these complexes.

If one of the two molecules containing a pnicogen atom is unsaturated, then a π-σ pnicogen bond can be formed. In these complexes, the first molecule donates π electrons to the second through its σ-hole, and that atom in turn donates a σ lone pair of electrons to the first through its π-hole. In general, complexes with π-σ pnicogen bonds are more stable than corresponding complexes with σ-σ pnicogen bonds. Charge-transfer also stabilizes π-σ pnicogen bonds. In complexes H_2C=(X)P:PXH_2, charge transfer from the π P=C bond of H_2C=PX to the σ* P-A orbital of PXH_2 is more stabilizing than charge transfer from the lone pair on P of PXH_2 to the π* P=C orbital. Similarly, for complexes H_2XP:PCX, charge-transfer from the π orbital of PCX to the σ* P-A orbital of PXH_2 is more stabilizing than the reverse transfer. It appears to be the nature of the σ* P-A orbital which is the dominant factor in determining the direction of charge transfer. PO_2X:NY and PO_2X:PY are also stabilized by π-σ pnicogen bonds. Since the molecules PO_2F and PO_2Cl have equilibrium planar structures, NY and PY act as σ electron-pair donors to P through its π-hole. In the more strongly bound complexes, PO_2F and PO_2Cl distort from their planar structures.

A second intermolecular interaction by a pnicogen-bonded binary complex can alter the strength of the pnicogen bond. Pnicogen-bonded trimers $(PH_2X)_3$ have greater total binding energies than the corresponding dimers, but the binding energies per pnicogen bond are reduced and the P-P distances are longer in the trimers compared to the corresponding dimers. The A-P...P alignment approaches linearity, and this facilitates charge transfer from the lone pair orbital of one P to the σ* P-A orbital of the P atom which is adjacent to the lone pair. The charge-transfer energies correlate with the trimer binding energies.

Dimers $(PH_2F)_2$ can also form ternary and higher-order complexes with FH, nFH:$(PH_2F)_2$, for $n = 1$–3, in which the FH molecules are hydrogen bonded to P-F, forming F-H...F hydrogen bonds. The addition of one, two, or three FH molecules at the same or different P-F bonds always increases the binding energies, with the greatest increase occurring in the complex with three FH molecules bonded to the same P-F bond. In this complex, the P...P bond acquires significant covalent character as it shortens and approaches the P-P distance in isolated P_2H_4. The pnicogen and hydrogen bond energies are synergistic. Charge-transfer stabilizes these complexes, and occurs from the molecule with fewer F-H...F hydrogen bonds to the molecule which forms the greater number of hydrogen bonds, and from that molecule to the FH molecules.

In contrast, hydrogen-bond formation may increase or decrease the energy of the P...N bond in nFH:(H_2FP:NFH_2) complexes depending on the number of FH molecules and their interaction sites. Hydrogen bonding at P-F decreases the P-N distance and increases the energy of the P...N bond, but not nearly to the extent seen for nFH:$(PH_2F)_2$ complexes. Hydrogen bonding at N-F increases the P-N distance and decreases the energy of the P...N bond. Thus, binding energies are synergistic in nFH:$(PH_2F)_2$ but may be either synergistic or diminutive in nFH:(H_2FP:NFH_2) depending on the hydrogen-bonding scheme. In nFH:(H_2FP:NFH_2) with hydrogen bonding at P-F, charge transfer occurs from the lone-pair orbital on N to the σ* P-F orbital, and then from the lone pair of that F to the σ* F-H orbital. These two charge-transfer interactions are in the same direction, and have a synergistic effect on the binding energies. In the nFH:(H_2FP:NFH_2) complexes with hydrogen bonding at N-F, the two charge transfers are in opposite directions, from the lone-pair orbital of N to the σ* P-F orbital of PH_2F, and from the lone pair on F of NH_2F to the σ* F-H orbital. This leads to a diminutive effect on binding energies.

Spin-spin coupling constants across pnicogen bonds provide further insight into pnicogen-bonded complexes. In general, ^{1p}J is dominated by the Fermi contact term, which is a very good approximation to total J. Coupling constants ^{1p}J(P-P) for complexes with σ-σ bonds are greater in absolute value than ^{1p}J(P-P) for corresponding complexes with π-σ bonds. This may be attributed to the dominance of the Fermi contact terms which depend on s electron densities in ground and excited states. For a related series of complexes, coupling constants usually increase in absolute value as the distance between the pnicogen-bonded atoms decreases. Thus, intermolecular distances could be extracted from these coupling constants. However, if the character of the pnicogen bond changes within a series of complexes, then the distance dependence of ^{1p}J also changes. At long distances, ^{1p}J may increase with decreasing distance, exhibit a maximum absolute value at an intermediate distance, and then decrease as the distance further decreases and approaches the covalent bond distance of a related isolated molecule.

Acknowledgments This work was supported by the Ministerio de Economía y Competitividad (Project No. CTQ2012–35513-C02–02) and the Comunidad Autónoma de Madrid (Project MADRISOLAR2, ref. S2009/PPQ1533). The authors are also grateful to the CTI of CSIC and the Ohio Supercomputer Center for computational support. Without all of this support, the work reported in this review would not have been possible.

References

1. Heyding RD, Calvert LD (1961) Arsenides of the transition metals. IV. A note on the platinum metal arsenides. Can J Chem 39:955–957
2. Girolami GS (2009) Origin of the terms pnictogen and pnictide. J Chem Educ 86:1200–1201
3. Cranton GE, Heyding RD (1965) Synthesis in aqueous solutions of binary compounds of the transition metals with group VA and VIA elements. Can J Chem 43:2027–2032
4. Arunan E, Desiraju GR, Klein RA, Sadlej J, Scheiner S, Alkorta I, Clary DC, Crabtree RH, Dannenberg JJ, Hobza P, Kjaergaard HG, Legon AC, Mennucci B, Nesbitt DJ (2011) Definition of the hydrogen bond (IUPAC Recommendations 2011). Pure Appl Chem 83:1637–1641
5. Desiraju GR, Ho PS, Kloo L, Legon AC, Marquardt R, Metrangolo P, Politzer P, Resnati G, Rissanen K (2013) Definition of the halogen bond (IUPAC Recommendations 2013). Pure Appl Chem 85:1711–1713
6. Politzer P, Murray JS, Clark T (2013) Halogen bonding and other σ-hole interactions: a perspective. Phys Chem Chem Phys 15:11178–11189
7. Mulliken RS, Person WB (1969) Molecular complexes. Wiley, New York
8. Metrangolo P, Resnati G, Pilati T, Biella S (2008) Halogen bonding in crystal engineering. Struct Bond 126:105–136

9. Zahn S, Frank R, Hey-Hawkins E, Kirchner B (2011) Pnicogen bonds: a new molecular linker? Chem Eur J 17:6034–6038
10. Klinkhammer KW, Pyykko P (1995) Ab Initio Interpretation of the closed-shell intermolecular E...E attraction in dipnicogen $(H_2E\text{-}EH_2)_2$ and dichalcogen $(HE\text{-}EH)_2$ hydride model dimers. Inorg Chem 34:4134–4138
11. Carré F, Chuit C, Corriu RJP, Monforte P, Nayyar NK, Reyé C (1995) Intramolecular coordination at phosphorus: donor-acceptor interaction in three- and four-coordinated phosphorus compounds. J Organomet Chem 499:147–154
12. Galasso V (2004) Theoretical study of the structure and bonding in phosphatrane molecules. J Phys Chem A 108:4497–4504
13. Murray JS, Lane P, Politzer P (2007) A predicted new type of directional noncovalent interaction. Int J Quantum Chem 107:2286–2292
14. Ganesamoorthy C, Balakrishna MS, Mague JT, Tuononen HM (2008) New tetraphosphane ligands $\{(X_2P)_2NC_6H_4N(PX_2)_2\}$ (X=Cl, F, OMe, OC_6H_4OMe-*o*): synthesis, derivatization, group 10 and 11 metal complexes and catalytic investigations. DFT calculations on intermolecular P...P interactions in halo-phosphines. Inorg Chem 47:7035–7047
15. Moilanen J, Ganesamoorthy C, Balakrishna MS, Tuononen HM (2009) Weak interactions between trivalent pnictogen centers: computational analysis of bonding in dimers $X_3E...EX_3$ (E=Pnictogen, X=Halogen). Inorg Chem 48:6740–6747
16. Scheiner S (2011) A new noncovalent force: comparison of P...N interaction with hydrogen and halogen bonds. J Chem Phys 134:094315
17. Bauer S, Tschirschwitz S, Lönnecke P, Frank R, Kirchner B, Clarke ML, Hey-Hawkins E (2009) Enantiomerically pure bis(phosphanyl)carbaborane(12) compounds. Eur J Inorg Chem 2009:2776–2788
18. Tschirschwitz S, Lonnecke P, Hey-Hawkins E (2007) Aminoalkylferrocenyldichlorophosphanes: facile synthesis of versatile chiral starting materials. Dalton Trans 1377–1382
19. Sundberg MR, Uggla R, Vias C, Teixidor F, Paavola S, Kivekäs R (2007) Nature of intramolecular interactions in hypercoordinate C-substituted 1,2-dicarba-cloro-dodecaboranes with short P...P distances. Inorg Chem Commun 10:713–716
20. LaBarge MS, Andrews AM, Taleb-Bendiab A, Hillig KW, Kuczkowski RL, Bohn RK (1991) Microwave spectrum, structure, and dipole moment of the phosphorus trifluoride-water complex. J Phys Chem 95:3523–3527
21. Politzer P, Murray J, Janjić G, Zarić S (2014) σ-Hole interactions of covalently-bonded nitrogen, phosphorus and arsenic: a survey of crystal structures. Crystals 4:12–31
22. Del Bene JE, Alkorta I, Sanchez-Sanz G, Elguero J (2011) ^{31}P–^{31}P spin–spin coupling constants for pnicogen homodimers. Chem Phys Lett 512:184–187
23. Del Bene JE, Alkorta I, Sanchez-Sanz G, Elguero J (2011) Structures, energies, bonding, and NMR properties of pnicogen Complexes $H_2XP{:}NXH_2$ (X=H, CH_3, NH_2, OH, F, Cl). J Phys Chem A 115:13724–13731
24. Del Bene JE, Alkorta I, Sanchez-Sanz G, Elguero J (2012) Structures, binding energies, and spin–spin coupling constants of geometric isomers of pnicogen homodimers $(PHFX)_2$, X=F, Cl, CN, CH_3, NC. J Phys Chem A 116:3056–3060
25. Del Bene JE, Sanchez-Sanz G, Alkorta I, Elguero J (2012) Homo- and heterochiral dimers $(PHFX)_2$, X=Cl, CN, CH_3, NC: to what extent do they differ? Chem Phys Lett 538:14–18
26. Alkorta I, Sánchez-Sanz G, Elguero J, Del Bene JE (2012) Influence of hydrogen bonds on the P...P pnicogen bond. J Chem Theory Comput 8:2320–2327
27. Del Bene JE, Alkorta I, Sánchez-Sanz G, Elguero J (2012) Interplay of F–H...F hydrogen bonds and P...N pnicogen bonds. J Phys Chem A 116:9205–9213
28. Blanco F, Alkorta I, Rozas I, Solimannejad M, Elguero J (2011) A theoretical study of the interactions of NF_3 with neutral ambidentate electron donor and acceptor molecules. Phys Chem Chem Phys 13:674–683
29. Alkorta I, Sánchez-Sanz G, Elguero J, Del Bene JE (2013) Exploring $(NH_2F)_2$, $H_2FP{:}NFH_2$, and $(PH_2F)_2$ potential surfaces: hydrogen bonds or pnicogen bonds? J Phys Chem A 117:183–191

30. Del Bene JE, Alkorta I, Sánchez-Sanz G, Elguero J (2013) Phosphorus as a simultaneous electron-pair acceptor in intermolecular P...N pnicogen bonds and electron-pair donor to lewis acids. J Phys Chem A 117:3133–3141
31. Alkorta I, Elguero J, Del Bene JE (2013) Pnicogen-bonded cyclic trimers $(PH_2X)_3$ with X=F, Cl, OH, NC, CN, CH_3, H, and BH_2. J Phys Chem A 117:4981–4987
32. Sánchez-Sanz G, Alkorta I, Trujillo C, Elguero J (2013) Intramolecular pnicogen interactions in PHF-$(CH_2)_n$-PHF (n=2–6) systems. ChemPhysChem 14:1656–1665
33. Alkorta I, Elguero J, Del Bene JE (2013) Pnicogen bonded complexes of PO_2X (X=F, Cl) with nitrogen BASES. J Phys Chem A 117:10497–10503
34. Del Bene JE, Alkorta I, Elguero J (2013) Characterizing complexes with pnicogen bonds involving sp^2 hybridized phosphorus atoms: $(H_2C{=}PX)_2$ with X=F, Cl, OH, CN, NC, CCH, H, CH_3, and BH_2. J Phys Chem A 117:6893–6903
35. Del Bene JE, Alkorta I, Elguero J (2013) Properties of complexes $H_2C{=}(X)P{:}PXH_2$, for X=F, Cl, OH, CN, NC, CCH, H, CH_3, and BH_2: P...P pnicogen bonding at σ-holes and π-holes. J Phys Chem A 117:11592–11604
36. Del Bene J, Alkorta I, Elguero J (2014) σ–σ and σ–π pnicogen bonds in complexes $H_2XP{:}PCX$, for X=F, Cl, OH, NC, CN, CCH, CH_3, and H. Theor Chem Acc 133:1–9
37. Alkorta I, Sánchez-Sanz G, Elguero J, Del Bene JE (2014) Pnicogen bonds between $X{=}PH_3$ (X=O, S, NH, CH_2) and phosphorus and nitrogen bases. J Phys Chem A 118:1527–1537
38. Del Bene JE, Alkorta I, Elguero J (2014) Influence of substituent effects on the formation of P...Cl pnicogen bonds or halogen bonds. J Phys Chem A 118:2360–2366
39. Del Bene JE, Alkorta I, Elguero J (2014) Pnicogen-bonded anionic complexes. J Phys Chem A 118:3386–3392
40. Azofra LM, Alkorta I, Elguero J (2014) Chiral discrimination in dimers of diphosphines (PH_2-PH_2 and PH_2-PHF). ChemPhysChem in press 15:33663–3670. doi:10.1002/cphc.201402086
41. Sánchez-Sanz G, Trujillo C, Alkorta I, Elguero J (2014) Intramolecular pnicogen interactions in phosphorus and arsenic analogues of proton sponges. Phys Chem Chem Phys in press 16:15900–15909. doi:10.1039/C4CP01072H
42. Grabowski SJ, Alkorta I, Elguero J (2013) Complexes between dihydrogen and amine, phosphine, and arsine derivatives. Hydrogen bond versus pnictogen interaction. J Phys Chem A 117:3243–3251
43. Alkorta I, Elguero J, Solimannejad M (2014) Single electron pnicogen bonded complexes. J Phys Chem A 118:947–953
44. Bauzá A, Alkorta I, Frontera A, Elguero J (2013) On the reliability of pure and hybrid DFT methods for the evaluation of halogen, chalcogen, and pnicogen bonds involving anionic and neutral electron donors. J Chem Theory Comput 9:5201–5210
45. Solimannejad M, Gharabaghi M, Scheiner S (2011) SH...N and SH...P blue-shifting H-bonds and N...P interactions in complexes pairing HSN with amines and phosphines. J Chem Phys 134:024312
46. Scheiner S (2011) Effects of substituents upon the P...N noncovalent interaction: the limits of Its STRENGTH. J Phys Chem A 115:11202–11209
47. Adhikari U, Scheiner S (2012) Substituent effects on Cl...N, S...N, and P...N noncovalent bonds. J Phys Chem A 116:3487–3497
48. Adhikari U, Scheiner S (2012) Sensitivity of pnicogen, chalcogen, halogen and H-bonds to angular distortions. Chem Phys Lett 532:31–35
49. Scheiner S (2011) Can two trivalent N atoms engage in a direct N...N noncovalent interaction? Chem Phys Lett 514:32–35
50. Adhikari U, Scheiner S (2011) Comparison of P...D (D=P, N) with other noncovalent bonds in molecular aggregates. J Chem Phys 135:184306
51. Scheiner S (2011) Weak H-bonds. Comparisons of CHO to NHO in proteins and PHN to direct PN interactions. Phys Chem Chem Phys 13:13860–13872
52. Scheiner S (2011) Effects of multiple substitution upon the P...N noncovalent interaction. Chem Phys 387:79–84

53. Scheiner S (2012) Extrapolation to the complete basis set limit for binding energies of noncovalent interactions. Comp Theor Chem 998:9–13
54. Scheiner S (2011) On the properties of X...N noncovalent interactions for first-, second-, and third-row X atoms. J Chem Phys 134:164313
55. Scheiner S (2013) Sensitivity of noncovalent bonds to intermolecular separation: hydrogen, halogen, chalcogen, and pnicogen bonds. CrystEngComm 15:3119–3124
56. Scheiner S (2012) The pnicogen bond: its relation to hydrogen, halogen, and other noncovalent bonds. Acc Chem Res 46:280–288
57. Scheiner S (2013) Detailed comparison of the pnicogen bond with chalcogen, halogen, and hydrogen bonds. Int J Quantum Chem 113:1609–1620
58. Bauza A, Quinonero D, Deya PM, Frontera A (2012) Pnicogen-π complexes: theoretical study and biological implications. Phys Chem Chem Phys 14:14061–14066
59. Bauzá A, Ramis R, Frontera A (2014) A combined theoretical and cambridge structural database study of π-hole pnicogen bonding complexes between electron rich molecules and both nitro compounds and inorganic bromides (YO_2Br, Y=N, P, and As). J Phys Chem A 118:2827–2834
60. Bauza A, Quinonero D, Deya PM, Frontera A (2013) Halogen bonding versus chalcogen and pnicogen bonding: a combined Cambridge structural database and theoretical study. CrystEngComm 15:3137–3144
61. Solimannejad M, Gholipour A (2013) Revealing substituent effects on the concerted interaction of pnicogen, chalcogen, and halogen bonds in substituted *s*-triazine ring. Struct Chem 24:17051711
62. Solimannejad M, Bayati E, Esrafili MD (2014) Enhancement effect of lithium bonding on the strength of pnicogen bonds: XH_2P...NCLi...NCY as a working model (X=F, Cl; Y=H, F, Cl, CN). Mol Phys 112:2058–2062
63. Solimannejad M, Ramezani V, Trujillo C, Alkorta I, Sánchez-Sanz G, Elguero J (2012) Competition and interplay between σ-Hole and π-Hole interactions: a computational study of 1:1 and 1:2 complexes of nitryl halides (O_2NX) with Ammonia. J Phys Chem A 116:5199–5206
64. Grabowski SJ (2013) σ-Hole bond versus hydrogen bond: from tetravalent to pentavalent N, P, and As atoms. Chem Eur J 19:14600–14611
65. Grabowski SJ (2014) Clusters of ammonium cation-hydrogen bond versus σ-hole bond. ChemPhysChem 15:876–884
66. Li Q-Z, Li R, Liu X-F, Li W-Z, Cheng J-B (2012) Pnicogen–hydride interaction between FH_2X (X=P and As) and HM (M=ZnH, BeH, MgH, Li, and Na). J Phys Chem A 116:2547–2553
67. Li Q-Z, Li R, Liu X-F, Li W-Z, Cheng J-B (2012) Concerted interaction between pnicogen and halogen bonds in XCl-FH_2P-NH_3 (X=F, OH, CN, NC, and FCC). ChemPhysChem 13:1205–1212
68. An X-L, Li R, Li Q-Z, Liu X-F, Li W-Z, Cheng J-B (2012) Substitution, cooperative, and solvent effects on π pnicogen bonds in the FH(2)P and FH(2)As complexes. J Mol Model 18:4325–4332
69. Li Q, Zhuo H, Yang X, Cheng J, Li W, Loffredo RE (2014) Cooperative and diminutive effects of pnicogen bonds and cation–π interactions. ChemPhysChem 15:500–506
70. Liu X, Cheng J, Li Q, Li W (2013) Competition of hydrogen, halogen, and pnicogen bonds in the complexes of HArF with XH_2P (X=F, Cl, and Br). Spectrochim Acta Part A Mol Biomol Spectros 101:172–177
71. Chen Y, Yao L, Lin X (2014) Theoretical study of $(FH_2X)_n \cdot Y$ (X=P and As, *n*=1–4, Y=F^-, Cl^-, Br^-, I^-, NO_3^- and SO_4^{2-}): the possibility of anion recognition based on pnicogen bonding. Comput Theor Chem 1036:44–50
72. Guan L, Mo Y (2014) Electron transfer in pnicogen bonds. J Phys Chem A 118:8911–8921.
73. Eskandari K, Mahmoodabadi N (2013) Pnicogen bonds: a theoretical study based on the laplacian of electron density. J Phys Chem A 117:13018–13024
74. Pople JA, Binkley JS, Seeger R (1976) Theoretical models incorporating electron correlation. Int J Quantum Chem Quantum Chem Symp 10:1–19

75. Krishnan R, Pople JA (1978) Approximate fourth-order perturbation theory of the electron correlation energy. Int J Quantum Chem 14:91–100
76. Bartlett RJ, Silver DM (1975) Many–body perturbation theory applied to electron pair correlation energies. I. closedâŁłshell first–row diatomic hydrides. J Chem Phys 62:3258–3268
77. Bartlett RJ, Purvis GD (1978) Many–body perturbation theory, coupled-pair many-electron theory, and the importance of quadruple excitations for the correlation problem. Int J Quantum Chem 14:561–581
78. Del Bene JE (1993) Proton affinities of ammonia, water, and hydrogen fluoride and their anions: a quest for the basis-set limit using the dunning augmented correlation-consistent basis sets. J Phys Chem 97:107–110
79. Dunning TH (1989) Gaussian basis sets for use in correlated molecular calculations. I. The atoms boron through neon and hydrogen. J Chem Phys 90:1007–1023
80. Woon DE, Dunning TH (1995) Gaussian basis sets for use in correlated molecular calculations. V. Core–valence basis sets for boron through neon. J Chem Phys 103:4572–4585
81. Frisch MJ, Trucks GW, Schlegel HB, Scuseria GE, Robb MA, Cheeseman, JR, Scalmani G, Barone V, Mennucci B, Petersson GA et al (2009) Gaussian, Inc.: Wallingford CT,. Gaussian-09, Revision A.01
82. Bader RFW (1991) A quantum theory of molecular structure and its applications. Chem Rev 91:893–928
83. Bader RFW (1990) Atoms in molecules, A quantum theory. Oxford University, Oxford
84. Popelier PLA (2000) Atoms In molecules. An introduction. Prentice Hall, Harlow, England
85. Matta CF, Boyd RJ (2007) The quantum theory of atoms in molecules: from solid state to DNA and drug design. Wiley-VCH, Weinheim
86. Silvi B, Savin A (1994) Classification of chemical bonds based on topological analysis of electron localization functions. Nature 371:683–686
87. Todd A, Keith TK (2011) Gristmill Software, Overland Park KS, USA. http://aim.tkgristmill.com. Accessed 1 Aug 2013 (AIMAll (Version 11.08.23))
88. Noury S, Krokidis X, Fuster F, Silvi B (1997) TopMod Package
89. Rozas I, Alkorta I, Elguero J (2000) Behavior of ylides containing N, O, and C atoms as hydrogen bond acceptors. J Am Chem Soc 122:11154–11161
90. Reed AE, Curtiss LA, Weinhold F (1988) Intermolecular interactions from a natural bond orbital, donor-acceptor viewpoint. Chem Rev 88:899–926
91. Glendening ED, Badenhoop JK, Reed AE, Carpenter JE, Bohmann JA, Morales CM, Weinhold F (2001) NBO 5.0. University of Wisconsin Press, Madison
92. Glendening ED, Badenhoop JK, Reed AE, Carpenter JE, Bohmann JA, Morales CM, Landis CR, Weinhold F (2013) NBO 6.0. University of Wisconsin Press. Madison
93. Becke AD (1993) DensityâŁłfunctional thermochemistry. III. The role of exact exchange. J Chem Phys 98:5648–5652
94. Lee C, Yang W, Parr RG (1988) Development of the colle-salvetti correlation-energy formula into a functional of the electron density. Phys Rev B 37:785–789
95. Jmol: an open-source Java viewer for chemical structures in 3D, version 13.0. http://www.jmol.org/. Accessed 26 Sept 2013
96. Patek M "Jmol NBO Visualization Helper" program. http://www.marcelpatek.com/nbo/nbo.html.Accessed 26 Sept 2013
97. Perera SA, Nooijen M, Bartlett RJ (1996) Electron correlation effects on the theoretical calculation of nuclear magnetic resonance spin–spin coupling constants. J Chem Phys 104:3290–3305
98. Perera SA, Sekino H, Bartlett RJ (1994) Coupled–cluster calculations of indirect nuclear coupling constants: the importance of non–fermi contact contributions. J Chem Phys 101:2186–2196
99. Schäfer A, Horn H, Ahlrichs R (1992) Fully optimized contracted gaussian basis sets for atoms Li to Kr. J Chem Phys 97:2571–2577
100. Stanton JF, Gauss J, Watts JD, Nooijen M, Oliphant N, Perera SA, Szalay PS, Lauderdale WJ, Gwaltney SR, Beck S, Balkova A, Bernholdt DE, Baeck KK, Tozyczko P, Sekino H, Huber

C, Bartlett RJ ACES II is a program product of the Quantum Theory Project, University of Florida. Integral packages included are VMOL (Almlöf J, Taylor PR); VPROPS (Taylor PR); ABACUS (Helgaker T, Jensen Aa HJ, Jørgensen P, Olsen J, Taylor PR). Brillouin-Wigner perturbation theory was implement by Pittner J
101. Knop O, Boyd RJ, Choi SC (1988) Sulfur-sulfur bond lengths, or can a bond length be estimated from a single parameter? J Am Chem Soc 110:7299–7301
102. Gibbs GV, Hill FC, Boisen MB, Downs RT (1998) Power law relationships between bond length, bond strength and electron density distributions. Phys Chem Miner 25:585–590
103. Alkorta I, Barrios L, Rozas I, Elguero J (2000) Comparison of models to correlate electron density at the bond critical point and bond distance. Theochem 496:131–137
104. Espinosa E, Alkorta I, Elguero J, Molins E (2002) From weak to strong interactions: a comprehensive analysis of the topological and energetic properties of the electron density distribution involving H...F systems. Part I: the transit region between pure closed-shell and shared-shell interactions. J Chem Phys 117:5529–5542
105. Alkorta I, Elguero J (2004) Fluorine-fluorine interactions: a NMR and AIM analysis. Struct Chem 15:117–120
106. Tang TH, Deretey E, Knak Jensen SJ, Csizmadia IG (2006) Hydrogen bonds: relation between lengths and electron densities at bond critical points. Eur Phys J D 37:217–222
107. Vener MV, Manaev AV, Egorova AN, Tsirelson VG (2007) QTAIM study of strong H-bonds with the O–H...A fragment (A=O, N) in three-dimensional periodical crystals. J Phys Chem A 111:1155–1162
108. Castillo N, Robertson KN, Choi SC, Boyd RJ, Knop O (2008) Bond length and the electron density at the bond critical point: X-X, Z-Z, and C-Z bonds (X=Li-F, Z=Na-Cl). J Comput Chem 29:367–379
109. Mata I, Alkorta I, Molins E, Espinosa E (2010) Universal features of the electron density distribution in hydrogen-bonding regions: a comprehensive study involving H...X (X=H, C, N, O, F, S, Cl, π) interactions. Chem Eur J 16:2442–2452
110. Zeng Y, Li X, Zhang X, Zheng S, Meng L (2011) Insight into the Nature of the interactions of furan and thiophene with hydrogen halides and lithium halides: Ab INITIO and QTAIM studies. J Mol Model 17:2907–2918
111. Cremer D, Kraka E (1984) A description of the chemical bond in terms of local properties of electron density and energy. Croat Chem Acta 57:1259–1281
112. Hankins D, Moskowitz JW, Stillinger FH (1970) Water molecule interactions. J Chem Phys 53:4544–4554
113. Xantheas S (1994) *Ab Initio* studies of cyclic water clusters $(H_2O)_N$, N=1–6. II. Analysis of many–body interactions. J Chem Phys 100:7523–7534

Chapter 9
Chalcogen Bonds in Protein Architecture

Michio Iwaoka

Abstract Nonbonded interactions between a divalent sulfur atom and polar functional groups, i.e., S$\cdots$X (X = O, N, and S) interactions, have recently been demonstrated to stabilize protein structures to some extent and play putative roles in their function and evolution, thanks to the interplay of statistical analysis of protein structure databases and theoretical calculation using simple molecular cluster models. The directional features observed between the interacting S and X atoms suggest that they can also be called S$\cdots$X chalcogen bonds in analogy to halogen bonds. While the existence of chalcogen bonds in proteins may be accepted today by the protein scientist community, there are still several issues that should be addressed clearly before going forward to applications of the interactions to protein engineering and the ligand design. Herein, the current status of the research on the S$\cdots$X chalcogen bonds in proteins is reviewed from several points of view, including the historical aspects and the analytical methods for mining chalcogen bonds in protein structures. Statistical, directional, and energetic features of four typical S$\cdots$X chalcogen bonds in proteins are presented. Possibility of analogous Se$\cdots$X chalcogen bonds in selenoproteins is also pointed out. Implications of S$\cdots$X chalcogen bonds in the structural control and in the function and evolution of some particular proteins are subsequently summarized from the recent literature. Finally, it is proposed that the concept of S$\cdots$X chalcogen bonds will be a useful tool for fully understanding not only protein structures but also the biological aspects. Thus, the chalcogen bonds, though their interaction energy is smaller than that of classical hydrogen bonds, can be an element of weak noncovalent interactions that control chemical and biological properties of protein architecture.

9.1 Introduction

Sulfur is involved in almost all proteins as cysteine (Cys) and methionine (Met) amino acid residues. In particular, it is contained at high concentrations in keratin [1, 2], a major component of wool, and metallothionein [3], a representative metal-

M. Iwaoka (✉)
Department of Chemistry, School of Science,
Tokai University, Kitakaname, Hiratsuka-shi, Kanagawa 259-1292, Japan
e-mail: miwaoka@tokai.ac.jp

S. Scheiner (ed.), *Noncovalent Forces*, Challenges and Advances in
Computational Chemistry and Physics 19, DOI 10.1007/978-3-319-14163-3_9

loprotein capable of binding many zinc, cadmium, and copper atoms per one protein molecule. Roles of sulfur atoms are diverse in proteins. For example, the sulfur atoms of Cys and Met residues can bind the prosthetic groups of several enzymes, like iron-sulfur clusters of ferredoxins [4] and zinc ions of zinc fingers [5], as soft ligands. The sulfanyl (SH) group of a Cys residue is sometimes utilized as an active center of oxidoreductases, such as thioredoxin (Trx) [6] and protein disulfide isomerase (PDI) [7, 8, 9], and cysteine proteases, such as papain and caspase [10]. Besides these roles, the disulfide (SS) bonds that are formed between two SH groups of Cys residues effectively stabilize protein structures [11]. Thus, sulfur atoms in proteins are important structural and functional elements for protein architecture. In the most cases, the sulfur atoms of Cys residues play more significant roles than those of Met residues. This is in accord with the observation that Cys residues are strongly conserved in the amino acid sequence during the evolution [11]. Indeed, site-directed mutation of Cys to other amino acids or the deletion should result in serious damage on the structure and function of a protein, while the consequence of similar modification on the Met residues would be much more limited.

Most sulfur atoms of proteins are present in a divalent state, having two covalent bonds and two lone pairs, and are contained in a SS bond formed between Cys residues or a methylsulfanyl group (MeS) of Met residues. These sulfur-containing functional groups, i.e., SS and MeS groups, are ubiquitous in proteins, but apart from the aforementioned distinct roles, they were usually considered as hydrophobic moieties, showing no apparent interaction with nearby polar functional groups. However, with accumulation of precise structural data of proteins [12], existence of weak noncovalent interactions around the divalent sulfur atoms has become evident thanks to the interplay of elaborate structural database analysis and sophisticated theoretical calculation. These interactions in protein architecture contains NH$\cdots$S and OH$\cdots$S hydrogen bonds [13, 14], S$\cdots\pi$ interactions [15, 16], and S$\cdots$X (X = O, N, and S) interactions [15, 17–19]. The last two can be categorized to "non-classical" interactions in the sense that the divalent sulfur atom is formally accepting electron donation from the π system or the X atom to the σ-hole, adapting a pseudo hypervalent coordination state [20]. The S$\cdots$X interactions that have pertinent directional propensity can also be called "chalcogen bonds" [21] in analogy to halogen bonds [22–24]. This contemporary terminology is adapted throughout this chapter, but the term, S$\cdots$X interaction, is also used to represent the interaction that has not yet been characterized by the directional propensity, and the term, chalcogen bond, is intentionally used for the interactions that have distinct directionalities.

While the concept of chalcogen bonds in proteins may be accepted today by the protein scientist community, there are still several issues that should be addressed clearly before going forward to applications of the interactions to protein engineering and the ligand design. The following questions would be particularly important to be understood with consensus.

Q1. Do the S$\cdots$X chalcogen bonds really exist in proteins?
Q2. Are the directional propensities similar to those observed for small organic molecules?

Q3. How do the potential surfaces look like?
Q4. Can the S$\cdots$X chalcogen bonds control protein structure?
Q5. Are the S$\cdots$X chalcogen bonds relevant to protein function and evolution?
Q6. Are there similar Se$\cdots$X chalcogen bonds in selenoproteins too?

In this chapter, to provide useful information as to above questions, the current status of the research on the S$\cdots$X chalcogen bonds in proteins is reviewed from several points of view, based mostly on the research carried out in the author's laboratory since 2001. First, the historical aspects including the discovery are overviewed briefly (Sect. 9.2). Then, the analytical methods for mining chalcogen bonds in protein structures are introduced (Sect. 9.3). Nonbonded S$\cdots$X interactions found in proteins are subsequently discussed in detail based on the probabilities, the directionalities, and the energetics in order to characterize four typical S$\cdots$X chalcogen bonds in proteins. Possibility of analogous Se$\cdots$X chalcogen bonds in selenoproteins is also pointed out (Sect. 9.4). Their implications in the structural control and in the function and evolution of some particular proteins are summarized from the recent literature in Sect. 9.5. Finally, the possible answers to the above six questions are addressed based on the data having been obtained so far (Sect. 9.6).

9.2 Historical Aspects

9.2.1 Nonbonded Interactions Involving a Divalent Sulfur Atom

Evidence of attractive S$\cdots$X interactions between a divalent sulfur atom (S) and nearby heteroatoms (X) in the crystals of small organic compounds [25, 26] was already known in late 1970s according to a pioneer work by Rosenfield and Parthasarathy [27], who analyzed close atomic contacts around S in Cambridge crystallographic database (CSD) [28] and found that X tends to approach S in the backside of one of the two covalent bonds (i.e., in the σ^*_S direction). Subsequently, Row and Parthasarathy [29] demonstrated that nonbonded S$\cdots$S interactions in organic crystals may be stabilized by the orbital interaction between the lone pair of one sulfur atom (n_S) and the anti-bonding orbital of the other sulfur atom (σ^*_S). This orbital model of the weak S$\cdots$S interactions was later supported by Desiraju and Nalini [30], whereas Dahaoui et al. [31] pointed out the importance of dispersion force for the S$\cdots$S interactions. On the other hand, the directional preference of S$\cdots$O interactions around the O atom was studied in detail by Kucsman and Kapovitz [32]. For intramolecular 1,4- and 1,5-S$\cdots$O$=$C interactions, the S atom tended to lie in the direction of the O lone pairs (i.e., the n_O direction) rather than the π electrons (i.e., the π_O direction), suggesting the importance of $n_O \rightarrow \sigma^*_S$ orbital interaction for the stability. The presence of S$\cdots\pi$ interactions between a divalent S atom and an aromatic ring in organic crystals was also suggested by Zauhar et al. [33].

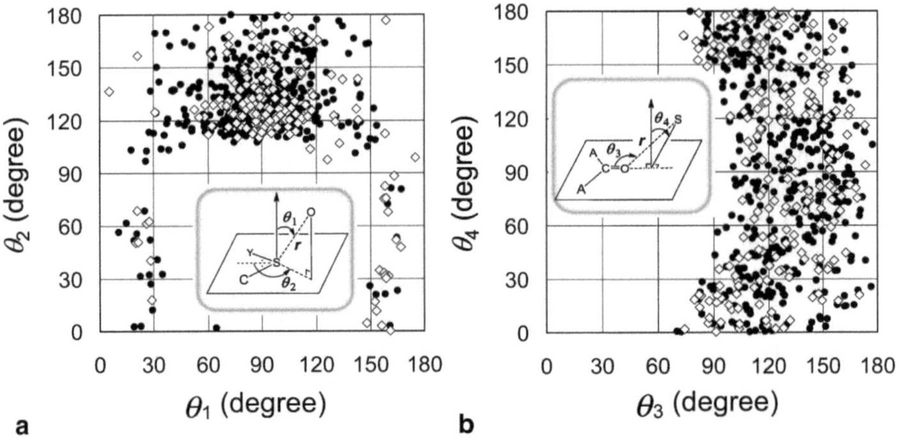

Fig. 9.1 Directionality of intermolecular S$\cdots$O=C interactions with $d \equiv r - vdw(\mathrm{S}) - vdw(\mathrm{O}) = r - 1.80\text{–}1.52 \leq 0.2$ Å. *Open diamonds* represent intermolecular S$\cdots$O(amide) interactions. *Filled circles* represent the other types of intermolecular S$\cdots$O interactions. **a** Spatial distribution of O relative to S by using angles θ_1 and θ_2. **b** Spatial distribution of S relative to O by using angles θ_3 and θ_4. This figure was modified from ref [34]

We [34] previously carried out statistical analysis of intermolecular S$\cdots$O=C interactions found in CSD in order to clearly characterize nonbonded S$\cdots$O interactions in organic crystals in comparison with those in proteins. Figure 9.1 shows the observed directionalities around S and O atoms for the intermolecular S$\cdots$O contacts. It is obvious that the O atom approaches to the S atom in the backside of the S–Y (Y = C or S) bond ($\theta_1 \sim 90°$ and $\theta_2 \sim 130°$ in Fig. 9.1a) in accord with previous observations for S$\cdots$O [27, 32] and S$\cdots$S [29, 30] interactions. However, no directional preference was observed around the O atom irrespective of whether it is involved in an amide group or other carbonyl groups (Fig. 9.1b). This is in strong contrast to the observation of Kucsman and Kapovitz [32] for the intramolecular 1,4- and 1,5-S$\cdots$O=C interactions, suggesting that the directional preference around the O atom in intermolecular S$\cdots$O=C interactions should be very week and hence it will be easily affected by other steric and energetic factors. On the other hand, the linearity of the Y–S$\cdots$O alignment was proved to be robust against the packing force of organic crystals.

As mentioned above, database analyses for various types of nonbonded S$\cdots$X (X = O, S, etc.) interactions have demonstrated that specific directional preferences are obvious for the S$\cdots$X interactions in organic crystals, although in the case of intermolecular S$\cdots$O=C interactions the preference is vanished. The directionalities strongly suggested the importance of orbital interactions, such as $n_{\mathrm{O}} \rightarrow \sigma^*_{\mathrm{S}}$ and $n_{\mathrm{S}} \rightarrow \sigma^*_{\mathrm{S}}$. However, the nonbonded distances between the two interacting atoms are sometimes only marginally shorter than the sum of the van der Waals radii, and, in such cases, the directionality becomes a little subtle. Therefore, there should be other electronic factors that are responsible for formation of the S$\cdots$X contacts.

Such factors would involve the dispersion force (or electron correlation effects) [31] and the electrostatic interaction between the positively charged S atom and the negatively charged O atom as suggested by Burling and Goldstein [35]. Recently, sophisticated theoretical methodologies have been applied for better understanding such noncovalent interactions. Nakanishi [36] attempted to describe S$\cdots$X interactions more exactly by using the total electron energy density and Laplacian of the electron density at the bond critical points of the interaction in the framework of atoms-in-molecules (AIM) theory [37, 38], while symmetry-adapted perturbation theory (SAPT) [39] was applied by Scheiner [40, 41] to decomposing the energetic components to electrostatic, induction, and dispersion terms.

Relevance of S$\cdots$O interactions to the biological functions of several organic molecules has been reported. For example, the 1,4-S$\cdots$O interaction of thiazole nucleoside analogues was found to be important for the antitumor activity [35]. Similarly, 1,5-S$\cdots$O interactions were shown to play important roles in the antagonism of (acylimino)thiadiazoline derivatives to an angiotensin II receptor [42] and in the antitumor activity of leinamycin [43]. Analogous interactions between a divalent selenium atom (Se) and X (i.e., Se$\cdots$X interactions) have also been well characterized in organic selenium compounds [44–46] and have been suggested to be relevant to the biological activity of selenium [47, 48].

9.2.2 Discovery of S$\cdots$X Chalcogen Bonds in Proteins

Weak noncovalent interactions are important physicochemical forces that control the structure of proteins [49]. Ionic interaction, hydrogen bond, van der Waals force, and hydrophobic interaction are mainly involved in this class of interactions, but some novel interaction patterns, such as C–H$\cdots$O hydrogen bond [50, 51, 52], cation-π interaction [53–56], and CH/π hydrogen bond [57–59], have also been characterized in folded protein structures. Importance of these new interactions for the stability and function of proteins has already been pointed out.

The SS and MeS groups of Cys and Met residues, however, were usually considered just as hydrophobic moieties in folded protein structures until recently, except for S$\cdots\pi$ interactions [16] and weak NH$\cdots$S and OH$\cdots$S hydrogen bonds [13, 14]. The S$\cdots\pi$ interactions in proteins were first pointed out by Morgan and co-workers [60], who analyzed the close atomic contact between S and a π-ring system in eight protein structures and found that the S atoms have a propensity to come over the π-plane. On the other hand, Reid et al. [61] suggested by using a larger set of protein structures that the close S$\cdots\pi$ contact in proteins can also be explained by C–H$\cdots$S interactions because the S atoms access to the π-plane from the side rather than the top. According to elaborate experimental and theoretical studies [15, 16, 62, 63], however, the nature of S$\cdots\pi$ interactions in proteins would be well rationalized by the interaction between the aromatic π electrons and the S atom [15, 16]. Meanwhile, NH$\cdots$S and OH$\cdots$S hydrogen bonds were suggested to play some roles in particular proteins [13, 14], but the interactions can be found with low frequency in protein structures. The S atoms of Cys and Met may have only a weak character as a hydrogen-bond acceptor.

Stereochemistry of the close S···X contacts involving a Met sulfur atom (a MeS group) was statistically analyzed in protein structures by Carugo in 1999 [17], but distinct directional preferences of the contacts were not observed. It was therefore suggested that nonbonded S···X interactions in proteins are either very weak or physicochemically different from those in small molecules. Two years after Carugo's first attempt, we [18] and Pal and Chakrabarti [19] reported more extensive database analysis of S···X interactions using a larger set of protein structures. These studies led us to the discovery of distinct directionality in the interactions. Thus, the presence of S···X chalcogen bonds in proteins was suggested for the first time. The chalcogen bonds can be characterized by two directional propensities. One is the linearity of C–S···X (or S–S···X) alignment, and the other is the directional preference of S around X. These features are discussed in detail in the following sections.

9.3 Analytical Methods

Characterization of weak noncovalent interactions is still a difficult task in any molecular systems. Usually, several experimental and theoretical techniques are collaboratively employed for the detection. For the study of S···X chalcogen bonds in proteins, only limited techniques may be applicable not only because the structure of a protein is very complicated but also because the interactions are somehow elusive. Therefore, the existence of S···X chalcogen bonds in proteins had not been recognized longtime. However, since high-resolution protein structural data [12] are now available, we can use the data for mining the chalcogen bonds.

Another useful technique available to characterize weak interactions is theoretical calculation. Sophisticated calculation techniques, such as ab initio calculation, cannot be directly applied for a whole protein molecule even today. However, the calculation for small model systems still provides us valuable information as shown later. Indeed, from such simple calculation we know the intrinsic interaction energy and directional preference of the interactions, which will be compared with what we have obtained from the statistical database analysis of protein structures. In this way, we can reasonably assume the nature of S···X chalcogen bonds and the structural perturbations in protein architecture. Other state-of-the-art spectroscopic techniques could also be applied for direct detection of the chalcogen bonds in proteins. Development of such techniques will have large impacts in versatile fields of protein research in future.

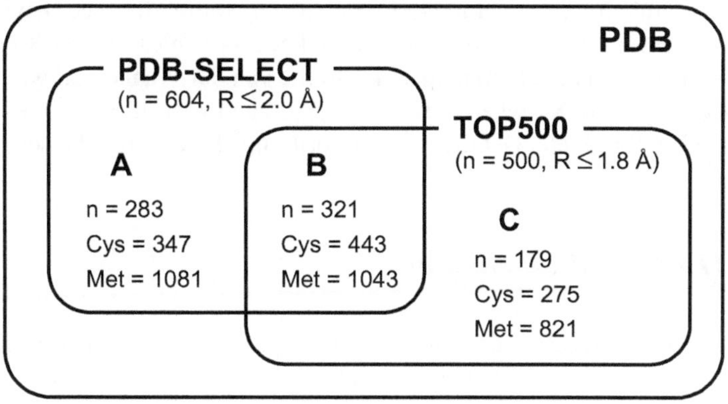

Fig. 9.2 The numbers of protein structures (*n*), cysteine residues (*Cys*), and methionine residues (*Met*) involved in PDB-SELECT and TOP500 databases. *R* represents the resolution. The numbers for each zone (***A***, ***B***, or ***C***) are shown

9.3.1 Protein Structure Database

All protein structures that have been determined to date are available from protein databank (PDB) [12], which we have utilized for characterization of S$\cdots$X chalcogen bonds in proteins [18, 34, 64–67]. However, PDB contains a lot of redundant structural data for one protein. Therefore, filtering of the data was necessary to obtain reliable statistical information about the relevant interactions from the database. Several filtering algorithms are known in order to choose nonredundant and heterogeneous protein structures from PDB. PDB_SELECT database [68, 69] collects 1085 nonredundant X-ray structures, which consist of 604 high resolution ($\leq$ 2.0 Å) and 481 low resolution (>2.0 Å) structures, from PDB with a criterion of the structural homology less than 25 %. The high-resolution data were employed in refs [18, 34, 64]. TOP500 database [70], which collects 500 heterogeneous protein structures with high resolution ($\leq$ 1.8 Å) from PDB for drawing Ramachandran-plot distributions, was applied in ref [67]. These two databases contain 321 common structures as illustrated in Fig. 9.2. Comparing the structural features of S$\cdots$X chalcogen bonds obtained from PDB-SELECT with those obtained from TOP500, we can guarantee reliability of the database analysis.

On the other hand, redundant structural data for one particular protein are useful for the analysis on the functional and evolutional aspects of specific S$\cdots$X chalcogen bonds because such data contain a variation of the ligands and mutations. By comparing the structural features of the S$\cdots$X chalcogen bonds among the redundant data, the plausible roles in the protein function and evolution can be inferred [65, 66].

A computer program was coded for extracting all non-hydrogen atoms (X) that have a close atomic contact in proteins to a divalent sulfur atom (S) of Cys (a SS group) or Met (a MeS group). The data of S$\cdots$X contacts obtained were sorted by the use of several structural parameters, such as an atomic distance between S and X (r), access angles of X to S (θ_1 and θ_2), and access angles of S to X (θ_3 and θ_4) (see Fig. 9.1). The relative atomic distance to the van der Waals radii was define by $d \equiv r - \mathrm{vdw(S)} - \mathrm{vdw(X)}$, where vdw(S) = 1.80 Å and vdw(X) = 1.70 Å for C, 1.55 Å for N, 1.52 Å for O, etc. Details of the algorithm were described elsewhere [64].

9.3.2 *Theoretical Calculation*

The purpose of theoretical calculation includes quantitative characterization of the structural features of S···X chalcogen bonds, estimation of the interaction energies, and identification of the energetic components of the interactions. To accomplish this mission, ab initio calculation has been carried out for simple cluster models, $CH_3SSCH_3 + CH_3CONHCH_3$ (cluster I) and $CH_3SCH_3 + CH_3CONHCH_3$ (cluster II), at various calculation levels, such as HF/6-31G(d), MP2/6-31G(d), and MP2/6-311G(d,p) [18, 34, 64]. The truncated models employed have an advantage that they will provide important information about the interactions in the absence of steric perturbations from the protein structure. Thus, intrinsic nature of the interactions can be deduced from the calculation. On the other hand, comparison of the results obtained at different calculation levels are applicable to addressing the energetic components because the second order Møller-Plesset perturbation theory (MP2) sufficiently incorporates electron correlation effects whereas the Hartree-Fock method (HF) does not [71]. The potential energy surfaces of the S···X chalcogen bonds were drawn by using the energies obtained as a function of access angles θ_1–θ_4 for the clusters at MP2/6-31(G) level [64].

9.4 Characterization of Chalcogen Bonds in Proteins

9.4.1 *Statistics of S···X Interactions*

Results of the statistical analysis of the nonbonded S···X contacts that were found in protein structure databases are summarized in Fig. 9.3. The S···X contacts of the SS group extracted from PDB-SELECT only (zone **A** of Fig. 9.3a) show that the probabilities of S···O and S···S contacts increase in a short distance range ($d \leq 0.0$ Å), while those of S···N and S···C contacts decrease contrarily. The same trends are observed for zone **B** (common in PDB-SELECT and TOP500) and zone **C** (TOP500 only), suggesting the presence of unique S–S···O and S–S···S interactions around the SS bonds in proteins. For the MeS group, on the other hand, the probabilities of S···O and S···N contacts increase in a short distance range, while that of S···C contacts decreases and S···S contacts are very rare from the beginning. Again, common trends were observed in zones **A-C**, suggesting the presence of unique C–S···O and C–S···N interactions around the MeS groups in proteins. It is important

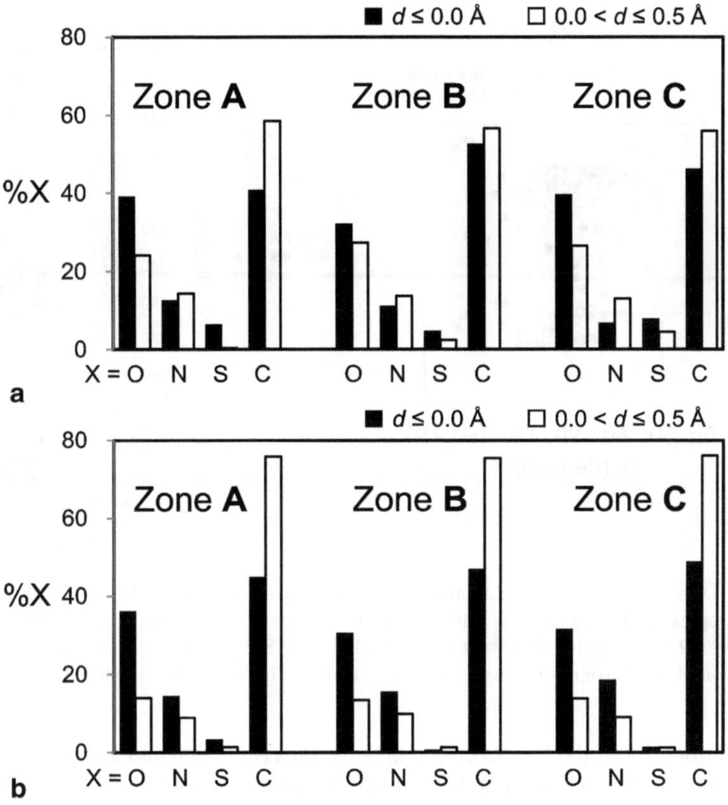

Fig. 9.3 Probabilities of S· · · X (X = O, N, S, and C) contacts observed for SS groups **a** and MeS groups **b** in the total S· · · X contacts extracted from the databases. Zones **A–C** correspond to PDB-SELECT only, common, and TOP500 only, respectively. See also Fig. 9.2

to notice that the probabilities do not change significantly depending on the databases employed. Thus, the existence of S· · · O, S· · · N, and S· · · S interactions should not be due to an artifact.

9.4.2 Directionality of S· · · X Interactions

Four types of S· · · X interactions were suggested by statistical analysis of the relative atomic distances (*d*) for the S· · · X contacts found in protein structure databases. These interactions were further analyzed by using access angles θ_1–θ_4 around the interacting atoms to characterize the directional propensity in proteins.

The S· · · O contacts observed for the SS group contained the interactions with various types of O, such as amide, hydroxyl, and water O atoms. Among these, the main-chain amide O atom was most frequently found (more than 50 %). Therefore,

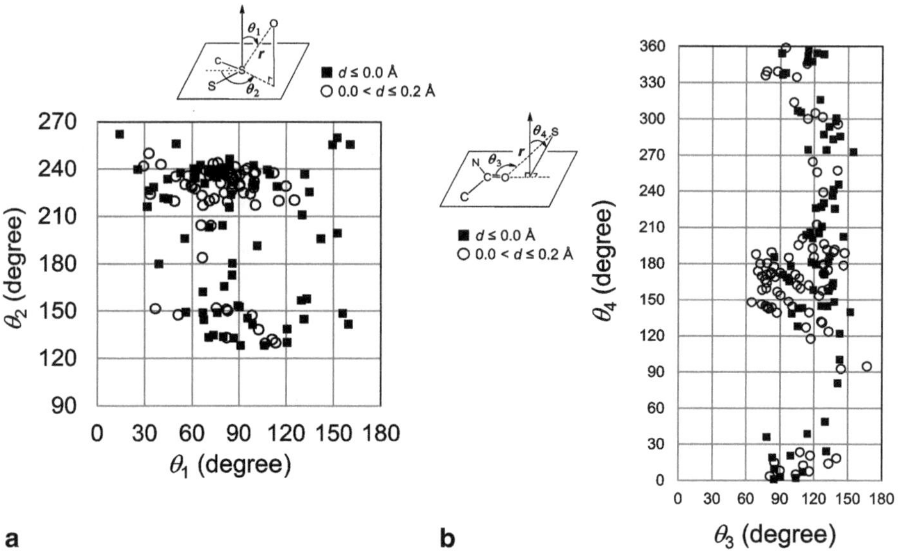

Fig. 9.4 Directionality of SS$\cdots$O(amide) interactions found in PDB-SELECT database. *Filled squares* represent the interactions with $d \leq 0.0$ Å. *Open circles* represent the interactions with $0.0 < d \leq 0.2$ Å. **a** Spatial distribution of O relative to S by using angles θ_1 and θ_2. **b** Spatial distribution of S relative to O by using angles θ_3 and θ_4. This figure was modified from ref [64]

the directionality of the S$\cdots$O contacts between a SS group and a main-chain amide O atom was analyzed. Figure 9.4 shows the directionality of SS$\cdots$O(amide) interactions found in PDB-SELECT database [64]. It is clearly seen that the O atom tends to approach the S atom in the backside of the S–S ($\theta_1 \sim 90°$ and $\theta_2 \sim 230°$) or S–C ($\theta_1 \sim 90°$ and $\theta_2 \sim 130°$) covalent bond. It should be noted that the definitions of the access angles are slightly different from those used in Fig. 9.1. On the other hand, the S atom tends to come above or below the amide plane ($\theta_3 \sim 90°$ and $\theta_4 \sim 180$ or 360°). The linearity of S–S$\cdots$O or C–S$\cdots$O alignment as well as the perpendicularity of S$\cdots$O$=$C alignment allows an effective charge transfer through the $\pi_O \rightarrow \sigma^*_S$ orbital interaction. Thus, the existence of two types of S$\cdots$X chalcogen bonds, i.e., S–S$\cdots$O$=$C (type 1) and C–S$\cdots$O$=$C (type 2), is strongly suggested. It is notable that not lone pairs but π electrons of the O atom participates in these chalcogen bonds. This feature is in significant difference from the trend observed in organic crystals (Fig. 9.1b) [34].

Similar but slightly less obvious directionalities were observed for the S$\cdots$O contacts between a MeS group and a main-chain amide O atom (Fig. 9.5). The directional preferences, i.e., linear C–S$\cdots$O alignment and perpendicular approach of S to the amide plane, provide a strong evidence for the existence of type-2 chalcogen bond (C–S$\cdots$O$=$C) in proteins.

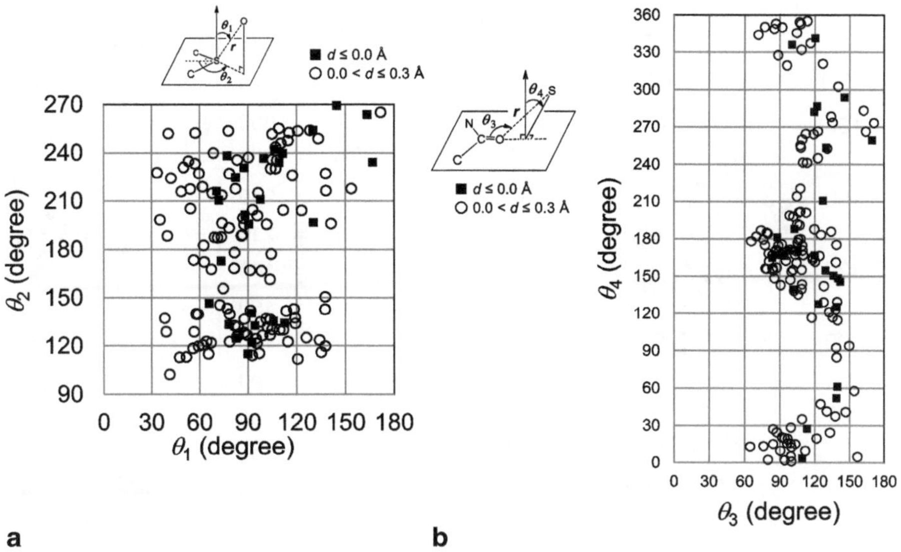

Fig. 9.5 Directionality of MeS$\cdots$O(amide) interactions found in PDB-SELECT database. *Filled squares* represent the interactions with $d \leq 0.0$ Å. *Open circles* represent the interactions with $0.0 < d \leq 0.3$ Å. **a** Spatial distribution of O relative to S by using angles θ_1 and θ_2. **b** Spatial distribution of S relative to O by using angles θ_3 and θ_4. This figure was modified from ref [64]

The S$\cdots$N contacts found around the MeS group of Met residues (Fig. 9.3b) mostly contained the N atoms of main-chain and side-chain amide groups. To characterize directional features of these S$\cdots$N contacts, the access angles between the S and the main-chain amide N atoms were analyzed in the range of $d \leq 0.3$ Å (Fig. 9.6). The scattergrams clearly indicate that the S$\cdots$N contacts contain two types of interactions.

One is a S$\cdots$N chalcogen bond (type 3) (open circles in Fig. 9.6) with linear C–S$\cdots$N alignment and perpendicular approach of S to the amide plane, allowing effective $\pi_N \rightarrow \sigma^*_S$ orbital interaction. The directionality around the nitrogen atom (Fig. 9.6b) was more explicit than that of type-1 and type-2 chalcogen bonds (Figs. 9.4b and 9.5b), whereas the directionality around the S atom (Fig. 9.6a) was less explicit. The other type of the S$\cdots$N interaction is a N–H$\cdots$S hydrogen bond (filled squares in Fig. 9.6) with linear N–H$\cdots$S alignment and roughly perpendicular approach of N to S. N–H$\cdots$S hydrogen bonds can be experimentally characterized by gas-phase IR/UV spectroscopy for short model peptides [14].

Statistical analysis of short S$\cdots$S contacts in PDB-SELECT suggested the presence of S$\cdots$S interactions around the SS groups in proteins although the frequency was low. Indeed, the directional analysis around the interacting S atoms (data no shown) revealed the unique direction preferences, which are similar to those of the nonbonded S$\cdots$S interactions found in organic crystals [29, 30]. Namely, one S atom is located in the backside of a S–S or S–C bond of another sulfur atom, and the latter

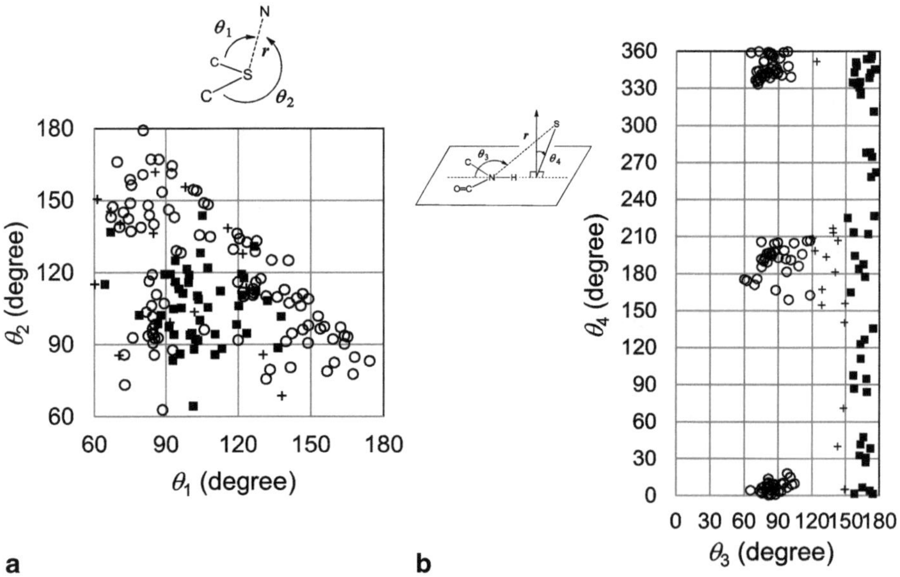

Fig. 9.6 Directionality of MeS$\cdots$N(amide) interactions ($d \leq 0.3$ Å) found in PDB-SELECT database. **a** Spatial distribution of the main-chain N atoms relative to the S atom. **b** Spatial distribution of the S atoms relative to the main-chain N atom. *Open circles* (○), *crosses* (+) and *filled squares* (■) represent the contacts with $\theta_3 \leq 120°$, $120 < \theta_3 \leq 150°$, and $150 < \theta_3 \leq 180°$, respectively. This figure was modified from ref [64]

S atom is approaching the former S atom in the lone pair directions. Thus, the S$\cdots$S interactions should have an advantage of an effective $n_S \rightarrow \sigma^*_S$ orbital interaction. This interaction can be categorized to type-4 chalcogen bond (S–S$\cdots$S–S).

According to the statistical and directional analysis of the S$\cdots$X contacts found in protein databases, existence of four typical S$\cdots$X chalcogen bonds (Fig. 9.7) was suggested in proteins. These chalcogen bonds are characterized by three structural features. (1) The nonbonded S$\cdots$X distance is only slightly less than or nearly equal to the sum of the van der Waals radii. (2) The X atom always comes to the backside of the S–S or S–C bond with preference of the extension of the S–S bond rather than the S–C bond, probably due to a more distinct σ-hole in the backside of the SS bond. These features are compatible with the definition of chalcogen bond [21]. (3) The S atom tends to approach the amide O or N atom in the directions perpendicular to the amide plane, while in the S$\cdots$S chalcogen bond the S atom tends to approach the other S atom in the direction of the lone pairs. Energetic aspects of these chalcogen bonds are shown in the subsequent section.

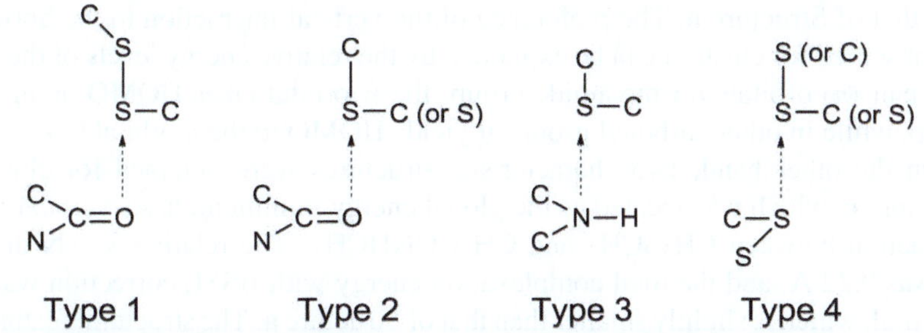

Fig. 9.7 Typical chalcogen bonds in proteins

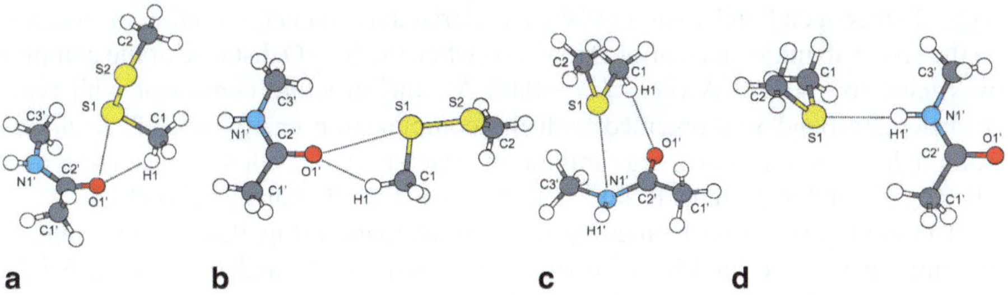

Fig. 9.8 Stable structures obtained for molecular clusters $CH_3SSCH_3 + CH_3CONHCH_3$ (**a** and **b**) and $CH_3SCH_3 + CH_3CONHCH_3$ (**c** and **d**) at MP2/6-31G(d). This figure was modified from ref [64]

9.4.3 Potential Surfaces of S··· X Interactions

In order to elucidate the intrinsic structural preferences as well as the strength of S···O chalcogen bonds, ab initio calculation was performed for molecular clusters of $CH_3SSCH_3 + CH_3CONHCH_3$ (cluster I) and $CH_3SCH_3 + CH_3CONHCH_3$ (cluster II) [18, 34, 64]. The representative structures obtained as a global or local energy minimum at MP2/6-31G(d) are shown in Fig. 9.8.

Structure **a** is the global energy minimum of cluster I, which has a vertical S···O interaction to the amide plane with keeping a linear S–S···O alignment. The relative S···O distance (*d*) was 0.02 Å. The structural features are in good agreement with those observed for type-1 chalcogen bond in proteins. The total stabilization energy due to the complexation with the correction for basis set superposition errors (BSSE) [72] was 3.21 kcal/mol including the contribution from a coexisting C–H···O hydrogen bond. The interaction distance was slightly decreased to $d = -0.02$ Å when a water molecule was hydrogen-bonded to the amide O atom [18]. Another stable structure located for cluster I (Structure **b**) has a horizontal S···O interaction to the amide plane. The relative S···O distance (*d*) was 0.06 Å, and the total complexation energy with BSSE correction was 2.18 kcal/mol, which is about 1 kcal/mol smaller than that of Structure **a**. The preference of the vertical interaction to the horizontal one observed for cluster I can be explained by the relative energy levels of the π and lone pair (*n*) orbitals for the amide group: the π orbital takes HOMO in an amide group, while in other carbonyl groups it yields HOMO to the *n* orbital [34].

On the other hand, two characteristic structures were obtained for cluster II. Structure **c**, which was located as the global energy minimum, has a vertical S··· N interaction between CH_3SCH_3 and $CH_3CONHCH_3$. The relative S··· N distance (*d*) was 0.22 Å, and the total complexation energy with BSSE correction was 2.88 kcal/mol, which is slightly smaller than that of Structure **a**. The structural features are consistent with type-3 S··· N chalcogen bond, supporting its existence in proteins. Structure **d** has a linear N–H··· S hydrogen bond with $d = 0.24$ Å. The relative energy of structure **d** to structure **c**, however, was + 0.78 kcal/mol. These calculation results are consonant with the observation for the S··· N contacts in proteins (Fig. 9.6). Type-2 chalcogen bond could not be characterized as an energy minimum structure in the above ab initio calculation. However, when the S··· O distance of the complex was fixed to $r = 3.30$ Å (i.e., $d = -0.02$ Å), the structure consonant with type-2 chalcogen bond was obtained with the complexation energy of 2.47 kcal/mol. Although it is not a local energy minimum structure, the result suggests that type-2 chalcogen bond in proteins is almost as strong as type-3 chalcogen bond.

The stable structures obtained by ab initio calculation (Fig. 9.8) strongly suggest the importance of orbital interactions, such as $\pi_O \rightarrow \sigma^*_S$ and $\pi_N \rightarrow \sigma^*_S$, for the chalcogen bonds in proteins. Indeed, such components were quantitatively evaluated by using the natural bond orbital (NBO) method [73]. However, the estimated value for structure **A** (0.64 kcal/mol) was not enough to explain the total interaction energy (3.21 kcal/mol) [18, 34]. On the other hand, a significant contribution of dispersion force was evident because the S··· O atomic distance for cluster I became larger when optimized at the HF/6-31G(d) level ($d = 0.37$ Å) and the stabilization energy reduced to 2.18 kcal/mol. Within the framework of the MP2 theory, the difference between MP2 and HF calculation results is assigned to the effects of high-order electrostatic interactions such as a dispersion force [71]. The S··· O chalcogen bonds in proteins may be primarily stabilized by the dispersion and/or long-range electrostatic forces but the directional propensity would be controlled by the HOMO–LUMO-type orbital interaction [64].

The potential surfaces of the type-1 S··· O chalcogen bond with a fixed S··· O distance ($d = 0.18$ Å) (Fig. 9.9) were then calculated for cluster I at MP2/6-31G(d) level [64]. The potential energy minimum is located at $\theta_1 \sim 90°$ and $\theta_2 \sim 240°$ around S and at $\theta_3 \sim 0°$ and $\theta_2 \sim 90°$ around O as indicated with x. The position corresponds nearly to the global energy minimum (i.e., structure **a**). The potential surface around O (Fig. 9.9b) is very flat compared to that around S (Fig. 9.9a), suggesting that the directionality around O is intrinsically weaker than that around S. It is more interesting to notice that the distribution patterns of the S··· O chalcogen bonds in proteins (Fig. 9.4) almost perfectly follow the potential surfaces calculated for the isolated model cluster. Thus, it is likely that the S··· O chalcogen bond is one of the important forces that determine protein structures. The attractive interaction was sustained at least in the range of $-0.22 < d < 0.58$ Å at the same calculation level,

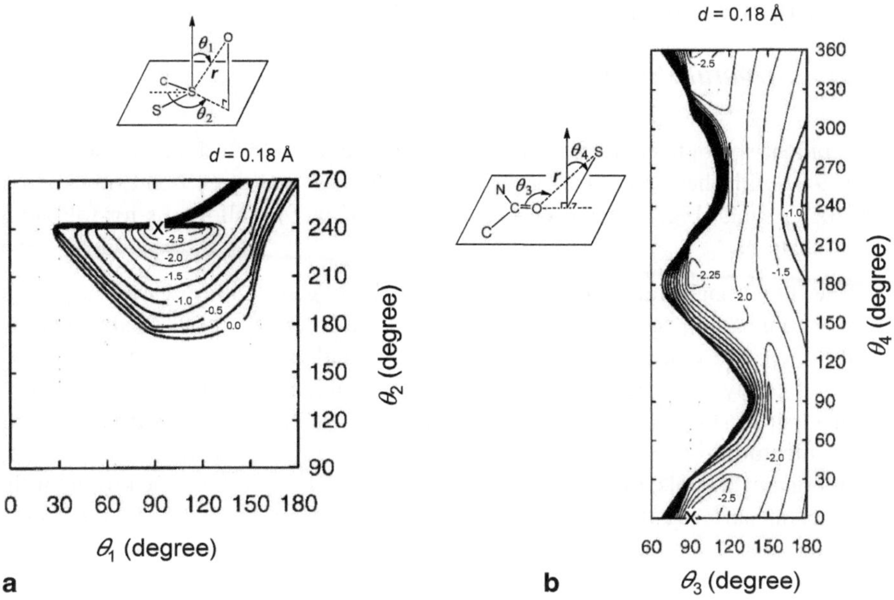

Fig. 9.9 Potential energy surfaces of type-1 S···O chalcogen bond calculated for molecular cluster $CH_3SSCH_3 + CH_3CONHCH_3$ at MP2/6-31G(d). The relative distance (*d*) was fixed at 0.18 Å. This figure was modified from ref [64]

suggesting that the potential energy surface of S···O chalcogen bond is also very flat along the axis of the interaction distance. Similar features were observed for the potential surfaces of type-2 chalcogen bond (C–S···O=C) [64].

9.4.4 Possibility of Se···X Chalcogen Bonds in Proteins

Selenium atoms are seldom found in protein structure databases. Therefore, statistical analysis of Se···X chalcogen bonds in proteins is not feasible at this moment. However, in some particular selenoproteins, such as glutathione peroxidase [74–77], thioredoxin reductase [78, 79], the chalcogen bonds may exist because a divalent selenium atom has a distinct σ-hole and nonbonded Se···X interactions (or Se···X chalcogen bonds) [44–46] are generally stronger than the S···X chalcogen bonds. Therefore, mining Se···X chalcogen bonds must be easier if high-resolution structures of selenoproteins will be accumulated. In the catalytic cycle of glutathione peroxidase, significant roles of Se···O and Se···N interactions at the active site have been suggested [47, 48]. It would be an interesting trial in future works to replace selenocysteine (Sec) for the Cys residue that is involved in a S···O chalcogen bond of a particular protein and see the change in the thermodynamic stability and the function.

9.5 Implications in Protein Science

9.5.1 Structural Control

The comparison of the potential energy surface calculated for a simple cluster model (Fig. 9.9) with the directional preferences of S$\cdots$O chalcogen bond observed in organic crystals (Fig. 9.1) and protein structures (Fig. 9.4) allows us for following considerations.

1. The directional preferences of the S$\cdots$O chalcogen bond observed in protein structures are in accord with the profiles of the potential surfaces calculated for the truncated cluster. The directionalities around both O and S atoms are not significantly perturbed in proteins. This remarkable agreement strongly suggests that the S$\cdots$O chalcogen bond is an important structural determinant for protein architecture.
2. In organic crystals, the directional preference around S follows the profile of the potential surface calculated for the isolated cluster, but the directional preference is not present around O. This suggests that crystal packing force can overwhelm the intrinsic directional preference around O, while it cannot disturb the intrinsic directionality around S.

Thus, the order of strength for the structural factors that are present in organic crystals and protein structures can be estimated to be: linearity of S–S$\cdots$O and C–S$\cdots$O $>$ crystal packing force $>$ verticality of S$\cdots$O$=$C $>$ protein structure [34, 66]. This order will be useful for protein engineering and molecular design of functional organic sulfur compounds.

9.5.2 Enzymatic Function

Importance of S$\cdots$X interactions in enzymatic functions has already been pointed out for some particular proteins [80–84]. For example, Taylor and Markham [80] reported that in the active site of *S*-adenosylmethionine synthetase the electrostatic S$\cdots$O interaction between the S atom of Met substrate and the carboxylate O atom of Asp118 plays a significant role. Brandt et al. [81] suggested that cleavage of a disulfide bond in the extracellular region of G-protein receptors is catalyzed by the S$\cdots$O interaction between Cys121 and the carboxylate side chain of Asp288. In most of these cases, the S$\cdots$O interaction is present between a divalent S atom and a negatively charged O atom. As for the electrically neutral S$\cdots$X interaction having pertinent directional features to S$\cdots$X chalcogen bonds, on the other hand, only a few examples [65, 66, 85] have been suggested to play roles in enzymatic functions of proteins.

After a success in structural characterization of S$\cdots$X chalcogen bonds in proteins, particular protein families or domains were employed for mining the specific S$\cdots$X chalcogen bonds that are commonly present in a wide range of homogeneous

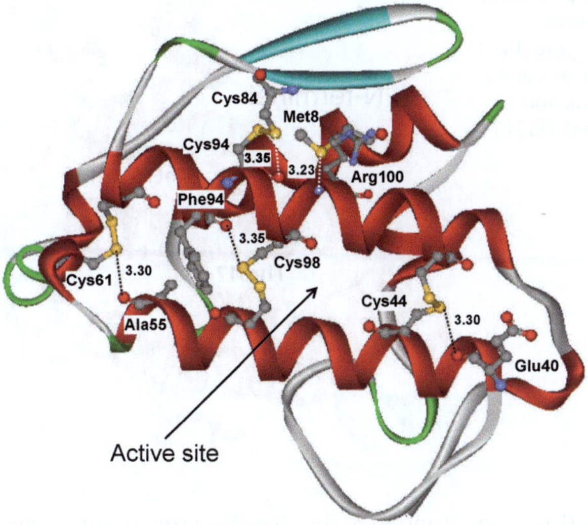

Fig. 9.10 Structure model of porcine phospholipase A_2 (PDB code, 1HN4) with indication of five S··· X interactions. The atomic distances are shown in Å

structures. For this purpose, redundant protein structures registered in PDB were very useful. Examples are shown below for the case of phospholipase A_2 (PLA_2) [65], ribonuclease A (RNase A) [66], insulin [66], and lysozyme [66]. A number of high-resolution structures can be found in PDB for these typical proteins

Phospholipase A_2 (PLA_2) [86, 87] is a small globular protein consisting of about 130 amino acid residues. This SS-rich protein catalyzes hydrolysis of the 2-acyl ester bond of phosphoglycerides in the presence of a calcium ion. Comprehensive search for S··· X chalcogen bonds in PLA_2 domain group revealed the presence of four common S··· O and one common S··· N interactions (Fig. 9.10). Among these, S(Cys61)··· O(Ala55), S(Cys84)··· O(Cys96), and S(Met8)··· N(Arg100) interactions can be clearly assigned to type-1, type-2, and type-3 chalcogen bonds, respectively, based on the directional characters. The other two have rather linear S··· O = C alignment. Most of the five S··· X interactions were present in vicinity of the active site, structurally supporting the conformational stability. They were not disturbed by the conformational changes caused by the binding of various substrates. Indeed, the enzymatic activity of porcine PLA_2 was significantly decreased in the Met8,20Leu mutant [88], which cannot hold the chalcogen bond of type-2 S(Met8)··· N(Arg100). Thus, this chalcogen bond was supposed to be important for the enzymatic activity of PLA_2 [65].

A similar functional role of S··· O interaction was proposed by Palenchar et al. [85] for *Bacillus subtilis* adenylosuccinate lyase, in which the MeS group of Met10 is located close to the main-chain amide O atom of Thr317 (Fig. 9.11). This S··· O interaction would effectively hold the N-terminal region close to the helix bundle. Indeed, the thermodynamic stability was significantly decreased upon substituting Met10 with leucine (Leu), resulting in a loss of the enzymatic activity to some extent.

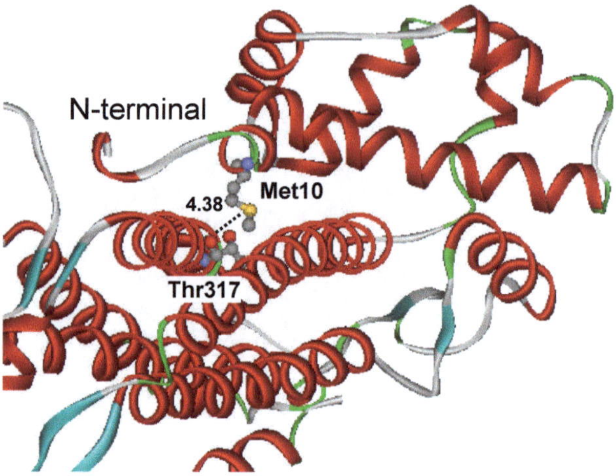

Fig. 9.11 Structure model of *Thermotoga maritima* adenylosuccinate lyase (PDB code, 1C3C) with indication of the S· · · O interaction between Met10 and Thr317

RNase A [89] is a small monomeric globular protein consisting of 124 amino acid residues having four SS bonds in the native state. For this domain, 43 high-resolution (≤ 2.0 Å) structures were extracted from PDB, and close S· · · X contacts were thoroughly sought. As a result, two common S· · · O and one common S· · · N interactions, i.e., S(Cys26)· · · Oγ(Thr99), S(Cys65)· · · O(Gln69), and S(Cys58)· · · N(Pro117), were characterized [66]. Their locations are shown in Fig. 9.12. The S(Cys26)· · · Oγ(Thr99) interaction has linear S–S· · · Oγ(side-chain) alignment. The other S(Cys65)· · · O(Gln69) and S(Cys58)· · · N(Pro117) interactions can be assigned to type-1 and type-3 chalcogen bonds, respectively, based on almost linear S–S· · · O and C–S· · · N alignment. The S(Cys65)· · · O(Gln69) chalcogen bond is located in the Cys65–Cys72 disulfide loop, which is one of the important sites that fold in the beginning of the oxidative folding [90] and gives significant thermodynamic stability to the native structure [90]. Therefore, structural and functional importance of the type-1 chalcogen bond was suggested [66].

Insulin [92] is a peptide hormone composed of two peptides (A and B chains), which are connected together by two SS bridges. Another SS bond is present in the A chain. For this protein domain, two common S· · · O interactions, i.e., S(Cys20)· · · O(Glu17) and S(Cys19)· · · O(Leu15), were detected within A and B chains, respectively, in the 23 high-resolution (≤ 2.0 Å) structures. Figure 9.13 shows their locations in the native structure. The two S· · · O interactions can be assigned to type-1 chalcogen bonds, forming a little bent O· · · S–S· · · O alignment. Interestingly, these chalcogen bonds were not disrupted by complexation with various ligands, while new S· · · O interactions were emerged in the A chain; S(Cys6)· · · O(Ile2), S(Cys7)· · · O(Val3), or S(Cys11)· · · O(Gln5). For the case of 1ZNI (Fig. 9.13), the former two S· · · O interactions were formed by binding zinc chloride. The allosteric effects may possibly contribute to the function of insulin [66].

Lysozyme [93] consists of 129 or 130 amino acid residues with four SS bonds in the native state. A large number of the structures with various ligand have been

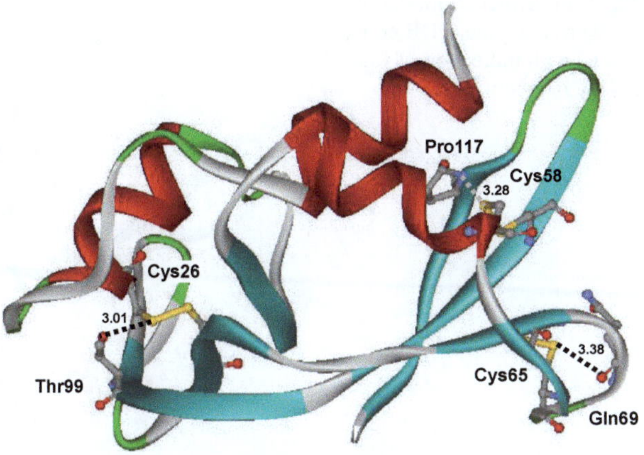

Fig. 9.12 Structure model of bovine ribonuclease A (PDB code, 9RAT) with indication of three S· · · X interactions. The atomic distances are shown in Å

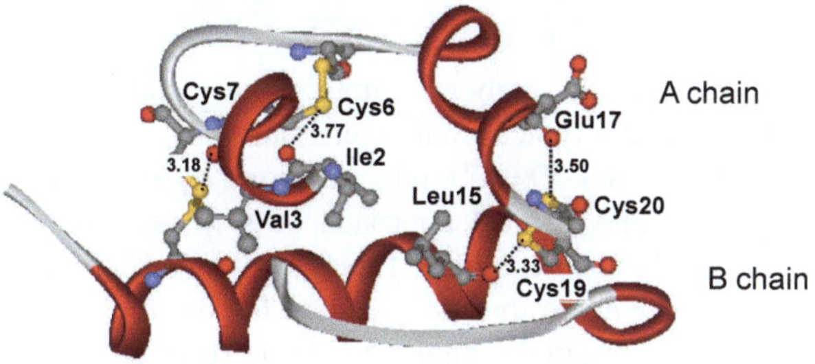

Fig. 9.13 Structure model of porcine insulin (PDB code, 1ZNI, A and B chains) with indication of four S· · · O interactions. $ZnCl_2$ is omitted for clarity. The atomic distances are shown in Å

registered in PDB. We [66] selected 28 high-resolution ($\leq$ 1.5 Å) structures of the lysozyme domain and analyzed close S· · · X contacts involved in the structures. One common S· · · O interaction, i.e., S(Cys127)· · · O(Ile124), and three common S· · · N interactions, i.e., S(Cys30)· · · Nε(Trp123), S(Cys80)· · · N(Asn65), and S(Cys127)· · · Nη(Arg5), were characterized as shown in Fig. 9.14. Among these, S(Cys127)· · · O(Ile124) and S(Cys127)· · · Nη(Arg5) make bifurcated interactions. However, functional roles of these interactions are not yet clear.

9.5.3 Evolution

The evolutional aspect of S· · · X interactions was analyzed for the snake PLA_2 domain [65]. For this domain group, the phylogenetic dendrogram was already obtained by Ohno et al. [94] using the amino acid sequences of various PLA_2 in venom of snakes inhabiting the southern islands of Japan. Mapping the S· · · O and S· · · N interactions commonly observed by the database analysis, it was found that most of the S· · · X interactions (including three chalcogen bonds shown in Fig. 9.10) make clusters on the dendrogram. The evolutional conservation suggests a possible role of S· · · X chalcogen bonds in molecular evolution of proteins [65].

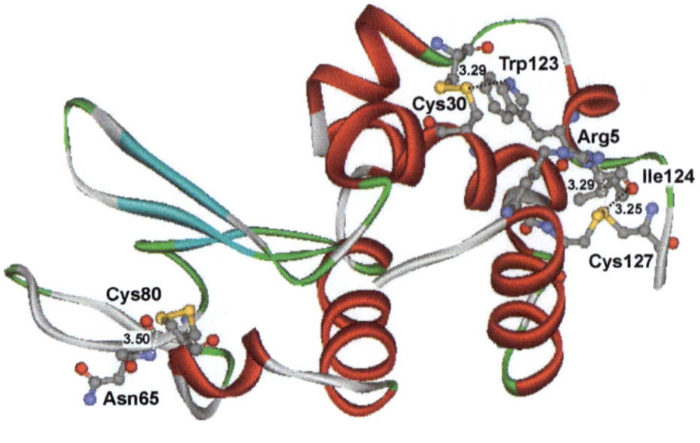

Fig. 9.14 Structure model of chicken lysozyme (PDB code, 194 L) with indication of four S$\cdots$X interactions

9.6 Outlook

According to the elaborate analysis of close S$\cdots$X contacts in protein structure databases, four typical chalcogen bonds have been characterized as shown in Fig. 9.7 [64]. The S–S$\cdots$O$=$C chalcogen bond (type 1) is characterized by linear S–S$\cdots$O alignment and vertical approach of S to the amide plane, allowing effective $\pi_O \rightarrow \sigma^*_S$ orbital interaction. The C–S$\cdots$O$=$C chalcogen bond (type 2) has similar structural features to type 1 except for the replacement of C–S for the S–S bond. The C–S$\cdots$N(amide) chalcogen bond (type 3) is characterized by linear C–S$\cdots$N alignment and vertical approach of S to the amide plane, allowing effective $\pi_N \rightarrow \sigma^*_S$ orbital interaction. The last S–S$\cdots$S–S chalcogen bond (type 4) has linear S–S$\cdots$S or C–S$\cdots$S alignment and roughly vertical approach of S to the other S atom, allowing $n_S \rightarrow \sigma^*_S$ orbital interaction. The strength of these chalcogen bonds would decrease in the order of type 1 > type2 $\approx$ type 3 > type 4, according to ab initio calculation for simple molecular cluster models. The maximum interaction energy was estimated to be 3.21 kcal/mol for type1. The potential energy surfaces are significantly flat around the O atom and along the interaction distance, while the directional preference around the S atom is distinct. For some particular proteins, putative roles of the chalcogen bonds in their enzymatic functions and evolution has been suggested. Thus, the chalcogen bonds, though their interaction energy is smaller than that of classical hydrogen bonds ($\sim$5 kcal/mol), can be an important element of weak noncovalent interactions that control chemical and biological properties of protein architecture.

On the basis of the data obtained from statistical database analysis and theoretical calculation on the S$\cdots$X chalcogen bonds in proteins, possible answers to the six questions raised earlier would be as follows.

Q1. Do S$\cdots$X chalcogen bonds really exist in proteins?

A1. Yes. According to the similarity between the statistical analysis results obtained by using two different structure databases (PDB-SELECT and TOP500), S$\cdots$X chalcogen bonds should exist widely in proteins.

Q2. Are the directional propensities similar to those observed for small organic molecules?

A2. No. In the S$\cdots$O$=$C interactions, the O atom approaches the S atom in the backside of a S–S or S–C bond (σ^*_S direction) in both organic crystals and proteins. However, the directionality around the O atom is in significant difference. There is no directional preference in organic crystals due probably to the diversity of the C$=$O functional groups as well as the crystal packing force, whereas in proteins the S atom comes over the carbonyl O atom in the π_O direction due to the elevated π_O orbital energy level in the amide group compared to other types of carbonyl groups, such as ketone and ester.

Q3. How do the potential surfaces look like?

A3. Profiles of the potential surfaces around the S and O atoms for both S–S$\cdots$O$=$C and C–S$\cdots$O$=$C interactions match well with the directional distributions of the S$\cdots$O interactions observed in proteins. There are distinct potential holes in the σ^*_S and π_O orbital directions. However, the potential surface around the O atom is rather flat compared to that around the S atom.

Q4. Can the S$\cdots$X chalcogen bonds control protein structure?

A4. Yes. Since the directionalities around S and O atoms of type-1 S$\cdots$O chalcogen bonds in proteins are not disturbed from their intrinsic propensities on the potential surfaces calculated for the truncated cluster models, S$\cdots$O chalcogen bonds should control the protein structures to some extent. Similar situation may be applied for the other S$\cdots$X chalcogen bonds in proteins.

Q5. Are the S$\cdots$X chalcogen bonds relevant to protein function and evolution?

A5. Probably yes. There are several examples in which specific S$\cdots$X chalcogen bonds are suggested to control protein functions (for PLA_2, RNase A, insulin, etc.) and evolution (for PLA_2).

Q6. Are there similar Se$\cdots$X chalcogen bonds in selenoproteins too?

A6. Probably yes. Se$\cdots$X chalcogen bonds should be stronger than S$\cdots$X chalcogen bonds because a Se atom has larger propensity than a S atom to adapt a hypervalent state. Therefore, Se$\cdots$X chalcogen bonds may possibly be found in selenoproteins if the high-resolution structures will be accumulated.

Thus, the concept of S$\cdots$X chalcogen bonds will be a useful tool for fully understanding the structure, function, and evolution of proteins. Their applications to protein engineering and the ligand design will be interesting and promising challenges in the fields of protein science.

Acknowledgment I thank Ms. Natsuki Babe for her assistance in preparation of graphical materials.

References

1. Tonin C, Aluigi A, Varesano A, Vineis, C (2010) Keratin-based nanofibers. In: Nanofibers KA (ed) InTech, chapter 7. pp 139–158
2. Fraser RDB, Parry DAD (2011) The structural basis of the filament-matrix texture in the avian/reptilian group of hard β-keratins. J Struct Biol 173:391–405
3. Coyle P, Philcox JC, Carey LC, Rofe AM (2002) Metallothionein: the multipurpose protein. Cell Mol Life Sci 59:627–647
4. Moura JJG, Macedo AL, Palma PN (1994) Ferredoxins. Methods Enzymol 243:165–188
5. Laity JH, Lee BM, Wright PE (2001) Zinc finger proteins: new insights into structural and functional diversity. Curr Opin Struct Biol 11:39–46
6. Nordberg J, Arner ESJ (2001) Reactive oxygen species, antioxidants, and the mammalian thioredoxin system. Free Rad Biol Med 31:1287–1312
7. Ellgaard L, Ruddock LW (2005) The human protein disulphide isomerase family: Substrate interactions and functional properties. EMBO Rep 6:28–32
8. Gruber CW, Čemažar M, Heras B, Martin JL, Craik DJ (2006) Protein disulfide isomerase: the structure of oxidative folding. Trends Biochem Sci 31:455–464
9. Hatahet F, Ruddock LW (2009) Protein disulfide isomerase: a critical evaluation of its function in disulfide bond formation. Antioxid Redox Signal 11:2807–2850
10. Rao MB, Tanksale AM, Ghatge MS, Deshpande VV (1998) Molecular and biotechnological aspects of microbial proteases. Microbiol Mol Biol Rev 62:597–635
11. Ventura S, Chang RJY (2011) Oxidative folding: coupling conformational folding and disulfide formation. In Folding of disulfide proteins, Chang RJY, Ventura S eds; Springer: New York, chapter 1, pp 1–22
12. Berman HM, Westbrook J, Feng Z, Gilliland G, Bhat TN, Weissig H, Shindyalov IN, Bourne PE (2000) The protein data bank. Nucl Acids Res 28:235–242
13. Ippolito JA, Alexander RS, Christianson DW (1990) Hydrogen bond stereochemistry in protein structure and function. J Mol Biol 215:457–471
14. Biswal H, Gloaguen E, Loquais Y, Tardivel B, Mons M (2012) Strength of NH$\cdots$S hydrogen bonds in methionine residues revealed by gas-phase IR/UV spectroscopy. J Phys Chem Lett 3:755–759
15. Bhattacharya R, Pal D, Chakrabarti P (2004) Disulfide bonds, their stereospecific environment and conservation in protein structures. Protein Eng Des Sel 17:795–808
16. Chakrabarti P, Bhattacharyya R (2007) Geometry of nonbonded interactions involving planar groups in proteins. Prog Biophys Mol Biol 95:83–137
17. Carugo O (1999) Stereochemistry of the interaction between methionine sulfur and the protein core. Biol Chem 380:495–498
18. Iwaoka M, Takemoto S, Okada M, Tomoda S (2001) Statistical characterization of nonbonded S$\cdots$O interactions in proteins. Chem Lett 132–133
19. Pal D, Chakrabarti P (2001) Non-hydrogen bond interactions involving the methionine sulfur atom. J Biomol Struct Dynam 19:115–128
20. Murray JS, Lane P, Politzer P (2008) Simultaneous σ-hole and hydrogen bonding by sulfur- and selenium-containing heterocycles. Int J Quant Chem 108:2770–2781
21. Wang W, Ji B, Zhang Y (2009) Chalcogen bond: a sister noncovalent bond to halogen bond. J Phys Chem A 113:8132–8135
22. Metrangolo P, Neukirch H, Pilati T, Resnati G (2005) Halogen bonding based recognition processes: a world parallel to hydrogen bonding. Acc Chem Res 38:386–395
23. Parisini E, Metrangolo P, Pilati T, Resnati G, Terraneo G (2011) Halogen bonding in halocarbon-protein complexes: a structural survey. Chem Soc Rev 40:2267–2278
24. Politzer P, Murray JS, Clark T (2010) Halogen bonding: an electrostatically-driven highly directional noncovalent interaction. Phys Chem Chem Phys 12:7748–7757
25. Ciuffarin E, Guaraldi G (1970) Nucleophilic substitution at dicoordinated sulfur. Effect of the leaving group on the reaction between triphenylmethyl sulfenyl derivatives and n-butylamine. J Org Chem 35:2006–2010

26. Lu J, Lu Y, Yang S, Zhu W (2011) Theoretical and crystallographic data investigations of noncovalent S··· O interactions. Struct Chem 22:757–763
27. Rosenfield Jr RE Parthasarathy R Dunitz JD (1977) Directional preferences of nonbonded atomic contacts with divalent sulfur. 1. Electrophiles and nucleophiles. J Am Chem Soc 99:4860–4862
28. Allen FH (2002) The Cambridge structural database: a quarter of a million crystal structures and rising. Acta Crystallogr Sect B 58:380–388
29. Row TNG, Parthasarathy R (1981) Directional preferences of nonbonded atomic contacts with divalent sulfur in terms of its orbital orientations. 2. S··· S interactions and nonspherical shape of sulfur in crystals. J Am Chem Soc 103:477–479
30. Desiraju GR, Nalini V (1991) Database analysis of crystal-structure-determining interactions involving sulphur: implications for the design of organic metals. J Mater Chem 1:201–203
31. Dahaoui S, Pichon-Pesme V, Howard JAK, Lecomte C (1999) CCD charge density study on crystals with large unit cell parameters: the case of hexagonal L-cystine. J Phys Chem A 103:6240–6250
32. Kucsman A, Kapovits I (1985) Nonbonded sulfur-oxygen interaction in organic sulfur compounds. In: Bernardi F, Csizmadia IG, Mangini A (eds) Organic sulfur chemistry: theoretical and experimental advances. Elsevier, Amsterdam, pp 191–245
33. Zauhar RJ, Colbert CL, Morgan RS, Welsh WJ (2000) Evidence for a strong sulfur-aromatic interaction derived from crystallographic data. Biopolymers 53:233–248
34. Iwaoka M, Takemoto S, Tomoda S (2002) Statistical and theoretical investigations on the directionality of nonbonded S··· O interactions. Implications for molecular design and protein engineering. J Am Chem Soc 124:10613–10620
35. Burling FT, Goldstein BM (1992) Computational studies of nonbonded sulfur-oxygen and selenium-oxygen interactions in the thiazole and selenazole nucleosides. J Am Chem Soc 114:2313–2320
36. Nakanishi W, Hayashi S, Narahara K (2008) Atoms-in-molecules dual parameter analysis of weak to strong interactions: behaviors of electronic energy densities versus Laplacian of electron densities at bond critical points. J Phys Chem A 112:13593–13599
37. Bader RFW (1990) Atoms in molecules: a quantum theory. Oxford University Press: New York
38. Gillespie RJ, Popelier PLA (2001) Chemical bonding and molecular geometry. Oxford University Press: New York
39. Jeziorski B, Moszynski R, Szalewicz K (1994) Perturbation theory approach to intermolecular potential energy surfaces of van der Waals complexes. Chem Rev 94:1887–1930
40. Scheiner S (2011) On the properties of X··· N noncovalent interactions for first-, second-, and third-row X atoms. J Chem Phys 134:164313
41. Adhikari U, Scheiner S (2011) The S··· N noncovalent interaction: comparison with hydrogen and halogen bonds. Chem Phys Lett 514:36–39
42. Nagao Y, Hirata T, Goto S, Sano S, Kakehi A, Iizuka K, Shiro M (1998) Intramolecular nonbonded S··· O interaction recognized in (acylimino)thiadiazoline derivatives as angiotensin II receptor antagonists and related compounds. J Am Chem Soc 120:3104–3110
43. Wu S, Greer A (2000) Attractive through-space S–O interaction in the DNA-cleaving antitumor antibiotic leinamycin. J Org Chem 65:4883–4887
44. Iwaoka M, Tomoda S (1996) Nature of the intramolecular Se··· N nonbonded interaction of 2-selenobenzylamine derivatives. An experimental evaluation by 1H, 77Se, and 15N NMR spectroscopy. J Am Chem Soc 118:8077–8084
45. Iwaoka M, Komatsu H, Katsuda T, Tomoda S (2004) Nature of nonbonded Se··· O interactions characterized by 17O NMR spectroscopy and NBO and AIM analyses. J Am Chem Soc 126:5309–5317
46. Iwaoka M, Komatsu H, Katsuda T, Tomoda S (2002) Quantitative evaluation of weak nonbonded Se··· F interactions and their remarkable nature as orbital interactions. J Am Chem Soc 124:1902–1909
47. Mugesh G, Panda A, Singh HB, Punekar NS, Butcher RJ (2001) Glutathione peroxidase-like antioxidant activity of diaryl diselenides: a mechanistic study. J Am Chem Soc 123:839–850

48. Takei T, Urabe Y, Asahina Y, Hojo H, Nomura T, Dedachi K, Arai K, Iwaoka M (2014) Model study using designed selenopeptides on the importance of the catalytic triad for the antioxidative functions of glutathione peroxidase. J Phys Chem B 118:492–500
49. Dobson CM, Šali A, Karplus M (1998) Protein folding: a perspective from theory and experiment. Angew Chem Int Ed 37:868–893
50. Derewenda ZS, Lee L, Derewenda U (1995) The occurrence of C–H···O hydrogen bonds in proteins. J Mol Biol 252:248–262
51. Bella J, Berman HM (1996) Crystallographic evidence for C(α)–H···O = C hydrogen bonds in a collagen triple helix. J Mol Biol 264:734–742
52. Vargas R, Garza J, Dixon DA, Hay BP (2000) How strong is the C(α)–H···O = C hydrogen bond? J Am Chem Soc 122:4750–4755
53. Burley SK, Petsko GA (1986) Amino-aromatic interactions in proteins. FEBS Lett 203:139–143
54. Flocco MM, Mowbray SL (1994) Planar stacking interactions of arginine and aromatic side-chains in proteins. J Mol Biol 235:709–717
55. Gallivan JP, Dougherty DA (1999) Cation-π interactions in structural biology. Proc Nat Acad Sci U S A 96:9459–9464
56. Gallivan JP, Dougherty DA (2000) A computational study of cation-π interactions vs salt bridges in aqueous media: Implications for protein engineering. J Am Chem Soc 122:870–874
57. Umezawa Y, Nishio M (1998) Ch/π interactions as demonstrated in the crystal structure of guanine- nucleotide binding proteins, src homology-2 domains and human growth hormone in complex with their specific ligands. Bioorg Med Chem 6:493–504
58. Takahashi O, Kohno Y, Nishio M (2010) Relevance of weak hydrogen bonds in the conformation of organic compounds and bioconjugates: evidence from recent experimental data and high-level ab initio MO calculations. Chem Rev 110:6049–6076
59. Brandl M, Weiss MS, Jabs A, Sühnel J, Hilgenfeld R (2001). C-H···π-interactions in proteins. J Mol Biol 307:357–377
60. Morgan RS, Tatsch CE, Gushard RH, McAdon J, Warme PK (1978) Chains of alternating sulfur and pi-bonded atoms in eight small proteins. Int J Peptide Protein Res 11:209–217
61. Reid KSC, Lindley PF, Thornton JM (1985) Sulphur-aromatic interactions in proteins. FEBS Lett 190:209–213
62. Viguera AR, Serrano L (1995) Side-chain interactions between sulfur-containing amino acids and phenylalanine in α-helices. Biochemistry 34:8771–8779
63. Cheney BV, Schulz MW, Cheney J (1989) Complexes of benzene with formamide and methanethiol as models for interactions of protein substructures. Biochim Biophys Acta 996:116–124
64. Iwaoka M, Takemoto S, Okada M, Tomoda S (2002) Weak nonbonded S···X (X = O, N, and S) interactions in proteins. Statistical and theoretical studies. Bull Chem Soc Jpn 75:1611–1625
65. Iwaoka M, Isozumi N (2006) Possible roles of S···O and S···N interactions in the functions and evolution of phospholipase A2. Biophysics 2:23–34
66. Iwaoka M, Isozumi N (2012) Hypervalent nonbonded interactions of a divalent sulfur atom. Implications in protein architecture and the functions. Molecules 17:7266–7283
67. Iwaoka M, Babe N (2015) Mining and structural characterization of S···X chalcogen bonds in protein database. Phosphorus Sulfur Silicon Relat Elem, in press.
68. Hobohm U, Scharf M, Schneider R, Sander C (1992) Selection of representative protein data sets. Protein Sci 1:409–417
69. Hobohm U, Sander C (1994) Enlarged representative set of protein structures. Protein Sci 3:522–524
70. Richardson Laboratory, 3D analysis:: 500 filtered structures. http://kinemage.biochem.duke.edu/databases/top500.php. Accessed Aug 2014
71. Jaffe RL, Smith GD (1996) A quantum chemistry study of benzene dimer. J Chem Phys 105:2780–2788
72. Schwenke DW, Truhlar DG (1985) Systematic study of basis set superposition errors in the calculated interaction energy of two HF molecules. J Chem Phys 82:2418–2426
73. Reed AE, Curtiss LA, Weinhold F (1988) Intermolecular interactions from a natural bond orbital, donor-acceptor viewpoint. Chem Rev 88:899–926

74. Flohé L, Günzler WA, Schock HH (1973) Glutathione peroxidase: a selfnoenzyme. FEBS Lett 32:132–134
75. Savaskan NE, Ufer C, Kühn H, Borchert A (2007) Molecular biology of glutathione peroxidase 4: from genomic structure to developmental expression and neural function. Biol Chem 388:1007–1017
76. Conrad M, Schneider M, Seiler A, Bornkamm GW (2007) Physiological role of phospholipid hydroperoxide glutathione peroxidase in mammals. Biol Chem 388:1019–1025
77. Toppo S, Flohé L, Ursini F, Vanin S, Maiorino M (2009) Catalytic mechanisms and specificities of glutathione peroxidases: variations of a basic scheme. Biochim Biophys Acta 1790:1486–1500
78. Holmgren A, Lu J (2010) Thioredoxin and thioredoxin reductase: current research with special reference to human disease. Biochem Biophys Res Commun 396:120–124
79. Prast-Nielsen S, Huang HH, Williams DL (2011) Thioredoxin glutathione reductase: its role in redox biology and potential as a target for drugs against neglected diseases. Biochim Biophys Acta 1810:1262–1271
80. Taylor JC, Markham GD (1999) The bifunctional active site of S-adenosylmethionine synthetase. Roles of the active site aspartates. J Biol Chem 274:32909–32914
81. Brandt W, Golbraikh A, Täger M, Lendeckel U (1999) A molecular mechanism for the cleavage of a disulfide bond as the primary function of agonist binding to G-protein-coupled receptors based on theoretical calculations supported by experiments. Eur J Biochem 261:89–97
82. Nagao Y, Honjo T, Iimori H, Goto S, Sano S, Shiro M, Yamaguchi K, Sei Y (2004) Intramolecular nonbonded S···O interaction in acetazolamide and thiadiazolinethione molecules in their dimeric crystalline structures and complex crystalline structures with enzymes. Tetrahedron Lett 45:8757–8761
83. Singh AK, Singh N, Sharma S, Shin K, Takase M, Kaur P, Srinivasan A, Singh TP (2009) Inhibition of lactoperoxidase by its own catalytic product: Crystal structure of the hypothiocyanate-inhibited bovine lactoperoxidase at 2.3-Å resolution. Biophys J 96:646–654
84. Nakamura T, Yamamoto T, Abe M, Matsumura H, Hagihara Y, Goto T, Yamaguchi T, Inoue T (2008) Oxidation of archaeal peroxiredoxin involves a hypervalent sulfur intermediate. Proc Nat Acad Sci U S A 105:6238–6242
85. Palenchar JB, Crocco JM, Colman RF (2003) The characterization of mutant Bacillus subtilis adenylosuccinate lyases corresponding to severe human adenylosuccinate lyase deficiencies. Protein Sci 12:1694–1705
86. Dennis EA (1994) Diversity of group types, regulation, and function of phospholipase A2. J Biol Chem 269:13057–13060
87. Jain MK, Gelb MH, Rogers J, Berg OG (1995) Kinetic basis for interfacial catalysis by phospholipase A2. Methods Enzymol 249:567–614
88. Yu BZ, Pan YH, Janssen MJW, Bahnson BJ, Jain MK (2005) Kinetic and structural properties of disulfide engineered phospholipase A2: insight into the role of disulfide bonding patterns. Biochemistry 44:3369–3379
89. Raines RT (1998) Ribonuclease A. Chem Rev 98:1045–1065
90. Xu X, Rothwarf DM, Scheraga HA (1996) Nonrandom distribution of the one-disulfide intermediates in the regeneration of ribonuclease A. Biochemistry 35:6406–6417
91. Li YJ, Rothwarf DM, Scheraga HA (1995) Mechanism of reductive protein unfolding. Nat Struct Biol 2:489–494
92. De Meyts P (2004) Insulin and its receptor: structure, function and evolution. BioEssays 26:1351–1362
93. Merlini G, Bellotti V (2005) Lysozyme: a paradigmatic molecule for the investigation of protein structure, function and misfolding. Clin Chim Acta 357:168–172
94. Ohno M, Chijiwa T, Oda-Ueda N, Ogawa T, Hattori S (2003) Molecular evolution of myotoxic phospholipases A2 from snake venom. Toxicon 42:841–854

Chapter 10
A Unified View of Halogen Bonding, Hydrogen Bonding and Other σ-Hole Interactions

Peter Politzer and Jane S. Murray

Abstract The term "σ-hole" refers to a region of diminished electronic density along the extension of a covalent single bond to a hydrogen or an atom of Groups IV—VII. This region often has a positive electrostatic potential through which the atom can interact attractively with a negative site (such as a lone pair of a Lewis base, π electrons or an anion) to form a noncovalent complex. Hydrogen bonding and halogen bonding are the most prominent examples of such σ-hole interactions, although they have long been known experimentally for Groups IV—VI as well (but without the σ-hole label). σ-Holes result from the anisotropic charge distributions of covalently-bonded atoms. It follows from the Hellmann-Feynman theorem that σ-hole interactions can be understood and described as Coulombic, which includes polarization and dispersion. In the context of noncovalent interactions, charge transfer is simply a mathematical formulation of polarization.

10.1 Noncovalent Interactions: The Chemistry of the Twenty-First Century

Recent years have seen a remarkable surge of research activity in the area of noncovalent interactions. One indication of this is the number of review and perspective papers that have appeared just since the year 2000; some of them (certainly not all) are cited here [1–17]. Indeed, an observation made by Schneider appears thus far to be borne out: "... one might assert that the chemistry of the last century was largely the chemistry of covalent bonding, whereas that of the present century is more likely to be the chemistry of noncovalent binding." [18].

Much of this recent interest has focused upon what is known as "halogen bonding," a highly-directional attractive interaction that is found to occur between many covalently-bonded halogen atoms and negative sites (lone pairs of Lewis bases, π electrons, anions, etc.). The interest is due in part to the applications of halogen bonding, existing and potential, in fields such as crystal engineering, molecular biology and pharmacology [5, 8, 10, 15], but it also reflects the intriguing fact that an

P. Politzer (✉) · J. S. Murray
Department of Chemistry, University of New Orleans, 70148 New Orleans, LA, USA
e-mail: ppolitze@uno.edu

S. Scheiner (ed.), *Noncovalent Forces,* Challenges and Advances in Computational Chemistry and Physics 19, DOI 10.1007/978-3-319-14163-3_10

electronegative halogen is attracted to a negative site. This was sometimes described as an enigma!

Our primary focus in this chapter will be upon halogen bonding. We emphasize, however, that this is simply a subset of a much larger category of noncovalent interactions, "σ-hole bonding," that can involve covalently-bonded atoms of Groups IV—VI as well as Group VII (the halogens) and also includes hydrogen bonding. The nature of the interaction is essentially the same in all of these cases, as shall be discussed. The formation of noncovalent complexes between Group IV—VII atoms and negative sites has been known to experimentalists for decades, although the unifying σ-hole interpretation is of more recent origin.

10.2 Some Historical Background

A very early example of what we now call halogen bonding was the report, in 1814, of an interaction between iodine and ammonia [19]. The product was later purified by Guthrie [20] and formulated as $I_2{\cdot}NH_3$. Additional complexes of Cl_2, Br_2 and I_2 with amines were subsequently observed [21, 22], as well as an adduct between iodoform and quinoline [23].

The twentieth century saw many more R-X—B systems being identified, where RX is a halide molecule and B is a Lewis base; a number of these studies are cited by Blackstock et al [24]. Of particular note was the spectral characterization of complexes of molecular iodine with benzene and other aromatic hydrocarbons by Benesi and Hildebrand [25, 26]. Their observations were a factor in Mulliken's development of his "charge-transfer" formalism [27], which is frequently invoked as a mathematical (but not physical) description of noncovalent bonding. However Mulliken incorrectly suggested that in the I_2—C_6H_6 complex, the I_2 axis is parallel to the benzene plane.

Important advances in elucidating the structure of halogen bonding have come through crystallography. In the 1960's, Hassel and his colleagues determined the crystal structures of a number of complexes between organic halides and oxygen/nitrogen Lewis bases; this work contributed to his receiving the Nobel Prize in 1969. These studies showed the crystals to be composed of chain-like structures *held together by weak X–B bonds*, e.g. **1** [28]. For a review, see Bent [29]. At about the same time, attractive Br—O intermolecular interactions were found in solid $POBr_3$ [30] and Cl—O in solid $POCl_3$ [31].

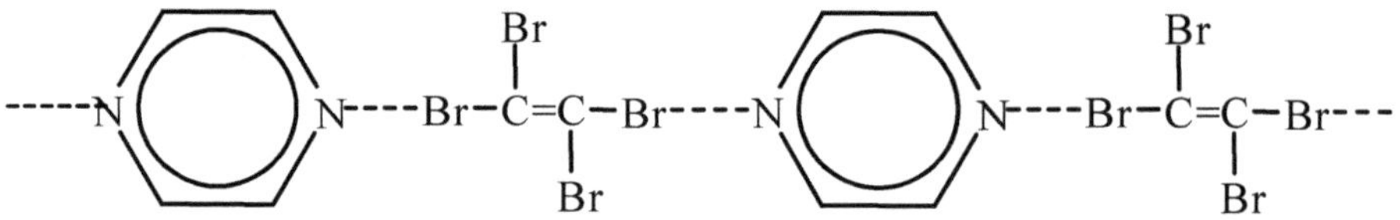

1

A few years later, Murray-Rust et al conducted extensive surveys of halide (RX; X = Cl, Br, I) crystal structures, using the Cambridge Structural Database [32–34]. Their objective was to identify close contacts between halogen atoms and atoms of neighboring molecules. Close contact was defined as being less than the sum of the respective van der Waals radii, and was taken to reflect an attractive interaction. A very significant pattern was revealed: Close contacts of halogen atoms with electrophilic sites, such as metal ions, were primarily at angles of 90–120° relative to the R–X bond (**2**); close contacts with nucleophilic sites, e.g. nitrogens and oxygens, tended to be nearly linear, the angles being 160–180° (**3**).

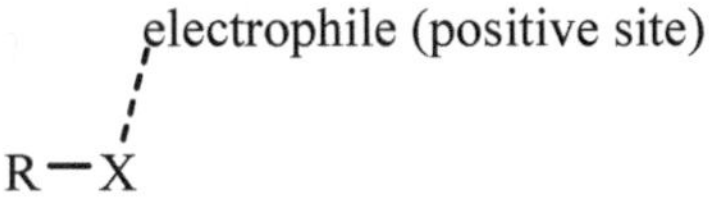

2

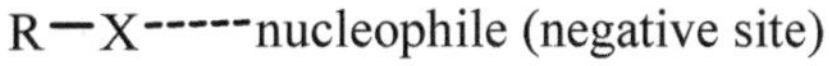

3

Thus a given halogen atom can interact favorably, but in different directions, with both positive and negative sites.

The near-linear interactions of covalently-bonded halogens with nucleophiles, as in **3**, are what we call halogen bonding, R-X—B. To our knowledge, this term was first used in 1976 [35, 36]. In the past, halogen bonding was sometimes viewed as very puzzling, since univalent halogens are generally regarded as being negative in character; how then can they be attracted to negative sites? This question was answered in 1992, as shall be discussed in Sect. 10.3.2.

During the last two decades, the practical significance and potential applications of halogen bonding have increasingly been recognized. Imakubo et al prepared semiconducting crystals and a superconductor by linking iodine-containing organic sulfides via negative ions (acting as the bases) [37, 38]. An important development was the realization that the role of halogen bonding in formulating new materials is enhanced if the halogen-containing molecule also contains strongly electron-attracting groups, e.g. perfluorination [39]. This helped to stimulate a great deal of activity in the area of crystal engineering, with applications in electronics, nonlinear optical activity, magnetic materials, liquid crystals, etc. [5, 8, 10, 40–42]. Catalysis through halogen bonding is also being explored [43, 44].

In chemical biology as well, there is a growing awareness of the importance of halogen bonding. A milestone was the survey of the Protein Data Bank by Auffinger et al [45], revealing C-X—O close contacts (X = Cl, Br, I); it was subsequently expanded by Lu et al to C-X—Y (X = Cl, Br, I; Y = O, N, S) [46]. Such interactions are now known to affect protein-ligand binding, recognition and assembly processes, docking, conformational stability, protein folding, etc. [5, 15, 41, 47], and are being exploited in drug design.

During roughly the same years as the surveys of halide crystal structures by Murray-Rust et al [32–34], discussed above, analogous studies of close contacts in organic sulfides and selenides were being carried out [48–50]. Very similar results were obtained. Electrophilic sites interacted with the lateral sides of the sulfur or selenium (**4**), while interactions with nucleophiles were close to linear, along the extensions of the R_1–S(Se) and/or R_2–S(Se) bonds (**5**). Some of the close contacts were between a sulfur in one molecule and the same sulfur in another identical molecule [49], i.e. "like attracting like" (**6**); the interactions were linear for one of the sulfurs and lateral for the other.

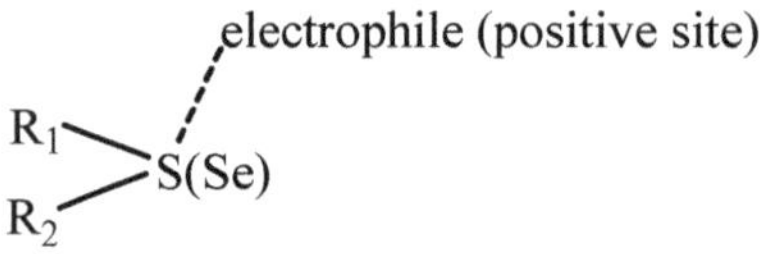

4

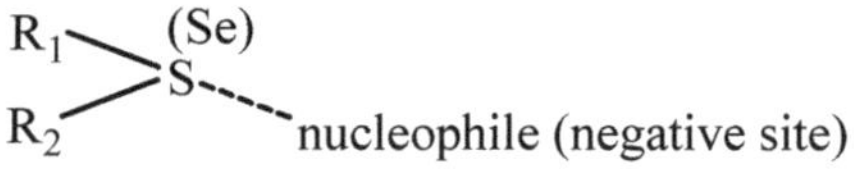

5

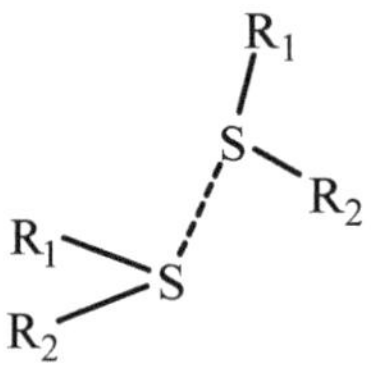

6

10.3 The Electrostatic Potential and the σ-Hole

All of these observations were explained and the "enigma" of halogen bonding was resolved in 1992—through the analysis of molecular electrostatic potentials. We will accordingly give some brief background concerning this property.

10.3.1 The Electrostatic Potential

The nuclei and electrons of a molecule (or other system) create an electrostatic potential $V(\mathbf{r})$ at each point $\mathbf{r}$ in the surrounding space:

$$V(\mathbf{r}) = \sum_{A} \frac{Z_A}{|\mathbf{R}_A - \mathbf{r}|} - \int \frac{\rho(\mathbf{r}')d\mathbf{r}'}{|\mathbf{r}' - \mathbf{r}|} \tag{10.1}$$

In Eq. (10.1), which is simply one form of Coulomb's Law, Z_A is the charge on nucleus A, located at $\mathbf{R}_A$, and $\rho(\mathbf{r})$ is the molecule's electronic density distribution. $V(\mathbf{r})$ is positive or negative in any given region depending upon whether the effects of the positive nuclei or the negative electrons are dominant there.

If a point charge Q is placed at $\mathbf{r}$, then its interaction energy ΔE with the molecule's nuclei and electrons is,

$$\Delta E = QV(\mathbf{r}) \tag{10.2}$$

If Q and V(**r**) have the same sign (positive or negative), then $\Delta E > 0$ and the interaction is repulsive; if they have opposite signs, then $\Delta E < 0$ and the interaction is attractive.

The electrostatic potential is accordingly an effective means of predicting close contacts and noncovalent interactions [51–53]. In general, regions of positive V(**r**) tend to interact favorably (at least initially) with negative sites and negative V(**r**) with positive sites. (A more detailed analysis needs to take polarization into account, as shall be discussed.)

We want to emphasize that the electrostatic potential is a real property of a system, a physical observable. It can be determined experimentally by diffraction methods [54–56], as well as computationally.

When V(**r**) is used for interpreting and predicting noncovalent interactions, it is typically computed on the molecular "surface" defined by the 0.001 au (electrons/bohr^3) contour of the molecule's electronic density ρ(**r**). This was suggested by Bader et al [57] as a reasonable representation of a molecular surface; it encompasses roughly 97 % of the electronic charge and reflects specific features such as lone pairs, π electrons and atomic anisotropies. V(**r**) on the 0.001 au surface is labeled V_S(**r**), and its locally most positive and most negative values (of which there may be several) are designated by $V_{S,max}$ and $V_{S,min}$, respectively.

10.3.2 The σ-Hole

In 1992, Brinck et al reported something quite unexpected [58]: covalently-bonded halogens with regions of positive electrostatic potential on their outer sides, on the extensions of the covalent bonds. The lateral sides of the halogens were negative, as anticipated. Additional studies in the next two years produced similar results [59, 60]. See, for instance, Fig. 10.1; the outer sides of the bromines, on the extensions of the C–Br bonds, have positive V_S(**r**) while the lateral sides of the bromines are negative (although less so after perfluorination).

Such positive outer regions appear to contradict the common view that univalent halogens are negative in character. However, as Brinck et al pointed out [58, 59], these findings do provide an immediate explanation of halogen bonding and of the patterns of crystallographic close contacts observed by Murray-Rust et al [32–34], i.e. structures **2** and **3**. Halogen bonding is the attractive interaction between the positive outer portion of the halogen and a negative site. Since the positive region on the halogen surface is on the extension of the covalent bond to the halogen, the same will be true of the interaction with the negative site; it will be near-linear, as shown in **3**. The halogen can also interact favorably with positive sites through its negative lateral sides, as in **2** (except for occasional instances in which the halogen's entire surface is positive). Some years later, Auffinger et al [45] and Awwadi et al [61] presented analogous intepretations of halogen bonding.

In 2007, the term "σ-hole" was introduced to denote the outer regions of positive potential on univalent halogens [62, 63]. The name reflects their being along the

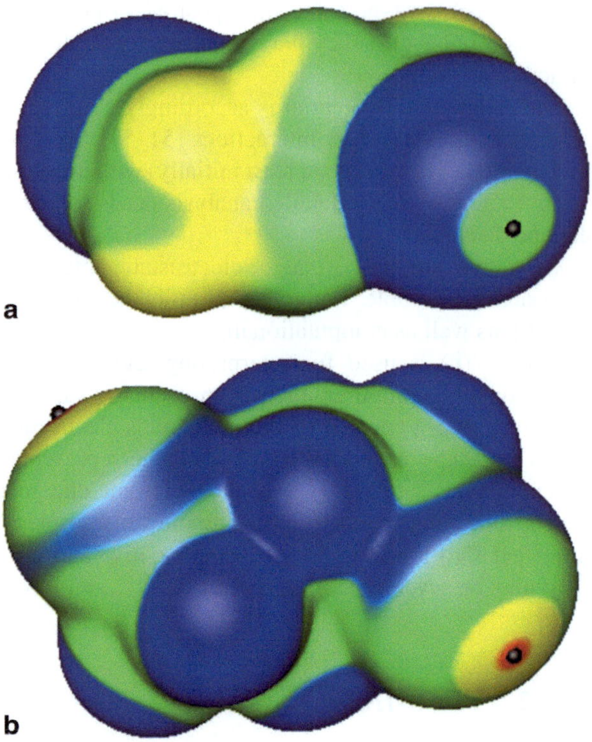

Fig. 10.1 Computed electrostatic potentials on 0.001 au molecular surfaces of **a** 1,4-dibromo-*n*-butane and **b** 1,4-dibromoperfluoro-*n*-butane. Bromines are at far *left* and far *right*. Black hemispheres indicate positive potential maxima on bromines, on extensions of C–Br bonds. Color ranges, in kcal/mol: *red*, greater than 25; *yellow*, between 25 and 15; *green*, between 15 and 0; *blue*, less than 0 (negative). Note that bromines become more positive after perfluorination. Computational level: M06-2X/6-311G(d, p)

extensions of σ bonds. As will be pointed out, however, σ-holes can sometimes also have negative potentials.

At about the same time that regions of positive potential were found on the extensions of covalent bonds to halogens [58–60], an analogous discovery was made for divalent sulfur. In a computational analysis of the heterocycle **7** [64], Burling and Goldstein showed that the sulfur has positive potentials on the extensions of the C–S bonds, with negative ones on its lateral sides. It was noted that this is consistent with the crystallographically-revealed patterns of close contacts in divalent sulfides and selenides [48–50], structures **4–6**; it also explains intramolecular S—O and Se—O close contacts that stabilize certain biologically-active thiazole and selenazole conformations [64]. (See also a later computational study involving other substituted sulfur—and selenium-containing heterocycles [65].) The basis for "like attracting like," **6**, now becomes apparent—for halides as well as for sulfides and selenides: The positive outer region on one of these atoms interacts favorably with a negative side of the same atom in another, identical molecule, as in **6**.

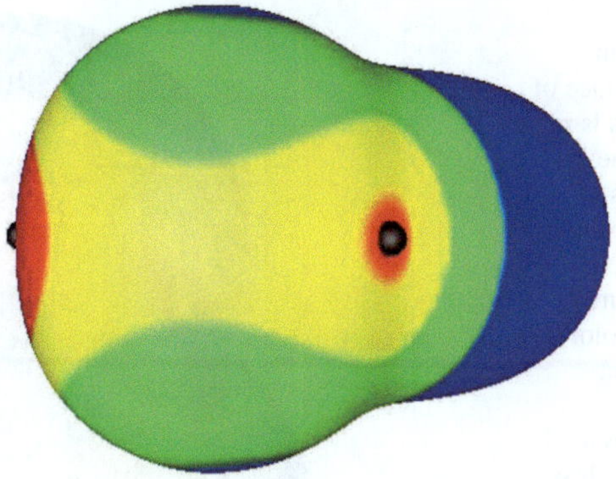

Fig. 10.2 Computed electrostatic potential on 0.001 au molecular surface of HSeCN. The selenium is in the foreground, cyano group at the *right*. *Black* hemispheres indicate selenium positive σ-holes on extensions of C–Se and H–Se bonds, the former being more positive. Color ranges, in kcal/mol: *red*, greater than 25; *yellow*, between 25 and 15; *green*, between 15 and 0; *blue*, less than 0 (negative). Computational level: M06-2X/6-311G(d, p)

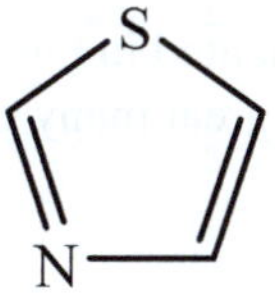

7

What Burling and Goldstein described are simply more examples of what are now called positive σ-holes: Regions of positive electrostatic potential on the outsides of singly-bonded atoms, along the extensions of the covalent bonds. Between 2007 and 2009, numerous positive σ-holes were found computationally on atoms of Groups IV—VI as well as Group VII (the halogens); this work in summarized in several reviews [11, 15, 16]. Through these regions, the atoms can interact favorably with negative sites (lone pairs of Lewis bases, π electrons, anions, etc.), forming noncovalent complexes. This is σ-hole bonding, of which halogen bonds are a subset. The interactions are highly directional, close to linear, along the extensions of the covalent single bonds to the atoms (see **3** and **5**).

Since Group IV—VI atoms can form four, three and two single bonds (or more if hypervalent [66, 67]), they can have the same numbers of σ-holes on the extensions of these bonds. This can be seen in Figs. 10.2, 10.3 and 10.4, which show that the σ-holes on a given atom can have different potentials, depending upon the partners in the bonds that gave rise to the σ-holes.

Some reports in the recent literature describe noncovalent interactions between Group IV—VI atoms and negative sites as new discoveries. This is incorrect (an error of which we have also been guilty [68]). Experimentalists have been familiar with

Fig. 10.3 Computed electrostatic potential on 0.001 au molecular surface of PF_2Br. The phosphorus is in the foreground, bromine at top. *Black* hemispheres indicate phosphorus positive σ-holes on extensions of Br–P and F–P bonds, the former being more positive. Color ranges, in kcal/mol: *red*, greater than 25; *yellow*, between 25 and 15; *green*, between 15 and 0; *blue*, less than 0 (negative). Computational level: M06-2X/6-311G(d, p)

them for decades, as is well documented in two reviews [16, 69]. What is new is the unifying σ-hole interpretation of a great many of these interactions [11,14–17].

10.4 Origins of σ-Holes

10.4.1 Anisotropies of Covalently-Bonded Atoms

The electronic density of a free neutral atom is, on the average, spherically-symmetrical [70]. The electrostatic potential V(**r**) due to its nucleus and electrons is positive for all $r < \infty$ [71], the effect of the nucleus dominating over that of the dispersed electrons. When two atoms begin to interact, at large separations, the electronic density of each of them becomes somewhat polarized toward the other [72], in response to the electric field of the latter. This results in the electronic density being less on the outer side of each atom (along the internuclear axis) than on its lateral sides. These diminished outer electron densities are incipient σ-holes!

As the atoms continue to approach and to interact in forming a covalent single bond, there is further redistribution of electronic densities, depending upon the atoms' relative polarizabilities and electronegativities. However the σ-holes continue to have lesser electronic densities than the surrounding portions of the atoms. It is indeed well established that covalently-bonded atoms have anisotropic charge distributions [14, 17, 49, 50, 61, 73–79]; their "radii" are less along the extensions of single bonds than perpendicular to them. For example, in the bromines in Fig. 10.1a, the distances to the 0.001 au surface are 2.06 Å along the extensions of the C–Br bonds and 2.28 Å in the perpendicular directions. In Fig. 10.1b, the same distances are 2.02 Å and

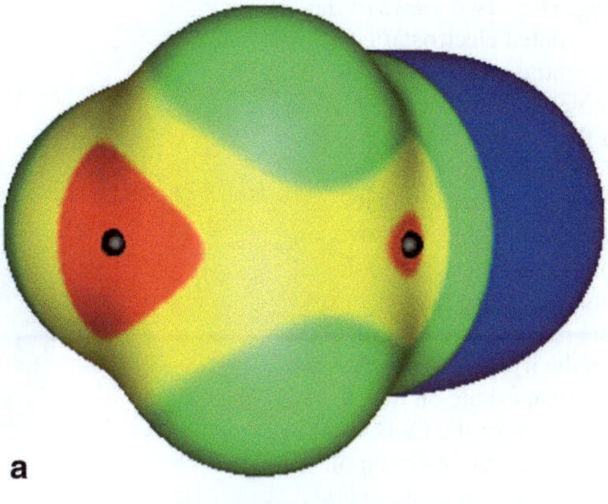

a

b

Fig. 10.4 Two views of the computed electrostatic potential on the 0.001 au molecular surface of SiH_3CN. In **a** the silicon is in the foreground, cyano group at *right*. In **b** a hydrogen is in the foreground, cyano group at *top*. *Black* hemispheres indicate silicon positive σ-holes on extensions of C–Si and H–Si bonds, the former being more positive. Color ranges, in kcal/mol: *red*, greater than 25; *yellow*, between 25 and 15; *green*, between 15 and 0; *blue*, less than 0 (negative). Computational level: M06-2X/6-311G(d, p)

2.24 Å. These values show the effect of the fluorines in attracting electronic density from the bromines.

The electronic density in a σ-hole is often sufficiently low that a positive electrostatic potential $V_S(\mathbf{r})$ results (Figs. 10.1, 10.2, 10.3 and 10.4). This is then a positive σ-hole, and favorable interactions with negative sites can be anticipated. In some instances, on the other hand, the electronic density in the σ-hole is high enough that its $V_S(\mathbf{r})$ is negative, i.e. a negative σ-hole, although less negative than its surroundings; see Fig. 10.5a. Examples of the $V_{S,max}$ of both positive and negative σ-holes are in Table 10.1 for the halogens [80, 81] and in Table 10.2 for Group IV—VI atoms [81]. The tables also give the most negative potentials ($V_{S,min}$) on these atoms.

It is very important to keep in mind, however, that the $V_S(\mathbf{r})$ computed for a free molecule in its unperturbed ground state, e.g. Figs. 10.1, 10.2, 10.3, 10.4 and 10.5, does not reflect the polarization induced by the electric field of another molecule

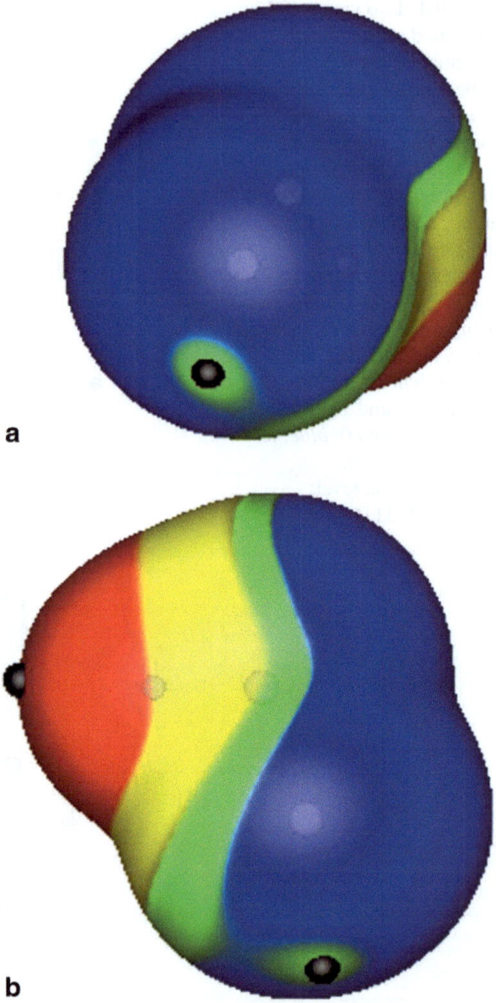

Fig. 10.5 Two views of the computed electrostatic potential on the 0.001 au molecular surface of FOH. In **a** the fluorine is in the foreground; the *black* hemisphere indicates its σ-hole, which is negative, on the extension of the O–F bond. In **b** the hydrogen is to the *left;* the *black* hemisphere on the hydrogen indicates the positive σ-hole on the extension of the O–H bond. Color ranges, in kcal/mol: *red*, greater than 30; *yellow*, between 30 and 0; *green*, between 0 and—8; *blue*, more negative than—8. Computational level: M06-2X/6-311G(d, p)

when they begin to interact. It has been demonstrated [82, 83] that polarization due to an external field can strengthen or weaken a σ-hole, and can sometimes convert it from negative (positive) to positive (negative). Thus one can occasionally find that a σ-hole complex forms even though the σ-hole may have been negative or near-neutral prior to the interaction; the negative site *induced* the positive σ-hole [84]. Some examples are H_3C-Cl—O = CH_2 [85] and H_3P—NSH [86], in which positive σ-holes were induced on the chlorine of H_3C–Cl and the phosphorus of H_3P.

Table 10.1 Most positive σ-hole potentials ($V_{S,max}$) and most negative potentials ($V_{S,min}$) on 0.001 au surfaces of the halogens shown in bold type, in kcal/mol. Computational method:M06–2X/6-311G(d, p)

Molecule	Bond producing σ-hole	$V_{S,max}$ (σ-hole)	$V_{S,min}$	Reference
HO–**F**	O–**F**	– 6.9	– 19.7	Present work
F–F	**F–F**	11.2	– 2.6	17
F–**Cl**	F–**Cl**	45.1	+0.3	80
F–**Br**	F–**Br**	53.2	– 0.1	80
Cl–Cl	**Cl–Cl**	25.4	– 2.7	80
Br–Br	**Br–Br**	29.0	– 4.1	80
H_3C–**F**	C–**F**	– 24.6	– 25.3	Present work
H_3C–**Cl**	C–**Cl**	– 0.9	– 15.6	Present work
H_3C–**Br**	C–**Br**	5.6	– 14.8	Present work
Br–$(CH_2)_4$–**Br**	C–**Br**	5.8	– 13.5	Present work
F_3C–**F**	C–**F**	– 1.3	– 2.8	81
F_3C–**Cl**	C–**Cl**	19.9	– 0.8	81
F_3C–**Br**	C–**Br**	25.3	– 2.0	81
Br–$(CF_2)_4$–**Br**	C–**Br**	26.3	– 1.7	Present work
F_3C–**I**	C–**I**	31.9	– 1.9	81
NC–**F**	C–**F**	12.8	10.9	17
NC–**Cl**	C–**Cl**	36.0	10.3	17
NC–**Br**	C–**Br**	42.7	8.6	17
NC–**I**	C–**I**	48.7	7.1	17
H_3Si–**Br**	Si–**Br**	0.2	– 11.9	Present work
H_3Ge–**Br**	Ge–**Br**	– 2.1	– 13.7	Present work
H_2P–**Br**	P–**Br**	3.5	– 14.3	Present work
F_2P–**Br**	P–**Br**	9.6	– 7.6	80
Cl–C≡C–**Cl**	C–**Cl**	23.7	– 0.6	Present work
Br–C≡C–**Br**	C–**Br**	30.1	– 2.1	17
C_6H_5–**Cl**	C–**Cl**	4.5	– 13.7	Present work
C_6H_5–**Br**	C–**Br**	10.2	– 13.7	Present work

10.4.2 Trends in σ-Hole Potentials

Since σ-holes are created by the shifting of electronic charge away from the outer portion of a singly-bonded atom, it is reasonable that the σ-hole electrostatic potential should become more positive as the atom is more polarizable and as it is less electronegative relative to the remainder of the molecule (particularly its bonding

Table 10.2 Most positive σ-hole potentials ($V_{S,max}$) and most negative potentials ($V_{S,min}$) on 0.001 au surfaces of the Group IV—VI atoms shown in bold type, in kcal/mol. Computational method: M06-2X/6-311G(d, p). The data are from the present work except as otherwise indicated

Molecule	Bond producing σ-hole	$V_{S,max}$ (σ-hole)	$V_{S,min}$
F**O**Br	F–**O**	10.5	− 11.7
	Br–**O**	−3.4	− 11.7
F**S**Br	F–**S**	35.9	− 4.0
	Br–**S**	27.2	− 4.0
F**Se**Br	F–**Se**	43.5	− 4.9
	Br–**Se**	35.2	− 4.9
H**Se**CN	H–**Se**	26.4	− 2.3
	C–**Se**	43.5	− 2.3
F_2**N**Br	F–**N**	13.2	− 9.4
	Br–**N**	11.8	− 9.4
H_2**P**Br	H–**P**	18.6	− 9.3
	Br–**P**	35.0	− 9.3
F_2**P**Br	F–**P**	32.3	11.4
	Br–**P**	35.1	11.4
F_2**As**Br	F–**As**	39.7	20.9
	Br–**As**	38.6	20.9
F_3**C**Br	F–**C**	16.0[a]	none[a]
	Br–**C**	21.6[a]	none[a]
F_3**Si**Br	F–**Si**	39.0[a]	none[a]
	Br–**Si**	45.0[a]	none[a]
F_3**Ge**Br	F–**Ge**	45.3[a]	none[a]
	Br–**Ge**	44.5[a]	none[a]
H_3**Si**CN	H–**Si**	26.8	none
	C–**Si**	37.7	none

[a]Reference 81

partner). Two common generalizations are [11,15–17]: (1) Within a given Group and for a particular molecular framework, the σ-hole potential ($V_{S,max}$) tends to become more positive in going from the lighter to the heavier (more polarizable and less electronegative) atoms, and (2) for a given atom, $V_{S,max}$ increases as the remainder of the molecule becomes more electron-attracting. These are useful empirical generalizations; supporting evidence can be seen in Tables 10.1 and 10.2. They explain, for example, why much of the early crystal engineering involving halogen bonding focused upon iodine, but in perfluorinated molecules [5, 8, 39]. Figure 10.1 and Table 10.1 show that perfluorination of the carbon chain of 1,4-dibromo-*n*-butane increases the bromine $V_{S,max}$ from 5.8 to 26.3 kcal/mol. The empirical generalizations also account for the fact that changing the bonding partner of bromine from the more electronegative carbon to silicon and germanium (in H_3C–Br, H_3Si–Br and H_3Ge–Br) causes the bromine σ-hole $V_{S,max}$ to become less positive and even negative (Table 10.1).

The first-row atoms N, O and F, being least polarizable and most electronegative, often have negative σ-holes (Fig. 10.5). It was indeed believed at one time that fluorine does not halogen bond. However it is well established since 2007 that fluorine *can* have a positive σ-hole and *can* participate in halogen bonding when it is in a sufficiently electron-attracting molecular environment [87–89]. In FCN, in fact, the portion of the molecular surface that is associated with the fluorine is entirely positive [87]; this is true as well of the other halocyanides (Table 10.1), demonstrating that positive σ-holes are not always surrounded by negative regions but sometimes simply by less positive ones. It has also been known since 2007–2009 that carbon [67], nitrogen [68] and oxygen [90] can also have positive σ-holes in appropriate molecular environments, and participate in σ-hole interactions. In the case of tetravalent Group IV atoms (which have no lone pairs), we have customarily found their entire exposed surfaces to be positive [67, 81].

It must be acknowledged that the generalizations concerning trends in σ-hole potentials, mentioned above, are somewhat oversimplified (although frequently valid). Consider, for instance, the molecule F_2PBr in Table 10.2 and Fig. 10.3. The phosphorus σ-hole produced by the bond to the bromine is more positive than that due to the bond to the fluorine, despite the fluorine's greater electronegativity. A detailed analysis [80] has shown that it is not only the polarizability and relative electronegativity of the σ-hole atom that are involved, but also the polarizability and electron-attracting power of the remainder of the molecule, plus factors such as overlapping with the electrostatic potentials of other parts of the molecule.

10.4.3 Covalently-Bonded Hydrogen

σ-Holes are not necessarily limited to singly-bonded atoms of Groups IV—VII. In particular, we must consider covalently-bonded hydrogen. Its electronic charge distribution, which involves just a single electron, is certainly anisotropic; it is centered in the internuclear region. This is why standard crystallographic techniques underestimate the lengths of covalent bonds to hydrogens [91, 92]. Accordingly, there is generally a positive σ-hole on the outer side of a hydrogen. This was in fact observed for the hydroxyl hydrogen in ethanol already in 1991 [93], although it was not labeled a σ-hole. Figure 10.5b clearly shows the hydrogen σ-hole in FOH. In some instances, as in Fig. 10.4b, a $V_{S,max}$ is not found on a hydrogen surface because the hydrogen's positive potential is overlapped by that of a larger neighboring atom. In Fig. 10.1a, the hydrogens do have $V_{S,max}$ but they are not shown explicitly. The interactions of such positive regions with negative sites readily explains hydrogen bonding as a σ-hole interaction [14, 15, 94, 95]; indeed, electrostatic interpretations of hydrogen bonding long preceded the discovery of σ-holes [55,93,96–100].

Since hydrogen has only the one valence electron, its lateral sides have relatively low electronic densities and the positive σ-hole usually covers a larger area than is typical of the Group V—VII atoms. Thus hydrogen σ-holes are less focused, as can be seen in Fig. 10.5b and in earlier reports [11, 15, 17, 94, 101]. It is accordingly not surprising that hydrogen bonds are overall less directional than are other σ-hole bonds [94, 102, 103].

10.5 σ-Hole Interactions

10.5.1 Interaction Energies

Attractive interactions between positive σ-holes and negative sites can give rise to numerous noncovalent complexes. Some examples are presented in Table 10.3 for the complexes R-X—B, R-Y—B and R-H—B, where X is a halogen, Y is a Group IV—VI atom and B represents the negative sites; in Table 10.3, these are the lone pairs of NH_3, HCN, $(H_3C)_2O$ and H_2S, the π electrons of C_6H_6, and the Br^- ion. The table includes the $V_{S,max}$ of the σ-holes, the $V_{S,min}$ of the negative sites, the interaction energies ΔE, the angles R-X—B, R-Y—B and R-H—B and the separations X—B, Y—B and H—B. The interaction energies were obtained with Eq. (10.3),

$$\Delta E = E(\text{complex}) - \sum_{i=1}^{2} E(\text{component } i) \qquad (10.3)$$

Table 10.3 confirms that the X—B, Y—B and H—B distances are less than or similar to the sums of the respective van der Waals radii [104, 105]. The interactions are approximately linear, i.e. along the extensions of the R–X, R–Y and R–H bonds. Deviations from linearity generally reflect secondary interactions involving neighboring atoms; this is illustrated in Figs. 10.6 and 10.7.

In view of the structural similarity between halogen and hydrogen bonds (both are due to positive σ-holes on univalent atoms), it is natural to compare them. For a given R and B, the R-H—B interaction is generally stronger than the R-X—B (ΔE more negative) when X = F or Cl, but they become competitive when X = Br or I, with halogen bonding sometimes being stronger [106–108]. Compare, for instance, the ΔE values for F_3C-H—NH_3 (−4.2 kcal/mol), F_3C-Cl—NH_3 (−2.5 kcal/mol) and F_3C-Br—NH_3 (−3.7 kcal/mol) in Table 10.3. Instances of halogen bonding dominating over hydrogen bonding have long been observed experimentally [109, 110].

Table 10.3 Computed data for various types of σ-hole complexes. Interaction energies ΔE and geometries were obtained by optimizations with the M06-2X/aug-cc-pVTZ procedure. σ-Hole $V_{S,max}$ and negative site $V_{S,min}$, *prior* to interaction, were calculated at the M06-2X/6-311G(d) level

Complex	ΔE	σ-Hole $V_{S,max}$	Negative site $V_{S,min}$	Separation[a] (Å)	Angle (degrees)
	(kcal/mol)				
Halogen-bonded:					
F-**F**—NH_3	− 1.5	11.2	− 46.7	**F**—N: 2.60 (3.05)	F-**F**—N: 177
F_3C-**Cl**—NCH	− 1.6	19.9	− 33.5	**Cl**—N: 3.13 (3.35)	C-**Cl**—N: 179
F-**F**—$O(CH_3)_2$	− 1.8	11.2	− 34.8	**F**—O: 2.48 (3.00)	F-**F**—O: 169
Br_2C= CBr(**Br**)—NCH	− 2.1	24.1	− 33.5	**Br**—N: 3.09 (3.46)	C-**Br**—N: 178
F_3C-**Cl**—NH_3	− 2.5	19.9	− 46.7	**Cl**—N: 3.03 (3.35)	C-**Cl**—N: 180
Br-C≡C-**Br**—NCH	− 2.7	30.1	− 33.5	**Br**—N: 3.05 (3.46)	C-**Br**—N: 180
Cl-**Cl**—SH_2	− 2.9	25.4	− 20.2	**Cl**—S: 3.13 (3.53)	Cl-**Cl**—S: 178
F_3C-**Br**—NH_3	− 3.7	25.3	− 46.7	**Br**—N: 3.07 (3.46)	C-**Br**—N: 180
N≡C-**Cl**—$O(CH_3)_2$	− 4.0	36.0	− 34.8	**Cl**—O: 2.77 (3.30)	C-**Cl**—O: 169
Br-C≡C-**Br**—$O(CH_3)_2$	− 4.1	30.1	− 34.8	**Br**—O: 2.86 (3.41)	C-**Br**—O: 172
Br-C≡C-**Br**—NH_3	− 4.2	30.1	− 46.7	**Br**—N: 2.99 (3.46)	C-**Br**—O: 180
F-**Br**—NCH	− 7.1	53.2	− 33.5	**Br**—N: 2.60 (3.46)	F-**Br**—N: 180
F_3C-**Cl**—Br^-	− 8.8	19.9	− 136.4	**Cl**—Br: 3.19 (3.59)	C-**Cl**—Br: 179
F_3C-**Br**—Br^-	− 12.9	25.3	− 136.4	**Br**—Br: 3.09 (3.70)	C-**Br**—Br: 180
Group IV-VI σ-hole-bonded:					
Cl_2**Se**—NCH	− 4.0	36.1	− 33.5	**Se**—N: 2.92 (3.51)	Cl-**Se**—N: 179
H_2F**P**—SH_2	− 4.1	38.8	− 20.2	**P**—S: 3.25 (3.59)	F-**P**—N: 166
H_3F**Si**—NCH	− 4.2	34.7	− 33.5	**F**—Si: 2.89 (3.54)	F-**Si**—N: 180
F_2**S**—NCH	− 4.5	40.2	− 33.5	**S**—N: 2.79 (3.40)	F-**S**—N: 174
H_2F**P**—NCH	− 4.7	38.8	− 33.5	**P**—N: 2.81 (3.41)	F-**P**—N: 164
H_3F**Ge**—NCH	− 4.9	43.0	− 33.5	**Ge**—N: 2.89 (–)	F-**Ge**—N: 180
HF**S**—SH_2	− 5.0	45.9	− 20.2	**S**—S: 3.08 (3.58)	F-**S**—S: 170
H_2F**As**—NCH	− 5.7	44.5	− 33.5	**As**—N: 2.82 (3.46)	F-**As**—N: 163
H_2F**P**—NH_3	− 7.2	38.8	− 46.7	**P**—N: 2.71 (3.41)	F-**P**—N: 167
HF**S**—NH_3	− 8.4	45.9	− 46.7	**S**—N: 2.48 (3.40)	F-**S**—N: 170
H_2F**As**—NH_3	− 8.7	44.5	− 46.7	**As**—N: 2.73 (3.46)	F-**As**—N: 165
ClF_3**Si**—NH_3	− 10.5	47.6	− 46.7	**Si**—N: 2.07 (3.71)	Cl-**Si**—N: 180
HF**Se**—NH_3	− 11.3	50.9	− 46.7	**Se**—N: 2.51 (3.51)	F-**Se**—N: 170

Table 10.3 Continued

Complex	ΔE (kcal/mol)	σ-Hole $V_{S,max}$	Negative site $V_{S,min}$	Separation[a] (Å)	Angle (degrees)
Hydrogen-bonded:					
NC-**H**—SH_2	– 2.6	51.3	– 20.2	**H**—S: 2.71 (2.88)	C-**H**—S: 178
F_3C-**H**—NCH	– 3.0	34.8	– 33.5	**H**—N: 2.39 (2.70)	C-**H**—N: 178
H_3CO-**H**—SH_2	– 3.1	44.4	– 20.2	**H**—S: 2.53 (2.88)	O-**H**—S: 168
H_3CO-**H**—NCH	– 4.1	44.4	– 33.5	**H**—N: 2.13 (2.70)	O-**H**—N: 160
F_3C-**H**—NH_3	– 4.2	34.8	– 46.7	**H**—N: 2.31 (2.70)	C-**H**—N: 178
NC-**H**—NCH	– 4.7	51.3	– 33.5	**H**—N: 2.21 (2.70)	C-**H**—N: 180
H_3CO-**H**—C_6H_6	– 4.8	44.4	– 19.4	b	b
NC-**H**—NH_3	– 6.6	51.3	– 46.7	**H**—N: 2.11 (2.70)	C-**H**—N: 180
H_3CO-**H**—NH_3	– 6.7	44.4	– 46.7	**H**—N: 1.96 (2.70)	O-**H**—N: 176
F_3C-**H**—Br^-	– 14.0	34.8	– 136.4	**H**—Br: 2.45 (2.94)	C-**H**—Br: 180
H_3CO-**H**—Br^-	– 14.5	44.4	– 136.4	**H**—Br: 2.34 (2.94)	O-**H**—Br: 160
NC-**H**—Br^-	– 19.5	51.3	– 136.4	**H**—Br: 2.29 (2.94)	C-**H**—Br: 180

[a]Values in parentheses are the sums of the van der Waals radii of the interacting atoms, references 104 and 105. No van der Waals radius was found for germanium
[b]The hydroxyl hydrogen is above the center of the benzene ring, and thus is not directly in line with any particular atom

10.5.2 Interaction Energy Relationships

Since we consider σ-hole interactions to be electrostatic, does it follow that the ΔE values should correlate directly with the magnitudes of the $V_{S,max}$ and $V_{S,min}$? Not necessarily, for several reasons. First, the attractive interactions are not simply between a single point on the σ-hole (the $V_{S,max}$) and a single point on the negative site (the $V_{S,min}$); the whole σ-hole and negative regions can participate, as was shown by Shields et al [94]. Furthermore, interactions involving the remaining portions of the molecules may be significant (Figs. 10.6 and 10.7) [65, 81, 111]. Second, as pointed out in Sect. 4.1, the $V_{S,max}$ and $V_{S,min}$ are for the unperturbed molecules, prior to interaction, and do not reflect their mutual polarizing effects; these can be quite important, as shall be seen.

Despite these potential complications, which can be real, ΔE has been related—perhaps surprisingly well—to $V_{S,max}$ and to combinations of $V_{S,max}$ and $V_{S,min}$. When the negative site is kept the same, good correlations have been obtained between ΔE and the $V_{S,max}$ of the σ-holes [15, 16, 83, 111, 112]; as $V_{S,max}$ is more positive, ΔE tends to become more negative (stronger interaction). For instance, plots of ΔE *vs.* $V_{S,max}$ for two series of Group IV—VII complexes with NH_3 and with HCN had R^2 of

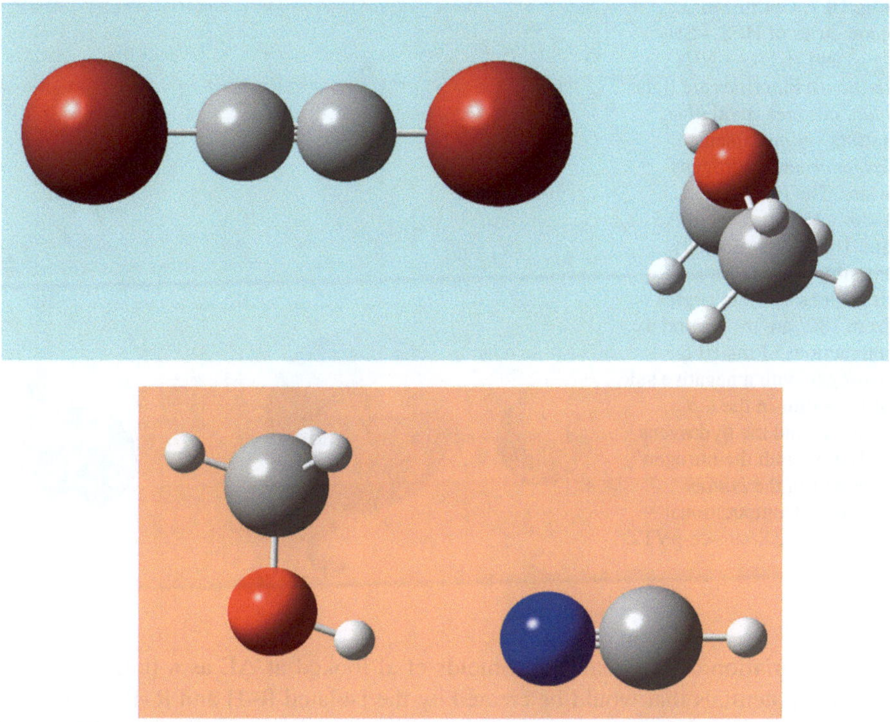

Fig. 10.6 Optimized geometries of Br-C≡C-Br—$O(CH_3)_2$ (*top*) and H_3CO-H—NCH (*bottom*). Carbons are *gray*, nitrogen *blue*, oxygens *red*, bromines *burgundy* and hydrogens *white*. The C-Br—O and O-H—N angles are 172.4 and 160.4°, respectively. Possible reasons for the deviations of these angles from 180° are the secondary interactions of methyl hydrogens with the negative sides of the bromine in the *top* structure and with the nitrogen lone pair in the *bottom* structure. Computational level: M06-2X/aug-cc-pVTZ

0.95 and 0.98, respectively [16]. When different negative sites are involved, then both the $V_{S,max}$ and the $V_{S,min}$ must be taken into account. Thus when the aforementioned complexes with NH_3 and HCN were taken together and ΔE was plotted against the product of the σ-hole $V_{S,max}$ and the negative site $V_{S,min}$, an R^2 of 0.96 was obtained.

We have now carried out a double regression analysis encompassing all 39 complexes in Table 10.3, thereby including σ-hole interactions of hydrogens, halogens and Groups IV—VI with six different negative sites. ΔE was expressed as,

$$\Delta E = c_1 V_{S,max} + c_2 V_{S,min} + c_3 \tag{10.4}$$

The relationship between the predicted and the computed ΔE is presented in Fig. 10.8; the R^2 is 0.91, which is noteworthy considering the varied nature of the database. (When the ΔE in Table 10.3 are correlated with the product of $V_{S,max}$ and $V_{S,min}$, R^2 is 0.88. While using the product may appear to be more consistent with Coulomb's Law, this is actually not the case, since $V_{S,max}$ and $V_{S,min}$ are potentials, not charges.)

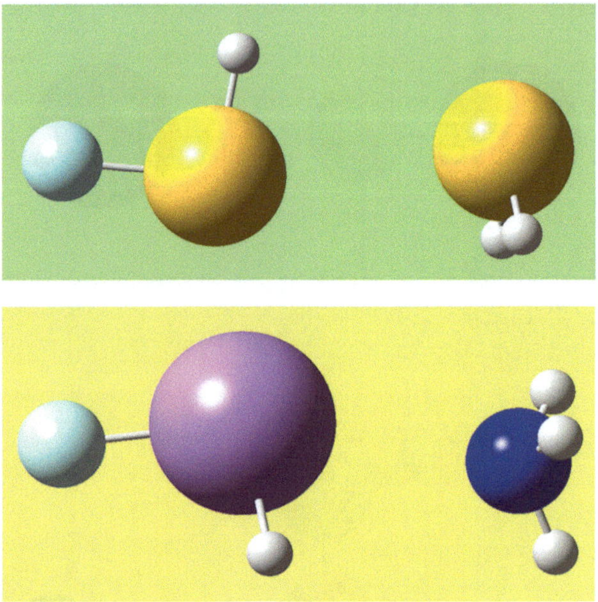

Fig. 10.7 Optimized geometries of HFS—SH_2 (*top*) and H_2FAs—NH_3 (*bottom*). Fluorines are light *blue*, nitrogen royal *blue*, sulfurs *yellow*, arsenic *lavender* and hydrogens *white*. The F-S—S and F-As—N angles are 169.5 and 165.0°, respectively. Possible reasons for the deviations of these angles from 180° are the secondary interactions of the HFS hydrogen with a negative side of the sulfur in the *top* structure and the hydrogens of H_2FAs with the nitrogen lone pair in the *bottom* structure. Computational level: M06-2X/aug-cc-pVTZ

In a variation of these studies, Shields et al looked at ΔE as a function of the positive potentials that would be created by the isolated R–H and R–Br molecules at the positions of the nitrogens in two series of complexes: (a) R-H—NCH and R-Br—NCH, and (b) R-H—NH_3 and R-Br—NH_3 [94]. This was done for R-H—N and R-Br—N angles ranging from 100 to 180°. For each series, ΔE correlated linearly and extremely well with the positive potentials at the nitrogen positions at the different angles; R^2 was 0.986 for the NH_3 complexes and 0.990 for the HCN. All of the interactions had negative ΔE, confirming that the entire σ-hole region can participate.

10.5.3 Thermodynamic Stability

We have discussed the energetics of σ-hole complex formation in terms of the interaction energies ΔE, as is customary. From a thermodynamic standpoint, however, stability is governed by the free energy change ΔG; at a given absolute temperature T,

$$\Delta G = \Delta H - T\Delta S \qquad (10.5)$$

where ΔH and ΔS are the changes in enthalpy and entropy that accompany the formation of the complex. For thermodynamic stability, ΔG must be negative.

For a noncovalent interaction $A + B \rightarrow A—B$ in the gas phase at $T = 298$ K, ΔH is normally within 1 kcal/mol of ΔE [16, 113], and if the interaction is attractive, as in Table 10.3, then $\Delta E < 0$ and $\Delta H < 0$. ΔS is also typically negative, because

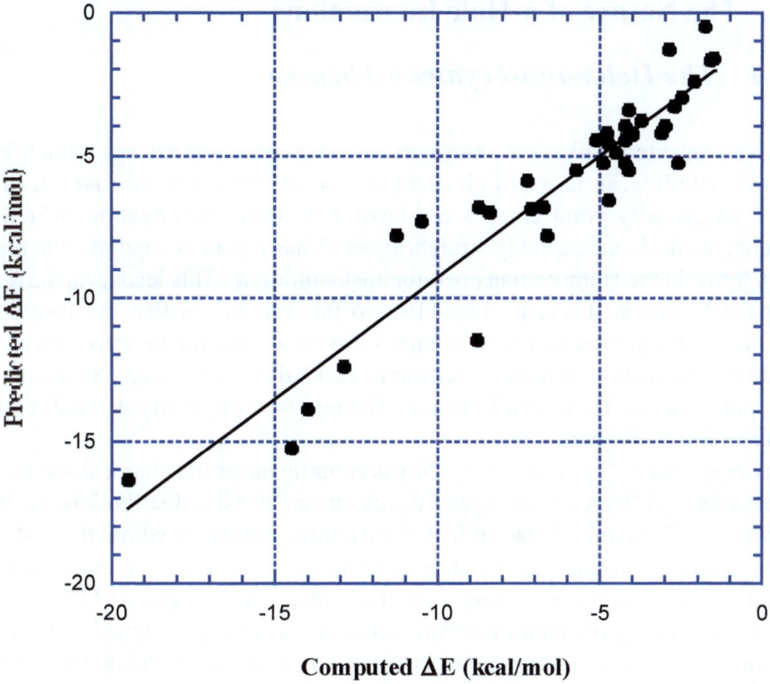

Fig. 10.8 Plot of ΔE predicted by double regression analysis, Eq. (10.4), *vs.* ΔE computed with Eq. (10.3) for the σ-hole complexes in Table 10.3. $R^2 = 0.91$

forming the complex diminishes the degrees of freedom of A and B. This can result in $|T\Delta S| > |\Delta H|$, so that $\Delta G > 0$ and the complex is thermodynamically unstable despite the interaction being attractive and having $\Delta E < 0$ and $\Delta H < 0$.

It has been found that $\Delta G > 0$ for many σ-hole complexes in the gas phase at 298 K [16, 113, 114]; it is only the relatively strong interactions that have $|\Delta H| > |T\Delta S|$ and therefore $\Delta G < 0$. It should be kept in mind, however, that $\Delta G > 0$ does not completely preclude an interaction; it simply means that the equilibrium constant is less than one.

When the interaction takes place in solution or in the solid state, additional factors are involved that may lead to $\Delta G < 0$ even when $\Delta G > 0$ for the gas phase complex. For example, the halogen-bonded solid —Cl-C(O)-C(O)-Cl—1,4-dioxane— was characterized crystallographically already in 1965 [115], although its gas phase ΔG(298 K) has been computed to be + 6.0 kcal/mol [16]. For further discussion of thermodynamic stability in relation to σ-hole interactions, see Politzer et al [16, 113].

10.6 The Nature of σ-Hole Interactions

10.6.1 The Hellmann-Feynman Theorem

We have presented a physical interpretation of σ-hole interactions, which include hydrogen bonding, in terms of electrostatics/polarization (the two are inextricably linked, unless only point charges are involved). This interpretation, which many theoreticians find unacceptably straightforward, has rigorous support: The potential energy terms in the Hamiltonian operator are Coulombic. This leads, via the rigorous Hellmann-Feynman theorem [116–119], to the conclusion that the forces acting upon the nuclei in a molecule or complex can be determined purely classically as Coulombic interactions with the electrons and with the other nuclei. (As an interesting aside, note that the Hellmann-Feynman theorem was originally derived neither by Hellmann nor by Feynman.)

As Levine put it, "... there are no 'mysterious quantum-mechanical forces' acting in molecules." [120] A knowledge of the electronic density distribution, the nuclear positions and Coulomb's Law suffices to determine the forces within the system—in fact the electronic density also locates the positions of the nuclei, and indeed all of the system's properties, according to the Hohenberg-Kohn theorem [121].

Notwithstanding all of the above, theoreticians have happily argued for years about the relative roles of covalency and electrostatics in hydrogen bonding [96, 122, 123]. The argument is unimpeded by the fact that covalent bonds are themselves Coulombic (Hellmann-Feynman theorem) nor by the absence of any rigorous definition of covalency. The latter point is indeed very advantageous: since covalency cannot be measured, no one can be proved wrong and the argument can (and doubtless will) continue indefinitely. To make matters even better, new victims have appeared—halogen bonding and other σ-hole interactions—and can be subjected to the same intense scrutiny!

How can the distressingly simple Coulombic explanation of noncovalent bonding (i.e. electrostatics/polarization) be reconciled with the formidable array of complexities that are typically invoked? These generally consist of some subset of electrostatics, exchange, Pauli repulsion, polarization (or induction), charge transfer (or donor-acceptor interaction), dispersion, orbital overlap, etc. (Different researchers emphasize different subsets.)

We will begin by separating exchange and Pauli repulsion from the others. They refer to mathematical requirements that must be satisfied by the wave function. Electrons are indistinguishable, which is handled by introducing exchange, and the wave function must be antisymmetric, which results in the Pauli exclusion principle (Pauli "repulsion"). These are mathematical effects [120, 124, 125], not physical forces. In the same context, orbitals are a valuable means of expressing wave functions but they have no physical existence [126]—nor does, therefore, orbital overlap.

Next, it should be recognized that analyses of noncovalent bonding commonly use the term "electrostatics" in a restricted and misleading manner, as referring to the Coulombic interaction between the *unperturbed* molecules *prior* to forming the complex. This is physically unrealistic [84]; it ignores the polarization of each molecule's electronic density distribution by the electric field of the other. This is shown schematically for the σ-hole complexes R-X—B, R-Y—B and R-H—B in **8–10** (X = halogen, Y = Group IV—VI atom). Such polarization is an intrinsic part of the Coulombic interaction [14–16, 85, 86, 95].

R-X---B	R-Y---B	R-H---B
← ←	← ←	← ←
8	**9**	**10**

The error incurred in treating the electrostatic interaction as involving only the unperturbed isolated molecules and ignoring their mutual polarization is graphically illustrated by computing the difference between the electron density of the complex and the sum of the electronic densities of the isolated molecules placed at the same separation as in the complex. Such density difference plots for σ-hole complexes show exactly the features depicted in **8–10** [14, 85, 127, 128]: The electric field of the σ-hole polarizes the electronic charge of B toward the σ-hole, while the electric field of the negative site on B polarizes the electronic charge near the σ-hole away from B.

Another demonstration of the importance of polarization was provided by Hennemann et al [82]. They showed computationally that the electric field of a point charge Q placed at a distance of 1.90 Å from one of the hydrogens of a water molecule, Q—H-OH, caused the σ-hole potential of that hydrogen to vary linearly as a function of the charge on Q. In the absence of the point charge, the hydrogen's $V_{S,max}$ was 57 kcal/mol. As Q was made increasingly negative, it repelled electronic charge from the hydrogen σ-hole and made it more positive; thus for $Q = -0.4$ au, the σ-hole $V_{S,max}$ was 75 kcal/mol. As Q was made increasingly positive, on the other hand, it attracted electronic charge to the hydrogen σ-hole and made it less positive; for $Q = 0.4$ au, $V_{S,max} = 38$ kcal/mol. This shows how polarization can affect the electronic density distribution of a molecule, and in particular, that it can strengthen, weaken or even induce a positive σ-hole. It explains why the complexes H_3C-Cl—O = CH_2 [85] and H_3P—NSH [86] were found to form despite the chlorine and the phosphorus σ-holes in H_3C–Cl and H_3P being near-neutral or negative; positive σ-holes were induced by the electric fields of O = CH_2 and NSH (Sect. 4.1).

The polarization depicted in **8–10** and clearly evident in density difference plots [14, 83, 127, 128] readily explains why the formation of an initial σ-hole bond may promote a second one ("cooperativity") or might inhibit it [11, 15, 79, 89, 129]. Thus, it can be anticipated that **11** will have an enhanced likelihood for further interactions through both the terminal nitrogen and the terminal hydrogen. In **12**, on the other hand, the terminal atoms should be less prone to forming additional σ-hole bonds.

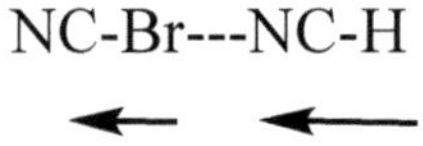

11

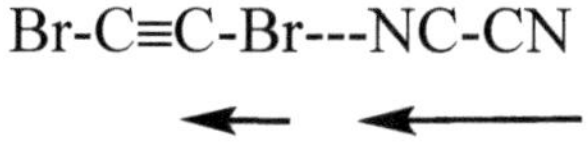

12

It was shown in Sect. 5.2 that the interaction energies of σ-hole complexes correlate quite well with the σ-hole $V_{S,max}$ and the negative site $V_{S,min}$, even though these are computed for the isolated molecules prior to interaction and therefore do not reflect their mutual polarization. An example of such a correlation is Fig. 10.8, which is based upon Table 10.3. Some σ-hole complexes, however, have been found to have unusually negative ΔE and short separations [130, 131], indicating atypically strong interactions. This simply means that polarization, particularly of the negative site, is very significant, and the polarizabilities of the σ-hole and especially of the negative site must be taken explicitly into account along with their $V_{S,max}$ and $V_{S,min}$. When this is done, via regression analyses, good correlations between predicted and computed ΔE are again obtained. For more extensive discussions of this, see Politzer et al [16, 17, 131].

Proceeding to dispersion, this is a very useful concept in interpreting noncovalent interactions since it can always be invoked when all else fails. The relative roles of dispersion and electrostatics are a favorite subject for debate, even though the Hellmann-Feynman theorem tells us that dispersion is part of the Coulombic interaction.

Dispersion is commonly linked to electronic correlation [101, 132, 133]. This refers to the instantaneous correlated movements of electrons as they respond to their mutual electrostatic repulsions. The usual view is that these movements create temporary dipoles, and it is the attraction between these dipoles that accounts for the stabilizing effect of dispersion.

Another view, widely overlooked, was proposed by Feynman [72]. He argued that the attraction is between the nuclei of each molecule and its own electronic charge, which has shifted to some extent into the intermolecular region. A study by Hirschfelder and Eliason [134] and a proof by Hunt [135] supported Feynman.

Whichever explanation one prefers, dipole-dipole or nuclear-electronic attractions, the point is that both are Coulombic. Thus dispersion is fully encompassed by the Hellmann-Feynman theorem.

10.6.2 Charge Transfer or Polarization?

Charge transfer from an "electron donor" to an "electron acceptor" has been widely invoked as an important factor in hydrogen and halogen bonding, and in σ-hole interactions in general. Some small fraction of an electron is supposedly transferred from the negative site B (the donor) into an antibonding orbital of the molecule R–X, R–Y or R–H that has the σ-hole (the acceptor). This weakens the R–X, R–Y or R–H bond and accounts for its stretching frequency often (but not always) being lower in the complex than in the isolated molecule (a red shift).

While this scenario can sometimes be effective as a purely mathematical model, it should not be confused with physical reality. Electrons are indivisible; a small portion of one cannot be plucked away. Orbitals—bonding, antibonding or otherwise—do not really exist [126], however useful and convenient the concept may be. It is increasingly being recognized [136–140] that charge transfer theory is a mathematical attempt to represent a physical process, which is the mutual polarization of the interacting molecules, the so-called "donor" and "acceptor." Thus it is redundant to cite both charge transfer and polarization as separate factors in noncovalent interactions [83, 84].

Hermansson [141] and Qian and Krimm [142] have derived formalisms that explain and predict both the red and the blue shifts in R–X, R–Y and R–H stretching frequencies in terms of just the electric field of B and the permanent and induced dipole moments of the R–X, R–Y and R–H molecules. These procedures have been extensively applied to σ-hole interactions [143, 144].

A simple demonstration of how polarization can produce either a red or a blue shift was provided by Hennemann et al [84]. They put a point charge Q near the hydroxyl hydrogen of methanol, Q—H-O-CH_3, and computed the O–H stretching frequency as a function of the charge on Q. Starting with $Q = 0$, as Q was made increasingly negative, the O–H frequency steadily decreased; when Q was made increasingly positive, the frequency increased until $Q = 0.3$ and then began to decrease. Thus both red and blue shifts could be produced by varying the charge and hence the electric field of Q, thereby polarizing the methanol molecule. The lower O–H frequencies cannot be due to the transfer of electronic charge into an antibonding orbital of the methanol because Q has no electronic charge to transfer. Both the red shifts and the blue shifts are purely polarization effects.

The mathematical rather than physical nature of the charge transfer concept is illustrated by an example due to Stone and Misquitta [145]. The usual quantum chemical description of a noncovalent complex, for example R-Y—B, is in terms of basis orbitals on both R–Y and B. However it could also be done quite satisfactorily using orbitals on only R–Y or only B, if enough of them were used. The polarization shown in **9** would be adequately described, by the computed electronic density distribution of the complex. However the charge transfer from an orbital of B into an orbital of R–Y would necessarily be zero, since either B or R–Y has no orbitals. The physical reality is maintained (by the electronic density distribution, which is an observable), but the mathematical model (charge transfer) fails.

To summarize, "charge transfer" in noncovalent interactions is, physically, polarization and polarization is Coulombic. The Hellmann-Feynman theorem lives on!

10.7 The Fallacy and Inadequacy of Atomic Charges

The preceding discussion is clearly linked to the question: Is it meaningful to assign quantitatively a charge to an atom in a molecule? In the laboratory, experience indicates the usefulness of viewing some atoms as having positive or negative characters relative to others. It is tempting to try to quantify this (a temptation to which one of the current authors yielded in his misspent youth [146]). However the concept of an atom in a molecule, while very useful in practice, does not have a rigorous basis—and therefore neither does giving it a numerical charge [53]. Since there is no correct way to do this, everyone is free to invent his/her own scheme, and many have done so. By 1994, more than 30 different procedures for assigning atomic charges had been proposed [147]. They sometimes produce remarkably varied results; for instance, Wiberg and Rablen cited calculated charges for the carbon in $H_3C–NO_2$ that ranged from—0.478 to + 0.564 [148].

Furthermore, attributing a single positive or negative charge to an atom in a molecule ignores the well-known anisotropies of its charge distribution and its electrostatic potential (Sect. 4.1). Price has referred to this rather bluntly as "a travesty of bonding theory" [149]. As was discussed in earlier sections, it is these anisotropies that account for the observed abilities of many covalently-bonded atoms to interact favorably with both positive and negative sites and for the phenomenon of "like attracting like." Calculated atomic charges cannot explain such behavior [45, 84, 150].

Auffinger et al [45] pointed out that since many force fields used in molecular mechanics and molecular dynamics do use atomic charges, they may miss some noncovalent interactions; an example was presented by Dobeš et al [151]. Accordingly several research groups have sought to develop more realistic force fields [152–155].

10.8 Concluding Comments

The avalanche of studies that has descended upon the area of noncovalent interactions in recent years has subjected them to minute dissection and extensive compartmentalization. For instance, the literature now mentions at least a dozen types of hydrogen bonding!

Our emphasis, however, has been upon the unifying principle that a great many noncovalent interactions—involving atoms of Groups IV—VII as well as hydrogen—fit under the umbrella of σ-hole bonding (illustrated on the front cover of Physical Chemistry Chemical Physics, volume 15, issue 27, 2013): A region of positive electrostatic potential, on the extension of a single covalent bond to the atom, interacts attractively with a negative site. Closely related to this is so-called "π-hole" bonding, in which the positive potential is perpendicular to an atom or region in a planar portion of a molecular framework. This is examined in detail elsewhere [16, 156, 157].

We have also invoked the rigorous Hellmann-Feynman theorem (which seems to often be overlooked) to point out that these noncovalent interactions can be fully understood and described as being Coulombic, which includes polarization and dispersion. If this seems simplistic to some, we argue that it is because of a tendency to confuse mathematical modeling with physical reality. The latter can be annoyingly straightforward. Newton said that, "Nature is pleased with simplicity" [158]. Einstein seemed to agree: "Nature is the realization of the simplest conceivable mathematical ideas" [158].

References

1. Müller-Dethlefs K, Hobza P (2000) Noncovalent interactions: a challenge for experiment and theory. Chem Rev 100:143–167
2. Engkvist O, Åstrand P-O, Karlström G (2000) Accurate intermolecular potentials obtained from molecular wave functions: bridging the gap between quantum chemistry and molecular simulations. Chem Rev 100:4087–4108
3. Chałasiński G, Szczêśniak (2000) State of the art and challenges of the *ab initio* theory of intermolecular interactions. Chem Rev 100:4227–4252
4. Braga D, Grepioni F (2000) Intermolecular interactions in nonorganic crystal engineering. Acc Chem Res 33:601–608
5. Metrangolo P, Neukirch H, Pilati T, Resnati G (2005) Halogen bonding based recognition processes: a world parallel to hydrogen bonding. Acc Chem Res 38:386–395
6. Belkova NV, Shubina ES, Epstein LM (2005) Diverse world of unconventional hydrogen bonds. Acc Chem Res 38:624–631
7. Hobza P, Zahradník R, Müller-Dethlefs K (2006) The world of non-covalent interactions: 2006. Collect Czech Chem Commun 71:443–531
8. Metrangolo P, Resnati G, eds (2008) Halogen bonding: fundamentals and applications. Springer, Berlin
9. Stone AJ (2008) Intermolecular potentials. Science 321:787–789
10. Legon AC (2010) The halogen bond: an interim perspective. Phys Chem Chem Phys 12:7736–7747
11. Politzer P, Murray JS, Clark T (2010) Halogen bonding: an electrostatically-driven highly directional noncovalent interaction. Phys Chem Chem Phys 12:7748–7757
12. Hobza P, Müller-Dethlefs K (2010) Non-Covalent Interactions: Theory and Experiment. Royal Society of Chemistry, Cambridge
13. Scheiner S (2011) Weak H-Bonds. comparisons of CH–O to NH–O in proteins and PH–N to direct P–N interactions. Phys Chem Chem Phys 13:13860–13872
14. Politzer P, Riley KE, Bulat FA, Murray JS (2012) Perspectives on halogen bonding and other σ-hole interactions: lex parsimoniae (Occam's Razor). Comput Theoret Chem 998:2–8
15. Politzer P, Murray JS (2013) Halogen bonding: an interim discussion. Chem Phys Chem 14:278–294
16. Politzer P, Murray JS, Clark T (2013) Halogen bonding and other σ-hole interactions: a perspective. Phys Chem Chem Phys 15:11178–11189
17. Politzer P, Murray JS, Clark T (2014) σ-Hole Bonding: A Physical Interpretation. Topics Curr Chem (in press).
18. Schneider H-J (2009) Binding mechanisms in supramolecular complexes. Angew Chem Int Ed 48:3924–3977
19. Colin M (1814) Ann Chim 91:252
20. Guthrie F (1863) XXVIII. On the Iodide of Iodammonium. J Chem Soc 16:239–244
21. Mallet JW (1881) Chem News 44:188
22. Remsen I, Norris JF (1896) The action of the halogens on the methylamines. Am Chem J 18:90–95

23. Rhoussopoulos O (1883) Einwirkung von Chinolin auf Chloroform and Jodoform. Berichte 16:202–203
24. Blackstock SC, Lorand JP, Kochi JK (1987) Charge-transfer interactions of amines with tetrahalomethanes. X-ray crystal structures of the donor-acceptor complexes of quinuclidine and diazabicyclo[2.2.2]octane with carbon tetrabromide. J Org Chem 52:1451–1460
25. Benesi HA, Hildebrand JH (1948) Ultraviolet absorption bands of iodine in aromatic hydrocarbons. J Am Chem Soc 70:2832–2833
26. Benesi HA, Hildebrand JH (1949) A Spectrophotometric investigation of the interaction of iodine with Aromatic Hydrocarbons. J Am Chem Soc 71:2703–2707
27. Mulliken RS (1952) Molecular compounds and their spectra. II. J Am Chem Soc 74:811–824
28. Dahl T, Hassel O (1966) A close relationship between the crystal structure of an acceptor and that of an addition compound. Acta Chem Scand 20:2009
29. Bent HA (1968) A structural chemistry of donor-acceptor interactions. Chem Rev 68:587–648
30. Olie K, Mijlhoff FC (1969) The crystal structure of $POBr_3$ and intermolecular bonding. Acta Cryst B 25:974–977
31. Olie K (1971) The crystal structure of $POCl_3$. Acta Cryst B 27:1459–1460
32. Murray-Rust P, Motherwell WDS (1979) Computer retrieval and analysis of molecular geometry. 4. intermolecular interactions. J Am Chem Soc 101:4374–4376
33. Murray-Rust P, Stallings WC, Monti CT, Preston RK, Glusker JP (1983) Intermolecular interactions of the carbon-fluorine bond: the crystallographic environment of fluorinated carboxylic acids and related structures. J Am Chem Soc 105:3206–3214
34. Ramasubbu N, Parthasarathy R, Murray-Rust P (1986) Angular preferences of intermolecular forces around halogen centers: preferred directions of approach of electrophiles and nucleophiles around carbon-halogen bond. J Am Chem Soc 108:4308–4314
35. Dumas J-M, Kern M, Janier-Dubry JL (1976) Cryoscopic and calorimetric study of mx_4-polar organic base interactions (M = C, Si, X = Cl, Br)—influence of element and of halogen. Bull Soc Chim Fr 11–1:1785–1790
36. Martire DE, Sheridan JP, King JW, O'Donnell SE (1976) Thermodynamics of molecular association. 9. An NMR study of hydrogen bonding of chloroform and bromoform to Di-n-octyl ether, Di-n-octyl thioether, and Di-n-Octylmethylamine. J Am Chem Soc 98:3101–3106
37. Imakubo T, Sawa H, Kato R (1995) Novel radical cation salts of organic π-donors containing iodine atom(s): the first application of strong intermolecular I–X (X = CN, halogen atom) interaction to molecular conductors. Synth Metals 73:117–122
38. Imakubo T, Tajima N, Tamura M, Kato R, Nishio Y, Kajita K (2003) Supramolecular organic conductor θ-$(DIETS)_2[Au(CN)_4]$: unique crystal structure and superconductivity under Uniaxial Strain (DIETS = Diodo(ethylenedithio) diselenadithiafulvalene). Synth Metals 133–134:181–183
39. Amico V, Meille SV, Corradi E, Messina MT, Resnati G (1998) Perfluorocarbon-hydrocarbon self-assembly. 1D infinite chain formation driven by nitrogen—iodine interactions. J Am Chem Soc 120:8261–8262
40. Rissanen K (2008) Halogen bonded supramolecular complexes and networks. Cryst Eng Comm 10:1107–1113
41. Metrangolo P, Meyer F, Pilati T, Resnati G, Terraneo G (2008) Halogen bonding in supramolecular chemistry. Angew Chem Int Ed 47:6114–6127
42. Priimagi A, Cavallo G, Metrangolo P, Resnati G (2013) The halogen bond in the design of functional supramolecular materials: recent advances. Acc Chem Res 46:2686–2695
43. Bruckmann A, Pena MA, Bolm C (2008) Organocatalysis through halogen-bond activation. Synlett 6:900–902
44. Kniep F, Jungbauer SH, Zhang Q, Walter SM, Schindler S, Schnapperelle I, Herdtweck E, Huber SM (2013) Organocatalysis by neutral multidentate halogen-bond donors. Angew Chemie Int Ed 52:7028–7032 (and references cited)
45. Auffinger P, Hays FA, Westhof E, Shing Ho P (2004) Halogen bonds in biological molecules. Proc Natl Acad Sci 48:16789–16794

46. Lu Y, Wang Y, Zhu W (2010) Nonbonding interactions of organic Halogens in biological systems: implications for drug discovery and Biomolecular design. Phys Chem Chem Phys 12:4543–4551
47. Murray JS, Riley KE, Politzer P, Clark T (2010) Directional weak intermolecular interactions: σ-Hole bonding. Aust J Chem 63:1598–1607.
48. Rosenfield RE Jr, Parthasarathy R, Dunitz JD (1977) Directional preferences of nonbonded atomic contacts with divalent sulfur. 1. Electrophiles and nucleophiles. J Am Chem Soc 99:4860–4862
49. Guru Row TN Parthasarthy R (1981) Directional preferences of nonbonded atomic contacts with divalent sulfur in terms of its orbital orientations. 2. S–S interactions and nonspherical shape of sulfur in crystals. J Am Chem Soc 103:477–479
50. Ramasubbu N, Parthasarathy R (1987) Stereochemistry of incipient electrophilic and nucleophilic reactions at divalent selenium center: electrophilic-nucleophilic pairing and anisotropic shape of Se in Se–Se interactions. Phosphorus Sulfur 31:221–229
51. Politzer P, Murray JS (1998) Statistical analysis of the molecular surface electrostatic potential: an approach to describing noncovalent interactions in condensed phases. J Mol Struct (Theochem) 425:107–114
52. Politzer P, Murray JS (2002) The fundamental nature and role of the electrostatic potential in atoms and molecules. Theor Chem Acc 108:134–142
53. Murray JS, Politzer P (2011) The electrostatic potential: an overview. WIREs Comp Mol Sci 1:153–163
54. Stewart RF (1979) On the mapping of electrostatic properties from bragg diffraction data. Chem Phys Lett 65:335–342
55. Politzer P, Truhlar DG, eds (1981) Chemical applications of atomic and molecular electrostatic potentials. Plenum Press, New York
56. Klein CL, Stevens ED (1988) Charge density studies of drug molecules. In Liebman JF, Goldberg A (eds) Structure and reactivity. VCH Publishers, New York, ch 2, pp 25–64
57. Bader RWF, Carroll MT, Cheeseman JR, Chang C (1987) Properties of atoms in molecules: atomic volumes. J Am Chem Soc 109:7968–7979
58. Brinck T, Murray JS, Politzer P (1992) Surface electrostatic potentials of halogenated methanes as indicators of directional intermolecular interactions. Int J Quantum Chem 44(Suppl 19):57–64
59. Brinck T, Murray JS, Politzer P (1993) Molecular electrostatic potentials and local ionization energies of group V—VII hydrides and their anions. Relationships for aqueous and gas-phase acidities. Int J Quantum Chem 48(Suppl 20):73–88
60. Murray JS, Paulsen K, Politzer P (1994) Molecular surface electrostatic potentials in the analysis of non-hydrogen-bonding noncovalent interactions. Proc Indian Acad Sci (Chem Sci) 106:267–275
61. Awwadi FF, Willett RD, Peterson KA, Twamley B (2006) The nature of halogen-halogen synthons: crystallographic and theoretical studies. Chem Eur J 12:8952–8960
62. Clark T, Hennemann M, Murray JS, Politzer P (2007) Halogen bonding: the σ-hole. J Mol Model 13:291–296
63. Politzer P, Lane P, Concha MC, Ma Y, Murray JS (2007) An overview of halogen bonding. J Mol Model 13:305–311
64. Burling FT, Goldstein BM (1992) Computational studies of nonbonded sulfur-oxygen and selenium-oxygen interactions in the thiazole and selenazole nucleosides. J Am Chem Soc 114:2313–2320
65. Murray JS, Lane P, Politzer P (2008) Simultaneous σ-hole and hydrogen bonding by sulfur- and selenium-containing heterocycles. Int J Quantum Chem 108:2770–2781
66. Clark T, Murray JS, Lane P, Politzer P (2008) Why are dimethyl sulfoxide and dimethyl sulfone such good solvents? J Mol Model 14:689–697
67. Murray JS, Lane P, Politzer (2009) Expansion of the σ-hole concept. J Mol Model 15:723–729
68. Murray JS, Lane P, Politzer P (2007) A predicted new type of directional noncovalent interaction. Int J Quantum Chem 107:2286–2292

69. Politzer P, Murray JS, Janjić GV, Zarić SD (2014) σ-hole interactions of covalently-bonded nitrogen, phosphorus and arsenic: a survey of crystal structures. Crystals 4:12–31
70. Delgado-Barrio G, Prat RF (1975) Deformed hartree-fock solutions for atoms. III. convergent iterative process and results for O^{--}. Phys Rev A 12:2288–2297.
71. Sen KD, Politzer P (1989) Characteristic features of the electrostatic potentials of singly-negative monoatomic ions. J Chem Phys 90:4370–4372
72. Feynman RP (1939) Forces in Molecules. Phys Rev 56:340–343
73. Pauling L (1960) The nature of the chemical bond, 3rd edn. Cornell University Press, Ithaca
74. Stevens ED (1979) Experimental electron density distributions of molecular chlorine. Mol Phys 37:27–45
75. Nyburg SC, Faerman CH (1985) A revision of van der Waals atomic Radii for molecular crystals: N, O, F, S, Cl, Se, Br and I bonded to carbon. Acta Cryst B 41:274–279
76. Ikuta S (1990) Anisotropy of electron density distribution around atoms in molecules: N, P, O and S atoms. J Mol Struct (Theochem) 205:191–201
77. Tsirelson VG, Zou PF, Tang T-H, Bader RFW (1995) Topological Definition of crystal structure determination of the bonded interactions in solid molecular chlorine. Acta Cryst A 51:143–153
78. Torii H (2003) The role of atomic quadrupoles in intermolecular electrostatic interactions of polar and nonpolar molecules. J Chem Phys 119:2192–2198
79. Bilewicz E, Rybarczyk-Pirek AJ, Dubis AT, Grabowski SJ (2007) Halogen bonding in crystal structures of 1-methylpyrrol-2-yl trichloromethyl ketone. J Mol Struct 829:208–211
80. Murray JS, Macaveiu L, Politzer P (2014) Factors affecting the strengths of σ-hole electrostatic potentials. J Comput Sci 5:590–596
81. Bundhun A, Ramasami P, Murray JS, Politzer P (2013) Trends in σ-hole strengths and interactions of F_3MX molecules (M = C, Si, Ge and X = F, Cl, Br, I). J Mol Model 19:2739–2746
82. Hennemann M, Murray JS, Politzer P, Riley KE, Clark T (2012) Polarization-induced σ-holes and hydrogen bonding. J Mol Model 18:2461–2469
83. Clark T, Murray JS, Politzer P (2014) Role of polarization in halogen bonds. Aust J Chem 67:451–456
84. Clark T, Murray JS, Politzer P (2015) Correct electrostatic treatment of non-covalent interactions: the importance of polarisation. WIREs Comp Mol Sci, DOI: 10.1002/wcms.1210
85. Riley KE, Hobza P (2008) Investigations into the nature of halogen bonding including symmetry adapted perturbation theory analyses. J Chem Theory Comp 4:232–242
86. Solimannejad M, Gharabaghi M, Scheiner S (2011) SH–N and SH–P blue-shifting H-bonds and N–P interactions in complexes pairing HSN with amines and phosphines. J Chem Phys 134:24312(1–6)
87. Politzer P, Murray JS, Concha MC (2007) Halogen bonding and the design of new materials: organic bromides, chlorides and even fluorides as donors. J Mol Model 13:643–650
88. Chopra D, Guru Row TN (2011) Role of organic fluorine in crystal engineering. Cryst Eng Comm 13:2175–2186
89. Metrangolo P, Murray JS, Pilati T, Politzer P, Resnati G, Terraneo G (2011) Fluorine-centered halogen bonding: a factor in recognition phenomena and reactivity. Cryst Growth Des 11:4238–4246
90. Murray JS, Lane P, Clark T, Politzer P (2007) σ-hole bonding: molecules containing group VI atoms. J Mol Model 13:1033–1038
91. Wilson CC (2000) Single crystal neutron diffraction from molecular materials.World Scientific, Singapore
92. Parkin A, Harte SM, Goeta AE, Wilson CC (2004) Imaging proton migration from X-rays and neutrons. New J Chem 28:718–721
93. Murray JS, Politzer P (1991). Correlations between the solvent hydrogen-bond-donating parameter α and the calculated molecular surface electrostatic potential. J Org Chem 56:6715–6717
94. Shields ZP-I, Murray JS, Politzer (2010) Directional tendencies of halogen and hydrogen bonding. Int J Quantum Chem 110:2823–2832

95. Clark T (2013) σ-holes. WIREs Comput Mol Sci 3:13–20
96. Kollman P, McKelvey J, Johansson A, Rothenberg S (1975) Theoretical studies of hydrogen-bonded dimers. Complexes involving HF, H_2O, NH_3, CH_1, H_2S, PH_3, HCN, HNC, HCP, CH_2NH, H_2CS, H_2CO, CH_4, CF_3H, C_2H_2, C_2H_4, C_6H_6, F^- and H_3O^+. J Am Chem Soc 97:955–965
97. Politzer P, Daiker KC (1981) Models for chemical reactivity. In: Deb BM (ed) The Force Concept in Chemistry. Van Nostrand Reinhold Co, New York, ch 6.
98. Legon AC, Millen DJ (1982) Determination of properties of hydrogen-bonded dimers by rotational spectroscopy and a classification of dimer geometries. Faraday Discuss Chem Soc 73:71–87
99. Buckingham AD, Fowler PW (1983) Do electrostatic interactions predict structures of Van der Waals complexes? J Chem Phys 79:6426–6428
100. Buckingham AD, Fowler PW (1985) A Model for the geometries of Van der Waals complexes. Can J Chem 63:2018–2025
101. Riley KE, Murray JS, Fanfrlík J, Řezáč J, Solá RJ, Concha MC, Ramos FM, Politzer P (2013) Halogen bond tunability II: the varying roles of electrostatic and dispersion contributions to attraction in halogen bonds. J Mol Model 19:4651–4659
102. Legon AC (1999) Prereactive complexes of dihalogens XY with Lewis bases B in the gas phase: a systematic case for the halogen analogue B–XY of the hydrogen bond B–HX. Angew Chem Int Ed 38:2686–2714
103. Legon AC (2010) The halogen bond: an interim perspective. Phys Chem Chem Phys 12:7736–7747
104. Bondi A (1964) van der Waals volumes and Radii. J Phys Chem 68:441–451
105. Rowland RS, Taylor R (1996) Intermolecular nonbonded contact distances in organic crystal structures: comparison with distances expected from van der Waals Radii. J Phys Chem 100:7384–7391
106. Politzer P, Murray JS, Lane P (2007) σ-hole bonding and hydrogen bonding: competitive interactions. Int J Quantum Chem 107:3046–3052.
107. Aakeröy CB, Fasulo M, Shultheiss N, Desper J, Moore C (2007) Structural competition between hydrogen bonds and halogen bonds. J Am Chem Soc 129:13772–13773
108. Alkorta I, Blanco F, Solimannejad M, Elguero J (2008) Competition of hydrogen bonds and halogen bonds in complexes of hypohalous acids with nitrogenated bases. J Phys Chem A 112:10856–10863
109. Di Paolo T Sandorfy C (1974) On the biological importance of the hydrogen bond breaking potency of fluorocarbons. Chem Phys Lett 26:466–469
110. Corradi E, Meille SV, Messina MT, Metrangolo P, Resnati G (2000) Halogen bonding versus hydrogen bonding in driving self-assembly processes. Angew Chem Int Ed 39:1782–1786
111. Riley KE, Murray JS, Fanfrlík J, Řez´aˇc J, Solá RJ, Concha MC, Ramos FM, Politzer P (2011) Halogen bond tunability I: the effects of aromatic fluorine substitution on the strengths of halogen-bonding interactions involving chlorine, bromine and iodine. J Mol Model 17:3309–3318
112. Riley KE, Murray JS, Politzer P, Concha MC, Hobza P (2009) Br–O complexes as probes of factors affecting halogen bonding: interactions of bromobenzenes and bromopyrimidines with acetone. J Chem Theory Comput 5:155–16
113. Politzer P, Murray JS (2013) Enthalpy and entropy factors in gas phase halogen bonding: compensation and competition. Cryst Eng Comm 15:3145–3150
114. Lu X, Li H, Zhu X, Zhu W, Liu H (2011) How does halogen bonding behave in solution? A Theoretical study using implicit solvation model. J Phys Chem A 115:4467–4475
115. Damm E, Hassel O, Romming C (1965) X-ray analysis of the (1:1) addition compounds of 1,4-Dioxan with oxalyl chloride resp. Oxalyl bromide. Acta Chem Scand 19:1159–1165
116. Güttinger P (1932) Das Verhalten von Atomen in magnetischen Drehfeld. Z Phys 73:169–184
117. Pauli W (1933) Principles of Wave Mechanics, Handbuch der Physik, 24. Springer, Berlin, p. 162

118. Hellmann H (1933) Zur Rolle der kinetischen Electronenenergie für die zwischenatomaren Kräfte. Z Phys 85:180–190
119. Feynman RP (1939) Forces in molecules. Phys Rev 56:340–343
120. Levine IN (2000) Quantum chemistry, 5th ed. Prentice-Hall, Upper Saddle River
121. Hohenberg P, Kohn W (1964) Inhomogeneous electron gas. Phys Rev 136:B864–B871
122. Ratajczak H, Orville-Thomas WJ (1976) Charge transfer theory of hydrogen bonds: relations between vibrational, spectra and energy of hydrogen bonds. Chem Phys 17:197–216 (and references cited)
123. Arunan E, Desiraju GR, Klein RA, Sadlej J, Scheiner S, Alkorta I, Clary DC, Crabtree RH, Dannenberg JJ, Hobza P, Kjaergaard HG, Legon AC, Mennucci B, Nesbitt DJ (2011) Defining the hydrogen bond: an account. Pure Appl Chem 83:1619–1636 (and references cited)
124. Berlin T (1951) Binding regions in diatomic molecules. J Chem Phys 19:208–213
125. Bader RFW (2006) Pauli repulsions exist only in the eye of the beholder. Chem Eur J 12:2896–2901
126. Scerri ER (2000) Have orbitals really been observed? J Chem Ed 77:1492–1494
127. Solimannejad M, Malekani M, Alkorta I (2010) Cooperative and dimunitive unusual weak bonding in F_3C–X–HMgH–Y and F_3C–X–Y–HMgH Trimers (X = Cl, Br; Y = HCN and HNC). J Phys Chem A 114:12106–12111
128. Scheiner S (2011) On the Properties of X–N Noncovalent Interactions for First-, Second-, and Third-Row X Atoms. J Chem Phys 134:164313(1–9)
129. Grabowski SJ, Bilewicz E (2006) Cooperative halogen bonding effect—Ab Initio calculations on H_2CO–$(ClF)_n$ complexes. Chem Phys Lett 427:51–55
130. Del Bene JE Alkorta I Elguero J (2010) Do traditional, chlorine-shared and ion-pair halogen bonds exist? An Ab Initio investigation of FCl:CNX complexes. J Phys Chem A 114:12958–12962
131. Politzer P, Murray JS (2012) Halogen bonding and beyond: factors influencing the nature of CN–R and SiN–R complexes with FCl and Cl_2. Theor Chem Acc 131:1114(1–10)
132. Hobza P, Zahradnik R (1992) An essay on the theory and calculations of intermolecular interactions. Int J Quantum Chem 42:581–590
133. Cramer CJ (2002) Essentials of computational chemistry. Wiley, Chichester
134. Hirschfelder JO, Eliason MA (1967) Electrostatic Hellmann-Feynman theorem applied to the long-range interaction of two hydrogen atoms. J Chem Phys 47:1164–1169
135. Hunt KLC (1990) Dispersion dipoles and dispersion forces: proof of Feynman's "Conjecture" and generalization to interacting molecules of arbitrary symmetry. J Chem Phys 92:1180–1187
136. Stone AJ, Price SL (1988) Some new ideas in the theory of intermolecular forces: anisotropic atom-atom potentials. J Phys Chem 92:3325–3335
137. Reed AE, Curtiss LA, Weinhold F (1988) Intermolecular interactions from a natural bond orbital, donor-acceptor viewpoint. Chem Rev 88:899–926
138. Sokalski WA, Roszak SM (1991) Efficient techniques for the decomposition of intermolecular interaction energy at SCF level and beyond. J Mol Struct (Theochem) 234:387–400
139. Chen J, Martínez TJ (2007) QTPIE: charge transfer with polarization current equalization: a fluctuating charge model with correct asymptotics. Chem Phys Lett 438:315–320
140. Clark T (2014) Directional electrostatic bonding. In: Frenking G (ed) The chemical bond: chemical bonding across the periodic table. Wiley-VCH, Weinheim, Germany, ch. 18.
141. Hermansson K (2002) Blue-shifting hydrogen bonds. J Phys Chem A 106:4695–4702
142. Qian W, Krimm S (2002) Vibrational spectroscopy of hydrogen bonding. Origin of the different behavior of the C–H–O hydrogen bond. J Phys Chem A 106:6628–6636
143. Wang W, Wang NB, Zheng W, Tian A (2004) Theoretical study on the blueshifting halogen bond. J Phys Chem A 108:1799–1805
144. Murray JS, Concha MC, Lane P, Hobza P, Politzer P (2008) Blue shifts vs. red shifts in σ-hole bonding. J Mol Model 14:699–704
145. Stone AJ, Misquitta AJ (2009) Charge-transfer in symmetry-adapted perturbation theory. Chem Phys Lett 473:201–205

146. Politzer P, Harris RR (1970) Properties of atoms in molecules. I. A proposed definition of the charge of an atom in a molecule. J Am Chem Soc 92:6451–6454
147. Meister J, Schwarz WHE (1994) Principal components of ionicity. J Phys Chem 98:8245–8252
148. Wiberg KB, Rablen PR (1993) Comparison of atomic charges derived via different procedures. J Comput Chem 14:1504–1518
149. Price SL (1996) Applications of realistic electrostatic modelling to molecules in complexes, solids and proteins. J Chem Soc Faraday Trans 92:2997–3008
150. Politzer P, Murray JS, Concha MC (2008) σ-hole bonding between like atoms. A fallacy of atomic charges. J Mol Model 14:659–665
151. Dobeš P, Řez´aˇc J, Fanfrlík J, Otyepka M, Hobza P (2011) Semiempirical quantum mechanical method PM6-DH2X describes the geometry and energetics of CK2-inhibitor complexes involving halogen bonds well, while the empirical potential fails. J Phys Chem B 115:8581–8589
152. Ibrahim MAA (2011) Molecular mechanical study of halogen bonding in drug discovery. J Comput Chem 32:2564–2574
153. Kolař M, Hobza P (2012) On extension of the current biomolecular empirical force field for the description of halogen bonds. J Chem Theory Comput 8:1325–1333
154. Carter M, Rappé AK, Shing Ho P (2012) Scalable anisotropic shape and electrostatic models for biological bromine halogen bonds. J Chem Theory Comput 8:2461–2473
155. Jorgensen WL, Schyman P (2012) Treatment of halogen bonding in the opls-aa force field: application to potent anti-hiv agents. J Chem Theory Comput 8:3895–3901
156. Murray JS, Lane P, Clark T, Riley KE, Politzer P (2012) σ-holes, π-holes and electrostatically-driven interactions. J Mol Model 18:541–548
157. Murray JS, Shields ZP-I, Seybold PG, Politzer P (2015) Intuitive and counterintuitive noncovalent interactions of aromatic π regions with the hydrogen and nitrogen of HCN. J Comput Sci DOI: 10.1006/j.jocs.02.001
158. Isaacson W (2007) Einstein: his life and universe. Simon and Schuster, New York, p. 549

Chapter 11
The X-C· · ·Y Carbon Bond

Devendra Mani and E. Arunan

Abstract The CH_3X (X more electronegative group/atom than C) molecules have their C atom partially positive which enables it to bond with another atom/group of atoms having an electron rich region, Y. This leads to a penta-coordinate carbon having X-C· · ·Y 'carbon bond' in addition to the four covalent bonds around the central carbon. The X-C· · ·Y carbon bond is very similar to the X-H· · ·Y hydrogen bond. The X-C bond lengthens and there is a red-shift in X-C stretching frequency. The ∠X-C· · ·Y is linear. There is an overlap between X-C σ* orbital and the donor orbital in Y, which leads to the weakening X-C bond and red-shift. Analysis of the electron density topology shows a bond critical point between C and Y. It also shows that there is mutual penetration of electron clouds of the C and Y atoms, leading to bond formation. The X-C· · ·Y carbon bond is an intermediate in S_N2 reactions, very much like the X-H· · ·Y hydrogen bond being an intermediate in proton transfer reactions. Moreover, the carbon bonded H_2O· · ·CH_3OH geometry is a textbook example for the ill-defined hydrophobic interactions, as the H_2O approaches methanol from the hydrophobic CH_3 side. Similar interactions involving Si and Ge containing molecules exist and these have all been named 'tetrel bonds', extending halogen bonding (Group 17), chalcogen bonding (Group 16) and pnicogen bonding (Group 15) to Group 14. While such interactions involving elements from all groups are likely, we expect that the carbon bonding would be next only to hydrogen bonding in its importance, as C and H are the elements chosen by nature and life as we know is largely based on these elements. This chapter summarizes the results on carbon bonding since it was proposed about a year ago.

11.1 Introduction

Non-covalent interactions are important binding forces utilized by nature to produce the condensed phases of neutral molecules [1, 2]. This term has been made popular

E. Arunan (✉) · D. Mani
Department of Inorganic and Physical Chemistry,
Indian Institute of Science, 560012
Bangalore, India
e-mail: arunan@ipc.iisc.ernet.in

S. Scheiner (ed.), *Noncovalent Forces*, Challenges and Advances in
Computational Chemistry and Physics 19, DOI 10.1007/978-3-319-14163-3_11

to denote intermolecular interactions, which were initially called van der Waals interactions following his well known equation. Without such interactions, there is no earth or life and hence the enormous interest in this field is not surprising. In the early part of twentieth century, stunning differences in intermolecular interactions between some molecules, such as H_2O, and other molecules such as H_2S, warranted a further classification of intermolecular interactions. Water is a high boiling liquid and H_2S is a gas at ambient conditions. In solid H_2O (ice), each molecule is surrounded by four molecules and in solid H_2S, each molecule is surrounded by twelve molecules. Only a sphere could accommodate 12 neighbours and it appeared that H_2S molecules were essentially spherical having isotropic interactions, implying that the H atoms in H_2S had no specific orientations. In ice, on the other hand each H_2O molecule has specific interactions with its four neighbours, with the O atom pointing towards two H atoms of two of the neighbouring molecules and each H atom pointing towards an O atom of a neighbouring molecule all with an O... H distance of $\approx$ 1.8 Å. This may be compared to the O–H covalent bond distance of $\approx$ 1.0 Å within the H_2O molecule. The longer intermolecular contact was called 'hydrogen bonding' and it makes the intermolecular interaction in ice very directional and anisotropic. It was decided that the anisotropic intermolecular interactions as in H_2O would be called 'hydrogen bonding' and isotropic intermolecular interactions as in H_2S would be left as van der Waals interaction. Pauling's classic book on *The Nature of Chemical Bond* was perhaps the first book to popularize hydrogen bonds [3] and it concluded that H_2O can form hydrogen bonds and H_2S cannot. Since then there have been numerous attempts to differentiate hydrogen bonding and van der Waals interactions in terms of the physical forces involved and it only led to a lot of discussion on semantics. Now we know that both H_2O and H_2S can form directional hydrogen bonds but at different conditions. Intermolecular forces in both these molecules originate from electrostatic interactions, induction/polarization, dispersion, exchange correlation and charge-transfer/covalency.

The recent IUPAC recommendation on the definition of hydrogen bonding states clearly that no unique physical forces can be attributed to hydrogen bonding and the forces acting between the interacting molecules should have contributions beyond dispersion [4]. Hydrogen bonding, often denoted as X-H$\cdots$Y, is clearly a subset of van der Waals interaction. Its unique features may be summarized as follows: (i) it is directional and the angle X-H$\cdots$Y tends to be linear, (ii) H is bonded to two (or more) atoms and is 'hypervalent' as it has a normal valency of 1 and (iii) X is more electronegative than H giving the H a partial positive charge. The uniqueness of hydrogen among all the elements of the periodic table made it easy to identify 'hydrogen bonding' as a well-defined intermolecular interaction. The fact that it controls the properties of the molecules of life such as H_2O and DNA ensured that enormous attention was paid to it resulting in almost one paper on it being published every hour in the last several years [4]. Its importance in biological sciences has been specifically addressed in some recent books [5, 6]. However, it was only a matter of time before similar interactions involving other elements in the periodic table were found.

Many specific non-covalent interactions, similar to hydrogen bonding, have now been recognised and classified. The obvious choices were interactions involving atoms in the alkali (Group 1) and halogen (Group 17) groups, as H has similarities to both these group elements. As shown in a recent comprehensive study, lithium bonding is dominated by electrostatics as LiX molecules are distinctly ionic as opposed to HX and XY molecules (X, Y halogens), both of which have significant covalent nature [7]. Indeed, halogen bonding has been characterized as a distinct intermolecular interaction similar to hydrogen bonding now [8]. In the last few years, chalcogen bonding (Group 16) [9] and pnicogen bonding (Group 15) [10] have also been characterized. Clearly all these interactions could be classified as van der Waals interactions. However, these new names are most certainly reflective of the detailed understanding of these interactions as observed in these specific groups.

Highly directional halogen bonding interactions have found great applications in supramolecular chemistry, crystal engineering and molecular recognition process [11]. The bonding in these interactions could be explained using the sigma-hole concept which was first introduced by Politzer and coworkers [12, 13] to explain halogen bonding. This was useful as there was some confusion about why the electronegative halogen atoms were interacting with other electronegative atoms. Even though, the presence of a sigma hole in the molecules containing Group 14 elements had also been predicted [13], work on such interactions involving carbon atom was particularly lagging. There had been quite a few studies for the complexes containing more polarisable Si and Ge atoms [13, 14] and in one of the studies complexes containing carbon atom were also considered [15]. This interaction involving C is much weaker compared to the interactions involving Si and Ge atoms, perhaps explaining the initial lack of interest. Another difficulty in identifying this interaction involving C atoms is that, in most molecules a tetravalent carbon is protected in all directions and penta-coordinate carbons have not been easy to identify. The fact that this interaction with C containing molecules is relatively weaker is important in nature, *vide infra*.

Our interest in carbon bonding similar to hydrogen bonding, arose following a microwave spectroscopic investigation on Ar···propargyl alcohol complex, which may not attract widespread interest [16, 17]. Propargyl alcohol could exist in gauche and trans conformations, gauche being more stable than trans. We attempted to observe the rotational spectra of the Ar complexes with both these forms, but could succeed in observing only the complex formed by the gauche conformer. Experimental microwave spectrum showed that the equilibrium structure had Ar positioned in between the OH and C≡C groups exhibiting both O–H···Ar and Ar···π interactions. Atoms in molecules (AIM) theoretical analysis on the ab initio optimized structure showed bond critical points confirming both these interactions. These resembled CH_3OH···Ar and HC≡CH···Ar complexes, as might be expected. While the experiment on Ar···trans-propargyl alcohol complex did not succeed, theory posed no such challenge. The trans conformer formed a complex with Ar exhibiting O–H···Ar interaction very similar to what is observed in the CH_3OH···Ar complex. The AIM calculations also revealed evidence for O–H···Ar interaction but showed another bond critical point connecting Ar to C! This was unexpected and given the contradicting claims of the relation between 'bonds' and 'bond critical points' [18, 19], it may have been prudent to ignore this result. However, the AIM analysis was repeated for CH_3OH···Ar complex, which also exhibited both OH···Ar and Ar···C bonds.

We had previously shown that the tetrahedron faces of CH_4 are electron rich, based on its electrostatic potential, and it could accept a hydrogen bond as in $H_4C\cdots HX$, leading to a penta-coordinate carbon [19]. How would the tetrahedron face of CH_3OH, opposite to the OH group, look? When we looked at it, see next section, carbon bond was born.

We showed that a specific interaction exists between the carbon atom of a CH_3X molecule, where X is an electron withdrawing atom or group, and electron rich centre (atom) of the other binding unit; like oxygen of H_2O [17]. Such interactions were recognised as unique X-C$\cdots$Y interactions and were called carbon bond after confirming their various bonding characteristics. Interactions with other atoms of the group IV have also been explored in detail recently and were found to be similar to the carbon bonding [20–23]. The interactions involving group 14 elements were collectively called tetrel bond by Grabowski and Bauza et al., almost simultaneously and independently [20, 24]. The very fact that nature prefers carbon over other tetrel atoms convinces us that carbon bonding would remain the most important subset of tetrel bonding.

In addition to 'hydrogen bonding' and van der Waals interactions, intermolecular interactions have often been classified as electrostatic interactions and hydrophobic interactions [2]. While the term 'electrostatic interactions' is easy to understand as the interactions between charges, dipoles, quadrupoles etc., the term hydrophobic interactions is not as well defined. A careful look at carbon bonding interactions revealed that they are a specific part of the ill-defined hydrophobic interaction [17].

Carbon bonding is quite young and is just a little older than 1 year. However, in this short span of time, there have been many studies on carbon bonds and much more is known about these now [20–23, 25–31]. Importance of carbon bonding in S_N2 reactions was pointed out in the original paper [17] and was later explored in detail by Grabowski [20]. Frontera et al. [27, 28] have shown that small substituted cycloalkane molecules are potential carbon bond donor. Alkorta and co-workers have pointed out the presence of intramolecular carbon bonding in carbohydrate molecules [26]. Varadwaj et al. [25] have studied complexes of CO with a number of CH_3X molecules. It was observed that both the ends of CO molecule can interact with the carbon atom of the CH_3X molecule leading to the formation of O$\cdots$C-X and C$\cdots$C-X carbon bonds, the latter can be called a dicarbon bond [25], in analogy to the 'dihydrogen bonds' [4b]. Moreover, carbon bonds have also been observed experimentally using charge density analysis [32].

In this chapter, we have tried to cover most of the important results on carbon bonding till date. The chapter begins with an introduction to the carbon bond interactions and then explores it in detail by using various theoretical tools. We summarize the results from Reference 17 and extend it for complexes with CH_3Cl and CH_3Br as carbon bond donors and also, π electrons, rare gases and sigma electrons in H_2 as carbon bond acceptors. We then summarize results from other groups on carbon bonds with unpaired electrons and anions as acceptors. Experimental results on carbon bonds are discussed next. This includes the experiments that were carried out after the carbon bond was proposed and also the earlier experiments that had given structures which could now be classified as 'carbon bonded complexes'.

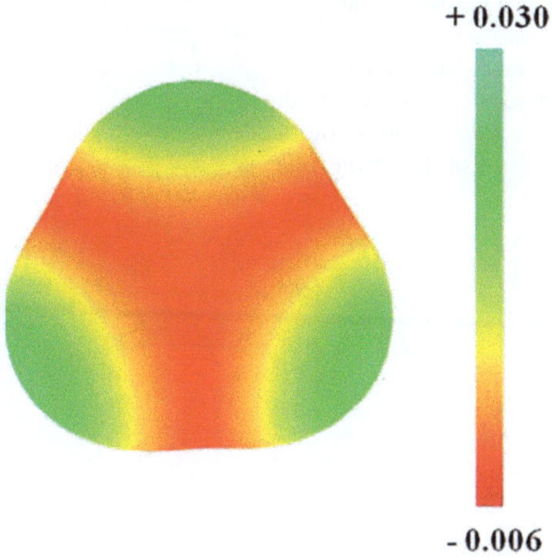

Fig. 11.1 Molecular electrostatic potential for methane molecule. Relative color code bar is shown on *right*

11.1.1 What is a Carbon Bond?

Carbon bond can be understood in terms of sigma hole concept which has already been used to explain other interactions like halogen bonding [8, 11, 12, 33]. Analysis of the molecular electrostatic potential (MESP) surfaces helps in locating regions of electronic charge depletion and accumulation in molecules [34, 35]. Regions with charge depletion can receive electronic charge from other electron rich species and may lead to bonding interactions. Similarly, electron rich regions interact with nucleophiles and may lead to bonding interactions. A sigma hole is one such electron depleted region along the direction opposite to a polar sigma bond.

We start our discussion here with the methane molecule. Its MESP is shown in Fig. 11.1. Red regions are electronegative and green regions are electropositive which represent regions of electron accumulation and electron depletion, respectively. The tetrahedral face of methane is electronegative and thus is nucleophilic in nature. The nucleophilic tetrahedral face interacts with molecules like HF, HCl, HCN, H_2O etc., in such a way that the hydrogen atom of these molecules points towards the tetrahedral face of methane. Such structures have been experimentally observed for the complexes of CH_4 with HF/HCl/HCN/H_2O using microwave spectroscopy [36–39]. Atoms in molecules (AIM) theoretical analysis showed a bond critical point between carbon atom of methane and the hydrogen atom of HX/H_2O molecule confirming the presence of a X-C···H hydrogen bond through the tetrahedral face [19].

What happens to the bonding nature of methane when one of its hydrogen atoms is replaced by some atom or group which is more electronegative than carbon? Is the CH_3 face of a molecule like CH_3OH hydrogen bond acceptor like that of CH_4? Let us have a look at the MESP of the CH_3OH molecule shown in Fig. 11.2.

In this molecule, the CH_3 tetrahedral face is electropositive in contrast to the electronegative tetrahedral face of methane. Being electropositive, it interacts with the electron rich sites like a lone-pair of electrons and leads to the stable geometries like the one shown for H_2O···CH_3OH complex Fig. 11.3.

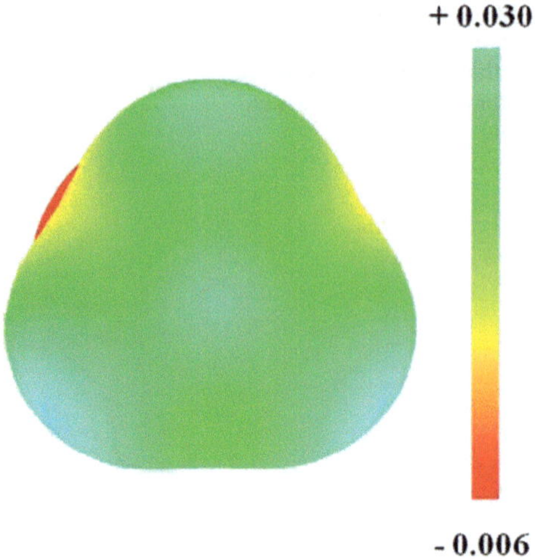

Fig. 11.2 Molecular electrostatic potential surfaces for CH_3OH molecule, viewed from the CH_3 side. Relative color code bar is also shown on *right*

The AIM analysis confirms the presence of a bond critical point between the electron rich atom and the carbon atom of CH_3X molecules as shown in Fig. 11.3b. This bond critical point follows several criteria used to classify hydrogen bonding and halogen bonding [17, 40] which confirms the formation of an X-C···Y carbon bond. Here H_2O is the carbon bond acceptor unit and methanol is the donor unit. Such interactions exist for a series of CH_3X complexes which are discussed in the next sections. It is worth pointing out again that both in $H_4C\cdots H_2O$ and in $H_2O\cdots CH_3OH$, we have a penta-coordinate carbon. In the former, CH_4 is the electron donor, bonding with the electron deficient H in H_2O and it is an O–H···C hydrogen bonded complex. In the later, H_2O is the electron donor, bonding with the electron deficient C in CH_3OH and it is an O–C···O carbon bonded complex. This convention in naming intermolecular bonding is needed today for two reasons. (1) The 'hydrogen bond' terminology is well established in science for nearly a century. (2) Such specific interactions involving other atoms have been widely recognized now. Without this convention, text book examples for hydrogen bonds, such as HF dimer might be classified as 'halogen bonded' and H_2O dimer might be classified as 'chalcogen bonded'.

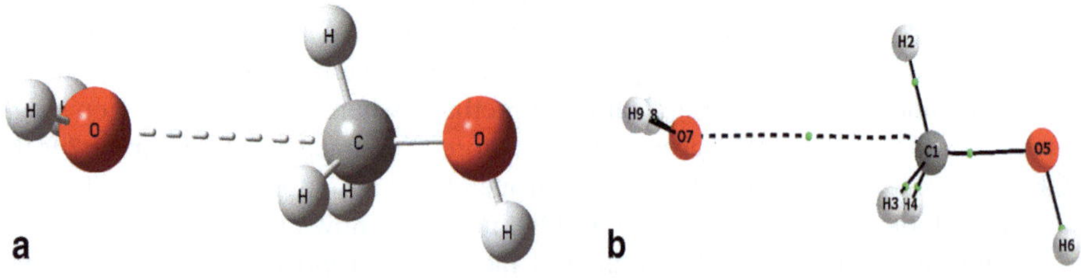

Fig. 11.3 **a** Optimized geometry of $H_2O\cdots CH_3OH$ complex, **b** Electron density topology for the complex; *small green dots* represent bond critical points

11.2 Computational Details

The geometry optimization was performed at MP2/6-311+G(3df,2p) as well as MP2/Aug-cc-pVTZ level of theories. Frequency calculations were performed on the optimized geometries to confirm that these geometries are minima on potential energy hypersurface as well as to find out the zero-point correction to the energies. Optimization as well as frequency calculations were performed using Gaussian 09 software suite [41]. Binding energies were corrected using the counterpoise method [42] inbuilt in Gaussian 09. The binding energies were also corrected for the zero-point energy. The MESP surfaces were plotted using GaussView software [43]. Electrostatic potential values were extracted using Multiwfn software [44]. The AIM analysis was performed on the optimized geometries using AIM-ALL [45] and AIM2000 [46] software suites. The NBO analysis was performed using NBO 6.0 software [47].

11.3 Lone Pairs as Carbon Bond Acceptor

11.3.1 Complexes with Methanol

As mentioned earlier, presence of the electron withdrawing hydroxyl group in methanol leads to electron depletion at its CH_3 face and facilitates the formation of a carbon bond. Hydrogen bonds were first observed with lone pairs of electrons as acceptors and not surprisingly the same is true for carbon bonds. This section focuses on lone pair acceptors.

Complexes of methanol with various lone pair donor molecules; H_2O, H_2S, HX (X = F/Cl/Br), LiX (X = F/Cl/Br), ClF, NH_3 and PH_3 were considered to explore these interactions in detail [17]. In each of these molecules, central atom bears one or more lone pair of electrons which can take part in intermolecular bonding. Geometry optimization was performed starting with guess geometries in which the atoms with lone-pair(s) of the carbon bond acceptor molecule faces the CH_3 face of methanol molecule. The optimized geometries for all these complexes are shown in Fig. 11.4. All these geometries were confirmed to be minima on the potential energy hypersurface by confirming the presence of real frequencies for all the normal modes of vibrations. Important geometrical parameters for the optimized geometries are given in Table 11.1.

Intermolecular bond lengths in all the complexes are shorter at MP2/Aug-cc-pVTZ level than at MP2/6-311+G(3df,2p) level. However, at both the levels of theory, the bond lengths for all the complexes appear long. The values are comparable to the sum of van der Waals radii of Y and C atoms in the case of complexes with H_2O, HF and ClF; are larger in case of complexes with H_2S, HCl, HBr, NH_3 and PH_3; and are much shorter in case of complexes with LiF, LiCl and LiBr. However, as pointed out in the IUPAC recommendation on hydrogen bonding [4], use of van der Waals 'radii' to confirm/rule out specific bonding interactions could be

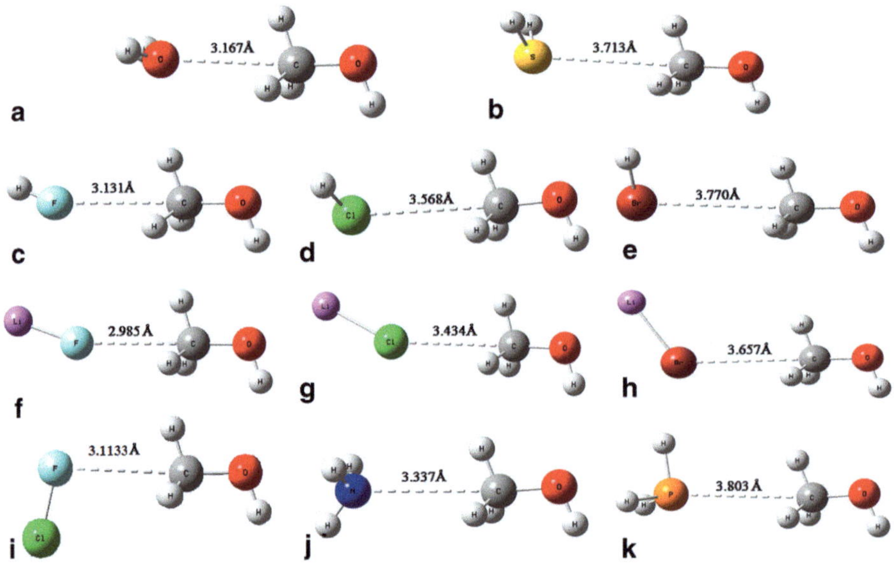

Fig. 11.4 Optimized geometries for **a** $H_2O\cdots CH_3OH$, **b** $H_2S\cdots CH_3OH$, **c** $HF\cdots CH_3OH$, **d** $HCl\cdots CH_3OH$, **e** $HBr\cdots CH_3OH$, **f** $LiF\cdots CH_3OH$, **g** $LiCl\cdots CH_3OH$, **h** $LiBr\cdots CH_3OH$, **i** $ClF\cdots CH_3OH$, **j** $H_3N\cdots CH_3OH$, **k** $H_3P\cdots CH_3OH$ complexes at MP2/6-311+G(3df,2p) level of theory. (Reproduced with permission from RSC, Reference 17)

Table 11.1 Intermolecular bonding parameters for various complexes of CH_3OH at MP2/6-311+G(3df,2P) and MP2/Aug-cc-pVTZ levels of theory. Distances are in angstroms and angles in degrees

Complex	MP2/6-311+G(3df,2p)		MP2/Aug-cc-pVTZ	
	d (Y···C)	Angle (Y···C-X)	d (Y···C)	Angle (Y···C-X)
$H_2O\cdots CH_3OH$	3.167	176.5	3.167	176.5
$H_2S\cdots CH_3OH$	3.713	177.6	3.692	177.6
$HF\cdots CH_3OH$	3.131	180.0	3.091	177.8
$HCl\cdots CH_3OH$	3.568	177.4	3.561	177.1
$HBr\cdots CH_3OH$	3.770	176.1	3.610	175.7
$ClF\cdots CH_3OH$	3.113	176.3	3.115	175.4
$LiF\cdots CH_3OH$	2.985	179.5	2.983	179.5
$LiCl\cdots CH_3OH$	3.434	179.2	3.434	179.2
$LiBr\cdots CH_3OH$	3.657	179.8	3.522	178.0
$H_3N\cdots CH_3OH$	3.377	175.7	3.384	174.9
$H_3P\cdots CH_3OH$	3.803	174.4	3.602	166.3

misleading. For all the complexes the Y···C-X angle is greater than 175° except for $H_3P\cdots CH_3OH$ complex. The atoms in molecules theoretical analysis, *vide infra*, does indicate that there is penetration of electron cloud between the two molecules

Table 11.2 Binding energies for the methanol complexes calculated at different levels of theory. The BSSE corrected, $\Delta E_{(BSSE)}$, and BSSE as well as zero-point corrected, $\Delta E_{(BSSE+ZPC)}$ values are given. All the values are given in kJ.mol^{-1}. (Reproduced with permission from RSC, Reference 17)

Complex	At MP2 /6-311+G(3df,2p)		At MP2 /Aug-cc-pVTZ		At CCSD-T /6-311+G(3df,2p)
	$\Delta E_{(BSSE)}$	$\Delta E_{(BSSE+ZPC)}$	$\Delta E_{(BSSE)}$	$\Delta E_{(BSSE+ZPC)}$	$\Delta E_{(BSSE)}$
$H_2O\cdots CH_3OH$	4.2	1.9	4.5	2.2	4.4
$H_2S\cdots CH_3OH$	2.5	1.0	3.5	2.3	2.7
$HF\cdots CH_3OH$	2.9	1.2	3.1	1.7	3.1
$HCl\cdots CH_3OH$	2.4	1.1	3.0	1.8	2.2
$HBr\cdots CH_3OH$	2.5	1.4	2.9	1.5	2.6
$ClF\cdots CH_3OH$	2.4	1.1	2.9	1.5	2.3
$LiF\cdots CH_3OH$	9.0	7.6	9.7	8.4	9.7
$LiCl\cdots CH_3OH$	7.2	6.0	8.1	7.0	7.0
$LiBr\cdots CH_3OH$	6.9	6.0	7.5	6.6	6.7
$H_3N\cdots CH_3OH$	4.3	2.2	4.7	2.7	4.5
$H_3P\cdots CH_3OH$	2.8	1.6	3.3	1.7	2.6

confirming the 'carbon bond' formation. Binding energies at both the levels of theory are given in Table 11.2. Values obtained from single point counterpoise calculations at CCSD(T)/6-311+G(3df,2p) level are also given.

Intermolecular bond distance pattern is reflected in the binding energies also for these complexes. Complexes are slightly more stable at the MP2/Aug-cc-pVTZ level as compared to the MP2/6-311+G(3df,2p) level. This is due to the shorter bond distances observed at the MP2/Aug-cc-pVTZ level. Values calculated at CCSD(T)/6-311+G(3df,2p) level are not very different from those obtained at MP2/6-311+G(3df,2p) level. Comparing the binding energies shows that the complexes with HY are much weaker than those with LiY (Y = F/Cl/Br). The LiY molecules have larger dipole moment than HY molecules and it can explain the difference in bond energies.

The Atoms in molecule theoretical analysis shows presence of a bond critical point between the Y atom and C atom in all the complexes. For the $H_2O\cdots CH_3OH$ complex the electron density topology is shown in Fig. 11.2. Intermolecular bond paths were found to be present in each of the complexes, which connect the intermolecular bond critical point to the Y and C atoms, suggesting bonding between the two. Electron density and Laplacian of electron density values at the intermolecular bond critical points are given in Table 11.3. The table also lists the values of mutual penetration between Y and C atoms. It is defined as the difference between the sum of the non-bonded 'radii' of the two atoms and the sum of their bonded 'radii'. Non-bonded 'radii' are calculated as the distance from the atom to the point where the electron density reduces to 0.001, *along the bond path.* Mutual penetration value for all the complexes is positive which confirms the bonding between the two atoms. We recall

Table 11.3 Electron density (ρ) and Laplacian of electron density ($\nabla^2\rho$) values at the intermolecular bond critical point for various complexes of methanol. Mutual penetration between the carbon and Y atoms are also given. (Reproduced with permission from RSC, Reference 17)

Complex	ρ(r) in a.u.	$\nabla^2\rho$ in a.u.	Mutual penetration (Å)
$H_2O\cdots CH_3OH$	0.0050	0.0248	0.41
$H_2S\cdots CH_3OH$	0.0037	0.0164	0.30
$HF\cdots CH_3OH$	0.0037	0.0230	0.28
$HCl\cdots CH_3OH$	0.0038	0.0174	0.34
$HBr\cdots CH_3OH$	0.0034	0.0140	0.31
$ClF\cdots CH_3OH$	0.0050	0.0299	0.39
$LiF\cdots CH_3OH$	0.0063	0.0350	0.54
$LiCl\cdots CH_3OH$	0.0049	0.0228	0.48
$LiBr\cdots CH_3OH$	0.0042	0.0181	0.42
$H_3N\cdots CH_3OH$	0.0044	0.0201	0.43
$H_3P\cdots CH_3OH$	0.0035	0.0143	0.34

that the distance between the two atoms is more than the sum of van der Waals radii in some cases, even though there is mutual penetration. This is because the van der Waals 'radii' assumes the atoms to be spherical, when they are not. The carbon bonds in these complexes follow most of the criteria proposed by Koch and Popelier [48] for C-H$\cdots$O hydrogen bonds and these will be discussed in the next section.

11.3.2 Complexes with Halomethanes (CH_3F, CH_3Cl and CH_3Br)

In this section, we consider the complexes between CH_3X (X = F/Cl/Br) molecules with the same 11 carbon bond acceptor molecules considered in the preceding section. The halogen atoms are more electron withdrawing than the OH group and therefore the sigma hole on the CH_3 face of halomethane molecules is more dominant in comparison to that of methanol. The ESP surface for CH_3F molecule is shown in Fig. 11.5. The sigma hole at the face centre is clearly visible in this case. The ESP values corresponding to maxima at the CH_3 face centres are +97.8, 82.3 and 78.6 kJ.mol^{-1} for CH_3F, CH_3Cl and CH_3Br respectively. These values are in accordance to the electronegativity order for F, Cl and Br.

Complexes of halomethanes were optimized following the methods as employed for the methanol complexes. Similar geometries, suggesting interaction between the atom with lone pair(s) of electrons and CH_3 face of halomethane molecules, were obtained on optimization. Binding energies for these complexes, obtained from different levels of theories are collected in Table 11.4. The values show that these complexes are stronger compared to the corresponding methanol complexes. Among

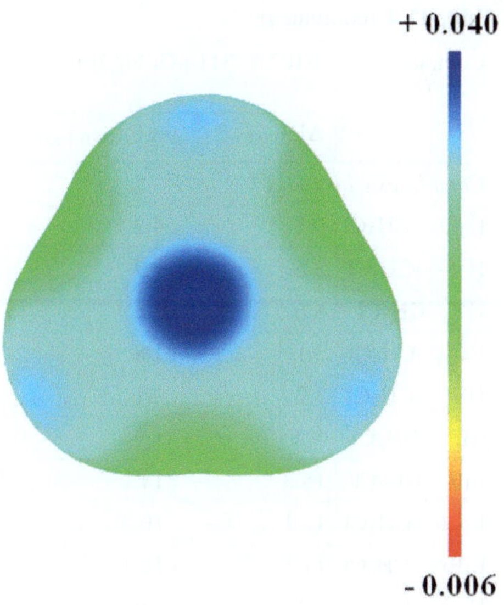

Fig. 11.5 Electrostatic potential surface for CH_3F molecule mapped on the 0.001 a.u. electron density surface

the halomethanes, the complexes with CH_3F are the strongest. The general trend of binding energy for a given acceptor is $CH_3F > CH_3Cl > CH_3Br$. For a particular halomethane, the complexes with LiY (Y = F/Cl/Br) molecules are the strongest ones as was the case with methanol complexes.

Table 11.4 Binding energies for the halomethane complexes calculated at different levels of theory. The BSSE corrected, $\Delta E_{(BSSE)}$, and BSSE as well as zero-point corrected, $\Delta E_{(BSSE+ZPC)}$ values are given. All the values are given in kJ.mol^{-1}

Complex	MP2 /6-311+G(3df,2p)		MP2 /Aug-cc-pVTZ		CCSD-T /6-311+G(3df,2p)
	$\Delta E_{(BSSE)}$	$\Delta E_{(BSSE+ZPC)}$	$\Delta E_{(BSSE)}$	$\Delta E_{(BSSE+ZPC)}$	$\Delta E_{(BSSE)}$
Complexes with CH_3F					
$H_2O\cdots CH_3F$	8.6	6.0	7.5	4.7	7.3
$H_2S\cdots CH_3F$	3.7	1.6	4.8	2.8	3.8
$HF\cdots CH_3F$	4.7	2.7	5.1	3.0	5.0
$HCl\cdots CH_3F$	5.0	3.5	3.7	2.2	2.8
$HBr\cdots CH_3F$	4.1	2.8	3.6	2.0	3.2
$ClF\cdots CH_3F$	2.6	1.0	3.7	2.2	3.1
$LiF\cdots CH_3F$	15.9	14.0	16.8	15.0	16.7
$LiCl\cdots CH_3F$	11.6	10.1	12.6	11.3	11.3
$LiBr\cdots CH_3F$	10.4	9.3	11.3	10.2	10.0
$H_3N\cdots CH_3F$	7.6	4.6	8.2	5.2	7.9
$H_3P\cdots CH_3F$	4.1	2.5	4.8	3.1	3.8

Table 11.4 (continued)

Complex	MP2 /6-311+G(3df,2p)		MP2 /Aug-cc-pVTZ		CCSD-T /6-311+G(3df,2p)
	$\Delta E_{(BSSE)}$	$\Delta E_{(BSSE+ZPC)}$	$\Delta E_{(BSSE)}$	$\Delta E_{(BSSE+ZPC)}$	$\Delta E_{(BSSE)}$
Complexes with CH_3Cl					
$H_2O\cdots CH_3Cl$	6.7	4.2	7.1	4.8	6.8
$H_2S\cdots CH_3Cl$	3.6	1.9	4.7	3.1	3.6
$HF\cdots CH_3Cl$	4.6	2.6	4.9	3.2	4.8
$HCl\cdots CH_3Cl$	3.1	1.8	3.7	2.4	2.7
$HBr\cdots CH_3Cl$	3.2	2.1	3.7	2.4	3.1
$ClF\cdots CH_3Cl$	3.2	1.9	3.8	2.5	3.0
$LiF\cdots CH_3Cl$	15.8	14.2	16.6	15.2	16.5
$LiCl\cdots CH_3Cl$	12.1	10.9	13.1	12.1	11.5
$LiBr\cdots CH_3Cl$	11.1	10.1	11.9	11.0	10.4
$H_3N\cdots CH_3Cl$	7.1	4.6	7.5	5.1	7.3
$H_3P\cdots CH_3Cl$	4.1	2.7	4.7	3.5	3.7
Complexes with CH_3Br					
$H_2O\cdots CH_3Br$	6.4	4.2	6.8	4.6	6.4
$H_2S\cdots CH_3Br$	3.5	1.5	4.6	3.1	3.3
$HF\cdots CH_3Br$	4.4	2.9	4.8	3.0	4.5
$HCl\cdots CH_3Br$	3.1	2.0	3.7	2.7	2.5
$HBr\cdots CH_3Br$	3.2	2.3	3.7	2.5	2.1
$ClF\cdots CH_3Br$	3.2	2.1	3.8	2.6	2.7
$LiF\cdots CH_3Br$	15.4	14.0	16.2	14.7	16.0
$LiCl\cdots CH_3Br$	12.0	10.9	13.0	12.2	11.3
$LiBr\cdots CH_3Br$	11.1	10.2	11.9	11.1	9.7
$H_3N\cdots CH_3Br$	6.8	4.4	7.2	5.0	6.8
$H_3P\cdots CH_3Br$	4.0	2.8	4.7	3.6	3.4

The AIM theoretical analysis was performed on each of these complexes as well. The analysis reveals the presence of an intermolecular bond critical point between the Y atom and C atom in all the complexes. Electron density (ρ), Laplacian of electron density ($\nabla^2\rho$) values, corresponding to intermolecular bond critical point connecting Y and C atoms, for all the complexes are given in Table 11.5. The table also lists values of mutual penetration between the Y and C atoms for each of the molecules. The positive mutual penetration values for all of these complexes confirm the intermolecular bonding.

Table 11.5 Electron density (ρ) and Laplacian of electron density ($\nabla^2\rho$) values at the intermolecular bond critical point for various complexes of methanol. Mutual penetration between the carbon and Y atoms are also given

Complex	ρ(r) (a.u.)	∇ (ρ) (a.u.)	Mutual penetration (Å)
Complexes with CH_3F			
$H_2O\cdots CH_3F$	0.0056	0.0307	0.50
$H_2S\cdots CH_3F$	0.0042	0.0184	0.45
$HF\cdots CH_3F$	0.0048	0.0298	0.38
$HCl\cdots CH_3F$	0.0041	0.0196	0.39
$HBr\cdots CH_3F$	0.0036	0.0164	0.32
$ClF\cdots CH_3F$	0.0051	0.0305	0.41
$LiF\cdots CH_3F$	0.0081	0.0459	0.67
$LiCl\cdots H_3F$	0.0058	0.0287	0.58
$LiBr\cdots CH_3F$	0.0050	0.0227	0.49
$H_3N\cdots H_3F$	0.0056	0.0262	0.55
$H_3P\cdots CH_3F$	0.0040	0.0163	0.42
Complexes with CH_3Cl			
$H_2O\cdots CH_3Cl$	0.0055	0.0282	0.52
$H_2S\cdots CH_3Cl$	0.0042	0.0172	0.46
$HF\cdots CH_3Cl$	0.0044	0.0260	0.37
$HCl\cdots CH_3Cl$	0.0041	0.0186	0.54
$HBr\cdots CH_3Cl$	0.0038	0.0157	0.36
$ClF\cdots CH_3Cl$	0.0049	0.0281	0.40
$LiF\cdots CH_3Cl$	0.0081	0.0442	0.69
$LiCl\cdots CH_3Cl$	0.0061	0.0287	0.46
$LiBr\cdots CH_3Cl$	0.0052	0.0227	0.53
$H_3N\cdots CH_3Cl$	0.0053	0.0232	0.53
$H_3P\cdots CH_3Cl$	0.0039	0.0150	0.40
Complexes with CH_3Br			
$H_2O\cdots CH_3Br$	0.0055	0.0264	0.51
$H_2S\cdots CH_3Br$	0.0042	0.0167	0.46
$HF\cdots CH_3Br$	0.0044	0.0252	0.39
$HCl\cdots CH_3Br$	0.0042	0.0185	0.43
$HBr\cdots CH_3Br$	0.0039	0.0153	0.37
$ClF\cdots CH_3Br$	0.0057	0.0321	0.41
$LiF\cdots CH_3Br$	0.0082	0.0435	0.70

Table 11.5 (continued)

Complex	ρ(r) (a.u.)	∇ (ρ) (a.u.)	Mutual penetration (Å)
$LiCl\cdots CH_3Br$	0.0062	0.0284	0.64
$LiBr\cdots H_3Br$	0.0054	0.0227	0.57
$H_3N\cdots H_3Br$	0.0052	0.0222	0.55
$H_3P\cdots CH_3Br$	0.0041	0.0148	0.43

11.4 Characteristics of Carbon bond

11.4.1 The AIM Descriptors for Carbon Bond

In the previous section, some of the most commonly used AIM descriptors for a bond, such as electron density at the BCP, its Laplacian and mutual penetration were presented along with the geometries and binding energies of the carbon bonded complexes with lone pair acceptors. In this section, we explore all the criteria based on AIM theoretical analysis given by Koch and Popelier [48] for the C-H$\cdots$O hydrogen bonds and discuss their application to carbon bonds. Complexes of CH_3OH and CH_3F were considered for this purpose and the results are summarized below:

1. The first criterion says that a bond critical point must be present between the two interacting atoms and also bond paths should connect the intermolecular bond critical point to both the interacting atoms. It is clear from the afore mentioned analysis that the CH_3OH as well as CH_3F complexes follow this criterion
2. The second criterion defines the range of acceptable values of electron density (ρ) at the bond critical point between the H and O atoms in case of C-H$\cdots$O hydrogen bonds. It suggests that the electron density value at the bond critical point should be within the range of 0.002–0.034 a.u. Note that small deviations are expected from this range due to the change in basis sets and level of theory used for the calculations. Values for the bond critical point corresponding to the carbon bond in CH_3OH as well as CH_3F complexes show that this criterion is followed by these complexes, Tables 11.3 and 11.5.
3. The third criterion defines the range for Laplacian of electron density ($\nabla^2\rho$) at the bond critical point between H and O atoms for C-H$\cdots$O hydrogen bonds. The acceptable range was given to be 0.024–0.139 a.u. for such hydrogen bonds. Note that again small deviations from this range are expected depending on the basis sets used. The Laplacian values for the carbon bond complexes of CH_3OH and CH_3F do lie within this range, see 11.3 and 11.5.
4. The fourth criterion says that mutual penetration between the two interacting atoms should be positive. This criterion is considered to be the necessary and sufficient criterion for intermolecular bonding. As mentioned before, the mutual penetration between the Y atom and C atom is positive for both the CH_3OH as well as CH_3F complexes. Thus, this most important criterion is followed by these complexes.

5. The criteria five to eight deal with the change in property of the hydrogen atom taking part in hydrogen bonding. The fifth criterion states that for hydrogen bonding, there should be a loss of charge on the hydrogen atom upon complex formation. This is calculated as the difference between the population on the hydrogen atom in monomer, N_H (monomer) and the population on the atom in the complex, N_H (complex) i.e. according to this criterion

$$\Delta N_H = N_H(complex) - N_H(monomer)\ should\ be\ negative$$

The corresponding ΔN_C values for the C atom were extracted from the AIM calculations and are complied in Table 11.6 for CH_3OH and CH_3F complexes. It is clear from the table that this criterion is followed by all the complexes except the $H_2O\cdots CH_3OH$ complex for which the ΔN_C value is nearly zero.

6. The sixth criterion deals with the energy of the hydrogen atom in the C-H···O hydrogen bonds and states that the hydrogen atom should be destabilized on complex formation i.e.

$$\Delta E_H = E_H(complex) - E_H(monomer)\ should\ be\ positive.$$

E_H (complex) and E_H (monomer) are the energies of the hydrogen atom in the hydrogen bonded complex and in monomer respectively. The ΔE values for the C atom (ΔE_C) extracted from the AIM calculations are given in Table 11.6 for the CH_3OH and the CH_3F complexes. This criterion is followed by all the complexes except the $H_2O\cdots CH_3OH$ and the $LiCl\cdots CH_3OH$ complexes for which ΔE value is negative.

7. The seventh criterion states that the magnitude of dipolar polarization (M) of hydrogen should decrease on complex formation i.e.

$$\Delta |M_H| = |M_H(complex)| - |M_H(monomer)|\ should\ be\ negative$$

M_H (complex) and M_H (monomer) are the dipolar polarization values of the hydrogen atom in the hydrogen bonded complex and in monomer respectively. The $\Delta|M_H|$ is the difference between the two. The $\Delta|M_C|$ values for the C atom are given in Table 11.6 for all the CH_3OH and CH_3F complexes. It is clear from the table that many of the CH_3OH complexes (with HF, HCl, HBr, ClF) do not follow this criterion and for them the $|\Delta M|$ values are positive. However, most of the CH_3F complexes follow this criterion except the $ClF\cdots CH_3F$ complex.

8. According to the eighth and final criterion, atomic volume of the H atom should decrease on complex formation i.e.

$$\Delta V_H = V_H(complex) - V_H(monomer)\ should\ be\ negative$$

Changes in the atomic volume of the carbon atom (ΔV_C) for all the complexes are given in Table 11.6. It can be noticed that all the complexes follow this criterion.

Table 11.6 Change in different properties of the carbon atom upon complex formation. Change in population, ΔN_C, energy ΔE_C, dipolar polarization $\Delta|M_C|$, and volume ΔV_C is given for the different complexes of CH_3OH and CH_3F. These properties were calculated using AIM theory. All the properties are given in atomic units. (Reproduced with permission from RSC, Reference 17)

CH_3OH complexes	ΔN_C	ΔE_C	$\Delta\|M_C\|$	ΔV_C	CH_3F complexes	ΔN_C	ΔE_C	$\Delta\|M_C\|$	ΔV_C
$H_2O\cdots CH_3OH$	0.002	−0.0270	−0.030	−3.44	$H_2O\cdots CH_3F$	−0.163	0.1906	−0.025	−8.92
$H_2S\cdots CH_3OH$	−0.204	0.2323	−0.011	−8.79	$H_2S\cdots CH_3F$	−0.163	0.2309	−0.029	−7.96
$HF\cdots CH_3OH$	−0.204	0.2323	0.006	−9.09	$HF\cdots CH_3F$	−0.164	0.2026	−0.006	−8.62
$HCl\cdots CH_3OH$	−0.204	0.2046	0.004	−8.22	$HCl\cdots CH_3F$	−0.164	0.2366	−0.009	−8.04
$HBr\cdots CH_3OH$	−0.205	0.2370	0.000	−8.56	$HBr\cdots CH_3F$	−0.163	0.2279	−0.013	−7.59
$ClF\cdots CH_3OH$	−0.205	0.2249	0.018	−9.89	$ClF\cdots CH_3F$	−0.168	0.2303	0.007	−9.37
$LiF\cdots CH_3OH$	−0.193	0.1919	−0.046	−9.74	$LiF\cdots CH_3F$	−0.168	0.1866	−0.036	−9.41
$LiCl\cdots CH_3OH$	−0.190	−0.4515	−0.040	−9.00	$LiCl\cdots CH_3F$	−0.151	0.2156	−0.050	−8.35
$LiBr\cdots CH_3OH$	−0.193	0.2259	−0.038	−9.16	$LiBr\cdots CH_3F$	−0.148	0.2184	−0.054	−8.11
$H_3N\cdots CH_3OH$	−0.203	0.1937	−0.020	−9.24	$H_3N\cdots CH_3F$	−0.162	0.1848	−0.043	−8.64
$H_3P\cdots CH_3OH$	−0.202	0.2325	−0.004	−7.57	$H_3P\cdots CH_3F$	−0.162	0.2315	−0.033	−7.56

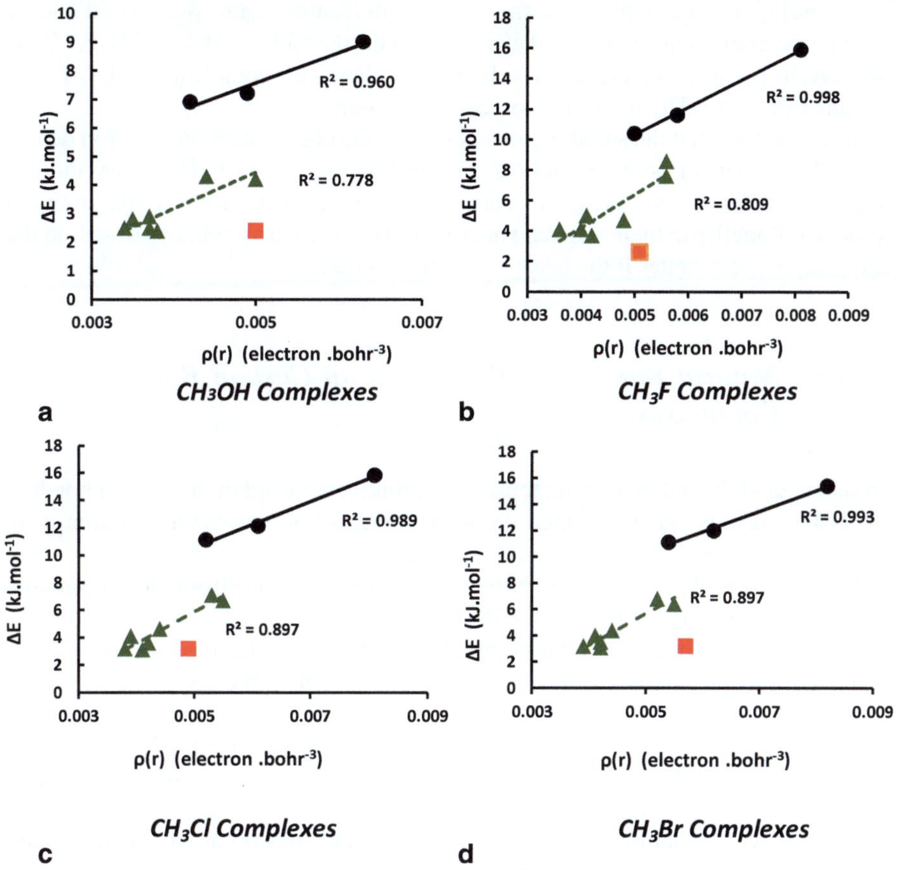

Fig. 11.6 Plot of stabilization energy, ΔE, against electron density, ρ, at the carbon bond critical point for **a** CH_3OH complexes, **b** CH_3F complexes, **c** CH_3Cl complexes and **d** CH_3Br complexes

11.4.2 Stabilization Energy vs. Electron Density

In the case of hydrogen bonding, there is a linear correlation between stabilization energy due to the hydrogen bonding and electron density at the bond critical point corresponding to the hydrogen bond. Are these two quantities correlated for the carbon bonds as well?

Plots between the stabilization energy and electron density at the intermolecular bond critical point are shown in Fig. 11.6 for the carbon bonded (a) CH_3OH, (b) CH_3F, (c) CH_3Cl and (d) CH_3Br complexes. In each of the correlation plots, two different fits are shown. The upper fits show the correlation for complexes with LiY (Y = F/Cl/Br) molecules. The lower fit shows the correlation for all other complexes except those with ClF molecule. In all the four cases, the stabilization energy is almost linearly correlated with the electron density for complexes with LiY molecules.

The correlation for complexes with the other molecules is also quite good and the correlation coefficient is 0.778, 0.809, 0.897 and 0.897 for CH_3OH CH_3F, CH_3Cl and CH_3Br complexes, respectively. In none of the cases, complexes with ClF were included in the fit. The fits were worsening in doing so.

It can be seen that the stabilization energy does correlate with the electron density at the bond critical point in case of carbon bonding as well. However, the correlation depends on the nature of carbon bond acceptor molecules. We note that Koch and Popelier in their original paper on C-H· · ·O hydrogen bonds observed that correlations were better if the acceptors were similar.

11.4.3 Natural Bond Orbital Analysis on Carbon Bonded Complexes

To understand the nature of interaction and orbitals involved in the carbon bonding, the NBO analysis was performed on all the complexes under consideration using wavefunctions for the optimized geometries at MP2/6-311+G(3df,2p) level as well as MP2/Aug-cc-pVTZ level. The NBO Charge transfer (ΔQ) from the carbon bond acceptor unit (Y) to the carbon bond donor unit (CH_3X) was calculated from the natural population analysis and is complied in Table 11.7. The negative value of ΔQ represents that the charge is transferred to the CH_3X unit which is the case in all the complexes.

The table also contains second order perturbation energies (E^2) resulting from the second order perturbation analysis of Fock matrix in the NBO basis at HF level of theory. The analysis shows a stabilization due to interaction between the lone pair orbital of Y with the anti-bonding σ^*(C-O/C-F/C-Cl/C-Br) orbital. In few cases, more than one lone pair of electrons contribute towards the stabilization and in such cases the reported value is sum of the energies from individual contributions, e.g. in case of LiBr· · ·CH_3F complex all three lone-pairs of Br interact with σ^*(C-F) orbital and the second order perturbation energies are 2.3, 1.8 and 0.8 kJ/mol and the reported value 4.9 kJ/mol is the sum of these three contributions. It is clear that the stabilization is due to the charge transfer from the lone pair orbital(s) of Y to the C-X anti-bonding orbital. It is similar to the lone-pair to anti-bonding (H-X) charge transfer found in the case of hydrogen bonding. The charge transfer from the lone-pair(s) of Y to the C-X anti-bonding orbital of CH_3X is also reflected in C-X vibrational frequency shift after complex formation as discussed below.

11.4.4 Red Shift in the C-X Vibrational Frequency

It is well known that in the case of hydrogen bonding generally there is a red shift in H-X stretching frequency. However, cases with blue shift hydrogen bonds [49] and 'no' shift hydrogen bonds [50] have also been reported.

Table 11.7 Charge transfer (ΔQ) from the carbon bond acceptor molecule to the donor CH_3X molecule. Second order perturbation energies for the intermolecular lone-pair and antibonding C-X orbital interactions are also given. Charges are given in atomic units and energies in kJ.mol^{-1}

Complex	MP2/6-311+G(3df,2p)		MP2/Aug-cc-pVTZ	
	ΔQ	E^2[l.p. → σ* (C-X)]	ΔQ	E^2[l.p. → s* (C-X)]
CH_3OH complexes				
$H_2O\cdots CH_3OH$	– 0.0014	2.6	– 0.0021	2.8
$H_2S\cdots CH_3OH$	– 0.0011	2.4	– 0.0016	3.0
$HF\cdots CH_3OH$	– 0.0010	1.5	– 0.0016	1.6
$HCl\cdots CH_3OH$	– 0.0007	2.0	– 0.0013	2.3
$HBr\cdots CH_3OH$	– 0.0007	2.0	– 0.0018	3.3
$ClF\cdots CH_3OH$	– 0.0018	2.3	– 0.0027	2.2
$LiF\cdots CH_3OH$	– 0.0022	3.6	– 0.0027	3.6
$LiCl\cdots CH_3OH$	– 0.0021	3.0	– 0.0036	3.1
$LiBr\cdots CH_3OH$	– 0.0019	2.9	– 0.0040	3.9
$H_3N\cdots CH_3OH$	– 0.0017	3.0	– 0.0019	2.8
$H_3P\cdots CH_3OH$	– 0.0009	1.9	– 0.0006	1.8
CH_3F complexes				
$H_2O\cdots CH_3F$	– 0.0020	3.6	– 0.0026	3.6
$H_2S\cdots CH_3F$	– 0.0017	3.3	– 0.0023	4.0
$HF\cdots CH_3F$	– 0.0014	2.3	– 0.0018	2.5
$HCl\cdots CH_3F$	– 0.0009	2.8	– 0.0015	3.1
$HBr\cdots CH_3F$	– 0.0010	2.5	– 0.0022	4.1
$ClF\cdots CH_3F$	– 0.0020	3.3	– 0.0026	2.9
$LiF\cdots CH_3F$	– 0.0035	5.9	– 0.0038	5.6
$LiCl\cdots CH_3F$	– 0.0029	4.1	– 0.0029	4.1
$LiBr\cdots CH_3F$	– 0.0033	3.6	– 0.0050	4.9
$H_3N\cdots CH_3F$	– 0.0027	5.1	– 0.0029	4.9
$H_3P\cdots CH_3F$	– 0.0016	2.8	– 0.0020	3.5
CH_3Cl complexes				
$H_2O\cdots CH_3Cl$	– 0.0016	2.5	– 0.0031	2.9
$H_2S\cdots CH_3Cl$	– 0.0011	2.6	– 0.0029	3.3
$HF\cdots CH_3Cl$	– 0.0015	1.5	– 0.0033	2.3
$HCl\cdots CH_3Cl$	– 0.0007	2.1	– 0.0027	2.8
$HBr\cdots CH_3Cl$	– 0.0006	2.0	– 0.0035	3.8
$ClF\cdots CH_3Cl$	– 0.0027	2.3	– 0.0066	3.1
$LiF\cdots CH_3Cl$	– 0.0027	4.3	– 0.0046	5.0

Table 11.7 (continued)

Complex	MP2/6-311+G(3df,2p)		MP2/Aug-cc-pVTZ	
	ΔQ	E^2[l.p. → σ* (C-X)]	ΔQ	E^2[l.p. → s* (C-X)]
$LiCl\cdots CH_3Cl$	– 0.0024	3.1	– 0.0061	4.0
$LiBr\cdots CH_3Cl$	– 0.0026	2.5	– 0.0073	4.0
$H_3N\cdots CH_3Cl$	– 0.0023	3.6	– 0.0032	4.3
$H_3P\cdots CH_3Cl$	– 0.0009	1.9	– 0.0024	2.6
CH_3Br complexes				
$H_2O\cdots CH_3Br$	– 0.0017	2.7	– 0.0034	3.1
$H_2S\cdots CH_3Br$	– 0.0016	3.0	– 0.0037	3.9
$HF\cdots CH_3Br$	– 0.0017	1.3	– 0.0032	1.6
$HCl\cdots CH_3Br$	– 0.0009	2.3	– 0.0030	2.8
$HBr\cdots CH_3Br$	– 0.0008	2.2	– 0.0039	4.0
$ClF\cdots CH_3Br$	– 0.0028	2.5	– 0.0068	2.6
$LiF\cdots CH_3Br$	– 0.0031	4.8	– 0.0051	5.6
$LiCl\cdots CH_3Br$	– 0.0029	3.4	– 0.0064	4.0
$LiBr\cdots CH_3Br$	– 0.0031	2.7	– 0.0079	4.1
$H_3N\cdots CH_3Br$	– 0.0026	3.9	– 0.0035	4.4
$H_3P\cdots CH_3Br$	– 0.0014	2.2	– 0.0029	2.9

To know the effect on C-X stretch frequency due to carbon bonding, frequency change corresponding to C-X stretching motion was calculated for all the complexes from the normal mode analysis. The frequency change (Δν) values at both the level of theories are given in Table 11.8. Negative value of Δν implies that the stretching frequency decreases on complex formation. In case of CH_3F, CH_3Cl and CH_3Br molecules, the C-X stretching mode is a local mode of vibration. On the other hand, in CH_3OH molecule, no mode, which is a pure C-O stretching mode, exists. However, there are two normal modes with significant contribution from C-O stretching. Frequency shifts on complex formation for both these modes were very similar and one of these values is reported in the Table 11.8. In all the cases, the Δν value is negative which shows that carbon bonding red shifts the C-X stretching frequency. This observation is again similar to hydrogen bonding. The frequency shift is largest in case of complexes with LiY which is in accordance with the maximum stabilization observed in these complexes.

11.4.5 Correlations Between the Frequency Shift and NBO Charge Transfer/Second Order Perturbation Energy

Hyperconjugation is considered to be one of the major reasons for the observed red shift in the X-H stretching frequency upon hydrogen bond formation. The molecular

Table 11.8 Frequency shifts, $\Delta\nu$, upon complex formation for CH_3OH, CH_3F, CH_3Cl and CH_3Br complexes. Values from both, MP2/6-311+G(3df,2p) and MP2/Aug-cc-pVTZ, levels are given

Complex	$\Delta\nu/cm^{-1}$		Complex	$\Delta\nu/cm^{-1}$	
	MP2 /6-311+G(3df,2p)	MP2 /Aug-cc-pVTZ		MP2 /6-311+G(3df,2p)	MP2 /Aug-cc-pVTZ
CH_3OH complexes			CH_3F complexes		
$H_2O\cdots CH_3OH$	– 9.1	– 7.8	$H_2O\cdots CH_3F$	– 16.3	– 14.8
$H_2S\cdots CH_3OH$	– 5.5	– 5.2	$H_2S\cdots CH_3F$	– 10.4	– 11.0
$HF\cdots CH_3OH$	– 6.3	– 6.5	$HF\cdots CH_3F$	– 10.2	– 7.8
$HCl\cdots CH_3OH$	– 4.0	– 3.2	$HCl\cdots CH_3F$	– 6.5	– 4.3
$HBr\cdots CH_3OH$	– 3.3	– 4.1	$HBr\cdots CH_3F$	– 6.3	– 5.7
$ClF\cdots CH_3OH$	– 2.8	– 3.1	$ClF\cdots CH_3F$	– 4.9	– 3.1
$LiF\cdots CH_3OH$	– 20.9	– 20.3	$LiF\cdots CH_3F$	– 40.4	– 39.0
$LiCl\cdots CH_3OH$	– 14.8	– 14.5	$LiCl\cdots CH_3F$	– 29.1	– 26.3
$LiBr\cdots CH_3OH$	– 12.5	– 14.3	$LiBr\cdots CH_3F$	– 24.3	– 24.8
$H_3N\cdots CH_3OH$	– 18.8	– 9.9	$H_3N\cdots CH_3F$	– 22.9	– 20.3
$H_3P\cdots CH_3OH$	– 3.9	– 1.0	$H_3P\cdots CH_3F$	– 9.9	– 8.9

Table 11.8 (continued)

Complex	$\Delta\nu/cm^{-1}$		Complex	$\Delta\nu/cm^{-1}$	
	MP2 /6-311+G(3df,2p)	MP2 /Aug-cc-pVTZ		MP2 /6-311+G(3df,2p)	MP2 /Aug-cc-pVTZ
CH_3Cl complexes			*CH_3Br complexes*		
$H_2O\cdots CH_3Cl$	− 9.2	− 9.9	$H_2O\cdots CH_3Br$	− 6.6	− 6.8
$H_2S\cdots CH_3Cl$	− 6.3	− 7.0	$H_2S\cdots CH_3Br$	− 5.5	− 5.5
$HF\cdots CH_3Cl$	− 5.3	− 5.6	$HF\cdots CH_3Br$	− 3.4	− 3.0
$HCl\cdots CH_3Cl$	− 2.4	− 2.8	$HCl\cdots CH_3Br$	− 2.0	− 1.7
$HBr\cdots CH_3Cl$	− 3.4	− 3.9	$HBr\cdots CH_3Br$	− 2.4	− 2.6
$ClF\cdots CH_3Cl$	− 1.3	− 1.3	$ClF\cdots CH_3Br$	− 0.2	0.3
$LiF\cdots CH_3Cl$	− 27.9	− 27.8	$LiF\cdots CH_3Br$	− 19.8	− 21.3
$LiCl\cdots CH_3Cl$	− 18.7	− 18.7	$LiCl\cdots CH_3Br$	− 13.6	− 13.9
$LiBr\cdots CH_3Cl$	− 16.6	− 17.6	$LiBr\cdots CH_3Br$	− 11.5	− 13.4
$H_3N\cdots CH_3Cl$	− 14.1	− 14.9	$H_3N\cdots CH_3Br$	− 10.5	− 11.5
$H_3P\cdots CH_3Cl$	− 6.4	− 7.1	$H_3P\cdots CH_3Br$	− 5.3	− 5.6

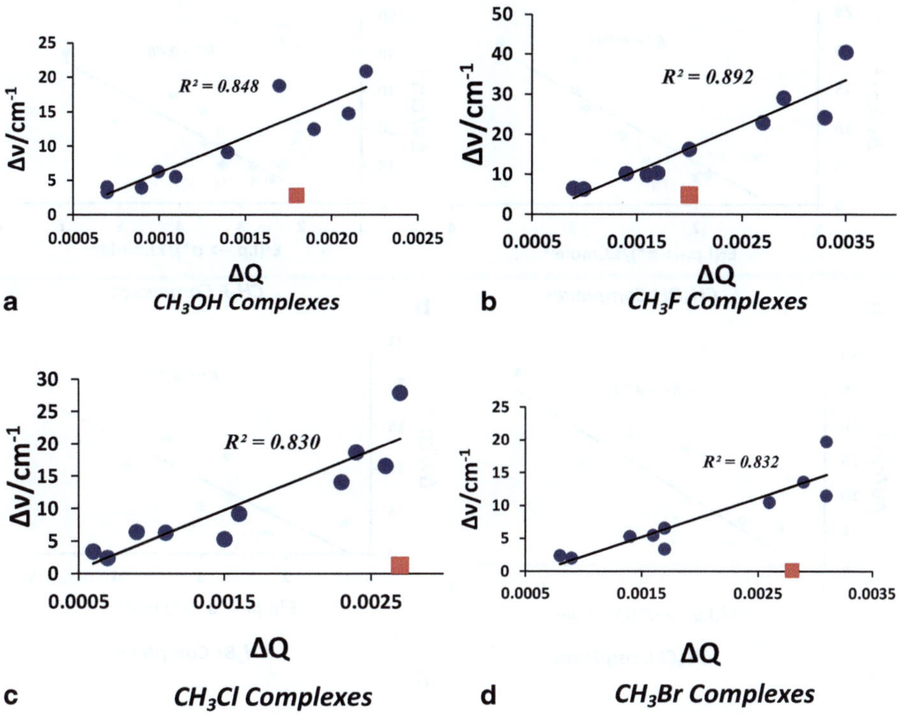

Fig. 11.7 Plots between NBO charge transfer (ΔQ) and C-X frequency shift (Δν) for the **a** CH_3OH, **b** CH_3F, **c** CH_3Cl and **d** CH_3Br complexes. Data points for ClF···CH_3X complexes are marked with brown squares and these were not included in the fit

orbitals involved in the hyperconjugation are usually lone pair orbitals of the hydrogen bond acceptor and the sigma anti-bonding orbital of the hydrogen bond donor [σ*(H-X)]. The electron transfer from the lone pair orbital(s) to the anti-bonding orbital results in the weakening of the H-X sigma bond and thus, results in the decrease of the stretching frequency. It is clear from the NBO data given in Table 11.7, the orbitals involved in the X-C···Y carbon bond interactions are l.p.(s) of Y and anti-bonding C-X sigma orbital (σ*(C-X)). Correlations between (i) total charge transfer from acceptor AY moiety to CH_3X moiety (ΔQ) and C-X frequency shift (Δν) and (ii) second order perturbation energy for the l.p.(s) – σ*(C-X) orbital interaction and the C-X frequency shift (Δν), are discussed below.

11.4.5.1 Charge Transfer (ΔQ) vs. C-X Stretching Frequency Shift (Δν)

The C-X stretching frequency shifts were plotted against the charge transfer (ΔQ) obtained from the NBO analysis. The plots are shown in Figs. 11.7a–11.7d for the different CH_3X (X = OH/F/Cl/Br) complexes. Correlation coefficients resulting from the linear fits are also given in the plots. In none of the plots, data for the

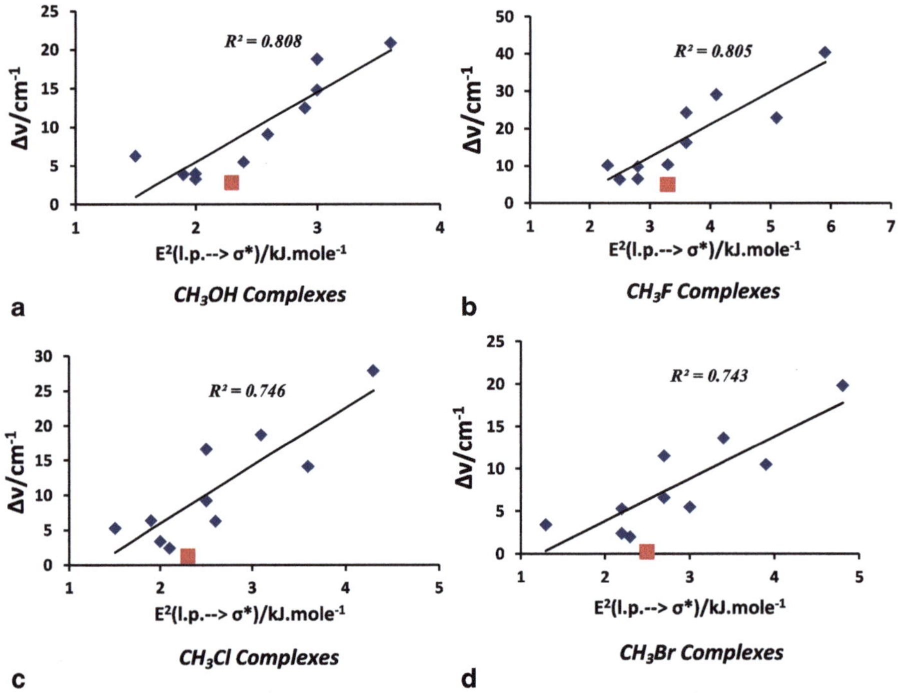

Fig. 11.8 Plots between second order perturbation energy, E^2(l.p. → σ*), and C-X frequency shift (Δν) for the **a** CH_3OH, **b** CH_3F, **c** CH_3Cl and **d** CH_3Br complexes. Data points for the ClF···CH_3X complexes are marked with brown squares and these were not included in the fit

ClF···CH_3X complexes was included since the frequency shift for these complexes are very small and the fit was worsening on including the data. In all the cases, correlation coefficient is more than 0.8 which shows the correlation between these two properties is reasonable.

11.4.5.2 Perturbation Energy vs. C-X Stretching Frequency Shift

Other set of correlation diagrams were plotted between the second order perturbation energy, E^2(l.p. → σ*(C-X)), and the C-X stretch frequency shift (Δν). The plots are shown in Fig. 11.8a–11.8d for the different CH_3X (X = OH/F/Cl/Br) complexes. Correlation coefficients are also given in each case. Again the data points for the ClF complexes were not included in the fit. The correlation coefficient for the CH_3OH and the CH_3F complexes is nearly 0.8 but is poorer for the CH_3Cl and the CH_3Br complexes, ~0.74.

11.4.6 The ESP Value at CH_3 Face Centre vs. Binding Energy

In this section, we discuss the correlation between the ESP values at CH_3 face centre of the CH_3X molecules and binding energies, keeping the carbon bond acceptor same. Values at the MP2/6-311+G(3df,2p) level were considered for these correlations. The correlation plots are given in Fig. 11.9a–11.9c. Figure 11.9a shows correlation fits for the $H_2O\cdots CH_3X$ (X = OH/F/Cl/Br) and $H_2S\cdots CH_3X$ (X = OH/F/Cl/Br) complexes. For both the sets the correlation is quite good and the correlation coefficients are 0.980 and 0.914, respectively. Figure 11.9b shows the correlation for the $HF\cdots CH_3X$ (X = OH/F/Cl/Br), $HCl\cdots CH_3X$ (X = OH/F/Cl/Br) and $HBr\cdots CH_3X$ (X = OH/F/Cl/Br) complexes. The fits are reasonable in the case of HF and HBr complexes, correlation coefficients being 0.905 and 0.920 respectively. However, in the case of the HCl complexes the correlation is poor and the correlation coefficient is 0.762. Figure 11.9c shows the correlation for the $LiF\cdots CH_3X$ (X = OH/F/Cl/Br), $LiCl\cdots CH_3X$ (X = OH/F/Cl/Br) and $LiBr\cdots CH_3X$ (X = OH/F/Cl/Br) complexes. The correlation coefficients in these cases are not very good and these are 0.860, 0.754 and 0.685, respectively. Figure 11.9d shows the correlation plots for the $H_3N\cdots CH_3X$ (X = OH/F/Cl/Br) and $H_3P\cdots CH_3X$ (X = OH/F/Cl/Br) complexes. In these cases the correlation coefficients are 0.954 and 0.857, respectively.

In general, there is some correlation between the ESP value at the CH_3 face centre and the binding energy if the acceptor is kept fixed. The correlation coefficients are reasonable in some cases and not very satisfactory in others. These observations indicate that the electrostatics does play an important role in carbon bond but it is not all.

It is clear from the above discussion that carbon bonds follow most of the established criteria for intermolecular bonding. Various properties show expected correlation in these complexes.

In the following sections we extend our discussion to carbon bonding with other types of acceptor molecules.

11.5 Other Acceptors/Donors for Carbon Bonds

11.5.1 Radicals as Carbon Bond Acceptor

Radicals can act as hydrogen bond acceptor and single electron hydrogen bonds [51, 52] have been reported in the literature. Li et al. [22] have recently shown that single electron carbon bonds also exist. They studied complexes of CH_3F, CH_3CN, CH_3NC and CH_3NO_2 with methyl radical and found a minimum geometry in which the methyl radical faces the CH_3 face of CH_3X molecule as shown in Fig. 11.10. The intermolecular $C\cdots C$ distance and binding energy for this complex were found to 3.486 Å and 2.8 $kJ.mol^{-1}$ respectively at UMP2/Aug-cc-pVTZ level. The AIM analysis confirms bonding between the two carbon atoms in this complex.

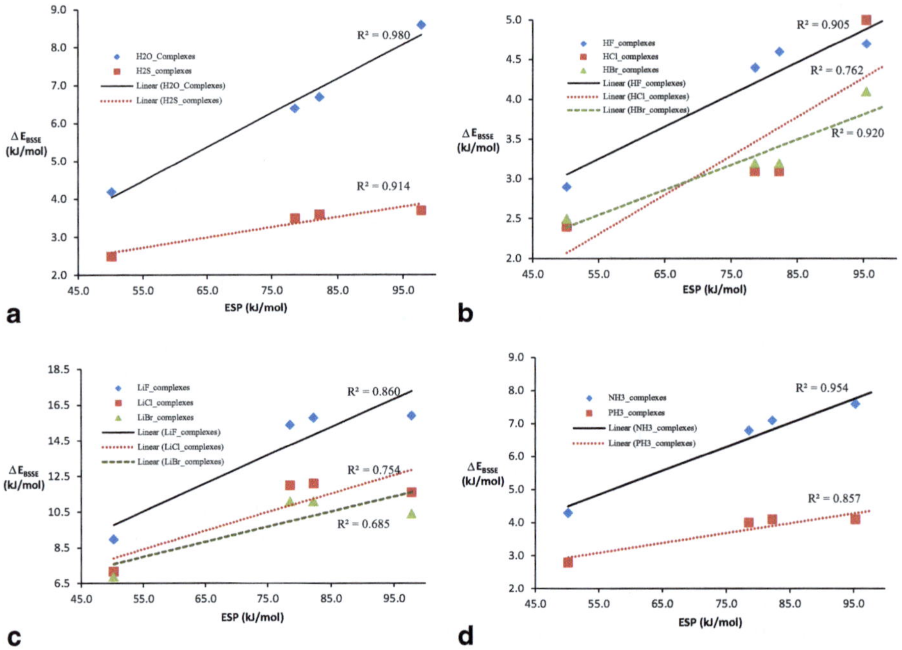

Fig. 11.9 **a** Correlation plot between ESP at the CH_3 face centre of CH_3X and binding energy for $H_2Y\cdots CH_3X$ (Y = O/S and X = OH/F/Cl/Br) complexes. **b** Correlation plot between ESP at the CH_3 face centre of CH_3X and binding energy for $HY\cdots CH_3X$ (Y = F/Cl/Br and X = OH/F/Cl/Br) complexes. **c** Correlation plot between ESP at the CH_3 face centre of CH_3X and binding energy for $LiY\cdots CH_3X$ (Y = F/Cl/Br and X = OH/F/Cl/Br) complexes. **d** Correlation plot between ESP at the CH_3 face centre of CH_3X and binding energy for $H_3Y\cdots CH_3X$ (Y = N/P and X = OH/F/Cl/Br) complexes

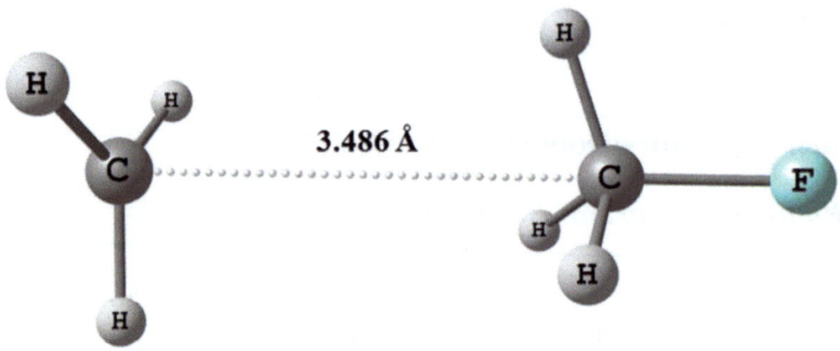

Fig. 11.10 Optimized geometry for $CH_3\cdots CH_3F$ complex at UMP2/Aug-cc-VTZ level (Reproduced with permission from RSC, reference 22)

It was found that subsequent substitutions of the hydrogen atoms of the acceptor methyl radical with electron donating CH_3 groups result in stronger bonding. The binding energies for the complexes of CH_3, CH_2CH_3, $CH(CH_3)_2$ and $C(CH3)_3$ radicals with CH_3F were found to be 2.8, 4.6, 6.1 and 7.3 kJ.mol^{-1} respectively.

11.5.2 Anions as Carbon Bond Acceptors

Grabowski considered complexes of carbon bond donors with atomic anions to explore S_N2 reactions [20]. McDowell and Joseph [21] studied these interactions in more detail. As expected the binding energies in these complexes were much higher than those for the complexes with the neutral species, e.g. binding energy for $F^{(-)} \cdots CH_3F$ complex was found to be 54.2 $kJ.mol^{-1}$ which is much higher in comparison to the values 5.1 and 16.8 $kJ.mol^{-1}$ for the $HF \cdots CH_3F$ and $LiF \cdots CH_3F$ complexes respectively. Interestingly, in case of carbon bonding with anions the interaction energy follows the pattern CH_3Br complexes > CH_3Cl complexes > CH_3F complexes. Note that this pattern is opposite to the pattern observed in case of carbon bonding with lone pair of electrons as acceptors. However, this pattern could be explained on the basis of the dipolar polarizability of the C-X bonds. The dipolar polarizability for these bonds decreases in the order C-Br > C-Cl > C-F. Thus, the polarization of the C-X bond by anion increases with increase in the size of the X atom, from F to Br. This results in stronger bonding.

More recently, Li et al. [31] have studied complexes of metal hydrides (LiH, BeH, NaH and MgH) with carbon bond acceptor CH_3F molecule as well as with other tetrel molecules. For these complexes, the binding energy increases in the order $BeH \cdots CH_3F < LiH \cdots CH_3F < NaH \cdots CH_3F$.

11.5.3 π-Electrons, σ-Electrons and Neutral Atoms as Carbon Bond Acceptor

The π-electrons are well known acceptors for hydrogen bonds [4–6]. In a recent publication [53], we have shown that the π-electrons can act as carbon bond acceptor too and lead to the formation of X-C··· π carbon bonds. Ethylene and acetylene were considered as the π-electron donors. Both the molecules, ethylene and acetylene, have no dipole moments and therefore the carbon bonded complexes with CH_3X were dominated by dispersion forces. All the complexes follow the AIM and NBO bonding criteria and also show red shifts in the C-X stretching frequencies on complexation. Complexes with molecules containing other tetrel atoms, Si and Ge, were also considered and it was found that the interaction energy increases in the order $CH_3X < SiH_3X < GeH_3X$, which is similar to that observed for other cases. However, electron density topology for most of the SiH_3X complexes was found to be very different than the CH_3X and GeH_3X complexes and no bond critical point was found connecting the π-electrons (C-atom) to the silicon atom.

Can σ-electrons and neutral atoms also act as the carbon bond acceptors? This was explored considering complex formation between H_2 molecule and Ar atom with CH_3F molecule. The $H_2 \cdots CH_3F$ complex was optimised taking initial geometry in which the σ-electrons of H_2 bond critical point towards the CH_3 face of CH_3F. Similarly, the $Ar \cdots CH_3F$ complex was optimised taking the initial geometry in which

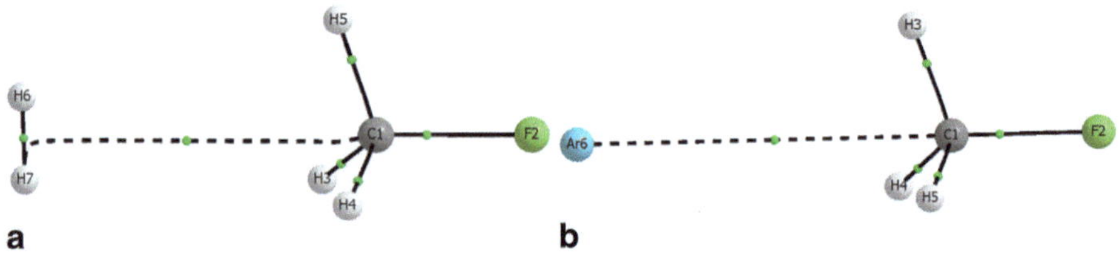

Fig. 11.11 Electron density topologies for the **a** H2· · ·CH_3F and **b** Ar· · ·CH_3F complexes

Ar atom points towards the CH_3 face of the CH_3F molecule. Geometry optimisation was performed at the MP2/Aug-cc-pVTZ level. The BSSE corrected binding energies for the H_2· · ·CH_3F and Ar· · ·CH_3F complexes are 1.3 and 1.9 kJ.mol^{-1},respectively. The AIM analysis was performed on both the complexes and the resulting electron density topologies are shown in Fig. 11.11a and 11.11b. The topologies confirm interaction between the σ-electrons (H-atom)/Ar and carbon atom in the H_2· · ·CH_3F and Ar· · ·CH_3F complexes, respectively. The (ρ, $\nabla^2\rho$) value at the intermolecular bond critical point are (0.0030, +0.0145) for the H_2· · ·CH_3F complex and (0.0037, +0.0190) for the Ar· · ·CH_3F complex.

11.5.4 Carbon Bonding in Cycloalkanes

Perfluoro cycloalkanes (C_n^F) and percyano cycloalkanes (C_n^{CN}) have been shown to be potential carbon bond donors by Bauzá et al. [27, 28]. Large cycloalkanes, like perfluorocyclohexane (C_6^F), possess very strong sigma hole at the centre of the ring in their planar conformation. It is similar to the sigma hole found in case of molecules like perfluorobenzene (C_6F_6). However, the planar conformations are not the most stable ones for the large ring cycloalkanes and the interaction with nucleophiles can not compensate for the conformational instability. The more stable non-planar conformers of these molecules, like chair form of perfluorocyclohexane, have much weaker sigma hole. Moreover, such sigma holes are not accessible to electron rich centres due to their proximity to the electron repelling F or CN groups. However, the smaller cycloalkane rings, like C3 and C4 are conformarily more rigid. Perfluro and percyano, cyclopropanes and cyclobutanes form multiple carbon bonds through the edges as well as ring faces.

11.6 Manifestations of Carbon Bonding

11.6.1 Carbon Bonding and Hydrophobic Interaction

Carbon bonding is very closely related to hydrophobic interactions. Let us consider methanol molecule as an example. The hydroxyl group of methanol is hydrophilic and attracts H_2O molecule to form a hydrogen bond. However, the methyl group in this molecule is hydrophobic in nature and any interaction such as the one shown in Fig. 11.3 will be considered as hydrophobic interaction. Hydrophobic effect is usually considered to be an entropic phenomenon, and the enthalpic contribution was considered as the generic van der Waals interaction. However, the discovery of carbon bond shows that there is a specific, small but non negligible enthalpic contribution from carbon bond to the hydrophobic effect.

11.6.2 Carbon Bonding and S_N2 Reactions

The carbon bonded complexes resemble the entrance channel complexes of an S_N2 reaction. We pointed out [17] that the carbon bonding could be the stabilizing factor of the intermediates to the S_N2 reactions. Grabowski [20] explored the role of carbon bonding in S_N2 reactions in detail. Carbon bonding interactions in which anions act as the acceptor are quite close to the S_N2 reaction intermediates. These interactions are quite strong as well, as in such complexes the CH_3 face is significantly deformed towards planarity. We recognize that the 'carbon bonded geometries' as stable minimum along the reaction coordinate for S_N2 reactions have been noted earlier by theoretical chemists exploring the potential surface and reaction dynamics of these reactions [54, 55]. However, their main focus was reaction dynamics and hence the similarity of these structures with that of hydrogen bonded complexes was not considered. The fact that S_N2 reactions happen offers the most convincing proof that there is electron communication between the Y and C atoms leading to the formation of the carbon bond. Such structures could have been misidentified in the past as 'trifurcated hydrogen bonds'.

11.6.3 Carbon Bonding in Biological Systems

Alkyl group bonded to electronegative atoms are present in abundance in the biological systems. Such abundance implies that the carbon bonding interactions could be present in these systems and could contribute to the functioning of the biological molecules. For example, in amino acids ($NH_2RCHCOOH$) the -RCH- group is sandwiched between two electron withdrawing groups, NH_2 and COOH. The middle carbon atom is positively charged and can participate in the carbon bonding interactions. This possibility was indeed checked in the glycine···FH complex. Calculations showed that the geometry in which fluorine of HF interacts with CH_2 carbon of glycine is a minimum. Alanine ($R = CH_3$) has a CH_3 group and thus is closer to the CH_3X molecules, already discussed. The ESP calculations show the presence of a maximum at the CH_3 face of alanine, with ESP value 19.6 kJ.mol^{-1}. It shows that the CH_3 face of alanine should form carbon bonds with molecules like HF, similar to the other CH_3X molecules. We optimized complex of alanine with HF at MP2/6-311+G(3df,2p) level of theory. The initial geometry was taken such that the F atom of HF faces the CH_3 face of alanine molecule. The optimized structure was confirmed to be a minimum by frequency calculation.

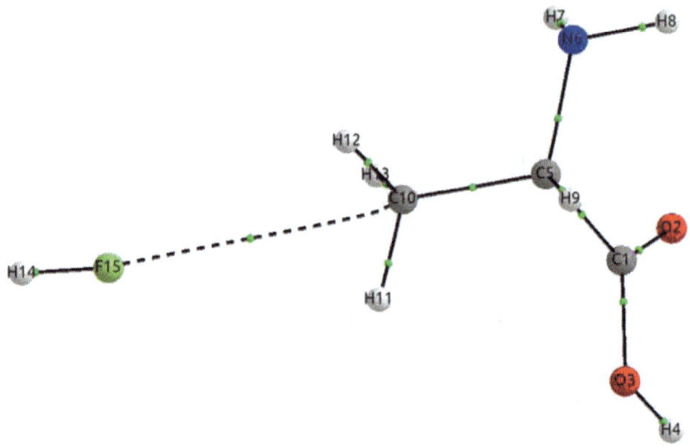

Fig. 11.12 X-C· · ·Y carbon bonding interaction in the HF· · ·alanine complex

The stabilization energy for this complex is 2.0 kJ.mol^{-1}. The AIM calculations show the presence of a BCP between F and C and a bond path connecting them, Fig. 11.12. The values of ρ(r) and $\nabla^2\rho(r)$ at the BCP are 0.0039 and +0.0205 a.u. respectively. This example shows that the X-C· · ·Y interactions could also be present in the biological systems and can play an important role in the biological processes such as protein folding.

Azofra et al. [26] have analysed carbohydrate molecules in gas phase using Density Functional Theory. They found intra-molecular O· · ·C-O carbon bonds in many of the structures, cyclic as well as open chain.

11.7 Tetrel Bond

Interactions similar to carbon bonding exist for all the elements of the IV group (Si, Ge, Sn). These interactions were named tetrel bond by Grabowski [20] and Bauzá et al. [24], simultaneously but independently. Tetrel bonds have been explored in detail now and like carbon bonds, tetrel bonds with radical [22] as well as anion acceptors [21, 23, 27, 29] have been reported in the literature. The interaction energy increases as one goes down the group from C to Sn. The reason follows: as the atomic size increases from C to Sn the atom becomes more polarizable and thus the sigma-hole becomes more prominent in the TH_3X (T = C, Si, Ge, Sn and X is an electron withdrawing atom or group) molecules. As an example, binding energies for the complexes of LiCN with CH_3F, SiH_3F and GeH_3F are 15.9, 32.2 and 35.9 kJ.mol^{-1}, respectively.

11.8 Experimental Evidence for Carbon Bonding

Following the first publication on carbon bond [17], Guru Row and co-workers carried out a systematic charge density analysis on specific crystals expected to show carbon bonding [32]. They chose two molecular systems in which carbon bonding

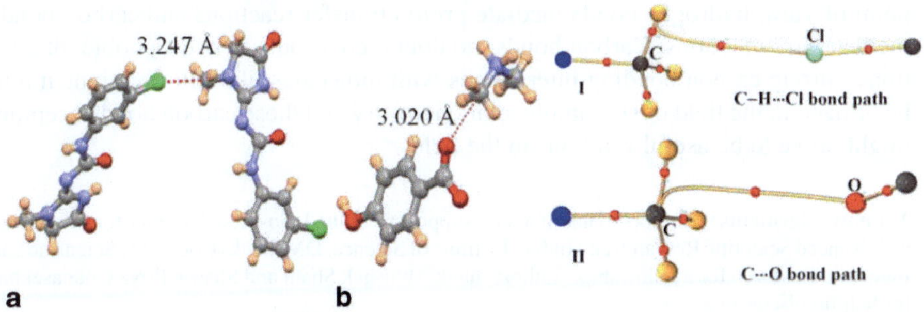

Fig. 11.13 Structure of **a** fenobam and **b** dimethylammonium-4-hydroxybenzoate, having potential for carbon bonds. Electron density analysis shows C-H· · ·Cl hydrogen bond in the former (**I**) and N-C· · ·O carbon bonding in the later (**II**). (Reproduced with permission from RSC, Reference 31)

could be found and solved their crystal structures, following it with charge density analysis. These were (i) the anxiolytic drug candidate fenobam (Fig. 11.13a) for a possible C· · ·Cl carbon bond, and (ii) dimethylammonium-4-hydroxybenzoate (Fig. 11.13b). Charge density analysis confirmed the N-C· · ·O carbon bond in the crystal of dimethylammonium-4-hydroxybenzoate (Fig. 11.13 **II**) but in the fenobam, it revealed the formation of C-H· · ·Cl hydrogen bond (Fig. 11.13 **I**). Indeed, experimental and theoretical charge density analyses are important in confirming the presence of carbon bond.

We note that a recent microwave spectroscopic investigation [56] on pyridine-CF_4 complex has led to a structure in which one of the tetrahedron face of CF_4 is pointing towards the N of pyridine. This could indeed be an example of an F-C· · ·N carbon bond. The difference in electron density distribution found between hydrocarbon and perfluorohydrocarbons have been noted for long. For example, benzene has an electron rich centre whereas the perfluorobenzene has its centre, electron depleted. This manifests in their interaction with other molecules. The H_2O interacts with benzene through O–H· · · π hydrogen bond(s) [57] but it interacts with C_6F_6 with the O in H_2O pointing towards the electron depleted π centre [58]. Indeed, this interaction would bear similarities to the 'carbon bond'. Bauzá et al. [59] have recently found similar interaction with $C_6H_5CF_3$ molecule using a combined CSD and ab intio analysis. The analysis showed many structures in which electron rich atoms interact with the carbon atom of CF_3 group through the face centre.

11.9 Conclusions

Carbon bond is an emerging field of research. Much is to be known about these interactions, especially in terms of their presence in nature as well as their utilization in synthetic approaches. Other tetrel bonds, with silicon and germanium, are much stronger than the carbon bonds. However, abundance of carbon in biological systems is likely to make carbon bonding as the most important tetrel bond. From a synthetic point of view, hydrogen bonds mediate proton transfer reactions and carbon bonds mediate S_N2 reactions. Carbon bonds, no doubt, contribute to hydrophobic interactions. Strong carbon bonding interactions with molecules like LiF or anions might be utilized in the field of supramolecular chemistry and these carbon bond acceptors might prove to be useful synthons in the field.

Acknowledgements Authors thank financial support from the Indo-French Center for Promotion of Advanced Scientific Research and Indian Institute of Science. DM thanks Council of Scientific and Industrial Research for a fellowship. Authors thank Abhishek Shahi and Sharon Priya Gnanasekhar for helpful discussions.

References

1. Hobza P, Müller-Dethlefs K (2010) Non-covalent interactions: theory and experiment. Royal Society of Chemistry, Cambridge
2. Karshikoff A (2006) Non-covalent interactions in proteins. World Scientific
3. Pauling L (1960) The nature of the chemical bond and the structure of molecules and crystals: an introduction to modern structural chemistry. Cornell University Press
4. Arunan E, Desiraju GR, Klein RA et al (2011) Definition of the hydrogen bond. Pure Appl Chem 83:1637–1641
5. Jeffrey GA, Saenger W (1994) Hydrogen bonding in biological structures. Springer-Verlag, Berlin
6. Desiraju GR, Steiner T (1999) The weak hydrogen bond in structural chemistry and biology. Oxford University Press, Oxford
7. Shahi A, Arunan E (2014) Hydrogen bonding, halogen bonding and lithium bonding: an atoms in molecules and natural bond orbital perspective towards conservation of total bond order, inter-and intra-molecular bonding. Phys Chem Chem Phys 16:22935–22952
8. Politzer P, Lane P, Concha M et al (2007) An overview of halogen bonding. J Mol Model 13:305–311
9. Wang W, Ji B, Zhang Y (2009) Chalcogen bond: a sister noncovalent bond to halogen bond. J Phys Chem A 113:8132–8135
10. Scheiner S (2011) A new noncovalent force: comparison of P [center-dot] [center-dot] [center-dot] N interaction with hydrogen and halogen bonds. J Chem Phys 134:094315–094319
11. Metrangolo P, Resnati G (eds) (2008) Halogen bonding: fundamentals and applications, of structure and bonding, vol 126. Springer, Berlin
12. Clark T, Hennemann M, Murray JS et al (2007) Halogen bonding: the σ-hole. J Mol Model 13:291–296
13. Murray J, Lane P, Politzer P (2009) Expansion of the σ-hole concept. J Mol Model 15:723–729
14. Politzer P, Murray JS (2013) Halogen bonding: an interim discussion. ChemPhysChem 14:278–294
15. Bundhun A, Ramasami P, Murray J et al (2013) Trends in σ-hole strengths and interactions of F_3MX molecules (M = C, Si, Ge and X = F, Cl, Br, I). J Mol Model 19:2739–2746
16. Mani D, Arunan E (2013) Microwave spectroscopic and atoms in molecules theoretical investigations on the Ar···Propargyl Alcohol Complex: Ar···H–O, Ar···π, and Ar···C interactions. ChemPhysChem 14:754–763
17. Mani D, Arunan E (2013) The X-CY (X = O/F, Y = O/S/F/Cl/Br/N/P) 'carbon bond' and hydrophobic interactions. Phys Chem Chem Phys 15:14377–14383

18. Farrugia LJ, Evans C, Tegel M (2006) Chemical bonds without "chemical bonding"? A combined experimental and theoretical charge density study on an iron trimethylenemethane complex. J Phys Chem A 110:7952–7961
19. Raghavendra B, Arunan E (2008) Hydrogen bonding with a hydrogen bond: the methane–water complex and the penta-coordinate carbon. Chem Phys Lett 467:37–40
20. Grabowski SJ (2014) Tetrel bond-σ-hole bond as a preliminary stage of the SN 2 reaction. Phys Chem Chem Phys 16:1824–1834
21. Mcdowell SA, Joseph JA (2014) The effect of atomic ions on model σ-hole bonded complexes of AH3Y (A = C, Si, Ge; Y = F, Cl, Br). Phys Chem Chem Phys 16:10854–10860
22. Li Q, Guo X, Yang X et al (2014) A σ-hole interaction with radical species as electron donors: does single-electron tetrel bonding exist? Phys Chem Chem Phys 16:11617–11625
23. Bauzá A, Ramis R, Frontera A (2014) Computational study of anion recognition based on tetrel and hydrogen bonding interaction by calix [4] pyrrole derivatives. Comput Theor Chem 1038:67–70
24. Bauzá A, Mooibroek TJ, Frontera A (2013) Tetrel-bonding interaction: rediscovered supramolecular force? Angew Chem Int Ed 52:12317–12321
25. Varadwaj PR, Varadwaj A, Jin B (2014) Significant evidence of CO and CC long-range contacts in several heterodimeric complexes of CO with CH_3-X, should one refer to them as carbon and dicarbon bonds! Phys Chem Chem Phys 16:17238–17252
26. Azofra LM, Quesada-Moreno MM, Alkorta I et al (2014) Carbohydrates in the gas phase: conformational preference of d-ribose and 2-deoxy-d-ribose. New J Chem 38:529–538
27. Bauza A, Mooibroek TJ, Frontera A (2014) Influence of ring size on the strength of carbon bonding complexes between anions and perfluorocycloalkanes. Phys Chem Chem Phys 16:19192–19197
28. Bauzá A, Mooibroek TJ, Frontera A (2014) Small cycloalkane (CN) 2C–C (CN) 2 structures are highly directional non-covalent carbon-bond donors. Chem-A Eur J 20:10245–10248
29. Mcdowell SA (2014) Sigma-hole cooperativity in anionic $[FX\cdots CH_3\cdots YF]^-$ (X, Y = Cl, Br) complexes. Chem Phys Lett 598:1–4
30. Arunan E (2013) Hydrogen bond seen, halogen bond defined and carbon bond proposed: intermolecular bonding, a field that is maturing! Curr Sci 105:892–894
31. Li Q-Z, Zhuo H-Y, Li H-B et al (2014) Tetrel–hydride interaction between XH_3F (X = C, Si, Ge, Sn) and HM (M = Li, Na, BeH, MgH). J Phys Chem A. doi:10.1021/jp503735u
32. Thomas SP, Pavan MS, Row TG (2014) Experimental evidence for 'carbon bonding'in the solid state from charge density analysis. Chem Commun 50:49–51
33. Politzer P, Murray JS, Clark T (2013) Halogen bonding and other [sigma]-hole interactions: a perspective. Phys Chem Chem Phys 15:11178–11189
34. Gadre SR, Shirsat RN (2000) Electrostatics of atoms and molecules. Universities Press, Hyderabad
35. Sjoberg P, Politzer P (1990) Use of the electrostatic potential at the molecular surface to interpret and predict nucleophilic processes. J Phys Chem 94:3959–3961
36. Suenram RD, Fraser GT, Lovas FJ et al (1994) The microwave spectrum of methane... water. J Chem Phys 101:7230–7240
37. Legon AC, Roberts BP, Wallwork AL (1990) Rotational spectra and geometries of the gas-phase dimers methane... HF and methane... HCl. Chem Phys Lett 173:107–114
38. Ohshima Y, Endo Y (1990) Rotational spectrum and internal rotation of a methane–HCl complex. J Chem Phys 93:6256–6265
39. Legon AC, Wallwork AL (1992) Methane as a proton acceptor: rotational spectrum and internal dynamics of a weakly bound dimer of methane and hydrogen cyanide. J Chem Soc Faraday Trans 88:1–9
40. Mani D (2013) Microwave spectroscopic and atoms in molecules theoretical investigations on weakly bound complexes: from Hydrogen Bond to 'Carbon Bond'. Ph.D. thesis. In: Department of Inorganic and Physical Chemistry Indian Institute of Science Bangalore

41. Frisch MJ, Trucks GW, Schlegel HB, Scuseria GE, Robb MA, Cheeseman JR, Scalmani G, Barone V, Mennucci B, Petersson GA, Nakatsuji H, Caricato M, Li X, Hratchian HP, Izmaylov AF, Bloino J, Zheng G, Sonnenberg JL, Hada M, Ehara M, Toyota K, Fukuda R, Hasegawa J, Ishida M, Nakajima T, Honda Y, Kitao O, Nakai H, Vreven T, Montgomery JA, Peralta JE, Ogliaro F, Bearpark M, Heyd JJ, Brothers E, Kudin KN, Staroverov VN, Kobayashi R, Normand J, Raghavachari K, Rendell A, Burant JC, Iyengar SS, Tomasi J, Cossi M, Rega N, Millam JM, Klene M, Knox JE, Cross JB, Bakken V, Adamo C, Jaramillo J, Gomperts R, Stratmann RE, Yazyev O, Austin AJ, Cammi R, Pomelli C, Ochterski JW, Martin RL, Morokuma K, Zakrzewski VG, Voth GA, Salvador P, Dannenberg JJ, Dapprich S, Daniels AD, Farkas, Foresman JB, Ortiz JV, Cioslowski J, Fox DJ (2009) Gaussian 09, Revision B.01. In: Wallingford CT
42. Boys SF, Bernardi F (1970) The calculation of small molecular interactions by the differences of separate total energies. Some procedures with reduced errors. Mol Phys 19:553
43. Dennington R, Keith T, Millam J (2009) GaussView, version 5. Semichem Inc., Shawnee Mission KS
44. Lu T, Chen F (2012) Multiwfn: a multifunctional wavefunction analyzer. J Comput Chem 33:580–592
45. Aimall K (Version 13.02.26), Todd A. AIMAll (Version 12.09.23). In: TK Gristmill Software, Overland Park KS, USA, 2012 (aim.tkgristmill.com)
46. Biegler-König F, Schönbohm J (2002) Update of the AIM2000-Program for atoms in molecules. J Comput Chem 23:1489–1494
47. Glendening ED, Landis CR, Weinhold F (2013) NBO 6.0: natural bond orbital analysis program. J Comput Chem 34:1429–1437
48. Koch U, Popelier PLA (1995) Characterization of C-H-O hydrogen bonds on the basis of the charge density. J Phys Chem 99:9747–9754
49. Hobza P, Havlas Z (2000) Blue-shifting hydrogen bonds. Chem Rev 100:4253–4264
50. Joseph J, Jemmis ED (2007) Red-, blue-, or no-shift in hydrogen bonds: a unified explanation. J Am Chem Soc 129:4620–4632
51. Alkorta I, Rozas I, Elguero J (1998) Radicals as hydrogen bond acceptors. Ber Bunsengesells Physikalische Chem 102:429–435
52. Raghavendra B, Arunan E (2007) Unpaired and σ bond electrons as H, Cl, and Li bond acceptors: an anomalous one-electron blue-shifting chlorine bond. J Phys Chem A 111:9699–9706
53. Mani D, Arunan E (2014) The X-C$\cdots\pi$ (X = F, Cl, Br, CN) carbon bond. J Phys Chem A 118:10081–10089
54. Manikandan P, Zhang J, Hase WL (2012) Chemical dynamics simulations of X– + CH3Y $\rightarrow$ XCH3 + Y– Gas-phase SN2 nucleophilic substitution reactions. Nonstatistical dynamics and nontraditional reaction mechanisms. J Phys Chem A 116:3061–3080
55. Gonzales JM, Cox RS, Brown ST et al (2001) Assessment of density functional theory for model SN2 Reactions: CH3X + F– (X = F, Cl, CN, OH, SH, NH2, PH2). J Phys Chem A 105:11327–11346
56. Maris A, Favero LB, Velino B et al (2013) Pyridine-CF4: a molecule with a rotating cap. J Phys Chem A 117:11289–11292
57. Gutowsky HS, Emilsson T, Arunan E (1993) LowâŁłJ rotational spectra, internal rotation, and structures of several benzene–water dimers. J Chem Phys 99:4883–4893
58. Mishra BK, Sathyamurthy N (2007) Van der Waals Complexes of Small Molecules with Benzenoid Rings: influence of Multipole Moments on Their Mutual Orientation. J Phys Chem A 111:2139–2147
59. Bauza A, Mooibroek TJ, Frontera A (2014) Non-covalent sp3 carbon bonding with ArCF3 is analogous to CH-[small pi] interactions. Chem Commun 50:12626–12629

Chapter 12
Interplay of Hydrogen, Halogen, Lithium and Beryllium Bonds in Complexes of Thiirane

Sean A. C. McDowell and Jerelle A. Joseph

Abstract Hydrogen, halogen, lithium and beryllium bonding are briefly surveyed as a prelude to a report of a computational study of the interplay between these various non-covalent interactions. Our study used model dimers and trimers involving the thiirane molecule, $(CH_2)_2S$, complexed with small molecules like HF, ClF, BrF, LiF and BeH_2 to assess and investigate the interplay between the different non-covalent interactions. The model trimer systems show positive cooperative effects when thiirane is one of the terminal molecules, whereas a negative cooperative effect is evident when it is at the center of the trimer. The changes in selected molecular properties, including the redistribution of charge densities obtained by the natural population analysis (NPA), implemented in the natural bond orbital (NBO) procedure, and an Atoms in Molecules (AIM) topological analysis, were useful in understanding these cooperative effects.

12.1 Introduction

12.1.1 Overview of Non-Covalent Interactions

Non-covalent interactions in molecular systems is a topic of much scientific interest as it spans a number of scientific subdisciplines. They are responsible for the existence of condensed phases, for which the forces operative between interacting neutral molecules are strong enough to limit the volume of samples of a particular material but weak enough to ensure the fluidity of such materials in bulk.

Such interactions are usually dominated by classical electrostatic forces, through the interaction between the permanent multipole moments of the interacting molecular species. These electrostatic forces are usually augmented by classical polarization and quantum-mechanical dispersion forces, and their long-range behavior dependent on some inverse power of their molecular separation (usually measured by the distance between the center of masses of the interacting molecules). At equilibrium

S. A. C. McDowell (✉) · J. A. Joseph
Department of Biological and Chemical Sciences,
The University of the West Indies, Cave Hill Campus, Wanstead, Barbados
e-mail: sacm@mail.com

S. Scheiner (ed.), *Noncovalent Forces,* Challenges and Advances in Computational Chemistry and Physics 19, DOI 10.1007/978-3-319-14163-3_12

separations, the electron-electron (Pauli) exchange repulsion plays a more prominent role and helps to modify the closeness of approach of the interacting subunits, as well as the shape of the resulting molecular complexes and clusters.

12.1.1.1 Hydrogen Bonding

There are a number of non-covalent interactions which are of scientific, as well as, technological interest. The most well-known and well-studied of these interactions is hydrogen bonding, which may be represented schematically by A-H. . . Y, with A being an atom or fragment which is more electron-withdrawing than the covalently-attached proton, which enjoys an attractive interaction with the electron-rich regions (e.g. lone pairs, π-electrons etc.) on a proton acceptor Y [1–7].

The hydrogen bond has been extensively studied and documented for more than a century now. This important concept continues to be relevant especially since considerable interest has developed in recent times for a variety of other non-covalent bonds, most prominently, for example, in halogen bonding [8–10], but also for other interactions with exotic names like chalcogen [11, 12], pnicogen [13] and tetrel bonding [14, 15]. The latter four interactions can all be considered as sub-sets of the more general concept of "σ-hole" bonding, with the differences between each type determined by the type of atom which gives rise to the σ-hole—Group IV atom (tetrel), Group V (pnicogen), Group VI (chalcogen) and Group VII (halogen).

A number of books have been written over the years about the hydrogen bond, with the seminal work by Pimentel and McClellan in 1960 [1] setting the stage for the future advances and current understanding of this concept. More recent monographs include *Theoretical Treatments of Hydrogen Bonding* [2]; *Hydrogen Bonding: A Theoretical Perspective* [4]; *An Introduction to Hydrogen Bonding* [3]; *The Weak Hydrogen Bond* [5]; a Frontiers article in 2008 also gives a future outlook on this topic [6].

Although the hydrogen bond derives its strength from the different relative contributions of the electrostatic, induction, charge transfer, dispersion and exchange repulsion energies to its overall energy, a good description of the attractive component of conventional hydrogen bonds (in gas and condensed phases) is possible with a combined electrostatic-plus-induction energy model [6]. The normal hydrogen bond is usually characterized by an increase in the A-H bond length due to the attraction of the electron-rich site of Y for the proton of A-H which is accompanied by a decrease of its vibrational frequency (a red shift) and increase in its IR intensity [4]. The transfer of electron density from the proton acceptor Y to the antibonding A-H σ^*orbital is also thought to occur and has the effect of weakening the A-H bond, resulting in its elongation.

However, in some instances there is an increase in the A-H stretching frequency on complex formation (a blue shift), usually accompanied by compression of the A-H bond and a decrease in IR intensity of the A-H stretching frequency. The reversal of the normal features of hydrogen bonding in the so-called blue-shifting hydrogen bonds received considerable attention about a decade ago, and provoked a lively and sometimes controversial discourse in the literature about its nature and origin and especially whether it was fundamentally different from the normal red-shifting analogue.

Numerous theoretical papers have been devoted to elucidation of the blue-shift phenomenon, with some researchers favouring an electrostatic model, while others advocate charge-transfer mechanisms [6]. A theoretical analysis using a perturbation theory model has been found useful in describing both red- and blue-shifting behavior in a set of model hydrogen-bonded complexes and this work suggests that both types of behavior can be described by the same theory [6]. Although there are many hydrogen-bonded complexes, the great majority are of the usual red-shifting variety and it would appear that the blue-shifting phenomenon is dependent on specific characteristics of the individual interacting molecular subunits which pre-dispose them to blue-shifting behavior. For example, the change in the permanent dipole moment of the proton donor when the A-H bond length increases and the chemical "hardness" of the atom of the proton acceptor to which the A-H molecule is directly bonded, appear to be two important parameters that determine whether the A-H. . . Y hydrogen bond is red- or blue-shifted [6].

12.1.1.2 Halogen Bonding

As mentioned earlier non-covalent interactions such as halogen bonds have been more extensively studied in recent years. Halogen bonds refer to the interaction between a covalently bonded halogen atom in a molecule A-X (A = electron-withdrawing atom or group; X = F, Cl, Br, I) and a nucleophilic site on Y (e.g a lone pair or anion) [8–10, 14, 16–22]. As is the case for hydrogen bonds, halogen bond formation normally leads to elongation of the A-X bond with a concomitant reduction in the A-X vibrational stretching frequency, though cases of 'blue-shifting' halogen bonds have also been reported in the literature [23, 24].

At first glance this interaction seems counterintuitive since halogen atoms are electronegative species and as such are not expected to participate in non-covalent bonding with electron-rich sites. Through the pioneering work of Brinck and others, it has been shown that when bonded to other electronegative species (A), an electron-deficient region develops on the surface of the halogen atom (adjacent to and collinear with the A-X molecular axis), termed a σ-hole, which allows for such interactions to occur [25–29].

A σ-hole on X is often characterized by computed molecular surface electrostatic potential plots for A-X and when A is a strong electron-withdrawing group, the σ-hole(s) on X acquires a positive electrostatic potential [25–29]. Several studies have reported good correlations between the σ-hole strength (i.e. the magnitude of the positive potential on X) and the stability of the resulting halogen-bonded complexes A-X. . . Y [30, 31], highlighting the electrostatic nature of halogen bonds. Since the polarizability of X increases as X gets larger (i.e. going from F to I), the σ-hole strength increases (for fixed A and Y) in the same order, generally leading to stronger halogen-bonded complexes [30, 31].

Halogen bonds are noted for their marked directionality (i.e. the halogen bond angle is invariably linear or very close to linearity) which is a result of the localized nature of σ-holes [8, 32, 33]. Another important feature of halogen bonding is the noted Lewis acid—Lewis base duality of X which allows for the simultaneous interaction of X with the Lewis base Y and with an electrophile [32, 34, 35]. This feature stems from the anisotropic charge distribution of X, in which regions along the A-X bonding axis are depleted of electronic charge (σ-hole), while those regions roughly perpendicular to the A-X bonding axis are electron-rich (lone pairs of X) [16, 36–38]. These and other interesting features of halogen bonds have stimulated the interest of researchers seeking to exploit the properties of halogen bonding in fields as diverse as biological modelling [39–42], crystal engineering and drug design [43–52], to name a few.

12.1.1.3 Lithium Bonding

The lithium bond is similar to the hydrogen bond and likewise may be denoted as A-Li. . . Y. The term 'lithium bond' is generally used to describe the interaction of a chemically bonded Li atom in a molecule A-Li with an electron-rich site in another molecule Y. One of the earliest theoretical studies of lithium bonding was undertaken by Kollman and coworkers who investigated a series of lithium-bonded dimers (HF. . . LiF, LiF. . . LiLi, LiF. . . LiF) and their hydrogen-bonded counterparts (HF. . . HF, LiF. . . HH, LiF. . . HF) [53]. Later, Ault et al obtained matrix isolation infrared spectra for A-Li. . . Y (A = Cl, Br; Y = $N(CH_3)_3$, NH_3, $O(CH_3)_2$, OH_2) complexes and provided the first experimental evidence of lithium halides participating in lithium bonding [54].

In a detailed review of lithium bonding, Sannigrahi et al described the lithium bond as a case where a lithium atom is di-coordinated [55]. For a fixed A, A-Li tends to be more ionic than A-H, since the Li atom is more electropositive than the H atom. Accordingly, the lithium bond (A-Li. . . Y) is stronger than the corresponding hydrogen bond (A-H. . . Y) [55, 56]. The relatively high electropositive character of the Li atom also accounts for the dominance of electrostatic forces in lithium bonds [56, 57].

Intermolecular charge transfer is thought to play a significant role in hydrogen bond formation, but is less significant for lithium bonding [56]. Vila and coworkers used Bader's Atoms in Molecules (AIM) theory [58] to perform topological analyses for several lithium- and hydrogen-bonded complexes. The relative electron density at bond critical points (ρ_{bcp}), for a given set of bonds, usually correlates well with the relative bond stability. Interestingly, it was found that the ρ_{bcp} for the lithium bonds was half that for the hydrogen bonds, although the lithium bond is stronger. This further emphasized the role of electrostatics in the formation of lithium bonds [57].

A recent study alluded to a difference in the mechanism for lithium and hydrogen bond formation. It was found that while the positive charge on the H atom of A-H. . . Y increases upon complex formation, with a concomitant decrease in H atomic volume, the Li atom of A-Li. . . Y showed a general reduction in positive charge and an increase in its volume [59]. It has also been reported that, unlike hydrogen bonds where the A-H bond extends (and its vibrational frequency decreases) or contracts (with its vibrational frequency increasing), the bond length changes and frequency shifts for A-Li in A-Li. . . Y do not conform to these 'rules' [60–62].

12.1.1.4 Beryllium Bonding

Beryllium bonds are formed when a Be atom in a AA′Be molecule (A′ may or may not be the same atom as A) interacts with an nucleophilic site in another molecule Y; i.e. AA′Be. . . Y [63–65]. The Be atom, like H and Li, is also highly electropositive and therefore the interaction energies of beryllium bonds tend to have a substantial electrostatic component and are generally comparable in strength to very strong hydrogen bonds.

However, due to the presence of vacant low-lying 2*p* orbitals on Be in AA′Be, considerable charge transfer into these orbitals may occur upon AA′Be. . . Y complex formation. Electronic charge may also be transferred into the A-Be/A′-Be antibonding orbital ($\sigma^*_{\text{A-Be/A}'\text{-Be}}$). The shift in charge densities greatly increase the covalent character of beryllium bonds [63, 64] and may result in relatively high ρ_{bcp} for these bonds compared to their hydrogen-bonded analogues [63].

Significant charge redistribution which occurs when beryllium bonds are formed usually causes bending of the linear AA′Be (A = A′) molecule, as well as elongation of the A-Be/A′-Be bond and large red or blue shifts in the stretching frequencies of the A-Be/A′-Be bond. It may also lead to substantial changes in the electron density of the Lewis base to the extent that the acidity of the base increases [63, 65].

12.1.2 Interplay Between Non-Covalent Interactions

The study of the interplay between the hydrogen bond and other less well-known interactions in molecular complexes is of widespread interest and the insight to be gained from such investigations is strong motivation for contemporary research on intermolecular forces, including the present contribution. Cooperativity is one of the most interesting features of non-covalent interactions and the study of this phenomenon may provide an understanding of cluster growth, aggregation and the formation of condensed phases.

Generally, cooperativity is manifested as an increase in the binding strength of a complex beyond the sum of its individual interactions taken a pair at a time. For example, a trimer system consisting of molecular subunits AX. . . Y. . . AX′ is said to exhibit positive cooperativity when the interaction energy of the trimer is greater than the sum of the separate AX. . . Y and Y. . . AX′ pair interaction energies. Conversely, a negative cooperative or a diminutive effect is associated with a decrease in the trimer interaction energy relative to sum of the pair interaction energies. Cooperativity in a trimer can be assessed from the relative contribution of the three-body pairwise non-additive energy to the total interaction energy.

Cooperativity results mainly from the mutual polarization of the constituent molecules of the cluster or complex—the energy of each pair of molecules in a trimer, for example, is modified by the presence of the third interacting molecule, especially if the molecules are both highly polar and polarizable. Hydrogen-bonded complexes commonly show positive cooperative effects, but negative cooperative effects are observed if one of the molecules in a trimer system acts as a double proton acceptor or a double proton donor [4].

Numerous studies have been devoted to the investigation of cooperative effects of hydrogen bonding and other non-covalent interactions. In particular, great emphasis has been placed on the interplay between multiple hydrogen bonds [66–77], hydrogen and dihydrogen bonds [78–80] and, most recently, hydrogen and halogen bonds [34, 81–91]. Parra et al investigated the cooperative effects of bifurcated hydrogen bonding. Interestingly, the hydrogen bond strength in the decamer was about 160 % higher than in the dimer, despite the hydrogen bonds being 3-center hydrogen bonds as opposed to being typical 2-center hydrogen bonds [70]. Arauju et al showed that the cooperative effect in hetero hydrogen-bonded chains was greater than that for homo chains for a series of $(HCN)_n$-HF and $(HCN)_n$ ($n = 1, 2, 3$) clusters [71]. Additionally, as the chains become longer there was a corresponding increase in the cooperative effects observed [71].

Cooperative effects in hydrogen-bonded systems may play a significant role in the self-assembly of certain complexes. In 2005, Planas and co-workers demonstrated this property for mercaptane-metallacarborane complexes, which contained both hydrogen and dihydrogen intermolecular bonds [78]. It appears that the self-assembly of these complexes is driven by the unusual cooperative effects between the aforementioned non-covalent bonds [78]. Li et al further studied the cooperativity between hydrogen and dihydrogen bonds for a set of LiH-$(HCN)_n$ ($n = 2, 3$) complexes and found that the non-additive energy in complexes with various types of hydrogen bonds was greater than those with the same type of hydrogen bonds [79]. Furthermore, they reported that charge transfer and polarization forces were major contributors to the cooperativity between these non-covalent bonds [79].

With regards to halogen bonding, it has been shown that halogen and hydrogen bonds can exhibit strong cooperative or diminutive effects on each other in molecular clusters. Zhou et al explored the cooperativity between these bonds for a series of ring-shaped complexes [A-(HF)X...NH_3 (X = Cl, Br and A = F, Cl, Br)] that contained two hydrogen bonds and one halogen bond [34]. A larger non-additive energy was reported for the AX = Br_2 cluster compared to the AX = Cl_2 cluster. This observation again showed the significant dependence of cooperative effects on the relative polarizabilities of the interacting species [34]. Grabowski et al conducted an investigation of Cl^-...HCCH...HF, Cl^-...ClCCH...HF, F^-...ClCCH...HF clusters and partitioned the total interaction energy for each cluster into various energy contributions. For all of the complexes which showed positive cooperative effects (i.e. $E_{\text{non-add}} < 0$), the most important attractive energy term was found to be the polarization term [88]. Additionally, it was reported that charge transfer forces were the most significant contributor to the formation of the hydrogen bonds [88].

Jing et al described a diminutive relationship between hydrogen and halogen bonds in model systems of HNC and HOBr when the HOBr molecule simultaneously acted as the hydrogen and halogen bond donor [83]. Conversely, in systems where HNC acted as both the proton donor and the halogen bond acceptor, the cooperative energy exceeded 20 % of the interaction energy [83]. Solimannejad and Malekani observed similar results for a series of complexes of HCCX (X = Cl, Br) with HCN and HNC [87].

More often than not, when both hydrogen and halogen bonds exist in molecular systems, one bond is strengthened at the expense of the other; usually the weaker bond experiences a positive cooperative effect. Zhao et al demonstrated this relationship for trimeric clusters of carbon tetrabromide, a halide, and a solvent molecule [91]. For clusters containing the smaller halide ions, the halogen bonds were elongated in the trimer relative to the dimer, while the hydrogen bond contracted [91]. Alkorta et al found that the hydrogen-bonded complexes were stronger for the hypohalous acids containing the smaller halogen atoms. The interaction energies for the dimers with the higher halogens, I and At, were similar or larger than that of their hydrogen-bonded counterparts [89].

Cooperative effects between lithium bonds and other non-covalent interactions have received increased attention in recent years. Studies on the interplay between lithium bonds and hydrogen [92–96], dihydrogen [95], halogen [93, 97, 98], chalcogen [99], pnicogen [100] and cation- π [101] bonds, as well as in lithium clusters [102–104] have been documented. Solimannejad showed that for a series of $F_3YLi...NCH...HMH$ and $F_3YLi...HMH...HCN$ trimers (Y = C, Si; M = Be, Mg), involving both lithium and dihydrogen bonds, the overall interaction was cooperative or diminutive, depending on whether HMH was located at the end or in the center of the trimer chains [95].

A similar study on the interplay between lithium and halogen bonds also found that the ordering of monomers had a notable impact on the cooperativity observed in the resulting trimers [98]. McDowell and Yarde, showed that for a series of model lithium-, halogen- and hydrogen-bonded trimers, the lithium bond was most significant in producing positive cooperative effects. They also reported that the non-additive energies were higher in the cyclic clusters than in their linear counterparts [93]. A study involving linear chains consisting of LiCN subunits bound by multiple lithium bonds revealed that electrostatic and polarization forces were the main contributors to the stability of the chains [102].

The mutual influence between beryllium bonds and other non-covalent bonds has also been investigated in recent years [105–107], though not as extensively as the preceding interactions. Mó et al carried out investigations on the effect of beryllium bonds on inter- and intra-hydrogen bonds [107]. The beryllium bonds and the inter-hydrogen bonds showed positive cooperative effects in the imidazole-BeX_2 (X = H, F) trimers. While in the malonaldehyde-BeX_2 trimers, the cooperative effects of the beryllium bonds and the intra-hydrogen bonds were heavily dependent on whether the Be atom was attached to the hydrogen bond donor or acceptor; the former being cooperative and the latter diminutive [107]. Another study involving beryllium bonds and multiple hydrogen bonds revealed that it was energetically more favorable to attach a BeX_2molecule to H_2O dimers or trimers than to solvate BeX_2 with the individual H_2O molecules [106].

12.2 Cooperative Effects in Unusual Thiirane Complexes

Our theoretical study of cooperativity is restricted to its manifestation in molecular trimers, with dimeric complexes of the thiirane molecule $(CH_2)_2S$ being the common motif throughout. A recent high-level computational study of the thiirane... ClF complex found an unusually strong interaction in this dimer accompanied by significant charge transfer from thiirane to the ClF molecule [108]. The author of this study suggested that this type of dimer be classified as a Mulliken "inner complex" of the form $[A\text{-}H]^+ \ldots B$, rather than that of a more weakly-bound "outer complex" of the usual A-H. . . Y form, where the charge transfer is smaller and the interaction more electrostatic in nature [108].

Mulliken and Person's work on molecular complexes [109] established this way of classifying donor-acceptor complexes, with specific reference to hydrogen-bonded complexes. The considerable interest in halogen bonding [18, 21] has also prompted some suggestion of classifying halogen-bonded complexes as either inner complexes $[A\text{-}X]^+ \ldots Y^-$ or outer complexes A-X. . . Y, where the former involve more substantial charge transfer between the interacting molecules than in the latter [108, 110].

The thiirane... ClF dimer has been investigated experimentally by gas-phase rotational spectroscopy [111] and its characterization through the explicitly correlated coupled cluster CCSD(T)-F12b/CBS method, established a relatively large value of -55.4 kJ mol^{-1} for this strong intermolecular interaction, compared with an interaction energy of -32.8 kJ mol^{-1} for the hydrogen-bonded analogue thiirane... HCl [108]. Furthermore, it was found that significantly more charge was transferred (by a factor of about four), mainly from the sulfur lone pair to the Cl-F antibonding orbital, than the corresponding charge transfer in thiirane... HCl.

The natural partial atomic (NPA) charges on the sulfur atom of thiirane... ClF, obtained from a natural bond orbital (NBO) analysis [112], found that remarkably the sulfur atom was positively charged, compared with the oxygen atom of oxirane... ClF, which has a negative partial charge. This finding, along with the large interaction energy and significant charge transfer, prompted Hill to suggest that thiirane... ClF be classified as the first "inner-type" halogen-bonded complex, in Mulliken's terminology [108].

Decomposition of the interaction energy of thiirane... ClF using symmetry-adapted perturbation theory (SAPT) [113] showed that the four main components of the interaction energy: the electrostatic (E_{elec}), exchange repulsion (E_{exch}), induction (E_{ind}), and dispersion (E_{disp}) energies, were all substantially larger than in the hydrogen-bonded thiirane... HCl and oxirane... HCl complexes, and in the halogen-bonded oxirane... HCl complex [108]. These SAPT results provide further support for the notion that thiirane... ClF should be considered as an "inner-type" complex.

12.2.1 Thiirane...XF/BeH$_2$ Dimers

We use Hill's computational study of thiirane...ClF as the starting point for this study of the cooperative effect of non-covalent interactions involving the familiar hydrogen bond and the less-familiar, but no less important, halogen bond, as well as two other unusual interactions—the lithium bond and the beryllium bond. Our approach is to optimize the thiirane...XF/BeH$_2$ (XF = HF, ClF, BrF, LiF) series of dimers using density functional theory (DFT) with the widely-used B3LYP functional as our computational method of choice. The large 6-311++G(3df,3pd) basis set, which is of triple-zeta quality and contains sets of diffuse functions and higher angular momentum functions, should be capable of producing reliable results with sufficiently high accuracy. The available experimental and computational data for thiirane...ClF is useful in gauging the accuracy of the present work.

The set of XF/BeH$_2$ molecules can attach to the S atom of thiirane via hydrogen, two types of halogen bonds (Cl, Br), lithium or beryllium bonds and this allows for an assessment of the relative strengths of these various intermolecular interactions. Computed parameters of the optimized dimer, which give useful insight into the nature of the operative non-covalent forces, include the interaction energy, the S...X intermolecular separation, the S...X-F bond angle, the change in the X-F bond length and the complexation-induced shift of the X-F harmonic vibrational stretching mode. The optimization and subsequent computation of these properties at the B3LYP/6-311++G(3df,3pd) level of theory for the thiirane...XF series was performed with the Gaussian 03 suite of programs [114].

The interplay between hydrogen and halogen bonding in sulfur-containing systems has potential relevance for thyroid chemistry [115, 116] and as such a study of the cooperative effects between these and other non-covalent interactions in model systems may provide a better understanding of the structure and properties of related biomolecular complexes, especially those that may govern important biological activities in living systems.

Table 12.1 shows the computed properties for the thiirane...XF dimer systems. Figures 12.1, 12.2 and 12.3 show the optimized geometries of the thiiriane...ClF, thiirane...LiF and thiirane...BeH$_2$ complexes and span the range of molecular

Table 12.1 Interaction energies (ΔE/kJ mol^{-1}), bonding distances (R/Å), bondlength changes (Δr/Å), frequency shifts ($\Delta\omega$/cm^{-1}) and bond angles (</°) for (CH$_2$)$_2$S...XF/BeH$_2$ (X = H, Cl, Br, Li) complexes at the B3LYP/6-311++G(3df,3pd) level of theory

Complex	ΔE	R(S...X/Be)	Δr(X-F)/ Δr(Be-H)	$\Delta\omega$(X-F)/ $\Delta\omega$(Be-H)	< S...X-F
(CH$_2$)$_2$S...HF	− 32.5	2.150	0.024	− 539.8	166.1
(CH$_2$)$_2$S...ClF	− 56.2	2.459	0.127	− 258.3	177.5
(CH$_2$)$_2$S...BrF	− 64.6	2.582	0.102	− 171.6	177.9
(CH$_2$)$_2$S...LiF	− 69.4	2.440	0.038	− 88.3	111.8
(CH$_2$)$_2$S...BeH$_2$	− 53.6	2.200	0.021 0.033	− 185.5^{asymm} − 122.0^{symm}	–

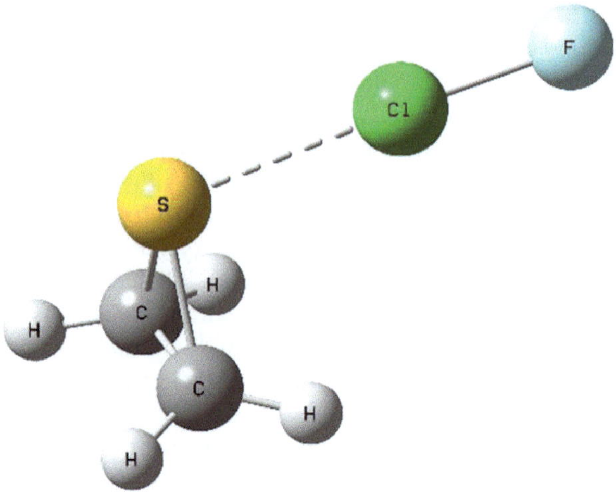

Fig. 12.1 Optimized geometry for $(CH_2)_2S...ClF$

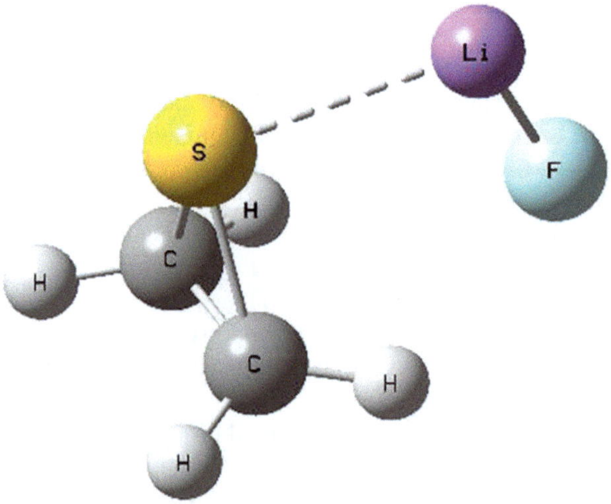

Fig. 12.2 Optimized geometry for $(CH_2)_2S...LiF$

structures that these dimers adopt. The S atom binds directly to the X atom (or Be atom) and sits at the apex of a 3-atom C-S-C cycle comprising the thiirane ring. The HF, ClF and BrF molecules adopt a similar orientation in the dimer structure, with the S... ClF/BrF orientation close to linearity (about 178°). The HF orientation, on the other hand, deviates by about 14° from linearity, consistent with its weaker bond strength relative to the halogen analogues. The LiF molecule is almost parallel to the C-S-C ring of thiirane (see Fig. 12.2) and the F atom appears to benefit from an attractive interaction with the H atoms of thiirane. The BeH_2 molecule is oriented with its hydridic H atoms bent away from the thiirane molecule (as shown in Fig. 12.3), which indicates a substantial perturbation of its structure due to the interaction with thiirane.

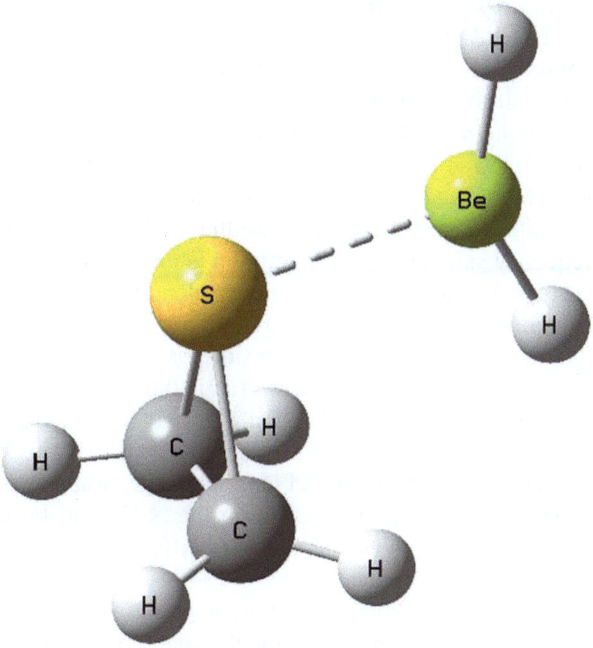

Fig. 12.3 Optimized geometry for $(CH_2)_2S\ldots BeH_2$

The interaction energies for the thiirane... XF/BeH_2 dimers are in the order LiF > BrF > ClF > BeH_2 > HF. The thiirane binds most strongly with LiF, presumably through an S...Li contact, as well as interaction of the F atom with the nearby protons of thiirane. This is inferred by the almost parallel orientation of LiF relative to thiirane. By contrast, the other XF molecules appear to maximize their interaction with the S atom by polarization of the thiirane molecule—this molecule has large dipole polarizabilities and the electric field arising from the XF molecules facilitates the shift of electron density from the sulfur lone pair towards the antibonding σ^* orbital of XF. The bonding in the BeH_2 complex is different from the XF complexes since the Be atom is oriented in such a way that the vacant $2p$ orbitals can overlap with the S lone pair thereby facilitating electron density transfer between thiirane and BeH_2. It is noteworthy that the hydrogen bond has the weakest non-covalent interaction with thiirane.

The order of the intermolecular S...X/Be separation in the complexes is HF < BeH_2 < LiF < ClF < BrF and does not correlate with the trend for interaction energies above. It seems that the intermolecular separation is dependent on the electron repulsion between S and X/Be, so that the H atom has the closest approach, while the large Br atom has the furthest, with the other atoms falling in between these limits. The order of the intermolecular separation roughly correlates with the number of electrons on the X atom. However, for the small Be atom, the two valence electrons are effectively localized on the terminal H atoms of BeH_2 allowing the tightly-held $1s$ core electrons to approach the S atom quite closely. It is noteworthy that the trend for the S...X/Be separation follows the order of the H/Be/Li ionic radius (H < Be < Li) and the Cl/Br atomic radius (Cl < Br); i.e. the respective values of the ionic or atomic radius for H^+, Be^{2+}, Li^+, Cl, Br are 0.00066, 0.34, 0.78, 0.99, 1.14 Å [117].

For all five dimers, there is an increase in the X-F (or Be-H) bond length and a corresponding red shift of the X-F (H-Be-H) stretching frequency. This elongation can be viewed as a consequence of the electrostatic attraction between the positive potential on X and the lone pair on S, as well as a consequence of the charge transfer from the S lone pair orbitals into the antibonding σ^* orbitals of X-F or Be-H. The frequency shift of the X-F vibration is dependent not only on the magnitude of the electrostatic forces acting on the covalent X-F bond or the extent of charge transfer from thiirane, but also on the strength (or stretching force constant) of the X-F bond.

The red shift and bond extension associated with the X-F molecule are usually correlated in a qualitative sense (as is the unconventional blue-shifting behaviour, which is usually correlated with a bond contraction in hydrogen-bonded complexes). However, a more quantitative (and proportionate) relationship between the frequency shift and the bond length change would depend on the relative sign and magnitude of the force exerted on the X-F bond by the electric field originating from the thiirane molecule and the "stiffness" of the X-F bond (as measured by its stretching force constant).

For example, the HF stretching frequency suffers the largest red shift (of $-540\,\text{cm}^{-1}$) even though the thiirane...HF complex is the most weakly-bound complex and has the smallest bond extension (0.024 Å). This large red shift is obviously due to the small size of the H atom, compared with the other X atoms. The largest bond extension (0.127 Å) is observed in thiirane...ClF, which also has the second highest computed red shift ($-258\,\text{cm}^{-1}$). These relatively large values appear to be consistent with the characteristics of a strongly-bound "inner type" complex, as noted by Hill in his recent study [108].

It is interesting to note that the second largest bond extension and red shift were obtained for thiirane...BrF, even though the complex has a larger interaction energy and BrF produces a larger σ-hole (and therefore, stronger halogen bond) than ClF. Do these results indicate that thiirane...BrF should also be considered as an inner complex? Comparison of the changes in electron density distribution, before and after complexation, in these two different halogen-bonded dimers should provide an answer to this intriguing question. These changes were investigated through the use of the electron density partitioning schemes implemented by the NBO in the Gaussian03 suite of programs and AIM analyses using the AIMAll software [118]—to be discussed later.

In the next sub-sections we shall see how the properties of the thiirane...XF/BeH_2 are modified when a third molecule interacts with the S atom or when it interacts with the negatively charged F atom of XF or the hydridic H atoms of BeH_2. The cooperative effects of the different non-covalent interactions can be assessed from these computational results.

Table 12.2 Interaction energies (ΔE/kJ mol^{-1}) and bonding distances (R/Å) for FX′/H_2Be... $(CH_2)_2$S...XF (X′ = H, Cl, Br, Li; X = H, Cl, Br) complexes at the B3LYP/6-311++G(3df,3pd) level. The $\sum \Delta E_{dimer}$ values represent the sum of the FX′/H_2Be... $(CH_2)_2$S and $(CH_2)_2$S...XF dimer interaction energies. The $(CH_2)_2$S...XF dimer results are shown for the sake of comparison

Complex	$\sum \Delta E_{dimer}$	ΔE	R(S...X)
(CH2)2S...HF			*2.150*
FH...$(CH_2)_2$S...HF	− 65.0	− 55.9	2.203
FCl...$(CH_2)_2$S...HF	− 88.7	− 70.2	2.278
FBr...$(CH_2)_2$S...HF	− 97.1	− 78.0	2.296
FLi...$(CH_2)_2$S...HF	− 101.9	− 92.3	2.199
H_2Be...$(CH_2)_2$S...HF	− 86.1	− 69.7	2.281
(CH2)2S...ClF			*2.459*
FH...$(CH_2)_2$S...ClF	− 88.7	− 70.2	2.529
FCl...$(CH_2)_2$S...ClF	− 112.4	− 81.1	2.593
FBr...$(CH_2)_2$S...ClF	− 120.8	− 87.5	2.612
FLi...$(CH_2)_2$S...ClF	− 125.6	− 106.5	2.517
H_2Be...$(CH_2)_2$S...ClF	− 109.8	− 77.9	2.620
(CH2)2S...BrF			*2.582*
FH...$(CH_2)_2$S...BrF	− 97.1	− 78.0	2.627
FCl...$(CH_2)_2$S...BrF	− 120.8	− 87.5	2.672
FBr...$(CH_2)_2$S...BrF	− 129.2	− 93.7	2.681
FLi...$(CH_2)_2$S...BrF	− 134.0	− 115.0	2.615
H_2Be...$(CH_2)_2$S...BrF	− 118.2	− 84.5	2.693

12.2.2 FX′/H_2Be...Thiirane...XF/BeH_2 Trimers

Table 12.2 shows selected properties for the FX′/H_2Be...thiirane...XF (X′ = H, Cl, Br, Li; X = H, Cl, Br) trimers which have the typical structure shown in Fig. 12.4. In these complexes the S atom acts as a double acceptor and all of the complexes show a negative cooperative or diminutive effect, as reflected by their reduced interaction energies, relative to the sum of the constituent dimer energies. This diminutive effect is also observed for hydrogen-bonded complexes involving a double proton acceptor [4].

The other computed parameters reflect this diminutive effect; i.e. all intermolecular S...X distances are increased, while all X-F bond extensions are smaller in the trimers, compared with the corresponding dimer values. The order of the diminutive effect due to the FX′ molecule as reflected by R(S...X/Be) is FBr > H_2Be > FCl > FH > FLi.

The FX′ molecule attaches to the free S lone pair of the thiirane...XF dimer to form the trimer and in so doing polarizes the S atom by pulling electron density towards itself, thereby reducing the lone pair density, both available for bonding to the X atom of XF, as well as reducing the charge transferred to XF, and vice versa. Consequently, the net binding weakens, the S...X separation increases, the XF bond

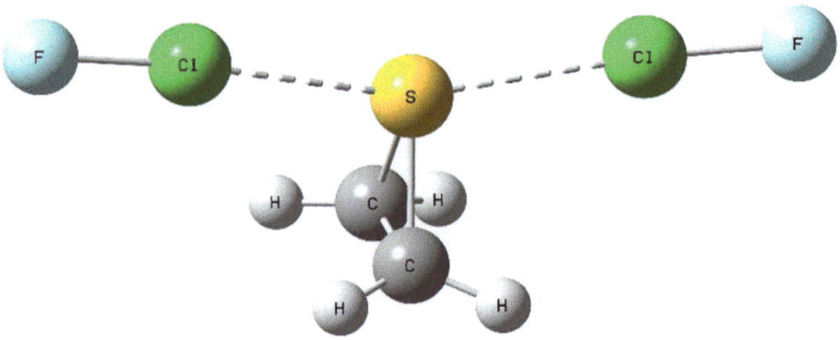

Fig. 12.4 Optimized geometry for FCl... $(CH_2)_2Si$... ClF

extension decreases and the XF red shift decreases, relative to the thiirane... XF dimer (note the bond extensions and frequency shifts are not shown in Table 12.2).

Despite the negative cooperative effect, all complexes are strongly-bound ($-\Delta E > 55$ kJ/mol) and all XF stretching frequencies are strongly red-shifted. The diminutive effect of FX′ on ΔE, R(S... X) and Δr(X-F) is always larger for FCl, FBr and BeH_2. This is because both halogen and beryllium bonding will shift electron density away from the central S atom, thereby diminishing the S... X interaction. This is manifested as an increase in the electron density transferred from S to the antibonding σ* orbital of the FX′molecule as a result of complex formation—see the NBO partial atomic charge results, to be discussed later. The least diminutive effect occurs for LiF and is probably because the repulsion between the Li 1*s* core electrons and the S lone pairs hinders charge transfer from S to the LiF molecule.

Table 12.3 shows selected properties for FX′/H_2Be... thiirane... LiF (Fig. 12.5) as well as for FX′/H_2Be... thiirane... BeH_2 (Fig. 12.6). Addition of FX′/H_2Be to form FX′/H_2Be... thiirane... LiF causes a net reduction of ΔE, increase in R(S... Li) and diminished LiF bond extension (values not shown in Table 12.3), as is the case for the other XF trimers, with FCl/FBr producing the largest diminutive effects.

A diminutive effect is also evident for FX′/H_2Be... thiirane... BeH_2; the net ΔE is smaller than the sum of the interaction energies of the constituent dyads, the S... Be separation is larger and the $\Delta r/\Delta\omega$ values (not shown in Table 12.3) are smaller than the corresponding dimer values, with the additional halogen-bonded and beryllium-bonded interactions yielding the largest diminutions. These results suggest that the displacement of the S lone pair electron density into the vacant *p* orbitals of the adjacent Be atom is reduced due to competition with the FX′ molecule.

In percentage terms, the largest diminutive effects in the FX′/H_2Be... thiirane... BeH_2trimers are for those involving ClF, BrF and BeH_2; this means that the halogen and beryllium bonds have the most significant diminutive effect. However, the most strongly-bound dimers and trimers are those which involve LiF, except FLi... thiirane... LiF, where the competition for the S lone pairs renders the whole complex unstable since no local minima were located. Perhaps these results indicate that the electron density transfer occurring in the ClF, BrF and BeH_2 trimers is the most significant cause of the negative cooperativity, whereas in the complexes

Table 12.3 Interaction energies (ΔE/kJ mol^{-1}) and bonding distances (R/Å) for FX′/H_2Be... $(CH_2)_2S$...LiF/BeH_2 (X′ = H, Cl, Br, Li) complexes at the B3LYP/6-311++G(3df,3pd) level. The $\sum \Delta E_{dimer}$ values represent the sum of the FX′/H_2Be...$(CH_2)_2S$ and $(CH_2)_2S$...LiF/BeH_2 dimer interaction energies. The $(CH_2)_2S$...XF dimer results are shown for the sake of comparison

Complex	$\sum \Delta E_{dimer}$	ΔE	R(S...X/Be)
(CH2)2S...LiF			*2.440*
FH...$(CH_2)_2S$...LiF	− 101.9	− 92.3	2.478
FCl...$(CH_2)_2S$...LiF	− 125.6	− 106.5	2.539
FBr...$(CH_2)_2S$...LiF	− 134.0	− 115.0	2.537
FLi...$(CH_2)_2S$...LiF	− 138.8	2 imag freq	
H_2Be...$(CH_2)_2S$...LiF	− 123.0	− 108.3	2.516
(CH2)2S... BeH2			*2.200*
FH...$(CH_2)_2S$... BeH_2	− 86.1	− 69.7	2.231
FCl...$(CH_2)_2S$... BeH_2	− 109.8	− 77.9	2.282
FBr...$(CH_2)_2S$... BeH_2	− 118.2	− 84.5	2.295
FLi...$(CH_2)_2S$... BeH_2	− 123.0	− 108.3	2.206
H_2Be...$(CH_2)_2S$...BeH_2	− 107.2	− 75.6	2.285

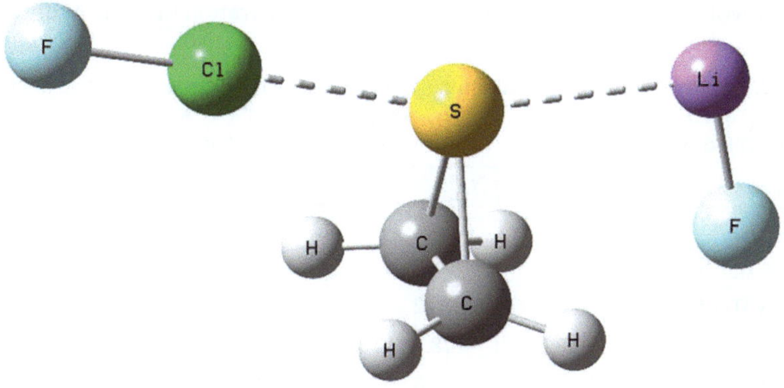

Fig. 12.5 Optimized geometry for FCl...$(CH_2)_2$Si...LiF

containing LiF, this charge transfer is reduced by the Li 1*s* core electrons, as well as its orientation relative to thiirane.

Table 12.4 shows the pairwise additive and non-additive energy contributions to the interaction energy. The non-additive energy ($E_{non\text{-}add}$) is a measure of cooperativity and is defined as $E_{non\text{-}add} = \Delta E - \sum \Delta E_{ij}$, where ΔE is the interaction energy of the trimer and $\Delta E_{ij}$is the pair interaction energy for a dimer subunit ij, computed at the geometry that this subunit adopts in the optimized trimer.

As expected, the pair interaction of the terminal FX′ and XF molecules is significantly smaller than the other pair interactions. The diminutive effect of FX′ is

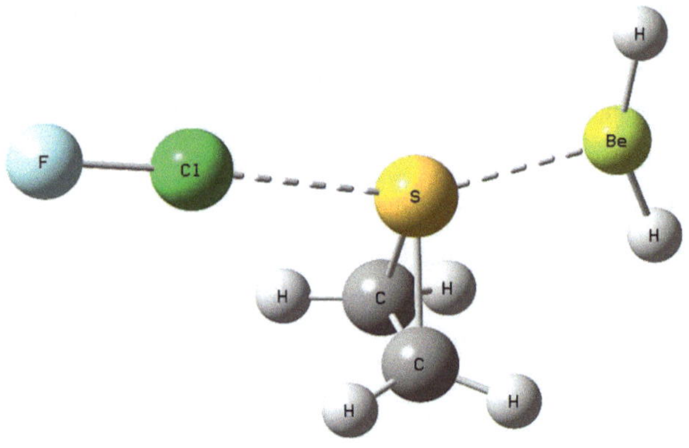

Fig. 12.6 Optimized geometry for FCl... $(CH_2)_2Si$... BeH_2

generally evident in the FX′... thiirane... XF complexes, except those involving either LiF or BeH_2. Recall that a diminutive effect was previously noted for all the trimers (relative to their respective thiirane... XF/BeH_2 dimers). The results in Table 12.4 does not contradict this observation. However, in Table 12.4 the interaction between the c... b pair (which is not accounted for in $\sum \Delta E_{dimer}$) is incorporated in the calculation of the non-additive energy. The c... b pair interaction is repulsive in all trimers and so modifies their estimated non-additive energies. Additionally, the trimers involving LiF or BeH_2 have a different configuration than the hydrogen- and

Table 12.4 Sum of the pairwise additive ($\sum \Delta E_{ij}$) and percentage non-additive contributions ($\%E_{non\text{-}add}$) of ΔE for $[FX'/H_2Be]^c$... $[(CH_2)_2S]^a$... $[XF/BeH_2]^b$ (X′, X = H, Cl, Br, Li) complexes at the B3LYP/6-311++G(3df,3pd) level. All values are measured in kJ/mol

Complexes	ΔE	$\sum \Delta E_{ij}$	$\%E_{non\text{-}add}$
c..........a..........b			
FH... $(CH_2)_2S$... HF	− 55.9	− 58.9	− 5.4
FCl... $(CH_2)_2S$... HF	− 70.2	− 72.0	− 2.6
FBr... $(CH_2)_2S$... HF	− 78.0	− 83.1	− 6.5
FLi... $(CH_2)_2S$... HF	− 92.3	− 91.1	1.3
H_2Be... $(CH_2)_2S$... HF	− 69.7	− 53.7	2.3
FCl... $(CH_2)_2S$... ClF	− 81.1	− 88.9	− 9.6
FBr... $(CH_2)_2S$... ClF	− 87.5	− 101.2	− 15.7
FLi... $(CH_2)_2S$... ClF	− 106.5	− 100.3	5.8
H_2Be... $(CH_2)_2S$... ClF	− 77.9	− 73.3	5.9
FBr... $(CH_2)_2S$... BrF	− 93.7	− 112.0	− 19.5
FLi... $(CH_2)_2S$... BrF	− 115.0	− 111.0	3.5
H_2Be... $(CH_2)_2S$... BrF	− 84.5	− 85.5	− 1.2
H_2Be... $(CH_2)_2S$... LiF	− 108.3	− 83.2	23.2
H_2Be... $(CH_2)_2S$... BeH_2	− 75.6	− 55.5	26.6

halogen-bonded counterparts and this may explain why there is a positive cooperative effect in these complexes—this configuration allows the negative F (of LiF) or H of (BeH_2) to interact favourably with the protons of the CH_2 subunits of thiirane.

12.2.2.1 NBO Analysis of Thiirane... XF/BeH_2 and FX′/H_2Be... Thiirane... XF/BeH_2

Table 12.5 shows that the positive charge on the S atom of the isolated thiirane molecule increases on complexation with HF, ClF and BrF and correlates with transfer of electron density from one of the S lone pairs into the antibonding σ^* orbital of XF, with the largest transfer to the ClF in accordance with Hill's findings [108]. Interestingly the EDT for thiirane... BrF is comparable to that of ClF. Therefore it is possible that this dimer is an inner complex as well.

The decrease in the positive charge on the S atom in the LiF dimer complex is due to the substantial polarization of the thiirane molecule by the large LiF dipole which is oriented in such a way that it forces electron density towards the S atom (its dipole is almost parallel to the C-S-C plane and in the opposite direction to the thiirane dipole). For the thiirane... BeH_2 dimer, there is evidence of electron transfer from the BeH_2bonding orbitals into the intermolecular S... Be region.

Going from dimer to trimer, there is a decrease in charge transfer from thiirane to XF due to competition between FX′ and XF for the S lone pairs. This supports the observed diminutive effect for the FX′/H_2Be... thiirane... XF/BeH_2complexes. Though diminished, the EDT for XF = Cl/Br is still considerably larger, relative to the trimers of the other XF molecules, and so these complexes may exhibit the characteristics of a Mulliken's inner complex [108].

12.2.3 Thiirane... XF/BeH2... X′F/BeH2 Trimers

Table 12.6 shows selected properties for the thiirane... XF/BeH_2... X′F/BeH_2 (X′ = H, Cl, Br, Li; X = H, Cl, Br, Li) trimers which span a range of various structures, some of which are shown in Figs. 12.7–12.10. In these complexes, the interaction of the S atom of thiirane with the X atom of XF (or Be atom of BeH_2) is modified by the attachment of the X′F/BeH_2 molecule to the negative F atom of X′F (or the negative H atom of BeH_2).

Generally, the interaction energy ΔE for each set of trimers (with fixed XF) seems to depend on the polarizing strength of the terminal X′F molecule. Hence, the lithium bond attached to the F of XF produces the most strongly-bound complexes, followed by the beryllium bond, then the hydrogen bond, with the halogen bonds either producing the most-weakly bound complexes or yielding no optimized structures at all. For the latter halogen-bonded complexes, the repulsion between the F lone pairs (of XF) and the lone pairs of the halogen atom X′, is likely to be the main destabilizing factor, especially for the Br atom. We shall now consider each set of trimers in turn.

Table 12.5 Changes in NPA partial atomic charges (Δq in e) w.r.t isolated monomers for $(CH_2)_2S...XF/BeH_2$, $FX'/H_2Be...(CH_2)_2S...XF/BeH_2$ (X′, X = H, Cl, Br, Li) complexes and electron density transfer (EDT in e) determined using the B3LYP/6-311++G(3df,3pd) procedure. Partial atomic charges for monomers are in parentheses

Complex	Δq (S)	EDT $(CH_2)_2S \rightarrow XF$	Complex	Δq (S)	EDT $(CH_2)_2S \rightarrow XF$
$(CH_2)_2S...HF$	0.003 (0.085)	0.070	$FH...(CH_2)_2S...BrF$	0.130	0.246
$(CH_2)_2S...ClF$	0.246	0.345	$FCl...(CH_2)_2S...BrF$	0.248	0.207
$(CH_2)_2S...BrF$	0.200	0.304	$FBr...(CH_2)_2S...BrF$	0.229	0.193
$(CH_2)_2S...LiF$	−0.108	0.037	$FLi...(CH_2)_2S...BrF$	0.053	0.256
$(CH_2)_2S...BeH_2$	−0.061	–	$H_2Be...(CH_2)_2S...BrF$	0.042	0.195
$FH...(CH_2)_2S...HF$	−0.026	0.048	$FH...(CH_2)_2S...LiF$	−0.130	–
$FCl...(CH_2)_2S...HF$	0.168	0.033	$FCl...(CH_2)_2S...LiF$	0.083	–
$FBr...(CH_2)_2S...HF$	0.130	0.029	$FBr...(CH_2)_2S...LiF$	0.053	–
$FLi...(CH_2)_2S...HF$	−0.130	0.047			
$H_2Be...(CH_2)_2S...HF$	−0.098	0.029	$H_2Be...(CH_2)_2S...LiF$	−0.196	–
$FH...(CH_2)_2S...ClF$	0.168	0.267	$FH...(CH_2)_2S...BeH_2$	−0.098	0.065
$FCl...(CH_2)_2S...ClF$	0.270	0.209	$FCl...(CH_2)_2S...BeH_2$	0.060	0.049
$FBr...(CH_2)_2S...ClF$	0.248	0.192	$FBr...(CH_2)_2S...BeH_2$	0.042	0.044
$FLi...(CH_2)_2S...ClF$	0.083	0.276	$FLi...(CH_2)_2S...BeH_2$	−0.196	0.063
$H_2Be...(CH_2)_2S...ClF$	0.060	0.198	$H_2Be...(CH_2)_2S...BeH_2$	−0.148	0.047

Table 12.6 Interaction energies (ΔE/kJ mol^{-1}) and bonding distances (R/Å) for $CH_2)_2S...XF/BeH_2...X'F/BeH_2$ (X, X′ = H, Cl, Br, Li) complexes at the B3LYP/6-311++G(3df,3pd) level

Complex	ΔE	R(S...X/Be)
$(CH_2)_2S...HF...HF$	− 72.0	2.033
$(CH_2)_2S...HF...ClF$	− 52.0	2.076
$(CH_2)_2S...HF...BrF$	− 58.2	2.053
$(CH_2)_2S...HF...LiF$	− 136.7	1.918
$(CH_2)_2S...HF...BeH_2$	− 97.8	1.857
$(CH_2)_2S...ClF...HF$	− 104.8	2.260
$(CH_2)_2S...ClF...ClF$	− 85.1	2.278
$(CH_2)_2S...ClF...LiF$	− 213.1	2.106
$(CH_2)_2S...ClF...BeH_2$	− 192.9	2.072
$(CH_2)_2S...BrF...HF$	− 110.8	2.455
$(CH_2)_2S...BrF...ClF$	− 91.9	2.470
$(CH_2)_2S...BrF...LiF$	− 199.2	2.325
$(CH_2)_2S...BrF...BeH_2$	− 165.0	2.325
$(CH_2)_2S...LiF...HF$	− 169.1	2.448
$(CH_2)_2S...LiF...BrF$	− 151.3	2.432
$(CH_2)_2S...LiF...LiF$	− 302.7	2.248
$(CH_2)_2S...LiF...BeH_2$	− 273.6	2.453
$(CH_2)_2S...BeH_2...LiF$	− 152.0	2.122
$(CH_2)_2S...BeH_2...BeH_2$	− 206.5	2.173

12.2.3.1 Thiirane... HF... X′F/BeH$_2$

1. All of these trimers are stable, with ΔE increasing as X′F = ClF < BrF < HF < BeH_2 < LiF, consistent with the order of decreasing R(S... H), which is ClF > BrF > HF > LiF > BeH_2, except that the last two molecules are switched around. The Be atom of BeH_2 makes the closest contact with the F atom of HF—in this complex, the interatomic electron repulsion is minimized since the 1s core electrons of Be are tightly held and its valence electrons are displaced towards the adjacent hydridic H atoms, which are bent away from the S atom. The F... X′ separation (not shown in Table 12.6) in these trimers increases accordingly with the interatomic repulsion in the order X′ = H < Li < Cl < Br.
2. The H-F bond extensions and vibrational red shifts are strongly correlated and generally scale with the binding strength.
3. The cooperativity between the S... H and the F... X′ non-covalent interactions is evident from the large non-additive energies displayed in Table 12.7, which are estimated to be more than 25 % of ΔE for those complexes. This (positive) cooperativity is also apparent in the shorter S... H distances for thiirane... HF... X′F shown in Table 12.6 compared with thiirane... HF (see Table 12.1).
4. For the thiirane... HF... X′F series, the cooperative effect of the S... H hydrogen bond and the non-covalent F... X′ interactions is significant, with the lithium and

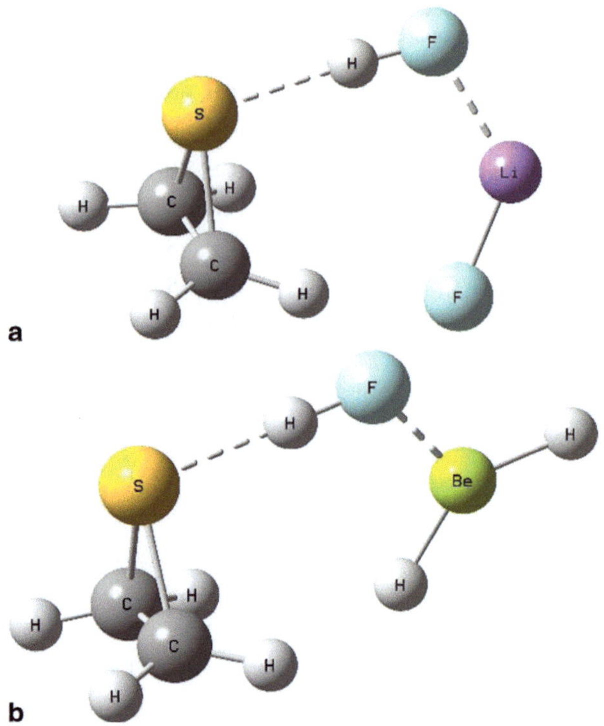

Fig. 12.7 Optimized geometries for **a** $(CH_2)_2S\ldots HF\ldots LiF$ and **b** $(CH_2)_2S\ldots HF\ldots BeH_2$

beryllium bonds having the largest, and the halogen bonds, the weakest effects. The optimized structures for thiirane... HF... LiF and thiirane... HF... BeH_2 (Fig. 12.7) show short contacts between the F of LiF (or one of the H atoms of BeH_2) and one of the nearby protons of thiirane, suggesting a cyclization involving the atoms of the three molecules constituting each trimer. This would suggest a substantial association involving all three molecules and is consistent with the large non-additive energies shown in Table 12.7.

12.2.3.2 Thiirane... ClF... X′F/BeH$_2$

1. A similar pattern to that observed for thiirane... HF... X′F is evident. The lithium and beryllium bonds produce more strongly-bound trimers (and shorter S... Cl contacts) than the halogen bonds, with the hydrogen bond falling in between. The largest ClF bond extensions and red shifts are also obtained with the lithium and beryllium bonds.
2. The substantial electron density transfer from the S atom of thiirane to ClF noted by Hill [108] is also evident from the NPA results in Table 12.8, which show an increase of 0.246*e* mainly into the antibonding σ^* orbital of ClF in the thiirane... ClF dimer. This transfer causes a significant increase in the ClF bondlength (by 0.127 Å) and appears to be further enhanced by the addition of the X′F or BeH_2 molecule to ClF.

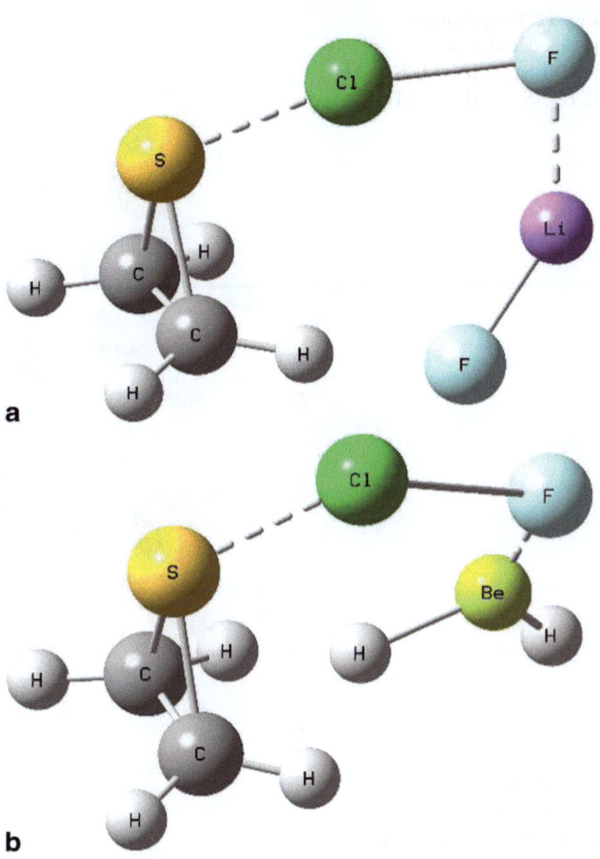

Fig. 12.8 Optimized geometries for **a** $(CH_2)_2S\ldots ClF\ldots LiF$ and **b** $(CH_2)_2S\ldots ClF\ldots BeH_2$

For example, Table 12.8 shows an increase in the positive charge on the S atom (corresponding to greater charge transfer to ClF) going from the less polarizing HF and ClF molecules to the more polarizing LiF and BeH_2 molecules. The S...Cl distance also decreases from 2.459 Å in the thiirane...ClF dimer to 2.278 Å for the most weakly-bound trimer (thiirane...ClF...ClF) to 2.072 Å for the strongly-bound thiirane...ClF...BeH_2, see Table 12.6.

3. Hence, the addition of a more polarizing X′F/BeH_2 molecule to the thiirane...ClF dimer leads to a stronger inner complex trimer. Table 12.7 also shows substantial cooperativity in the thiirane...ClF...X′F series, to a much greater extent than is evident for the thiirane...HF...X′F set. The F...Li and F...Be distances in the optimized structures of thiirane...ClF...LiF and thiirane...ClF...BeH_2 are short (1.757 and 1.444 Å, respectively). These complexes are more strongly-bound than the other members of this set of trimers; there is evidence of the cyclization of atoms from each of the three molecules constituting the trimer—the F atom (of LiF) and the H atom (of BeH_2) make close contacts (1.680 and 1.456 Å, respectively) with one of the nearby protons of the CH_2 fragment of thiirane. There is little evidence of strong association between the terminal thiirane and X′F molecules in the other members of this set.

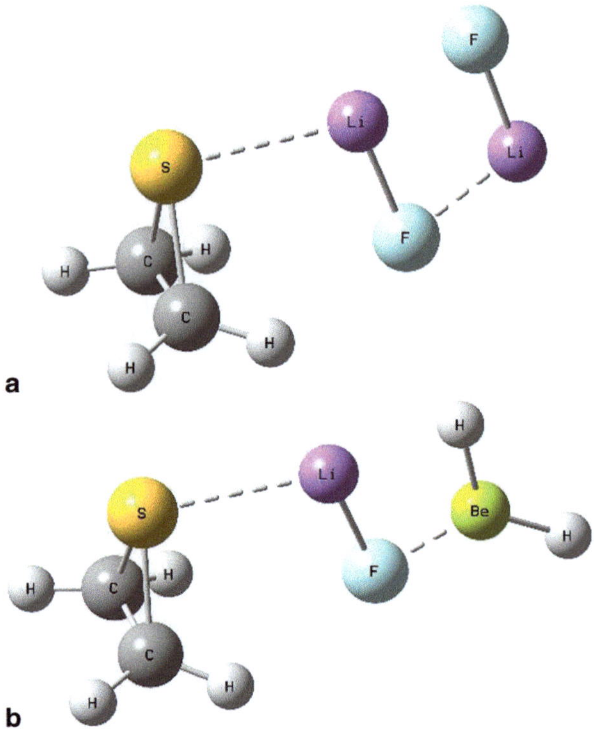

Fig. 12.9 Optimized geometries for **a** $(CH_2)_2S...LiF...LiF$ and **b** $(CH_2)_2S...LiF...BeH_2$

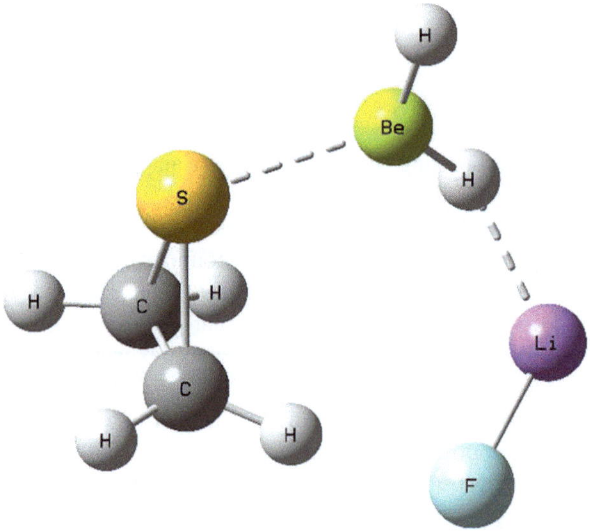

Fig. 12.10 Optimized geometry for $(CH_2)_2S...BeH_2...LiF$

Table 12.7 Sum of pairwise additive ($\sum \Delta E_{ij}$) and percentage non-additive contributions ($\% E_{non-add}$) of ΔE for $[(CH_2)_2S]^a \ldots [XF/BeH_2]^b \ldots [X'F/BeH_2]^c$ (X, X′ = H, Cl, Br, Li) complexes at the B3LYP/6-311++G(3df,3pd) level. All values are measured in kJ/mol

Complexes	ΔE	$\sum \Delta E_{ij}$	$\% E_{non-add}$
a..........b..........c			
$(CH_2)_2S \ldots HF \ldots HF$	− 72.0	− 50.1	30.4
$(CH_2)_2S \ldots HF \ldots ClF$	− 52.0	− 38.2	26.5
$(CH_2)_2S \ldots HF \ldots BrF$	− 58.2	− 40.1	31.1
$(CH_2)_2S \ldots HF \ldots LiF$	− 136.7	− 86.3	36.9
$(CH_2)_2S \ldots HF \ldots BeH_2$	− 97.8	4.0	104.1
$(CH_2)_2S \ldots ClF \ldots HF$	− 104.8	0.3	100.3
$(CH_2)_2S \ldots ClF \ldots ClF$	− 85.1	9.8	111.5
$(CH_2)_2S \ldots ClF \ldots LiF$	− 213.1	90.9	142.7
$(CH_2)_2S \ldots ClF \ldots BeH_2$	− 192.9	299.9	255.5
$(CH_2)_2S \ldots BrF \ldots HF$	− 110.8	− 41.7	62.4
$(CH_2)_2S \ldots BrF \ldots ClF$	− 91.9	− 34.0	63.0
$(CH_2)_2S \ldots BrF \ldots LiF$	− 199.2	10.1	105.1
$(CH_2)_2S \ldots BrF \ldots BeH_2$	− 165.0	21.3	112.9
$(CH_2)_2S \ldots LiF \ldots HF$	− 169.1	− 106.7	36.9
$(CH_2)_2S \ldots LiF \ldots BrF$	− 151.3	− 139.6	7.7
$(CH_2)_2S \ldots LiF \ldots LiF$	− 302.7	− 279.3	7.7
$(CH_2)_2S \ldots LiF \ldots BeH_2$	− 273.6	− 197.8	27.7
$(CH_2)_2S \ldots BeH_2 \ldots LiF$	− 152.0	− 56.3	63.0
$(CH_2)_2S \ldots BeH_2 \ldots BeH_2$	− 206.5	− 44.3	78.5

12.2.3.3 Thiirane... BrF... X′F/BeH$_2$

1. The trends observed for thiirane... ClF... X′F/BeH_2 are mirrored by this set of trimers, except that there is less charge transfer from S to Br (Table 12.8), smaller non-additive energies (Table 12.7), longer S... Br distances (Table 12.6) and F... X′/Be distances, and smaller X-F/BeH bond extensions/red shifts (not shown). The larger Br atom probably introduces more electron density in the intermolecular S... Br region (relative to the S... Cl region of thiirane... ClF... X′F) and thereby reduces the electron density transferred from thiirane.
2. The lithium and beryllium bonds due to LiF and BeH_2 produce the most strongly-bound trimers and the close interatomic contacts in these complexes suggest an association between the terminal molecules (as was the case for thiirane... ClF... X′F/BeH_2) which leads to relatively large interaction and non-additive energies, compared with the other members of this set.

Table 12.8 Changes in NPA partial atomic charges (Δq in e) w.r.t isolated monomers for $(CH_2)_2S...XF/BeH_2...X'F/BeH_2$ (X′, X = H, Cl, Br, Li) complexes and electron density transfer (EDT in e) determined using the B3LYP/6-311++G(3df,3pd) procedure. Partial atomic charges for monomers are in parentheses

Complex	Δq (S)	Complex	Δq (S)
$(CH_2)_2S...HF$	0.003 (0.085)	$(CH_2)_2S...BrF...HF$	0.276
$(CH_2)_2S...ClF$	0.246	$(CH_2)_2S...BrF...ClF$	0.270
$(CH_2)_2S...BrF$	0.200		
$(CH_2)_2S...LiF$	− 0.108	$(CH_2)_2S...BrF...LiF$	− 0.336
$(CH_2)_2S...BeH_2$	− 0.061	$(CH_2)_2S...BrF...BeH_2$	0.372
$(CH_2)_2S...HF...HF$	0.009	$(CH_2)_2S...LiF...HF$	− 0.078
$(CH_2)_2S...HF...ClF$	0.016		
$(CH_2)_2S...HF...BrF$	0.021	$(CH_2)_2S...LiF...BrF$	− 0.073
$(CH_2)_2S...HF...LiF$	0.009	$(CH_2)_2S...LiF...LiF$	− 0.059
$(CH_2)_2S...HF...BeH_2$	− 0.278	$(CH_2)_2S...LiF...BeH_2$	− 0.057
$(CH_2)_2S...ClF...HF$	0.378		
$(CH_2)_2S...ClF...ClF$	0.350		
$(CH_2)_2S...ClF...LiF$	0.473	$(CH_2)_2S...BeH_2...LiF$	− 0.118
$(CH_2)_2S...ClF...BeH_2$	0.578	$(CH_2)_2S...BeH_2...BeH_2$	− 0.012

12.2.3.4 Thiirane. . . LiF. . . X′F/BeH$_2$

1. The thiirane. . . LiF dimer shows the least electron density transfer from thiirane to XF (Table 12.7) probably because this charge shift is likely to be retarded by repulsion from the tightly held core electrons on Li. On the other hand, the binding in this most strongly-bound dimer is dominated by the polarization induced in thiirane by the large LiF dipole moment.
2. The thiirane. . . LiF. . . X′F/BeH$_2$ complexes are the most strongly-bound of all the trimer sets and their structures suggest that the LiF and X′F/BeH$_2$ molecular subunits form strong aggregates or clusters, which interact with thiirane through a S. . . Li contact. This idea is supported by the fact that the LiF. . . X′F/BeH$_2$ pair energies were found to be the most dominant attractive component of the interaction, followed by the thiirane. . . LiF pair energy.
3. The non-additive energies are relatively low for these trimers, with the largest cooperative effect estimated for the X′F = HF species (representing 37 % of ΔE). The thiirane. . . LiF. . . LiF is the most strongly bound trimer ($\Delta E = -303$ kJ/mol) and has the closest contact between the F atom of LiF and one of the protons of thiirane.

Table 12.9 B3LYP/6-311++G(3df,3pd) QTAIM electron density (ρ/ a.u) and negative Laplacian of electron density (*L*/a.u) at the S...X/Be and X-F bond critical points for $(CH_2)_2S...XF/BeH_2$, $FX'/H_2Be...(CH_2)_2S...XF/BeH_2$ (X′, X = H, Cl, Br, Li) complexes

Complex	(ρ,*L*)S...X or (ρ,*L*)S...Be
$(CH_2)_2S...HF$	(0.034, – 0.045)
$(CH_2)_2S...ClF$	(0.064, – 0.090)
$(CH_2)_2S...BrF$	(0.058, – 0.072)
$(CH_2)_2S...LiF$	(0.020, – 0.083)
$(CH_2)_2S...BeH_2$	(0.042, – 0.134)
$FH...(CH_2)_2S...HF$	(0.029, – 0.048)
$FH...(CH_2)_2S...ClF$	(0.055, – 0.092)
$FH...(CH_2)_2S...BrF$	(0.052, – 0.075)
$FH...(CH_2)_2S...LiF$	(0.018, – 0.075)
$FH...(CH_2)_2S...BeH_2$	(0.038, – 0.126)

12.2.3.5 Thiirane...BeH_2...X′F/BeH_2

1. These complexes are the most unstable set of trimers. In the thiirane...BeH_2 dimer, the small and highly positive Be atom can approach the S atom quite closely. However, this close approach causes the hydridic H atoms to be repelled by the S lone pair thereby pushing the H atoms away from the S atom and bending the H-Be-H angle. This creates a dipole moment which can then polarize the S atom, shifting electron density onto the S atom and making it less positive and the Be atom more positive (Table 12.8).
   Attachment of a hydrogen bond or halogen bonds to one of the H atoms of BeH_2 apparently destabilizes the complex, possibly because the electron cloud of the H atom is too diffuse to form a strong attachment to the proton of HF or repels the electron density on the halogen atom (Cl or Br).
2. The thiirane...BeH_2...LiF trimer shows cyclization of the atoms from each of the three monomers, with a short contact of 1.829 Å between the H (of thiirane) and the F atom of LiF joining the two terminal molecules together. The non-additive energy that results from this interaction is estimated to be about 63 % of ΔE (Table 12.7) and indicates significant positive cooperativity.
3. By contrast, the thiirane...BeH_2...BeH_2 trimer is more strongly bound and benefits from a strong aggregation between the two BeH_2 molecules. The BeH_2...BeH_2 pair energy and the non-additive energy dominate the interaction and suggest the existence of a sizeable Be...Be interaction.

12.2.4 *AIM Analysis of Thiirane...XF/BeH_2, FX′/H_2Be...Thiirane...XF/BeH_2, and Thiirane...XF/BeH_2...X′F/BeH_2*

For the thiirane...XF/BeH_2 dimer, ρ varies between 0.020 and 0.064 a.u and *L* between – 0.045 and – 0.134 a.u (Table 12.9) for the S...X/Be bcp, indicating

Table 12.10 B3LYP/6-311++G(3df,3pd) QTAIM electron density (ρ) and negative Laplacian of electron density (*L*) at the S…X/Be and X-F bond critical points for $(CH_2)_2S\ldots XF/BeH_2$ and $(CH_2)_2S\ldots XF/BeH_2\ldots X'F/BeH_2$ (X′, X = H, Cl, Br, Li) complexes

Complex	(ρ,*L*)S…X or (ρ,*L*)S…Be	Complex	(ρ,*L*)S…X or (ρ,*L*)S…Be
$(CH_2)_2S\ldots HF$	(0.034, – 0.045)	$(CH_2)_2S\ldots BrF\ldots HF$	(0.075, – 0.061)
$(CH_2)_2S\ldots ClF$	(0.064, – 0.090)	$(CH_2)_2S\ldots BrF\ldots ClF$	(0.073, – 0.063)
$(CH_2)_2S\ldots BrF$	(0.058, – 0.072)		
$(CH_2)_2S\ldots LiF$	(0.020, – 0.083)	$(CH_2)_2S\ldots BrF\ldots LiF$	(0.099, – 0.023)
$(CH_2)_2S\ldots BeH_2$	(0.042, – 0.134)	$(CH_2)_2S\ldots BrF\ldots BeH_2$	(0.098, – 0.029)
$(CH_2)_2S\ldots HF\ldots HF$	(0.045, – 0.029)	$(CH_2)_2S\ldots LiF\ldots HF$	(0.020, – 0.081)
$(CH_2)_2S\ldots HF\ldots ClF$	(0.040, – 0.036)		
$(CH_2)_2S\ldots HF\ldots BrF$	(0.043, – 0.031)	$(CH_2)_2S\ldots LiF\ldots BrF$	(0.021, – 0.083)
$(CH_2)_2S\ldots HF\ldots LiF$	(0.059, – 0.005)	$(CH_2)_2S\ldots LiF\ldots LiF$	(0.018, – 0.072)
$(CH_2)_2S\ldots HF\ldots BeH_2$	(0.068, – 0.032)	$(CH_2)_2S\ldots LiF\ldots BeH_2$	(0.020, – 0.080)
$(CH_2)_2S\ldots ClF\ldots HF$	(0.096, – 0.079)		
$(CH_2)_2S\ldots ClF\ldots ClF$	(0.094, – 0.069)		
$(CH_2)_2S\ldots ClF\ldots LiF$	(0.136, – 0.029)	$(CH_2)_2S\ldots BeH_2\ldots LiF$	(0.052, – 0.145)
$(CH_2)_2S\ldots ClF\ldots BeH_2$	(0.143, – 0.061)	$(CH_2)_2S\ldots BeH_2\ldots BeH_2$	(0.044, – 0.149)

non-covalent interactions. In the $FX'/H_2Be\ldots$ thiirane… XF/BeH_2trimers, for fixed XF/BeH_2, there is little variation in ρ and *L* values for S…X/Be bonds. In going from dimer to trimer, there is a decrease in ρ due to the FX′ interaction, consistent with the competition for the S lone pair electron density by both FX′ and XF. This decrease in ρ is also consistent with the previously discussed negative cooperative effect.

Compared with FX′…thiirane…XF, there is a greater variation in ρ and *L* for the S…X bcp for the thiirane…$XF/BeH_2\ldots X'F/BeH_2$ trimers. Going from dimer to trimer, ρ(S…X/Be) increases, which is consistent with the positive cooperative effect due to FX′. Generally, the most polarizing molecules, LiF and BeH_2, produce the largest ρ(S…X/Be) values since these molecules would shift more electron density into the S…X intermolecular region. (Table 12.10)

12.3 Conclusions

a. Our study supports and extends Hill's work on thiirane…ClF, where he proposed that this species be considered as an example of a Mulliken inner complex.
b. The binding strength of the thiirane…XF/BeH_2 dimers increases in the order: lithium bond > halogen bond > beryllium bond > hydrogen bond.

c. For the $FX'/H_2Be...(CH_2)_2S...XF/BeH_2$ trimers, where the thiirane acts as a double acceptor, negative cooperativity is evident, with the largest effects due to the halogen and beryllium bonds.
d. For the $(CH_2)_2S...XF/BeH_2...X'F/BeH_2$trimers, the binding strength depends on the polarizing power of the terminal $X'F/BeH_2$molecules and large cooperative effects are evident, with some complexes showing cyclization involving the terminal molecules.
e. The diminutive effect in $FX'/H_2Be...(CH_2)_2S...XF/BeH_2$and the cooperative effect in $(CH_2)_2S...XF/BeH_2...X'F/BeH_2$are mirrored by the respective decrease and increase in the electron density in the intermolecular S...X region.
f. Generally, the complexes with relatively small S to XF charge transfer are dominated by strong electrostatic interactions (e.g. XF = LiF and HF), whereas greater charge transfer occurs in the halogen-bonded systems, where the electrostatic contribution to the binding is presumably relatively smaller than in the more strongly-bound analogues.

Acknowledgement We would like to thank the School for Graduate Studies and Research of the University of the West Indies, Cave Hill Campus, for financial support.

References

1. Pimentel G, McClellan A (1960) The hydrogen bond. W. H. Freeman & Co., San Francisco
2. Henri-Rousseau O, Blaise P, Hadzi D (1997) Theoretical treatments of hydrogen bonding. Wiley, New York
3. Jeffrey GA (1997) An introduction to hydrogen bonding, vol 12. Oxford University Press, New York
4. Scheiner S (1997) Hydrogen bonding: a theoretical perspective. Oxford University Press, New York
5. Desiraju GR (1999) Weak hydrogen bond. Oxford University Press, New York
6. Buckingham AD, Del Bene JE, McDowell SAC (2008) The hydrogen bond. Chem Phys Lett 463(1–3):1-10. doi:10.1016/j.cplett.2008.06.060
7. Arunan E, Desiraju GR, Klein RA, Sadlej J, Scheiner S, Alkorta I, Clary DC, Crabtree RH, Dannenberg JJ, Hobza P, Kjaergaard HG, Legon AC, Mennucci B, Nesbitt DJ (2011) Definition of the hydrogen bond (iupac recommendations 2011). Pure Appl Chem 83(8):1637–1641. doi:10.1351/pac-rec-10–01-02
8. Metrangolo P, Resnati G (2001) Halogen bonding: a paradigm in supramolecular chemistry. Chem-a Eur J 7(12):2511–2519. doi:10.1002/1521-3765(20010618)7:12<2511::aid-chem25110>3.0.co;2-t
9. Clark T, Hennemann M, Murray JS, Politzer P (2007) Halogen bonding: the sigma-hole. J Mol Model 13(2):291–296. doi:10.1007/s00894-006-0130-2
10. Politzer P, Murray JS (2013) Halogen bonding: an interim discussion. Chemphyschem 14(2):278–294. doi:10.1002/cphc.201200799
11. Bleiholder C, Werz DB, Koppel H, Gleiter R (2006) Theoretical investigations on chalcogen-chalcogen interactions: what makes these nonbonded interactions bonding? J Am Chem Soc 128(8):2666–2674. doi:10.1021/ja056827g
12. Murray JS, Lane P, Clark T, Politzer P (2007) Sigma-hole bonding: molecules containing group vi atoms. J Mol Model 13(10):1033–1038. doi:10.1007/s00894-007-0225-4

13. Murray JS, Lane P, Politzer P (2007) A predicted new type of directional noncovalent interaction. Int J Quantum Chem 107(12):2286–2292. doi:10.1002/qua.21352
14. Murray JS, Lane P, Politzer P (2009) Expansion of the sigma-hole concept. J Mol Model 15(6):723–729. doi:10.1007/s00894-008-0386-9
15. Bauza A, Mooibroek TJ, Frontera A (2013) Tetrel-bonding interaction: re-discovered supramolecular force? Angew Chem-Int Ed 52(47):12317–12321. doi:10.1002/anie.201306501
16. Auffinger P, Hays FA, Westhof E, Ho PS (2004) Halogen bonds in biological molecules. Proc Natl Acad Sci U S A 101(48):16789–16794. doi:10.1073/pnas.0407607101
17. Politzer P, Lane P, Concha MC, Ma YG, Murray JS (2007) An overview of halogen bonding. J Mol Model 13(2):305–311. doi:10.1007/s00894-006-0154-7
18. Metrangolo P, Resnati G (2008) Halogen bonding: fundamentals and applications, vol 126. Springer,
19. Legon AC (2010) The halogen bond: an interim perspective. Phys Chem Chem Phys 12(28):7736–7747. doi:10.1039/c002129f
20. Politzer P, Murray JS, Clark T (2010) Halogen bonding: an electrostatically-driven highly directional noncovalent interaction. Phys Chem Chem Phys 12(28):7748–7757. doi:10.1039/c004189k
21. Desiraju GR, Ho PS, Kloo L, Legon AC, Marquardt R, Metrangolo P, Politzer P, Resnati G, Rissanen K (2013) Definition of the halogen bond (iupac recommendations 2013). Pure Appl Chem 85(8):1711–1713. doi:10.1351/pac-rec-12-05-10
22. Politzer P, Murray JS, Clark T (2013) Halogen bonding and other sigma-hole interactions: a perspective. Phys Chem Chem Phys 15(27):11178–11189. doi:10.1039/c3cp00054k
23. Wang WZ, Wong NB, Zheng WX, Tian AM (2004) Theoretical study on the blueshifting halogen bond. J Phys Chem A 108(10):1799–1805. doi:10.1021/jp036769q
24. Murray JS, Concha MC, Lane P, Hobza P, Politzer P (2008) Blue shifts vs red shifts in sigma-hole bonding. J Mol Model 14(8):699–704. doi:10.1007/s00894-008-0307-y
25. Brinck T, Murray JS, Politzer P (1992) Surface electrostatic potentials of halogenated methanes as indicators of directional intermolecular interactions. Int J Quantum Chem 57–64
26. Murray JS, Paulsen K, Politzer P (1994) Molecular-surface electrostatic potentials in the analysis of non-hydrogen-bonding noncovalent interactions. Proc Indian Acad Sci-Chem Sci 106(2):267–275
27. Murray JS, Politzer P (1998) Statistical analysis of the molecular surface electrostatic potential: an approach to describing noncovalent interactions in condensed phases. Theochem-J Mol Struct 425(1–2):107–114
28. Politzer P, Murray JS (2002) The fundamental nature and role of the electrostatic potential in atoms and molecules. Theor Chem Acc 108(3):134–142. doi:10.1007/s00214-002-0363-9
29. Murray JS, Politzer P (2011) The electrostatic potential: an overview. Wiley Interdiscip Rev-Comput Mol Sci 1(2):153–163. doi:10.1002/wcms.19
30. Riley KE, Murray JS, Fanfrlik J, Rezac J, Sola RJ, Concha MC, Ramos FM, Politzer P (2011) Halogen bond tunability i: the effects of aromatic fluorine substitution on the strengths of halogen-bonding interactions involving chlorine, bromine, and iodine. J Mol Model 17(12):3309–3318. doi:10.1007/s00894-011-1015-6
31. Bundhun A, Ramasami P, Murray JS, Politzer P (2013) Trends in sigma-hole strengths and interactions of F3MX molecules (M = C, Si, Ge and X = F, Cl, Br, I). J Mol Model 19(7):2739–2746. doi:10.1007/s00894-012-1571-4
32. Murray-Rust P, Motherwell WDS (1979) Computer retrieval and analysis of molecular geometry. 4. Intermolecular interactions. J Am Chem Soc 101(15):4374–4376
33. Hassel O, Hvoslef J (1954) The structure of bromine 1, 4-dioxanate. Acta Chem Scand 8(5):873–873
34. Zhou P-P, Qiu W-Y, Liu S, Jin N-Z (2011) Halogen as halogen-bonding donor and hydrogen-bonding acceptor simultaneously in ring-shaped H3NX(Y)HF ($X = Cl, Br$ and $Y = F, Cl, Br$) complexes. Phys Chem Chem Phys 13(16):7408–7418. doi:10.1039/c1cp00025j

35. McDowell SAC, Joseph JA (2012) Communication: an unusual halogen-bonding motif: the LiBr...BrF dimer as a model system. J Chem Phys 137(17). doi:10.1063/1.4766932
36. Bilewicz E, Rybarczyk-Pirek AJ, Dubis AT, Grabowski SJ (2007) Halogen bonding in crystal structure of 1-methylpyrrol-2-yl trichloromethyl ketone. J Mol Struct 829(1–3):208–211. doi:10.1016/j.molstruc.2006.06.032
37. Politzer P, Murray JS, Concha MC (2007) Halogen bonding and the design of new materials: organic bromides, chlorides and perhaps even fluorides as donors. J Mol Model 13(6–7):643–650. doi:10.1007/s00894-007-0176-9
38. Price SL, Stone AJ, Lucas J, Rowland RS, Thornley AE (1994) The nature of -Cl...Cl- intermolecular interactions. J Am Chem Soc 116(11):4910–4918. doi:10.1021/ja00090a041
39. Voth AR, Hays FA, Ho PS (2007) Directing macromolecular conformation through halogen bonds. Proc Natl Acad Sci U S A 104(15):6188–6193. doi:10.1073/pnas.0610531104
40. Lu Y, Shi T, Wang Y, Yang H, Yan X, Luo X, Jiang H, Zhu W (2009) Halogen bonding-a novel interaction for rational drug design? J Med Chem 52(9):2854–2862. doi:10.1021/jm9000133
41. Voth AR, Khuu P, Oishi K, Ho PS (2009) Halogen bonds as orthogonal molecular interactions to hydrogen bonds. Nat Chem 1(1):74–79. doi:10.1038/nchem.112
42. Hardegger LA, Kuhn B, Spinnler B, Anselm L, Ecabert R, Stihle M, Gsell B, Thoma R, Diez J, Benz J, Plancher J-M, Hartmann G, Banner DW, Haap W, Diederich F (2011) Systematic investigation of halogen bonding in protein-ligand interactions. Angew Chem-Int Ed 50(1):314–318. doi:10.1002/anie.201006781
43. Imakubo T, Sawa H, Kato R (1995) Novel radical-cation salts of organic pi-donors containing iodine atom(s)—the first application of strong intermolecular -I-X- (X = CN, halogen atom) interaction to molecular conductors. Synth Met 73(2):117–122. doi:10.1016/0379-6779(95)03322-x
44. Amico V, Meille SV, Corradi E, Mesina MT, Resnati G (1998) Perfluorocarbon–hydrocarbon self-assembling. 1d infinite chain formation driven by nitrogen···iodine interactions. J Am Chem Soc 120(32):8261–8262. doi:10.1021/ja9810686
45. Iwaoka M, Takemoto S, Tomoda S (2002) Statistical and theoretical investigations on the directionality of nonbonded (SO)-O- Interactions. Implications for molecular design and protein engineering. J Am Chem Soc 124(35):10613–10620. doi:10.1021/ja026472q
46. Metrangolo P, Neukirch H, Pilati T, Resnati G (2005) Halogen bonding based recognition processes: a world parallel to hydrogen bonding. Acc Chem Res 38(5):386–395. doi:10.1021/ar0400995
47. Saha BK, Nangia A, Jaskolski M (2005) Crystal engineering with hydrogen bonds and halogen bonds. Crystengcomm 7:355–358. doi:10.1039/b501693b
48. Syssa-Magale JL, Boubekeur K, Schollhorn B (2005) First molecular self-assembly of 1,4-diiodo-tetrafluoro-benzene and a ketone via (o...) non-covalent halogen bonds. J Mol Struct 737(2–3):103–107. doi:10.1016/j.molstruc.2004.10.008
49. Lucassen ACB, Karton A, Leitus G, Shimon LJW, Martin JML, van der Boom ME (2007) Co-crystallization of sym-triiodo-trifluorobenzene with bipyridyl donors: consistent formation of two instead of anticipated three N...I halogen bonds. Cryst Growth Des 7(2):386–392. doi:10.1021/cg0607250
50. Metrangolo P, Meyer F, Pilati T, Resnati G, Terraneo G (2008) Halogen bonding in supramolecular chemistry. Angew Chem-Int Ed 47(33):6114–6127. doi:10.1002/anie.200800128
51. Rissanen K (2008) Halogen bonded supramolecular complexes and networks. Crystengcomm 10(9):1107–1113. doi:10.1039/b803329n
52. Moilanen J, Ganesamoorthy C, Balakrishna MS, Tuononen HM (2009) Weak interactions between trivalent pnictogen centers: computational analysis of bonding in dimers X3E...EX3 (E = pnictogen, X = halogen). Inorg Chem 48(14):6740–6747. doi:10.1021/ic900635f
53. Kollman PA, Liebman JF, Allen LC (1970) Lithium bond. J Am Chem Soc 92(5):1142–1150
54. Ault BS, Pimentel GC (1975) Matrix isolation infrared studies of lithium bonding. J Phys Chem 79(6):621–626

55. Sannigrahi A, Kar T, Niyogi BG, Hobza P, Schleyer PvR (1990) The lithium bond reexamined. Chem Rev 90(6):1061–1076
56. Ammal SSC, Venuvanalingam P (1998) Ab initio and dft investigations of lithium/hydrogen bonded complexes of trimethylamine, dimethyl ether and dimethyl sulfide. J Chem Soc-Faraday Trans 94(18):2669–2674
57. Vila A, Vila E, Mosquera RA (2006) Topological characterisation of intermolecular lithium bonding. Chem Phys 326(2–3):401–408. doi:10.1016/j.chemphys.2006.02.032
58. Bader RFW (1990) Atoms in molecules—a quantum theory. Oxford University Press, Oxford
59. Lipkowski P, Grabowski SJ (2014) Could the lithium bond be classified as the sigma-hole bond?—qtaim and nbo analysis. Chem Phys Lett 591:113–118. doi:10.1016/j.cplett.2013.11.017
60. Feng Y, Liu L, Wang JT, Li XS, Guo QX (2004) Blue-shifted lithium bonds. Chem Commun (1):88–89. doi:10.1039/b310723j
61. Li QZ, Wang YF, Li WZ, Cheng JB, Gong BA, Sun JZ (2009) Prediction and characterization of the HMgHLiX (X = H, OH, F, CCH, CN, and NC) complexes: a lithium–hydride lithium bond. Phys Chem Chem Phys 11(14):2402–2407. doi:10.1039/b820309a
62. McDowell SAC, St Hill JAS (2011) A theoretical study of hydrogen- and lithium-bonded complexes of F-H/Li and Cl-H/Li with NF3, NH3, and NH2(CH3). J Chem Phys 135(16). doi:10.1063/1.3653476
63. Yanez M, Sanz P, Mo O, Alkorta I, Elguero J (2009) Beryllium bonds, do they exist? J Chem Theory Comput 5(10):2763–2771. doi:10.1021/ct900364y
64. Eskandari K (2012) Characteristics of beryllium bonds; a qtaim study. J Mol Model 18(8):3481–3487. doi:10.1007/s00894-012-1360-0
65. Yanez M, Mo O, Alkorta I, Elguero J (2013) Can conventional bases and unsaturated hydrocarbons be converted into gas-phase superacids that are stronger than most of the known oxyacids? The role of beryllium bonds. Chem-a Eur J 19(35):11637–11643. doi:10.1002/chem.201300808
66. Suhai S (1994) Cooperative effects in hydrogen-bonding—4th-order many-body perturbation-theory studies of water oligomers and of an infinite water chain as a model for ice. J Chem Phys 101(11):9766–9782. doi:10.1063/1.467942
67. Tsuzuki S, Houjou H, Nagawa Y, Goto M, Hiratani K (2001) Cooperative enhancement of water binding to crownophane by multiple hydrogen bonds: analysis by high level ab initio calculations. J Am Chem Soc 123(18):4255–4258. doi:10.1021/ja0037264
68. Szczesna B, Urbanczyk-Lipkowska Z (2002) Cooperative effect of multiple hydrogen bonding involving the nitro group: Solid state dimeric self-assembly of o-, m- and p-hydroxyphenyl-2,4-dinitrophenylhydrazones. New J Chem 26(2):243–249. doi:10.1039/b105498h
69. Thallapally PK, Katz AK, Carrell HL, Desiraju GR (2002) Unusually long cooperative chain of seven hydrogen bonds. An alternative packing type for symmetrical phenols. Chem Commun (4):344–345. doi:10.1039/b110036j
70. Parra RD, Bulusu S, Zeng XC (2003) Cooperative effects in one-dimensional chains of three-center hydrogen bonding interactions. J Chem Phys 118(8):3499–3509. doi:10.1063/1.1535441
71. Araujo RCMU, Soares VM, Oliveira BG, Lopes KC, Ventura E, Do Monte SA, Santana OL, Carvalho AB, Ramos MN (2006) Theoretical study of cooperative effects in the homo- and heteromeric hydrogen bond chains (HCN)(n)–HF. with n = 1, 2, and 3. Int J Quantum Chem 106(13):2714–2722. doi:10.1002/qua.21132
72. Larsen RW, Suhm MA (2006) Cooperative organic hydrogen bonds: the librational modes of cyclic methanol clusters. J Chem Phys 125(15). doi:10.1063/1.2358349
73. Song H-J, Xiao H-M, Dong H-S (2006) Cooperative effects, strengths of hydrogen bonds, and intermolecular interactions in circular cis, trans-cyclotriazane clusters (n = 3-8). J Chem Phys 125(7). doi:10.1063/1.2336209

74. Znamenskiy VS, Green ME (2007) Quantum calculations on hydrogen bonds in certain water clusters show cooperative effects. J Chem Theory Comput 3(1):103–114. doi:10.1021/ct600139d
75. Fradelos G, Kaminski JW, Wesolowski TA, Leutwyler S (2009) Cooperative effect of hydrogen-bonded chains in the environment of pi → pi* chromophore. J Phys Chem A 113(36):9766–9771. doi:10.1021/jp906483z
76. Rodziewicz P, Rutkowski KS, Melikova SM, Koll A (2009) Cooperative effects in blue-shifted hydrogen bonded cluster of CF3H... HF)(1 < = n < = 3) from first principles simulations. Chem Phys 361 (3):129–136. doi:10.1016/j.chemphys.2009.05.017
77. Roztoczynska A, Kozlowska J, Lipkowski P, Bartkowiak W (2014) Does the spatial confinement influence the electric properties and cooperative effects of the hydrogen bonded systems? Hcn chains as a case study. Chem Phys Lett 608:264–268. doi:10.1016/j.cplett.2014.05.102
78. Planas JG, Vinas C, Teixidor F, Comas-Vives A, Ujaque G, Lledos A, Light ME, Hursthouse MB (2005) Self-assembly of mercaptane–metallacarborane complexes by an unconventional cooperative effect: a C–H...S–H...H–B hydrogen/dihydrogen bond interaction. J Am Chem Soc 127(45):15976–15982. doi:10.1021/ja055210w
79. Li Q-Z, Hu T, An X-L, Gong B-A, Cheng J-B (2008) Cooperativity between the dihydrogen bond and the N...HC hydrogen bond in lih–(hcn)n complexes. Chemphyschem 9(13):1942–1946. doi:10.1002/cphc.200800320
80. Solimannejad M, Rabbani M, Ahmadi A, Esrafili MD (2014) Cooperative and diminutive interplay between the sodium bonding with hydrogen and dihydrogen bondings in ternary complexes of NaC3N with HMgH and HCN (HNC). Mol Phys 112(15):2017–2022. doi:10.1080/00268976.2013.879496
81. Lankau T, Wu Y-C, Zou J-W, Yu C-H (2008) The cooperativity between hydrogen and halogen bonds. J Theor Comput Chem 7(1):13–35. doi:10.1142/s0219633608003563
82. Li Q, Lin Q, Li W, Cheng J, Gong B, Suo J (2008) Cooperativity between the halogen bond and the hydrogen bond in H3N...XY...HF complexes (X, Y = F, Cl, Br). Chemphyschem 9(15):2265–2269. doi:10.1002/cphc.200800467
83. Jing B, Li Q, Li R, Gong B, Liu Z, Li W, Cheng J, Sun J (2011) Competition and cooperativity between hydrogen bond and halogen bond in HNC...(HOBr)n and (HNC)n...HOBr (n = 1 and 2) systems. Comput Theor Chem 963(2–3):417–421. doi:10.1016/j.comptc.2010.11.006
84. Solimannejad M, Malekani M, Alkorta I (2011) Cooperativity between the hydrogen bonding and halogen bonding in F3CX ... NCH(CNH) ... NCH(CNH) complexes (X = Cl, Br). Mol Phys 109(13):1641–1648. doi:10.1080/00268976.2011.582050
85. Zhao Q, Feng D, Hao J (2011) The cooperativity between hydrogen and halogen bond in the XY...HNC...XY (X, Y = F, Cl, Br) complexes. J Mol Model 17(11):2817–2823. doi:10.1007/s00894-011-0974-y
86. Li Q, Li R, Zhou Z, Li W, Cheng J (2012) S...X halogen bonds and H...X hydrogen bonds in H2CS–XY (XY = FF, ClF, ClCl, BrF, BrCl, and BrBr) complexes: cooperativity and solvent effect. J Chem Phys 136(1). doi:10.1063/1.3673540
87. Solimannejad M, Malekani M (2012) Cooperative and diminutive interplay between the hydrogen bonding and halogen bonding in ternary complexes of HCCX (X = Cl, Br) with hcn and hnc. Comput Theor Chem 998:34–38. doi:10.1016/j.comptc.2012.05.021
88. Grabowski SJ (2013) Cooperativity of hydrogen and halogen bond interactions. Theor Chem Acc 132(4). doi:10.1007/s00214-013-1347-7
89. Alkorta I, Blanco F, Deya PM, Elguero J, Estarellas C, Frontera A, Quinonero D (2010) Cooperativity in multiple unusual weak bonds. Theor Chem Acc 126(1–2):1–14. doi:10.1007/s00214-009-0690-1
90. McDowell SAC, Joseph JA (2012) Cooperative effects of noncovalent bonds to the Br atom of halogen-bonded H3N... BrZ and HCN... BrZ (Z = F, Br). J Chem Phys 137(7). doi:10.1063/1.4745838

91. Zhao XR, Wu YJ, Han J, Shen QJ, Jin WJ (2013) Theoretical study of the triangular bonding complex formed by carbon tetrabromide, a halide, and a solvent molecule in the gas phase. J Mol Model 19(1):299–304. doi:10.1007/s00894-012-1518-9
92. Li Q, Hu T, An X, Li W, Cheng J, Gong B, Sun J (2009) Theoretical study of the interplay between lithium bond and hydrogen bond in complexes involved with hli and hcn. Chemphyschem 10(18):3310–3315. doi:10.1002/cphc.200900549
93. McDowell SAC, Yarde HK (2012) Cooperative effects of hydrogen, lithium and halogen bonding on f–h/lioh2 complexes. Phys Chem Chem Phys 14(19):6883–6888. doi:10.1039/c2cp40203c
94. Solimannejad M, Rezaei Z, Esrafili MD (2013) Competition and interplay between the lithium bonding and hydrogen bonding: R3c center dot center dot center dot hy center dot center dot center dot liy and r3c center dot center dot center dot liy center dot center dot center dot hy triads as a working model (r = h, ch3; y = cn, nc). J Mol Model 19(11):5031–5035. doi:10.1007/s00894-013-2006-6
95. Solimannejad M (2012) Cooperative and diminutive interplay between lithium and dihydrogen bonding in f3yli center dot center dot center dot nch center dot center dot center dot center dot hmh and f3yli center dot center dot center dot hmh center dot center dot center dot hcn triads (y = c, si; m = be, mg). Chemphyschem 13(13):3158–3162. doi:10.1002/cphc.201200333
96. McDowell SAC, Joseph JA (2013) Cooperative effects in novel lif/hf···lif···xf (x = f, cl, br) clusters. J Chem Phys 138(16). doi:10.1063/1.4801863
97. Li Q, Li R, Liu Z, Li W, Cheng J (2011) Interplay between halogen bond and lithium bond in mcn-licn-xcch (m = h, li, and na; x = cl, br, and i) complex: The enhancement of halogen bond by a lithium bond. J Comput Chem 32(15):3296–3303. doi:10.1002/jcc.21916
98. Esrafili MD, Esmailpour P, Mohammadian-Sabet F, Solimannejad M (2013) Theoretical study of the interplay between halogen bond and lithium-pi interactions: cooperative and diminutive effects. Chem Phys Lett 588:47–50. doi:10.1016/j.cplett.2013.10.009
99. Esrafili MD, Mohammadian-Sabet F, Solimannejad M (2014) A theoretical evidence for mutual influence between s···n(c) and hydrogen/lithium/halogen bonds: Competition and interplay between π-hole and σ-hole interactions. Struct Chem 25(4):1197–1205. doi:10.1007/s11224-014-0392-8
100. Solimannejad M, Bayati E, Esrafili MD (2014) Enhancement effect of lithium bonding on the strength of pnicogen bonds: Xh2p···ncli···ncy as a working model (x = f, cl; y = h, f, cl, cn). Mol Phys 112(15):2058–2062. doi:10.1080/00268976.2014.884250
101. Li Q, Li R, Liu X, Cheng J, Li W (2012) Ab initio study of synergetic effects of two strong interactions of cation–π interaction and lithium bond in m+···phenyl lithium···n (m = li, na, k; n = h2o and nh3) complex. Mol Phys 110(8):457–465. doi:10.1080/00268976.2012.655793
102. Esrafili MD, Fatehi P, Solimannejad M (2013) Cooperative effects in cyclic licn and hcn clusters: a comparative study. Comput Theor Chem 1022:115–120. doi:10.1016/j.comptc.2013.08.011
103. Solimannejad M, Ghafari S, Esrafili MD (2013) Theoretical insight into cooperativity in lithium-bonded complexes: linear clusters of licn and linc. Chem Phys Lett 577:6–10. doi:10.1016/j.cplett.2013.05.023
104. Esrafili MD, Esmailpour P, Mohammadian-Sabet F, Solimannejad M (2014) Substituent effects on cooperativity between lithium bonds. Int J Quantum Chem 114(4):295–301. doi:10.1002/qua.24560
105. Albrecht L, Boyd RJ, Mo O, Yanez M (2014) Changing weak halogen bonds into strong ones through cooperativity with beryllium bonds. J Phys Chem A 118(23):4205–4213. doi:10.1021/jp503229u
106. Albrecht L, Boyd RJ, Mo O, Yanez M (2012) Cooperativity between hydrogen bonds and beryllium bonds in (h2o)(n)bex2 (n = 1-3, x = h, f) complexes. A new perspective. Phys Chem Chem Phys 14(42):14540–14547. doi:10.1039/c2cp42534c
107. Mo O, Yanez M, Alkorta I, Elguero J (2012) Modulating the strength of hydrogen bonds through beryllium bonds. J Chem Theory Comput 8(7):2293–2300. doi:10.1021/ct300243b

108. Hill JG (2014) The halogen bond in thiirane... Clf: an example of a mulliken inner complex. Phys Chem Chem Phys 16(36):19137–19140. doi:10.1039/c4cp03412k
109. Mulliken RS, Person WB (1969) Molecular complexes: a lecture and reprint volume. Wiley-Interscience, New York
110. Legon AC (1999) Prereactive complexes of dihalogens xy with lewis bases b in the gas phase: a systematic case for the halogen analogue b—xy of the hydrogen bond b—hx. Angew Chem-Int Ed 38(18):2687–2714
111. Evans CM, Holloway JH, Legon AC (1996) Nature and angular geometry of the pre-reactive complex thiirane-chlorine monofluoride from its rotational spectrum. Chem Phys Lett 255(1–3):119–128. doi:10.1016/0009-2614(96)00340-5
112. Reed AE, Curtiss LA, Weinhold F (1988) Intermolecular interactions from a natural bond orbital, donor-acceptor viewpoint. Chem Rev 88(6):899–926
113. Jeziorski B, Moszynski R, Szalewicz K (1994) Perturbation-theory approach to intermolecular potential-energy surfaces of van-der-waals complexes. Chem Rev 94(7):1887–1930. doi:10.1021/cr00031a008
114. Frisch MJ, Trucks GW, Schlegel HB, Scuseria GE, Robb MA, Cheeseman JR, Montgomery JA, Vreven T, Kudin KN, Burant JC, Millam JM, Iyengar SS, Tomasi J, Barone V, Mennucci B, Cossi M, Scalmani G, Rega N, Petersson GA, Nakatsuji H, Hada M, Ehara M, Toyota K, Fukuda R, Hasegawa J, Ishida M, Nakajima T, Honda Y, Kitao O, Nakai H, Klene M, Li X, Knox JE, Hratchian HP, Cross JB, Adamo C, Jaramillo J, Gomperts R, Stratmann RE, Yazyev O, Austin AJ, Cammi R, Pomelli C, Ochterski JW, Ayala PY, Morokuma K, Voth GA, Salvador P, Dannenberg JJ, Zakrzewski VG, Dapprich S, Daniels AD, Strain MC, Farkas O, Malick DK, Rabuck AD, Raghavachari K, Foresman JB, Ortiz JV, Cui Q, Baboul AG, Clifford S, Cioslowski J, Stefanov BB, Liu G, Liashenko A, Piskorz P, Komaromi I, Martin RL, Fox DJ, Keith T, Al-Laham MA, Peng CY, Nanayakkara A, Challacombe M, Gill PMW, Johnson B, Chen W, Wong MW, Gonzalez C, Pople JA (2003) Gaussian 03, revision b.05. Gaussian, Inc., Pittsburgh
115. Arman HD, Gieseking RL, Hanks TW, Pennington WT (2010) Complementary halogen and hydrogen bonding: sulfur center dot center dot center dot iodine interactions and thioamide ribbons. Chem Commun 46(11):1854–1856. doi:10.1039/b925710a
116. Li QZ, Jing B, Li R, Liu ZB, Li WZ, Luan F, Cheng JB, Gong BA, Sun JZ (2011) Some measures for making halogen bonds stronger than hydrogen bonds in h2cs-hox (x = f, cl, and br) complexes. Phys Chem Chem Phys 13(6):2266–2271. doi:10.1039/c0cp01543a
117. Emsley J (1998) The elements, 3rd edn. Clarendon Press, Oxford
118. Keith TA (2012) Aimall (version 12.11.09). TK Gristmill Software, Overland Park

Chapter 13
Understanding Lone Pair-π Interactions from Electrostatic Viewpoint

Shridhar R. Gadre and Anmol Kumar

Abstract Over the last two decades, studies on lone pair-π interaction have attracted lot of attention of experimental as well as theoretical chemists due to its intriguing nature and its suspected presence in biological systems. The present Chapter begins with a brief overview of the earlier theoretical and experimental work done in this area. This is followed by exploration of the nuances of bonding in lone pair-π interaction, employing the tool of molecular electrostatic potential (MESP) since such weak interactions are mainly dominated by electrostatic features of host and guest molecules. The critical points associated with the scalar field of MESP are exploited for scrutinizing the directionality and bonding sites involved in the lone pair-π complexes. Furthermore, the electrostatic potential for intermolecular complexation (EPIC) model developed by Gadre et al., has been employed for finding out the electrostatically optimized structures and interaction energies of these complexes. The outcomes of EPIC model are compared with the results obtained from quantum chemical calculations of the complexes employing M06L/6-311++G(d,p) level of theory. The present study details out four different cases of lone pair-π complexes, which are currently in vogue. Hexafluorobenzene, one of the most explored π-deficient host in the present context, is initially taken up to demonstrate various facets of MESP for gaining insights into this interaction. This is followed by the scrutiny of special classes of recently synthesized highly π-deficient molecules, viz. tetraoxacalix [2]arene[2]triazine and naphthalenediimide, which are known to have specificity and large affinity, respectively, towards the electron rich species. The chapter ends with the description of lone pair-π interaction in the case of urate oxidase, an enzyme present in biological systems.

S. R. Gadre (✉) · A. Kumar
Department of Chemistry, Indian Institute of Technology Kanpur, Kanpur 208016, India
e-mail: gadre@iitk.ac.in

A. Kumar
e-mail: anmol@iitk.ac.in

S. Scheiner (ed.), *Noncovalent Forces,* Challenges and Advances in Computational Chemistry and Physics 19, DOI 10.1007/978-3-319-14163-3_13

13.1 Introduction

Supramolecular chemistry is generally governed by non-covalent forces. The non-covalent interactions are known to play an indispensable role in ensuring functionality of most of the chemical, physical and especially biological processes. Their relevance is brought out by several examples, e.g. the existence of water in liquid form is solely due to non-covalent interaction [1] viz. the hydrogen bond. Among the various types of non-covalent interactions, there exists a category, which involves interactions with aromatic rings such as in π-π, CH-π, cation-π interactions. Several biochemical phenomenon, either mechanical or cognitive such as protein folding, DNA/RNA stacking, drug-receptor interactions etc. are the consequence of the harmonic interplay between many such non-covalent interactions. Synthetic chemists exploit such weak interactions in the field of biomimetics [2], crystal engineering, medicine [3, 4], catalysts [5], material science etc. A recent addition to the aforementioned list of non-covalent aromatic interactions is the interaction of π-deficient aromatic rings with lone pair-containing species, popularly known as anion-π interaction.

A glimpse at this interaction may leave one amused as the aromatic ring is generally perceived to be endowed with a π electron cloud and the very idea of interaction between lone pair and π cloud may seem antithetical. However, theoretical as well as experimental investigations are suggestive of the attractive nature of this interaction. Due to its rather mysterious nature and its possibility of existence in biological systems, lone pair-π interaction has recently captured the attention of many theoretical and experimental chemists. Being neutral in nature, π-deficient aromatic ring also offers a promising alternative to cationic anion-receptors. Although some active research on the anion-π interaction is being pursued, its cationic counterpart viz. cation-π interaction is widely recognized as an important non-covalent interaction by the scientific community since long [6–13]. The significant role of the latter in biological systems can be depicted through many examples such as protein crystal structures, enzyme catalysis, molecular signaling etc. to name a few. It has been verified that the nucleic acid bases in double-stranded DNA exhibit cation-π interaction with amino acid side chains of proteins at the interface between protein and DNA.

In the recent past, several reviews on the anion-π interaction have been published [14–21], providing quite comprehensive details of research done in this field. The present Chapter is aimed at providing a brief summary of the earlier research done on lone pair-π interactions and then scrutinize the nature of such interactions employing the tools of molecular electrostatics and *ab initio* calculations to provide mechanistic insights.

Before proceeding further, we would like to clarify the usage of the terms, "lone pair-π interaction" and "anion-π interaction" in the present Chapter. The term lone pair-π interaction seems more generalized than anion-π interaction due to three reasons:

1. Every anion has at least one lone pair and the interaction with π-deficient aromatic system generally occurs through the lone pair of anion.
2. Even the neutral molecules possessing lone pairs can interact with π-deficient aromatic system.
3. The usage of the term "lone pair" brings in the sense of directionality of the interaction.

In view of this, the present Chapter will employ both the terms viz. "lone pair-π" and "anion-π", depending on the nature of the species involved.

Although the term "anion-π interaction" was first coined by Quiñonero et al. [22] in 2002, its existence was already noted long back in 1973 by Boden et al. [23] in X ray analysis of solid hexafluorobenzene (HFB) at 120 K. This showed fluorine atom of one HFB molecule lying above the centroid of the other one in a unit cell of monoclinic lattice. Later in 1980, Vrbancich et al. [24] experimentally verified the quadrupole moment of hexafluorobenzene, benzene and other non-dipolar aromatic molecules using electric field-gradient birefringence method. Unlike benzene, which was found to possess a large negative quadrupole moment of $(-33.3 \pm 2.1) \times 10^{-40}$ Cm^2, hexafluorobenzene showed a highly positive quadrupole moment of $(+31.7 \pm 1.7) \times 10^{-40}$ Cm^2 explaining the possible cause of interaction between the fluorine atom of one HFB and the centroid of the other. Interestingly, the actual experimental verification of existence of lone pair-π interaction was achieved by Schneider et al. [25–27] in the early 90's. NMR shifts indicated attractive interaction between anions and organic neutral π-deficient arene systems, the change in the free energy for which, was estimated (through force field calculations) to be about 2 kJ mol^{-1}.

Research in this area till early nineties had been sporadic and accelerated only after the simultaneous pioneering theoretical studies done by the groups of Alkorta [28, 29] and Dougherty [30] in the late 1990's. In their works, Alkorta and coworkers [28, 29, 31] studied the ability of perfluoro aromatic rings to bind to electron donor molecules/anions such as HF, :CH^{-2}, F^-, Cl^-, CN^-. Since all the latter systems exhibit diffuse electronic distribution, Alkorta et al. performed quantum chemical calculations on them at HF, MP2 and DFT levels of theory employing Pople basis sets with diffuse and polarization functions. The results confirmed the presence of local minimum in PES for complexes formed between perfluoro aromatic rings and electron donor molecules. For the systems studied by them, the interaction energy was found to vary widely in the range of -1 to -20 kcal/mol depending on electron donating tendency of anions/lone pair-containing molecule and π deficiency of the aromatic ring. In a similar quest, Gallivan et al. [30] determined the attractive nature of interaction between lone pair of water and HFB ring employing quantum chemical calculations and the interaction energy was found to be quite comparable with that of global minimum of water-benzene complex.

Significant contributions into understanding the nature of anion-π interaction, the role of anions and π systems were made by Frontera et al. [22, 32–34]. Molecular potential with polarization model (MIPp) was employed by them for separating out components of interaction energy into electrostatic, classical dispersion-repulsion and perturbationally derived polarization terms. Critical analysis of weak interactions may also be carried out by employing symmetry adapted perturbation theory (SAPT) [35, 36] which separates out physically meaningful components of a given interaction viz. electrostatics, induction and dispersion interaction. To achieve this decomposition, the corresponding Hamiltonian is partitioned into monomer Fock operators, Møller-Plesset fluctuation operators and the intermolecular interaction operators. It was first employed by Kim et al. [37] for scrutinizing the separate contributions of interaction energy components in anion-π interaction. The optimized geometries of weak complexes between anions (F^-, Cl^-, Br^-, CN^-, NC^-, $NO_3{}^-$ and $CO_3{}^{-2}$) and π systems (tetrafluoroethene, hexafluorobenzene and 1,3,5-triazine) were obtained [37] using supermolecular calculations. SAPT was then employed on those optimized geometries using hybrid basis viz. 6-31G** for fluorine atoms in perfluoroaromatic ring and 6-31++G* for rest of the atoms. The results showed interaction energies of anion-π systems to be nearly comparable to those in cation-π systems and the contributions from dispersion energies was found to be higher in former, especially in the case of complexes with organic anions. However, in the case of halide anions, the interactions were observed to be dominated by electrostatic and induction energies. Hence both MIPp and SAPT studies are indicative of electrostatic dominance in anion-π interaction.

The pioneering work of specifically synthesizing complexes exhibiting anion-π interaction was reported by Meyer and coworkers [38] in 2004. They synthesized stable crystals of a complex wherein the chloride ions were found to attractively interact with ligand containing electron deficient triazine ring. Crystallographic studies showed that in a single molecule of the above mentioned compound, there exist two triazine-containing ligands stacked parallely and held together by three Cu^{+2} ions forming a carousel structure. Interestingly, the counteranions Cl^- and $CuCl_4^{-2}$ were seen to lie $\sim$ 3.15 Å above the centroid of the two parallely stacked triazine rings. Electron spray ionization (ESI) mass spectrum also showed the presence of two Cl^- ions interacting with the two triazine rings of the cationic part in methanol solution. Later in the same year, Reedijk et al. [39] synthesized an octadentate dendritic ligand, (N,N$'$,N$''$,N$'''$-tetrakis{2,4-bis(di-2-pyridylamino)-1,3,5-triazinyl}-1,4,8,11-tetraazacyclotetradecane (azadendtriz)) the cationic tetranuclear copper moiety of which could bind Cl^- in a cavity created by four pyridine rings. The distance/s between Cl^- and pyridine ring/s varied in the range of 3.5 Å to 3.7 Å.

Interestingly, Cambridge Structural Database (CSD) search of many previously synthesized compounds and biological systems reveals [22] the evidence of lone pair-π interactions. As pointed out by Quiñonero et al. [22], a search of CSD shows HOHJUJ structure originally synthesized by Frohn et al. [40] possessing tetrafluoroborate anion ($BF_4{}^-$) with fluorine atoms facing aromatic ring of three pentafluorophenyl moieties. It may be noted that Meyer and Reedijk [38, 39] succeeded in their above described work to synthesize compounds that distinctly show anion-π interactions, but the species interacting with anions were cationic in nature. The surplus positive charge could enhance the π deficiency of the π acidic rings making the potential more positive, ensuring the high binding energy with anions.

The hope that a π-deficient aromatic ring could prove to be a promising anion receptor has prompted the experimental chemists to practically achieve hosts that are specific to anion reception. Designing and synthesizing anion receptors has been a challenging field of research due to larger size of anions as compared to their isoelectronic cationic counterparts, their wide range of shape and size and their pH sensitivity [41]. The anion receptors generally bind anions through salt bridges, hydrogen bonds, ion pair or hydrophobic effects. The π aromatic ring adds to the category of neutral anion receptors which ensures lesser sensitivity of host towards pH and thus provides a wider pH window to remain operational for anion reception. Although there has been increasing experimental evidence of neutral π-deficient anion receptors in solid-state and solution, such an evidence for anion-π interaction in gas phase seems to be conspicuous by its absence from the literature. In a study carried out in solution phase, Rosokha et al. [42] have clearly illustrated anion-π complex formation through their work on electron deficient olefinic and aromatic centers interacting with halides in acetonitrile solution. Job's method based on continuous variation of absorbance with respect to mole fraction ensured 1:1 complexation between tetracyanopyrazine ring (TCP) and Br^-. New absorbance band in UV-vis spectra at about 400 nm in addition to characteristic absorption band of TCP ($\lambda = 220$–300 nm) confirmed the charge transfer in the complex. X-ray analysis showed short halide-C contacts (~ 0.4 Å shorter than the sum of respective van der Waals radii) in crystal structure of salt-admixtured complexes formed by simple electron deficient molecules viz. TCP with halides. All these analyses confirmed the presence of anion-π interactions and the concomitant charge-transfer.

Due to the low ΔG^0 value and weak nature of interaction, anion-π interaction is a poor candidate to bring specificity in anion receptors on its own, but can pave a way to challenges of designing anion receptors assisted by the cooperative effects of other non-covalent interactions. However, the additive nature of anion-π interactions with an increase in number of π-deficient rings may allow the synthetic chemists to design hosts which could be specific and bind mainly through anion-π interaction [43]. Li et al. [44] recently synthesized a triazine-based host (Bis(phenylcarbamoyl)-functionalized tetraoxacalix[2]arene[2]triazine) bearing H-bonding motifs and anion-π domains. This host could bind selectively to $H_2PO_4^-$ using both H-bonding and anion-π interaction. Spectroscopic titration of the host with variety of tetrabutylammonium salt of anions was performed in acetonitrile solution at the room temperature. The use of solvents with high dielectric constant may increase the ΔG^0 value mainly due to ion pair formation with complex [45–48]. $H_2PO_4^-$ showed selective quenching of fluorescence spectrum at ~ 308 nm (signature of host system) and a formation of a new fluorescence emission band at nearly 412 nm, whereas all other anions such as Cl^-, Br^-, SCN^-, $NO_3{}^-$, $BF_4{}^-$ could not vary either UV or fluorescence spectrum of the host.

Recently, synthetic chemists are constantly engaged in employing anion-π interaction to tune the fragile chemical pathways. One such example is the determination of particular structural motif of Ag(I) complexes by modulating interaction between anion and π-acidic aromatic rings during self-assembly reactions [49]. Matile and Mareda [50] have come up with synthesis of a highly electron deficient class of molecule, i.e. substituted naphthalenediimide (NDI) with the quadrupole moment as high as +55.5 B. A whole new synthetic branch has emerged with the varied use of NDI and its derivatives in terms of application to anions [50–54]. An example is the use of oligomers of NDIs for making anion-π slides where the host possesses a gradient of electron deficiency along its length and they have been synthetically exploited to transfer anions through membrane [50]. Another application of NDI shows that the catalysis of the Kemp elimination is achieved by utilizing anion-π interactions [51]. The features of molecular electrostatic potential (MESP) for NDI have been briefly explored by Matile and coworkers [50]. In the later part of the present Chapter, this unique set of molecules will be scrutinized through the topographical features of MESP.

We conclude by quoting an example of anion-π interaction in biological systems which has put it at the center stage in the realm of non-covalent interactions. In their recent work, Estarellas et al. [34] have reported experimental findings retrieved from Protein Data Bank (PDB) [55] unveiling relevant anion-π interaction operative in urate oxidase enzyme (UOX). UOX is an antioxidant enzyme (absent in humans) which oxidizes uric acid (π-deficient ring) into (S)-allantoin in the presence of molecular oxygen. Although attempts to crystallize UOX with molecular oxygen have been unsuccessful, its probable position is guessed by the location acquired by CN^- which inhibits UOX functionality in solution. Experimental results revealed that CN^- is located above the plane of uric acid (URC) in UOX/URC/CN^- complex wherein the distances of the carbon and nitrogen atoms of CN^- from the mean plane of the URC molecule are 3.2 Å and 3.0 Å respectively. All these experimental results were supported by quantum chemical calculations at a high level of theory and sufficiently large basis, viz. RI-MP2/CBS. The interaction energy of uric acid and CN^- was found to be of the order of -17 kcal/mol. This case has been elaborated in the forthcoming parts of the present Chapter in terms of justification to location and orientation of anions above uric acid ring.

A critical understanding of the energetic and geometric features of lone pair-π interactions is essential to the development of this field, particularly the science of biomimetics and the development of new synthetic applications in catalysis, materials science, and medicine. The present chapter mainly focuses on the aspects governing lone pair-π interactions. It seems from the search of literature that there is a need to acquire better understanding of location and orientation of anions or lone pair-containing molecules above the π-deficient ring. This Chapter is aimed at offering an electrostatic view of the lone pair π interactions. The following subsection summarizes the features of molecular electrostatic potential (MESP) and its critical points.

13.2 Use of Molecular Electrostatic Potential for Probing Lone Pair-π Interactions

We first elaborate the basic features of the scalar field of molecular electrostatic potential. Molecular electrostatic potential (MESP) and its topography [56–65] is a well-established tool for studying molecular reactivity [59, 66–73] (especially weak interactions) and other chemical phenomena such as existence of lone pairs [74, 75], aromatic π cloud [76], resonance and inductive effect [77] etc. The research in this area was pioneered by Tomasi and Pullman where the MESP was exploited to understand sites of electrophilic attack [56–58, 78]. The works of Politzer et al. [59, 66–71] popularized the use of electrostatics in understanding various interactions between molecules. Extensive work by Gadre et al. on electrostatics and its topography has made the use of MESP more quantitative. MESP can be mathematically written as sum of the nuclear and electronic potential, viz.

$$V(\mathbf{r}) = \sum_{A}^{N} \frac{Z_A}{|\mathbf{r} - \mathbf{R}_A|} - \int \frac{\rho(\mathbf{r}')d^3r'}{|\mathbf{r} - \mathbf{r}'|} \tag{13.1}$$

where $\{Z_A\}$ are the nuclear charges centered at position $\{R_A\}$ and $\rho(r')$ is the continuous electron density. Thus MESP can attain both positive and negative values which ensure richer topographical features than electron density. The values of MESP can be estimated theoretically using standard ab initio packages. Since MESP is electron density derived quantity, its experimental determination is possible using X-ray crystallography. Lecomte et al. have reported experimental techniques of determining MESP in molecular crystals [79–82]. The features of topography of three dimensional scalar field such as MED, MESP etc. is succinctly brought out in terms of its critical points (CPs), an area of research extensively explored by Gadre et al. [60–65, 83]. Critical points are locations, P in space, where all the first order partial derivatives of MESP with respect to the coordinates vanish, i.e.

$$|\nabla V(\mathbf{r})|_P = \left| \mathbf{i}\frac{\partial V}{\partial x} + \mathbf{j}\frac{\partial V}{\partial y} + \mathbf{k}\frac{\partial V}{\partial z} \right|_P = 0 \tag{13.2}$$

Classification of these CP's is done on the basis of signs of the eigenvalues of the corresponding Hessian matrix at the CP. Hessian matrix is formed by second order partial derivatives, the terms of which can be given as

$$H_{ij} = \left(\frac{\partial^2 V(\mathbf{r})}{\partial r_i \partial r_j} \right)_{r=r_{CP}} \tag{13.3}$$

where V is the electrostatic field, r_i and r_j denote Cartesian coordinate i.e. x, y or z and the Hessian is evaluated at any given critical point (CP). A CP is represented in terms of R and S viz. (R, S) where R denotes the rank i.e. the number of non-zero eigenvalues of the Hessian matrix and S denotes the signature i.e. algebraic sum of the signs of these eigenvalues. If all the eigenvalues at a CP are non-zero, then such a CP is called non-degenerate CP. On the other hand, if any one or more eigenvalue is zero, the associated CP is called degenerate CP. Degenerate CPs are unstable and either vanish with slightest conformational change in nuclear position or bifurcate into non-degenerate CPs. There exist four distinct type of non-degenerate CPs for MESP viz. $(3, -3)$, $(3, -1)$, $(3, +1)$ and $(3, +3)$. A $(3, +3)$ CP is referred to as local minimum as all three eigenvalues are positive. Negative-valued $(3, +3)$ CP corresponds to a point where the negative potential due to the electrons, dominates maximally over the positive potential generated by the nuclei, thus indicating electron-dense region in a molecule. The $(3, +1)$ CP is a saddle point with two positive eigenvalues. All chemical structures possessing topograhical rings show $(3, +1)$ type of CP as their signature. A $(3, +1)$ CP also appears between two $(3, +3)$ CPs as a topographical artifact. The $(3, -1)$ CP is another type of non-degenerate saddle, with two negative eigenvalues. Two bonded atoms are marked by the presence of a $(3, -1)$ CP. Lastly a $(3, -3)$ CP is a local maximum which explicitly represents a nucleus. Gadre et al. [84] proved that non-nuclear maximum in MESP cannot exist. The mathematical argument put for this condition is following: In the case of MESP, the eigenvalues of the Hessian matrix of MESP satisfy the Poisson relation, given as Eq. 13.4

$$\nabla^2 V(\mathbf{r}) = (\lambda_1 + \lambda_2 + \lambda_3) = 4\pi\rho(\mathbf{r}) \tag{13.4}$$

However, for a non-nuclear maximum to exist, all the three eigenvalues and thus their sum should be negative, whereas the right hand side of the above equation is always non-negative. Due to non-existence of non-nuclear MESP maxima, Politzer came up with the method of locating most positive potential on van der Waals surface of the the molecule [67] so as to justify the nucleophillic attack.

MESP isosurfaces of a typical π-rich aromatic ring, viz. benzene and a π-deficient system viz. hexafluorobenzene (HFB) are illustrated in Fig. 13.1, along with corresponding MESP CPs. Benzene ring clearly shows negative regions of π electrons spread above as well as below the ring plane (represented by blue MESP isosurface in Fig. 13.1). On each side of the π cloud, a total of six $(3, +3)$ CP's (red dots) mediated by six $(3, +1)$ CP's (green dots) forms a ring of CPs with a $(3, -1)$ CP located at the center of the ring formed by these CPs. The typical MESP value at $(3, +3)$ CPs in aromatic π cloud of benzene is -18.7 kcal/mol. Hexafluorobenzene, on the other hand, does not possess any such negative region on either side of the ring plane. Rather, the lone pairs on fluorine atoms in HFB are disposed in the plane of the ring, forming a zone of negative MESP around HFB. However, the MESP value associated with the CP signifying the lone pair of fluorine is unusually low (-6.6 kcal/mol) as compared to lone pair strength of fluorine in HF (-29.0 kcal/mol). Other commonly used π-deficient systems [75] viz. s-triazine (STZ), trinitrobenzene (TNB), tetracyanobenzene (TCB) and cyanuric acid (CUA) are displayed in Fig. 13.2. Similar to HFB, absence of π electrons and presence of negative MESP only around the periphery of the ring, are common features of these systems too. The complete absence of the electrostatic π cloud in these molecules (such as HFB) is in conflict with the usual connotation of the term "π" in "lone pair-π interaction". This electrostatic picture brings out the reason why electron rich species interact attractively with π-deficient aromatic systems.

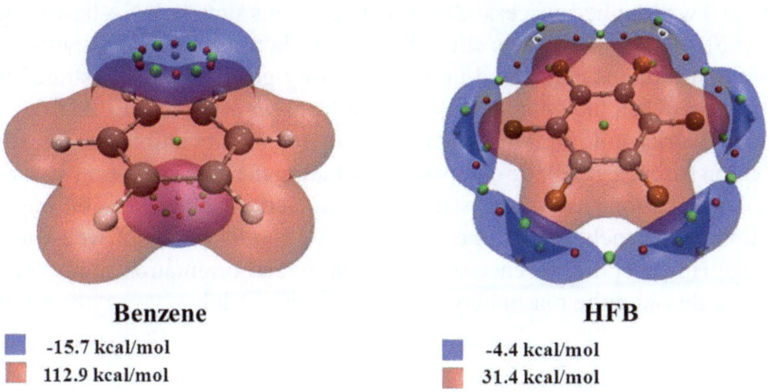

Fig. 13.1 Positive (*red*) and negative (*blue*) molecular electrostatic potential (MESP) isosurfaces (isovalue indicated under the molecules) of benzene and hexafluorobenzene (HFB). Small *red*, *green* and *gray spheres* denote (3,+3), (3,+1) and (3, −1) CPs respectively. See text for details

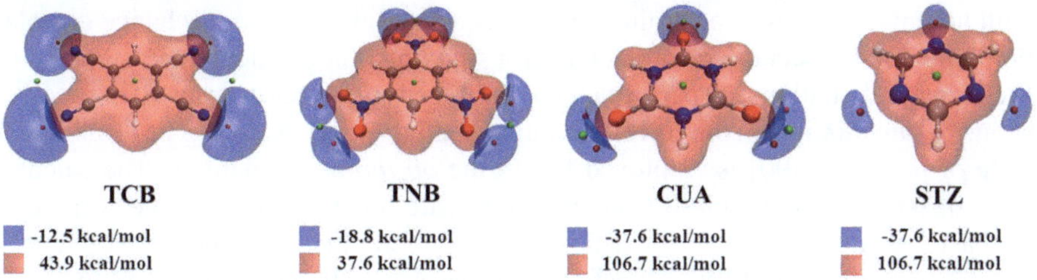

Fig. 13.2 Positive (*red*) and negative (*blue*) molecular electrostatic potential (MESP) isosurfaces (isovalue indicated under corresponding molecules) for π-deficient host systems viz. tetracyanobenzene (TCB), trinitrobenzene (TNB), cyanuric acid (CUA) and s-triazine (STZ). Color coding of CPs identical to that in Fig. 13.1. See text for details

Based on electrostatics and its topographical features, Gadre et al. [85, 86] developed a model several years' ago, named as electrostatic potential for intermolecular complexation (EPIC), which provides a simple tool for estimating the interaction energy and geometry of the weakly interacting molecular species. In this model, the interaction energies are calculated by summing the MESP driven point charges $q_{A,i}$ of the guest molecule multiplied with the electrostatic potential value of the host at the location of the point charge and vice versa.

$$E_{EPIC} = \frac{1}{2}\left\{\sum_i V_{A,i} q_{B,i} + \sum_j V_{B,j} q_{A,j}\right\} \tag{13.5}$$

where V_A, V_B, q_A and q_B are the MESP and the MESP-derived charges on the interacting moieties A and B, respectively.

This interaction energy is minimized by rotating and translating the guest molecule such that the van der Waals surfaces of the guest and the host do not interpenetrate. The

EPIC model was applied successfully to many systems such as DNA base dimers and trimers [86], hydration of crown ether [87] and stacked and H-bonded cytosine dimer [88]. Previous studies performed in our laboratory generally demonstrated striking similarities between the geometries obtained from *ab initio* and EPIC calculations. This model works quite well for the systems where the interaction energy has a large contribution from electrostatic component including the cases of cation-π and anion-π complexes. In the case of a system where induction and dispersion interactions are predominant (such as benzene dimer), EPIC model does not well-capture the energetics. However, the essence of directionality and orientation of host and guest is still brought out quite reasonably by electrostatic model.

13.3 Methodology

The quantum chemical calculations of the systems reported in this Chapter are executed by employing M06L DFT functional subjected to 6-311++G(d,p) basis set without any symmetry constraints. M06L is a local meta-GGA exchange correlation functional especially recommended for studying non-covalent interactions [89]. Frequency calculations are performed for optimized structures to ensure if they correspond to true minima on the corresponding potential energy surface. The Gaussian09 suite of programs [90] is employed for all the *ab initio* calculations. The calculations of interaction energies of the complexes are done considering the correction for basis set superposition error (BSSE) by employing the Boys and Bernardi counterpoise technique [91]. Wavefunctions and density matrices of all the systems are extracted for topographical analysis at M06L(DFT) level of theory employing large Pople basis viz. 6-311++G(d,p). However, it is an established fact that the nature of the critical points, their location and the corresponding MESP value are nearly constant beyond Hartree-Fock level of theory applied to reasonably large basis viz. 6-31G(d,p) [92]. The topographical analysis of MESP has been carried out on a large variety of molecular systems comprising neutral molecules and anions. The location and characterization of the MESP CPs is carried out employing the rapid topography mapping Fortran code developed recently by Yeole et al. [93]. This code uses a hierarchy of the scalar field viz. bare nuclear potential (BNP), molecular electron density (MED) and MESP for building the topography i.e. critical points of BNP, serve as guess points for generating MED CPs, which in-turn serve as guess points for generating MESP CPs. To ensure that no CP is left unidentified in a given region, the guess points are enriched with randomly generated points inside a cube around each nucleus. All these guess points are optimized using the L-BFGS code. The topography of MESP is much richer than that of BNP and MED in terms of the number of CPs. The underlying code for evaluation of MED, MESP and the corresponding gradients is based on the deformed atoms in molecules (DAM) procedure proposed by Rico et al. [94]. DAM method, based on partitioning of scalar field into atomic contributions, provides a rapid and sufficiently accurate function and gradient calculation.

All the EPIC calculations are performed using the integrated EPIC program recently put together by Yeole and Gadre [95] which optimizes the starting geometry into electrostatically most favorable geometry, such that the van der Waals surfaces of the interacting species do not interpenetrate. Method of interaction energy calculation mentioned in the Eq. 13.5 is replaced by a simpler equation, as described in the Eq. 13.6.

$$E_{EPIC} = \sum_i V_{A,i} q_{B,i} \tag{13.6}$$

where V_A is MESP of one of the interacting species (preferably electron rich species) and q_B are the MESP-derived point charges of the other. In the case of anion-π complexes, the initial guess geometry is provided to EPIC model by placing the $(3, +3)$ CP of electron-rich anion/molecule over the π-deficient region of host molecule.

13.4 Results and Discussion

Several examples showing lone pair-π interactions have been briefly mentioned in the Introduction Section. The results of four representative case studies of lone pair-π interaction are discussed below from the electrostatic perspective. For this purpose, we select hosts which are widely studied in the literature on this subject.

13.4.1 Hexafluorobenzene

HFB is one of the most commonly employed π-deficient systems in theoretical as well as experimental studies for understanding lone pair-π interactions. As previously discussed in Introduction Section, one of the reasons behind this molecule being in vogue is its quadrupole moment, which is roughly equal to that of the benzene, but opposite in sign. We limit ourselves to the investigation of complexation between HFB and lone pair-containing neutral species and anions. In one of our previous works [75] we also studied interaction between radicals and the π-deficient hosts. Topographical features of hexafluorobenzene (cf. Fig. 13.1), reveal that the ring centroid of HFB is electron deficient due to the electron withdrawing property of fluorine atoms. This is the physical reason why the lone pair of a guest molecule orients along the C_6 axis of HFB.

We have analyzed the topographical features of several molecules including neutral species and anions. These molecules/anions interact with π-deficient system mostly through their electron rich regions, which are brought out by $(3, +3)$ MESP CPs. It has been observed in our earlier works [75] that the value of MESP at a $(3, +3)$ CP reflects the strength of the lone pair and it correlates remarkably well with the interaction energy of the lone pair-π complex. This primarily indicates the large electrostatic contribution to the energetics of such complexes.

Table 13.1 Relevant MESP topographical parameters and quantum chemically calculated interaction energies of neutral species and anions with Hexafluorobenzene. Dist implies distance of CP (MESP minima) from lone pair-containing atom. D R and D S denotes degenerate ring and degenerate spheres respectively. For explanation of the angles θ, θ' and $\Delta\theta$, see text and Fig. 13.4

System	Nature and no. of CPs	Dist (Å)	MESP (kcal/mol)	E_{int} (kcal/mol)	θ (^{o})	θ' (^{o})	$\Delta\theta$ (^{o})
PH_3	(3,+3), 1	1.88	−23.8	−2.1	123.3	123.9	0.6
Pyridine	(3,+3), 1	1.28	−61.2	−4.3	121.6	121.5	−0.1
H_2O	(3,+3), 2	1.24	−54.0	−3.1	119.8	124.3	4.5
H_2CO	(3,+3), 2	1.29	−39.2	−2.7	131.3	174.0	42.7
O_2	D R, 2	2.09	−1.8	−1.9[a]	–	–	–
HCl	D R, 1	1.90	−12.3	−1.7	–	–	–
NO_3^-	(3,+3), 6	1.20	−167.7	−14.3	107.1	86.4	−20.7
$H_2PO_4^-$	(3,+3), 4	1.19	−168.4	−15.0	124.2	93.7	−30.5
CN^-	(3,+3), 2	1.23	−186.9	−14.5	180.0	85.8	−94.2
N_3^-	D R, 2	1.25	−169.9	−15.9	–	–	–
F^-	D S, 1	1.08	−240.3	−21.1	–	–	–
Cl^-	D S, 1	1.57	−171.0	−14.5	–	–	–

[a]E_{int} for O_2 is calculated without BSSE correction. See text for more details

Topographical features of some prototype examples of lone pair-containing molecules and anions are summarized in Table 13.1 and Fig. 13.3. The interaction energies reported in Table 13.1 refer to the complexes of corresponding molecules/anions with hexafluorobenzene ring. As an example, we elaborate the case of H_2O wherein the two lone pairs on oxygen are characterized by two (3, +3) CP, disposed tetrahedrally with respect to the hydrogen atoms in H_2O molecule. The MESP value at both the CPs is −54.0 kcal/mol showing the electron rich nature of the oxygen atoms. These two (3, +3) CPs are joined by an intervening (3, +1) CP possessing comparable MESP value. It is a case where all the CPs are non-degenerate in nature. However, there are certain molecular systems exhibiting degenerate CPs in the electron rich region. This suggests a delocalized distribution of lone pair electrons. For instance, Fig. 13.3 shows O_2, HCl and N_3^- each endowed with degenerate ring/s of the negative potential. Similarly, MESP topography of Cl^- reveals a sphere of negative minimum around the ion. Further, as may be expected, it turns out that the MESP minima in anions are much deeper as compared to those in the neutral species (see Table 13.1), making them better guests for lone pair-π interaction than the latter ones. Among the anions presented in Table 13.1, fluoride ion (F^-) exhibits the deepest minimum of value −240.3 kcal/mol.

A closer look at the nature of these CPs reveals the fact that the one of the three eigenvalues associated with the (3, +3) CP is significantly larger compared to the other two. For instance, in the case of H_2O, the three eigenvalues of the Hessian matrix of second derivative of MESP are 0.0064, 0.0327 and 0.1304. In a recent

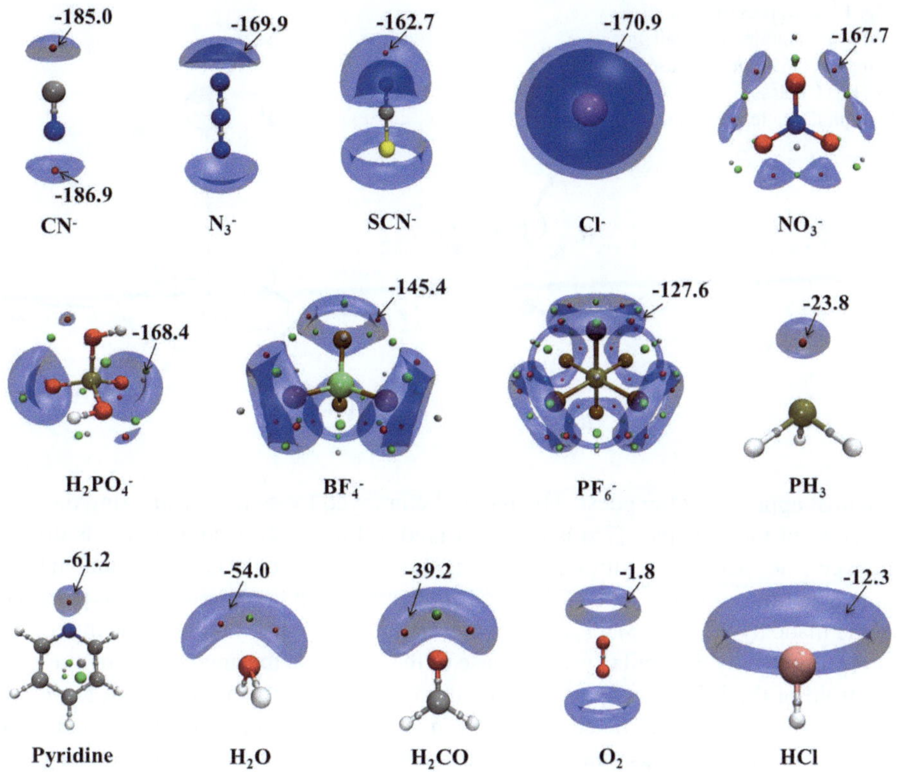

Fig. 13.3 Representation of neutral species and anions with the corresponding MESP CPs wrapped around by MESP isosurfaces. Small *red*, *green* and *gray* spheres denote (3,+3), (3,+1) and (3, −1) CPs respectively. See text for details

work [74], the authors have proposed that the largest among the three eigenvalues enables one to distinguish a CP associated with lone pair from other CPs associated with aromatic ring. Thus, the characterization of lone pairs using MESP topography reveals the location and strength of the lone pairs. This information indeed provides insights into lone pair-π interactions wherein the electron rich lone pair of the guest molecule interacts with the electron deficient aromatic rings.

The complexation between HFB and the lone pair-containing species is scrutinized below using the results obtained from MESP topography, EPIC as well as *ab initio* calculations. Topographical analysis is performed on the *ab initio* optimized structures of HFB:X complex (X = lone pair-containing species). MESP topographical features of these complexes may be analyzed from the following two perspectives.

Directionality of Lone Pair During Complexation During lone pair π complexation, it is expected that the lone pair gets directed towards the electron deficient region of the HFB ring. But there exists no signature of the interacting lone pair in the complex, as the nature of (3, +3) CP present in isolated species changes when

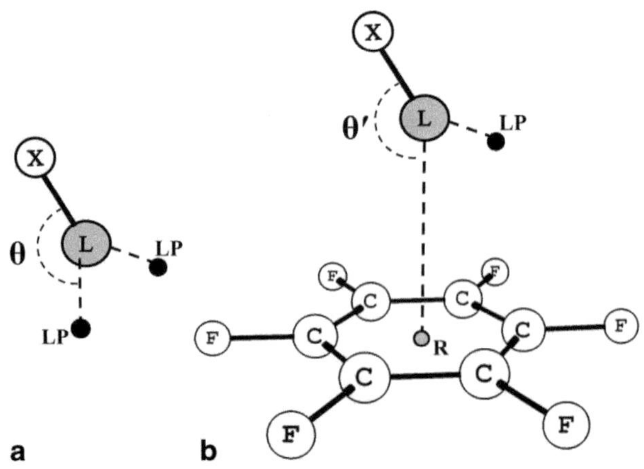

Fig. 13.4 Representation of θ and θ' in a model lone pair-π complex. *Black dot* represents (3, +3) CP associated with lone pair. See text for details

the host approaches the guest. The method employed by us for scrutinizing the orientation of the lone pair [75] is demonstrated in Fig. 13.4. A comparison is drawn between the spatial disposition of lone pair in the free molecules/anions to that in lone pair-π complex. In the case of isolated molecule/anion containing a lone pair, the angle made by the (3, +3) CP, atom possessing it and a nearby atom in the monomer is represented by θ. Similarly, in the case of the complex, the angle made by the ring centroid of the HFB, lone pair bearing atom and the nearby atom chosen to define θ, is represented by θ'. The change in orientation of lone pair upon the formation of the complex is thus measured by calculating the difference between these two angles $(\theta' - \theta)$ designated as $\Delta\theta$. Table 13.1 lists the parameters θ, θ' and $\Delta\theta$ for all the guest molecules studied here.

It is evident from above study that the molecules bearing only one lone pair, such as pyridine and PH_3, orient their lone pair almost perfectly above the ring centroid of HFB. Similarly, the monoatomic anions like F^- and Cl^- possessing spherically degenerate ring CP and molecules like HCl and O_2 with degenerate ring of CPs also align themselves such that the region of lone pair falls directly on top of the centroid of HFB ring. However, the cases wherein the guest molecule bears two or more competing minima such as H_2CO and $H_2PO_4^-$, prefer off-center geometry over the HFB ring so as to maximize the electrostatic interaction of minima with the positive potential. The angle $\Delta\theta$ for such systems is found to be as large as 20–40° displaying the fact the concerned species binds closer to the periphery of the ring of HFB. The angle of 94.2^o in case of CN^- is due to two competing minima on the C-N axis, with the latter lying parallel to the HFB plane. The off-center geometry is also preferred in molecules which have hydrogen atoms such that it can form weak H-bond with the fluorine atoms of HFB, thus distorting the usual expected orientation and maximizing the attractive forces.

Since the lone pair plays a central role in lone pair-π interaction, its strength, as provided by the MESP value, is expected to decide the interaction energy of the complex. This is in affirmation of the fact that MESP value at the (3, +3) CP

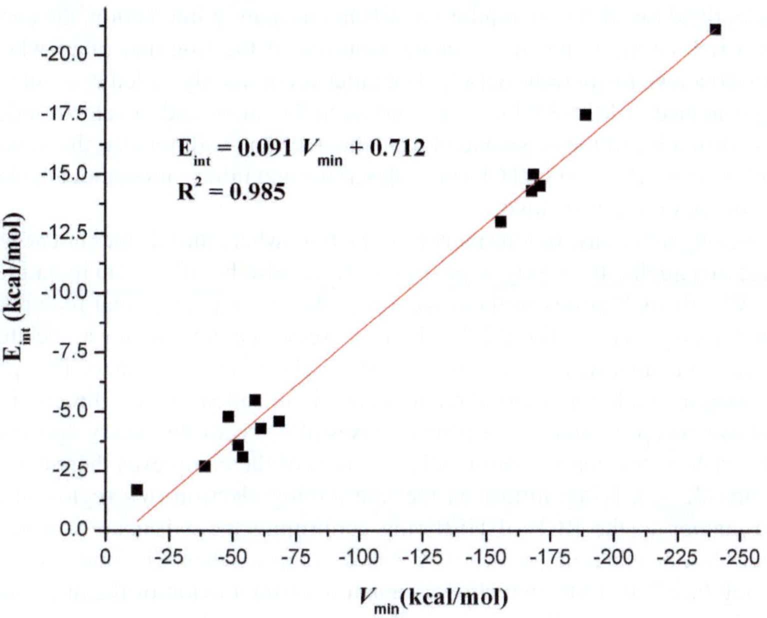

Fig. 13.5 Plot of the interaction energies (E_{int}) of the HFB:X complexes versus the MESP value (V_{min}) at the (3, +3) CP which interacts with HFB. Set of points on upper end are related to anion-π complexes, while the points located in the lower region of plot are associated with neutral species-π complexes. See text for details

correlates extremely well with the interaction energy of the complex with correlation coefficient $R^2 = 0.985$. The species included in this plot are those given in Table 13.1, augmented by few more molecules such as CH_3OH, $O(CH_3)_2$, $N(CH_3)_2$, imidazole, Br^- and $CH_3CO_2^-$. Figure 13.5 depicts the correlation between interaction energy (E_{int}) of HFB:X complex and the MESP value V_{min} at the most negative lone pair CP of the neutral molecules/anions. It is incisive from the plot that higher is the MESP value, larger is the interaction energy. This suggests that the directionality and interaction energy in the lone pair-π interaction is determined by the number, location and strength of the lone pairs.

Appearance of New CPs Between the Host and the Guest The lone pair-π interaction is marked by the appearance of new CP/s in the region connecting the van der Waals extreme of the two interacting species. As pointed out in the Introduction Section, a (3, −1) CP located between two interacting nuclei may be indicative of the bond between them. Elaborating further, one of the eigenvector of this (3, −1) CP corresponding to positive eigenvalue (directional minimum), points in the direction of nuclei connected through that CP. Let us designate the eigenvector associated with positive eigenvalue as PEV. Interaction between the lone pair and π-deficient system can be rather viewed as an interaction of a MESP minimum of electron rich species with a region of diffuse positive potential of electron deficient host. Experimentalists have identified two different binding motifs in lone pair-π interaction, the one where electron rich center is located near the centroid of the ring and other where it is positioned above the periphery [42]. The latter is commonly called as σ interaction. The appearance of MESP CPs located between the anion and the electron deficient host is affirmative of the existence of weak interaction. Additionally, the direction of PEV of these newly appeared CPs provides a succinct information about the bonding motifs of the two species involved.

Especially in the case of lone pair-π interaction, where the CP does not necessarily connect two nuclei, the newly appearing CPs can also be of (3, +1) in nature. Still, the PEV at this CP points in the direction of the two topographical features of the two interacting species. The RCP in HFB possesses an eigenvector associated with negative eigenvalue which is aligned perpendicular to the ring plane. This provides it the topographical property of directional maximum atleast in one direction. Figure 13.6 shows a representative set of complexes along with the newly appeared CPs and the PEV of the corresponding CPs. In most of the complexes there exists a CP of nature (3, −1), lying almost on the line joining electron rich region of neutral species/anions to the RCP of HFB ring confirming the existence of lone pair-π interaction. For example, a (3, −1) CP is developed when Cl^- interacts with the HFB ring (see Fig. 13.6), which is a signature of interaction of the anion with the π-deficient host. The PEV points in the direction of Cl^- and the RCP of HFB, indicating that Cl^- interacts with the ring centroid of HFB. On the other hand, the cases where there are more than one competing minima, the CPs can be generated elsewhere in the region connecting the two species e.g. HFB:CN^- complex, where interaction is marked by two (3, −1) CPs mediated by a (3, +1) CP (see Fig. 13.6). The PEVs of the two CPs indicate that the nitrogen atom of CN^- is interacting with the RCP of HFB, whereas the carbon end of CN^- interacts with peripheral carbon atom of HFB. The introduction of negative charge on complexation with the anion, especially monoatomic anion viz. Cl^- atom brings out yet another intriguing feature i.e. appearance of symmetric pattern of CPs resembling those in the π cloud of benzene (see Fig. 13.6). These CPs are located on one side of the HFB ring opposite to approach of anion, indicating the phenomenon of charge transfer in the complex. The case of HFB:HCl is interesting because geometrical alignment of HCl over the HFB ring puts an impression as if the Cl atom is directly interacting with the ring centroid. However, the newly appeared CPs show that chlorine interacts with the two oppositely located peripheral carbons of HFB. Thus the topographical analysis of these complexes succinctly brings out the nature as well as different modes of interaction between electron rich species and π-deficient system.

The partitioning between positive and negative electrostatic potential is indicated by zero isosurface. Gadre et al. [83] have shown that the negative potential region of the anions lies outside the zero isosurface encapsulating the positive nuclear potential. This theorem can be further applied to HFB:anion complexes, as they carry extra negative charge/s. Indeed the pattern of zero isosurface in HFB:anion complex can be easily distinguished from that in HFB:neutral species complex (cf. Fig. 13.7), as the former possesses zero isosurface enclosing both the anionic nuclear framework and π-deficient host separately, leaving all the negative potential to lie outside this surface.

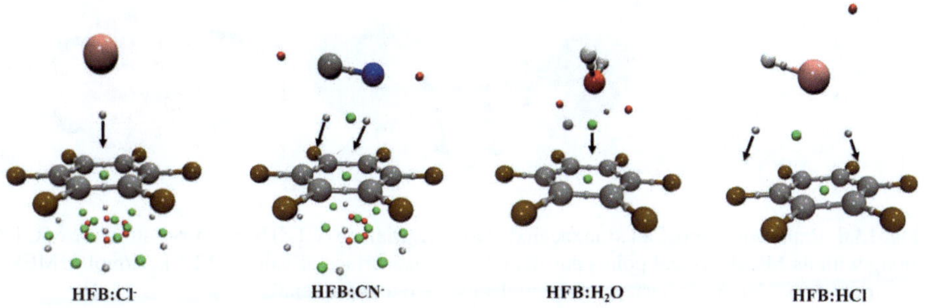

Fig. 13.6 Representation of HFB:X (X = lone pair-containing species) complexes along with relevant MESP CPs. Small *red*, *green* and *gray spheres* denote (3,+3), (3,+1) and (3, −1) CPs respectively. CPs associated with fluorine lone pairs are removed for clarity. See text for details. The arrows indicate the directions of eigenvectors (PEV)

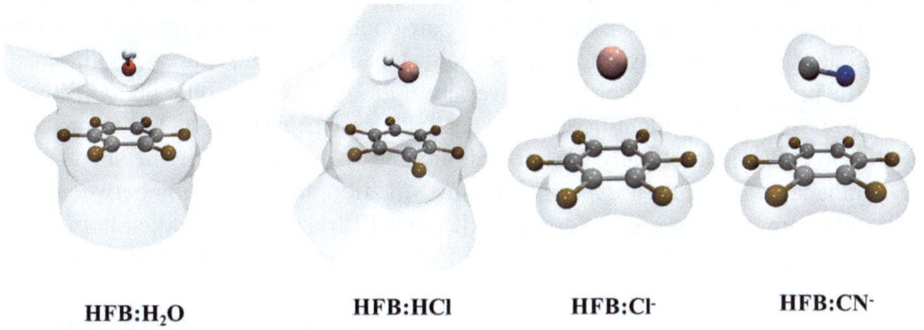

Fig. 13.7 Representation of HFB:X (X = lone pair-containing species) complexes wrapped around by zero valued isosurfaces of MESP. See text for details.

It is evident from above discussion that electrostatic forces mainly govern the energetics and geometries the lone pair-π complexes. In view of this, EPIC calculations are performed to obtain the electrostatically favored conformations and their interaction energies. The EPIC optimized geometries serve as very good initial guess for the follow-up *ab initio* calculations. Our previous work [75] on lone pair-π interaction has demonstrated an excellent correlation of EPIC energies with the corresponding *ab initio* energies. The comparison of magnitude of E_{EPIC} with respect to E_{int} shows that electrostatic component in HFB:X complexation is nearly 65 % of the total interaction energy. To this end, it can be incisively concluded that electrostatic effects play a dominant role in governing the energetics and structural orientation of such complexes. However, the difference in EPIC and *ab initio* calculations is indicative of the fact that other forces such as dispersion and induction forces may play a significant role in lone pair-π interaction.

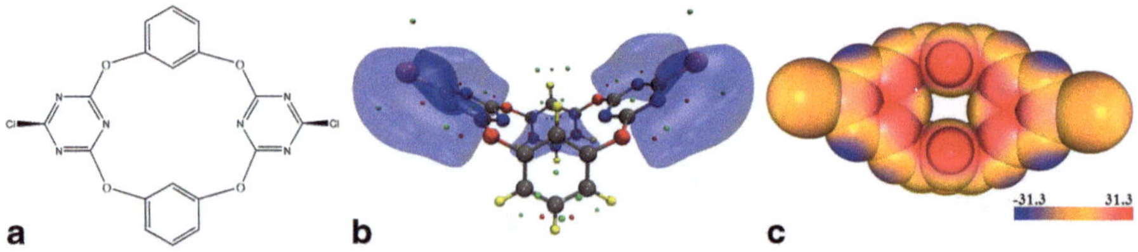

Fig. 13.8 Representation of **a** tetraoxacalix[2]arene[2]triazine (CTZ) **b** optimized structure of CTZ along with its MESP critical points enveloped by an isosurface of value −12.5 kcal/mol. **c** MESP textured on van der Waals surface of the molecule. See text for details

13.4.2 Tetraoxacalix[2]arene[2]triazine

A myriad of cationic anion receptors already exist, but experimental chemists have always thrived for alternatives, especially for achieving selective recognition of polyatomic anions. The fact that the lone pair-containing species interacts with the π-deficient system through a weak non-covalent interaction, empowers the hosts with the property of specificity and reversibility towards the anionic guest. Wang et al. [96] have recently identified tetraoxacalix[2]arene[2]triazine (CTZ) to have the ability to chelate structurally diverse anionic guests, by self tuning the shape of its π-deficient cavity. In their experimental investigation, the interaction between various polyatomic anions and tetraoxacalix[2]arene[2]triazine was studied both in the solution and in solid state employing the tool of spectrometric titrations and X-ray crystallography, respectively.

The present subsection provides insights into the bonding motifs involved in the recognition of four different polyatomic anions of varying dimensionality, employing the tool of MESP. The host system i.e. tetraoxacalix[2]arene[2]triazine (CTZ) and anionic systems viz. SCN^-, BF_4^-, PF_6^- and NO_3^- are optimized using M06L DFT functional with a sufficiently large basis i.e. 6-311++G(d,p). Moreover, the coordinates of the complexes, obtained from crystal structure [96] are used for *ab initio* calculation with a view to explain the nature of bonding therein. The optimized structure of CTZ, along with its MESP CPs enveloped by negative valued MESP isosurface is illustrated in Fig. 13.8. Figure shows the three dimensional orientation of the benzene and triazine rings in the host molecule. The two benzene rings are positioned parallel to each other, whereas the two oppositely located triazine rings tend to form a V-shaped cleft. An intriguing feature of MESP is revealed by projecting the negative MESP isosurface along with the CTZ molecule, showing an isolated region of π electrons sandwiched between the two benzene rings. The π-deficient nature of this V-shaped cleft is prominently brought out by mapping the MESP on van der Waals surface of the molecule (see Fig. 13.8 c). The red region on the van der Waals surface spread all over the large cleft shows highly positive surface potential. This brings out the fact that unlike simpler systems such as HFB, the present system can readily bind larger anions. Upon complexation with polyatomic anions, the orientation of the rings adjust so as to accommodate varying sizes of the anions.

Figure 13.9 displays the topographical features of the four complexes viz. CTZ:SCN^-, CTZ:NO_3^-, CTZ:BF_4^- and CTZ:PF_6^-. A large number of newly generated CPs, in addition to individual CPs of host and the guest are quite noticeable in this Figure. These newly generated CPs are indicative of the existence of non-covalent bonding between CTZ and the anions. The dispersed distribution of these CPs is suggestive of the fact that anions do not bind merely due to lone pair-π interaction, rather the whole cincture of the positive region in the cleft offers binding sites to the anions. The discussion on the nature of bonding within the complexes is taken up individually. In the case of CTZ:SCN^- complex, the SCN^- is aligned over CTZ such that the nitrogen atom in SCN^- lies 3.0 Å away from the centroid of one of the triazine rings, while the sulphur atom at the other end is located 3.6 Å away from another triazine moiety. Amongst the newly appeared CPs of CTZ:SCN^- complex, there exists a $(3, -1)$ CP between the nitrogen and the triazine ring, indicative of the lone pair-π interaction between them. One of the eigenvector of this CP, corresponding to positive eigenvalue, reveals the off-center weak σ interaction of nitrogen with the π-deficient ring. Another $(3, -1)$ CP can be observed between the sulphur atom of SCN^- and the second triazine ring, the eigenvector of which, indicates interaction with a peripheral carbon of the triazine ring. Nonetheless the positive eigenvalue associated with the eigenvector is very low ($\sim$ 0.06 au) indicating this interaction to be extremely weak. Another significant attractive interaction is provided by the two hydrogen atoms located on the two benzene rings.

CTZ:NO_3^- complex distinctly shows a $(3, -1)$ CP connecting one of the oxygen's of NO_3^- and centroid of the triazine ring. However, the other two oxygen atoms of NO_3^- do not show any signature of anion-π interaction with the second triazine ring. Nevertheless the other newly appeared saddle points show a dispersed electrostatic interaction with the bed of positive potential lying underneath the negative potential of NO_3^-.

The lone pair-π interaction is exquisitely brought out in CTZ:BF_4^- complex, where the interaction between two fluorine of BF_4^- and the two triazine rings are represented by the two $(3, -1)$ CPs connecting them. A closer look at the eigenvectors associated with the two CPs reveals that one of the fluorine atom interacts with the centroid of the triazine ring, while the other one possesses an off-center σ interaction with another triazine ring. Weaker interaction with hydrogen atoms of benzene ring is also brought out by the mediating $(3, -1)$ CPs.

In the case of CTZ:PF_6^- complex, the PF_6^- is positioned symmetrically in the cavity formed by two triazine rings. Interestingly, the significant interaction to hold the anion is provided only by the two hydrogen atoms of the benzene ring. However, the anion-π interaction is very weak and does not contribute significantly to the total interaction energy.

The discussion on bonding motifs present in CTZ:anion complex can be concluded by stating that anions bind to the cavity of tetraoxacalix[2]arene[2]triazine molecule, not just through anion-π interaction but also utilizing complementary positive potential stretched along the cavity.

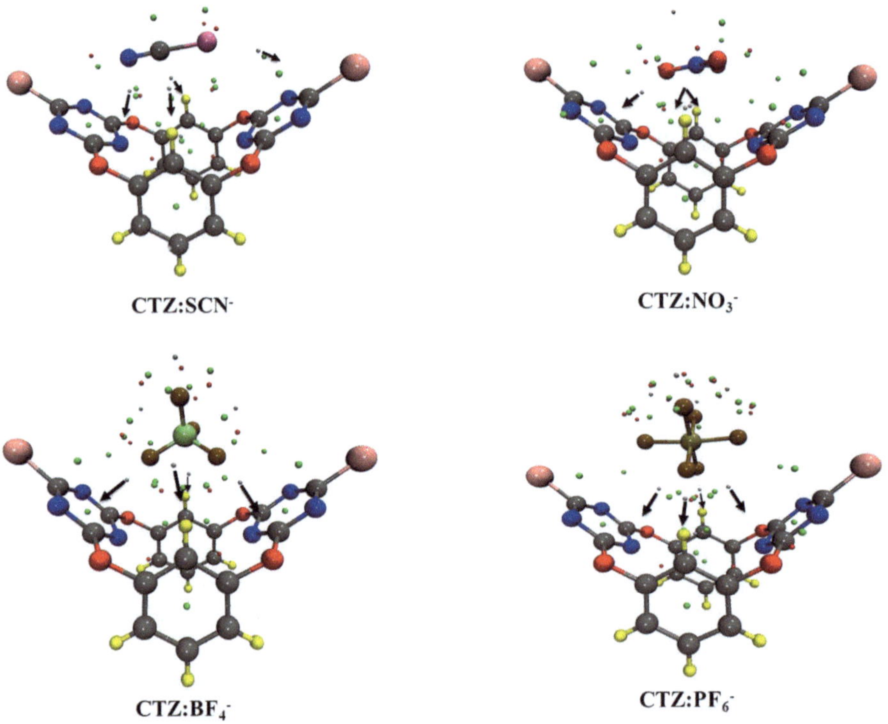

Fig. 13.9 Optimized structures of CTZ:X complexes along with its MESP critical points. Small *red*, *green* and *gray spheres* denote (3,+3), (3,+1) and (3, −1) CPs respectively. Only those CPs indicative of interaction are retained for clarity. See text for details. The arrows indicate the directions of eigenvectors (PEV)

13.4.3 Naphthalenediimide

Naphthalenediimide (NDI) and its derivatives are currently getting widespread attention of theoretical as well as experimental chemists due to the exceptional π-deficient nature of the former. Matile and Mareda [50–54] have recently reported the synthesis of few molecules of this class such as N-tetramethylphenyl 2,6-dicyano-NDI and N-tetramethylphenyl −2,3,6,7-tetracyano-NDI, with spectacular quadrupole moment values of +39.2 B and +55.5B, respectively. They suggested in their work, the use of linear chain of these molecules to design anion-π slides, which can selectively transfer the anions. Use of such anion-π slide to create electroneutral photosystems have also been explored by them.

The large quadrupole moments of these molecules can be attributed to cyano groups attached to NDI scaffold which drastically pulls out the π electrons. Additionally, the tetramethylphenyl ring attached in plane perpendicular to the plane of NDI contributes locally to the positive quadrupole moment. Figure 13.10 shows the structure and the highly positive electrostatic potential textured on the van der Waals

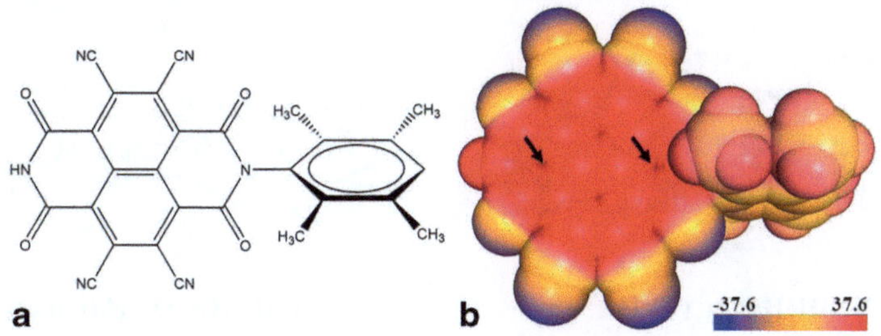

Fig. 13.10 **a** Structure of N-tetramethylphenyl −2,3,6,7-tetracyano-Naphthalenediimide (TNDI), **b** MESP values textured on van der Waals surface of the molecule. *Arrows* indicate the location of most positive potential on the surface. See text for details

surface of N-tetramethylphenyl −2, 3, 6, 7-tetracyano-NDI (TNDI). The potential at the two most positive locations on the surface is as high as ~66kcal/mol, marked by black arrows in the Figure.

The quantum chemical calculations have been performed herein at similar level of theory and basis as discussed in Methodology Section. The initial guess geometries for the *ab initio* optimization are obtained from EPIC calculations, which provide electrostatically most favorable geometries of the complexes. Following discussion brings out the accuracy of EPIC method achieved in predicting the structure and energetics of the complex formed between TNDI and lone pair-containing species. The structures obtained from both EPIC and *ab initio* calculations are illustrated in Fig. 13.11. The gross similarity of the structures obtained by the two methods is indicative of the electrostatic dominance of energy in the interaction of lone pair-containing species with TNDI. Most of the species studied here can be observed to align their lone pair towards either of the two positive MESP location over the π-deficient surface. Figure 13.12 shows the correlation between the interaction energies of TNDI:X (X = lone pair-containing species) obtained from EPIC and *ab initio* calculations. Anions like HCO_2^-, NO_3^- and $H_2PO_4^-$ show numerically large interaction energy of −40 kcal/mol, higher than strongest H-bond. Nonetheless these polyatomic anions interact through more than one lone pair, thus additive effect of lone pair-π interaction comes into play. The average electrostatic contribution to the total interaction energy is nearly 70 %, which justifies the use electrostatics as a tool for studying anion-π interaction.

13.4.4 Anion-π Interaction in Urate Oxidase

To the best of our knowledge, present example is the first clear experimental verification of presence of lone pair-π interaction in biological systems. The active center of urate oxidase possibly carries out oxidation of uric acid employing lone pair-π interaction with molecular oxygen as suggested by Estarellas et al. [34]. This example has been briefly introduced in previous part of Chapter highlighting the essential

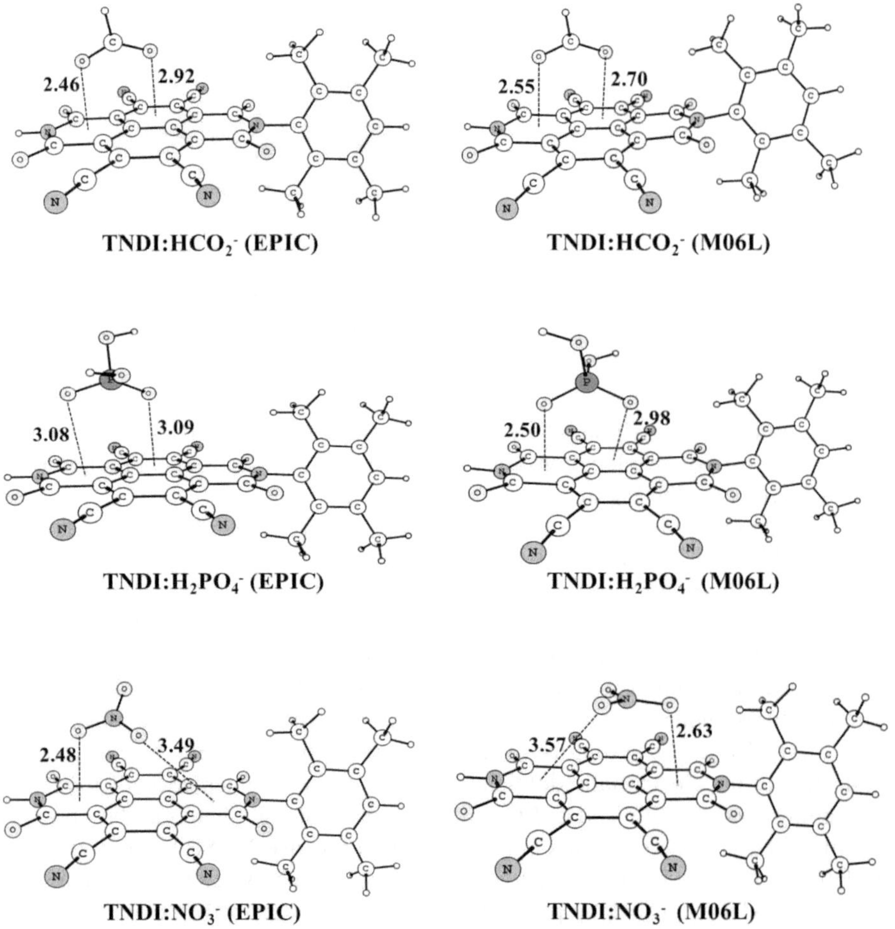

Fig. 13.11 Comparative representation of EPIC and *ab initio* optimized structures of TNDI:anion complexes. See text for details

features. Figure 13.13 shows the enzymatic cascade reaction for uric acid oxidation in UOX, as proposed in the literature.

Uniqueness of UOX activity lies in the fact that unlike other oxidase enzymes, it does not require any metal atom or organic co-factor for catalysis. The substrate (molecular oxygen) is expected to be aligned over the π-deficient ring of URC through anion-π interaction. As mentioned earlier, the inference of conformation of UOX/URC/O_2 is drawn indirectly [34] using the X-ray crystal structure of UOX/URC/CN^-.

In what follows, we investigate the bonding motifs and the nature of interaction involved in uric acid oxidation inside UOX by essentially carrying out MESP analysis of the host, guest and complex. Although the main substrate which binds to uric acid

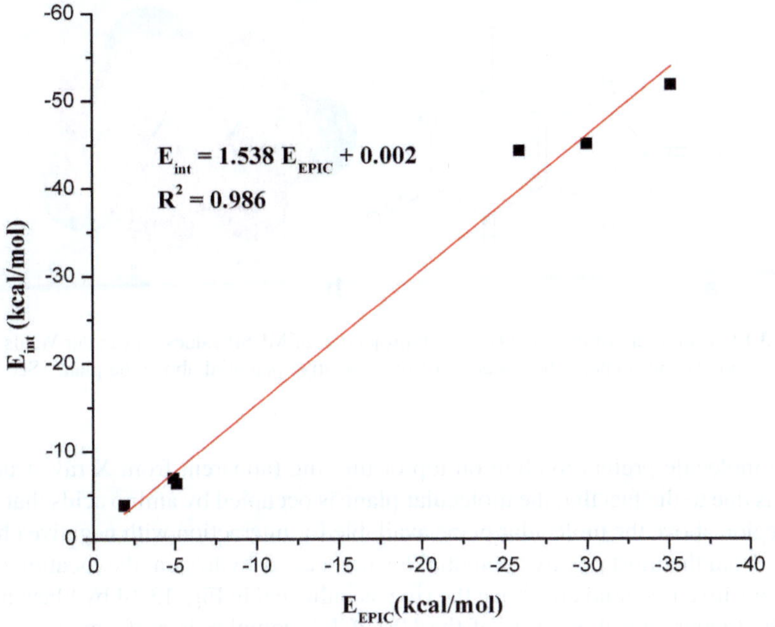

Fig. 13.12 Correlation between the quantum chemically calculated interaction energy (E_{int}) and EPIC (E_{EPIC}) energy for TNDI:X complexes. See text for details

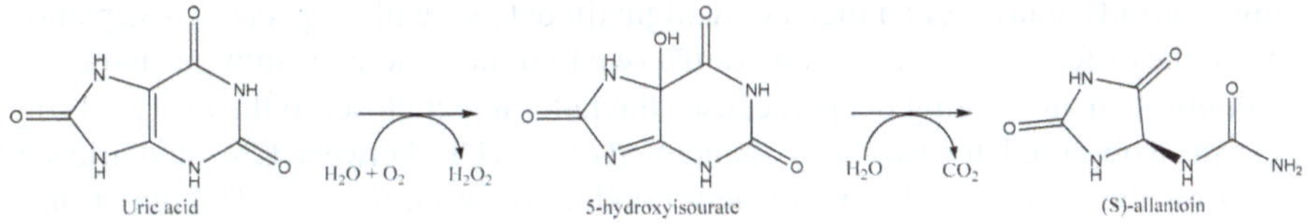

Fig. 13.13 Enzyme cascade for uric acid oxidation in UOX

in UOX is molecular oxygen, the other lone pair-containing anions viz. CN^- and N_3^- are known to inhibit the activity of UOX. As evident from Fig. 13.3, all these three systems do possess similar topographical features, i.e. existence of two competing minima at both ends, on the internuclear axis. The O_2 molecule shows two rings of degenerate CPs with a very low MESP value of −1.8 kcal/mol, whereas the anions, CN^- and N_3^- possess very deep MESP minimum of value −186.9 kcal/mol and −169.9 kcal/mol respectively. The relative strength of negative ESP of latter species provides enough reason why they could possibly bind to π-deficient surface of URC replacing O_2 and thus inhibit the UOX activity.

The MESP features of electron deficient counterpart of the complex i.e. uric acid are reflected by mapping the MESP values on the van der Waals surface of URC as illustrated in Fig. 13.14. The electrostatic potential mapping of URC shows that the most positive values lie in the molecular plane near the α hydrogen of URC, still the

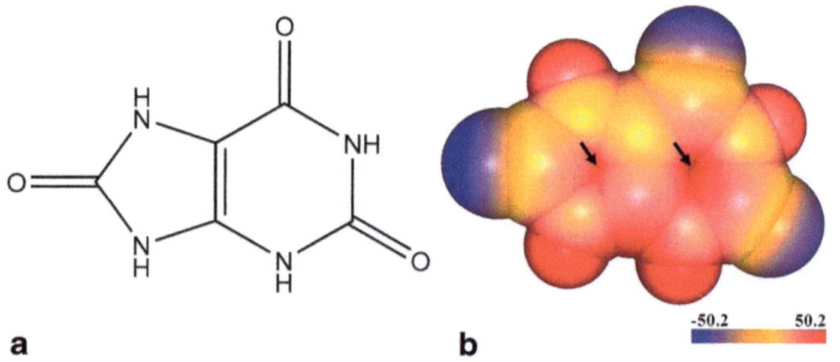

Fig. 13.14 **a** Structure of uric acid (URC), **b** projection of MESP values on van der Waals surface of uric acid. *Arrows* indicate the location of most positive potential above the plane. See text for details

guest molecule prefers to align on top of the ring (apparent from X ray structure). This is due to the fact that the molecular plane is occupied by amino acids that leaves the region above the molecular plane available for interaction with negative charges. Apart from the most positive potential located near α hydrogen, the location of next most positive potential are above the ring as indicated in Fig. 13.14 by black arrows.

The topographical analysis of the URC/CN^- complex is performed in order to probe the bonding motifs involved in the complexation. Figure 13.15 illlustrates the geometrical alignment of CN^- over the uric acid along with MESP CPs of URC/CN^- complex. Both nitrogen and carbon are roughly positioned above the two ring centroids of uric acid molecule. A slight tilt of CN^- with respect to URC plane is due to the difference in the strength of the two minima, where the minima associated with nitrogen atom being deeper, causes the latter to get closer to the uric acid ring. As can be observed, the two newly generated (3,+1) CPs between CN^- and uric acid confirms the existence of non-covalent bonding between the two. The direction of PEVs at the corresponding (3, +1) CPs reveals that both carbon and nitrogen strongly interact with the two RCPs of the uric acid molecule. Since the topographical features of O_2 molecule bears compelling qualitative similarity with CN^-, it is expected to bind to uric acid in a similar fashion, but with a weaker interaction energy. This is in accordance with the fact that O_2 helps in enzymatic process of the enzyme cascade, while CN^- ion is the inhibitor. Thus, the electrostatic features incisively emphasizes the evidence of existence of lone pair π interaction operational in the active center of urate oxidase.

13.5 Concluding Remarks

For more than a decade now, the intriguing nature of lone pair-π interaction has captivated experimental as well as theoretical chemists. Several theoretical studies have employed quantum chemical calculations on complexes formed between electron rich species and model π-deficient hosts such as hexafluorobenzene and s-triazine

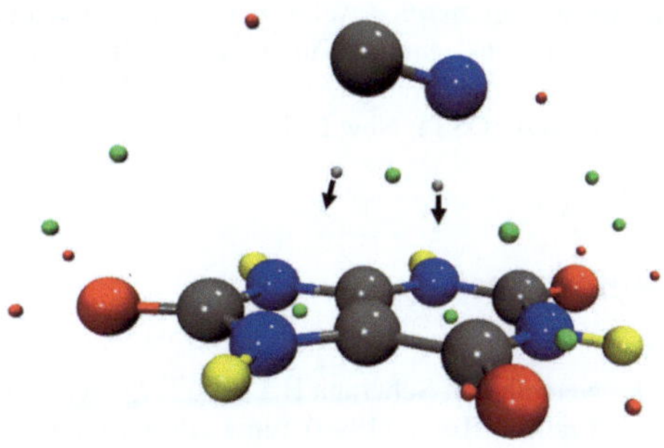

Fig. 13.15 The optimized geometry of URC/CN^- complex along with its MESP CPs. Small *red*, *green* and *gray spheres* denote (3,+3), (3,+1) and (3, −1) CPs respectively. The *arrows* indicate the direction of eigenvectors (PEV). Only those CPs indicative of interaction are retained for clarity. See text for details

in order to verify the attractive nature of the interaction. Experimental chemists have been successful in synthesizing stable complexes which exhibit lone pair-π interaction. Even the long existing skepticism on its biological role [42, 50] was cleared with the study on urate oxidase enzyme, which distinctly reveals the importance of such interaction in carrying out the cascade reaction at its active center. The methods employed to understand such interaction have been based on simpler criteria such as contact distances between the atoms of the interacting species, mapping of electron density followed up by atoms in molecules (AIM) analysis. Attempts using the tool of electrostatic potential have also been explored in the literature, but their use is mainly limited to depict the complementarity of the species involved. In one of our recent works [75],we have used electrostatics in wider sense i.e. to understand MESP features of host and guest in terms of CPs and their role in determining the conformation of the complexes.

In the present chapter, we have elaborated the role of electrostatics for scrutinizing the nature of interaction and the bonding motifs involved in the interaction. We could successfully explain the scenario of lone pair-π interaction in certain experimentally determined complexes, by analyzing the MESP CPs generated due to complexation. Literature reports that the electron rich species binds either to the ring centroid or the periphery of π-deficient rings, the latter disposition being known in the literature as σ interaction. This fact is succinctly brought by the eigenvector (PEV) associated with the CP located between the interacting species. Molecules possessing one lone pair are usually observed to bind to RCP of the host, as evident from the PEV at the intermediate (3, −1) CP, whereas cases where more than one competing minima exist in the molecule, tend to lie at peripheral ends so as to minimize the interaction energies. EPIC model has been employed to predict the electrostatically favorable structures as well as to estimate the electrostatic contribution to total interaction energies in such complexes. The resemblance of structures obtained from EPIC model and *ab initio* calculations clearly brings out the dominance of electrostatics in the formation of lone pair-π complexes. However, subtle differences are indicative of the role of dispersion and inductive forces in tuning the final conformation of the complexes. In summary, the essence of lone pair-π interaction is vividly brought out by the molecular electrostatic potential and its critical points, proving it to be a valuable tool for this purpose.

Acknowledgements Authors are thankful to Dr. C. H. Suresh and Dr. P. Balanarayan for fruitful discussions. Anmol Kumar thanks the Council of Scientific and Industrial Research (CSIR) for research fellowship. Professor Shridhar Gadre is grateful to the Department of Science and Technology (DST), New Delhi for the award of J. C. Bose National Fellowship.

References

1. Némethy G, Scheraga HA (1962) J Chem Phys 36:3382–3400
2. Tzalis D, Tor Y (1996) Tetrahedron Lett 37:8293–8296
3. Müller-Dethlefs K, Hobza P (2000) Chem Rev 100:143–168
4. Duncan R, Kopeček J (1984) Soluble synthetic polymers as potential drug carriers. In: Polymers in medicine, vol. 57. Springer, Berlin
5. Huck WTS, Prins LJ, Fokkens RH, Nibbering NMM, van Veggel FCJM, Reinhoudt DN (1998) J Am Chem Soc 120:6240–6246
6. Zhong W, Gallivan JP, Zhang Y, Li L, Lester HA, Dougherty DA (1998) Proc Nat Acad Sci USA 95:12088–12093
7. Mecozzi S, West AP, Dougherty DA (1996) J Am Chem Soc 118:2307–2308
8. McCurdy A, Jimenez L, Stauffer DA, Dougherty DA (1992) J Am Chem Soc 114:10314–10321
9. Wheeler SE, Houk KN (2009) J Am Chem Soc 131:3126–3127
10. Sayyed FB, Suresh CH (2012) J Phys Chem A 116:5723–5732
11. Sayyed FB, Suresh CH (2012) Chem Phys Lett 523:11–14
12. Sayyed FB, Suresh CH (2011) J Phys Chem A 115:9300–9307
13. Mahadevi AS, Sastry GN (2013) Chem Rev 113:2100–2138
14. Egli M, Sarkhel S (2007)Acc Chem Res 40:197–205
15. Caltagirone C, Gale PA (2009) Chem Soc Rev 38:520–563
16. Wenzel M, Hiscock JR, Gale PA (2012) Chem Soc Rev 41:480–520
17. Gale PA, Busschaert N, Haynes CJE, Karagiannidis LE, Kirby IL (2014) Chem Soc Rev 43:205–241
18. Schottel BL, Chifotides HT, Dunbar KR (2008) Chem Soc Rev 37:68–83
19. Chifotides HT, Dunbar KR (2013) Acc Chem Res 46:894–906
20. Frontera A, Gamez P, Mascal M, Mooibroek TJ, Reedijk J (2011) Angew Chem Int Ed 50: 9564–9583
21. Mooibroek TJ, Black CA, Gamez P, Reedijk J (2008) Cryst Growth Des 8:1082–1093
22. Quiñonero D, Garau C, Rotger C, Frontera A, Ballester P, Costa A, Deyà PM (2002) Angew Chem Int Ed 41:3389–3392
23. Boden N, Davis P, Stam C, Wesselink G (1973) Mol Phys 25:81–86
24. Vrbancich J, Ritchie GLD (1980) J Chem Soc Faraday Trans 2 76:648–659
25. Schneider H-J, Werner F, Blatter T (1993) J Phys Org Chem 6:590–594
26. Schneider H-J, Blatter T, Palm B, Pfingstag U, Ruediger V, Theis I (1992) J Am Chem Soc 114:7704–7708
27. Schneider H-J (1991) Angew Chem Int Ed 30:1417–1436
28. Alkorta I, Rozas I, Elguero J (1997) J Org Chem 62:4687–4691
29. Alkorta I, Rozas I, Elguero J (2002) J Am Chem Soc 124:8593–8598
30. Gallivan JP, Dougherty DA (1999) Org Lett 1:103–106
31. Alkorta I, Elguero J (2003) J Phys Chem A 107:9428–9433
32. Garau C, Frontera A, Quiñonero D, Ballester P, Costa A, Deyà PM (2003) Chem Phys Chem 4:1344–1348
33. Estarellas C, Frontera A, Quiñonero D, Deyà PM (2011) Chem Asian J 6:2316–2318
34. Estarellas C, Frontera A, Quiñonero D, Deyà PM (2011) Angew Chem Int Ed 50:415–418
35. Jeziorski B, Moszynski R, Szalewicz K (1994) Chem Rev 94:1887–1930
36. Lao KU, Herbert JM (2014) J Chem Phys 140:044108–8

37. Kim D, Tarakeshwar P, Kim KS (2004) J Phys Chem A 108:1250–1258
38. Demeshko S, Decher S, Meyer F (2004) J Am Chem Soc 126:4508–4509
39. de Hoog P, Gamez P, Mutikainen I, Turpeinen U, Reedijk J (2004) Angew Chem Int Ed 43:5815–5817
40. Frohn H, Giesen M, Welting D, Henkel G (1996) Eur J Solid State Inorg Chem 33:841–853
41. Sessler JL, Gale PA, Cho W-S (2006) Anion receptor chemistry, vol 8. Royal Society of Chemistry
42. Rosokha YS, Lindeman SV, Rosokha SV, Kochi JK (2004) Angew Chem Int Ed 43:4650–4652
43. Müller M, Albrecht M, Gossen V, Peters T, Hoffmann A, Raabe G, Valkonen A, Rissanen K (2010) Chem Eur J 16:12446–12453
44. Li S, Wang D-X, Wang M-X (2012) Tetrahedron Lett 53:6226–6229
45. Arranz-Mascarós P, Bazzicalupi C, Bianchi A, Giorgi C, Godino-Salido M-L, Gutiérrez-Valero M-D, Lopez-Garzón R, Savastano M (2013) J Am Chem S 135:102–105
46. Giese M, Albrecht M, Krappitz T, Peters M, Gossen V, Raabe G, Valkonen A, Rissanen K (2012) Chem Commun 48:9983–9985
47. Ballester P (2013) Acc Chem Res 46:874–884
48. Gil-Ramírez G, Escudero-Adán EC, Benet-Buchholz J, Ballester P (2008) Angew Chem 120:4182–4186
49. Schottel BL, Chifotides HT, Shatruk M, Chouai A, Pérez LM, Bacsa J, Dunbar KR (2006) J Am Chem Soc 128:5895–5912
50. Mareda J, Matile S (2009) Chem Eur J 15:28–37
51. Zhao Y, Domoto Y, Orentas E, Beuchat C, Emery D, Mareda J, Sakai N, Matile S (2013) Angew Chem Int Ed 52:9940–9943
52. Gorteau V, Julliard MD, Matile S (2008) J Membr Sci 321:37–42
53. Vargas Jentzsch A, Emery D, Mareda J, Metrangolo P, Resnati G, Matile S (2011) Angew Chem Int Ed 50:11675–11678
54. Lin N-T, Vargas Jentzsch A, Guenee L, Neudorfl J-M, Aziz S, Berkessel A, Orentas E, Sakai N, Matile S (2012) Chem Sci 3:1121–112
55. Berman HM, Westbrook J, Feng Z, Gilliland G, Bhat TN, Weissig H, Shindyalov IN, Bourne PE (2000) Nucleic Acids Res 28:235–242
56. Scrocco E, Tomasi J (1973) The electrostatic molecular potential as a tool for the interpretation of molecular properties. In New concepts II, vol 42. Springer, Berlin, pp 95–170
57. Tomasi J, Mennucci B, Cammy M (1996) Molecular electrostatic potentials: concepts and applications. Elsevier, Amsterdam
58. Tomasi J, Mennucci B, Cammi R (1996) Theor Comp Chem 3:1–103
59. Politzer P, Murray JS (2002) Theor Chem Acc 108:134–142
60. Yeole SD, Gadre SR (2011) J Phys Chem A 115:12769–12779
61. Balanarayan P, Kavathekar R, Gadre SR (2007) J Phys Chem A 111:2733–2738
62. Gadre SR, Shirsat RN (2000) Electrostatics of atoms and molecules. Universities Press, Hyderabad
63. Gadre SR, Kulkarni SA, Shrivastava IH (1992) J Chem Phys 96:5253–5260
64. Balanarayan P, Gadre SR (2003) J Chem Phys 119:5037–5043
65. Shirsat RN, Bapat SV, Gadre SR (1992) Chem Phys Lett 200:373–378
66. Politzer P, Murray JS, Peralta-Inga Z (2001) Int J Quantum Chem 85:676–684
67. Politzer P, Landry SJ, Waernheim T (1982) J Phys Chem 86:4767–4771
68. Murray JS, Lane P, Politzer P (2007) Int J Quantum Chem 107:2286–2292
69. Politzer P, Murray JS, Lane P (2007) Int J Quantum Chem 107:3046–3052
70. Shields ZP, Murray JS, Politzer P (2010) Int J Quantum Chem 110:2823–2832
71. Politzer P, Murray JS, Clark T (2010) Phys Chem Chem Phys 12:7748–7757
72. Suresh CH, Koga N, Gadre SR (2000) Organometallics 19:3008–3015
73. Elango M, Subramanian V, Rahalkar AP, Gadre SR, Sathyamurthy N (2008) J Phys Chem A 112:7699–7704
74. Kumar A, Gadre SR, Mohan N, Suresh CH (2014) J Phys Chem A 118:526–532
75. Mohan N, Suresh CH, Kumar A, Gadre SR (2013) Phys Chem Chem Phys 15:18401–18409

76. Suresh CH, Gadre SR (2007) J Phys Chem A 111:710–714
77. Suresh CH, Alexander P, Vijayalakshmi KP, Sajith P, Gadre SR (2008) Phys Chem Chem Phys 10:6492–6499
78. Pullman B (1990) Int J Quantum Chem 38:81–92
79. Pichon-Pesme V, Lecomte C (1998) Acta Crystallogr Sect B 54:485–493
80. Benabicha F, Pichon-Pesme V, Jelsch C, Lecomte C, Khmou A (2000) Acta Crystallogr Sect B 56:155–165
81. Bouhmaida N, Ghermani N-E, Lecomte C, Thalal A (1997) Acta Crystallogr Sect A 53:556–563
82. Bouhmaida N, Thalal A, Ghermani Ne, Lecomte C (1999) Acta Crystallogr Sect A 55:729–738
83. Gadre SR, Shrivastava IH (1991) J Chem Phys 94:4384–4390
84. Gadre SR, Pathak RK (1990) Proc Ind Acad Sci (Chem Sci) 102:189–192
85. Gadre SR, Pundlik SS (1997) J Phys Chem B 101:3298–3303
86. Pundlik SS, Gadre SR (1997) J Phys Chem B 101:9657–9662
87. Pingale SS, Gadre SR, Bartolotti LJ (1998) J Phys Chem A 102:9987–9992
88. Sivanesan D, Babu K, Gadre SR, Subramanian V, Ramasami T (2000) J Phys Chem A 104:10887–10894
89. Zhao Y, Truhlar DG (2006) J Chem Phys 125:194101–194118
90. Frisch MJ et al. (2009) Gaussian09 Revision D.01. Gaussian Inc. Wallingford
91. Boys S, Bernardi F (1970) Mol Phys 19:553–566
92. Gadre SR, Kulkarni SA, Suresh C, Shrivastava IH (1995) Chem Phys Lett 239:273–281
93. Yeole SD, López R, Gadre SR (2012) J Chem Phys 137:074116–074117
94. Rico JF, López R, Ramírez G, Ema I, Ludeña EV (2004) J Comput Chem 25:1355–1363
95. Yeole SD, Gadre SR (2011) J Chem Phys 134:084111–084118
96. Wang D-X, Wang M-X (2013) J Am Chem Soc 135:892–897

Part III
Aromatic Systems

Chapter 14
Unraveling the Origin of Substituents Effects in π-Stacking Interactions

Steven E. Wheeler

Abstract Non-covalent interactions involving aromatic rings, which include π-stacking, anion-π, and cation-π interactions, among others, are central to modern chemical research. These interactions play vital roles in everything from protein-DNA interactions and the properties of organic electronic materials to stereoselective organocatalyzed reactions. We discuss recent efforts to understand the impact of substituents on the strength of π-stacking interactions through the application of modern tools of computational quantum chemistry. We first provide an account of previous efforts to develop qualitative physical models of these interactions, followed by apparent flaws in these previous models. We then present our local, direct interaction model of substituent effects in π-stacking interactions, and discuss recent extensions of this model based on the examination of electric fields of arenes. We also discuss related misconceptions regarding molecular electrostatic potentials (ESPs), and offer a simpler view of the origin of ESP differences arising from the incorporation of heteroatoms or substituents into aromatic rings. Finally, we conclude with an outlook for future advances in our understanding of π-stacking interactions.

14.1 Introduction

Non-covalent interactions involving aromatic rings play vital roles in a myriad of contexts ranging from supramolecular chemistry and crystal engineering to organocatalysis [1–5]. Understanding the factors that impact both the strength and geometry of these non-covalent interactions has proved vital to harnessing their power in practical applications. π-stacking interactions are traditionally defined as attractive non-covalent interactions between aromatic rings (see Fig. 14.1), and their existence has been known for many years [6]. These interactions play key roles in many areas of chemistry and molecular biology. For example, π-stacking interactions between DNA bases are the primary stabilizing factor in double-helical DNA, [7–10] and are also responsible for the intercalation of drugs and various carcinogens into the DNA

S. E. Wheeler (✉)
Department of Chemistry, Texas A&M University,
College Station, TX 77843, USA
e-mail: wheeler@chem.tamu.edu

S. Scheiner (ed.), *Noncovalent Forces*, Challenges and Advances in Computational Chemistry and Physics 19, DOI 10.1007/978-3-319-14163-3_14

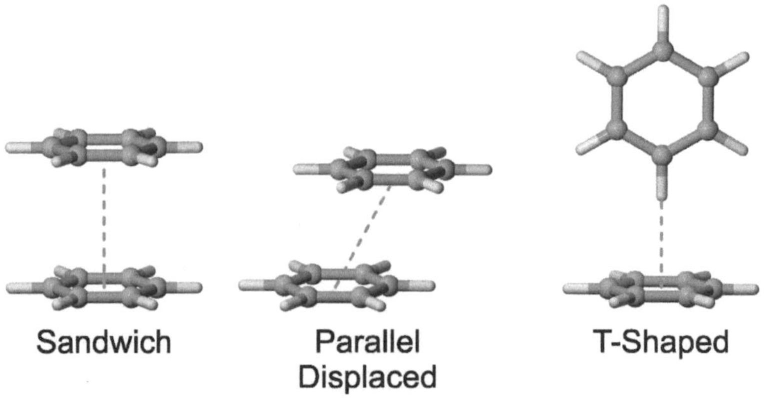

Fig. 14.1 Prototypical configurations of the benzene dimer. Of these, only the parallel-displaced and T-shaped dimers are energy minima. Both the parallel displaced and T-shaped dimers are bound by about 2.8 kcal mol^{-1} in the gas phase, whereas the sandwich dimer is bound by 1.8 kcal mol^{-1} [17–18]

double helix [9, 11, 12]. These interactions play central roles in many other areas of chemistry, including the packing of polycyclic systems in organic electronic materials. Finally, in the last few years, there has been increasing attention afforded to organocatalytic transformations in which π-stacking and other non-covalent interactions play key roles in both catalytic activity and selectivity [13–16].

There has been some discussion recently [19–21] regarding the preferred name for π-stacking interactions, which are alternatively referred to as "π-π interactions", "π-π stacking interactions", "aromatic stacking interactions", "aromatic donor-acceptor interactions", "charge-transfer interactions", or simply "stacking interactions". Each of these terms is loaded with connotations, and the existence and widespread use of these disparate terms reflects the storied history of our understanding of these non-covalent interactions. Along these lines, we note that even though π-stacking interactions are often *defined* as occurring between aromatic molecules, there is growing evidence that aromaticity plays no direct role in these interactions. For example, in 2008, Grimme [19] showed that, at least for aromatic systems smaller than anthracene, there does not appear to be anything "special" about aromatic π-π interactions. More recently, Bloom and Wheeler [20] showed more directly that aromatic π-electron delocalization actually hinders π-stacking interactions in model complexes. Finally, Martinez and Iverson [21] recently reviewed the experimental and theoretical literature on π-stacking interactions, concluding that the term "π-stacking interaction" does not accurately describe the forces responsible for the association of most aromatic molecules. Ultimately, they suggested that the terms "π-stacking interaction" and "π-π stacking interaction" should be avoided altogether!

We view "stacking interactions" as the safest choice of terminology for these non-covalent interactions, since this term can be used purely as a geometric descriptor. That is, "stacking" invokes a relatively clear physical picture of planar molecules in roughly parallel, overlapping orientations, devoid of the added implications of the other widely-used terms. However, in the present work we will stick with the more popular moniker "π-stacking interaction", with the caveat that the "π-" refers to the fact that the stacking interactions we are discussing involve aromatic "π-systems."

14.2 Computational Methods for Studying π-Stacking Interactions

The study of non-covalent interactions involving aromatic rings, including π-stacking interactions, provides a number of challenges for standard electronic structure methods. In particular, gas-phase π-stacking interactions are dominated by dispersion effects whose description requires robust correlated *ab initio* methods paired with large basis sets [17, 18, 22, 23]. Previously, this requirement limited computational studies of non-covalent interactions involving aromatic rings to a small model complexes containing 10–20 non-hydrogen atoms [4, 23]. However, recent years have witnessed an explosion in the development of density functional theory (DFT) methods capable of describing dispersion-dominated interactions [24–43]. As a result, methods are now readily available that enable routine studies of π-stacking interactions in relatively large systems (100–200 atoms), or studies of massive numbers of non-covalent complexes involving relatively small systems. Perhaps the simplest, yet most popular, of the various means of treating dispersion-bound complexes using DFT is through the inclusion of empirical dispersion corrections. This was popularized by Grimme, [25, 26] who introduced so-called DFT-D2 (and later DFT-D3) approaches [25–27, 29–31]. We have relied heavily on the B97-D approach, [26, 44] which, when paired with triple-ζ quality basis sets, provides accurate π-stacking interaction energies at a very modest computational cost [45, 46].

Of course, any application of DFT, particularly highly-paramterized functionals or those including empirical corrections, must be done with care. To this end, our strategy has been to apply B97-D/TZV(2*d*,2*p*) to a large collection of π-stacked dimers, and then also compute robust coupled cluster energies using large basis sets [*e.g.* CCSD(T)/aug-cc-pVTZ] for selected systems. In this way, we have been able to confirm the performance of B97-D for a subset of the systems under consideration while still examining a broad collection of π-stacked dimers. In other words, we can use B97-D to study a large enough collection of systems to draw general and convincing conclusions, while anchoring these computations to robust *ab initio* results for selected systems.

Another powerful tool for the study of non-covalent interactions is symmetry-adapted perturbation theory (SAPT) [47, 48]. SAPT not only provides rigorous interaction energies for non-bonded dimers, but simultaneously decomposes these interaction energies into electrostatic, exchange-repulsion, induction, and dispersion contributions. By exploiting density-fitting techniques, Sherrill and co-workers have implemented [23, 49–52] an incredibly efficient SAPT code in the open-source software package Psi4 [53]. At lowest-order (SAPT0), the SAPT code in Psi4 can routinely be applied to systems of several hundred atoms, enabling detailed studies of the various components underlying non-covalent interactions in quite large systems. We have relied heavily on SAPT0/jun-cc-pVDZ energies to study substituent effects in π-stacking interactions in diverse model stacked sandwich dimers [54]. Through error-cancelation, this level of theory provides surprisingly robust predictions at very low computational cost [52].

14.3 Models of π-Stacking Interactions

There have been a number of efforts to develop predictive physical models of π-stacking interactions over the years. Chief among these is the seminal work by Hunter and Sanders, published in 1990 [6]. Models of substituent effects in π-stacking interactions have been around nearly as long [55–63]. However, starting in the early 2000s, apparent flaws in these models were identified based on gas-phase computations, and, in recent years, these models have been challenged by new views of substituent effects in simple π-stacked systems [18, 64–77]. Below, we provide a brief overview of the most popular "conventional" views of substituent effects in π-stacking interactions, followed by the most recent developments in this area.

The benzene dimer is typically used as a model system for π-stacking interactions, for which three limiting configurations are typical considered (see Fig. 14.1). Among these configurations, the sandwich dimer is a saddle-point on the potential energy surface, whereas both the parallel-displaced and T-shaped dimers are energy minima with roughly equal interaction energies. However, we do not consider the T-shaped interactions to be an example of π-stacking interactions. First, this dimer does not resemble a "stacked" geometry (we challenge the reader to try "stacking" plates in this configuration!). Perhaps more importantly, the non-covalent interactions operative in the T-shaped and edge-to-face dimers are qualitatively different from those in either the sandwich or parallel displaced benzene dimer [18].

We also note that, for simplicity, most models of substituent effects in π-stacking interactions are cast in terms of the benzene sandwich dimer, even though this configuration is not a stable minimum in the unsubstituted case. This raises some concern that conclusions drawn from studies of model sandwich dimers are not necessarily applicable to "real world" π-stacking interactions. However, we have shown [73] that substituent effects in the parallel-displaced dimer are strongly correlated with those in the sandwich dimer, meaning that results arising from studies of substituent effects in the benzene sandwich dimer are directly transferable to more realistic parallel-displaced configurations.

14.3.1 Conventional Views of Substituent Effects in π-Stacking Interactions

The pioneering work of Hunter and Sanders [6] provided an intuitive physical model of π-stacking interactions. In particular, Hunter and Sanders [6] showed that the strength and preferred orientations of π-stacking interactions between disparate aromatic rings could be understood and predicted based on a simple electrostatic model, in which the aromatic rings are viewed as a collection of atom-centered local quadrupole moments. That is, the basic behavior of stacked aromatic rings is recovered by representing each non-hydrogen atom as a positive charge of $+2q$ flanked by charges of $-q$ located at distances $\pm d$ above and below the molecular plane. This simple model provided simple and intuitive explanations of the preference for aromatic rings to form parallel-displaced and T-shaped configurations.

Among other things, Hunter and Sanders [6] also emphasized that the "donor-acceptor concept can be misleading when used to describe π-π interactions: it is the properties of the atoms in the regions of intermolecular contact that control the strength and geometry of interactions, rather than the overall molecular oxidation or reduction potentials." In other words, the still all too common view of π-stacking interactions as donor-acceptor complexes is, in many cases, incorrect. Furthermore, Hunter and Sanders emphasized [6] that the charge-transfer transitions in the UV spectra of π-stacked complexes are a consequence of the π-π interaction, not the cause!

The Hunter-Sanders model [6] also laid the groundwork for the development of an intuitive physical model of the impact of substituent effects in π-stacking interactions. This was carried out primarily by Hunter, who published a series of papers on this subject over the ensuing decades [55–57, 78–80]. The substituent effect model of Hunter and co-workers, depicted in Fig. 14.2, is purely electrostatic in nature. Intriguingly, whereas the original Hunter-Sanders model represented the π-electron system using atom-centered local quadrupoles, in describing substituent effects Hunter cast his model in terms of the overall "π-electron system." In particular, Hunter proposed that electron-withdrawing substituents (*e.g.* –CN, $-NO_2$, *etc.*) enhance π-stacking interactions by depleting the π-electron density of the substituted ring and relieving the π-π repulsion between the two rings. On the other hand, electron-donating groups (e.g. $-CH_3$, $-OCH_3$, *etc.*) hinder π-stacking interactions through the opposite mechanism. More generally, Hunter and co-workers [55, 57] emphasized that the strongest π-stacking interactions will occur between aromatic rings with complementary electrostatic character. That is, "electron-rich" rings stack most favorably with "electron-poor" rings. In terms of computed electrostatic potentials (ESPs), this means that strong π-stacking interactions occur when one arene has a negative ESP above the center and the other a largely positive ESP. This intuitive model is widespread in the literature, and seems to jive with the basic concept that, when it comes to electrostatic interactions, opposites attract.

This intuitive model of substituent effects in π-stacking interactions is entrenched in the chemical literature, and has ruled over the field for decades. Moreover, this view of π-stacking interactions has enabled the rational exploitation of π-stacking interactions in many contexts and powered the growing fields of supramolecular chemistry and crystal engineering. However, in recent years, high-accuracy computational studies of model π-stacking interactions have pinpointed what appear to be deep-rooted flaws in this venerable model [18, 72, 74, 81].

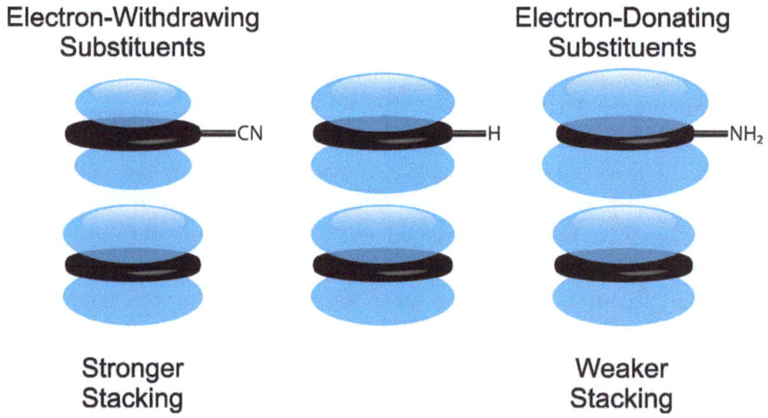

Fig. 14.2 Depiction of the electrostatic model of substituent effects in π-stacking interactions from Hunter and co-workers [55–57]

14.3.2 Unexpected Substituent Effects in the Benzene Dimer

Starting in the early 2000s, Sherrill and co-workers, [17, 18, 22, 23, 45, 46, 64–69, 82–85] as well as others, [5, 75–77, 86] began publishing systematic, high-accuracy gas-phase interaction energies for prototypical stacked dimers of benzene and substituted benzenes. The data quickly revealed unexpected substituent effects for these gas-phase dimers, [18, 64–67] kicking off a frenzy of computational and experimental studies of substituent effects in π-stacking interactions. In particular, these revelations engendered lively debate regarding the relative importance of electrostatic and dispersion effects in substituent effects in π-stacking interactions [18, 57, 65–69, 77, 79].

As noted above, the conventional view is that electron-withdrawing substituents enhance π-stacking interactions, whereas electron-donating groups hinder these interactions. Sherrill and co-workers [18, 64–67] found qualitatively different behavior for model benzene dimers. In particular, they predicted enhanced π-stacking interactions for all substituted dimers, regardless of the nature of the substituent. Moreover, SAPT computations revealed that these substituent effects are not driven solely by electrostatic effects, but result from a mélange of other effects (dispersion, induction, *etc.*). Indeed, Sherrill found no correlation between computed interaction energies and Hammett σ-constants. This flew in the face of conventional views of substituent effects in π-stacking interactions, [55–57] as well as a vast body of experimental data showing a correlation of π-stacking interactions with Hammett σ-constants!

14.3.3 *The Importance of Direct Interactions*

In 2008, Wheeler and Houk [70] entered the fray with a short communication on the substituted benzene sandwich dimer. Unlike Sherrill et al. [18, 64–67] who utilized high-level *ab initio* methods and were therefore limited in the number of substituted benzene dimers that could reasonably be examined, Wheeler and Houk [70] used the then recently developed M05-2X DFT functional [32]. This allowed the examination of a far broader range of substituents. In particular, in contrast to the four substituents examined by Sherrill and co-workers, [18, 65, 66] (OH, CH_3, F, and CN), Wheeler and Houk [70] considered a diverse set of 25 substituents. This broader view afforded a much clearer picture of the overall trend in substituent effects in gas-phase benzene sandwich dimers. In particular, Wheeler and Houk [70] showed that there is a correlation between computed gas-phase interaction energies and σ_m (see Fig. 14.3a). However, there was one outlier—the unsbubstituted benzene dimer. In particular, when X = H was excluded, computed gas-phase interaction energies are correlated with σ_m ($r = 0.91$), but the best-fit line had a y-intercept of -0.4 kcal mol^{-1}. That is, the substituted sandwich dimers interact more favorably by 0.4 kcal mol^{-1} compared to the value predicted simply based on the Hammett σ_m constant. This non-zero y-intercept was attributed to dispersion interactions, which were postulated to stabilize all substituted dimers compared to the unsubstituted case. In other words, Wheeler and Houk showed that the origin of Sherrill's observation that even electron-donating substituents enhance π-stacking interactions is due to the stabilizing effects of dispersion interactions on all substituted benzene sandwich dimers.

More importantly, Wheeler and Houk [70] showed that substituent effects in the benzene sandwich dimer could be captured without the substituted benzene! More precisely, they showed that interaction energies of H-X... C_6H_6 dimers, in which the substituent (X) was positioned exactly as it is in the corresponding C_6H_5X... C_6H_6 dimer, are also correlated with σ_m. Moreover, the best-fit line through the HX... C_6H_6 data has the same slope and intercept as the C_6H_5X... C_6H_6 results (see Fig. 14.3b). This provided compelling evidence that the factors that are responsible for the substituent effects in the HX... C_6H_6 complexes are the same as those in the C_6H_5X... C_6H_6 sandwich dimers. Moreover, since hydrogen has no π-electrons, these substituent effects in both complexes must not be the result of the π-resonance effects that are central to Hunter's model! Instead, Wheeler and Houk proposed that substituent effects in the benzene sandwich dimer were due primarily to direct interactions between the local dipole moment associated with the substituent (and captured by H–X) and the other ring. Similarly, the dispersion interactions responsible for the 0.4 kcal mol^{-1} "extra" stabilization of the substituted dimers arose from dispersion interactions between the substituents and the other ring.

Thus, the overall picture that emerged was one in which the substituent effect trend is due to electrostatic effects, but *all substituted dimers* are stabilized by dispersion interactions by an "extra" 0.4 kcal mol^{-1}, on average. However, the electrostatic effects giving rise to the substituent effect trend were not the ones proposed by

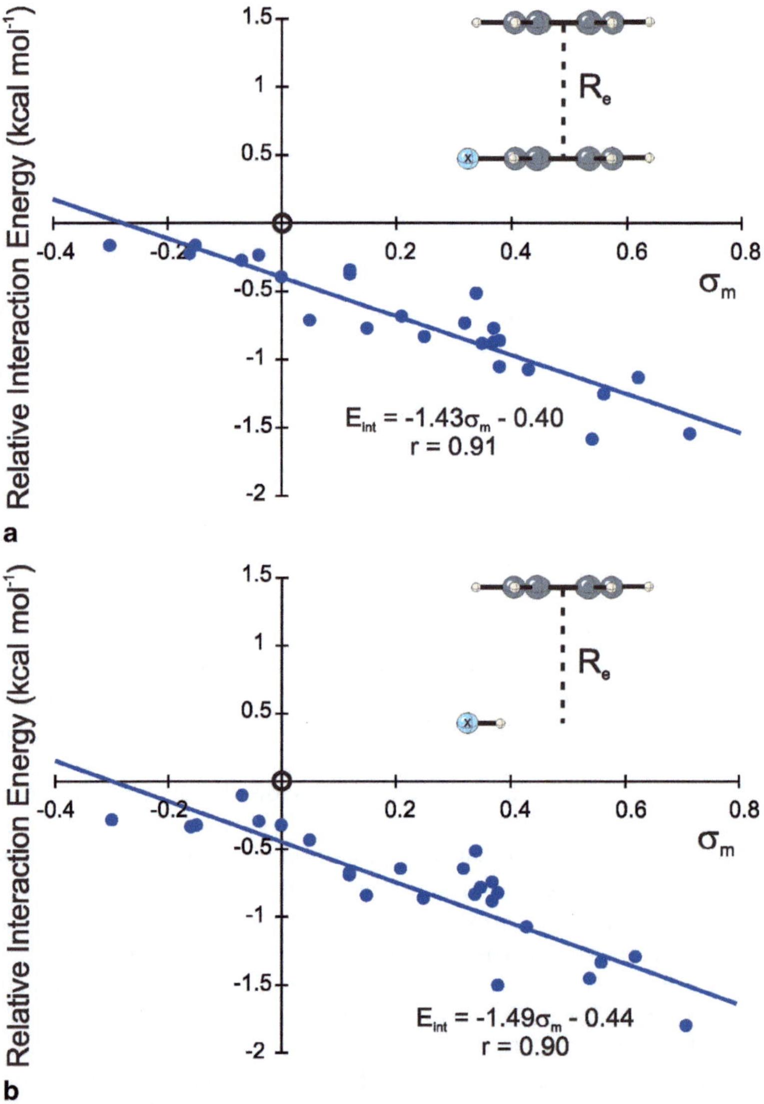

Fig. 14.3 Interaction energies (kcal mol^{-1}), relative to X = H, for **a** $C_6H_5X \ldots C_6H_6$ sandwich dimers and **b** $HX \ldots C_6H_6$ dimers versus Hammett σ_m constants for the substituents X. The open circles represent the X = H case. (Reprinted with permission from [70]. Copyright 2008 American Chemical Society)

Hunter and co-workers, [55–57] but were instead due to *direct interactions* between the substituents and the other ring!

In subsequent work, Sherrill and co-workers [68] and Lewis et al. [77] continued to debate the relative role of dispersion and electrostatic effects in substituent effects in the benzene sandwich dimer. For example, Sherrill and co-workers [68] examined polysubstituted benzene sandwich dimers, for which the correlation of interaction energies with Hammett σ_m constants again breaks down. This was explained shortly thereafter in a remarkable paper by Sherrill and co-workers, [69] in which they showed that at the close intermolecular distances characteristic of these highly-substituted benzene sandwich dimers, electrostatic effects were always stabilizing because of interpenetration effects. At larger inter-ring distances, the correlation with Hammett σ_m constants reappeared. Meanwhile, Lewis and co-workers [77] also examined a broad range of mono- and polysubstituted benzene sandwich dimers using SAPT. Their results revealed that even though the impact of dispersion interactions varies greatly among substituted dimers, these variations are almost exactly cancelled by complementary variations in the exchange-repulsion and induction components of these interactions. Consequently, the trends in substituent effects in diverse substituted stacked dimers are almost completely due to electrostatic effects; changes in the other components of the interaction tend to always cancel. Thus, the overall picture that emerged is one in which electrostatic interactions are the dominant source of substituent effects in π-stacking interactions, but these electrostatic effects are often counterintuitive due to penetration effects.

14.3.4 The Local, Direct Interaction Model

In 2011, we introduced the local, direct interaction model of substituent effects in π-stacking interactions, [73] which can be viewed as an extension of the direct interaction model from Wheeler and Houk [70]. The underlying concept is that substituent effects in π-stacking interactions are due primarily to the direct, through-space interactions of the substituent with the proximal vertex of the other ring (see Fig. 14.4). In other words, the impact of a given substituent will depend only on what is in its local environment. This model was motivated by the observation that substituent effects in the benzene sandwich dimer could be reproduced by only considering HX interacting with propene (with all conserved atoms in the same position as in the intact sandwich dimer). That is, not only is the substituted phenyl ring unnecessary to capture the substituent effects, but neither is half of the unsubstituted benzene.

This local nature of substituent effects in π-stacking interactions has a number of important implications for understanding substituent effects in general π-stacking interactions. For example, substituent effects in π-stacking interactions of substituted benzenes with 1,2,3-trifluorobenzene are drastically different depending on whether the substituent sits over the fluorinated or non-fluorinated vertex of the trifluorobenzene. If the substituent is over the fluorinated end of trifluorobenzene, then the substituent effects are essentially identical to those in the $C_6H_5X\ldots C_6F_6$ dimer. On the other hand, when the substituent is positioned over the non-fluorinated end of trifluorobenzene, the substituent effects are simply those of the benzene dimer. That is, the presence of the fluorines has no impact on the effect of substituent X unless *they are in the local environment* of the substituent. Similarly, the local, direct

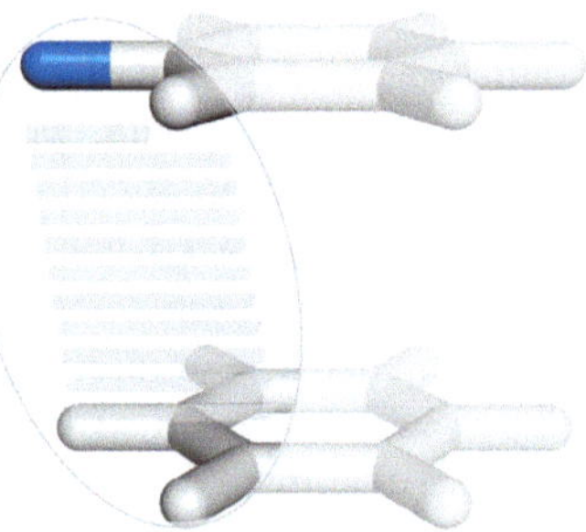

Fig. 14.4 Depiction of the local, direct interaction model of Wheeler and co-workers [73]. (Reprinted with permission from. Copyright 2011 American Chemical Society)

interaction model also explains why 1,3,5-trifluorobenzene engages in p-stacking interactions (-5.9 kcal mol^{-1}) that are nearly as strong as those between benzene and perfluorobenzene (-6.3 kcal mol^{-1}).

This local nature of substituent effects also provides a simple means of understanding substituent effects in π-stacking interactions involving *N*-heterocycles. In fact, unless the heteroatom is located in the local environment of the substituent, it does not change the substituent effect at all. This means that the impact of substituents on pyridine-benzene and pyridine-pyridine dimers is the same as those in the benzene dimer as long as the substituent is not located in the local environment of one of the nitrogens. As seen below, the insensitivity of substituent effects to the presence of heteroatoms extends well beyond a single nitrogen atom.

Overall, the local, direct interaction model implies that substituent effects in π-stacking interactions will be additive, orientationally dependent, and transferable. Many of these features have been previously observed. For example, Sherrill and co-workers [66] showed the additivity of substituent effects in several configurations of the benzene dimer. However, at the time, there was no straightforward explanation of this additivity, since this behavior is not expected based on Hunter's view of substituent effects. The local, direct interaction model provides a simple and physically sound explanation of the additivity of substituent effects in π-stacking interactions.

Similarly, it has been observed experimentally that substituent effects are sensitive to relative orientations in cases where both rings are substituted [87, 88]. This is not consistent with a straightforward application of Hunter's substituent effect model, [55–57] which formally depends only on the π-electron density. This dependence on relative orientation is a direct result of the local nature of substituent effects. This orientational dependence also provides a potentially powerful tool for controlling the local orientation of stacked discotic systems, which have potential applications in organic electronic materials. For example, we recently showed [89] that the local orientations of model stacked coronenes and circumcoronenes can be controlled by exploiting the local, direct interactions of substituents on the stacked rings.

Finally, the transferability of substituent effects in π-stacking interactions means that the impact of a given substituent will be the same across diverse π-stacked dimers as long as the local environment is unchanged. This was demonstrated above for π-stacking interactions with partially-fluorinated rings, as well as *N*-heterocycles. In 2013, we tested the limits of this transferability by considering monosubstituted sandwich dimers involving benzene, borazine, and triazine [54]. Ultimately, substituent effects were independent of the identity of the substituted ring. Moreover, substituent effects were the same across any pair of dimers in which the unsubstituted ring was unchanged. For example, in Fig. 14.5a, relative interaction energies for sandwich dimers of boron-substituted borazine are plotted again substituted benzene sandwich dimers. These interaction energies are strongly correlated, and the slope of the best-fit line is very close to unity. This confirms that substituent effects in π-stacking interactions are independent of the identity of the substituted ring.

However, there were also instances of correlations among substituent effects in π-stacked dimers in which the unsubstituted ring was not the same [54]. For example, interaction energies for stacked dimers of substituted triazine with borazine are correlated with those in stacked dimers of boron-substituted borazine with benzene. Explaining unexpected correlations such as this, which were not predicted by the local, direct interaction model, required an extension of the model.

14.3.5 Substituent Effects and Electric Fields

The unexpected correlations between substituent effects in π-stacked dimers in which the unsubstituted ring is different (e.g. Fig. 14.5b) can be explained by examining the electric fields of the unsubstituted rings. This is because the local, direct interactions responsible for substituent effects in π-stacking interactions are dominated by the electrostatic interaction of the local dipole moment associated with the substituent and the electric field of the other ring.

The electric field of benzene in the plane bisecting the ring is shown in Fig. 14.6a, superimposed over the ESP in this same plane. Also shown is a typical electron density isosurface of benzene ($r = 0.001\ e/\text{au}^3$), along with the silhouette of a stacked benzene at 3.65 Å. The electric field at the position of the substituent on this stacked ring will interact with the parallel component of the local dipole associated with the substituent. The electric field surrounding borazine is shown in Fig. 14.6b. Even though the electric fields of benzene and borazine are different globally, the two electric fields are parallel at the position of the substituents on the other ring. This is demonstrated more quantitatively in Fig. 14.6c, where we plot the dot product of the normalized electric fields of benzene and borazine. As seen in Fig. 14.6c, the electric fields of benzene and borazine are parallel across much of this region, particularly at the position of substituents above either the boron or nitrogen vertices of borazine. Consequently, the electrostatic interaction with a substituent will be similar, and π-stacked dimers involving unsubstituted benzenes and borazine will exhibit the same substituent trends. Overall, substituent effect trends are the same in two stacked dimers not only in cases in which the unsubstituted ring is unchanged, but in any case where the direction of the electric field is parallel at the position of the substituent. This explains the expected correlation in Fig. 14.5b, as well as all other sandwich dimers considered in which either benzene or borazine was the unsubstituted ring.

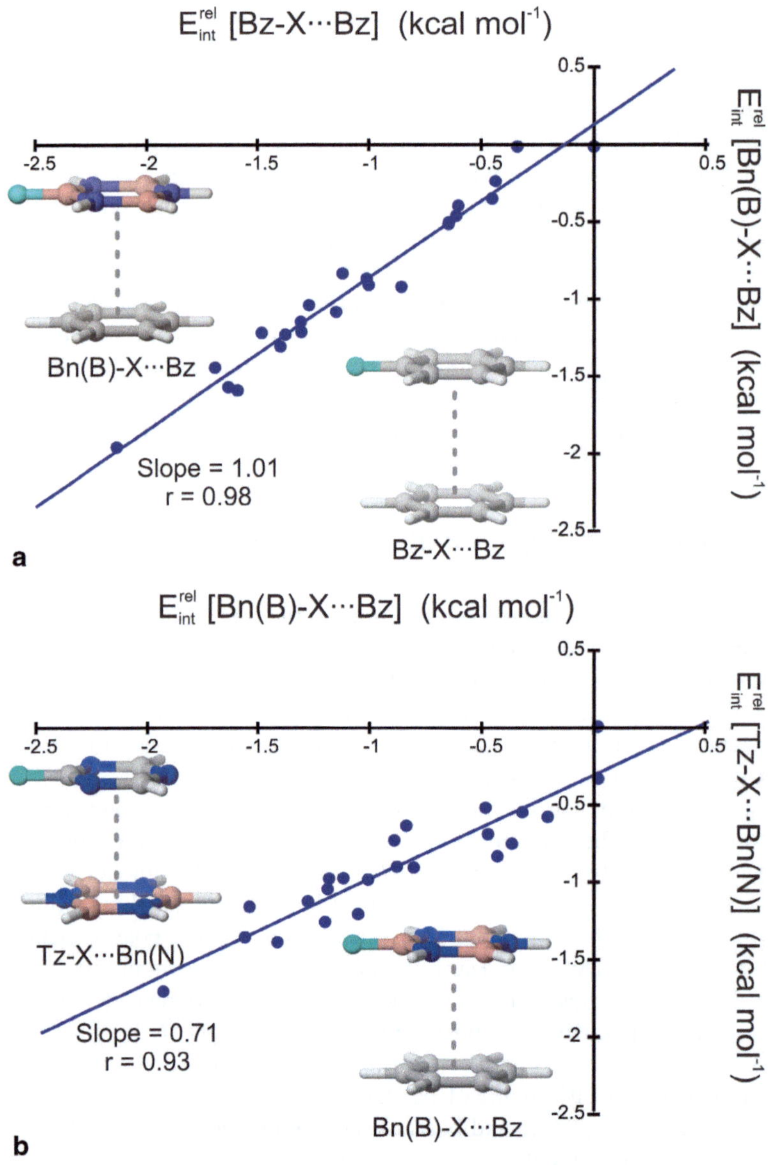

Fig. 14.5 SAPT0/jun-cc-pVDZ interaction energies, relative to the unsubstituted case for sandwich dimers of **a** boron-substituted borazine with benzene vs. substituted benzene with benzene and **b** substituted triazine with borazine and boron-substituted borazine with benzene. (Reprinted with permission from [73]. Copyright 2011 American Chemical Society)

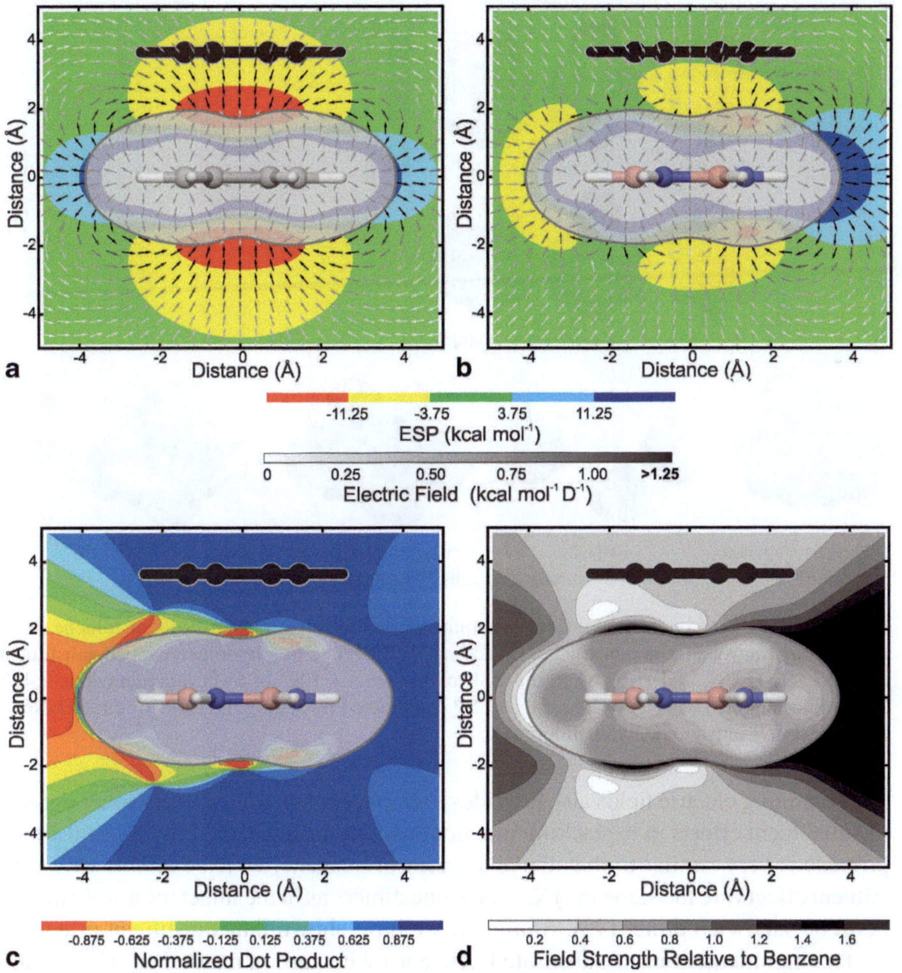

Fig. 14.6 Electric fields (*arrows*) and electrostatic potentials (*solid colors*) of **a** benzene and **b** borazine in a plane bisecting the two rings. **c** Dot product of the normalized electric fields surrounding benzene and borazine in the plane bisecting the rings. At a given point, blue indicates that the two electric fields are parallel, while *red* signifies anti-parallel electric fields. *Green* indicates regions where the two electric fields are perpendicular. **d** Strength of the electric field of borazine relative to benzene in the plane bisecting the rings. *Darker colors* indicate that the electric field of borazine is stronger than that of benzene at that point, while *lighter colors* indicate that the electric field of borazine is weaker than that of benzene. In all four panels, the *gray* shaded region is an electron density isosurface ($\rho = 0.001$ *e*/bohr^3) and the black silhouette is of a sandwich stacked benzene at $R = 3.65$ Å. (Reprinted with permission from [54]. Copyright 2013 American Chemical Society)

However, even if the electric fields above two rings are parallel, their strengths are not necessarily the same. This is true for borazine and benzene, which underlies the non-unit slope of the best-fit line in Fig. 14.5b. In particular, the electric field above the vertex of borazine is weaker than that above benzene, as indicated in Fig. 14.6d.

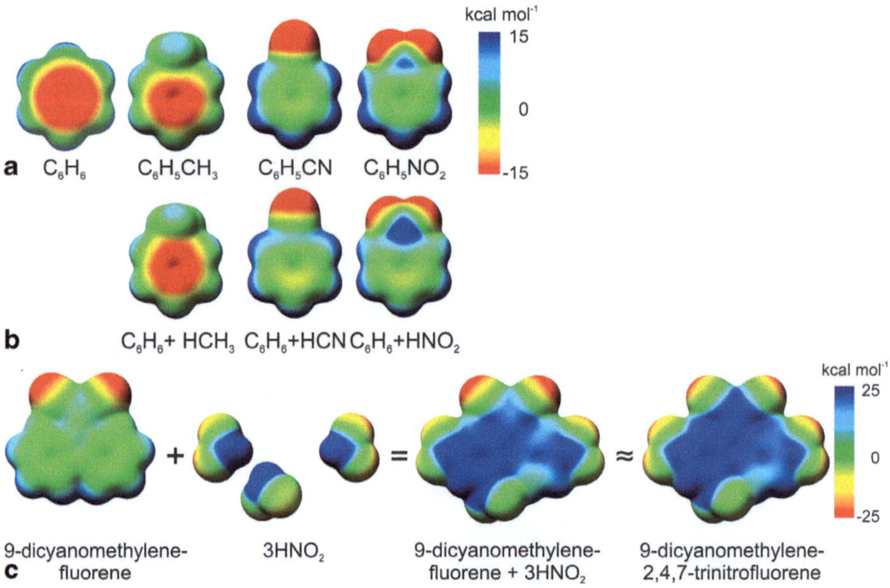

Fig. 14.7 **a** ESPs of benzene, toluene, benzonitrile, and nitrobenzene. **b** ESPs of toluene, benzonitrile, and nitrobenzene constructed by adding the ESP due to a hydrogen capped substituent (i.e. H-CH_3, H-CN, and H-NO_2) to the ESP of benzene. **c** ESP of 9-dicyanomethylene-2,4,7-trinitrofluorine (*far right*) as well as the an ESP constructed by adding the ESPs of three H-NO_2s to the ESP of 9-dicyanomethylenefluorene

Examining electric fields also provides a simple explanation of many other aspects of substituent effects in π-stacking interactions. For instance, the electric field above propene is very similar to that above a vertex of benzene, which explains why substituent effects are the same in HX... propene dimers as in the intact monosubstituted benzene sandwich dimer [73]. Similarly, we have shown that substituent effects in π-stacking interactions of substituted benzenes with carbon nanotubes (CNTs) and a small graphene model depend only on the polarizability of the substituents, not any electrostatic character [90]. This is a consequence of the weak strength of the electric field above the center of CNTs and graphene.

14.4 What About Molecular Electrostatic Potentials?

Central to discussions of π-stacking and many other non-covalent interactions are molecular electrostatic potentials (ESPs, e.g. see Fig. 14.7a). These colorful plots can provide key insights into both the strength and preferred orientation of non-covalent complexes, and have emerged as powerful tools for the analysis and prediction of non-covalent interactions. Unfortunately, these ESP plots are widely misinterpreted among chemists.

At a given point in space in the vicinity of a molecule, the ESP is the electrostatic interaction that a positive test charge would experience at that position. In other words, the ESP reflects the balance between the repulsion of a positive test charge by the nuclei and the attraction of a test charge by the electron density. As such, differences in the ESPs between two molecules indicate differences in both the electron density distribution *and* the nuclear charge distribution. Furthermore, the ESP at a given point depends on the electron density everywhere in space. Consequently, even a remote change in the electron density distribution will impact the ESP at a given point. Practically, this means that a change in ESP in a given region is *not necessarily an indication of a local change* in the electron density.

However, it is common among chemists to associate differences in ESPs between two molecules with differences in the *local* electron density. This is seemingly the basis of popular terms such as "electron-rich" and "electron-poor' in describing aromatic systems in particular. For example, the ESPs above benzonitrile and nitrobenzene are significantly more positive than the ESP above benzene (see Fig. 14.7a). This difference is widely explained as arising from the π-resonance-based delocalization of a formal positive charge throughout the π-system of the aromatic rings bearing the electron-withdrawing substituents, since the presence of this positive charge in the π-system of the ring would lead to repulsion of a positive test charge. That is, the positive ESPs above these rings seem to reinforce the classic view of benzonitrile and nitrobenzene as having an "electron-deficient π-system." This is not actually the case.

In 2009, Wheeler and Houk [91] showed that the ESPs above the faces of a diverse collection of substituted arenes can be reproduce by simply adding the through-space effects of the substituents to the ESPs of the corresponding unsubstituted arenes. For example, in Fig. 14.7b we show the ESPs of a fictitious versions of toluene, benzonitrile, and nitrobenzene in which we simply added the ESP arising from hydrogen capped-substituents to the ESP of benzene. These "additive" ESPs, in which both the σ- and π-electron density of the arene itself are *identical to that of benzene*, are strikingly similar to the ESP of the intact substituted systems. In other words, the difference in the ESPs between these substituted phenyl rings and benzene can be reproduced without making any changes in either the σ- or π-electron distribution of the ring! Instead, the changes in the ESPs above these substituted arenes primarily reflect the through-space effects of the local dipole associated with the substituents. Indeed, Wheeler and Houk [91] found for nearly all substituted arenes that substituent-induced changes in the ESPs are primarily the result of through-space effects of the substituents, not changes in the π-electron-density. This is shown explicitly for 9-dicyanomethylene-2,4,7-trinitrofluorene in Fig. 14.7c.

These results underscore the local nature of substituent effects in π-stacking interactions. That is, even though substituents result in drastic changes in the ESP above aromatic systems, these changes do not reflect changes in the π-electron density. Instead, these changes result from the electrostatic potential associated with the substituent, which will typically be dominated by the local dipole moment associated with that substituent.

Similar misconceptions underlie common discussion of the ESPs of *N*-heterocycles. For example, the ESP of benzene and five azines are shown in Fig. 14.8. Generally, the positive ESPs above these "electron-deficient" *N*-heterocycles are

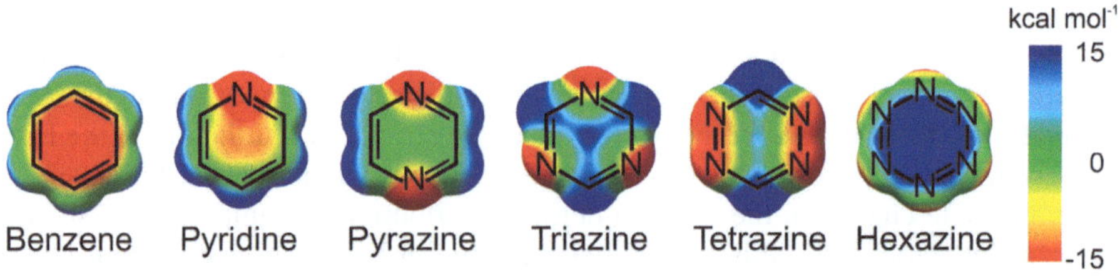

Fig. 14.8 Electrostatic potentials of benzene and five azines—pyridine, pyrazine, s-triazine, s-tetrazine, and the hypothetical molecule hexazine

attributed to the depletion of π-electron density above the carbon atoms due to the presence of the electronegative nitrogens. However, this is not the case! We recently examined the differences in ESPs between benzene and these five azines. These *N*-heterocycles exhibit ESPs that are drastically different from that of benzene, yet are all both σ- and π-isoelectronic with benzene. This allowed direct comparisons among the various contributions to the ESP differences.

In particular, we showed that the positive ESP above the center of the azines does not arise from any change in the π-electron system. Instead, it simply reflects the greater proximity of nuclear charge to the ring center in the azines, compared to benzene. For example, the ESPs of benzene and triazine in the plane bisecting these rings are shown in Fig. 14.9a. Figure 14.9b shows the difference in these ESPs in this plane, as well as the contribution of the nuclei, σ-electrons, and π-electrons to this difference. Notably, the reason that the ESP above triazine is positive is the position of the nuclei—there simply is a greater amount of positive nuclear charge near the center of tetrazine than the center of benzene. Moreover, the impact of both the σ- and π-electron densities of triazine depress the ESP value above the ring, compared to benzene. That is, compared to benzene, there is a more favorable electrostatic interaction between the σ- and π-electron density and a positive test charge above this so-called "electron-deficient" *N*-heterocycles! Similar results were found for the other azines.

Although the implications of these most recent findings have only been examined for anion-π interactions, [92] they have important implications for our understanding of the impact of heteroatoms on all non-covalent interactions in which electrostatic interactions play a role. This includes π-stacking interactions, and we think that this is an area poised for a substantial shift in our understanding.

14.5 Conclusions and Outlook

In the late 1990s, π-stacking interactions seemed relatively well-understood [6, 55, 93, 94]. These interactions were abundant in chemical and biochemical systems, and there were well-trodden physical models to describe π-stacking interactions in both unsubstituted and substituted π-stacked dimers [55–57]. This all changed

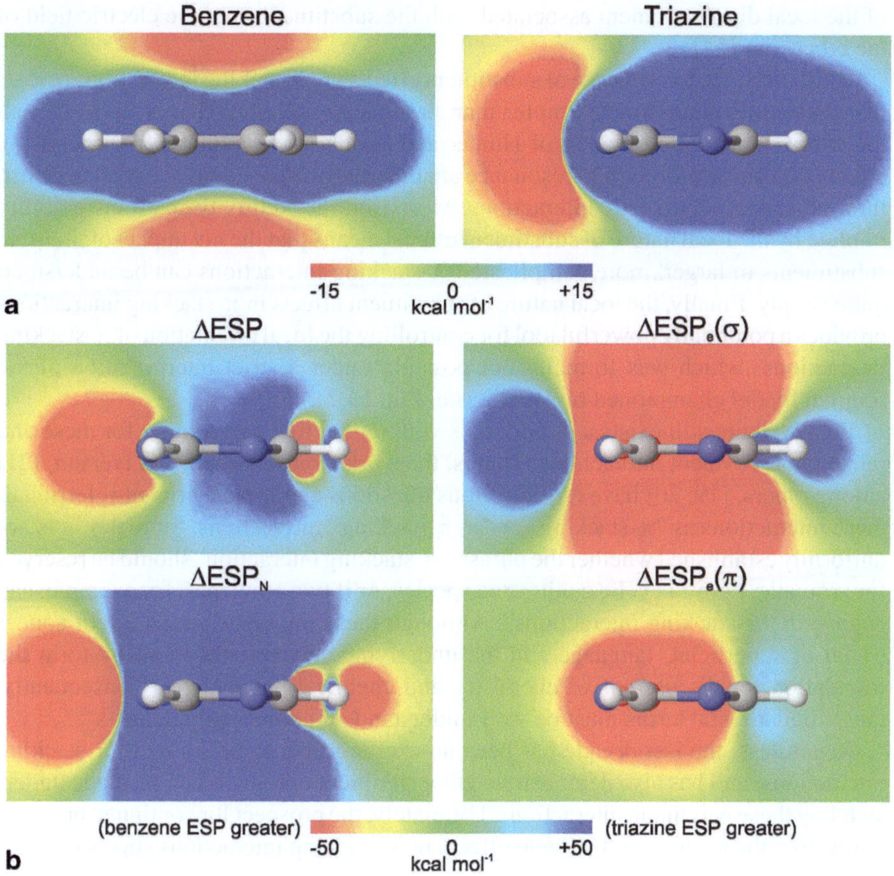

Fig. 14.9 **a** Electrostatic potential in the planes bisecting benzene and triazine. **b** The difference in ESPs between triazine and benzene (ΔESP) in this plane, as well as the nuclear (ΔESP_N), σ-electronic [$\Delta ESP_e(\sigma)$], and π-electronic [$\Delta ESP_e(\pi)$] contributions to ΔESP

in the early 2000s, when high-accuracy gas-phase computations revealed a much richer collection of forces at play in even simple π-stacked systems (i.e. the benzene dimer). Moreover, studies of substituted π-stacked dimers revealed what seemed to be significant weaknesses in the prevailing models of these interactions. Although some of these weaknesses stemmed from differences in gas-phase versus solution-phase interactions, the door had been opened for a dramatic shift in our understanding of these interactions.

Consequently, there have been significant advances over the last 10 years in our understanding of substituent effects in π-stacking interactions. Although it is still not fully settled, there is mounting evidence that local, direct interactions between the substituents and the other rings play a significant, and previously unrecognized, role in tuning the strength of π-stacking interactions [73, 74, 81, 89]. We now understand these local, direct interactions as arising primarily from the interaction of the local dipole moment associated with the substituents and the electric field of the other ring [54].

Ultimately, the local nature of substituent effects in π-stacking interactions reveals that these effects are much simpler than previously thought. That is, even though the π-resonance-based picture of Hunter and Sanders was intuitive (for chemists), effects transmitted through π-resonance are inherently more complicated than direct, through-space electrostatic interactions. Moreover, the additivity and transferability implied by the local nature of substituent effects means that the net impact of multiple substituents in larger, more complicated π-stacking interactions can be understood quite simply. Finally, the local nature of substituent effects in π-stacking interactions provides a potentially powerful tool for controlling the local orientation of π-stacking interactions, which was formally not possible under a strict interpretation of the original model championed by Hunter (i.e. Fig. 14.2).

Finally, there still exists a minor crisis with regard to nomenclature for these and related non-covalent interactions. That is, the work from Martinez and Iverson, [21] among others, [19, 20] have raised serious questions about the merits of referring to these interactions as "π-stacking" or "π-π stacking" interactions. Similarly, it is not uniformly established whether the phrase "π-stacking interaction" should be reserved for aromatic, or even cyclic conjugated species, or if two ethylenes, for example, can engage in "π-stacking interactions". Although these linguistic issues might appear trivial or superficial, language and the underlying concepts they convey form the foundation for discussions of chemistry and chemical phenomena. Consequently, these fundamental terms must be well-understand and universally defined.

Regardless, the last decade has been an exciting time to be studying π-stacking interactions, and has also demonstrated that significant progress can be made understanding these enigmatic interactions. Ultimately, the prospect for continued progress unraveling the origin of substituent effects in π-stacking interactions appears bright.

Acknowledgment This work was supported by the National Science Foundation (Grant CHE-1254897) and the Welch Foundation (Grant A-1776). We also thank the Texas A&M Supercomputing Center for providing computational resources and J. W. G. Bloom, R. K. Raju, K. N. Houk, D. A. Dougherty, C. Corminboeuf, H. M. Jaeger, F. A. Evangelista, B. L. Iverson, and J. S. Siegel for many fruitful discussions about π-stacking interactions.

References

1. Meyer EA, Castellano RK, Diederich F (2003) Interactions with aromatic rings in chemical and biological recognition. Angew Chem Int Ed 42:1210–1250
2. Salonen LM, Ellermann M, Diederich F (2011) Aromatic rings in chemical and biological recognition: energetics and structures. Angew Chem Int Ed 50:4808–4842
3. Schneider H-J (2009) Binding mechanisms in supramolecular complexes. Angew Chem Int Ed 48:3924–3977
4. Müller-Dethlefs K, Hobza P (2000) Noncovalent interactions: a challenge for experiment and theory. Chem Rev 100:143–168
5. Kim KS, Tarakashwar P, Lee JY (2000) Molecular clusters of π-systems: theoretical studies of structures, spectra, and origin of interaction energies. Chem Rev 100:4145–4186

6. Hunter CA, Sanders JKM (1990) The nature of π-π interactions. J Am Chem Soc 112: 5525–5534
7. Yakovchuk P, Protozanova E, Frank-Kamenetskii MD (2006) Base-stacking and base-pairing contributions into thermal stability of the DNA double helix. Nucleic Acids Res 34:564–574
8. Kool ET (2002) Replacing the nucleobases in DNA with designer molecules. Acc Chem Res 35:936–943
9. Riley KE, Hobza P (2013) On the importance and origin of aromatic interactions in chemistry and biodisciplines. Acc Chem Res 46:927–936
10. Cooper VR, Thonhauser T, Puzder A, Schröder E, Lundqvist BI, Langreth DC (2008) Stacking interactions and the twist of DNA. J Am Chem Soc 130:1304–1308
11. Riley KE, Hobza P (2011) Noncovalent interactions in biochemistry. WIREs Comp Mol Sci 1:3–17
12. Hargis JC, Schaefer HF, Houk KN, Wheeler SE (2010) Non-covalent interactions of a Benzo[a]pyrene diol epoxide with DNA base pairs: insight into the formation of adducts of (+)-BaP DE-2 with DNA. J Phys Chem A 114:2038–2044
13. Knowles RR, Jacobsen EN (2010) Attractive noncovalent interactions in asymmetric catalysis: links between enzymes and small molecule catalysts. Proc Natl Acad Sci U S A 107: 20678–20685]
14. Takenaka N, Chen JS, Captain B (2011) Helical chiral 2,2'-Bipyridine N-monoxides as catalysts in the enantioselective propargylation of aldehydes with allenyltrichlorosilane. Org Lett 13:1654–1657
15. Lu T, Zhu R, An Y, Wheeler SE (2012) Origin of enantioselectivity in the propargylation of aromatic aldehydes catalyzed by helical N-Oxides. J Am Chem Soc 134:3095–3102
16. Krenske EH, Houk KN (2013) Aromatic interactions as control elements in stereoselective organic reactions. Acc Chem Res 46:979–989
17. Sinnokrot MO, Valeev EF, Sherrill CD (2002) Estimates of the Ab initio limit for π–π interactions: the benzene dimer. J Am Chem Soc 124:10887–10893
18. Sinnokrot MO, Sherrill CD (2006) High-accuracy quantum mechanical studies of π-π interactions in benzene dimers. J Phys Chem A 110:10656–10668
19. Grimme S (2008) Do special noncovalent π-π stacking interactions really exist? Angew Chem Int Ed 47:3430–3434
20. Bloom JWG, Wheeler SE (2011) Taking aromaticity out of aromatic interactions. Angew Chem Int Ed 50:7847–7849
21. Martinez CR, Iverson BL (2012) Rethinking the term "π-stacking". Chem Sci 3:2191–2201
22. Sherrill CD, Takatani T, Hohenstein EG (2009) An assessment of theoretical methods for nonbonded interactions: comparison to complete basis set limit coupled-cluster potential energy curves for the benzene dimer, the methane dimer, benzene–methane, and benzene–H_2S. J Phys Chem A 113:10146–10159
23. Hohenstein EG, Sherrill CD (2012) Wavefunction methods for noncovalent interactions. WIREs Comp Mol Sci 2:304–326
24. Dion M, Rydberg H, Schröder E, Langreth DC, Lundqvist BI (2004) Van der Waals density functional for general geometries. Phys Rev Lett 92:246401
25. Grimme S (2004) Accurate description of van der Waals complexes by density functional theory including empirical corrections. J Comp Chem 25:1463–1473
26. Grimme S (2006) Semiempirical GGA-type density functional constructed with a long-range dispersion correction. J Comp Chem 27:1787–1799
27. Grimme S, Antony J, Schwabe T, Mück-Lichtenfeld C (2007) Density functional theory with dispersion corrections for supramolecular structures, aggregates, and complexes of (bio)organic molecules. Org Biomol Chem 5:741–758
28. Schwabe T, Grimme S (2008) Theoretical thermodynamics for large molecules: walking the thin line between accuracy and computational cost. Acc Chem Res 41:569–579
29. Grimme S, Antony J, Ehrlich S, Krieg H (2010) A consistent and accurate ab initio parametrization of density functional dispersion correction (DFT-D) for the 94 elements H-Pu. J Chem Phys 132:154104
30. Grimme S (2011) Density functional theory with London dispersion corrections. WIREs Comp Mol Sci 1:211–228

31. Ehrlich S, Moellmann J, Grimme S (2013) Dispersion-corrected density functional theory for aromatic interactions in complex systems. Acc Chem Res 46:916–926
32. Zhao Y, Schultz NE, Truhlar DG (2006) Design of density functionals by combining the method of constraint satisfaction with parametrization for thermochemistry, thermochemical kinetics, and noncovalent interactions. J Chem Theory Comput 2:364–382
33. Zhao Y, Truhlar DG (2006) A new local density functional for main-group thermochemistry, transition metal bonding, thermochemical kinetics, and noncovalent interactions. J Chem Phys 125:194101
34. Zhao Y, Truhlar DG (2007) Density functionals for noncovalent interaction energies of biological importance. J Chem Theory Comput 3:289–300
35. Zhao Y, Truhlar DG (2008) The M06 suite of density functionals for main group thermochemistry, thermochemical kinetics, noncovalent interactions, excited states, and transition elements: two new functionals and systematic testing of four m06 functionals and twelve other functionals. Theor Chem Acc 120:215–241
36. Zhao Y, Truhlar DG (2008) Density functionals with broad applicability in chemistry. Acc Chem Res 41:157–167
37. Vydrov OA, Van Voorhis T (2009) Nonlocal van der Waals density functional made simple. Phys Rev Lett 103:063004
38. Vydrov OA, Van Voorhis T (2012) Benchmark assessment of the accuracy of several van der Waals density functionals. J Chem Theory Comput 8:1929–1934
39. Vydrov OA, Van Voorhis T (2010) Dispersion interactions from a local polarizability model. Phys Rev B 81:062708
40. Becke A; Johnson ER (2005) Exchange-hole dipole moment and the dispersion interaction. J Chem Phys 122:154104
41. Johnson ER, Mackie ID, DiLabio GA (2009) Dispersion interactions in density-functional theory. J Phys Org Chem 22:1127–1135
42. Steinmann SN, Corminboeuf C (2010) A system-dependent density-based dispersion correction. J Chem Theory Comput 6:1990–2001
43. Tkatchenko A, Scheffler M (2009) Accurate molecular van der Waals interactions from ground-state electron density and free-atom reference data. Phys Rev Lett 102:073005
44. Becke A (1997) Density-functional thermochemistry. V. systematic optimization of exchange-correlation functionals. J Chem Phys 107:8554–8560
45. Vázquez-Mayagoitia Á, Sherrill CD, Apra E, Sumpter BG (2010) An assessment of density functional methods for potential energy curves of nonbonded interactions: the XYG3 and B97-D approximations. J Chem Theory Comput 6:727–734
46. Burns LA, Vázquez- Mayagoitia Á, Sumpter BG, Sherrill CD (2011) Density-functional approaches to noncovalent interactions: a comparison of dispersion corrections (DFT-D), exchange-hole dipole moment (XDM) theory, and specialized functionals. J Chem Phys 134:084107
47. Jeziorski B, Moszyński R, Szalewicz K (1994) Perturbation theory approach to intermolecular potential energy surfaces of van der Waals complexes. Chem Rev 94:1887–1930
48. Szalewicz K (2012) Symmetry-adapted perturbation theory of intermolecular forces. WIREs Comp Mol Sci 2:254–272
49. Hohenstein EG, Sherrill CD (2010) Density fitting and Cholesky decomposition approximations in symmetry-adapted perturbation theory: implementation and application to probe the nature of π-π interactions in linear acenes. J Chem Phys 132:184111
50. Hohenstein EG, Sherrill CD (2010) Efficient evaluation of triple excitations in symmetry-adapted perturbation theory via second-order Møller-Plesset perturbation theory natural orbitals. J Chem Phys 133:104107
51. Hohenstein EG, Sherrill CD (2010) Density fitting of intramonomer correlation effects in symmetry-adapted perturbation theory. J Chem Phys 133:014101
52. Parker TM, Burns LA, Parrish RM, Ryno AG, Sherrill CD (2014) Levels of symmetry adapted perturbation theory (SAPT). I. Efficiency and performance for interaction energies. J Chem Phys 140:094106

53. Turney JM, Simmonett AC, Parrish RM, Hohenstein EG, Evangelista FA, Fermann JT, Mintz BJ, Wilke JJ, Abrams ML, Russ NJ et al (2012) Psi4: an open-source ab *initio* electronic structure program. WIREs Comp Mol Sci 2:556–565
54. Raju RK, Bloom JWG, Wheeler SE (2013) Broad transferability of substituent effects in π-stacking interactions provides new insights into their origin. J Chem Theory Comput 9:3479–3490
55. Hunter CA, Lawson KR, Perkins J, Urch CJ (2001) Aromatic interactions. J Chem Soc Perkin Trans 2:651–669
56. Cockroft SL, Hunter CA, Lawson KR, Perkins J, Urch CJ (2005) Electrostatic control of aromatic stacking interactions. J Am Chem Soc 127:8594–8595
57. Cockroft SL, Perkins J, Zonta C, Adams H, Spey SE, Low CMR, Vinter JG, Lawson KR, Urch CJ, Hunter CA (2007) Substituent effects on aromatic stacking interactions. Org Biomol Chem 5:1062–1080
58. Cozzi F, Cinquini M, Annunziata R, Dwyer T, Siegel JS (1992) Polar/π interactions between stacked aryls in 1,8-Diarylnaphthalenes. J Am Chem Soc 114:5729–5733
59. Cozzi F, Cinquini M, Annunziata R, Siegel JS (1993) Dominance of polar/π over charge-transfer effects in stacked phenyl interactions. J Am Chem Soc 115:5330–5331
60. Cozzi F, Ponzini F, Annunziata R, Cinquini M, Siegel JS (1995) Polar interactions between stacked π systems in fluorinated 1,8-Diarylnaphthalenes: importance of quadrupole moments in molecular recognition. Angew Chem Int Ed 34:1019–1020
61. Cozzi F, Siegel JS (1995) Interaction between stacked aryl groups in 1,8-Diarylnaphthalenes: dominance of polar/π over charge-transfer effects. Pure Appl Chem 67:683–689
62. Cozzi F, Annunziata R, Benaglia M, Cinquini M, Raimondi L, Baldridge KK, Siegel JS (2003) Through-space interactions between face-to-face, center-to-edge oriented arenes: importance of polar-π effects. Org Biomol Chem 1:157–162
63. Cozzi F, Annunziata R, Benaglia M, Baldridge KK, Aguirre G, Estrada J, Sritana-Anant Y, Siegel JS (2008) Through-space interactions between parallel-offset arenes at the van der Waals distance: 1,8-diarylbiphenylene syntheses, structure and QM computations. Phys Chem Chem Phys 10:2686–2694
64. Sinnokrot MO, Sherrill CD (2003) Unexpected substituent effects in face-to-face π-stacking interactions. J Phys Chem A 107:8377–8379
65. Sinnokrot MO, Sherrill CD (2004) Substituent effects in π-π interactions: sandwich and T-shaped configurations. J Am Chem Soc 126:7690–7697
66. Ringer AL, Sinnokrot MO, Lively RP, Sherrill CD (2006) The effect of multiple substituents on sandwich and T-Shaped π-π interactions. Chem Eur J 12:3821–3828
67. Arnstein SA, Sherrill CD (2008) Substituent effects in parallel-displaced π-π interactions. Phys Chem Chem Phys 10:2646–2655
68. Ringer AL, Sherrill CD (2009) Substituent effects in sandwich configurations of multiply substituted benzene dimers are not solely governed by electrostatic control. J Am Chem Soc 131:4574–4575
69. Hohenstein EG, Duan J, Sherrill CD (2011) Origin of the surprising enhancement of electrostatic energies by electron-donating substituents in substituted benzene sandwich dimers. J Am Chem Soc 133:13244–13247
70. Wheeler SE, Houk KN (2008) Substituent effects in the benzene dimer are due to direct interactions of the substituents with the unsubstituted benzene. J Am Chem Soc 130:10854–10855
71. Wheeler SE, McNeil AJ, Müller P, Swager TM, Houk KN (2010) Probing substituent effects in aryl–aryl interactions using stereoselective Diels–Alder cycloadditions. J Am Chem Soc 132:3304–3311
72. Raju RK, Bloom JWG, An Y, Wheeler SE (2011) Substituent effects in non-covalent interactions with aromatic rings: insights from computational chemistry. ChemPhysChem 12:3116–3130
73. Wheeler SE (2011) Local nature of substituent effects in stacking interactions. J Am Chem Soc 133:10262–10274

74. Wheeler SE (2013) Understanding substituent effects in non-covalent interactions involving aromatic rings. Acc Chem Res 46:1029–1038
75. Lee EC, Kim D, Jurečka P, Tarakeshwar P, Hobza P, Kim KS (2007) Understanding of assembly phemona by aromatic-aromatic interactions: benzene dimer and the substituted systems. J Phys Chem A 111:3446–3457
76. Seo J-I, Kim I, Lee YS (2009) π-π interaction energies in monosubstituted-benzene dimers in parallel- and antiparallel-displaced conformations. Chem Phys Lett 474:101–106
77. Watt M, Hardebeck LKE, Kirkpatrick CC, Lewis M (2011) Face-to-face arene-arene bidning energies: dominated by dispersion but predicted by electrostatic and dispersion/polarizability substituent constants. J Am Chem Soc 133:3854–3862
78. Hunter CA (2004) Quantifying intermolecular interactions: guidelines for the molecular recognition toolbox. Angew Chem Int Ed 43:5310–5324
79. Cockroft SL, Hunter CA (2007) Chemical double-mutant cycles: dissecting non-covalent interactions. Chem Soc Rev 36:172–188
80. Cockroft SL, Hunter CA (2009) Desolvation and substituent effects in edge-to-face aromatic interactions. Chem Commun 3961–3963
81. Wheeler SE, Bloom JWG (2014) Toward a more complete understanding of non-covalent interactions involving aromatic rings. J Phys Chem A 118:6133–6147
82. Tauer T, Sherrill CD (2005) Beyond the benzene dimer: an investigation of the additivity of π–π interactions. J Phys Chem A 109:10475–10478
83. Hohenstein EG, Sherrill CD (2009) Effects of heteroatoms on aromatic π-π interactions: benzene-pyridine and pyridine dimer. J Phys Chem A 113:878–886
84. Sherrill CD (2009) Computations of noncovalent pi interactions. In: Lipkowitz KB Cundari TR (eds) Reviews computational chemistry, vol 26. Wiley-VCH, New York, pp 1–38
85. Marshall MS, Burns LA, Sherrill CD (2011) Basis set convergence of the coupled-cluster correction, δ(MP2)(CCSD(T)): best practices for benchmarking non-covalent interactions and the attendant revision of the S22, NBC10, HBC6, and HSG databases. J Chem Phys 135:194102
86. Beg S, Waggoner K, Ahmad Y, Watt M, Lewis M (2008) Predicting face-to-face arene-arene binding energies. Chem Phys Lett 455:98–102
87. Gung BW, Emenike BU, Alvereza CN, Rakovan J, Kirschbaum K, Jain N (2010) Relative substituent position on the strength of π-π stacking interactions. Tetrahedron Lett 51:1648–1650
88. Benitex Y, Baranger AM (2011) Control of the stability of a protein-RNA Complex by the position of fluorine in a base analog. J Am Chem Soc 133:3687–3689
89. Wheeler SE (2012) Controlling the local arrangements of π-stacked polycyclic aromatic hydrocarbons through substituent effects. CrystEngComm 14:6140–6145
90. Munusamy E, Wheeler SE (2013) Endohedral and exohedral complexes of cyclohexane and substituted benzenes with carbon nanotubes and graphene. J Chem Phys 139:094703
91. Wheeler SE, Houk KN (2009) Through-space effects of substituents dominate molecular electrostatic potentials of substituted arenes. J Chem Theory Comput 5:2301–2312
92. Wheeler SE, Bloom JWG (2014) Anion-π interactions and positive electrostatic potentials of N-heterocycles arise from the positions of the nuclei, not changes in the π-electron distribution. Chem Commun 50:11118–11121
93. Hunter CA (1994) Meldola lecture. The role of aromatic interactions in molecular recognition. Chem Soc Rev 23:101
94. Carver FJ, Hunter CA, Seward EM (1998) Structure-activity relationship for quantifying aromatic interactions. Chem Commun 775–776

Chapter 15
Noncovalent Interactions of Organic Ions with Polar Molecules in the Gas Phase

M. Samy El-Shall, Isaac K. Attah and Sean P. Platt

Abstract The chapter is focused on noncovalent interactions of organic ions with small polar molecules in the gas phase. The organic ions studied include cyclic $C_3H_3^+$ and the radical cations of benzene ($C_6H_6^{\bullet+}$), pyridine ($C_5NH_5^{\bullet+}$), pyrimidine ($C_5N_2H_4^{\bullet+}$), fluorobenzene ($C_6H_5F^{\bullet+}$), phenylacetylene ($C_8H_6^{\bullet+}$), benzonitrile ($C_7NH_5^{\bullet+}$) and naphthalene ($C_{10}H_8^{\bullet+}$). In addition, protonated pyridine (pyridine.H^+) and protonated pyrimidine (pyrimidine.H^+) are also included for comparison with the radical cations. The solvent molecules include water (H_2O), hydrogen cyanide (HCN) and acetonitrile (CH_3CN). The results presented include experimental thermochemical data ($\Delta H°$ and $\Delta S°$) for the stepwise association of the solvent molecules with the organic ions and theoretically calculated binding energies and structures. The four major topics discussed are: (1) Trends in binding energies and entropy changes, (2) Ionic hydrogen bonds with organic ions, (3) Internal vs. external solvation of the organic ions, and (4) Intracluster proton transfer and deprotonation of the organic ions.

15.1 Introduction

Intermolecular forces, including hydrogen bonds and ion-molecule interactions, [1–5] are important in many biological, chemical, and astrochemical processes such as the conformation and folding of proteins, base pair stacking in DNA, drug design, macromolecular assemblies, molecular recognition, clathrate hydrate formation, and the formation of complex organics and ices in interstellar space [1–7].

A special class of hydrogen bonding interactions, usually referred to as ionic hydrogen bonds (IHBs), involves hydrogen bonding between radical ions or protonated molecules and neutral polar molecules such as water, methanol, ammonia, and hydrogen cyanide [1]. IHBs have bond strengths higher than the typical

M. S. El-Shall (✉) · I. K. Attah · S. P. Platt
Department of Chemistry, Virginia Commonwealth University, 23284–2006 Richmond, VA, USA
e-mail: mselshal@vcu.edu

M. S. El-Shall
Department of Chemistry, Faculty of Science, King Abdulaziz University,
Jeddah 21589, Saudi Arabia

S. Scheiner (ed.), *Noncovalent Forces*, Challenges and Advances in
Computational Chemistry and Physics 19, DOI 10.1007/978-3-319-14163-3_15

conventional hydrogen bond in neutral systems and could reach up to 35 kcal/mol, nearly a third of the strength of covalent bonds [1]. These strong interactions are critical in many fields such as ion induced nucleation, ionic clusters, ion solvation, radiation chemistry, electrochemistry, acid-base chemistry, and self-assembly in supramolecular chemistry [1–5]. IHBs are also important in biological systems including protein folding, proton transport, membranes, enzyme active centers, and molecular recognition [8, 9] Organic ions can form hydrogen bonds with solvents in nature, for example, in icy grains doped with polycyclic aromatic hydrocarbons that are subjected to ionizing radiation in interstellar dust grains [10–13].

Unconventional carbon-based IHBs are formed when the hydrogen donors are ionized hydrocarbons containing CH groups and the hydrogen acceptors are electron lone pairs on hetero atoms such as O and N or π electrons in olefin double bonds and aromatic systems [1]. For example, carbon-based $CH^{\delta+}$–O IHBs appear in the hydration of ionized aromatics such as benzene ($C_6H_6^{\bullet+}$), cyclic $C_3H_3^+$, cyclobutadiene ($C_4H_4^{\bullet+}$), phenylacetylene ($C_8H_6^{\bullet+}$), and naphthalene ($C_{10}H_8^{\bullet+}$) [14–19]. Organic ions can also interact with water molecules by stronger conventional hydrogen bonds, such as in protonated pyridine or protonated pyrimidine where the NH^+–O hydrogen bonds are formed [1, 20, 21].

In addition to water, other polar molecules containing lone pairs of electrons such as hydrogen cyanide, methanol, and acetonitrile can participate in hydrogen bonding interactions with the ring hydrogen atoms ($CH^{\delta+}$) of ionized aromatics. Hydrogen cyanide is a useful probe of non-covalent interactions because it is highly polar (dipole moment = 2.98 D), [22] and it can serve both as a hydrogen donor and as a lone-pair hydrogen acceptor in hydrogen bonds [1]. HCN has received considerable attention because of its role in atmospheric chemistry as a result of its release by biomass burning, [23] and it is also one of the main interstellar/nebula molecules [13]. HCN can be produced in space from the reactions of ammonia and methane, and could play a very important role in the formation of nitrogen-containing polycyclic aromatic hydrocarbons (NPAHs) [24]. Since molecules in outer space are subjected to ionizing radiation from stars, radioactive decay and cosmic rays, ionized aromatics such as benzene and PAHs may act as nucleation centers for the condensation of astrochemical molecules such as hydrogen cyanide to form clusters that could undergo intracluster reactions leading to nitrogen containing PAHs [6, 24]. Nitrogen-containing aromatics are of interest because biologically significant molecules such as DNA, RNA, and certain amino acids and proteins contain nitrogen-substituted rings. As a result, a fundamental understanding of the reactions and structures of nitrogen-containing species in space may provide insights into the origin of life [24–26]. In fact, HCN polymers have been shown to exist in meteorites, comets, planets, moons, and in circumstellar envelopes [24, 27, 28].

Insight into the basic molecular interactions can be obtained from the energies and structures of the key species involved in the stepwise association of polar molecules with organic ions. These data can be obtained experimentally by measuring sequential binding energies of solvent molecules to organic ions in the gas phase, and computationally by calculating the structures and binding energies of the hydrated and solvated organic ions [1,14–21, 29–32]. One of the most established methods for measuring binding energies of solvent molecules to organic ions in the gas phase is the mass-selected ion mobility (MSIM) method which allows measurements of thermochemical equilibria between a mass-selected ion and several solvent molecules (typically 1–6) [14–21, 30–32]. These measurements, when conducted at different temperatures, yield enthalpy and entropy changes associated with the stepwise association of solvent molecules with the organic ions. A good deal of work over the last decade or two has been conducted using equilibrium measurements to obtain thermochemical data on the stepwise solvation of organic ions with a variety of solvent molecules including water [1, 33–42]. This work continues to this day, with a focus on trends in binding energies and correlations with the properties of the solvent molecule such as dipole moment and proton affinity, and the nature of the organic ion including size, charge distribution and degree of charge delocalization. Our group has been active in studying the stepwise hydration and solvation of different types of organic ions, and this chapter provides an overall review of these recent studies [14–21, 43–48]. In particular, we focus on the interactions of organic ions with three solvent molecules: water, hydrogen cyanide and acetonitrile. We specifically address three issues: (1) variation of binding energies with the structures and properties of the solvent molecules, (2) the onset of ion solvation involving a small number (four to six) of solvent molecules and whether internal or external solvation is preferred, and (3) the degree of intracluster proton transfer within the solvated ions and the ability of a number of solvent molecules to abstract a proton from the organic ion to form a more stable protonated solvent cluster and covert the organic radical cation into a radical which could be more reactive in addition reactions with small organic molecules.

The review begins with a brief description of the MSIM technique to carry out the thermochemical equilibrium measurements to determine the sequential binding energies and entropy changes of the stepwise solvation of organic ions. The next section describes the main experimental and computational results for the solvation of organic ions. Attention is focused on: (1) Trends in Binding Energies & Entropy Changes, (2) Structures of the Hydrated/Solvated Ions, (3) Internal vs. External Solvation, and (4) Intracluster Proton Transfer & Deprotonation Reactions.

15.2 Experimental Measurements of Sequential Enthalpy and Entropy Changes

Figure 15.1 illustrates the essential components of the ion mobility system at Virginia Commonwealth University (VCU) [17, 18, 21]. In the experiments, mass-selected ions (generated by electron impact ionization of the neutral molecules or clusters) are injected (in 20–30 μs pulses) into the drift cell containing a helium carrier gas typically at 1 Torr containing a known partial pressure of the solvent vapor. Flow controllers are used to maintain a constant pressure inside the drift cell. The drift cell temperature can be set within the range of 78–773 K, and can be controlled to $\pm$ 1 K using four temperature controllers, using liquid nitrogen flowing through actuated solenoid valves to cool down the drift cell for temperature experiments below room

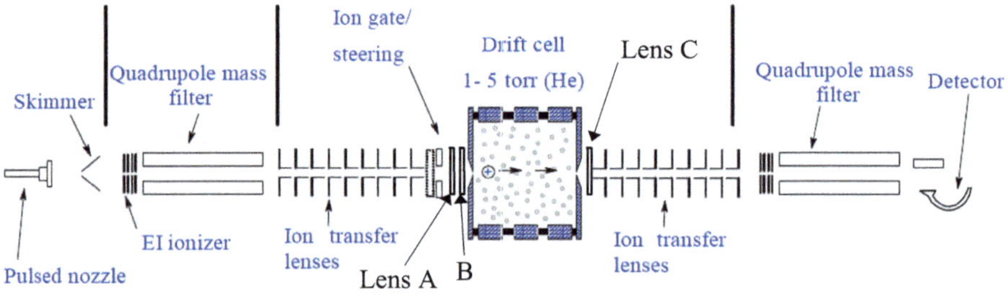

Fig. 15.1 VCU mass-selected ion mobility system [18]

temperature. The reaction products can be identified by scanning a second quadrupole mass filter located coaxially after the drift cell. The arrival time distributions (ATDs) are collected by monitoring the intensity of each ion as a function of time. The reaction time can be varied by varying the drift voltage. Most of the ion thermalization occurs outside the cell entrance by collisions with the solvent vapor escaping from the cell entrance orifice. At a cell pressure of 0.2 Torr, the number of collisions that the ion encounters with the solvent molecules within the 1.5 ms residence time inside the cell is about 10^4 collisions, which is sufficient to ensure efficient thermalization of most of the organic ions.

Figures 15.2a and 15.2b show typical examples of mass spectra obtained following the injection of the mass-selected pyrimidine and benzene radical cations into the drift cell containing water and HCN vapors, respectively [21, 45]. In the absence of water in the drift cell (Fig. 15.2a), only the mass-selected pyrimidine radical cation is observed as shown in Fig. 15.2a (top). In the presence of water, both the $(C_4N_2H_4^{\bullet+})(H_2O)_n$ with $n = 1$–5 and the protonated water series $H^+(H_2O)_n$ with $n = 4$–9 are observed [21]. They consistently shift towards higher n as the temperature of the drift cell decreases. For example, at 238 K the observed ions are $(C_4N_2H_4^{\bullet+})(H_2O)_n$ with $n = 2-5$ and $H^+(H_2O)_n$ with $n = 5-9$ as shown in Fig. 15.2a (bottom). The formation of protonated water clusters $H^+(H_2O)_n$ with $n \geq 4$ is attributed to the dissociative proton transfer reactions represented by Reaction 1 below Eq. (15.1):

$$[C_4H_4N_2^{\bullet+}(H_2O)_{n-1}] + H_2O \rightarrow C_4H_3N_2^{\bullet} + [H^+(H_2O)_n]; n \geq 4 \qquad (15.1)$$

Similar dissociative proton transfer reactions have been observed in the hydration of benzene and the cyclic C_3H_3 cations [14–16]. These reactions are further discussed in Sect. 15.2.4.

In Fig. 15.2b, the injection of the mass-selected $C_6H_6^{\bullet+}$ion into the drift cell containing 0.1 Torr HCN vapor at 298 K, results in the formation of the association products $C_6H_6^{\bullet+}(HCN)_n$ with $n = 1$ and 2 [45]. At 178 K (the lowest achievable temperature before the condensation of HCN), the cluster population is dominated by the $C_6H_6^{\bullet+}(HCN)_n$ series from $n = 2$ to 6 with $n = 3$ and 4 being the major ions observed as shown in Fig. 15.2b (bottom) [45].

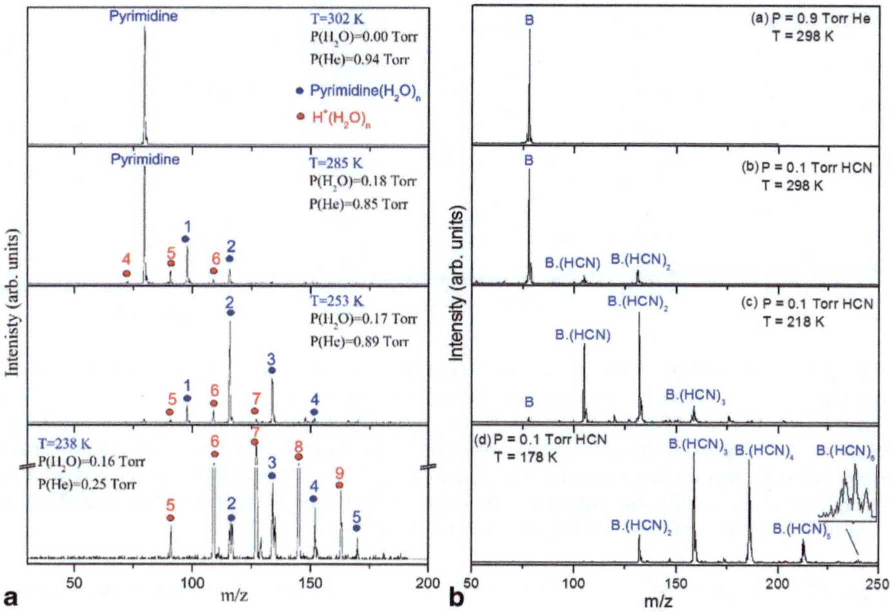

Fig. 15.2 **a** Mass spectra resulting from the injection of the mass-selected pyrimidine radical cation ($C_4N_2H_4^{\bullet+}$) into a helium (He)—water (*W*) vapor mixture at different temperatures using 12.7 eV injection energy (*laboratory frame*) and 2.5 V/cm drift field [21]. **b** Mass spectra obtained following the injection of the mass-selected $C_6H_6^{\bullet+}$(B) into the drift cell containing He and HCN vapor at different pressures and temperatures as indicated [45]

A good test of equilibrium is the observation of identical ATDs of the reactant and product ions. If the $I^{\bullet+}(S)_{n-1}$ and $I^{\bullet+}(S)_n$ ions are in equilibrium, their ATDs must be identical [15–17, 22]. This is evident from the ATDs of the $(C_4N_2H_4^{\bullet+})(H_2O)_n$ ions with $n = 1$–5 as shown in Fig. 15.3a. The ion intensity ratio $I^{\bullet+}(S)_n/I^{\bullet+}(S)_{n-1}$ is measured from the integrated peak areas of the ATDs as a function of decreasing cell drift field corresponding to increasing reaction times, and equilibrium is achieved when a constant ratio is obtained. The equilibrium constants are then obtained using Eq. (15.2).

$$K_{eq} = [I^{\bullet+}(S)_n]/[I^{\bullet+}(S)_{n-1}]P(S) \tag{15.2}$$

Here, $[I^{\bullet+}(S)_n]$ and $[I^{\bullet+}(S)_{n-1}]$ are the intensities of the peaks taken from the integrated ATDs, and P(S) is the partial pressure of the solvent (S) in atm. The equilibrium constants measured as a function of temperature yield ΔH° and ΔS° from the slopes and intercepts, respectively of the van't Hoff plots as illustrated by the typical examples shown in Fig. 15.3b.

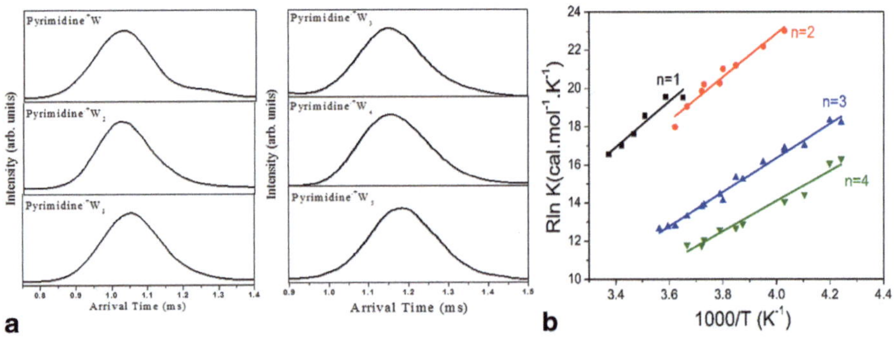

Fig. 15.3 **a** (*Left*) ATDs of the pyrimidine$^{\bullet+}(H_2O)_n$ ions with $n = 1$–3 collected following the injection of mass-selected pyrimidine$^{\bullet+}$into the drift cell containing 0.21 Torr water vapor + 0.22 Torr helium buffer gas at 268 K. (*Right*) ATDs of the pyrimidine$^{\bullet+}(H_2O)_n$ ions with $n = 3$–5 collected following the injection of mass-selected pyrimidine^{+} into the drift cell containing 0.19 Torr water vapor and 0.22 Torr of helium buffer gas at 243 K [21]. **b** van't Hoff plots of the temperature dependence of the equilibrium constant of the stepwise hydration of the pyrimidine radical cation and formation of pyrimidine$^{\bullet+}(H_2O)_n$ with $n = 1$–4 [21]

Table 15.1 Ionization energies and proton affinities of the studied aromatic and polar molecules [22]. The dipole moments (μ) of the polar molecules are also included

Compound	Ionization energy (eV)	Proton affinity (kcal/mol)
Benzene	9.2	179.3
Phenylacetylene	8.8	198.9
Fluorobenzene	9.2	180.7
1,4 di-Fluorobenzene	9.2	171.8
Benzonitrile	9.7	194.0
Naphthalene	8.1	191.9
Pyridine	9.3	222.0
Pyrimidine	9.3	211.7
H_2O ($\mu = 1.9$ D)	12.6	165.0
CH_3OH ($\mu = 1.7$ D)	10.8	180.3
HCN ($\mu = 2.98$ D)	13.6	179.4
CH_3CN ($\mu = 3.9$ D)	12.2	186.2

15.2.1 Trends in Binding Energies & Entropy Changes

Table 15.1 lists molecular properties relevant to the organic ions and solvent molecules discussed in this chapter [22]. Tables 15.2, 15.3, 15.4 list the enthalpy and entropy changes corresponding to the association of water, hydrogen cyanide, and acetonitrile, respectively with different organic ions.

Table 15.2 Thermochemistry of hydrated organic cations

Cation (Ref)	$-\Delta H^o$(kcal/mol)	$-\Delta S^o$(cal mol^{-1} K^{-1})	ΔE (Calc) (kcal/mol)
Cyclic $C_3H_3^+$ [16]	11.7	18.8	11.8
Benzene$^{\bullet+}$ [15]	9.0	19.5	8.5
Phenylacetylene$^{\bullet+}$ [17]	8.0	17.8	7.8
Benzonitrile$^{\bullet+a}$	11.5	29.4	7.7
Naphthalene$^{\bullet+}$ [19]	7.8	19.5	7.7
Pyridine$^{\bullet+}$ [20]	15.2	33.1	15.4 (distonic ion) 10.2 (conventional)
Protonated pyridine [20]	15.6	27.0	14.5
Pyrimidine$^{\bullet+}$ [21]	11.9	23.6	10.8
Protonated pyrimidine [21]	16.7	38.6	16.9

[a]Unpublished results

Table 15.3 Thermochemistry of HCN complexes with organic cations

Cation (Ref)	$-\Delta H^o$ (kcal/mol)	$-\Delta S^o$ (cal mol^{-1} K^{-1})	ΔE (Calc) (kcal/mol)
Benzene$^{\bullet+}$ [45]	9.2	19.1	9.4
Phenylacetylene$^{\bullet+}$ [46]	10.5	24.6	9.5
F-benzene$^{\bullet+}$ [47]	11.2	24.4	9.6
1,4-di F-benzene$^{\bullet+}$ [47]	11.2	22.5	9.8
Benzonitrile$^{\bullet+}$ [47]	9.6	19.1	9.5
Naphthalene$^{\bullet+a}$	6.8	15.3	7.8
Pyridine$^{\bullet+}$ [48]	11.4	21.8	11.9
Protonated pyridine [48]	14.0	26.6	16.0
Pyrimidine$^{\bullet+}$ [48]	12.0	23.3	12.7

[a]Unpublished results

Figure 15.4a compares the hydration energies and the sequential binding energies of several water molecules to the *c*-$C_3H_3^+$, benzene ($C_6H_6^{\bullet+}$), phenylacetylene ($C_6H_5CCH^{\bullet+}$) and naphthalene ($C_{10}H_8^{\bullet+}$) ions. The hydration energy is the largest for the *c*-$C_3H_3^+$(11.7 kcal/mol) followed by the benzene and the phenylacetylene cations (9.0 and 8.0 kcal/mol, respectively), and then the naphthalene cation (7.8 kcal/mol). The range of the hydration energies is consistent with the usual strength of ionic hydrogen bonds of the type $CH^{\delta+}\cdots OH_2$ which are typically about 9 kcal/mol [1]. The IHB in the hydrated *c*-$C_3H_3^+$ is nearly 50 % stronger than in the respective hydration of the naphthalene cation due to the higher charge density on the $C_3H_3^+$

Table 15.4 Thermochemistry of CH_3CN complexes with organic cations

Cation (Ref)	$-\Delta H^o$ (kcal/mol)	$-\Delta S^o$ (cal mol^{-1} K^{-1})	ΔE (Calc) (kcal/mol)
$C_3H_3^+$ [16]	15.3	21.5	–
Benzene$^{\bullet+}$ [45]	14.0	19.9	13.0
Protonated Pyrimidine [21]	–	–	23.7
Naphthalene$^{\bullet+a}$	11.2	20.3	10.8

[a]Unpublished results

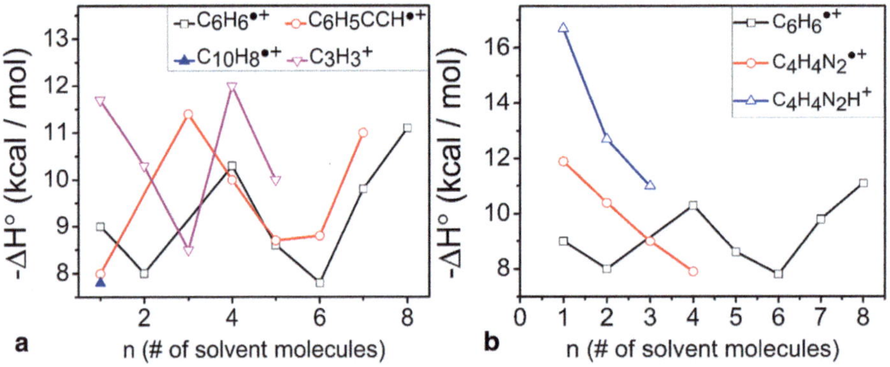

Fig. 15.4 **a** Sequential binding energies of water molecules to the *c*-$C_3H_3^+$, benzene ($C_6H_6^{\bullet+}$), phenylacetylene ($C_6H_5CCH^{\bullet+}$) and naphthalene ($C_{10}H_8^{\bullet+}$) cations [15–17, 19]. **b** Sequential binding energies of water molecules to the benzene ($C_6H_6^{\bullet+}$) and pyrimidine ($C_4H_4N_2^{\bullet+}$) radical cations and to protonated pyrimidine ($C_4H_4N_2H^+$) [15, 21]

ion as compared to the $C_{10}H_8^{\bullet+}$ ion. The charge delocalization over the bicyclic naphthalene cation decreases the charge-dipole interaction of the $CH^{\delta+}\cdots OH_2$ bond in the naphthalene$^{\bullet+}$(H_2O) complex.

For both the *c*-$C_3H_3^+$ and the benzene cation, an increase in the binding energy of the fourth water molecule is observed and for the phenylacetylene cation this increase is observed for the addition of the third water molecule. This could suggest the formation of geometrically stable structures of the first solvent shell around the ion as shown from the DFT calculations discussed in Sect. 15.2.3.

Another interesting result is the apparent increase in the sequential binding energies associated with the formation of the benzene$^{\bullet+}$$(H_2O)_7$, benzene$^{\bullet+}$$(H_2O)_8$, and phenylacetylene$^{\bullet+}$$(H_2O)_7$ clusters. The formation of these large clusters is also associated with large entropy changes (−25.5, −32.6 and −32 cal/mol K, respectively) as compared to the nearly constant $-\Delta S^o$ value of 20 ± 3 cal/mol K for the $n = 1$–5 clusters [15, 17]. This could suggest strong orientational restraint of water in the larger clusters [49]. In fact, three-dimensional cage-like structures involving multiple rings sharing edges are the lowest energy conformers of the water heptamer and octamer [49, 50]. The observed large negative entropy of the phenylacetylene$^{\bullet+}$$(H_2O)_7$ and benzene$^{\bullet+}(H_2O)_8$ clusters could be explained by the formation of cage-like structures by 7–8 H_2O molecules similar to neutral water clusters [49, 50]. This observation could be related to the common bulk view of hydrophobic hydration where water molecules tend to organize around small hydrophobic units without sacrificing hydrogen bonds [51]. The entropy loss associated with this organization is the major reason for the low solubility of nonpolar organic molecules in water [51]. However, in the hydration of large organic ions such as benzene and phenylacetylene, it is not theoretically confirmed that the organization of water molecules into cage-like structures would incorporate such organic ions within the cages.

Figure 15.4b compares the sequential hydration energies of the pyrimidine radical cation ($C_4H_4N_2^{\bullet+}$) and protonated pyrimidine ($C_4H_4N_2H^+$) with those of the benzene radical cation. With the protonated pyrimidine, water is bonded by a conventional $NH^+\cdots OH_2$ bond which is significantly stronger (16.7 kcal/mol) than the unconventional $CH^{\delta+}\cdots OH_2$ bond (8–9 kcal/mol) [21]. Also, the hydration energy of the pyrimidine cation (11.9 kcal/mol) is higher than that of the benzene cation (9.0 kcal/mol) although in both cases the water binds to a $CH^{\delta+}$ site by an unconventional IHB ($CH^{\delta+}\cdots OH_2$) as indicated by the DFT calculations of the lowest energy structures [15, 21]. The sequential binding energies of the pyrimidine$H^+(H_2O)_n$ clusters decrease with n and approach the limiting macroscopic value of 10.5 kcal/mol, the condensation energy of water at $n = 3$ [21]. This follows the usual trend in systems with conventional ionic hydrogen bonds [1]. However, this trend is clearly different from the hydration of the benzene cation as shown in Fig. 15.4b. In relation to these data, the C-H hydrogens of the pyrimidine radical cation ($C_5H_4N_2^{\bullet+}$) can only form carbon-based $CH^{\delta+}\cdot\cdot OH_2$ bonds to water, similar to those formed by the benzene$^{\bullet+}$ ion [14, 15]. In fact, DFT calculations below also show that the hydrogen bond strength of the pyrimidine$^{\bullet+}$ ion to water is 10.8 kcal/mol, in good agreement with the measured value of 11.9 kcal/mol [21].

The hydration energy of protonated pyrimidine measured by the MSIM method (16.7 kcal/mol) is consistent with the hydration energies of other aromatic ions recently measured using energy-resolved collision-induced dissociation (CID) [29]. For example; the CID hydration energies of protonated aniline, acetophenone and phenol were measured as 14.4, 15.6 and 17.5 kcal/mol, respectively [29]. The hydration energy of protonated pyrimidine is higher than that of aniline and slightly lower than of that phenol. The measured hydration energies appear to correlate well with the proton affinity of the aromatic molecules as shown in Tables 15.1 and 15.2 [1].

Figure 15.5a compares the sequential binding energies of HCN molecules with protonated pyridine and the radical cations of pyrimidine, pyridine, phenylacetylene, and benzene. The binding of multiple HCN molecules to the benzene cation is mostly due to unconventional $CH^{\delta+}$–NCH hydrogen bonds directly connected to the $CH^{\delta+}$ sites of the benzene cation and also hydrogen bonding chains (HCN–HCN) among the HCN molecules [45]. The small difference in the bond strength of the two types of interactions ($CH^{\delta+}$–NCH and HCN–HCN) results in relatively small changes of $(-\Delta H^\circ_{n-1,n})$ for $n = 1$–4 as shown in Fig. 15.5a. HCN binds more strongly to the pyridine and pyrimidine radical cations due to ion-dipole interactions in addition to the $CH^{\delta+}$–NCH hydrogen bonding interactions [48]. The strongest binding of HCN

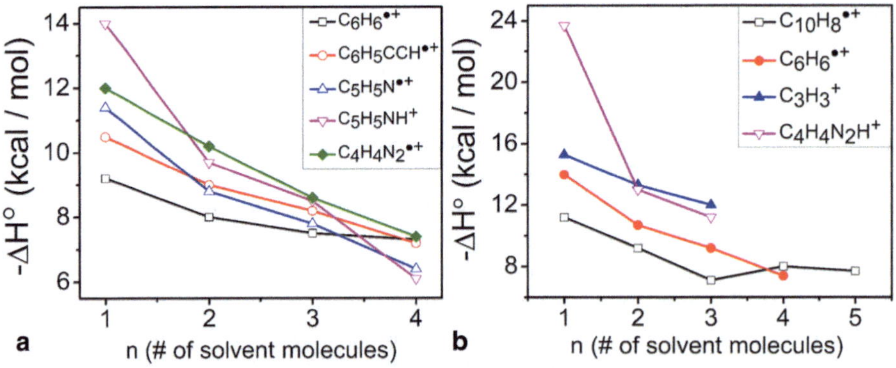

Fig. 15.5 **a** Sequential binding energies of HCN molecules to the benzene ($C_6H_6^{\bullet+}$), phenylacetylene ($C_6H_5CCH^{\bullet+}$), pyridine ($C_5H_5N^{\bullet+}$), and pyrimidine ($C_4H_4N_2^{\bullet+}$) radical cations, and protonated pyridine ($C_5H_5NH^+$) [45–48]. **b** Sequential binding energies of CH_3CN molecules to c-$C_3H_3^+$, benzene ($C_6H_6^{\bullet+}$), naphthalene ($C_{10}H_8^{\bullet+}$) and protonated pyrimidine ($C_4H_4N_2H^+$) [16, 45, 48]

is observed with protonated pyridine as shown in Fig. 15.5a. This bonding interaction is mainly due to an IHB that forms between the NH^+ group of the protonated pyridine and the N atom of HCN (NH^+–NCH bond). A significant drop in the binding energy (31 %) is observed upon the addition of the second HCN molecule to the protonated pyridine in contrast to the smaller changes observed upon the addition of the second HCN molecule to the pyridine or pyrimidine radical cations (23, and 15 %, respectively) as shown in Fig. 15.5a [48]. Despite the strong bonding of HCN to protonated pyridine, the interaction decreases sharply by further addition of HCN molecules and the binding of the fourth HCN molecule is only 6.1 kcal/mol [48]. In fact, in all of the studied HCN clusters around ionized or protonated aromatics, [45–48] the binding energies converge with the addition of 4–5 HCN molecules to the enthalpy of vaporization of HCN liquid (ΔH^{o}_{vap}), which is 6.0 kcal/mol at 298 K [22].

Figure 15.5b compares the sequential binding energies of acetonitrile molecules to protonated pyrimidine, c-C_3H_3 cation and benzene and naphthalene radical cations. DFT calculations indicate that acetonitrile forms a strong proton-bound dimer with protonated pyrimidine (calculated binding energy = 23.7 kcal/mol) [46]. This is a consequence of the large dipole moment of acetonitrile (3.9 D) as compared to 1.6 D for water. As a result, the ion-dipole interaction term is stronger in the case of acetonitrile, and the overall binding energy is significantly higher in pyrimidine$H^+(NCCH_3)$ (23.7 kcal/mol) than in pyrimidine$H^+(OH_2)$ (16.7 kcal/mol) [21, 48].

Figure 15.5b also shows a significant drop in the binding energy (43 %) upon the addition of the second acetonitrile molecule to the proton-bound dimer $(C_4N_2H_4)H^+(NCCH_3)$ as compared to the water interactions where the corresponding drop in binding energy is 24 % as shown in Fig. 15.4b. The transition from strong ionic hydrogen bonding in the proton-bound dimer to weaker $CH^{\delta+}$–N type of bonds is responsible for the sharp drop in the binding energy of the $(C_4N_2H_4)H^+(CH_3CN)_2$ cluster. However, in the case of water, extended hydrogen bonding networks can be formed as shown in the calculated low energy structures of the $(C_4N_2H_4)H^+(H_2O)_2$ clusters discussed in Sect. 15.2.3.

The effect of charge localization on the organic ion is clearly observed by comparing the binding energies of $c\text{-}C_3H_3^+(CH_3CN)_n$ and naphthalene$^{\bullet+}(CH_3CN)_n$ clusters shown in Fig. 15.5b. The naphthalene$^{\bullet+}(CH_3CN)_n$ clusters show the weakest sequential binding energies among the ions shown in Fig. 15.5b. This is again a result of charge delocalization on the naphthalene radical cation which leads to weaker charge-dipole interactions in comparison with the $c\text{-}C_3H_3^+(CH_3CN)_n$ clusters.

15.2.2 *CH$^{\delta+}$....O and CH$^{\delta+}$....N Ionic Hydrogen Bonds with Organic Ions*

The calculated lowest energy structures of several hydrated ions along with their calculated binding energies (kcal/mol) are shown in Fig. 15.6. Except for the protonated pyrimidine, protonated pyridine and distonic pyridine ion, all the ions show unconventional IHBs of the type $CH^{\delta+}\cdots OH_2$. For the $c\text{-}C_3H_3^+(H_2O)$, a single hydrogen bond between the water oxygen atom and one of the $C_3H_3^+$ hydrogen atoms is formed. The charges on $c\text{-}C_3H_3^+$ are equally distributed and localized among the carbon atoms (90 % of the charge) [16]. Upon clustering with one water molecule, the charge distribution changes with significant localizations on the hydrogen atoms of the $c\text{-}C_3H_3^+$, and one C-H bond shows elongation resulting from the linear H-bond formation with the oxygen atom of the water molecule [16].

For the benzene$^{\bullet+}$, phenylacetylene$^{\bullet+}$, naphthalene$^{\bullet+}$, and pyrimidine$^{\bullet+}$ radical cations, the lowest energy hydrated ion has a bifurcated structure with H_2O bonding to two CH hydrogens as shown in Fig. 15.6. This structure was also found in several *ab initio* studies of the benzene$^{\cdot+}(H_2O)$ cluster [52–57]. The MP2 corrected binding energy of 8.5 kcal/mol (Table 15.1) matches well the experimental value of 9.0 ± 1 kcal/mol, [15] and that reported from the IR photodissociation experiment of 9.4 ± 0.3 kcal/mol of the benzene$^{\bullet+}(H_2O)$ cluster [58].

For the hydrated phenylacetylene cation, $C_8H_6^{\bullet+}(H_2O)$, another structure where the water molecule is attached to the hydrogen of the acetylene group is only 1 kcal/mol higher in energy than the lowest energy isomer (at the MP2//ROHF/6-31+G** level) [17]. This indicates that the hydration of the organic ions could involve several structures with energetically similar binding sites. Experimentally, the observed equilibrium could involve different structural isomers depending on the relative stability of these isomers and the temperature of the experiment. As a result, the measured binding energy may represent an average value for water binding to different hydrogen atoms on the ion.

For the hydrated naphthalene cation, the two lowest energy isomers (at the B3LYP/6-311++G** level) have bifurcated structures with H_2O bonding to two CH hydrogens as shown in Fig. 15.6d [19]. The bifurcated structures have significantly larger binding energies than the ion-dipole structure where the water molecule lies above the plane of the naphthalene cation [19]. Also the calculated binding energies

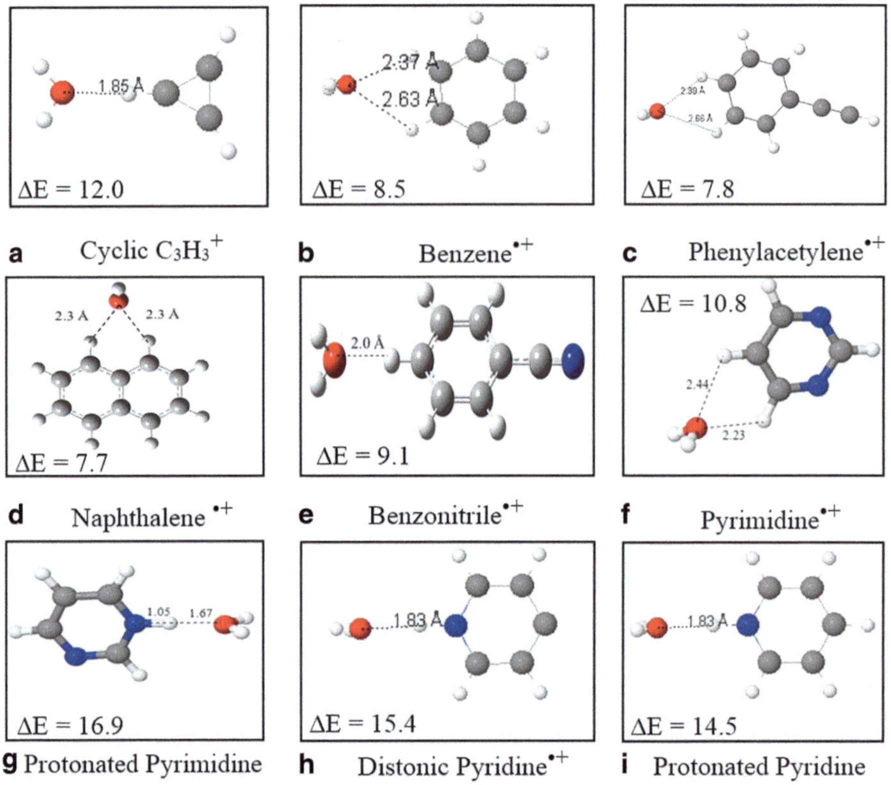

Fig. 15.6 Lowest energy structures of hydrated organic cations. ΔE is the calculated binding energy in kcal/mol. The calculated structures were obtained at the MP2//ROHF/6-31+G(d, p) level for the c-$C_3H_3^+$, benzene$^{\bullet+}$, phenylacetylene$^{\bullet+}$, pyridine$^{\bullet+}$, distonic pyridine$^{\bullet+}$ and protonated pyridine cations, [15–17,20] at the B3LYP/6-311++G(d, p) level for the naphthalene$^{\bullet+}$ and benzonitrile$^{\bullet+}$, [19] and at the M06-2X/6-311++G(d, p) level for the pyrimidine$^{\bullet+}$ and protonated pyrimidine [21]

of the bifurcated structures (7.7–6.8 kcal/mol) are in good agreement with the experimentally determined value (7.8 ± 1 kcal/mol) [19].

For the hydration of the pyridine radical cation, the measured binding energy of the pyridine$^{\bullet+}$(H_2O) cluster (15.2 kcal/mol) was similar to that of the protonated pyridine-water cluster ($C_5H_5NH^+$)(H_2O) (15.6 kcal/mol) that involves a $NH^+ \cdot\cdot OH_2$ bond, and different from those of the hydrated benzene radical cation (9.0 kcal/mol) (Table 15.2) [20]. These relationships indicated that the hydrated pyridine$^{\bullet+}$ ions have the distonic $^{\bullet}C_5H_4NH^+$ structure that can form $NH^+ \cdot\cdot OH_2$ bonds [20]. The calculated $CH^{\delta+} \cdot\cdot OH_2$ binding energy for the conventional pyridine ion to one water molecule is 10.2 kcal/mol, which does not agree with the experimental measured value of 15.2 kcal/mol, while the calculated $NH^+ \cdot\cdot OH_2$ binding energy of the distonic pyridine ion to one water molecule is 15.4 kcal/mol, in excellent agreement with the experimental measured values of 15.2 and 15.6 kcal/mol for the($C_5H_5N^{\bullet+}$)(H_2O) and ($C_5H_5NH^+$)(H_2O) clusters, respectively [20]. However, the hydrated pyrimidine

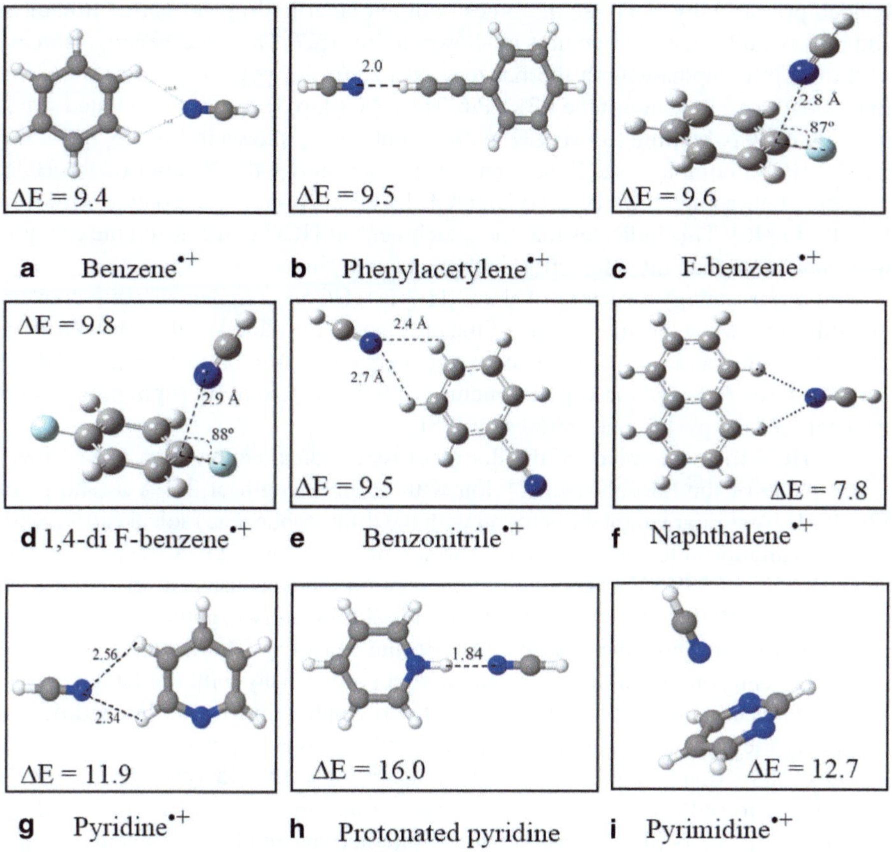

Fig. 15.7 Lowest energy structures of the organic cation –(HCN) clusters. ΔE is the calculated binding energy in kcal/mol. The calculated structures were obtained at the B3LYP/6-311++G(d, p) level for the benzene$^{\bullet+}$, phenylacetylene$^{\bullet+}$, F-benzene$^{\bullet+}$, 1,4-di F-benzene$^{\bullet+}$, benzonitrile$^{\bullet+}$, and naphthalene$^{\bullet+}$ radical cations [45–47], and at the M06-2X/6-311++G(d, p) level for the pyridine$^{\bullet+}$, pyrimidine$^{\bullet+}$ and protonated pyridine [48]

ions were found to have the conventional radical cation structures similar to the hydrated benzene cation. For the hydrated pyrimidine radical cation $(C_4N_2H_4^{\bullet+})(H_2O)_n$, the lowest energy isomer has a bifurcated structure with H_2O bonding to two CH hydrogens with relatively larger distances (2.4 Å and 2.2 Å) than typical H-bonds as shown in Fig. 15.6f. However, the lowest energy isomer of the hydrated protonated pyrimidine has the water molecule directly attached to the NH^+ center via a NH^+–O hydrogen bond of 1.67 Å (Fig. 15.6g), similar to the structure of the hydrated protonated pyridine [21]. This is mediated by the stronger interaction that involves a $NH^+ \cdot\cdot OH_2$ bond similar to other ionic hydrogen bonds where the proton is shared between two centers containing lone pairs of electrons.

The calculated lowest energy structures of HCN complexes with several organic ions are shown in Fig. 15.7. The complexes of HCN with ionized aromatics are bonded primarily by electrostatic forces both in planar hydrogen-bonded structures and in vertical L-shaped structures as shown in Fig. 15.7. The lowest energy isomers of the HCN complexes with the benzene, benzonitrile, pyridine and naphthalene radical cations calculated at the B3LYP/6-311++G(d, p) level have bifurcated structures with HCN binding to two CH hydrogen atoms as shown in Fig. 15.7. For the $C_8H_6^{\bullet+}$(HCN) complex, the lowest energy isomer shows the N-atom of the HCN molecule interacting with the acetylenic CH by a 2.0 Å hydrogen bond as shown in Fig. 15.7b [46]. This indicates that the attachment of HCN to the acetylene group is more favorable than the interaction with the phenyl ring.

The lowest energy structure of the $C_5H_5NH^+$(HCN) complex shows a conventional IHB between the NH^+ group of the protonated pyridine and the nitrogen atom of HCN as shown in Fig. 15.7h. The lowest energy structure of the $C_4H_4N_2^{\bullet+}$(HCN) complex has a T-shaped ion-dipole structure with the N atom of HCN pointing toward the center of the pyrimidine cation ring [48].

The HCN molecule binds to the fluorobenzene radical cation at an angle of 87° to the plane of the fluorobenzene$^{\bullet+}$ ion with a bond length of 2.8 Å as shown in Fig. 15.7c. A similar geometry is found with the difluorobenzene radical cation, with the ion plane-to-molecule angle and bond length of 88° and 2.9 Å, respectively as shown in Fig. 15.7d [47]. The atomic charges on the ions are also of interest. In fluorobenzene$^{\bullet+}$, the electronegative F atom is significantly negative and its carbon has a high partial positive charge (0.60) in the ion [47]. The negative N end of the HCN neutral molecule (− 0.31) can interact attractively with the large positive charge at this carbon. On the other hand, in hydrogen bonding to the ring hydrogens, especially the para hydrogen, HCN interacts with a smaller positive charge (0.24), but this interaction is not destabilized by nearby negative charges. These effects balance out fortuitously to lead to practically equal binding energies in the planar and vertical isomers [47]. Similar charges, interactions and binding energies apply in 1,4-difluorobenzene$^{\bullet+}$. However, in benzonitrile$^{\bullet+}$, although the CN group is electron withdrawing, there are no large positive charges on the ring carbons, and a planar bifurcated hydrogen-bonded structure has the lowest energy similar to the benzene$^{\bullet+}$(HCN) complex [47].

15.2.3 "Internally" & "Externally" Solvated Ions

It is of interest to compare "*internally solvated*" structures where the solvent molecules are bonded directly to the ion with "*externally solvated*" structures where the solvent molecules are bonded to each other and the ion is hydrogen bonded to the exterior of a solvent cluster. Figure 15.8 displays the lowest energy structures of the hydrated organic ions by two water molecules. It is interesting to note that all the structures shown in Fig. 15.8, with the exception of the *c*-$C_3H_3^+$ and pyrimidine ions, form "externally hydrated" structures where the two water molecules are bonded to each other. This indicates that the formation of "*internally hydrated*" structures in

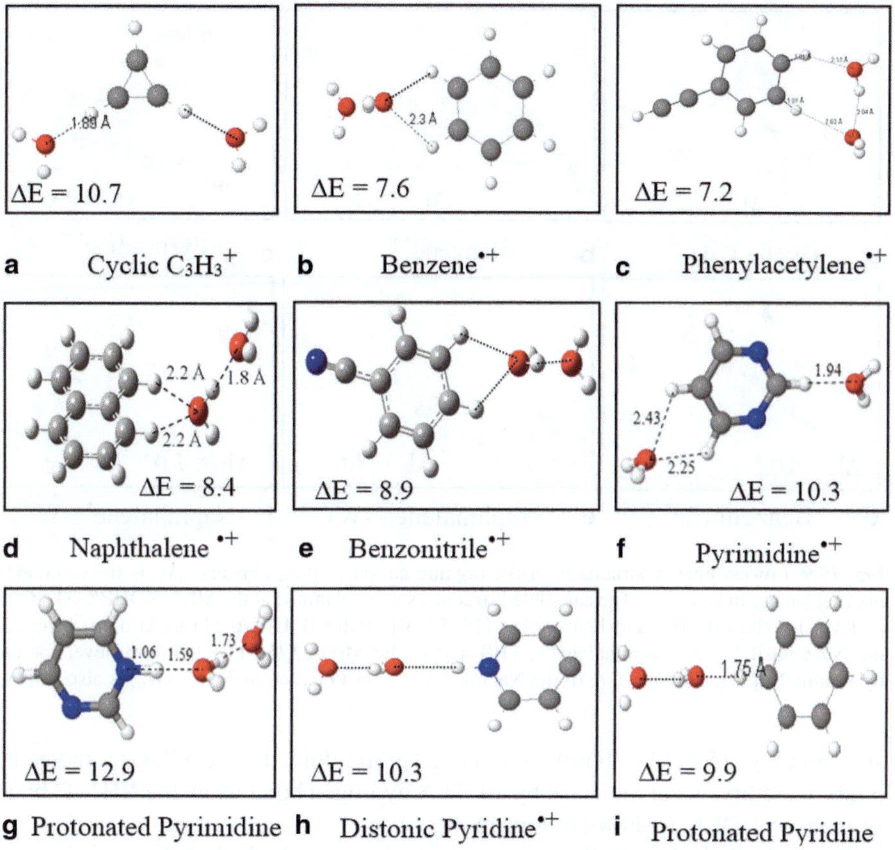

Fig. 15.8 Lowest energy structures of the organic cation-$(H_2O)_2$ clusters. ΔE is the calculated binding energy in kcal/mol. The calculated structures were obtained at the MP2//ROHF/6-31+G(d, p) level for the c-$C_3H_3^+$, benzene$^{\bullet+}$, phenylacetylene$^{\bullet+}$, pyridine$^{\bullet+}$, distonic pyridine$^{\bullet+}$ and protonated pyridine, [15–17,20] at the B3LYP/6-311++G(d, p) level for the naphthalene$^{\bullet+}$ and benzonitrile$^{\bullet+}$ radical cations, [19] and at the M06-2X/6-311++G(d, p) level for the pyrimidine$^{\bullet+}$ and protonated pyrimidine [21]

which the water molecules are bonded to the organic ion may require more than two water molecules.

Clustering of the second water molecule on the $C_3H_3^+(H_2O)$ ion gives a symmetric *internally* hydrated structure as shown in Fig. 15.8a [16]. A similar structure is found for the addition of the second water molecule to the pyrimidine ($C_4N_2H_4^{\bullet+}$) ion as shown in Fig. 15.8f [21]. However, the calculated structures of the hydrated protonated pyrimidine $(C_4N_2H_5^+)(H_2O)_n$ clusters are significantly different from those of the hydrated radical cation [21]. The lowest energy pyrimidineH$^+(H_2O)$ cluster, has the water molecule directly attached to the NH$^+$ center via a NH$^+$–O hydrogen bond of 1.67 Å (Fig. 15.6g) similar to the structure of the hydrated protonated pyridine

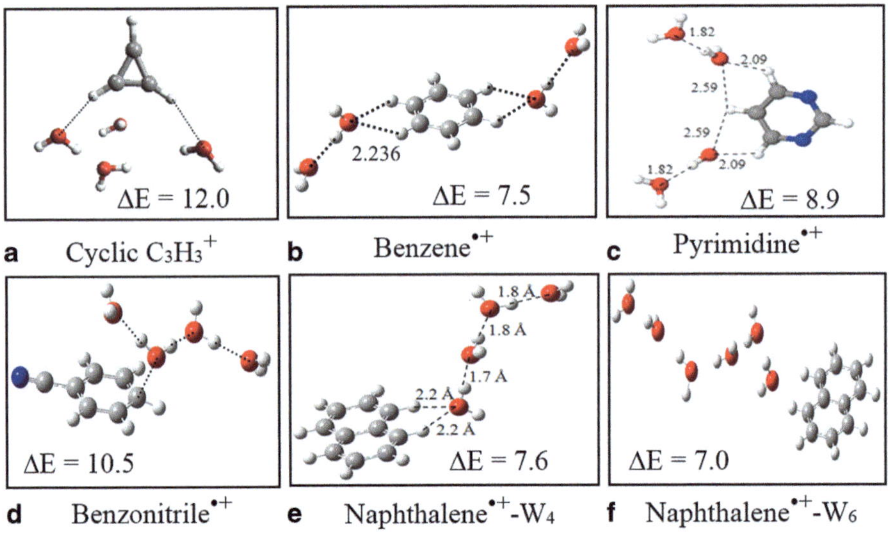

Fig. 15.9 Lowest energy structures of the organic cation-$(H_2O)_4$ clusters. ΔE is the calculated binding energy in kcal/mol. The calculated structures were obtained at the MP2//ROHF/6-31+G(d, p) level for the c-$C_3H_3^+$ and benzene$^{\bullet+}$, [15–17,20] at the B3LYP/6-311++G(d, p) level for the benzonitrile$^{\bullet+}$ and naphthalene$^{\bullet+}$, [19] and at the M06-2X/6-311++G(d, p) level for the pyrimidine$^{\bullet+}$ [21]. The structure of the Naphthalene$^{\bullet+}$-$(H_2O)_6$ (Naphtalene$^{\bullet+}$-W_6) is also shown

shown in Fig. 15.6i [21]. For the protonated pyrimidine, the second water molecule binds to the first water molecule by a 1.73 Å hydrogen bond while the NH^+–O bond shortens to 1.59 Å as shown in Fig. 15.8g.

Figure 15.9 displays the lowest energy structures of the hydrated organic ions by four water molecules. The sequential addition of four water molecules to the $C_3H_3^+$ ion results in a cyclic water tetramer bound with two long hydrogen bonds to the $C_3H_3^+$ hydrogens [16]. It is interesting that the lowest energy isomer of the $C_3H_3^+(H_2O)_3$ cluster involves the formation of a cyclic water trimer [16]. This propensity of water molecules to form cyclic structures in the presence of the c-$C_3H_3^+$ ion is consistent with the structural behavior of neutral water clusters [49, 59].

The lowest energy structure of $C_6H_6^{\bullet+}(H_2O)_4$ is symmetrical with a water dimer on each side of the ion, both forming bifurcated bonds with the ion [15]. IR spectroscopy and *ab initio* calculations indicate that the first water molecule forms two IHBs with two adjacent hydrogen atoms of the benzene cation as shown in Figs 15.6b [55, 60]. For the $C_6H_6^{\bullet+}(H_2O)_n$ with n = 2–4 both "*internally solvated*" isomers where the water molecules are bonded directly to the benzene cation and "*externally solvated*" isomers where the benzene cation is attached to a cluster of water molecules have been predicted [54, 55, 60].

The lowest energy isomers of the pyrimidine$^{\bullet+}(H_2O)_4$ cluster show both "*internally solvated*" and "*externally solvated*" structures with similar binding energies [21]. However, the calculated structures of the hydrated protonated pyrimidine $(C_4N_2H_5^+)(H_2O)_n$ clusters are different from those of the hydrated radical cation

[21]. The pyrimidine$H^+(H_2O)_3$ shows a symmetric structure where a central water molecule binds by 1.75 Å hydrogen bonds to two other water molecules; and to protonated pyrimidine by a shorter NH^+–O bond of 1.43 Å [21]. This structure is consistent with a growth pattern that could lead to dissociative proton transfer within the pyrimidine$H^+(H_2O)_4$ cluster to form the closed shell solvated hydronium ion $H_3O^+(H_2O)_3$ as observed experimentally [21].

Unlike the hydration of the benzene radical cation where the $C_6H_6^{+\bullet}(H_2O)_n$ clusters with n up to 4 exhibit mainly "*internally*" solvated structures with the water molecules forming bifurcated hydrogen bonds with the $CH^{\delta+}$ groups of the benzene cation, [15] the sequential addition of water molecules onto the naphthalene cation favors *external* solvation where the naphthalene cation remains exterior with respect to the water sub-cluster [19]. The calculated structures of the naphthalene$^{\bullet+}(H_2O)_4$ and naphthalene$^{\bullet+}(H_2O)_6$ are shown in Figs. 15.9e and 15.9f, respectively. Similar structures are also obtained for naphthalene$^{\bullet+}(CH_3OH)_4$ and naphthalene$^{\bullet+}(CH_3OH)_6$ [19].

This indicates that hydrogen bonding interactions among water molecules are more favorable than the unconventional hydrogen bonding formed between the $CH^{\delta+}$ sites of the naphthalene cation and the water molecules. This observation suggests that the solvation of larger PAH cations would also favor the "*externally*" solvated structures where the large organic species remain accessible at the surface of the water nanodroplets. This could have an important implication to the reactivity of large PAH ions in interstellar medium since the ions can interact with incoming small molecules under UV irradiation where both ion-molecule and photochemical processes can lead to the formation of complex organics on the surface of the ice grains [10, 19].

The above results and discussions indicate that unlike the hydration of metal ions, where the ion-water interaction is significantly stronger than the water-water interaction, water-water interaction can be very similar or even stronger than ion-water interaction in the hydration of organic ions [1]. Therefore, it appears that the common ion hydration picture where the ion is surrounded by 4–6 water molecules may not be valid for the hydration of organic ions. The observed trend of a large entropy loss with the higher order hydration steps as a result of forming cyclic or bicyclic hydrogen bonded water clusters supports this view [15, 17]. In other words, unless the organic ion is very small (e.g. $C_3H_3^+$), "internally hydrated" structures may not form and the large organic ions remain outside the cage-like hydrogen bonded water structures [19].

15.2.3.1 Solvation of Organic Ions by HCN Molecules

Solvation of the organic ions by HCN molecules show a tendency to form extended HCN chains associated with the benzene and phenylacetylene radical cations and protonated pyridine. Figure 15.10 displays the calculated lowest energy structures of the benzene$^{\bullet+}(HCN)_n$, phenylacetylene$^{\bullet+}(HCN)_n$ and protonated pyridine$H^+(HCN)_n$ clusters with $n = 2$ and 4. It is clear that the addition of the second HCN molecule

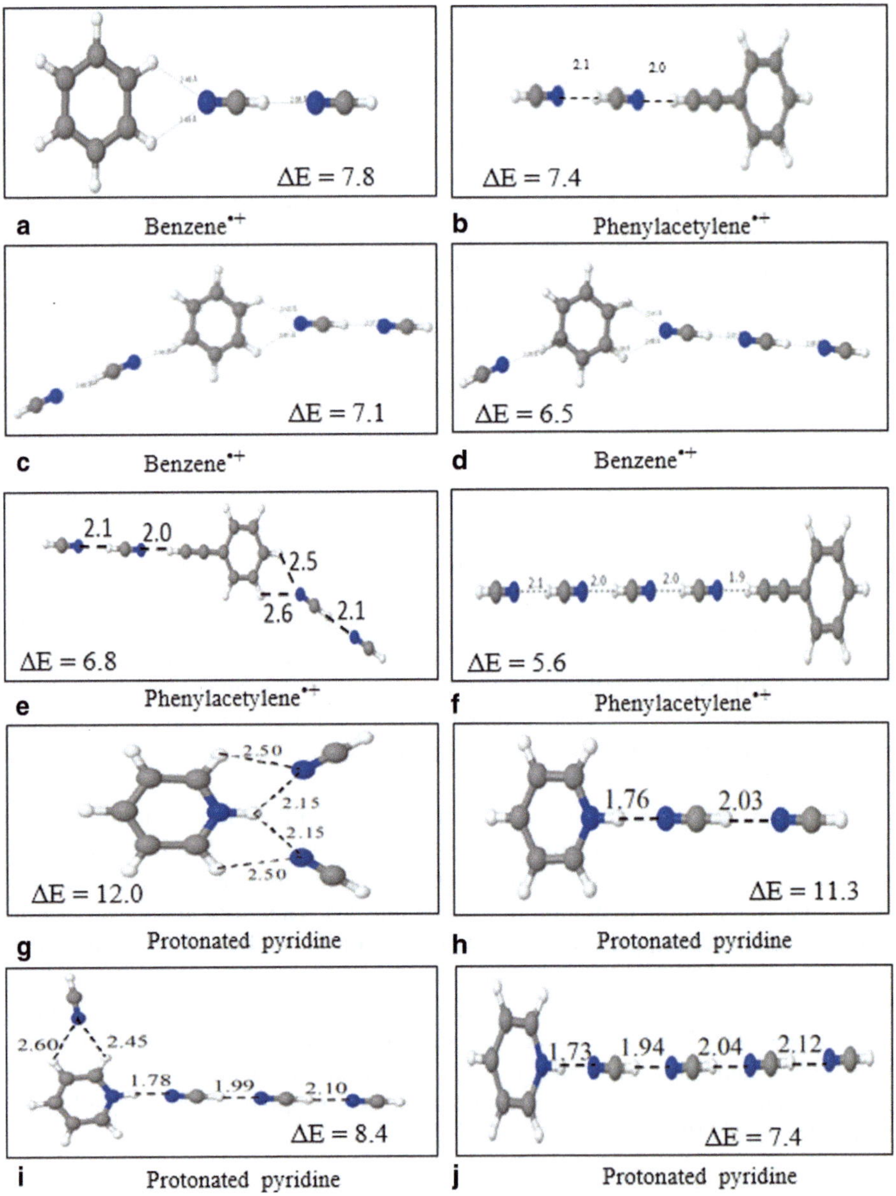

Fig. 15.10 Lowest energy structures of organic cation-$(HCN)_n$ clusters with $n = 2$ and 4. ΔE is the calculated binding energy in kcal/mol. The calculated structures were obtained at the B3LYP/6-311++G(d, p) level for the benzene$^{\bullet+}$ and phenylacetylene$^{\bullet+}$ radical cations, [45, 46] and at the M06-2X/6-311++G(d, p) level for protonated pyridine [48]

is more likely to form HCN... HCN hydrogen bond than forming a second C-H^{δ}.... NCH bond with the organic ion. The structures of the ions solvated by four

HCN molecules represent a competition between the assembly of a linear hydrogen bonding HCN chain and internal solvation of the cation. Interestingly, structures with linear hydrogen bonded chains extended from the phenylacetylene cation are only 1 kcal/mol higher in energy than the lowest energy isomer. For protonated pyridine, the extended chain structure with four HCN molecules (Fig. 15.10j) is 1.0 kcal/mol higher in energy than the lowest energy isomer with an extended chain of three HCN molecules shown in Fig. 15.10i. These extended structures are good models for the design of molecular wires formed by hydrogen bonding chains that could be utilized for low temperature molecular devices [61–63]. For example, the lengths of these molecular wires (the distance from the carbon atom of the phenyl ring connected to the C≡C group to the hydrogen atom of the terminal HCN molecule in the one-sided chain structures) can vary from 0.8 to 2.1 nm depending on the number of HCN molecules. For the two sided chain structures where the phenyl ring can be considered as part of the wire, the distance between the hydrogen atoms of the terminal HCN molecules on both sides of the ring can reach 2.5 nm in the $C_8H_6^{\bullet+}(HCN)_4$ cluster. These wire assemblies connected by hydrogen bonds could provide unique opportunities for systematic spectroscopic and structure studies of charge distributions and separations in these systems.

It should be noted that extended linear chain structures involving HCN and acetylene units are common in the solid state molecular crystals of these molecules. For example, in the solid state, both hydrogen cyanide and cyanoacetylene (H-C≡C-C≡N) crystals consist of extended, linear hydrogen-bonded chains [64, 65]. However, in the gas phase both linear and cyclic structures are known to exist based on several spectroscopic and theoretical studies [66–69]. Interestingly, clusters of linear chains have been reported to exist in superfluid helium expansion [70]. In cluster ions, the measured thermochemistry of protonated HCN clusters $[H^+(HCN)_n]$ suggests the formation of linear hydrogen-bonded chains extended from a strongly bound core consisting of the proton-bound HCN dimer ($HCN.H^+.NCH$) [71, 72]. In the present system, the core ion is the phenylacetylene cation which allows the formation of HCN extended chains by hydrogen bonding to the acetylene hydrogen or to the hydrogen atoms of the phenyl ring. It appears that the formation of a one-sided chain by hydrogen bonding of the HCN molecules to the acetylene hydrogen, and two-sided chains by additional hydrogen bonding to the para hydrogen atom (with respect to the acetylene group) of the phenyl ring represents the most favorable modes for the growth of the hydrogen cyanide molecular wires attached to the phenylacetylene cation.

The second group of organic ions which includes pyridine$^{\bullet+}$, pyrimidine$^{\bullet+}$, F-benzene$^{\bullet+}$, 1,4-di-F-benzene$^{\bullet+}$, benzonitrile$^{\bullet+}$, and naphthalene$^{\bullet+}$ show more tendency for internal solvation by four HCN molecules than for the formation of extended HCN chains. The calculated lowest energy structures of these solvated ions are shown in Fig. 15.11.

The lowest energy structures of the $C_5H_5N^{\bullet+}(HCN)_4$ and $C_4H_4N_2^{\bullet+}(HCN)_4$ clusters show bifurcated structures involving multiple hydrogen bonding sites with the ring hydrogen atoms as shown in Figs. 15.11a and 15.11b, respectively. However, no chain structures involving four HCN molecules have been found in these clusters

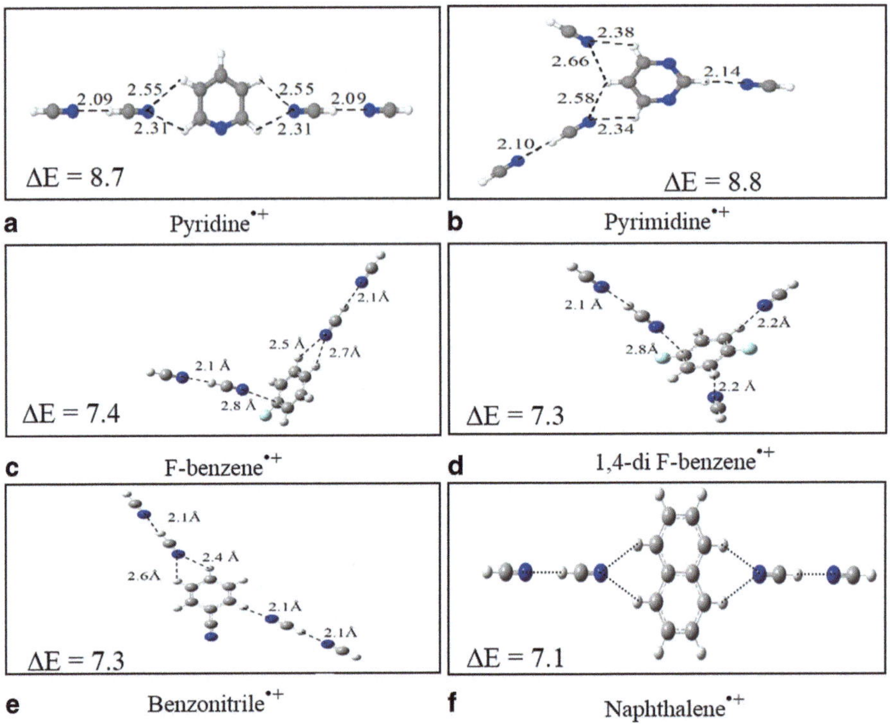

Fig. 15.11 Lowest energy structures of organic radical cation-$(HCN)_4$ clusters. ΔE is the calculated binding energy in kcal/mol. The calculated structures were obtained at the B3LYP/6-311++G(d, p) level for the F-benzene$^{\bullet+}$, 1,4-di F-benzene$^{\bullet+}$ and benzonitrile$^{\bullet+}$, [47] and at the M06-2X/6-311++G(d, p) level for pyridine$^{\bullet+}$ and pyrimidine$^{\bullet+}$ radical cations [48]

indicating that HCN interactions with the pyridine or pyrimidine ring cations are more favorable than the interactions within the HCN chains.

In the $C_6H_5F^{\bullet+}(HCN)_4$ cluster, the fourth HCN binds externally to the third bifurcated HCN with a bond length of 2.1 Å as shown in Fig. 15.11c. The measured binding energy for this addition is 7.8 kcal/mol in good agreement with the DFT calculated value of 7.4 kcal/mol. The fourth HCN in $C_6H_4F_2^{\bullet+}(HCN)_4$ binds externally to the second meta position on the aromatic ring as shown in Fig. 15.11d, with a bond length of 2.2 Å. The measured binding energy of 7.7 kcal/mol agrees well with the theoretical value of 7.3 kcal/mol. In the benzonitrile cluster ion $C_6H_5CN^{\bullet+}(HCN)_4$, the fourth HCN ligand binds externally to the HCN at the ortho position, with a bond length of 2.1 Å (Fig. 15.11e). The measured binding energy for this addition is 6.1 kcal/mol, which is lower than the DFT binding energy of the lowest energy isomer (Fig. 15.11e) of 7.3 kcal/mol, but within the experimental uncertainty range of $\pm$ 1 kcal/mol for clustering equilibrium temperature studies.

The lowest energy structure of the naphthalene$^{\bullet+}(HCN)_4$ is very similar to that of the benzene$^{\bullet+}(HCN)_4$ cluster where two HCN molecules are attached to the cation ring from two opposite sides. This structure provides efficient solvation of the organic ion. It is interesting to note that the naphthalene cation is *externally* solvated by four water molecules (Fig. 15.9e) but it can be *internally* solvated by four HCN molecules (Fig. 15.11f). As in all such clusters, the binding energies, with the addition of 4–6 ligands, converge to the enthalpy of vaporization of the ligand (ΔH^{o}_{vap}), which is 6.0 kcal/mol for HCN at 298 K [1, 22, 48].

15.2.4 Intracluster Proton Transfer and Deprotonation Reactions

Intracluster proton transfer (IPT) coupled with cluster's dissociation reactions have been observed in several solvated organic radical cations $R\text{-}H^{\bullet+}(S)_n$ with specific numbers of solvent molecules (S_n). These reactions, also known as Dissociative Proton Transfer (DPT) reactions, generate a protonated solvent cluster $S_nH^+$and an organic radical ($R^{\bullet}$). Intracluster DPT reactions can occur within the $R\text{-}H^{\bullet+}(S)_n$ clusters if the proton transfer from the radical cation to the solvent sub-cluster $(S)_n$ becomes exothermic and the resulting energy is sufficient to dissociate the $R^{\bullet}\cdots H^+(S)_n$ cluster. The cluster size (n) at which the PT occurs can be predicted by comparing the proton affinities of the organic radical ($R^{\bullet}$) and the solvent sub-cluster $(S)_n$. These reactions have been observed in $C_6H_6^{\bullet+}(H_2O)_n$, $C_3H_3^+(H_2O)_n$, $C_4N_2H_4^{\bullet+}(H_2O)_n$, and $C_4N_2H_5^+(H_2O)_n$ with $n \geq 4$ in all cases [15, 16, 20, 21]. The IPT and DPT reactions are represented by Eqs. (15.3) and (15.4) for the benzene$^{\bullet+}(H_2O)_n$ clusters.

Intracluster Proton Transfer (IPT)

$$C_6H_6^{\bullet+}(H_2O)_n \rightarrow C_6H_5^{\bullet+}(H_2O)_nH^+ \quad (15.3)$$

Intracluster Dissociative Proton Transfer (IDPT)

$$C_6H_6^{\bullet+}(H_2O)_n \rightarrow (H_2O)_nH^+ + C_6H_5^{\bullet} \quad (15.4)$$

The driving force for the deprotonation of the organic ion is the formation of strong ionic hydrogen bonds in the solvent clusters. The solvent clusters, e.g. $(H_2O)_nH^+$ can be attached to a radical carbon site through a relatively weak $(H_2O)_nH^+\cdots C$ hydrogen bond, or it can form a stronger $(H_2O)_nH^+\cdots N$ bond to a basic site of the radical such as in pyridine or pyrimidine. These are a new class of solvated clusters where a protonated water cluster is hydrogen bonded to an organic radical. In other words, in these clusters the deprotonated radicals are solvated externally, i.e., the radicals remain outside, and bonded to the surface, of the solvent clusters.

The thermochemistry of the overall reaction (15.3 or 15.4) is relevant if the reaction proceeds effectively in one step, i.e., if the cluster is assembled without stabilization, retaining the exothermicity of all the association steps as internal energy for deprotonation/dissociation to form products. This may be possible in the experiments that use neat water (solvent) vapor, where collision with every successive H_2O molecule can lead to association with the release of the binding energy into internal energy, rather than removing energy from the cluster. However, if a third body is present, it would stabilize the growing cluster and no internal energy would be available for deprotonation/dissociation. In fact, the third-body effect was observed in the benzene$^+$/water system where the addition of He buffer gas quenched the deprotonation reaction [15].

The observation of IPT reactions in the benzene$^{\bullet+}$/water system is consistent with the IR dissociation spectra of the $C_6H_6^{\bullet+}(H_2O)_n$ clusters which were very similar to the spectra of $(H_2O)_nH^+$ after $n = 4$, indicating that intracluster proton transfer resulted in the formation of a protonated water cluster attached to the phenyl radical [54]. However, H/D-exchange experiments under thermal ion mobility conditions showed that $C_6H_6^{\bullet+}(D_2O)_n$ clusters with $n = 2$–8 did not exchange a proton with D_2O to yield a $C_6H_5^{\bullet}(D_2O)_nD^+$ ion, although this would be expected for the $C_6H_5^{\bullet}(D_2O)_nH^+$ structure [15]. It was, therefore, suggested that the intracluster proton transfer has an energy barrier that cannot be overcome in thermal systems below 280 K in the ion mobility experiments [15].

The deprotonation of organic ions such as $C_3H_3^+$ and $C_6H_6^{\bullet+}$ requires the assembly of several solvent molecules (at least 4 for H_2O) which is strongly facilitated at lower temperatures. Therefore, the rate coefficients of these reactions showed uniquely large negative temperature coefficients, with pseudo second-order coefficients varying as $k = cT^{-63}$ and $k = cT^{-67}$- in the hydrated $C_3H_3^+$ and $C_6H_6^{\bullet+}$ ions, respectively [15, 16]. The large negative temperature coefficients result from a multi-body mechanism in which five or more components need to be assembled in the activated complex [for example, $C_6H_6^{\bullet+}(H_2O)_4]^*$ for the reaction to proceed. This means that only the fraction of the total cluster population that is in the $n \geq 4$ clusters is reactive. Therefore, only this fraction among all the collisions of the clusters with H_2O molecules are reactive, and the small collision efficiency of 10^{-5}–10^{-6} reflects this effect [15, 16]. The population of the large reactive clusters increases rapidly with decreasing temperature, and this leads to the large negative temperature coefficient.

IPT reactions were not observed in the naphthalene $^{\bullet+}(H_2O)_n$ clusters because of the higher proton affinity (PA) of the naphthalene radical ($C_{10}H_7^{\bullet}$, 234 kcal/mol) [73] as compared to that of the phenyl radical ($C_6H_5^{\bullet}$, 212 kcal/mol) [15]. This means that more than seven or eight water molecules will be required to associate with the naphthalene radical cation for the IPT to occur. However, IPT reactions can occur within the $C_{10}H_8^{\bullet+}(CH_3OH)_n$ clusters if the proton transfer from the $C_{10}H_8^{\bullet+}$ radical cation to the methanol sub-cluster $(CH_3OH)_n$ becomes exothermic and the resulting energy is sufficient to dissociate the $C_{10}H_7^{\bullet}\cdots H^+(CH_3OH)_n$ cluster. Based on the estimates of the PAs of the naphthalene radical ($C_{10}H_7^{\bullet}$) [73] and the methanol clusters [74], the PT reaction from $C_{10}H_8^{\bullet+}$ to the methanol sub-cluster in $C_{10}H_8^{\bullet+}(CH_3OH)_n$ is expected to becomes exothermic at $n > 5$. This is consistent with the changes in the mass spectra of the $C_{10}H_8^{\bullet+}(CH_3OH)_n$ clusters at 238 K where the ion signal of the naphthalene-containing ions disappeared and only the $H^+(CH_3OH)_n$ clusters with $n \geq 5$ were observed [19].

The calculated the structures of the clusters formed by the association of the naphthalene radical ($C_{10}H_7^{\bullet}$) with protonated methanol $[H^+(CH_3OH)_n]$ clusters with $n = 4$–6 and their binding energies are shown in Fig. 15.12.

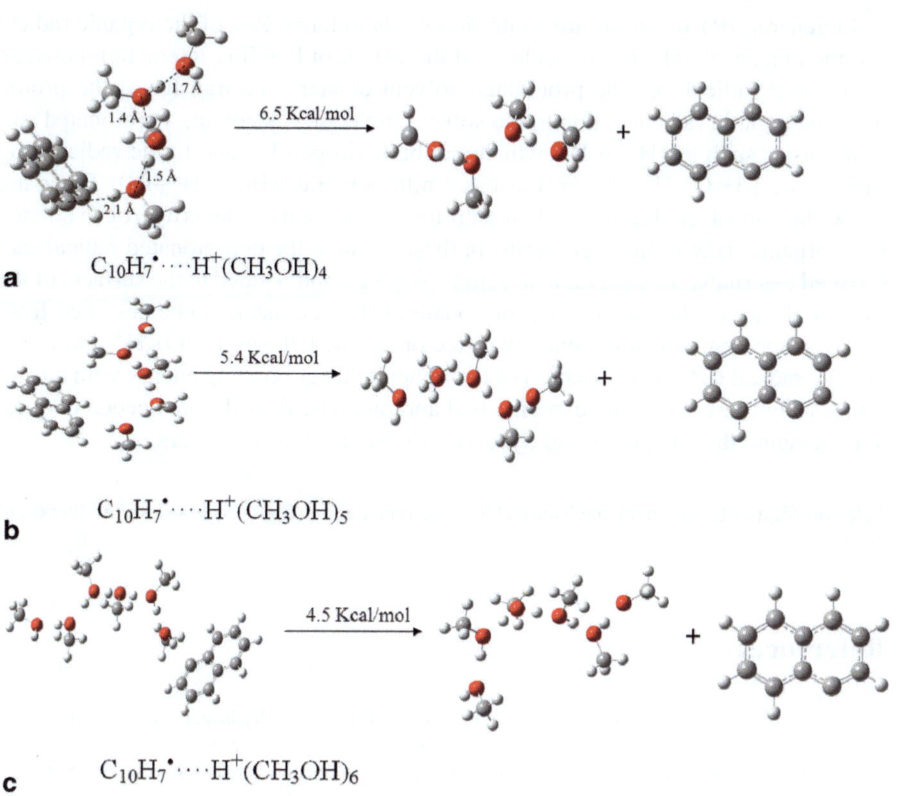

Fig. 15.12 Structures and binding energies of the lowest energy structures of the $C_{10}H_7\bullet$. $H^+(CH_3OH)_n$ clusters with $n = 4 - 6$ obtained at the B3LYP/6-311^{++}G(d, p) level [19]

It is clear that the binding energy of the protonated methanol cluster to the naphthalene radical decreases as the size of the protonated methanol cluster (n) increases. For example, the binding energies of the $H^+(CH_3OH)_n$ clusters to the $C_{10}H_7^\bullet$ radical decrease from 6.5 to 5.4 to 4.5 for $n = 4, 5$ and 6, respectively according to the B3LYP/6-311^{++}G(d, p) calaculations [19]. This can be explained by the decrease in the charge-induced dipole interaction between the protonated methanol cluster and the naphthalene radical as the effective charge becomes more delocalized on the methanol cluster and further away from the radical. Eventually, the effect of the charge would disappear and the binding energy would converge to that of the neutral system. Since both the intracluster PT from the naphthalene radical cation to the methanol sub-cluster in the $C_{10}H_8^{\bullet+}(CH_3OH)_n$ clusters becomes more exothermic as n increases and the binding energies of the resulting $C_{10}H_7^\bullet \cdots H^+(CH_3OH)_n$ clusters decrease as n increases, dissociation of the larger $C_{10}H_8^{+\bullet}(CH_3OH)_n$ clusters into the $C_{10}H_7^\bullet$ radical and protonated methanol clusters is expected to be the predominant process at low temperatures. This is consistent with the experimental observation mentioned above [19].

In general, IPT reactions are controlled by the relative PAs of the organic radical and the cluster of solvent molecules and the extent of bonding interaction between the organic radical and the protonated solvent cluster. The transfer of the proton from the organic radical cation to the solvent molecules generates a protonated solvent cluster such as $(H_2O)_nH^+$ which can be hydrogen bonded to the radical ring carbon site $[(H_2O)_nH^+ \cdots C]$ or at a basic nitrogen site $[(H_2O)_nH^+ \cdots N]$. These are a new class of solvated clusters where a protonated solvent cluster is hydrogen bonded to an organic radical. In other words, in these clusters the deprotonated radicals are solvated externally, i.e., the radicals remain outside, and bonded to the surface, of the solvent clusters. Alternatively, the protonated solvent cluster can be detached from the deprotonated radical to generate a free protonated cluster $(H_2O)_nH^+$ and a free organic radical ($R^\bullet$). The variation of the radical/cluster bonding energy with cluster size is of interest, and more experimental and theoretical works are needed in order to understand the energetics and dynamics of these solvated clusters.

Acknowledgment We thank the National Science Foundation (CHE-0911146) for the support of this work.

References

1. Meot-Ner (Mautner), M (2012) Update 1 of: strong ionic hydrogen bonds. Chem Rev 112:PR22–103; (2005) The Ionic hydrogen bond. Chem Rev 105:213–284
2. Tsuzuki S (2005) In: Wales DJ (ed) Intermolecular Forces and Clusters I, vol. 115. Springer, Berlin, p 149
3. Cerny J, Hobza P (2007) Non-covalent interactions in biomacromolecules. Phys Chem Chem Phys 9:5291
4. Conway BE (1981) Ionic hydration in chemistry and biophysics. Elsevier, Amsterdam
5. Sloan EDMD (2008) Clathrate hydrates of natural gases, vol 3. CRC Press, Florida
6. Nuevo M, Milam SN, Sandford SA, Elsila J, Dworkin JP (2009) Formation of uracil from the ultraviolet photo-irradiation of pyrimidine in pure H_2O ices. Astrobiol 9:683–695
7. Menor-Salvan C, Marin-Yaseli MR (2012) Prebiotic chemistry in eutectic solutions at the water–ice matrix. Chem Soc Rev 41:5404–5415
8. Ben-Naim A (1980) Hydrophobic interactions. Plenum, New York
9. Tanford C (1980) The hydrophobic effect: formation of micelles and biological membranes, 2nd edn. Wiley, New York
10. Gudipati RS, Allamandola LJ (2006) Unusual stability of polycyclic aromatic hydrocarbon radical cations in amorphous water ices up to 120 K: astronomical implications. Astrophys J 638:286–292
11. Rhee YM, Lee TJ, Gudipati MS, Allamandola LJ, Head-Gordon, M (2007) Charged polycyclic aromatic hydrocarbon clusters and the galactic extended red emission. Proc Natl Acad Sci U S A 104:5274–5278
12. Snow TP, Bierbaum VM (2008) Ion chemistry in the interstellar medium. Annul Rev Anal Chem 1:229
13. Herbst E, van Dishoeck EF (2009) Complex organic interstellar molecules. Annul Rev Astron Astrophys 47:427–480
14. Ibrahim, Y, Alsharaeh E, Dias K, Meot-Ner M, El-Shall MS (2004) Stepwise hydration, and multi-body deprotonation with steep negative temperature dependence in the Benzene∞^{+}– water system. J Am Chem Soc 127:12766–12767

15. Ibrahim Y; Alsharaeh E, Meot-Ner M, El-Shall MS, Scheiner S (2005) Stepwise hydration of ionized aromatics, energies, structures of the hydrated benzene cation, and the mechanism of deprotonaion reactions. J Am Chem Soc 127:7053–7064
16. Mabrouki R, Ibrahim Y, Xie E, Meot-Ner M, El-Shall MS (2006) Gas phase hydration and deprotonation of the cyclic $C_3H_3^+$cation. Solvation with acetonitrile and comparison with the benzene radical cation. J Phys Chem A 110:7334–7344
17. Momoh PO, El-Shall MS (2008) Gas phase hydration of organic ions. Phys Chem Chem Phys 10:4827–4834
18. Momoh PO, Hamid AM, Abrash SA, El-Shall MS (2011) Structure and hydration of the $C_4H_4^{\bullet+}$ Ion formed by electron impact ionization of acetylene clusters. J Chem Phys 134:204315
19. Attah IK, Platt SP, Meot-Ner (Mautner) M, Aziz SG, Alyoubi AO, El-Shall MS (2014) Hydrogen bonding of the naphthalene radical cation to water and methanol and attachment of the naphthalene ion to extended hydrogen bonding chains. Chem Phys Lett 613:45–53
20. Ibrahim Y, Mabrouki R, Meot-Ner M, El-Shall MS (2007) Hydrogen bonding interactions of pyridine$^{\bullet+}$ with water: stepwise solvation of distonic cations. J Phys Chem A 111:1006–1014
21. Hamid AM, Sharma P, Hilal R, Elroby S, Aziz SG, Alyoubi AO, El-Shall MS (2013) Hydration of the pyrimidine radical cation and stepwise solvation of protonated pyrimidine with water, methanol and acetonitrile. J Chem Phys 139:084304
22. NIST Chemistry Web Book NIST Standard Reference Database Number 69; National Institute of Standards and Technology: Gaithersburg MD, 20899 (http://webbook.nist.gov)
23. Li QB, Jacob DJ, Yantosca RM, Heald CL, Singh HB, Koike M, Zho YJ, Sachse GW, Streets DG (2003) A global 3-D model evaluation of the atmospheric budgets of HCN and CH_3CN: constraints from aircraft measurements over the western pacific. J Geophys Atmos 108:8827
24. Matthews CN, Ludicky RA (1992) Hydrogen cyanide polymers on comets. Adv Space Res 12:21–32
25. Matthews CN, Minard RD (2006) Hydrogen cyanide polymers, comets and the origin of life. Faraday Discuss 133:393–401
26. Pizzarello S, Holmes W (2009) Nitrogen-containing compounds in two CR2 meteorites: 15N composition, molecular distribution and precursor molecules. Geochim Cosmochim Acta 73:2150–2162
27. Mumma MJ, DiSanti MA, Magee-Sauer K, Bonev BP, Villanueva GL, Kawakita H, Russo ND, Gibb EL, Blake GA, Lyke JE, Campbell RD, Aycock J, Conrad A, Hill GM (2005) Parent volatiles in comet 9P/Tempel 1: before and after impact. Science 310:270
28. Lahuis F, van Dishoeck EF, Boogert ACA, Pontopiddan KM, Blake GA, Dullemond CP, Evans II NJ, Hogerheijde MR, Jørgensen JK, Kessler-Silacci JE, Knez C (2006) Hot organic molecules toward a young low-mass star: a look at inner disk chemistry. Astrophys J Lett 636:L145
29. Haupert LJ, Enthold PG (2013) Hydration energies of aromatic ions in the gas phase. J Phys Chem A 117:1164–1170
30. Wyttenbach T, Bowers MT (2009) Hydration of biomolecules. Chem Phys Lett 480:1–16
31. Gao B, Wyttenbach T, Bowers MT (2009) Hydration of protonated aromatic amino acids: phenylalanine, tryptophan, and tyrosine. J Am Chem Soc 131:4695–4701
32. Liu D, Wyttenbach T, Bowers MT (2006) Hydration of mononucleotides. J Am Chem Soc 128:15155–15163
33. Kebarle P (1997) Ion thermochemistry and solvation from gas phase ion equilibria. Annu Rev Phys Chem 28:445
34. Keesee RG, Castleman AW (1986) Thermochemical data on gasâŁłphase ionâŁłmolecule association and clustering reactions. J Phys Chem Ref Data 15:1011
35. Larson JW, McMahon TB (1987) Hydrogen bonding in gas-phase anions. The energetics of interaction between cyanide ion and Broensted acids determined from ion cyclotron resonance cyanide exchange equilibria. J Am Chem Soc 109:6230
36. Gorman GS, Amster JI (1993) Gas-phase basicity measurements of dipeptides that contain valine. J Am Chem Soc 115:5729
37. Gross DS, Schnier PD, Rodriguez-Cruz SE, Fagerquist CK, Williams ER (1996) Conformations and folding of lysozyme ions in vacuo. Proc Natl Acad Sci U S A 93:3143

38. Deakyne CA (1997) In: Scheiner S (ed) Molecular interactions: from van der Waals to strongly bound complexes. Wiley, New York, p 217
39. Meot-Ner (Mautner) M (2003) The proton affinity scale, and effects of ion structure and solvation. Int J Mass Spectrom 227:525
40. Meot-Ner (Mautner) M, Lias SG (2003) Binding energies between ions and molecules, and the thermochemistry of cluster ions. In: Linstrom PJ, Mallard WG (eds) NIST chemistry webbook. NIST Standard Reference Database Number 69; National Institute of Standards and Technology: Gaithersburg, MD, March 2003 (http://webbook.nist.gov)
41. Wyttenbach T, Bowers MT (2007) Intermolecular interactions in biomolecular systems examined by mass spectrometry. Annu Rev Phys Chem 58:511
42. Kemper PR, Dupuis NF, Bowers MT (2009) A new, higher resolution, ion mobility mass spectrometer. Int J Mass Spectrom 287:46
43. Momoh PO, El-Shall MS (2007) Stepwise hydration of ionized acetylene trimer. Further evidence for the formation of benzene radical cation. Chem Phys Letters 436:25–29
44. Momoh PO, Xie E, Abrash SA Meot-Ner M, El-Shall MS (2008) Gas phase reactions between acetylene radical cation and water. Energies, structures and formation mechanism of $C_2H_3O^+$ and $C_2H_4O^+$ ions. J Phys Chem A 112:6066–6073
45. Hamid AM, Soliman AR, El-Shall MS (2013) Stepwise association of hydrogen cyanide and acetonitrile with the benzene radical cation: structures and binding energies of $(C_6H_6^{\bullet+})(HCN)_n$, $n = 1$–6, and $(C_6H_6^{\bullet+})(CH_3CN)_n$, $n = 1$–4, clusters. J Phys Chem A 117:1069–1078
46. Hamid AM, Soliman AR, El-Shall MS (2012) Assembly of HCN hydrogen bonding chains in the gas phase. binding energies and structures of phenylacetylene$^+$$(HCN)_n$ clusters, $n = 1$–4. Chem Phys Lett 543:23–27
47. Attah IK, Hamid AM, Meot-Ner (Mautner) M, Aziz SG, Alyoubi AO, El-Shall MS (2013) Substituent effects on non-covalent bonds: complexes of ionized benzene derivatives with hydrogen cyanide. J Phys Chem A 117:10588–10597
48. Hamid AM, El-Shall MS, Hilal R, Elroby S, Aziz SG (2014) Unconventional hydrogen bonding to organic ions in the gas phase. Stepwise association of hydrogen cyanide with the pyridine and pyrimidine radical cations and protonated pyridine. J Chem Phys 141:054305–054315
49. Ludwig R (2001) Water: from clusters to the bulk. Angew Chem Int Ed Engl 40:1808
50. Gruenloh CJ, Carney JR, Hagemeister FC, Arrington CA, Zwier TS, Fredericks SY, Wood JT III, Jordan KD (1998) Resonant ion-dip infrared spectroscopy of the S_4 and D_{2d} water octamers in benzene-$(water)_8$ and benzene$_2$-$(water)_8$. J Chem Phys 109:6601
51. Lum K, Chandler D, Weeks JD (1999) Hydrophobicity at small and large length scales. J Phys Chem B 103:4570
52. Solca N, Dopfer O (2001) IR spectrum of the benzene–water cation: direct evidence for a hydrogen-bonded charge–dipole complex. Chem Phys Lett 347:59–64
53. Miyazaki M, Fujii A, Ebata T, Mikami N (2001) Infrared spectroscopy of the benzene–H_2O cluster cation: experimental study on the drastic structural change upon photoionization. Chem Phys Lett 349:431–436
54. Miyazaki M, Fujii A, Ebata T, Mikami N (2003) Infrared spectroscopy of hydrated benzene cluster cations, $[C_6H_6\text{-}(H_2O)_n]^+ (n = 1\text{–}6)$: Structural changes upon photoionization and proton transfer reactions. Phys Chem Chem Phys 5:1137–1148
55. Solca N, Dopfer O (2003) Prototype microsolvation of aromatic hydrocarbon cations by polar ligands: ir spectra of benzene$^+$-L_n clusters (L) H_2O, CH_3OH). J Phys Chem A 107:4046–4055
56. Tachikawa H, Igarashi M, Ishibashi T (2001) Vibrational frequency-shifts of H2O caused by complex formation with a molecular cation: a density functional study. Phys Chem Chem Phys 3:3052–3056
57. Tachikawa H, Igarashi M (1998) Dynamics of the ionization processes of benzene−H2O clusters: a direct ab initio dynamics study. J Phys Chem A 102:8648–8656
58. Miyazaki M, Fujii A, Mikami N (2004) Binding energy of the benzene−water cluster cation: an ar-mediated ir photodissociation study. J Phys Chem A 108:8269–8272
59. Liu K, Cruzan JD, Saykally RJ (1996) Water clusters. Science 271:929–933

60. Fujii A, Patwari GN, Ebata T, Mikami N (2002) Vibrational spectroscopic evidence of unconventional hydrogen bonds. Inter J Mass Spetrom 220:289
61. Mehata MS (2008) Proton translocation and electronic relaxation along a hydrogen-bonded molecular wire in a 6-hydroxyquinoline/acetic acid complex. J Phys Chem B 112:8383
62. Wang C, Batsanov AS, Bryce MR, Martin S, Nichols RJ, Higgins SJ, Garcia-Suarez VM, Lambert CJ (2009) Oligoyne single molecule wires J Am Chem Soc 131:15647
63. Puigmarti-Luis J, Minoia A, Lei S, Geskin V, Li B, Lazzaroni R, De Feyter S, Amabilino DB (2011) Self-assembly of supramolecular wires and cross-junctions and efficient electron tunnelling across them. Chem Sci 2:1945
64. Dulmage WJ, Lipscomb WN (1951) The crystal structures of hydrogen cyanide, HCN. Acta Crystallogr 4:330
65. Shallcross FV, Carpenter GB (1958) The crystal structure of cyanoacetylene. Acta Crystallogr 11:490
66. Jucks KW, Miller REJ (1988) Near infrared spectroscopic observation of the linear and cyclic isomers of the hydrogen cyanide trimer. Chem Phys 88:6059
67. Ruoff RS, Emilsson T, Klots TD, Chuang C, Gutowsky HS (1988) Rotational spectrum and structure of the linear HCN trimer. J Chem Phys 89:138
68. Karpfen A (1996) Linear and cyclic clusters of hydrogen cyanide and cyanoacetylene: a comparative ab initio and density functional study on cooperative hydrogen bonding. J Phys Chem 100:13474
69. Sanchez M, Provasi PF, Aucar GA, Alkorta I, Elguero J (2005) Theoretical study of HCN and HNC neutral and charged clusters. J Phys Chem B 109:18189
70. Nauta K, Miller RE (1999) Nonequilibrium self-assembly of long chains of polar molecules in superfluid helium. Science 283:1895
71. Meot-Ner (Mautner) M (1978) Solvation of the proton by hydrogen cyanide and acetonitrile. Condensation of hydrogen cyanide with ions in the gas phase. J Am Chem Soc 100:4694
72. Meot-Ner (Mautner) M, Speller CV (1989) Multicomponent cluster ions. 2. Comparative stabilities of cationic and anionic hydrogen-bonded networks. Mixed clusters of water and hydrogen cyanide. J Phys Chem 93:3663
73. Feng WY, Lifshitz C (1996) Ion/molecule reactions of naphthalene and 1-chloronaphthalene radical cations. Inter J Mass Spectrom Ion Proc 152:157
74. Knochenmuss R, Cheshnovsky O, Leutwyler S (1988) Proton transfer reactions in neutral gas—phase clusters: 1-Naphthol with H2O, D2O, CH3OH, NH3 and piperidine. Chem Phys Lett 144:317

Chapter 16
Anion-π Interactions in Supramolecular Chemistry and Catalysis

Antonio Bauzá, Pere M. Deyà and Antonio Frontera

Abstract Non-covalent interactions play a major role in supramolecular chemistry and biochemistry by dominating the central parts of living systems since they dictate the functionality of many biological and host-guest systems. A good comprehension of the different non-covalent forces is necessary for the rational design of new drugs and developing improved synthetic receptors capable to function in competitive media. Interactions involving aromatic rings (or π-systems in general) are very relevant in supramolecular chemistry, exemplified by the cation–π interaction and its importance in protein structure and enzyme catalysis. From a traditional point of view, the π-system is usually considered as electron rich (π-basic). The naissance of the counterintuitive anion–π interaction –the attractive interaction between an anion and an electron poor π-system (π-acid)– was somewhat controversially discussed by the scientific community. However, in the last decade a great deal of theoretical and experimental investigations has time-honored the anion–π interaction as an important supramolecular bond. Herein we describe the physical nature of this noncovalent interaction and the different strategies that can be used to modulate its strength. Finally, selected state-of-the-art reports illustrating the rational utilization of the anion–π interaction in supramolecular chemistry (anion receptors), biological applications and catalysis are described in this chapter.

16.1 Introduction

Nowadays supramolecular chemistry is probably the most multidisciplinary field of research. It is exponentially growing as indicated by the large number of articles, reviews, and books published since the beginning of this millennium. The rapid development of supramolecular chemistry has a profound effect on the increasing efficiency for preparation of structures of different sizes, shapes and functionalities. Supramolecular chemists rely on the comprehension of the non-covalent forces, which form the basis of highly specific recognition, transport, and regulation

A. Frontera (✉) · A. Bauzá · P. M. Deyà
Departament de Química, Universitat de les Illes Balears, 07122 Palma de Mallorca, Baleares, Spain
e-mail: toni.frontera@uib.es

S. Scheiner (ed.), *Noncovalent Forces*, Challenges and Advances in Computational Chemistry and Physics 19, DOI 10.1007/978-3-319-14163-3_16

mechanisms. The orchestration of many chemical and biological processes is often dominated by an intricate combination of non-covalent interactions [1, 2], which are the foundation of the life process itself, the ultimate expression of function. In general host-guest chemistry, interactions between targeted guests and rationally designed receptors drive the formation of assemblies of different sizes, shapes and affinities [3–6]. The correct description of interactions between molecules is needed for the understanding and progress of the supramolecular chemistry that usually relies on strong, directional interactions such as hydrogen bonding and halogen bonding, and less directional forces like ion pairing. More recently, general σ-hole interactions are also considered as an important addition to the family of well-established directional non-covalent interactions [7–9]. In addition, non-covalent interactions involving aromatic rings are enormously significant in this field [10, 11]. They play a crucial role in chemistry and biology [12], in particular drug–receptor interactions, crystal engineering, enzyme inhibition and protein folding [12]. A clear example is the role of π–stacking interactions in DNA, RNA [14, 15] and protein-DNA aromatic interactions [16]. An important and widely recognized non-covalent attractive force that involves aromatic rings is the cation–π interaction [17, 18], of great significance in biology [19–21] and supramolecular chemistry [22–25]. Other weak interactions involving π-systems are also at the forefront of interdisciplinary research. For example, CH–π [26, 27], lone pair–π [28, 29] and salt-bridge–π [30] interactions have been recently used in several supramolecular chemistry fields, especially in crystal engineering and protein-ligand interactions.

In the last decade, the anion–π interaction [31–33], i.e. the attractive noncovalent force between an electron-deficient π–system and an anionic moiety [34, 35], has also been recognized as a non-covalent bonding interaction. Its nature has been described by a plethora of theoretical studies [36–39] in addition to an increasing amount of experimental investigations [40–44]. Anion–π interactions are expected to become noticeable players in fields as diverse as medicine, environmental chemistry and biochemical processes [31, 45–47]. Moreover, their application to the design of highly selective anion receptors and transport channels definitively confirms their significance in the field of supramolecular chemistry [48, 49].

As a matter of fact, the design and synthesis of selective receptors designed to bind anions is a topic of continuous interest [50–55]. The main reason is related to the vital function played by anions, which are ubiquitous throughout biological systems [56, 57]. In addition, some anions are increasingly recognized as problematic environmental contaminants. For example, the overuse of anionic pollutants (phosphate and nitrate) as fertilizers causes the disruption of aquatic ecosystems. The eutrophication caused by the fertilizing properties of these oxy-anions limits the populations of algae and phytoplankton in rivers and lakes [58]. Other anions, such as pertechnetate are generated from the re-processing of nuclear fuel [59] and their release to the sea is strictly controlled or prohibited. Another harmful anion is arsenate that caused the largest mass poisoning in human history (200 million) due to the dissolution in the groundwater of the Bengal basin [60]. Interestingly, a new class of anion receptors based on the anion–π interaction is emerging in the literature [31, 61]. For instance, an interesting receptor, that combines hydrogen bonding and anion–π interaction

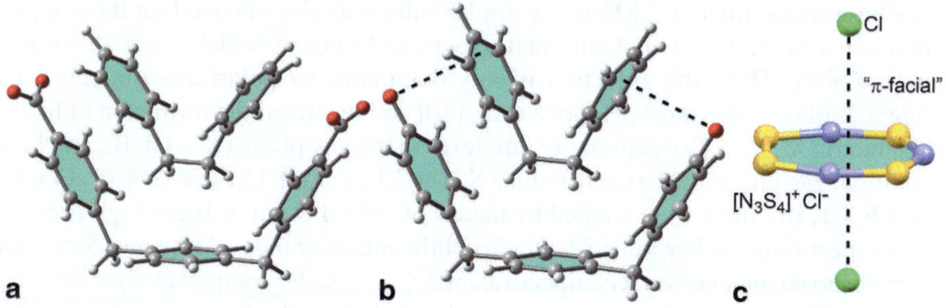

Fig. 16.1 **a** and **b** Host–Guest complexes reported by Schneider and coworkers. **c** Partial view of the X-ray structure of thiotrithiazylium chloride

for the binding of anions with neutral π-acceptors, has been recently published by Albrecht and collaborators [62] and its ability to trap anions both in solution and in the solid state. Similarly, Johnson and collaborators have developed receptors for selective nitrate binding in competitive hydrogen bonding solvents where anion–π interactions facilitate the selectivity [63].

This chapter is not intended to be a bibliographic survey of the literature related to anion–π interactions since several excellent reviews have been written for that purpose [31, 35, 64–66]. Instead, we intend to emphasize the bonding relationship between anions and π systems, at the experimental and theoretical forefront of this noncovalent interaction. The chapter is developed under five headings. First, we describe early publications on this topic, which have not been properly cited and adequately placed into perspective. Second, the physical nature of the interaction is explained for assimilation by a wide readership. Third, we highlight selected examples from the literature where molecular recognition driven by anion–π bonding is very relevant. Next, attractive applications and wonderful examples of anion–π bonding to catalysis are described. Finally, clear evidence that illustrates the increasing interest on the study of anion–π interactions in biological systems is provided.

16.2 Early Publications

Early reports on weak attractive interactions involving negatively charged residues and polarizable aryl groups in host–guest systems [67–69] were reported before the publication of a series of computational studies in 2002 [36–38] supporting the existence of attractive forces between anions and the positive electrostatic potential on the ring edge of electron-deficient aromatic groups.

In 1993 Schneider and coworkers demonstrated attractive interactions between simple organic host and guest molecules by means of NMR measurements [67]. Since their experiments were carried out in a highly competitive media (water) and the aryl groups that participate in the anion–π interaction were not electron deficient (see Fig. 16.1a and 16.1b), the calculated interactions energies were very modest, reaching approximately 2 kJ/mol. A similar value was also obtained for the complex between a calixarene with 4-sulfonato groups and toluene, which is not an electron poor moiety. Thus, the attraction is only dominated by polarization effects (*vide infra*). Earlier to this work, Hiraoka et al. [70] demonstrated the formation of highly symmetric $X^- \cdots C_6F_6$ anion–π complexes in the gas-phase (X = CI, Br, and I). In addition, gas-phase clustering reactions $X^-@(C_6F_6)_{n-1} + C_6F_6 = X^-@(C_6F_6)_n$ for X = F, C1, Br, and I were studied by means of pulsed electron-beam high-pressure mass spectrometry. For fluoride a nucleophilic attack and covalent bond formation was observed and, conversely, noncovalent $X^- \cdots C_6F_6$ complexes were detected for X = CI, Br, and I where the anion was located along the C_6 main symmetry axis of C_6F_6.

Another early progress in this direction [71] was reported in 1996 by Woollins and collaborators. They named the anion–π interaction as "π-facial" [72] in their description of the close contact between chloride and the seven-membered aromatic $[S_4N_3]^+$ ring that they observe in the X-ray structure of thiotrithiazylium chloride (see Fig. 16.1c).

16.3 Physical Nature

The physical nature of the anion–π interaction has been widely studied using high level *ab initio* and Density Functional Theory (DFT) calculations. Moreover, several partition energy schemes have been used in order to decompose the total interaction energy into individual components [73, 74]. The general conclusion is that electrostatic forces and ion-induced polarization are the main forces that contribute to the anion–π interaction [75–77]. The electrostatic term is explained by means of the permanent quadrupole moment (Q) of the arene, which is the first non-zero multipole moment in symmetric arenes. Moreover, in most asymmetric arenes, where the dipole moment (μ) is non-zero, the μ_z component is approximately zero and, likewise symmetric arenes, the electrostatic attraction is basically due to the existence of the adequate component (perpendicular to the ring plane) of the quadrupole moment (Q_{zz}), which describes the charge distribution on both sides of the aromatic plane. The Q_{zz} of benzene is negative, but can be turned into positive by attaching electron withdrawing substituents to the ring (see Fig. 16.2a). Similarly, the Q_{zz} of pyridine is negative but can be turned into a positive value by attaching a coordination metal to the nitrogen atom [66]. Therefore, the electrostatic charge-quadrupole interaction between an anion and an aromatic ring can become attractive either attaching electron-withdrawing substituents or coordinating metal ions in case of heteroaromatic rings. The ion-induced polarization of the π-electron system by the anion is significant, inducing a dipole (see Fig. 16.2b). Therefore, a polarization contribution to the total interaction energy is derived from the interaction of the anion with the induced dipole [73–76]. An alternative explanation on the nature of anion–π interactions involving benzene rings has been proposed by Wheeler and Houk [78], who examined substituent effects in $Cl^- \cdots C_6H_{6-n}X_n$ complexes. In contrast

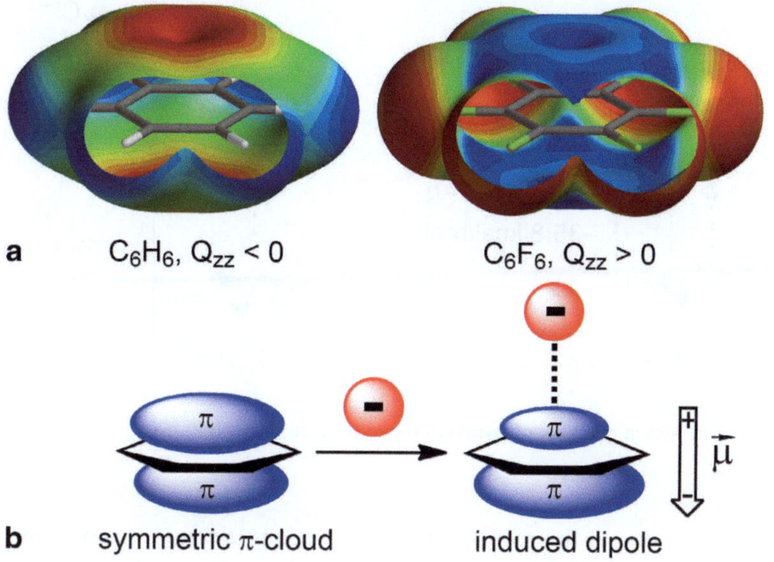

Fig. 16.2 a Molecular electrostatic potential of benzene and hexafluorobenzene. **b** Schematic representation of the ion-induced dipole

to the intuitive view where the substituent induces changes in the aryl π system, Wheeler and Houk propose a model where substituent effects in these systems can be attributed mainly to direct interactions between the anion and local C–X dipoles.

A proper understanding of the physical nature (*vide supra*) is necessary to explain, on one hand, the dual binding mode exhibited by arenes with negligible quadrupole moments [79, 80]. That is, since electrostatic and polarization terms are the main contributions to both anion–π and cation–π interactions, molecules with very small Q_{zz} values such as 1,3,5-trifluorobenzene ($Q_{zz} = 0.57$ B) and *s*-triazine ($Q_{zz} = 0.90$ B) are able to interact with both anions and cations because the electrostatic term is negligible and the interaction is dominated by the polarization term, which is always attractive. On the other hand, it is useful to rationalize the interaction of anions with electron-rich aromatic rings, such as benzene, which are not strongly repulsive. This is due to a compensating effect between the electrostatic (unfavorable) and ion-induced polarization (favorable) forces that roughly cancel each other out. For this reason, the anion–π interaction energy of benzene with chloride is very small, but favorable [81, 82].

The electrostatic term depends on the magnitude of the Q_{zz} and the polarization term on the magnitude of the molecular polarizability parallel to the main symmetry axis (denoted as $\alpha_{||}$) in symmetric arenes (or perpendicular to the ring plane in asymmetric arenes, denoted as α_{zz}), which are intrinsic properties of the π-system. Therefore, it is clear that in order to design an efficient anion receptor based on the anion–π interaction, the π–binding units should have a large and positive quadrupole moment and a large molecular polarizability. However, there is a limitation for the

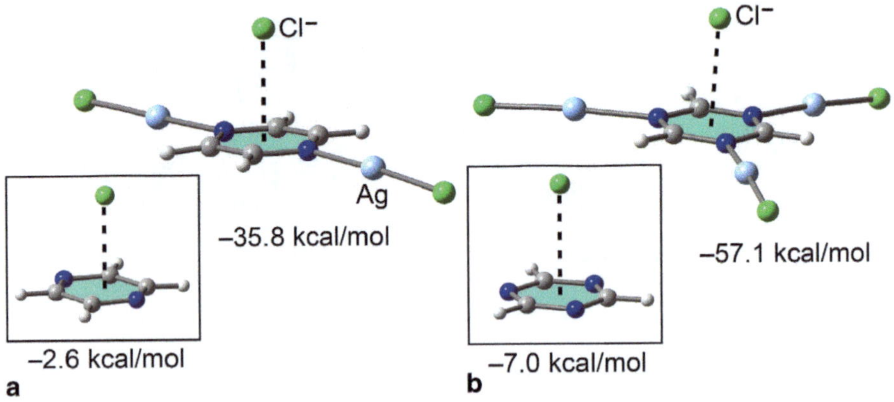

Fig. 16.3 Interaction energies of pyrazine **a** and triazine **b** anion–π complexes from references [84, 85]

first condition due to the reduced number of strong electron withdrawing groups available for constructing the binding blocks. To have a large value of Q_{zz}, the use of $-NO_2$ and -CN groups is required. However, it is synthetically complicated to attach more than three strong electron withdrawing groups and the spacer to the aromatic ring to build the receptor. An intelligent solution to this limitation is the utilization of heteroaromatic rings, especially di-, tri- and tetrazines. The metal coordination to heteroaromatic rings strongly increases the π-acidity of the ring, thus favoring the anion–π interactions [83, 84]. It has been demonstrated that the coordination of both pyrazine (Fig. 16.3a) and *s*-triazine (Fig. 16.3b) to Ag^I dramatically enhances their ability to establish anion–π binding [83, 85].

The same behavior has been demonstrated [84] for *s*-tetrazine using both theory and experiment. As a matter of fact, *s*-tetrazine coordinated to four Ag^I atoms is the most powerful anion–π acceptor binding block reported to date. Theoretically, the interaction energy of *s*-tetrazine with nitrate ion strengthens from − 9.6 kcal/mol to − 62.4 kcal/mol when the arene is tetracoordinated to Ag^I. Experimentally, several X-ray crystal structures of *s*-tetrazine μ_4–coordinated to Ag^I have been reported, exhibiting very close contacts between the anion and the *s*-tetrazine ring (see Fig. 16.4) indicating strong anion–π interactions, in agreement with theoretical predictions. Interestingly the anion–π distance for the perchlorate anion in the X-ray structure shown in Fig. 16.3a (2.61 Å) is the shortest reported to date.

The anion–π interaction has been also observed in the solid state of novel hybrid inorganic–organic assemblies generated from $H_4SiW_{12}O_{40}$ as Keggin-type polyoxometalates (POM) and several trinuclear lanthanide clusters of type $\{Na(H_2O)_3[Ln(HCAM)(H_2O)_3]_3\}^{4+}$ (Ln = La, Ce, Eu and H_3CAM = chelidamic acid or 2,6-dicarboxy-4-hydroxypyridine) [86]. These unprecedented anion–π interactions between the POM (a tetra-anion) and the coordinated aromatic ligand rings play a crucial role in the crystal packing formation (see Fig. 16.5). This investigation of the interaction in Keggin-type POM-based inorganic–organic frameworks also

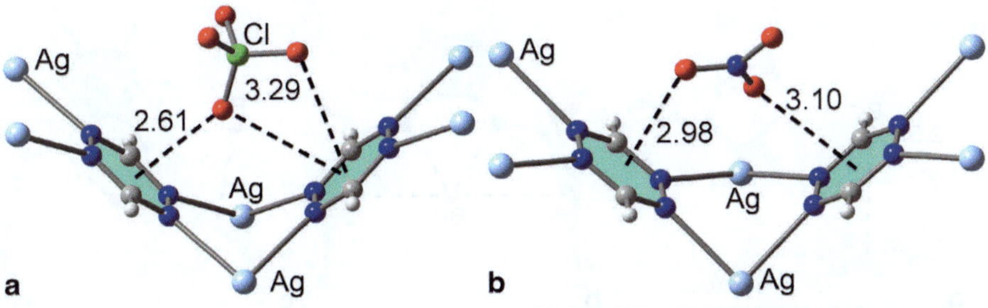

Fig. 16.4 Fragments of the X-ray crystal structures containing μ_4-coordination of 1,2,4,5-tetrazine from ref [84]. The relevant anion–π interactions are indicated by dashed lines (distances in Å)

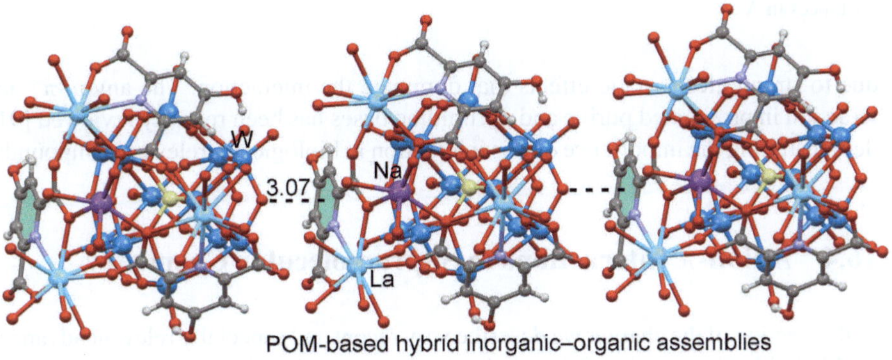

Fig. 16.5 Partial view of the crystal packing of compound $\{Na(H_2O)_3[La(C_7H_3NO_5)(H2O)_3]_3\}$ $[SiW_{12}O_{40}]$ published by Mirzaei et al. [86] with indication of the intermolecular anion–π interaction established between the organic ligand and the POM (distance in Å)

includes a theoretical study devoted to analyze the effect of the ligand coordination to the metal center that increases the π-acidity of the aromatic ring, and consequently enhances its ability to establish anion–π interactions.

The coordination of the heteroaromatic ring to a transition metal is similar to the effect of protonation. For instance, pyridine, diazines, etc. can be easily protonated by simply adjusting the pH of the medium increasing the anion binding ability of the ring (anion–π^+ interactions). The geometric and energetic features of anion–π^+ complexes of several aromatic cations (tropylium, quinolizinylium) and various anions have been reported along with crystallographic structures [87, 88]. This field of research has recently attracted attention and several works have appeared in the literature [89, 91]. For instance, the anion–π^+ interaction participates in the formation of a robust recognition motif in the transition metal malonate complexes using protonated 2-amino-4-picoline and 2-aminopyridine as the auxiliary ligands [89, 90]. As expected, these complexes exhibit very large (>80 kcal/mol) interaction energies

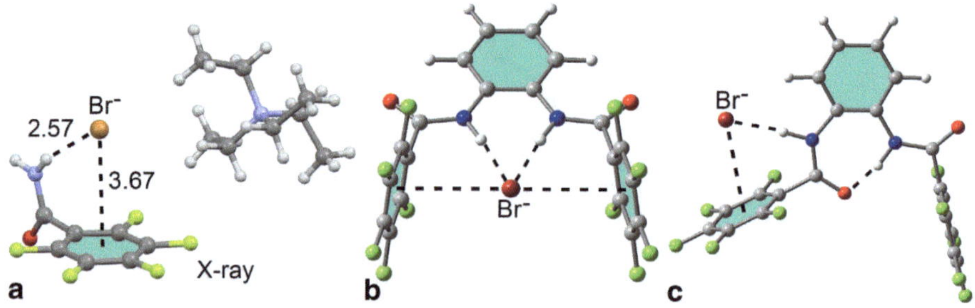

Fig. 16.6 **a** Partial view of the X-ray crystal structure reported by Giese et al. [62] showing the pentafluorobenzamide, the bromide and the tetraethylammonium cation. **b** and **c** Modeled binding motifs for the interaction of bromide with a receptor containing two pentafluorobenzamide moieties. Distances in Å

due to strong electrostatic effects that dominate the interaction. The anion–π^+ interaction in protonated purine and pyrimidine bases has been recently reviewed [91] demonstrating the importance of this interaction in biologically relevant compounds.

16.4 Anion-π Interactions in Supramolecular Chemistry

In this section of the chapter we describe very recent and especially relevant advances in this field. The quality of the works demonstrates that there is a continuous and increasing interest for investigating and developing novel anion-binding hosts and transporters based on electron deficient aromatic rings.

Giese et al. [62] have studied the binding of a series of anions with neutral π-acceptors by means of concurrent hydrogen bonding and anion-π interaction. Interestingly, latter interaction is demonstrated both in the solid state and in solution, and further evidenced by a computational study. The receptors are based on pentafluorobenzamides (see Fig. 16.6), which were found appropriate systems for studying anion-π interactions. In case of bromide, the anion–π complex was characterized by X-ray spectroscopy and it is the first solid state structure where anion-π interactions between an uncharged pentafluorophenyl derivative and an anion are observed. Moreover, the investigation in solution showed differences between electron rich and poor systems, which are explained by a cooperative effect of N–H · · · anion and anion–π interactions and the enhanced acidity of the amide proton by the electron-withdrawing C_6F_5-unit.

Remarkably, Watt et al. [63] have demonstrated that the selective nitrate binding in competitive media by a tripodal urea receptor (see Fig. 16.7) is facilitated by anion–π interactions. Using ^{1}H–NMR titrations they show that the higher affinity observed for nitrate over the halides for the fluorinated receptor is lost when the fluorine atoms are absent. An anion–π interaction between the nitrate and the π-system of the ethynyl-substituted arene is proposed as the source of this selectivity.

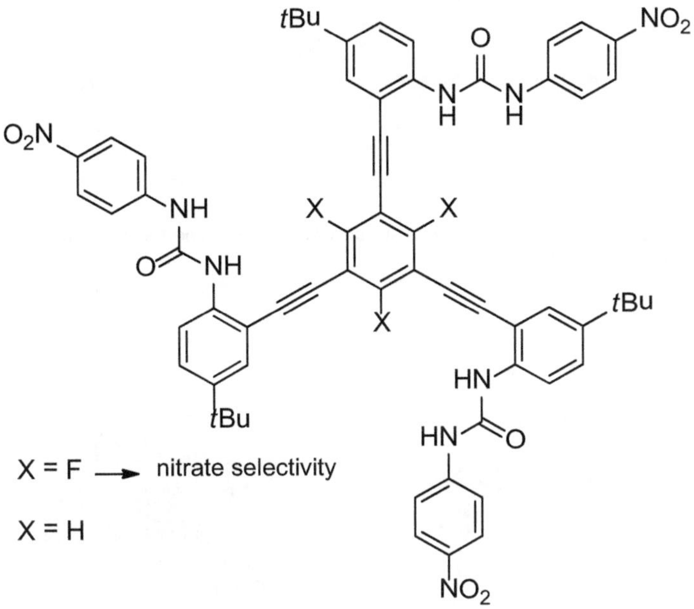

Fig. 16.7 Tripodal urea receptors synthesized by Watt et al. [63]

The fluorinated receptor trends: $NO_3^- > Cl^- > Br^- > I^-$, with a moderate selectivity for nitrate. The existence of the anion–π interaction is demonstrated using an indirect prove. That is, binding studies of the non-fluorinated receptor showed three important issues: first, the nitrate selectivity is lost; second, the association constant of nitrate is diminished compared to the fluorinated receptor; and, third, the binding mode of nitrate is different than that for halides. These data support a model in which nitrate binds to the receptors through a combination of hydrogen bonds with the urea moieties and an anion–π-type interaction regardless of the electronic nature of the central core, thus raising the association constant and selectivity for nitrate in the fluorinated receptor and the opposite in the non-fluorinated. Latter receptor is more selective for chloride since it binds the receptors via hydrogen bonding interactions, exclusively.

Anion–π interactions have been systematically studied by Wang and Wang [92] using tetraoxacalix[2]arene[2]triazine (see Fig. 16.8), an electron-deficient and neutral macrocyclic host. Using electrospray ionization mass spectrometry (ESI-MS), fluorescence titration and X-ray crystallography, the authors demonstrate the formation of 1:1 host-guest complexes with four typical polyatomic anions of different geometries and shapes in the gaseous phase, in solution, and in the solid state. The association constants for the formation of anion–π complexes in acetonitrile are impressive in some cases, ranging from 239 to 16950 M^{-1}, ($NO_3^- > BF_4^- > PF_6^- > SCN^-$). The X-ray molecular structures of the complexes show that two opposed triazine rings of the host interact with the anionic guests through cooperative anion–π and lp–π

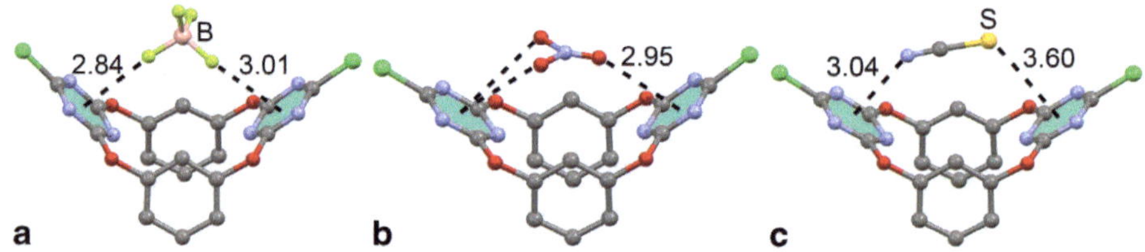

Fig. 16.8 Crystal structures of the 1:1 host-guest complexes reported by Wang and Wang [92] Distances in Å

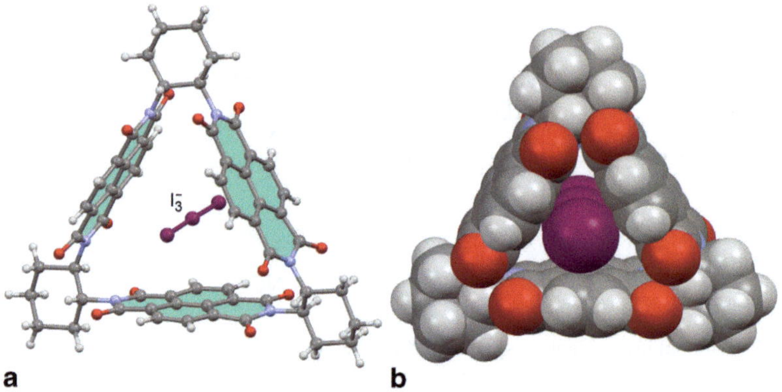

Fig. 16.9 Two views in ball and stick **a** and space filling **b** of the crystal structure of the triangular prism complexed to I_3^- reported by Stoddart and coworkers [93]

interactions. In this comprehensive study, the generality and diversity of anion–π interaction motifs certainly provide a new dimension to molecular recognition of anions and anion-governed self-assembly processes.

Stoddart et al. [93] have reported an interesting anion–π recognition study in molecular triangular prisms formed by naphthalenediimide redox centers. They have demonstrated, both experimentally and by the application of theory, through-space orbital interactions and electron sharing phenomena in their synthesized triangular, redox-active naphthalenediimide prisms (see Fig. 16.9). The resulting electronic communication among the naphthalenediimide units leads to an unusually large number of individually accessible redox states in the triangular prisms, opening the door to potential applications in the field of molecular electronics. The electron deficient cavities of the molecular prisms are ideal for studying anion–π interactions. This ability is demonstrated by the encapsulation of linear I_3^- anions inside the prismatic cavities, causing a profound change in the packing of the prisms in the extended solid-state architecture. The inclusion of I_3^- anions induces π–π stacking of the chiral prisms into supramolecular helices, providing a unique example of anion-induced self-assembly with potential applications as ion-channels. In addition, the chirality endowed by the six stereogenic centers in the occupied prisms dictates the either right- or left-handed associated with their packing in the solid state.

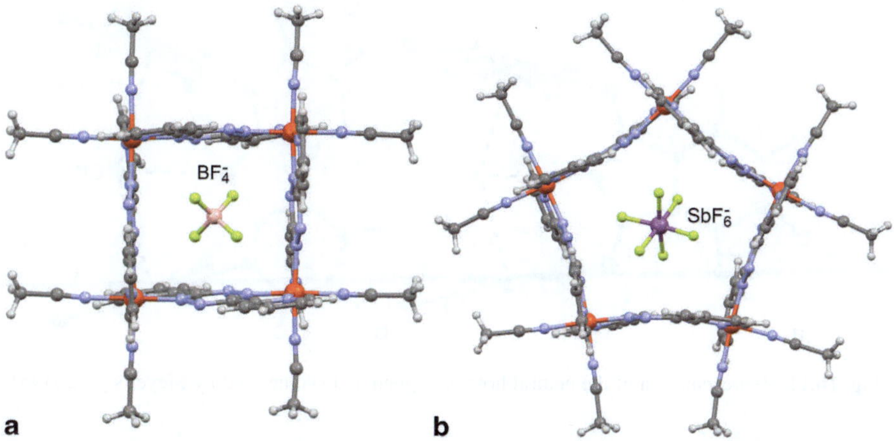

Fig. 16.10 Anion-templated X-ray structures of a molecular square **a** and a molecular pentagon **b** reported by Chifotides et al. [94]

Taking advantage of the enhanced π-acidity of organic ligands upon coordination to transition metals, Chifotides et al. [94] have reported supramolecular architectures with 3,6-bis(2-pyridyl)-*s*-tetrazine cavities where anion–π interactions participate in the remarkable stability of Fe(II) metallacycles in solution. This comprehensive investigation provides convincing evidence that anion–π interactions are the main driving force in the formation (in high yields) of self-assembled Fe(II)-templated metallacycles. Combining several technics, like X-ray crystallography, ^{1}H NMR, solution and solid-state^{19}F NMR spectroscopies, cyclic voltammetry and mass spectroscopy, they demonstrated that the anion acts as a template occupying the π–acidic cavities and controlling the nuclearity of the cages. That is, $[BF_4]^-$ and $[ClO_4]^-$ anions template molecular squares (see Fig. 16.10a) and $[SbF_6]^-$, $[AsF_6]^-$ and $[PF_6]^-$anions template molecular pentagons (see Fig. 16.10b) establishing close directional F· · ·C *s*-tetrazine contacts with the *s*-tetrazine rings that are up to 0.4 Å shorter than the sum of the F· · ·C van der Waals radii (3.17 Å). The number and strength of F· · ·C tetrazine contacts are maximized. They have also performed unprecedented solid-state^{19}F MAS NMR studies, where the templating anions showed downfield chemical shifts $\Delta\delta(^{19}F)$ ranging 3.5–4.0 ppm (compared to peripheral anions) due to their participation in anion–π interactions. NMR, cyclic voltammetry and mass spectroscopy studies also establish that the molecular squares and pentagons remain intact in solution. This study provides unambiguous evidence that anion–π interactions are the main driving force in the templation process leading to the formation of Fe(II) metallacycles with π-acidic cavities. The F atoms of the encapsulated anions are directly located over the more π-acidic *s*-tetrazine C atoms, establishing six simultaneous anion–π contacts with the metallacycle edges. More importantly, they evidenced the instrumental role of the templating anions in solution, thus favoring and stabilizing the Fe(II) metallacycles of specific nuclearities.

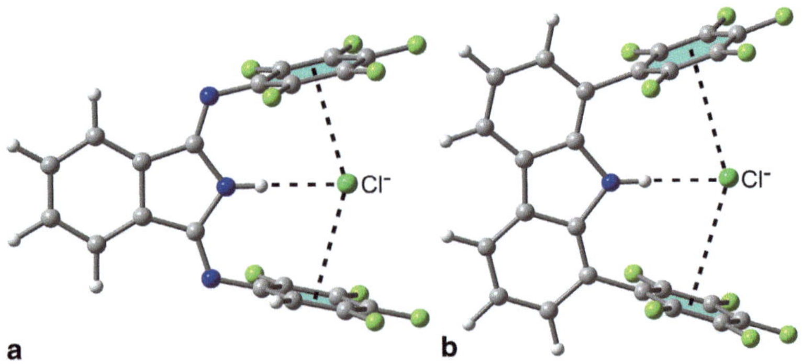

Fig. 16.11 Representation of the neutral hosts designed and synthesized by Meyer's group [95]

Meyer's group [95] has designed and synthesized pre-organized anion traps (Cl^- and Br^-) for exploiting anion–π interactions based on 1,3-bis(pentafluorophenyl-imino) isoindoline (Fig. 16.11a) and 3,6-di-*t*-butyl-1,8-bis(pentafluorophenyl)-9H–carbazole (Fig. 16.11b) in various solvents. Both neutral receptors provide a central N–H $\cdots$ X^- hydrogen bond that directs the halide anion into a pre-organized clamp formed by the two electron deficient arenes. Crystal structures of host–guest complexes reveal that in all cases the guest is located in the cleft between the perfluorinated flaps. In solution, association constants up to 960 M^{-1} have been determined depending on the solvent by NMR spectroscopy. Their study also includes a detailed computational analysis of the host–guest complexes and an energetic decomposition of the ring–anion interactions that confirm the contribution of the anion–π interactions to the stabilization of these complexes (~50 % of the total energy). These receptors contribute to increasing the relative low number of examples of neutral receptors that are well pre-organized for exploiting anion–π interactions.

Ballester and his group have dedicated much effort to the experimental quantification of anion–π interactions in solution using neutral host–guest model systems [96]. The quantification of anion–π interactions is commonly provided by computational studies of simple models that are useful to estimate the binding energy. The scientific community has no doubt about the existence of attractive anion-π interactions in the gas phase and in the solid state. However, there are still few examples of attractive anion-π interactions in solution. Ballester's group has reported several examples of neutral molecular receptors that bind anions in solution as a combination of anion–π interactions and hydrogen bonding [97, 98]. The strength of the anion–π interaction is indirectly detected as a modulation of the stronger hydrogen bonding interaction. The dissection of the energy contribution of the anion–π interaction to the overall binding is complex and requires the use of appropriate reference systems. Ballester and coworkers have designed a model system based on a series of "four wall" aryl-extended calix[4]pyrrole receptors. They contain deep aromatic cavities with fixed walls (see Fig. 16.12a, for the nitro-derivative as example). The formation of four concurrent H-bonds between the anion and the NH groups of the calix[4]pyrrole

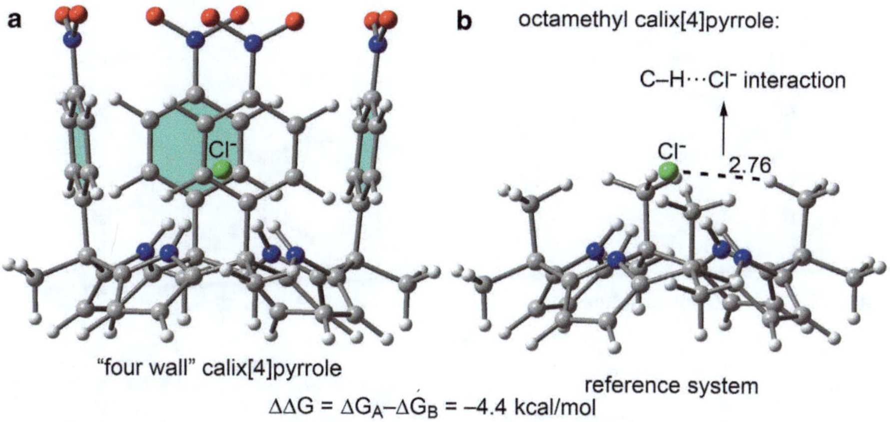

Fig. 16.12 Structures of a four wall calix[4]pyrrole **a** and the receptor used by Ballester's group [97] as reference **b** for the estimation of chloride–π interactions. Distance in Å

scaffold fixes the halide in the aromatic cavity, above the planes of the π-systems of the four meso-aryl substituents, as probed using ^{1}H NMR spectroscopy. Different *para* substituents in the aromatic walls were used to tune the electronic density of the aromatic rings and Cl^- as the interacting anion. Interestingly, the magnitude of the association constant was increased with the electron-withdrawing character of the *para*-substituent in the meso-aryl groups. The difference in free energy ($\Delta\Delta G$, see Fig. 16.12) of binding between different complexes provides a direct measurement of the relative interaction energy of the halide with the different aromatic systems. Using this approach, they determined a maximum contribution of −4.4 kcal/mol for the four chloride–π interactions to the overall binding free energy in the *para*-nitroaryl substituted receptor. This is likely an underestimation of the anion–π interaction energy because the reference system used by Ballester's group was octamethylcalix[4]pyrrole, which provides four C–H· · · Cl interactions (see Fig. 16.12b). Therefore, the estimated contribution of –4.4 kcal/mol likely means that each anion–π interaction is 1.1 kcal/mol more favorable than each C–H· · · Cl^- interaction that is established in the reference complex and, consequently, it cannot be used as an absolute estimate of the anion–π interaction in solution.

Calix[4]pyrrole based receptors featuring two additional pyrrole side arms have been also used by Chang et al. [99] for the molecular recognition of anions. This hexapyrroliccalix[4]pyrrole (see Fig. 16.13) has two additional pyrrole suspended above or below the calix[4]pyrrole core and presents *cis*/*trans* isomerism. Anion binding experiments revealed interesting differences in the binding mode depending on the isomer. That is, whilst the *trans* isomer displays only hydrogen bonding interactions, the *cis* isomer displays a mixed binding mode featuring a combination of hydrogen bonding and anion–π interactions resulting in an unexpected strong binding (see Fig. 16.13c). In fact, UV spectrophotometry and NMR titrations reveal

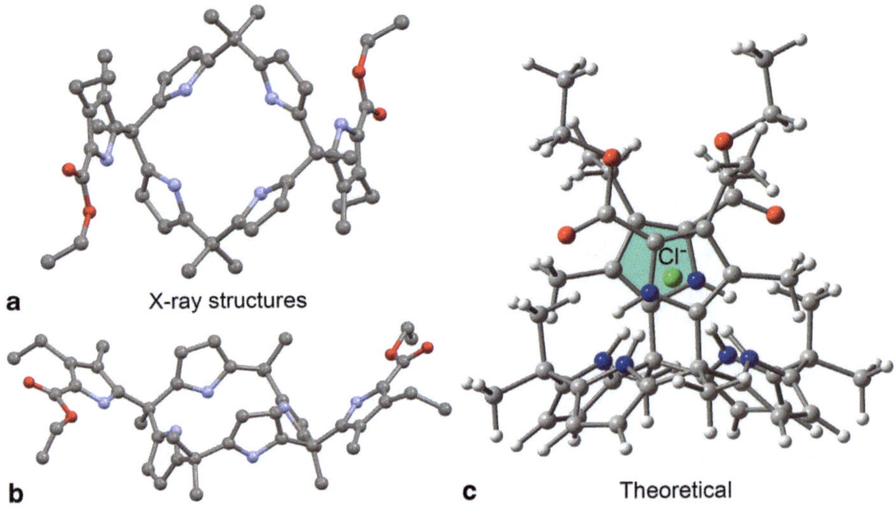

Fig. 16.13 X-ray structures of the trans **a** and cis **b** isomers of *meso*-substituted hexapyrrolic-calix[4]pyrrolereceptors reported by Chang et al. [99] and the modeled complex **c** between chloride and the *cis* isomer

that *cis* isomer displays higher affinity (105–106 M^{-1}) for anions while the *trans* isomer is more selective.

An interesting research has been recently published by He et al. [100] devoted to the study of the stability of size-regulable vesicles based on anion–π interactions Taking tetraoxacalix[2]arene[2]triazine as a functionalization platform (see Fig. 16.14), He et al. synthesized a series of new amphiphilic anion receptors that self-assemble into stable vesicles in a mixture of THF and water, with the surface of the vesicles engineered by electron-deficient cavities. Strikingly, several anions are able to influence the size of self-assembled vesicles selectively, following the order of $F^{\mu} < ClO_4^{\mu} < SCN^{\mu} < BF_4^{\mu} < Br^{\mu} < Cl^{\mu} < NO_3^{\mu}$, as revealed by dynamic light scattering (DLS) experiments and independently with the hydration cost. This order of selectivity agrees with the binding strength of anions with tetraoxacalix[2]arene[2]triazine receptor, demonstrating that the anion–π interaction most probably competed over other possible weak interactions and is responsible for this interesting selectivity. Furthermore, the chloride permeation process across the membrane of the vesicles was also studied by He et al. using fluorescent experiments. This investigation shows the potentiality of heteracalix aromatics as new models to construct functional vesicles and gives a new dimensionality to the anion–π interaction in aqueous medium and, potentially, in living systems.

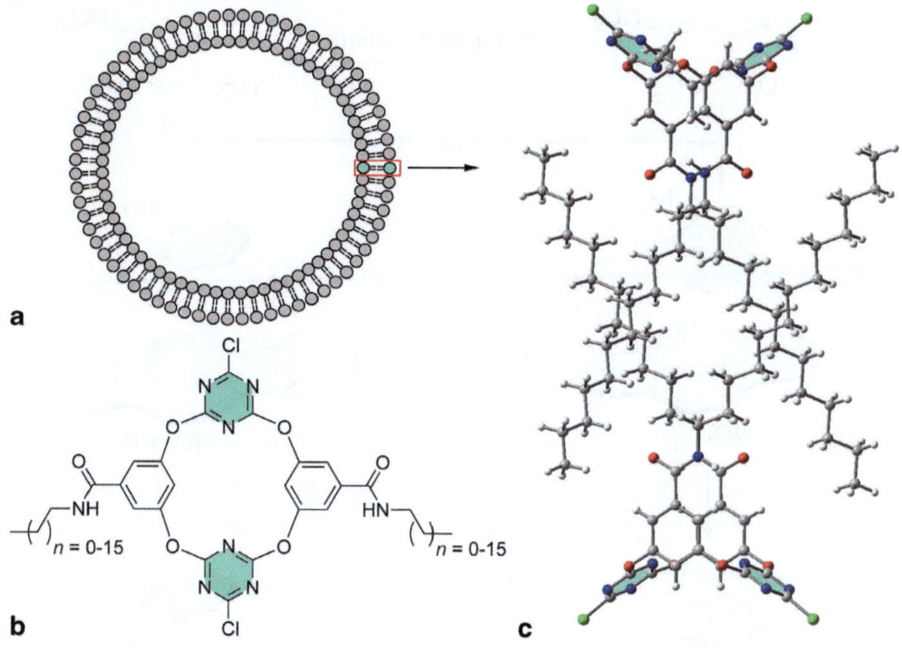

Fig. 16.14 Schematic illustration of the vesicle (**a** and **b**) and long alkyl chain derivatives of tetraoxacalix[2]arene[2]triazine compounds (**c**) reported by He et al. [100]

16.5 Catalysis

Zhao et al. reported in 2013 for the first time experimental evidence suggesting that anion–π interactions contribute to the catalysis of the Kemp elimination reaction (see Fig. 16.15) by π-acidic naphthalene diimides (NDI) leading to conceptually innovative design strategies to stabilize anionic transition states [101]. Subsequent studies [102] with modified sulfur-containing NDI catalysts confirmed the general validity of increasing transition-state stabilization while increasing π-acidity with regard to the Kemp elimination. Moreover, computational simulations are in excellent agreement with experimental results, confirming that the stabilization of the anionic transition states (but not the neutral ground states) increases with the π-acidity of the catalysts, i.e., the existence of anion–π catalysis. The proposed catalytic cycle is shown in Fig. 16.15 and the key point is the location of the carboxylate base on the π-acidic surface of catalyst NDI. The initial substrate-NDI complex (NDI + S) is likely stabilized by a combination of π–π and hydrogen bonding interactions. In the transition state (TS), the negative charge flows over the π-acidic surface from the carboxylate base over the carbanion of the conjugate base to the phenolate oxygen. The TS is stabilized by the π-acidic surface of the NDI. The proton transfer from the carboxylic acid to the phenolate in the intermediate (I) prevents product inhibition and regenerates the catalyst NDI. Therefore, in this catalytic cycle, the NDI stabilizes the

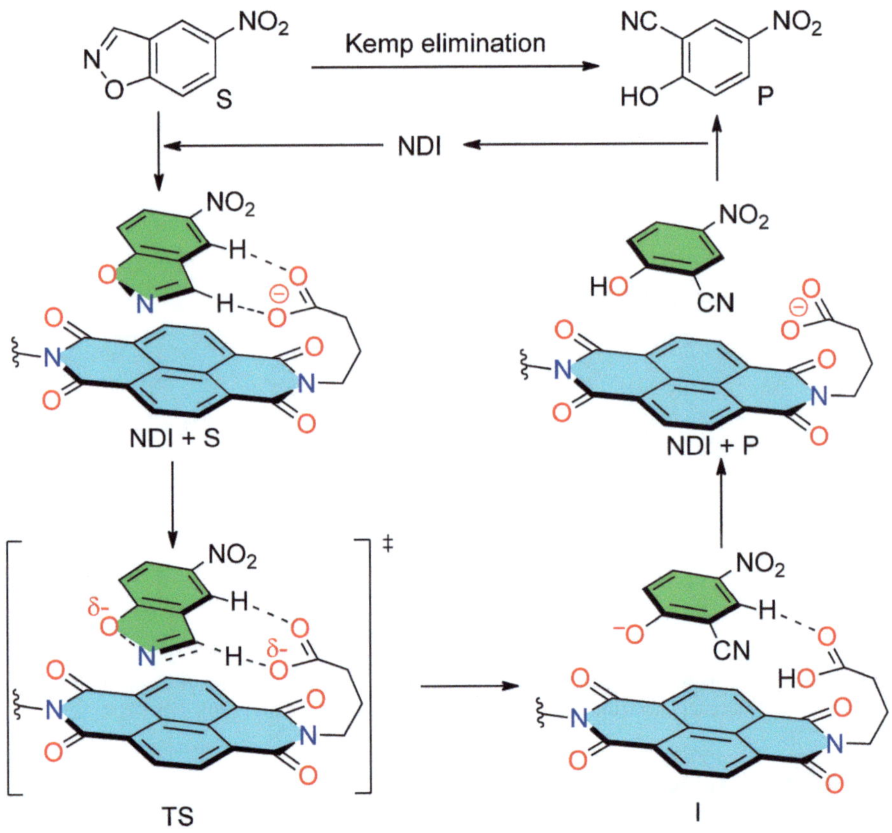

Fig. 16.15 Catalytic cycle proposed by Zhao et al. [101] (NDI + S = catalyst–substrate complex, I = reactive intermediate, NDI + P = catalyst–product complex)

anionic TS by means of anion–π interactions, which are proved by the acceleration of the Kemp elimination observed experimentally. Both pioneering works [101, 102] on catalysis with anion–π interactions are extremely important since clearly open a new avenue and move beyond the grand principles operating in nature.

The catalytic cycle shown in Fig. 16.15 has been analyzed theoretically to examine the most important aspects of the anion–π catalyzed Kemp elimination considering solvent effects in the calculations [102]. Already in the early stage of the reaction (NDI + S), the carboxylate anion is positioned so that an efficient intramolecular anion–π interaction with the most π-acidic part of the NDI surface takes place. The carboxylate group also anchors the benzisoxazole substrate above the NDI surface via two C–H· · · O interactions (see Fig. 16.15), therefore favoring the π–π interaction between the two parallel disposed aromatic systems. This facilitates the proton transfer between the catalyst and the substrate, leading to the transition state TS with the activation barrier of 15.05 kcal/mol (Fig. 16.16a). At this stage the electron

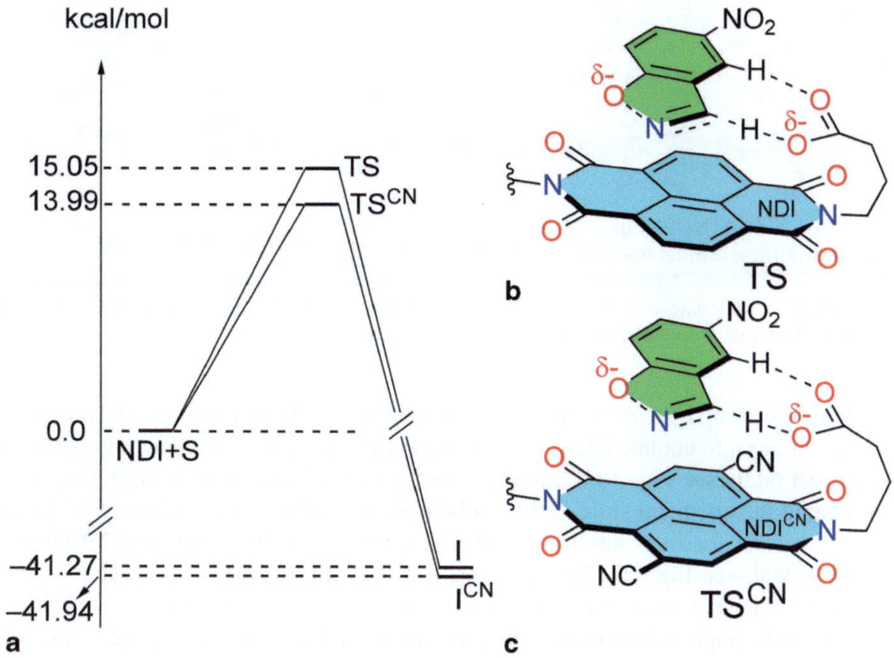

Fig. 16.16 **a** Free energy diagram (IEFPCM/M06-2X/def2-TZVP//B97-D/6-311G**) for the Kemp elimination with anion–π catalyst NDI and NDI^{CN}. **b** Transition state (TS) for the reaction catalyzed by catalyst NDI, negative charge transfer is highlighted in red. **c** Transition state (TS^{CN}) for the reaction catalyzed by catalyst NDI^{CN} [101]

transfer, which occurs over several atoms from carboxylate anion to phenolate oxygen, is efficiently stabilized by the π-acidic surface of NDI. The reaction progresses toward the anionic intermediate I, while the negative charge is fully transferred to the benzisoxazole substrate. The benzisoxazole oxygen accumulates most of the charge and its distance from the NDI surface decreases to 2.995 Å. The conformation of this complex once again favors anion–π interactions by placing the anionic oxygen right above the preferential binding site of NDI. Interestingly, the Kemp elimination in the presence of the 3,7-dicyano-substituted catalyst (NDI^{CN}, see Fig. 16.16c) follows a similar pathway. However, the increased π-acidity enhances the TS^{CN} transition-state stabilization by 1.06 kcal/mol when compared to TS (Fig. 16.16a). This stabilization enhancement of TS^{CN} by the more π-acidic NDI surface of NDI^{CN} confirms that anion–π interactions contribute significantly to this reaction. The comparison of certain geometric parameters during the reaction mechanism involving catalysts NDI and NDI^{CN} also reflects enhanced anion–π implication. For instance the more pronounced decrease of the distance between benzisoxazole oxygen and catalyst plane on going from the initial complex to the TS is correlated with the higher strength of the anion–π interaction in the NDI^{CN}.

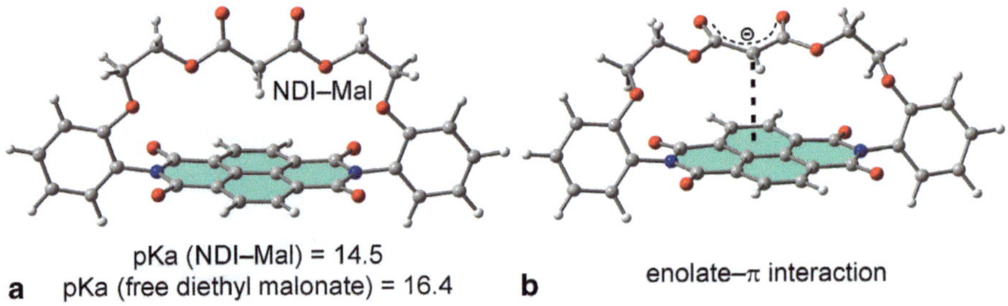

Fig. 16.17 **a** Malonate covalently bonded to NDI reported by Matile's group [103]. **b** DFT-optimized geometry obtained for the enolate–π interaction

Matile's group has taken one step further the research on anion–π catalysis [103] by extending it to enolate chemistry. They have covalently attached a malonate moiety to an NDI (see Fig. 16.17) and by means of ^{1}H NMR spectroscopy they have compared the chemical shift of the acidic hydrogen atoms of free diethyl malonate with the corresponding ones upon its attachment to the naphthalene diimide (denoted as NDI–Mal, see Fig. 16.17a). As a result, the chemical shift of 3.38 ppm for the acidic hydrogen atoms in free diethylmalonate changes to 1.78 ppm due to their exposure to the naphthalene ring current in NDI–Mal that causes the upfield shift (see Fig. 16.17a). This direct experimental evidence for the fixed covalent positioning of the malonate above the π–acidic surface is very important because it assures that any changes in acidity can be unambiguously attributed to the stabilization of the enolate by anion–π interactions (see Fig. 16.17b). In fact, they have demonstrated using ^{1}H NMR titrations that anion–π interactions stabilize the enolate by almost two pK_a units. Remarkably, the addition of these anion–π-stabilized reactive enolate intermediates to enones and nitroolefins occurs with significant transition-state stabilizations (up to 11 kJ/mol) [103]. Moreover, anionic cascade reactions that cover aldol condensation, elimination and transesterification also accelerate on π-acidic surfaces. These findings are very significant because enolate chemistry is fundamental in chemistry and biology. This research is expected to stimulate the use of anion–π interactions in catalysis in the broadest sense.

16.6 Biologically Relevant Anion–π Interactions

Clear evidence that illustrates the growing interest in the study anion–π interactions in biological systems is the development of software capable to search anion–π contacts in the PDB and related biological databases. To this respect, the STAAR (statistical analysis of aromatic rings) program [47] can identify anion–π interactions in a large structural database of biomolecules. The program is freely available for download through the web (http://staar.bio.utk.edu) and has been tested in a recent version of the PDB demonstrating the high prevalence and relatively strong anion–π energies involving side-chain/side-chain interactions in biomolecules. The program is currently limited to phenylalanine residues and is expected to include tryptophan and tyrosine residues in the new version. The program also provides pairwise interaction energies with Asp and Glu as anions. Other future project for improving the program includes the possibility to investigate anion–π interaction in protein/ligand complexes.

A pioneering work on the study of anion-π interaction in biomolecular recognition was published in 2012 by Chakravarty et al. [46]. Combining theoretical and convincing experimental evidences they demonstrated that the interaction between an anion and an aromatic π system plays an important role in the formation and recognition of biomolecular structures. Though less frequent than its counterpart cation–π interactions, the examination of high-resolution structures of proteins and nucleic acids indicated the presence anion–π interactions. Interestingly, they have been observed unambiguously, occurring in protein/nucleic acid loops and often involving conserved/coevolving sites in proteins. These findings suggest that this interaction plays an important role in macromolecular folding and function. Two examples are highlighted in Fig. 16.18. In the first one, the anion–π interaction is observed in the nucleic acid backbone (tetraloop hairpin, 1m5o structure), where the guanine-nucleotide (G75) establishes a clear anion–π interaction with the oxygen atom of the phosphate anion. Apart from this interacting phosphate, the rest of the phosphate anionic oxygen atoms in the nucleic acid backbone face outward. Remarkably, the phosphate anion that faces inward is poorly solvated, facilitating the anion–π interaction (Fig. 16.18a, top). The nucleic acid base likely compensates for the energetic cost of the desolvation of the anion. In many RNA hairpins, even in structures with low resolutions the same interacting pattern is observed [46] suggesting that anion–π interactions may be a regular feature of nucleic acid loops. The second example involves an RNA-protein interaction that corresponds to a crucial biological process. It is well-known [104] that the RRM (RNA recognition motif) of the U1A protein binds to U1 snRNA hairpin II. The structure of the complex reveals that RRM Asp92 stacks in parallel to the Cytosine-12 residue of the hairpin loop, establishing an anion–π interaction. Amazingly, the same feature is observed in the structure of the RRM of U2B" bound to hairpin IV of U2 snRNP (Fig. 16.18b). U2B" Asp92 stacks on Guanine-12 of the hairpin loop of U2 and this arrangement is a conserved feature of spliceosomal RNA hairpins involving a C/G12 and Asp92-carboxylates.

Frontera and coworkers [45] have studied the important role of anion–π interactions in the active site of the urate oxidase enzyme (see Fig. 16.19). The inhibition of this enzyme by cyanide ions is caused by the existence of an anion–π interaction between the anionic inhibitor and the π-electron deficient enzymatic substrate (uric acid). They demonstrated using high level *ab initio* calculations that the anion–π interaction between the inhibitor and the substrate in a model of the active site is energetically favorable. In addition, a favorable cooperativity effect between a π–π stacking interaction involving the substrate and a phenylalanine residue (PHE159, see Fig. 16.19) and the anion–π interaction also contributes to the binding energy. This was the first example where the presence of an anion–π interaction between an inhibitor and an enzymatic substrate was proposed to be crucial in the inhibition of

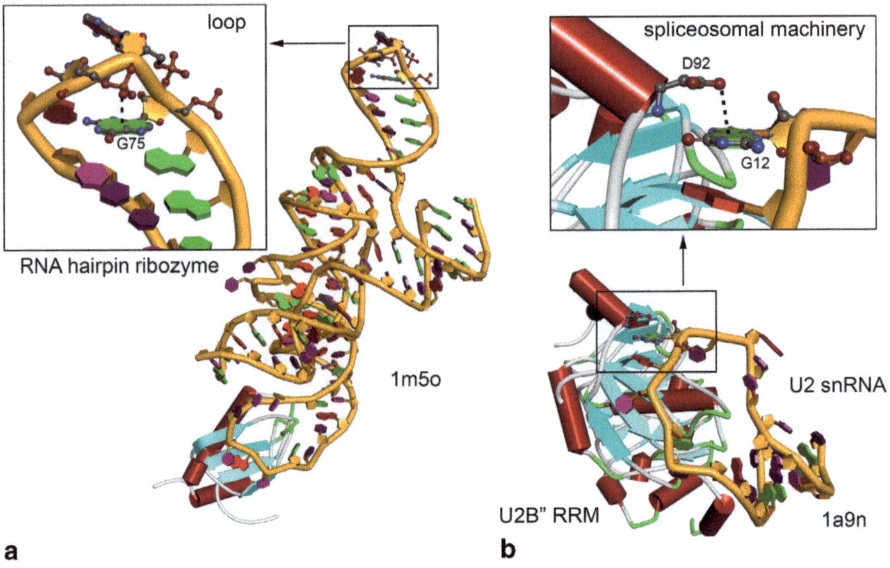

Fig. 16.18 **a** Anion–π interaction in nucleic acids between the base of the ith nucleotide with the oxygen atoms of an $(i+2)^{th}$ phosphate of a tetraloop hairpin. **b** Anion–π in RRM (split-pea *green*, 1a9n) of U2B″ (spliceosomalmachinary) in complex with U2 snRNA

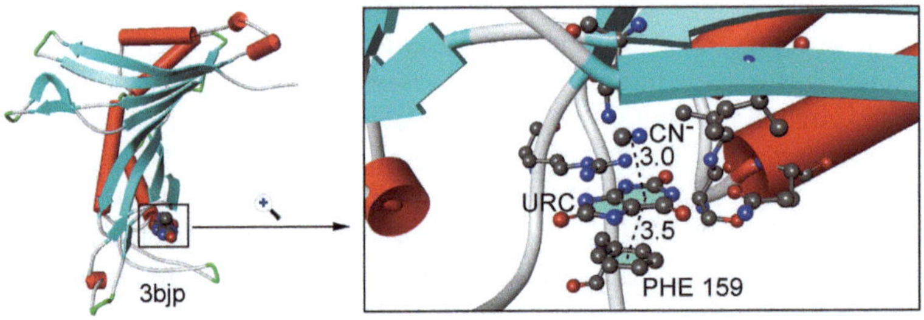

Fig. 16.19 The anion–π interaction between the CN^- inhibitor and the uric acid substrate (URC) found in the X-ray structure 3bjp studied theoretically by Frontera's group [45] is shown. Distances in Å

an enzyme. This investigation extended the relevance of anion–π interactions to an important field such as enzyme chemistry.

The relevance of anion–π interactions in flavoproteins has been analyzed by Estarellas et al. [105]. An initial search in the PDB provided evidence of anion–π interactions in a large number of proteins. Many of them are reductases that have affinity for small negatively charged ions such as acetate, chloride and thiocyanate, which are competitive inhibitors. Two interesting examples were described in their investigation. In the first one (2ar8, tryptophan 7-halogenase), the anion binds the FAD at the enzymatic center by means of an anion–π interaction and participates in the enzymatic process [106]. This enzyme catalyzes the regioselective chlorination at the 7 position of tryptophan. The Cl^- is bound near to the entrance of the tunnel leading to the tryptophan and positioned to make a nucleophilic attack on the flavin peroxide resulting in the formation of hydroxylated FAD and HClO that is the real chlorination agent [106]. A close examination of the active site reveals that the anion is located above the most electron deficient ring of the FAD establishing an anion–π interaction. In the ternary complex (see Fig. 16.19a), the distance between the Cl^- anion and the ring centroid of the pyrimidinic ring of FAD is 3.3 Å in 2ar8. The second example is a FMN-dependent nitroreductase (1ylU) [107] that plays a prominent role in the reduction of the antibiotics nitrofuranone and nitrofurantoin to the hydroxylamine derivatives, which are the active antibacterial agents. Inhibition studies of nitroreductase (NTR) have demonstrated that acetate displayed competitive inhibition with respect to both the substrate and NADH [107]. More interestingly, it has been demonstrated that acetate binds only to the oxidized form of the enzyme. The crystal form of the enzyme complexed to the acetate and the FMN coenzyme shows one acetate molecule in each active site (see Fig. 16.20b). The acetate anion is bonded to the active site by means of a bifurcated hydrogen bond, a $-CH_3 \cdots \pi$ interaction and an anion–π interaction with the pyrimidinic ring of FMN. The distance between the acetate anion and the ring centroid is short (3.0 Å) indicating a strong interaction. This investigation further confirms the relevance of anion–π interactions in the scientific field of enzyme chemistry.

Bauzá et al. [108] have studied the importance of anion-π interactions in the mechanism of sulfide:quinone oxidoreductase. It is a flavin-dependent enzyme that plays a physiological role in two important processes. First, it is responsible for sulfide detoxification by oxidizing sulfide ions to elementary sulfur and the electrons are first transferred to flavin adenine dinucleotide (FAD), which in turn passes them to the quinone pool in the membrane [109]. Second, this enzyme plays a key role in the sulfide-dependent respiration and anaerobic photosynthesis, deriving energy for their growth from reduced sulfur in sulfidotrophic bacteria [110]. Two mechanisms of action for this enzyme have been proposed [108] that involve a common anionic intermediate that it is stabilized by a relevant anion–π interaction (INT–FAD, see Fig. 16.21). The formation of the intermediate is facilitated by reducing the transition state barrier, owing to an anion–π interaction that involves the π system of FAD. By analyzing the X-ray structures of SQRs available in the Protein Data Bank (PDB) and using DFT calculations, they have demonstrated the relevance of the anion–π interaction in the enzymatic mechanism.

In particular, Bauzá et al. [108], have demonstrated theoretically using DFT calculations (B3LYP/6-311++G**) that the energy barrier is reduced when the reaction occurs in the present of FAD, using cyclo-L-cysteine (see Figure 16.21b) as model of disulfide bond. In Fig. 16.22a the reaction coordinate diagram is represented where the relative energies of the different species involved in the mechanism are also indicated. It can be observed that the presence of FAD (in red) stabilizes all species. However, the most important result is that the stabilization of the transition state is higher than the stabilization of the intermediates, thus supporting the crucial role of FAD not only as an electron sink in subsequent enzymatic steps of

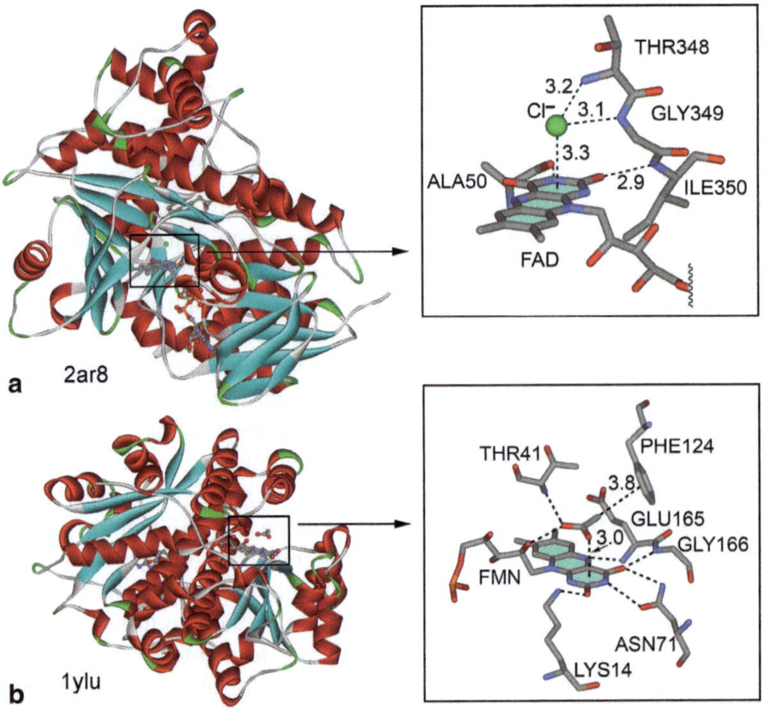

Fig. 16.20 3D X-ray structures of 2ar8 (**a**) and 1ylu (**b**) are shown and the anion–π interactions observed in the FAD-Cl^- (**a**) and FMN-acetate (**b**) complexes are highlighted. Distances in Å

the mechanism but also facilitating the initial addition-elimination reaction to take place. Moreover, the presence of FAD stabilizes the reaction product placing it in an ideal position either to attack or transfer an electron to the isoalloxazine ring. The theoretical study also includes the computation of the NCI (Non Covalent Interaction) plot to study the anion–π interactions observed in the structures retrieved from the PDB. This visualization index based on the electron density and its derivatives enables the identification and visualization of non-covalent interactions efficiently [111]. The isosurfaces correspond to both favorable and unfavorable interactions, as differentiated by the sign of the second density Hessian eigenvalue and defined by the isosurface color. In Fig. 16.22b the representation of the NCI plot computed around the FAD (within 4.0 Å) cofactor in 3SX6 protein is shown. Obviously, several non-covalent regions clearly appear between the FAD and the aminoacids. For instance, several green isosurfaces are found around the methyl groups of FAD, which are characteristic of hydrophobic interactions. Interestingly the trisulfide group (represented in ball and stick) originates an extended isosurface that covers two rings of FAD and confirms the importance of the anion–π interaction.

Finally, Bauzá et al. [112] have also studied long-range effects in anion–π interactions and their crucial role in the inhibition mechanism of mycobacterium

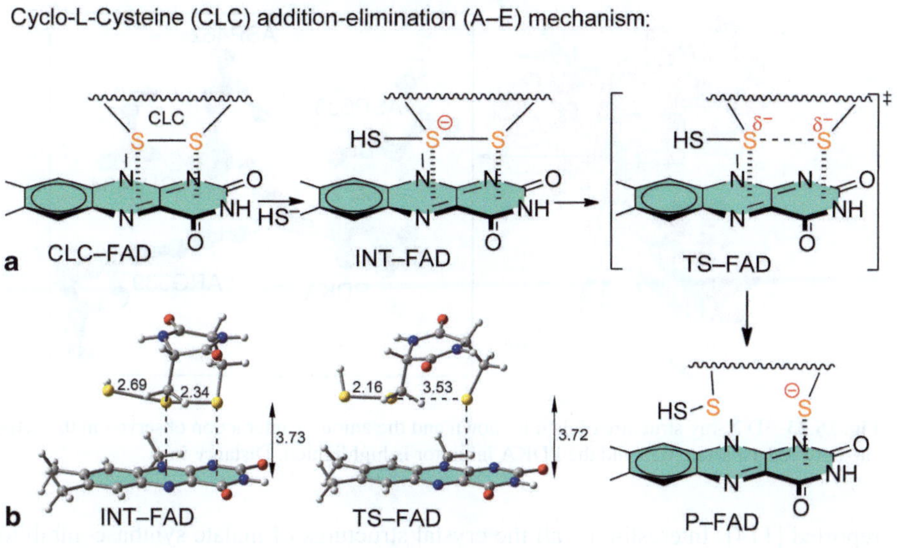

Fig. 16.21 **a** Addition-elimination mechanism in the presence of FAD. **b** B3LYP/6-311 + G* optimized complexes of INT-FAD and TS-FAD. Distances in Å

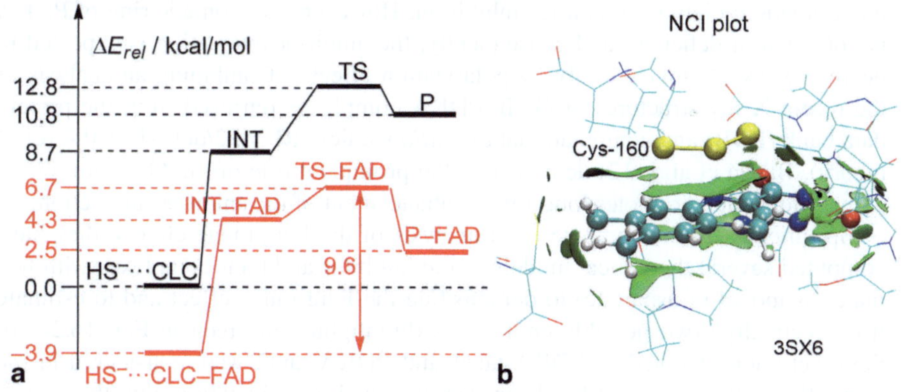

Fig. 16.22 **a** Addition-elimination mechanism in the presence of FAD. **b** B3LYP/6-311+G* optimized complexes of INT-FAD and TS-FAD. Distances in Å

tuberculosis malate synthase. This enzyme together with isocitrate lyase forms the glyoxylate shunt that is an anaplerotic bypass of the traditional Krebs cycle. It plays a prominent role in Mycobacterium tuberculosis virulence, so it can be exploited for the development of antitubercular therapeutics [113]. The shunt bypasses two steps of the tricarboxylic acid cycle, allowing the incorporation of carbon, and thus, refilling oxaloacetate under carbon-limiting conditions. A catalytic Mg^{2+} unit is located at the bottom of the cavity, and plays a very important role. Recently, the development of effective antituberculosis drugs based on phenyldiketo acids (PDKAs) has been

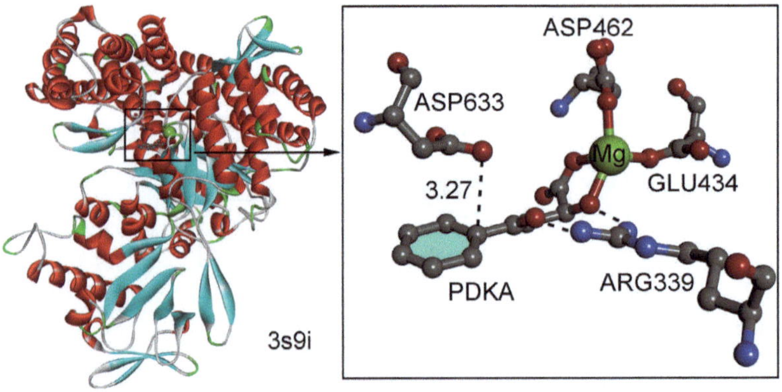

Fig. 16.23 3D X-ray structure of 3s9i is shown and the anion–π interaction observed in the active site between aspartate-633 and the PDKA inhibitor is highlighted. Distance in Å

reported [114]. Interestingly, all the crystal structures of malate synthase–inhibitor complexes exhibit close contact between the carboxylate of Asp633 and the face of the aromatic ring of the inhibitor (see Fig. 16.23). Remarkably, the replacement of the phenyl ring in PDKA by aliphatic moieties yields inactive inhibitors, suggesting that the aromatic moiety is crucial for inhibition. However, the aromatic ring of PDKA is not electron-deficient, and consequently, the anion–π interaction is expected to be very weak (dominated only by polarization effects). Combining an analysis of the recent X-ray structures of GlcB–PDKA complexes retrieved from the protein data bank (PDB) and computational *ab initio* studies (RI-MP2/def2-TZVP level of theory), Bauzá et al. [112] demonstrate the prominent role of the Mg^{2+} ion in the active site, which promotes long-range enhancement of the anion–π interaction.

Specifically, in the computational analysis of the long range effects, they have computed several theoretical models of the inhibitor and their complexes with formate (as model of aspartate) to demonstrate the long range effect and to estimate it energetically. Two possible enol forms (highlighted in green in Fig. 16.24) of the diketo acid denoted as PDKA-taut1 and PDKA-taut2 were considered in the energetic analysis, since it has been demonstrated experimentally that the PDKA chemical structure is in its enol tautomer by solution-phase ^{1}H– and 13C–NMR data [114]. In addition, a theoretical model of the inhibitor coordinated to the Mg^{2+} ion was also considered, denoted as PDKA-Mg. In this model (see Fig. 16.23c), a formate ligand was added in order to keep the model neutral and to exemplify the active site where the Mg^{2+} is coordinated to ASP462 (see Fig. 16.23). Interestingly, the interaction energy of the anion–π complex of formate with PDKA-taut1 is −4.5 kcal/mol and with PDKA-taut2 is almost negligible (−0.2 kcal/mol) indicating that the intramolecular hydrogen bonding is very important since it increases the electron withdrawing ability of the carbonyl group attached to the ring. Interestingly, in the PDKA-Mg complex, the interaction energy is large and negative (−10.6 kcal/mol), indicating a strong binding. Therefore, the theoretical study clearly demonstrates

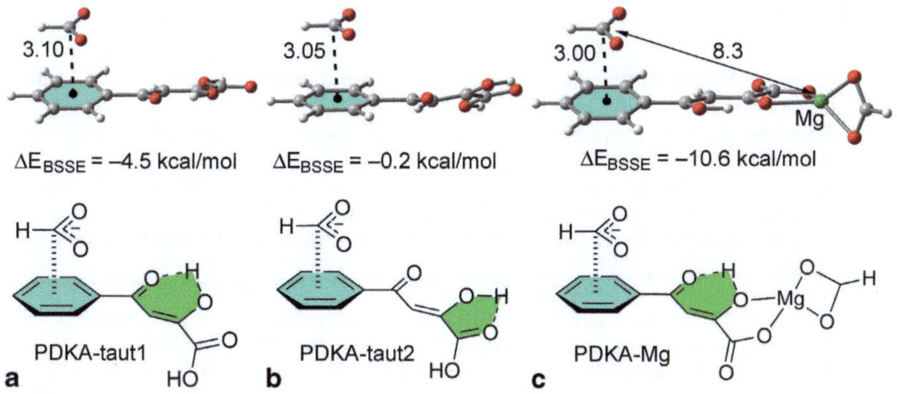

Fig. 16.24 Optimized formate complexes with the PDKA considering both tautomers (**a** and **b**) and with the PDKA coordinated to Mg^{2+} from reference [112]. The interaction energies (ΔE_{BSSE}) are indicated. Distances in Å

that the presence of the Mg ion coordinated to the inhibitor enhances the anion–π interaction. This long-range effect should be emphasized since the distance from the Mg ion to the interacting anion is 8.3 Å (see Fig. 16.24c).

16.7 Conclusions

The purpose of this chapter has been to expose the bonding relationship between anions and π systems by describing high quality work and demonstrating the extraordinary potential of this relatively new interaction to impact the field of supramolecular science, including catalysis and enzyme chemistry. It is now evident for the scientific community that anion–π interactions are prominent in a wide range of systems and should be considered as an important and general noncovalent binding force. The potential of the anion–π interaction is not limited to the design of novel hosts and sensors. Its important role in RNA recognition motifs, enzymatic chemistry, catalysts, and crystal engineering has provided a new dimension to this interaction. The effect of anion–π interactions should not be overlooked in chemical and biological systems that involve anion and electron-deficient aromatic species.

Acknowledgment We are grateful to Carol Garau, Xavier Lucas, Daniel Escudero and David Quiñonero with whom we have had the good fortune to work and their names are contained within the pertinent references. We thank CONSOLIDER-Ingenio 2010 (project CSD2010-0065) and the MICINN of Spain (project CTQ2011-27512 FEDER funds) for financial support. We thank the Direcció General de Recerca, Desenvolupament Tecnològic i Innovació del Govern Balear (project 23/2011, FEDER funds) for financial support.

References

1. Schneider HJ (2009) Binding mechanisms in supramolecular complexes. Angew Chem Int Ed 48:3924–3977
2. Schneider HJ, Yatsimirski A (2000) Principles and methods in supramolecular chemistry. Wiley, Chichester
3. Lehn JM (1995) Supramolecular chemistry concepts and perspectives. Wiley-VCH, Weinheim
4. Vögtle F (1993) Supramolecular chemistry: an introduction. Wiley, New York
5. Beer PD, Gale PA, Smith DK (1999) Supramolecular chemistry. Oxford University Press, Oxford
6. Steed JW, Atwood JL (2000) Supramolecular chemistry. Wiley, Chichester
7. Politzer P, Murray JS, Clark T (2010) Halogen bonding: an electrostatically driven highly direccional noncovalent interaction. Phys Chem Chem Phys 12:7748–7757.
8. Bundhun A, Ramasami P, Murray JS, Politzer P (2013) Trends in σ-hole strenghts and interactions of F_3MX molecules (M = C, Si, Ge and X = F, Cl, Br, I). J Mol Model 19:2739–2746
9. Bauzá A, Mooibroek TJ, Frontera A (2013) Tetrel–bonding: rediscovered supramolecular force? Angew Chem Int Ed 52:12317–12321
10. Meyer EA, Castellano RK, Diederich F (2003) Interactions with aromatic rings in chemical and biological recognition. Angew Chem Int Ed 42:1210–1250
11. Schneider HJ, Yatsimirsky A (2008) Selectivity in supramolecular host-guest complexes. Chem Soc Rev 37:263–277
12. Salonen LM, Ellermann M, Diederich F (2011) Aromatic rings in chemical and biological recognition: energetics and structures. Angew Chem Int Ed 50:4808–4842
13. Müller-Dethlefs K, Hobza P (2000) Noncovalent interactions: a challenge for experiment and theory. Chem Rev 100:143–167
14. Burley SK, Petsko GA (1985) Aromatic-aromatic interaction: a mechanism of protein structure stabilization. Science 229:23–28
15. Li S, Cooper VR, Thonhauser T, Lundqvist BI, Langreth DC (2009) Stacking interactions and DNA intercalation. J Phys Chem B 113:11166–11172
16. Rutledge LR, Campbell-Verduyn LS, Wetmore SD (2007) Characterization of the stacking interactions between DNA or RNA nucleobases and the aromatic amino acids. Chem Phys Lett 444:67–175
17. Tsuzuki S, Yoshida M, Uchimaru T, Mikami M (2001) The origin of the cation/π interaction: the significant importance of the induction in Li^+ and Na^+ complexes. J Phys Chem A 105:769–773
18. Ma JC, Dougherty DA (1997) The cation-π interaction. Chem Rev 97:1303–1324
19. Crowley PB, Golovin A (2005) Cation-π interactions in protein-protein interfaces. Proteins Struct Funct Genet 59:231–239
20. Gallivan JP, Dougherty DA (2000) A computational study of cation-π interactions vs salt bridges in aqueous media: implications for protein engineering. J Am Chem Soc 122:870–874
21. Lamoureux JS, Maynes JT, Glover JMN (2004) Recognition of 5′-YpG-3′ sequences by coupled stacking/hydrogen bonding interactions with amino acid residues. J Mol Biol 335:399–408
22. Frey J, Kraus T, Heitz V, Sauvage JP (2005) A catenane consisting of a large ring threaded through both cyclic units of a handcuff-like compound. Chem Commun 49:5310–5312
23. Masu H, Sakai M, Kishikawa K, Yamamoto M, Yamaguchi K, Kohmoto S (2005) Aromatic foldamers with iminodicarbonyl linkers: their structures and optical properties. J Org Chem 70 1423–1431
24. Bao X, Isaacsohn I, Drew AF, Smithrud DB (2007) Determining the binding and intracellular transporting abilities of a host-[3]rotaxane. J Org Chem 72:3988–4000
25. Heemstra JM, Moore JS (2004) Helix stabilization through pyridinium-π interactions. Chem Commun 40:1480–1481

26. Nishio M, Hirota M, Umezawa Y (1998) In the C–H/π interaction: evidence, nature, consequences. Wiley-VCH, New York
27. Nishio M (2004) CH/π hydrogen bonds in organic and organometallic chemistry. Cryst Eng Comm 6:130–156
28. Egli M, Sarkhel S (2007) Lone pair-aromatic interactions: to stabilize or not to stabilize. Acc Chem Res 40:197–205
29. Mooibroek TJ Gamez P Reedijk J (2008) Lone pair–π interactions: a new supramolecular bond? CrystEngComm 10:1501–1515
30. Mitra M, Manna P, Seth SK, Das A, Meredith J, Helliwell M, Bauzá A, Choudhury SR, Frontera A, Mukhopadhyay S (2013) Salt-bridge-π (sb-π) interactions at work: associative interactions of sb-π, π-π and anion-π in Cu(II)-malonate-2-aminopyridine-hexafluoridophosphate ternary system. CrystEngComm 15:686–696
31. Frontera A, Gamez P, Mascal M, Mooibroek TJ, Reedijk J (2011) Putting anion-π interactions into perspective. Angew Chem Int Ed 50:9564–9583
32. Schottel BL, Chifotides HT, Dunbar KR (2008) Anion-π interactions. Chem Soc Rev 37:68–83
33. Caltagirone C, Gale PA (2009) Anion receptor chemistry: highlights from 2007. Chem Soc Rev 38:520–563
34. Gamez P, Mooibroek TJ, Teat SJ, Reedijk J (2007) Anion binding involving π-acidic heteroaromatic rings. Acc Chem Res 40 435–444
35. Hay BP, Bryantsev VS (2008) Anion-arene adducts: C–H hydrogen bonding, anion-π interaction, and carbon bonding motifs. Chem Commun 44:2417–2428
36. Quiñonero D, Garau C, Rotger C, Frontera A, Ballester P, Costa A, Deyà PM (2002) Anion-π interactions: do they exist? Angew Chem Int Ed 41:3389–3392
37. Mascal M, Armstrong A, Bartberger MD (2002) Anion-aromatic bonding: a case for anion recognition by π-acidic rings. J Am Chem Soc 124:6274–6276
38. Alkorta I, Rozas I, Elguero J (2002) Interaction of anions with perfluoro aromatic compounds. J Am Chem Soc 124:8593–8598
39. Kim DY, Singh NJ, Kim KS (2008) Cyameluric acid as anion-π type receptor for ClO_4^- and NO_3^-: π-Stacked and edge-to-face structures. J Chem Theory Comput 4:1401–1407
40. Schottel BL, Bacsa J, Dunbar KR (2005) Anion dependence of Ag(I) reactions with 3,6-bis(2-pyridyl)-1,2,4,5-tetrazine (bptz): isolation of the molecular propeller compound $[Ag_2(bptz)_3][AsF_6]_2$. Chem Commun 41:46–47
41. Han B, Lu JJ, Kochi JK (2008) Anion recognitions via cocrystallizations with organic π-acids in the efficient self-assembly of nanoscopic one-dimensional molecular chains. Cryst Growth Des 8:1327–1334
42. Mascal M, Yakovlev I, Nikitin EB, Fettinger JC (2007) Fluoride-selective host based on anion-π interactions, ion pairing, and hydrogen bonding: synthesis and fluoride-ion sandwich complex. Angew Chem Int Ed 46:8782–8784
43. Götz RJ, Robertazzi A, Mutikainen I, Turpeinen U, Gamez P, Reedijk J (2008) Concurrent anion-π interactions between a perchlorate ion and two π-acidic aromatic rings, namely pentafluorophenol and 1,3,5-triazine. Chem Commun 44:3384–3386
44. Albrecht M, Müller M, Mergel O, Rissanen K, Valkonen A (2010) CH-directed anion-π interactions in the crystals of pentafluorobenzyl-substituted ammonium and pyridinium salts. Chem Eur J 16:5062–5069
45. Estarellas C, Frontera A, Quiñonero D, Deyà PM (2011) Relevant anion-π interactions in biological systems: the case of urate oxidase. Angew Chem Int Ed 50:415–418
46. Chakravarty S, Sheng ZZ, Iverson B, Moore B (2012) "η6"-Type anion-π in biomolecular recognition. FEBS Lett 586:4180–4185
47. Jenkins DD, Harris JB, Howell EF, Hinde RJ, Baudry J (2013) STAAR: Statistical Analysis of Aromatic Rings. J Comput Chem 34:518–522
48. Dawson RE, Hennig A, Weimann DP, Emery D, Ravikumar V, Montenegro J, Takeuchi T, Gabutti S, Mayor M, Mareda J, Schalley CA, Matile S (2010) Experimental evidence for the functional relevance of anion-π interactions. Nat Chem 2:533–538

49. Sakai N, Mareda J, Vauthey E, Matile S (2010) Core-substituted naphthalenediimides. Chem Commun 46:4225–4237
50. Gale PA, Caltagirone C (2014) Anion sensing by small molecules and molecular ensembles. Chem Soc Rev. doi:10.1039/c4cs00179f
51. Caballero A, Zapata F, Beer PD (2013) Interlocked host molecules for anion recognition and sensing. Coord Chem Rev 257:2434–2455
52. Gale PA (2011) Anion receptor chemistry. Chem Commun 47:82–86
53. Wenzel M, Hiscock JR, Gale PA (2012) Anion receptor chemistry: highlights from 2010. Chem Soc Rev 41:480–520
54. Gale PA, Quesada R (2006) Anion coordination and anion-templated assembly: highlights from 2002 to 2004. Coord Chem Rev 250:3219–3244
55. Bowman-James K, Bianchi A, García-Espaa E (2011) Anion coordination chemistry. Wiley-VCH, Weinheim
56. Berg JM (1995) Zinc finger domains: from predictions to design Acc Chem Res 28:14–19
57. Ashcroft FM (2000) Ion channels and disease. Academic, San Diego
58. Moss B (1996) A land awash with nutrients-the problem of eutrophication. Chem Ind 11:407–411
59. Asakura T, Kim SY, Morita Y, Ozawa M (2005) Reduction of pertechnetate in nitric extraction of rare metals for future reprocessing. J Nucl Radiochem Sci 6:267–269
60. Benner S (2010) Hydrology: anthropogenic arsenic. Nat Geosci 3:5–6
61. Frontera A, Quiñonero D, Deyà PM (2011) Cation-π and anion-π interactions. WIREs Comput Mol Sci 1:440–459
62. Giese M, Albrecht M, Krappitz T, Peters M, Gossen V, Raabe G, Valkonen A, Rissanen K (2012) Cooperativity of H-bonding and anion-π interaction in the binding of anions with neutral π-acceptors. Chem Commun 48:9983–9985
63. Watt MM, Zakharov LN, Haley MM, Johnson DW (2013) Selective nitrate binding in competitive hydrogen bonding solvents: do anion-π interactions facilitate nitrate selectivity? Angew Chem Int Ed 52:10275–10280
64. Gamez P, Mooibroek TJ, Teat SJ, Reedijk J (2007) Anion-π interactions involving π-acidic rings. Acc Chem Res 40:435–444
65. Ballester P (2008) Recognition of anions. Springer Berlin
66. Frontera A (2013) Encapsulation of anions: macrocyclic receptors based on metal coordination and anion-π interactions. Coord Chem Rev 257:1716–1727
67. Schneider HJ, Werner F, Blatter T (1993) Attractive interactions between negative charges and polarizable aryl parts of host-guest systems. J Phys Org Chem 6:590–594
68. Schneider HJ (1991) Mechanisms of molecular recognition—investigations of organic host guest complexes. Angew Chem Int Ed Engl 30:1417–1436
69. Hiraoka K, Mizuse S, Yamabe S (1987) High-symmetric structure of the gas-phase cluster ions $X^- \cdots C_6F_6$ (X = Cl, Br and I). J Phys Chem 91:5294–5297
70. Hiraoka K, Mizuse S, Yamabe S (1987) A determination of the stabilities and structures of F-(C_6H_6) and F-(C_6F_6) clusters. J Chem Phys 86:4102–4105
71. Bauzá A, Quiñonero D, Deyà PM, Frontera A (2013) Anion-π interactions in $[S_4N_3]^+$ rings. New J Chem 37:2636–2641
72. Galán-Mascarós JR, Slawin AMZ, Woollins JD, Williams DJ (1996) π-Facial interactions between Cl^- and $[S_4N_3]^+$. Polyhedron 15:4603–4605
73. Kim D, Tarakeshwar P, Kim KS (2004) Theoretical investigations of anion-π interactions: the role of anions and the nature of π systems. J Phys Chem A 108:1250–1258
74. Kim DY, Singh NJ, Lee JW, Kim KS (2008) Cyameluric acid as anion-π type receptor for ClO_4^- and NO_3^-: π-Stacked and edge-to-face structures. J Chem Theory Comput 4:1162–1169
75. Garau C, Frontera A, Quiñonero D, Ballester P, Costa A, Deyà PM (2003) A topological analysis of the electron density in anion-π interactions. Chem Phys Chem 4:1344–1348
76. Quiñonero D, Garau C, Frontera A, Ballester P, Costa A, Deyà PM (2002) Counterintuitive interaction of anions with benzene derivatives. Chem Phys Lett 359:486–492

77. Bauzá A, Deyà PM, Frontera A, Quiñonero D (2014) Substituent effects in cation-π interactions revisited: a general approach based on intrinsic properties of the arenes. Phys Chem Chem Phys 16:1322–1326
78. Wheeler SE, Houk KN (2010) Are anion/π interactions actually a case of simple charge–dipole interactions? J Phys Chem A 114:8658–8664
79. Garau C, Quiñonero D, Frontera A, Ballester P, Costa A, Deyà PM (2003) Dual binding mode of s-triazine to anions and cations. Org Lett 5:2227–2229
80. Garau C, Frontera A, Quiñonero D, Ballester P, Costa A, Deyà PM (2004) Cation-π versus anion-π interactions: energetic, charge transfer, and aromatic aspects. J Phys Chem A 108:9423–9427
81. Garau C, Quiñonero D, Frontera A, Ballester P, Costa A, Deyà PM (2003) Anion-π interactions: must the aromatic ring be electron deficient? New J Chem 27:211–214
82. Quiñonero D, Frontera A, Garau C, Ballester P, Costa A, Deyà PM (2006) Interplay between cation-π, anion-π and π-π interactions. Chem Phys Chem 7:2487–2491
83. Quiñonero D, Frontera A, Deyà PM (2008) High-level ab initio study of anion-π interactions in pyridine and pyrazine rings coordinated to Ag(I). Chem Phys Chem 9:397–399
84. Gural'skiy IA, Escudero D, Frontera A, Solntsev PV, Rusanov EB, Chernega AN, Krautscheid H, Domasevitch KV (2009) 1,2,4,5-Tetrazine: an unprecedented μ_4-coordination that enhances ability for anion-π interactions. Dalton Trans 38:2856–2864
85. Bauzá A, Quiñonero D, Deyà PM, Frontera A (2012) Tuning of the anion-π interaction. Theor Chem Acc 131:1219–1230
86. Mirzaei M, Eshtiagh-Hosseini H, Lotfian N, Salimi A, Bauzá A, Van Deun R, Decadt R, Barceló-Oliver M., Frontera A (2014) Syntheses, structures, properties and DFT study of hybrid inorganic-organic architectures constructed from trinuclear lanthanide frameworks and Keggin-type polyoxometalates. Dalton Trans 43:1906–1916
87. Estarellas C, Frontera A, Quiñonero D, Deyà PM (2008) Theoretical and crystallographic study of the dual σ/π binding affinity of quinolizilynium cation. J Chem Theory Comput 4:1981–1989
88. Quiñonero D, Frontera A, Escudero D, Ballester P, Costa A, Deyà PM (2007) Theoretical study of anion-π interactions in seven membered rings. Chem Phys Chem 8:1182–1187
89. Das A, Choudhury SR, Dey B, Yalamanchili SK, Helliwell M, Gamez P, Mukhopadhyay S, Estarellas C, Frontera A (2010) Supramolecular assembly of Mg(II) complexes directed by asociative lone pair-π/ π-π/ π-anion-π/ π-lone pair interactions. J Phys Chem B 114:4998–5009
90. Das A, Choudhury SR, Estarellas C, Dey B, Frontera A, Hemming J, Helliwell M, Gamez P, Mukhopadhyay S (2011) Supramolecular assemblies involving anion-π and lone pair-π interactions: experimental observation and theoretical analysis. Cryst Eng Comm 13:4519–4527
91. Fiol JJ, Barceló-Oliver M, Tasada A, Frontera A, Terrón A, García-Raso A (2013) Structural characterization, recognition patterns and theoretical calculations of longchain Nalkyl substituted purine and pyrimidine bases as ligands: on the importance of anion-π interactions. Coord Chem Rev 257:2705–2715
92. Wang DX, Wang MX (2013) Anion–π interactions: generality, binding strength, and structure. J Am Chem Soc 135:892–897
93. Schneebeli ST, Frasconi M, Liu Z, Wu Y, Gardner DM, Strutt NL, Cheng C, Carmieli R, Wasielewski MR, Stoddart JR (2013) Electron sharing and anion-π recognition in molecular triangular prisms. Angew Chem Int Ed 52:13100–13104
94. Chifotides HT, Giles ID, Dunbar KR (2013) Supramolecular architectures with π-acidic 3,6-bis(2-pyridyl)-1,2,4,5- tetrazine cavities: role of anion–π interactions in the remarkable stability of Fe(II) metallacycles in solution. J Am Chem Soc 135:3039–3055.
95. Bretschneider A, Andrada DM, Dechert S, Meyer S, Mata RA, Meyer F (2013) Preorganized anion traps for exploiting anion-π interactions: an experimental and computational study. Chem Eur J 19:16988–17000

96. Ballester P (2013) Experimental quantification of anion-π interactions in solution using neutral host-guest model systems. Acc Chem Res 46:874–884
97. Gil-Ramírez G, Escudero-Adán EC, Benet-Buchholz J, Ballester P (2008) Quantitative evaluation of anion-π interactions in solution. Angew Chem Int Ed 47:4114–4118
98. Adriaenssens L, Gil-Ramírez G, Frontera A, Quiñonero D, Escudero-Adán EC, Ballester P (2014) Thermodynamic characterization of halide-π interactions in solution using "two wall" aryl extended calix[4] pyrroles as model system. J Am Chem Soc 136:3208–3218
99. Chang KC, Minami T, Koutník P, Savechenkov PY, Liu Y, Anzenbacher P (2014) Anion binding modes in meso-substituted hexapyrrolic calix[4]pyrrole isomers. J Am Chem Soc 136:1520–1525
100. He Q, Han Y, Wang Y, Huang ZT, Wang DX (2014) Size-regulable vesicles based on anion-π interactions. Chem Eur J 20:7486–7491
101. Zhao Y, Domoto Y, Orentas E, Beuchat C, Emery D, Mareda J, Sakai N, Matile S (2013) Catalysis with anion-π interactions. Angew Chem Int Ed 52:9940–9943
102. Zhao Y, Beuchat C, Domoto Y, Gajewy J, Wilson A, Mareda J, Sakai N, Matile S (2014) Anion-π catalysis. J Am Chem Soc 136:2101–2111
103. Zhao Y, Sakai N, Matile S (2014) Enolate chemistry with anion-π interactions. Nature Comm 5:3911
104. Oubridge C, Ito N, Evans PR, Teo CH, Nagai K (1994) Crystal-structure at 1.92 Angstrom resolution of the RNA-binding domain of the U1a spliceosomal protein complexed with an RNA hairpin. Nature 372:432–438
105. Estarellas C, Frontera A, Quiñonero D, Deyà PM (2011) Anion-π interactions in Flavoproteins. Chem Asian J 6:2316–2318
106. Dong C, Flecks S, Unversucht S, Haupt C, van Pee KH, Naismith JH (2005) The structure of tryptophan 7-halogenase (PrnA) suggests a mechanism for regioselective chlorination. Science 309:2216–2219
107. Race PR, Lovering AL, Green RM, Ossor A, White SA, Searle PF, Wrighton CJ, Hyde EI (2005) Structural and mechanistic studies of *Escherichia Coli* nitroreductase with the antibiotic nitrofurazone. Reserved binding orientations in different redox states of the enzyme J Biol Chem 280:13256–13264
108. Bauzá A, Quiñonero D, Deyà PM, Frontera A (2013) On the importance of anion-π interactions in the mechanism of sulfide: quinone oxidoreductase. Chem Asian J 8:2708–2713
109. Cherney M, Zhang Y, Solomonson M, Weiner JH, James MNG (2010) Crystal structure of sulfide:quinone oxidoreductase from *Acidithiobacillus Ferrooxidans*: insights into sulfidotrophic respiration and detoxification. J Mol Biol 398:292–305
110. Pham VH, Yong JJ, Park SJ, Yoon DN, Chung WH, Rhee SK (2008) Molecular analysis of the diversity of the sulfide: quinone reductase (sqr) gene in sediment environments. Microbiol 154:3112–3121
111. Johnson E, Keinan S, Mori-Sánchez P, Contreras-García J, Cohen A, Yang W (2010) Revealing non-covalent interactions. J Am Chem Soc 132:6498–6506
112. Bauzá A, Quiñonero D, Deyà PM, Frontera A (2014) Long-range effects in anion-π interactions: their crucial role in the inhibition mechanism of *Mycobacterium Tuberculosis* malate synthase. Chem Eur J 20:6985–6990
113. McKinney JD, Höner zu Bentrup K, Muoz-Elías EJ, Miczak A, Chen B, Chan WT, Swenson D, Sacchettini JC, Jacobs WR Jr, Russell DG (2000) Persistence of *Mycobacterium Tuberculosis* in macrophages and mice requires the glyoxylate shunt enzyme isocitrate lyase. Nature 406:735–738
114. Krieger IV, Freundlich JS, Gawandi VB, Roberts JP, Gawandi VB, Sun Q, Owen JL, Fraile MT, Huss SI, Lavandera JL, Ioerger TR, Sacchettini JC (2012) Structure-guided discovery of phenyl-diketo acids as potent inhibitors of *M. Tuberculosis* malate synthase. Chem Biol 19:1556–1567

Chapter 17
A Survey of DNA–Protein π–Interactions: A Comparison of Natural Occurrences and Structures, and Computationally Predicted Structures and Strengths

Katie A. Wilson and Stacey D. Wetmore

Abstract Non-covalent interactions between DNA and proteins play critical roles in cellular functions, including DNA replication and repair. To gain an appreciation of the biomolecular components involved, several bioinformatics studies have data mined experimental X-ray crystal structures to identify close contacts between DNA and protein building blocks. These critical studies have revealed that DNA–protein non-covalent interactions include π–π, C–H · · · π, O–H · · · π, N–H · · · π or lone pair–π (X · · · π, X = O, N or S) contacts. Unfortunately, however, experimental structural data cannot provide information about the relative strength of biologically-relevant non-covalent interactions. Therefore, quantum mechanical calculations have been used to determine the stability of DNA–protein π–heterodimers, as well as the dependence of the interaction energy on changes in relative monomer orientations. In this light, the present review summarizes work done in the literature to characterize π–interactions between the DNA nucleobases (A, C, T and G) or deoxyribose moiety and cyclic (His, Phe, Tyr and Trp) or acyclic (Arg, Glu and Asp) amino acid side chains. Collectively, this body of work emphasizes the importance of DNA–protein π–interactions for providing stability to biomolecular complexes and driving key cellular functions.

17.1 Introduction

Non-covalent interactions are important in a wide variety of applications. For example, Tokay geckos use non-covalent interactions to walk up walls [1–3]. As a result, the Defense Advanced Research Project Agency in the United States of America has used the Tokay geckos as a model to develop paddles that allow a person to climb an eight-meter glass wall [4, 5]. Additionally, non-covalent interactions play microscopic roles in many biological processes, including those that involve interactions between DNA and proteins such as DNA repair and DNA replication.

S. D. Wetmore (✉) · K. A. Wilson
Department of Chemistry and Biochemistry, University of Lethbridge, 4401 University Drive West, T1K 3M4 Lethbridge, Alberta, Canada
e-mail: stacey.wetmore@uleth.ca

S. Scheiner (ed.), *Noncovalent Forces*, Challenges and Advances in Computational Chemistry and Physics 19, DOI 10.1007/978-3-319-14163-3_17

These non-covalent DNA–protein interactions can occur between two π–containing systems (denoted as π–π interactions) or between one π–system and another (conjugated or non-conjugated) molecular component (including C–H· · · π, O–H· · · π, N–H· · · π or lone pair–π (X· · · π, X = O, N or S) interactions).

DNA repair provides a textbook example of the wide function of DNA–protein π–π interactions. For example, the DNA glycosylases, which are the first enzymes involved in the base excision repair (BER) of non-bulky nucleobase damage [6], are known to use π–π interactions between DNA nucleotide substrates and the aromatic (or cyclic) amino acids (Phe, His, Trp and Tyr; Fig. 17.1). The best example of the contribution of π–π interactions to BER is provided by alkyladenine DNA glycosylase (AAG). Indeed, several π–interactions afforded by the AAG active site allow this enzyme to excise a wide variety of substrates (including both cationic and neutral nucleobases). Specifically, upon substrate binding, AAG inserts Tyr162 into the helix 3′ with respect to the site of the damaged nucleobase, and this residue forms π–π interactions that stabilize the DNA helix when the damaged base is flipped into the enzyme active site (Fig. 17.2a) [7]. In addition, simultaneous interactions between Tyr127, His136 and Tyr159 and the damaged nucleobase bind the substrate in the AAG active site (Fig. 17.2b) [8]. Notably, binding of labile cationic substrates in the AAG active site occurs due to stronger π_{cation}–π interactions than the π–π contacts formed with the neutral lesions. As a result, the π–π interactions between AAG and the substrate have been shown to be catalytic for the excision of neutral lesions, but anti-catalytic or non-catalytic for the removal of cationic lesions [8]. Although AAG solely relies on π–π interactions to facilitate DNA repair, other DNA repair enzymes (for example, uracil DNA glycosylase [9] and alpha-ketoglutarate-dependent dioxygenase [10]) use π–π interactions in conjunction with other types of non-covalent contacts to bind and excise damaged DNA. Furthermore, other enzymes are known to implement DNA-protein π–π interactions with π–containing (cyclic or acyclic) amino acids (Fig. 17.1) for catalysis (for example, aldehyde dehydrogenase [11], antibody interactions [12], tropinone reductase [13] and ketosteroid isomerase [14]).

DNA replication also exploits DNA–protein π–interactions. However, unlike the π–π contacts between two aromatic rings used by DNA repair enzymes, all classes of DNA polymerases invoke interactions between an amino acid π–system and deoxyribose. Specifically, C–H· · · π interactions between the sugar of the incoming deoxynucleoside triphosphate (dNTP) and a conserved Glu (A-family) or Phe/Tyr (B−, X−, Y− and RT-families) residue occurs during DNA replication (Fig. 17.2c) [15–17]. In fact, these so-called "steric gate" interactions are responsible for the selective incorporation of dNTP (DNA) over RNA nucleoside triphosphate (NTP) by a 1,000,000 [18] to 100 [19] ratio. Nevertheless, this conserved interaction between the deoxyribose and a π–containing amino acid was only recently acknowledged in the literature to be more than simply a steric interaction [20, 21]. This example illustrates that there are still many aspect of non-covalent DNA–protein contacts that remain to be explored.

DNA–protein π–π and sugar–π interactions are ideal for binding between biomolecules and catalysis of biological processes due to their significant strength, yet ability to readily degrade in order to continuously afford function to the protein.

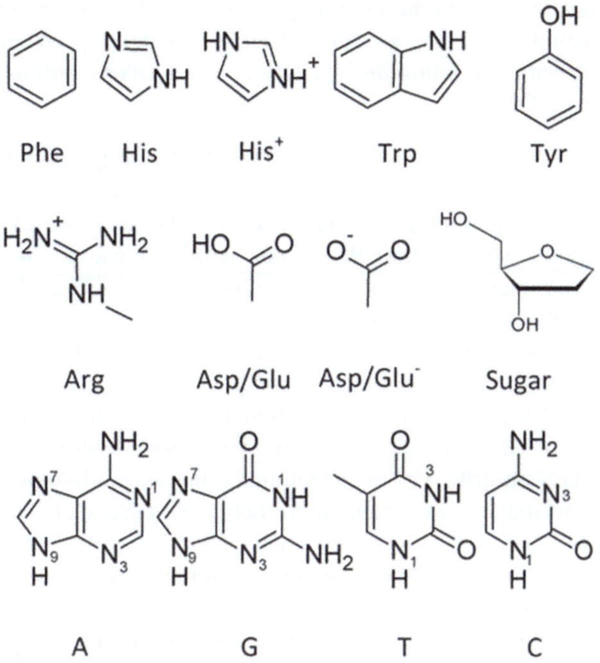

Fig. 17.1 Models of π–containing amino acids, the natural DNA nucleobases and DNA deoxyribose moiety (sugar) for which non-covalent interactions are discussed in the present review

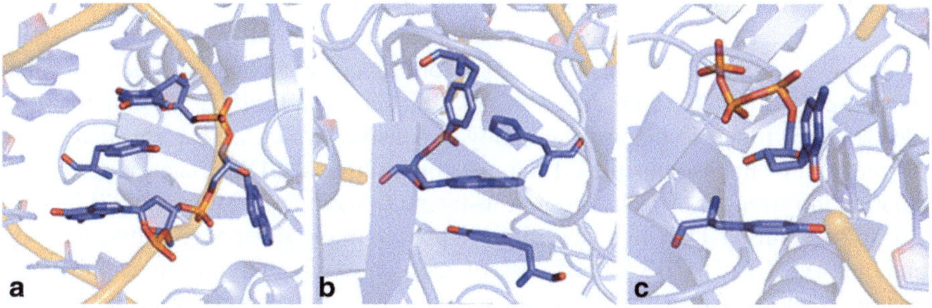

Fig. 17.2 AAG active site depicting the **a** interaction between Y162 and T8, and **b** π–π interactions with the (ethenoadenine) substrate, as well as **c** a representative sugar–π interaction between deoxyribose and Tyr12 in a polymerase active site

Experimental methods can provide information about the abundance and structure (NMR and X-ray crystallography), as well as catalytic contribution (mutagenesis and kinetic studies), of DNA–protein contacts. Unfortunately, however, it is difficult to gain information about the relative strength of discrete DNA–protein interactions from experimental data. In contrast, computational chemistry is an ideal tool for studying these non-covalent interactions since this approach can provide atomic level details about the interactions, including the strength of intermolecular forces acting between the monomers and the effects of geometrical variables on the interaction energy. As a result, quantum mechanical calculations are commonly used to determine the binding strength (ΔE) of DNA–protein π–interactions according to Eq. 17.1

$$\Delta E = E^{dimer} - E^{aa} - E^{nt} \tag{17.1}$$

In this equation, E^{dimer} stands for the electronic energy of the interaction, while E^{aa} and E^{nt} stand for the electronic energy of the amino acid and deoxyribose or nucleobase moiety of the nucleotide, respectively. Indeed, although experimental data tells us about the structures of π–π interactions in nature, quantum mechanical calculations are the only method that can be used to determine their energetic contribution to DNA–protein binding, stability and function.

There are many computational methods that can be used to gain information about DNA–protein π–interactions. Unlike hydrogen bonding, which can be accurately described by many computational methodologies due to the large electrostatic contribution, the interaction energy of π–contacts has a large dispersion component and therefore requires methods that recover a large portion of the total electron correlation. As a result, early studies on π–interactions implemented high-level quantum mechanical techniques, which require an abundance of computer resources. The most widely used and successful *ab initio* approach was MP2/6-31G*(0.25), which replaces the standard d-exponent (0.8) with a value of 0.25 to obtain accurate interaction energies in comparison to gold-standard CCSD(T)/CBS results at a reduced computational cost. Nevertheless, the accuracy of MP2/6-31G*(0.25) arises due to a cancelation of errors [22, 23]. More recently, advances in density functional theory (DFT) have produced methods that account for dispersion interactions [24], but are much less computationally expensive, which allows for analysis of π–interactions in biomolecules. The most commonly used DFT functionals for calculating the strength of DNA–protein π–interactions include M06-2X, B3LYP-D3, B97-B3 and ωB97-D, which have all been proven to provide energies that are comparable to CCSD(T)/CBS [20, 25–29]. In addition to quantum mechanical methods that allow the structure and energy of π–contacts to be characterized, the electron density obtained from Quantum Theory of Atoms in Molecules (QTAIM) can provide complementary information about the nature of non-covalent interactions [30].

The goal of this review is to summarize recent work on biologically important DNA protein π–interactions. Specifically, the occurrence, composition, structure and strength of π–interactions between π–containing amino acids (Trp, Phe, His, Tyr, Arg, Glu and Asp) and the natural DNA nucleobases (A, T, G and C) or deoxyribose sugar as determined according to experimental crystal structures of DNA–protein complexes will be examined. Additionally, the effects of geometrical parameters that dictate relative monomer orientations and the optimal dimer arrangements that have been determined using quantum mechanical techniques will be reviewed for the most well studied interactions to date, including contacts between the aromatic amino acids (Trp, Phe, His, and Tyr) and the DNA bases, and the acyclic amino acids (Arg, Glu and Asp) and cytosine. These 7 distinct π–containing amino acids also

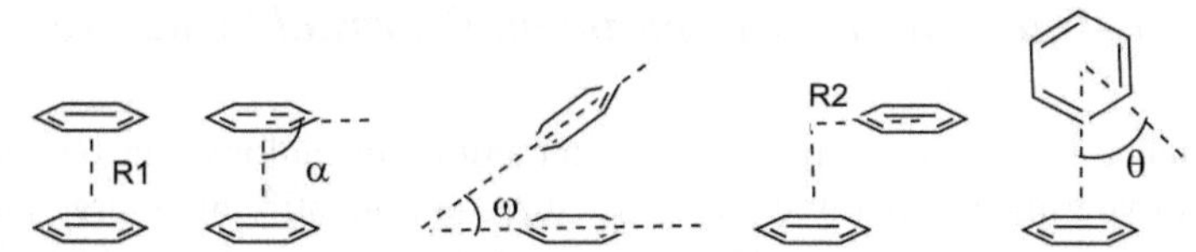

Fig. 17.3 Definitions of key geometrical parameters that define the relative monomer orientations in DNA–protein π–π interactions

reveal the effects of charge (anionic, cationic and neutral), cyclic versus acyclic π–system, and π–system size on the frequency, structure and strength of DNA–protein π–interactions in nature, as well as on the structure and strength of the optimum orientation of DNA–protein π–interactions. The results from purely computational studies will be critically compared to interactions found in nature. Finally, the effects of the biological environment, including biological backbones, solvent and additional non-covalent interactions, will be presented, and future perspectives for the field will be summarized.

17.2 Interactions Between Aromatic Amino Acids and Nucleobases

As discussed in the Introduction, the presence of π–π interactions between the aromatic amino acids (Trp, Tyr, Phe and His) and the DNA nucleobases has been accepted to be important in critical bioprocesses, including DNA repair. Furthermore, the function of interactions between other non-aromatic π–containing amino acids (Arg, Glu, and Asp) and the DNA nucleobases have also been more recently acknowledged. In general, regardless of the nature of the amino acid involved, the structural differences between nucleobase–amino acid π–π interactions are defined on the basis of four variables, namely R1, α, R2, ω and θ (Fig. 17.3). R1 represents the vertical separation between the center of mass of the amino acid and the center of mass of the nucleobase, α is the twist of the amino acid with respect to the nucleobase (defined according to the center of mass), R2 symbolizes the horizontal displacement between the center of masses of the two π–systems, and ω is the angle between the planes of the amino acid and nucleobase rings. Notably, ω defines whether the monomer relative orientation corresponds to a stacking (when the amino acid and nucleobase are parallel or $\omega \approx 0°$) or T-shaped (when the amino acid and nucleobase are perpendicular or $\omega \approx 90°$) binding arrangement. A final variable is also important for defining T-shaped interactions, namely θ or the edge directed toward the π–system. Specifically, either the edge of the amino acid can interact with the face (π–system) of the nucleobase or the edge of the nucleobase can interact with the face of the amino acid. The upcoming discussion will first highlight the occurrence, structure and strength of π–interactions identified in experimental crystal structures of DNA–protein complexes, and then describe how the calculated strength of these important π–π interactions depends on the R1, α, R2, ω and θ variables.

17.2.1 π–Interactions in Experimental Crystal Structures

The existence of specific non-covalent interactions in biology can be verified by searching experimental X-ray crystal structures and identifying a large number of contacts prior to undertaking in-depth computational investigations of specific interactions. In order to appreciate the persistence of DNA–protein π–π interactions in nature, several groups have searched the protein data bank (PDB) to determine the frequency of different types of non-covalent DNA–protein contacts, including π–π interactions between amino acids and DNA nucleobases. The first studies of this kind were restricted by the absence of accurate X-ray crystal structures due to limitations in the available experimental technologies at the time [31–34]. With recent advances in experimental methods and equipment, a large number of high-resolution crystal structures of DNA–protein complexes became available for analysis [35–38]. However, due to differences in search criteria used and relatively undefined geometric definitions of van der Waals interactions, these studies yielded starkly different conclusions. For example, one paper reports that van der Waals interactions are more common than hydrogen bonding [35], while another study determined that hydrogen bonding is more common than van der Waals interactions [37]. Nevertheless, these early works agree that van der Waals interactions are common and compose more than 30 % of all non-covalent contacts identified.

Aside from broadly determining the relative frequency of different types of non-covalent interactions, other studies have conducted PDB searches to specifically investigate π–π interactions between the aromatic amino acids and DNA nucleobases. The first such study considered π–π interactions between adenosine 5′-triphosphate (ATP) and the aromatic amino acids, and found that, although hydrogen bonding occurs 2.7 times more often than π–π interactions, π–π interactions are vital for substrate binding [39]. A more recent investigation considered the π–π interactions between the aromatic amino acids (Trp, Tyr, Phe and His) and all four natural nucleobases (A, T, G and C) in 141 DNA–protein complexes [40]. Unfortunately, these two studies lead to different overall trends in terms of which amino acid preferentially interacts with A. Although this discrepancy may be an artefact of searching different experimental structures, distinctive (distance only) search criteria were also used in each study. Furthermore, many other parameters besides the separation distance contribute to the characteristics of π–π interactions. In fact, due to the distance only search criteria, not all previously identified non-covalent interactions correspond to DNA–protein π–π interactions. Indeed, closer examination reveals some of the monomer separations that fall within the distance cut-off represent hydrogen-bonding or sugar–π interactions [20, 21]. Therefore, despite the usefulness of previous work for verifying the presence of close contacts between DNA and protein π–residues, none of these studies truly determined the frequency or clarified the structure of these π–π interactions.

Although database tools have been more recently created to identify possible DNA–protein π–interactions in experimental crystal structures [41–47] quantum chemical calculations have also been used to gain information about the strength of

DNA–protein π–π interactions in nature. For example, Mao et al. used data mining and MP2/6-311+G* calculations to investigate aromatic amino acid π–π contacts with ATP in 68 crystal structures of adenylate-binding proteins [39]. The corresponding interaction energies were determined to be substantial, and the work thereby established the molecular basis for adenine recognition in proteins. Subsequently, the Rooman group used MP2 to considered 89 interactions between His and Phe, Tyr, Trp or A observed in X-ray structures of proteins and protein-ligand complexes in several environments (gas phase, water, acetone, THF and CCl_4) [48]. This analysis concluded that the strength and frequent occurrence of His interactions in active sites has implications for their importance in biological processes. In 2007, Baker and Grant investigated 141 DNA–protein complexes for interactions between the DNA nucleobases and Tyr, Phe, His and Trp [40]. Subsequent MP2/6-31G calculations helped verify the importance of Phe interactions in transcription. The Tschumper group further considered A:Phe binding by building 26 9-methyladenine–toluene dimers from protein/ligand crystal structures and determining the binding strengths with highly correlated methods (CCSD(T) and RI-MP2) [49]. Uniquely, MP2 optimizations of the nucleobase–amino acid complexes were also implemented, which lead to six distinctive dimers. In a follow-up study by the same group, 20 model A:Tyr systems were fully optimized to 11 unique dimers, which were found to have binding strengths nearly equivalent to the A:Phe systems [50]. These representative examples of combined bioinformatics and quantum chemical calculations were critical contributions to the literature.

Despite the large number and strength of interactions between the DNA nucleobases and the aromatic amino acids in nature, other amino acid side chains also have π–systems and form π–contacts with nucleobases. Indeed, small distances between the non-aromatic π–side chains of arginine, aspartic/glutamic acid or aspartate/glutamate and the DNA nucleobases have been identified in experimental X-ray crystal structures [51–56]. More specifically, the structure of stair motifs, which are formed upon binding of a charged amino acid side chain to two neighbouring nucleobases, have been assessed by critical analysis of 52 X-ray crystal structures, and most complexes were determined to involve G and Arg [55]. This study was followed by MP2/6-31G*(0.2) calculations on a wide variety of stair motifs [57, 58], which verified their stability. Alternatively, while attempting to quantify the role of cation–π interactions at the interfaces of DNA–protein complexes, Wintjens et al. identified a large number of Arg contacts, comprising more than half of the cation–π interactions found [54]. Furthermore, the most frequent nucleobase–amino acid pairing was determine to occur between G and Arg, which was also the most stable pair according to MP2 calculations [51, 52]. In contrast, a search of cation–π interactions in proteins bound to ligands revealed a number of contacts between DNA nucleobases and positively charged side chains, with the most frequent pairing occurring between A and Arg [53]. Similarly, a study of 55 cation/amino–π motifs found in X-ray structures that involve A moieties and charged (Arg, Lys, Asn, or Gln) side chains reveals that the Arg:A pair is the most favourable according to MP2 binding strengths and the most commonly observed [56]. Finally, PDB data mining and molecular dynamics simulations verify that anionic interactions, including those with Asp and Glu side chains, mediate protein–nucleic acid interactions, albeit through an abundance of hydrogen–bonding interactions [59]. It should be noted that the Arg, Glu and Asp series is also of interest from a fundamental perspective since the variety of charges these side chains adopt depending on the biological environment permits evaluation of differences in the nature of π–π, π_{cation}–π and π_{anion}–π interactions.

Overall, although previous studies have been useful for verifying the existence and determining the importance of π–interactions, differences in the crystal structures considered, types of contacts investigated, search methods implemented and quantum chemical calculations used for analysis makes it difficult to compare between the studies. Therefore, in the discussion below, we will primarily summarize the most recent work from our group that systematically considered high-resolution X-ray crystal structures of DNA–protein complexes published before January 2014 in the PDB [20, 21]. A unique aspect of our work is the inclusion of a visual inspection step to definitively characterize the identified contacts between π–containing amino acids (Phe, Trp, Tyr, His, Glu, Asp or Arg) and the natural DNA nucleobases (A, T, G or C).

17.2.1.1 Occurrence of π–Interactions in Experimental Crystal Structures

952 DNA–protein π–π interactions were found in the 672 crystal structures searched by our group. It has been determined that 50 % of these 672 structures contain a π–π interaction involving Trp, Tyr, His or Phe, while 25 % of the searched PDB structures contained a π–π interaction involving Arg, Glu or Asp. Among the 331 structures with a π–π interaction involving Trp, Tyr, His or Phe, 61 % contain only one contact. Similarly, of the 166 structures with a π–π interaction involving Arg, Glu or Asp, 73 % only contain one contact. These DNA–protein interactions occur in a wide variety of protein complexes, supporting proposals that these non-covalent contacts are important for many biological processes. However, the percentage distribution in the types of proteins the interactions were found in is equal to that for the types of proteins searched; therefore, π–π interactions are unlikely to preferentially occur in a particular protein class.

17.2.1.2 Composition of π–Interactions in Experimental Crystal Structures

Among the interactions identified in nature, 75 % involve a cyclic aromatic amino acid (Fig. 17.4a). When the amino acids involved in the contacts are considered, the majority of interactions occur with Phe (33 %) followed by Tyr (22 %) and Arg (21 %), while all other amino acids investigated (Trp, His, Glu and Asp) make up less than 15 % of the contacts found. When the interacting nucleobase is considered, the majority of contacts occur with T (30 %), while A, C, and G have a similar number of interactions (23 %, Fig. 17.4b). However, when the composition with respect to the nucleobase is divided according to whether the interaction occurs with a cyclic or acyclic amino acid, T forms the most contacts with Phe, Tyr, Trp or His, but the

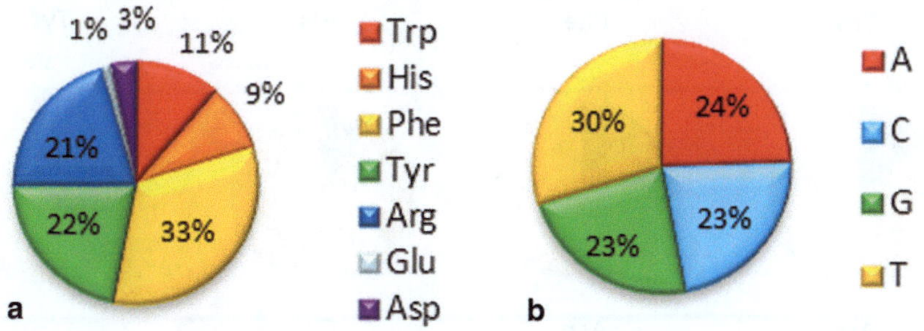

Fig. 17.4 Proportions of **a** amino acids and **b** nucleobases observed in nucleobase-amino acid π–π interactions found in X-ray crystal structures of DNA–protein complexes

least with Arg, Glu or Asp, while G has the largest number of contacts with Arg, Glu or Asp, but the least with Phe, Tyr, Trp or His. Notably, A and C have similar trends with respect to the cyclic and acyclic amino acids, each forming about 20–30 % of the interactions.

17.2.1.3 Structure of π–Interactions in Experimental Crystal Structures

The interactions found in nature adopt many different relative orientations of the two residues involved. When the closest heavy atom distance between the monomers is considered, Arg adopts the largest distances among the amino acids (up to 4.8 Å), while Asp forms the closest interactions, with separation distances <2.8 Å. Overall, the majority of contacts have separation distances between 3.0 and 4.0 Å, with 45 to 90 % of the interactions for each amino acid ranging between 3.0 and 3.4 Å (Fig. 17.5). Interestingly Glu has the smallest proportion of structures between 3.0 and 3.4 Å (45 %), and the largest number of structures with a separation distance >4.0 Å (22 %). Notably, Arg, His and Glu are the only amino acids that never adopt a separation distance greater than 4.0 Å. Conversely, Glu and Phe never adopt a separation distance less than 3.0 Å. Regardless, there is no particular trend for the separation distance with respect to the nucleobase. The range of distances adopted in the crystal structures is indicative of possible interaction strengths, where interactions occurring closer than the van der Waals radius of the atoms involved will be repulsive, while interactions that are too far away will not be stabilizing.

The DNA–protein π–π interactions found in nature adopt a variety of tilt angles, which correspond to stacked ($\omega = 0$–$20°$), inclined (20–70°) or T-shaped ($\omega = 70$–$90°$) orientations. In terms of the DNA nucleobases, all preferentially adopt the stacked orientation. Additionally, Phe, Tyr, Trp, His and Arg are more likely to appear in a stacked orientation (47–89 %) than an inclined or T-shaped orientation with respect to a DNA nucleobase (less than 31 %, Fig. 17.6). Conversely, potentially anionic amino acids (Glu and Arg), which form relatively few interactions, never

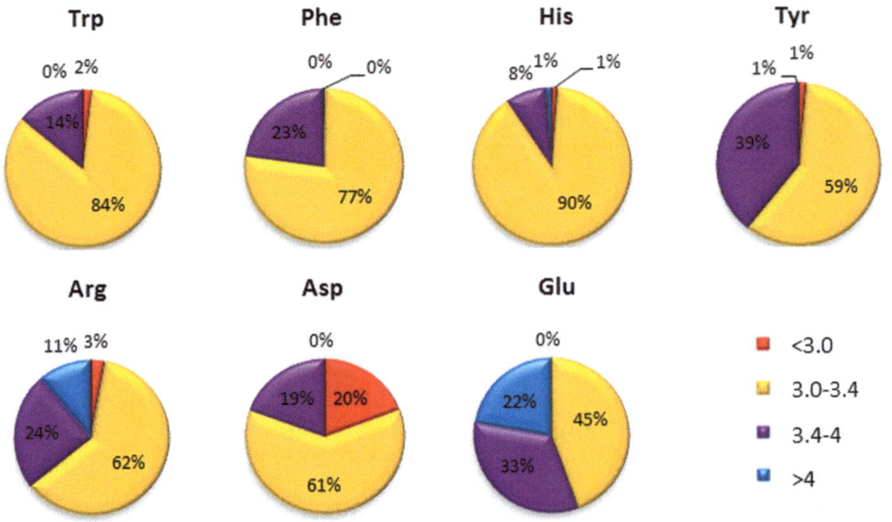

Fig. 17.5 Occupancies of different distances (Å) observed in nucleobase–amino acid π–π interactions found in X-ray crystal structures of DNA–protein complexes

adopt a tilt angle less than 50°, and therefore are most commonly found in a T-shaped orientation (56–70 %). These features thereby highlight the distinct nature of π_{anion}–π and π_{cation}–π interactions [40, 60].

Each inclined and T-shaped contact was visually inspected to identify the interacting edge (atoms interacting with the π–system). For Trp, His, Phe and Tyr side chains, amino acid–edge dimers [170] are more common than nucleobase–edge dimers [84]. Among the amino acid–edge dimers, 56 % involve a bridged (two H atoms directed towards the π–ring) edge of Phe (Fig. 17.7a). Conversely, the nucleobase edge–dimers are more structurally diverse, with the most prevalent cytosine–edge occurring when two CH groups (C5 and C6) are directed towards the π–ring (18 %, Fig. 17.7b). Conversely there are less amino acid–edge dimers (31) than nucleobase–edge dimers (114) for the acyclic amino acids. The most common amino acid–edge for the acyclic amino acids involves two NH groups (Nϵ and Nη) of Arg (26 %, Fig. 17.7c), which are the most acidic hydrogens. The most common nucleobase–edge interactions with the acyclic amino acids involve the G edge that contains CH (C8) and N (N7) groups (16 %, Fig. 17.7d). Notably, there is larger diversity in the nucleobase edges employed in DNA–protein T-shaped interactions, but a preference for particular edges in the amino acid edge dimers, partly due to the greater chemical diversity of nucleobase edges available for π–interactions.

17.2.1.4 Strength of π–Interactions in Experimental Crystal Structures

High-level quantum mechanical calculations (M06-2X/6-31G+(d,p)//MP2/6-31G(d)) show that DNA–protein π–π interactions adopt a wide variety of interaction

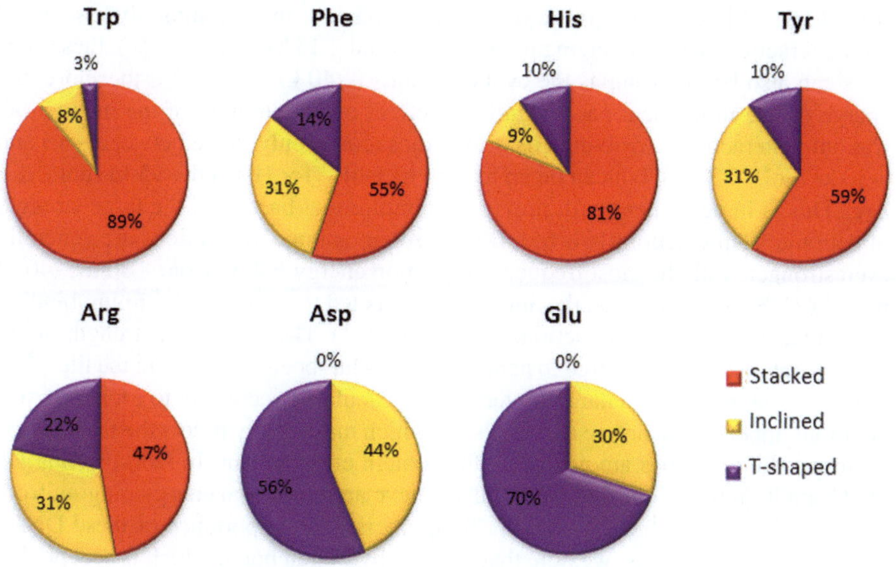

Fig. 17.6 Occupancies of different binding orientations of nucleobase–amino acid π–π interactions found in X-ray crystal structures of DNA–protein complexes

Fig. 17.7 Most common T-shaped edges for **a** Phe (3I0X; F179:A9) **b** C (3LDY; Tyr35:C5) **c** Arg (3PX6; Arg472:C203), and **d** G (3U6Q; Arg264:G9) found in experimental X-ray crystal structures

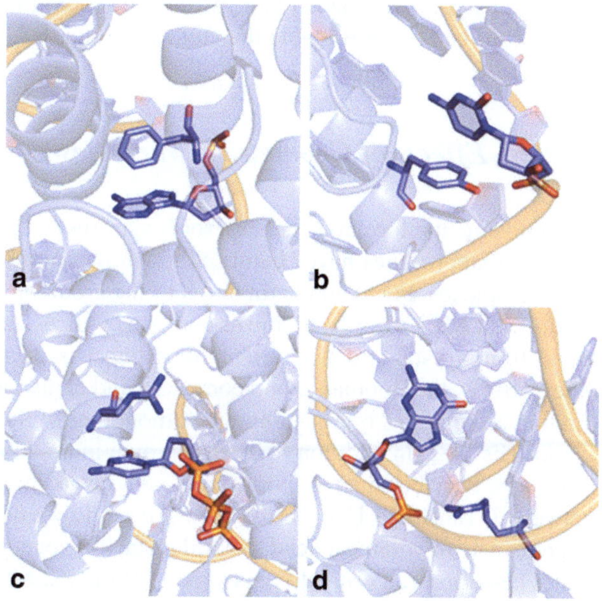

energies and can provide stability to DNA–protein complexes. Specifically, contacts with Trp, (neutral) His, Phe and Tyr most commonly have binding strengths between -20 and -25 kJ mol^{-1} (25 %), but can have interaction energies up to −40 kJ/mol

(Fig. 17.8). While the neutral acyclic interactions are most commonly associated with interaction energies ranging between -10 and -15 kJ mol^{-1} (26 %), these contacts can also be as strong as the cyclic variants (-40 kJ mol^{-1}). Furthermore, the charge of the amino acid has a significant effect on the strength of the interaction. Cationic interactions involving His or Arg most commonly range between -15 and -30 kJ mol^{-1} (32 %), but can be up to -50 kJ mol^{-1} for His and -95 kJ mol^{-1} for Arg. This is in agreement with previous literature that has noted the large strength of cationic π–interactions [61–65]. DNA–protein anionic interactions are generally even stronger, with the most frequent interaction energy falling from -30 to -40 kJ mol^{-1} (21 %). Nevertheless, the anionic contacts led to a similar maximum binding strength as the cationic interactions (-95 kJ mol^{-1}). The significant strength of anionic π–interactions compared to neutral contacts has been discussed in the literature [64, 66]. Notably, many charged interactions calculated according to monomer orientations appearing in nature are repulsive, which may simply reflect the fact that the corresponding residues are neutral in the protein environment. In fact, the charged His/Asp/Glu interactions that are repulsive have an interaction energy ranging from -10 to -20 kJ mol^{-1} when neutral. To appreciate the importance of these DNA–protein π–π interactions, we note that the A:T hydrogen bond, which is accepted to be an important interaction in nature for the stabilization of DNA duplexes, has been calculated to be approximately -70 kJ/mol at the gold-standard CCSD(T)/CBS level of theory [67]. Thus, neutral DNA–protein π–π interactions approach the strength of the A:T hydrogen-bonded pair, while even more impressively, the charged π–π interactions can be more stable than the A:T dimer.

17.2.2 Effects of Geometrical Variables

To complement information gained from experimental X-ray crystal structures, computational approaches have been used to study a number of isolated DNA–protein π–π complexes. For example, in terms of the aromatic amino acids, the Tschumper group used MP2 to fully optimize complexes of A stacked with Phe and Tyr initiated from crystal structure orientations (discussed in the Introduction) [49, 50]. Alternatively, the Rooman group compared and contrasted the stacked and T-shaped binding orientations by scanning the A monomer of fixed geometry with respect to Phe or His [48]. In addition to A, pyrimidine interactions with the aromatic amino acids have been considered. For example, Cysewski investigated stacked complexes between C or U and His, Phe, Tyr or Trp, and revealed that considerable degrees of geometric freedom can lead to similar stacking energies [68]. The U:Phe stacked dimer has also been critically analyzed using MP2, M05-2X and M06-2X coupled with 6-31G** and full optimizations, as well as AIM, by Ebrahimi and co-workers [69]. In terms of nucleobase contacts with the acyclic π–containing amino acids, relatively little computational work has been dedicated to characterizing the dimer potential energy surfaces, with Cysewski having considered the dimers between C or U and Arg [68]. Aside from looking at interactions with the natural bases, contacts

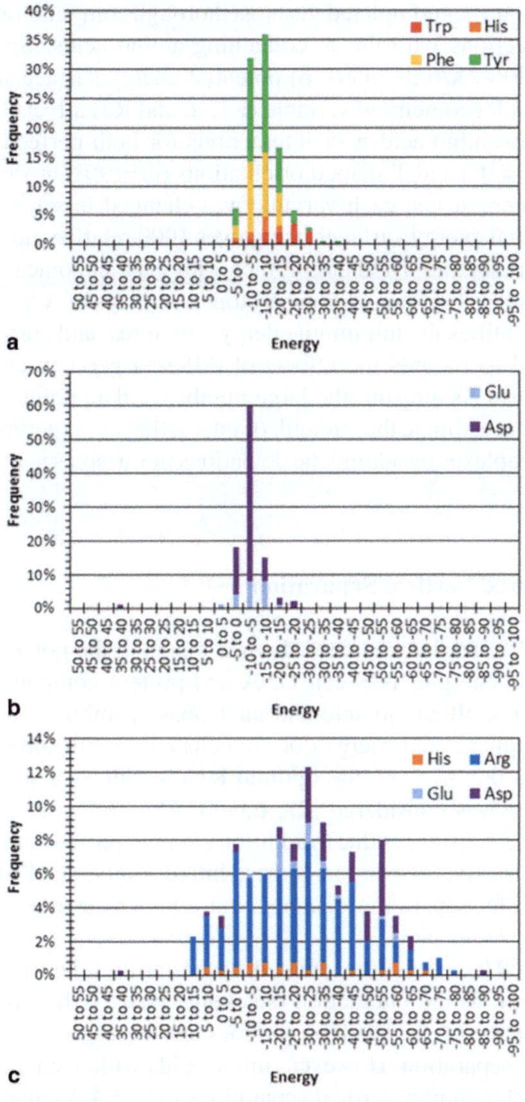

Fig. 17.8 Binding strengths for π–π interactions involving the **a** neutral (cyclic) aromatic, **b** neutral acyclic, and **c** charged amino acids in experimental X-ray crystal structure geometries

between non-canonical bases and the aromatic amino acids have also been studied using computational chemistry. For example, interactions involving damaged nucleobases have been investigated, such as inosine paired with Phe [70] and methylated bases with all four aromatic amino acids [71–73].

To date, our group has completed the most thorough computational studies of DNA nucleobase interactions with the π–containing amino acids. Specifically, MP2/6-31G*(0.25) or M06-2X/6-31+G(d, p) potential energy surface scans were used to determine how each geometrical variable (R1, α, and R2) affects the binding energy of the nucleobase–amino acid π–π interactions for both perfectly stacked binding arrangements ($\omega = 0°$) and T-shaped orientations ($\omega = 90°$) at various θ values [28, 63, 74, 75]. In these scans, each variable was changed in set intervals to map the interaction potential energy surfaces using over 1000 relative monomer orientations for each Phe, Trp, Tyr, or His heterodimer with each canonical (DNA and RNA) nucleobase, as well as between various protonation states of Arg, Glu, or Asp and C. This approach identifies the minimum energy structures, and corresponding binding strengths, as well as records the effects of different geometrical variables on the interaction energies. As a result, the large number of data points gained from these scans can be used to estimate the strength of interactions in experimental structures of DNA–protein complexes by adding the deviations in variables to the global minimum values [75].

17.2.2.1 Preferred Vertical Separation

The first geometric variable considered when mapping the potential energy surface for the interaction energies between DNA and protein components is the vertical separation (R1). For all amino acid and nucleobase combinations investigated, the magnitude of the interaction energy does not change significantly with small spatial deviations (0.1 to 0.2 Å) from the optimal R1 regardless of whether a stacked or T-shaped orientation is considered [28, 63, 74, 75]. However, the binding energy decreases at short R1 distances due to additional repulsion and at large R1 separations due decreased attractive dispersion forces. Interestingly, stacked dimers typically have a shorter prefer separation distance, but a larger dependence on R1, than T-shaped dimers.

The preferred R1 can also be divided based on the identity of the interacting amino acid, as well as the interaction (stacking versus T-shaped) classification. In contrast, the identity of the nucleobase does not have a consistent effect on the preferred vertical separation. However, amino acids with a smaller π–system (Arg, His, Asp/Glu) prefer shorter vertical separations (3.2–3.4 Å) than amino acids with larger, cyclic π–systems (Phe, Tyr or Trp; 3.4–3.6 Å). Interestingly, cationic His consistently leads to shorter vertical separation distances than neutral His [63], while there is no specific trend in the effect of a negative charge on Asp/Glu on the preferred vertical separation distance [28].

17.2.2.2 Preferred Angle of Rotation

The angle of rotation (α) influences the interaction energy to a greater extent than R1 and R2 for all nucleobase–amino acid dimers expect those involving cationic

His [63]. For stacking interactions, changing the angle of rotation increases the electrostatic interactions in the system by anti-aligning the monomer dipole moment vectors and/or decreases the steric repulsion by removing close contacts between the monomers [28, 63, 74, 75]. Indeed, the dependence of the binding strength on the angle of rotation increases as A < T < G < C for the nucleobases and as Phe < Tyr < Trp < His for the neutral cyclic amino acids, which correlates with the dipole moments of the monomers. Furthermore, the dependence on α is largest for the anionic systems, with the dependence being up to 20, 15 and 32 kJ/mol for neutral, cationic and anionic stacked systems, respectively.

For T-shaped DNA–protein dimers, the dependence of the interaction energy on α is related to the strength of secondary intermolecular hydrogen bonds, the alignment of acid/base sites of the monomers and/or steric repulsions between monomers. Specifically, when secondary hydrogen bonding is weaker (stronger) than the dipole–dipole stacking interaction, then the dependence on α is less (more) than for the stacked arrangement between the same amino acid and nucleobase. For the neutral cyclic amino acids, the dependence on α ranges between 1 and 33 kJ/mol regardless of whether an amino acid edge or nucleobase edge is involved. Conversely, the interaction energy changes more with α (by up to 13 kJ/mol) for cationic His T-shaped than stacking contacts, while cationic Arg dimers exhibit the same α dependence regardless of the binding orientation. Notably, due to greater symmetry, rotation about α does not have as large of an effect on Arg compared to His. As for the stacking interactions, the dependence on α is generally stronger for anionic C:Asp/Glu dimers (up to 32 kJ/mol).

17.2.2.3 Preferred Horizontal Displacement

For all neutral systems, the original horizontal displacement aligns the center of masses of the monomers in both the T-shaped and stacked orientations. Upon calculating the preferred R2, only small changes relative to the initial position occur in most dimers. Furthermore, for the neutral dimers, the R2 shift leads to very small changes in the interaction energy (generally less than 2 kJ/mol). In cases where the R2 shift affects the interaction energy, the corresponding change in the relative monomer alignment generally reduces steric repulsion, increases the π–π overlap or aligns the acidic/basic regions in the monomers to maximize the electrostatic contribution, with the latter being especially true for T-shaped dimers. For charged interactions, the horizontal displacement tends to increase the interaction energy to a greater extent (by up to approximately 15 kJ/mol). Furthermore, for charged dimers, the monomers shift to align the electropositive and electronegative regions of the residues. Most importantly, R2 can affect the binding energy of heterodimers involving cationic systems more than any other geometrical variable.

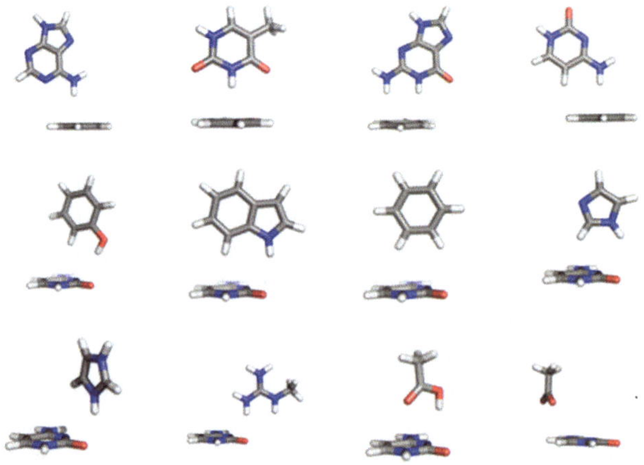

Fig. 17.9 Binding arrangements corresponding to the strongest DNA–protein T-shaped edge interactions as determined through quantum mechanical calculations

17.2.2.4 Preferred T-Shaped Edge

The structure and interaction energy of DNA–protein T-shaped dimers exhibit a large dependence on the interacting monomer edge. However, some overall trends in the data are evident. For example, monomers with large dipole moments, and therefore a large variation in the acid/base properties of the edges, show the greatest dependence on θ. Furthermore, the strongest neutral and anionic T-shaped interactions involve the monomer edge with the largest acidity. However, for the cationic systems, the most stable T-shaped interactions occur with the monomer edge with the greatest basicity. Due to greater overlap between π–systems, stronger interactions generally occur when the monomer edge is a bridge, which involves a bond or two atoms directed toward the π–ring, rather than single atom interactions. An exception to this rule occurs for neutral Asp/Glu heterodimers, which have the smallest π–systems of the amino acids considered. Notably, edges that give rise to lone pair–π interactions with a carbonyl oxygen atom are unstable, while lone pair–π interactions with nitrogen atoms are stable by up to -23 kJ/mol. For nucleobase–edge dimers, the glycosidic bond is generally the preferred edge, but this is not biologically relevant for most proteins that process DNA strands rather than nucleobase components. Therefore, the amino group of A (N6) or C (N4), the NH groups at N1 and N2 for G, and the NH group at N3 for T are the most preferred edges (Fig. 17.9). For amino acid–edge dimers, the preferred edges contains a CH (Cζ) and N or NH group (Nϵ or Nδ) (neutral) His, NH (Nϵ) or NH (Nϵ) and CH (Cζ) for (cationic) His, two CH groups for Phe, and OH and CH group for Tyr and a NH and CH group for Trp, a OH group for neutral Asp/Glu and the edge of the carboxylic group for anionic Asp/Glu and two NH groups (Nϵ and Nη) for Arg (Fig. 17.9).

17.2.2.5 Energetics of Preferred Orientation

Although the above discussion provides information about the dependence of the binding strength on the relative monomer orientation, an important remaining question is: what is the maximum interaction energy for a particular nucleobase–amino acid pairing? The data obtained from the potential energy surface scans gives information about the maximum stability a particular pairing can provide to a DNA–protein complex. The interaction strengths for the neutral cyclic amino acids range from −20 to −43 kJ/mol when in the stacked orientation, with two thirds of the contacts possessing interaction energies between −25 and −35 kJ/mol. Furthermore, the T-shaped interactions for the neutral cyclic amino acids range from −12 to −50 kJ/mol, with the T-shaped interactions involving a nucleobase edge being stronger than interactions involving an amino acid edge. Furthermore, the most favourable nucleobase–edge interactions are just as strong as or stronger than the stacking interactions between the same nucleobase–amino acid combination.

When the contacts are divided according to the interacting nucleobase, both stacking and T-shaped contacts are similar in magnitude. As a function of the amino acid, the strongest stacking and T-shaped interactions increase as Phe $<<$ Tyr $\sim$ His $<$ Trp due to the relative magnitude of the dipole moments and π–systems, as well as the strength of secondary intermolecular contacts between the amino acid or nucleobase edge in the T-shaped interactions. When cationic His is considered, the stacked dimers have binding strengths ranging from −35 to −69 kJ/mol, which is an increase of 130–223 % compared to the neutral His dimers. When cationic His adopts a T-shaped orientation, the interaction energies range from −20 kJ/mol to −105 kJ/mol, which is an increase of 186–332 % from neutral His dimers and larger than the increase for stacked dimers. Notably, the increase in the interaction energy upon protonation of His is due to increased electrostatics and is the greatest for His edge dimers.

The energetics for the acyclic amino acid π–systems with C are similar in magnitude to those for the cyclic systems with the same charge. In particular, the stacking interactions are up to −26 kJ/mol, −57 kJ/mol and −48 kJ/mol for the neutral, cationic and anionic acyclic systems, respectively. The interaction energies for these contacts are within 4 kJ/mol of those for contacts between C and His regardless of (neutral or cationic) charge. For the T-shaped dimers, the interactions are up to −40 kJ/mol, −91 kJ/mol and −99 kJ/mol for the neutral, cationic and anionic acyclic systems, respectively. Therefore, the T-shaped interactions are also similar in magnitude to the neutral and cationic His dimers (−33 and −92 kJ/mol, respectively). As for the cyclic protein components, T-shaped interactions involving the acyclic amino acids are stronger than the stacked arrangements, and the cytosine–edge dimers are stronger than amino acid–edge counterparts (by up to 40 kJ/mol). Despite differences between neutral and charged binding strengths, both systems adopt distinct geometries in order to maximize the interaction energy. Specifically, when the preferred geometry of the neutral Asp/Glu:C dimer (corresponding binding strength of −25.5 kJ/mol) is used to calculated the interaction energy with anionic Asp/Glu, the binding strength decreases to only −9 kJ/mol. Similarly, when the preferred geometry for the anionic Asp/Glu:C dimer (corresponding binding strength of −47.9 kJ/mol) is used to calculated the binding strength for the corresponding neutral Asp/Glu interaction, the stability decreases to only −5 kJ/mol.

Overall, analysis of DNA–protein interactions involving 7 amino acids of varying size and charge reveals that charge is more important than the size of the π–system for determining the stability of the contact. Additionally, these π–π interactions are strong in the optimal binding orientations, approaching the stability provided to DNA by the A:T base pair (-70 kJ mol^{-1} calculated with CCSD(T)/CBS) [67]. Therefore, when coupled with the abundance of such contacts in DNA–protein complexes (discussed above), we can conclude that many types of nucleobase–amino acid π–interactions must play an important role in biology. However, since similar interaction strengths are found for many monomer combinations and geometries, these interactions likely most often contribute to binding without selectivity. Nevertheless, examples do exist in the literature of the potential selectivity of DNA–protein π–π interactions [12]. Regardless, since very stable interactions occur in many non-optimal binding orientations, there is great flexibility in the contributions these interactions can make to DNA–protein binding.

17.2.2.6 Effect of the Biological Environment

For all calculations discussed thus far, the amino acid and nucleobase systems were treated with truncated models such that only the π–system of interest was considered (without the DNA or protein backbone). However, it is possible that the biological backbone(s) affects the structure or magnitude of the π–interaction through, for example, polarizing the π–system. Therefore, the validity of this approximation has been considered for all π–π (neutral) stacked nucleobase–aromatic amino acid combinations [76, 77], while T-shaped interactions were investigated for the His:A dimer [78], with either the DNA or protein backbone. As anticipated, the addition of the protein or DNA backbone affects the preferred geometry of the dimer due to steric and/or electrostatic interactions between the π–system and the backbone. Although R1 changes very little when the preferred α for the extended system is considered, the preferred angle of rotation changes by more than 30° in many dimers upon inclusion of the DNA backbone (for example, 38 of the 105 stacked dimers investigated exhibit larger deviations in α). Nevertheless, the α increment used in the previous potential energy surface scans was 30°, and the protein backbone negligibly affects the preferred α. Finally, although there is not a consistent trend in the effect of the biological backbone on the horizontal displacement (R2), the binding energy of few structures increases more than 5 kJ/mol upon accounting for the R2 shift, which generally decreases the steric repulsion or increases the π–overlap between the monomers, as well as allows secondary interactions between the backbone and the π–system.

The inclusion of the biological backbone can change the strength of the interaction between a nucleobase and amino acid. Specifically, the protein backbone strengthens the interactions (by up to approximately 10 kJ/mol), while the DNA backbone can strengthen or weaken the interactions (by up to approximately 15 kJ/mol). Nevertheless, the binding energy of some dimer pairs decreases upon including the backbone due to changes in the relative ring orientations that move the dimer away from its optimal geometry. However, changes in the interaction energy are not due to a modification in the π–π contact, but rather additional attractive backbone–π contacts with the aromatic system, the presence of which was verified with QTAIM [77, 78]. These additional backbone–π interactions can increase the total DNA–protein binding energy by up to 18 kJ/mol. Furthermore, the backbone and π–π interactions are additive [77], and therefore the energy of the π–π interactions can be accurately calculated without the inclusion of the protein or DNA backbone. This key finding justifies the small models used to estimate the stability of DNA–protein interactions in nature.

In addition to the majority of previous computational studies on DNA–protein π–π interactions employing truncated models, most calculations have been performed in the gas phase and the results are therefore relevant to active sites of low polarity. However, due to the presence of water in biological systems (i.e., filling cavities in solvent exposed active sites, mediating interactions between biomolecules, as a catalytic residue or as a product of an enzymatic reaction), solvent effects on the magnitude of the calculated interactions energies must be considered. In this light, select works have studied the effect of the polarity of the surrounding environment using implicit solvation models [39, 48, 73, 74, 79]. Unfortunately, the accuracy of this approach for monomers separated by large distances is unclear. Therefore, to the best of our knowledge, one study in the literature to date has considered microsolvation effects by examining discrete interactions between water and (neutral or cationic) His when in a stacked or (nucleobase or amino acid–edge) T-shaped orientation with respect to A [80]. In this contribution, up to four water molecules were added to His:A dimers in binding arrangements that represent (N–H $\cdots$ O, N $\cdots$ H–O or C–H $\cdots$ O) hydrogen bonding, O–H $\cdots$ π T-shaped and O $\cdots$ π lone pair–π interactions. The water contacts affect the (M06-2X/6-31+G(d, p)) calculated strength of the neutral His:A dimer by less than 3 kJ/mol (15 %) regardless of the number of water molecules, water binding orientation or nucleobase–amino acid relative monomer orientations considered. This supports the use of gas–phase calculations to determine the magnitude of DNA–protein π–π interactions and suggests that the significant strength of such π–π binding arrangements likely applies to many biological environments (i.e., low polarity active sites and environments with solvent or even side chain interactions with the nucleobases). However, the presence of water drastically decreases the stability of cationic His:A dimers (by up to 15 kJ/mol or 30 %) for all (water and nucleobase–amino acid) binding arrangements considered. This decrease arises from a reduction in the positive charge on the amino acid upon interacting with water, which leads to an interaction energy between that of the fully protonated and charge neutral His dimers. Nevertheless, even in the presence of explicit water, the binding strength of the cationic His:A dimers are significant in both stacked and T-shaped arrangements and therefore may be important for many cellular functions.

Although studying dimers improves our fundamental understanding of DNA–protein interactions, the effects of additional contacts on the preferred geometry and binding strength must be determined. For example, the AAG active site discussed in the Introduction (Fig. 17.2b) provides a prime example of DNA–protein interactions in nature, which seldom appear as an isolated dimer. Indeed, there are three simultaneous nucleobase–amino acid π–π interactions in the AAG active site upon binding ethenoadenine [7]. As a result, six test trimers composed of adenine or (cationic) 3-methyladenine sandwiched between two (neutral and/or cationic) His residues were considered using MP2/6-31G(0.25) and QTAIM analysis [81]. Potential energy surface scans analogous to those performed for dimers determined that the presence of an additional nucleobase–amino acid π–π contact does not affect the preferred monomer orientation in the first interaction, which is true regardless of the charge of the monomers involved. Furthermore, the two nucleobase–amino acid interactions in the trimer, as well as the amino acid–amino acid 1–3 term, sum to yield the overall trimer stabilization energy. Thus, the interaction energies reported for dimer contacts that appear in nature, and for isolated nucleobase–amino acid dimers, likely extend to more complicated biological environments.

17.2.3 Comparisons Between Natural Interactions and Preferred Geometric Parameters

Comparisons can be made between the extensive studies of the structure and strength of DNA–protein π–π interaction in the natural state [20, 21] and the optimal calculated parameters [28, 63, 74, 75]. Of the variables considered in studies of the optimal DNA–protein interactions, similarities and differences between the optimal and the natural structure can be discussed for ω, R_1 and θ. However, since the center of mass was not added to interactions found in experimental crystal structures, slight differences arise in the way in which these variables are measured between the two classes of studies. Additionally, R2 and α were not recorded for the interactions identified in experimental crystal structures. Starting with ω, studies of the optimal interactions only considered two ω values, namely 0 (stacking) and 90° (T-shaped), while interactions in nature were found to adopt all ω values between 0 and 90°. Although only two orientations were investigated using scan calculations, the interactions in nature that deviate significantly from $\omega = 0$ or 90° have a significant binding strength. Therefore, future work should consider the effect of rotating ω on the interaction energy. Notably, the optimal interactions were found to be stronger for the T-shaped interactions; however, although this trend is maintained in nature, the stacking interactions are overall much more prevalent than T-shaped interactions.

In terms of R1, the optimal separation distance was found to be 3.2–3.6 Å and natural interactions commonly adopt distances within this range (most commonly being from 3–4 Å). However, these numbers are difficult to compare since the method used to measure the distances was different for the natural and optimal interactions (center of mass for the optimal distances versus closest heavy atom contact for the natural interactions). Additionally, although amino acids with smaller π–systems adopt closer optimal separation distances according to the calculations, this trend is not observed for the natural interactions.

Finally, when the edge interacting in the T-shaped interactions is considered, it was determined to be more favourable to have a bridged edge (a bond or two atoms) interacting with the π–ring rather than a single proton edge, which corresponds to the greater number of bridged edges found in nature. However, the edges that lead to the strongest interaction energy according to the calculations are not the edges that occur most frequently in nature. Nevertheless, it is unreasonable to expect this type of agreement since the studies of the optimal interactions only considered perfectly T-shaped interactions, while the edge interacting in nature is highly dependant on the geometric constraints imposed by the protein. Nevertheless, DNA–protein π–π interactions in nature still have significant strength and can contribute to DNA–protein binding and other biological functions.

17.3 Interactions Between Aromatic Amino Acids and Deoxyribose

As mentioned in the Introduction, in addition to the π–π interactions between the DNA nucleobases and π–containing amino acids, interactions can occur between a biological π–ring and a non-conjugated molecular component, which leads to C–H$\cdots$π, O–H$\cdots$π, N–H$\cdots$π or lone pair–π (X$\cdots$π, X = O, N or S) interactions. Indeed, examples of such non-covalent interactions between the aromatic amino acids and carbohydrates have been identified in experimental crystal structures and the strengths of these contacts have been verified using quantum mechanical methods [82–85]. These so-called carbohydrate–π interactions have also been shown to contribute to many bioprocesses including bacterial cell wall recognition, fertilization, immune response, cell–cell communication and post-translational modifications [86–89]. Perhaps the most detailed computer modeling has been performed on interactions between various pyranoses and Phe, Tyr or His, which has shown that the amino acids interact favourably with either carbohydrate face to form interactions that are stable by up to −50 kJ/mol [90–96].

Analogous to the carbohydrate–π interactions believed to be important in glycobiology, contacts between π–containing amino acids and DNA deoxyribose have been implicated in the mechanism of action of DNA polymerases (discussed in the Introduction) [15–19]. Indeed, early searches of experimental X-ray crystal structures available in the PDB revealed close contacts between the π–containing amino acids and the DNA deoxyribose moiety [35, 37]. Although some of these interactions likely correspond to hydrogen bonding, including with the phosphate backbone, many relative DNA–protein orientations correspond to π–contacts (sugar–π). Similar to the carbohydrate–π interactions, these DNA–protein sugar–π contacts can likely involve many different sides of the sugar moiety and lead to several classifications of non-covalent interactions, including C–H$\cdots$π and lone pair–π contacts. In the coming

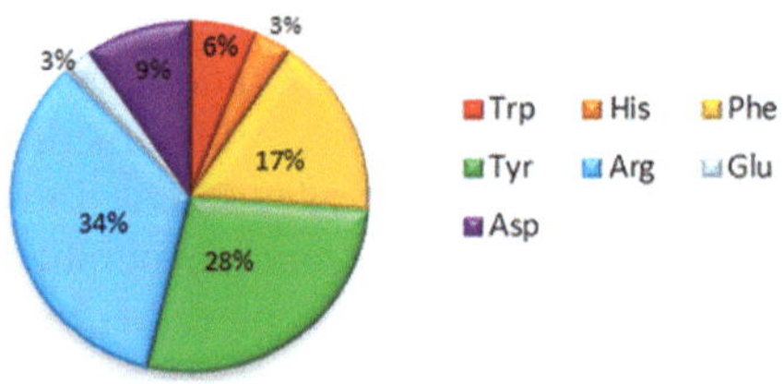

Fig. 17.10 Frequency of particular amino acids involved in sugar–π interactions with the DNA deoxyribose

sections, the frequency, structure and stability of DNA–protein sugar–π interactions in nature will be presented, and compared to the observed nucleobase–amino acid π–π interactions. Since the only work on sugar–π interactions in DNA–protein complexes in the literature to date has been performed by our group, the discussion will be solely based on our studies [20, 21].

17.3.1 Natural Frequency of Sugar–π Interactions

A search of 672 DNA–protein complexes available in the PDB identified 813 sugar–π interactions between deoxyribose and the cyclic or acyclic amino acids. 31 % of the structures searched contain a contact with the acyclic amino acids and 38 % with the aromatic amino acids. As discussed for the nucleobase–amino acid π–π interactions, the sugar–π contacts occur in a wide variety of proteins. Just over half of the structures with a single sugar–π contact contained only one such contact (57 or 53 % for cyclic or acyclic amino acids, respectively). Notably, sugar–π contacts are as common as nucleobase–amino acid π–π contacts, comprising 46 % of all DNA–protein π–interactions. Among the structures searched, 80 % contain either a sugar–π or π–π interaction. These statistics suggest that sugar–π and nucleobase π–π interactions are both important components of DNA–protein binding.

17.3.2 Composition of Sugar–π Interactions

Sugar–π interactions occur with all π–containing amino acids in nature and are almost as frequent with the cyclic (54 %) and acyclic (46 %) protein residues (Fig. 17.10). This contrasts the nucleobase–amino acid π–π contacts, where 75 % of interactions occur with the cyclic amino acids. Most sugar–π interactions invoke Arg (34 %). In fact, Arg contacts comprise 73 % of all acyclic sugar–π interactions. Nevertheless, more sugar–π interactions occur with Asp [78] than nucleobase π–π interactions [10]. As found for the nucleobase π–π interactions, most cyclic sugar–π interactions occur with Tyr (28 %) and Phe (17 %).

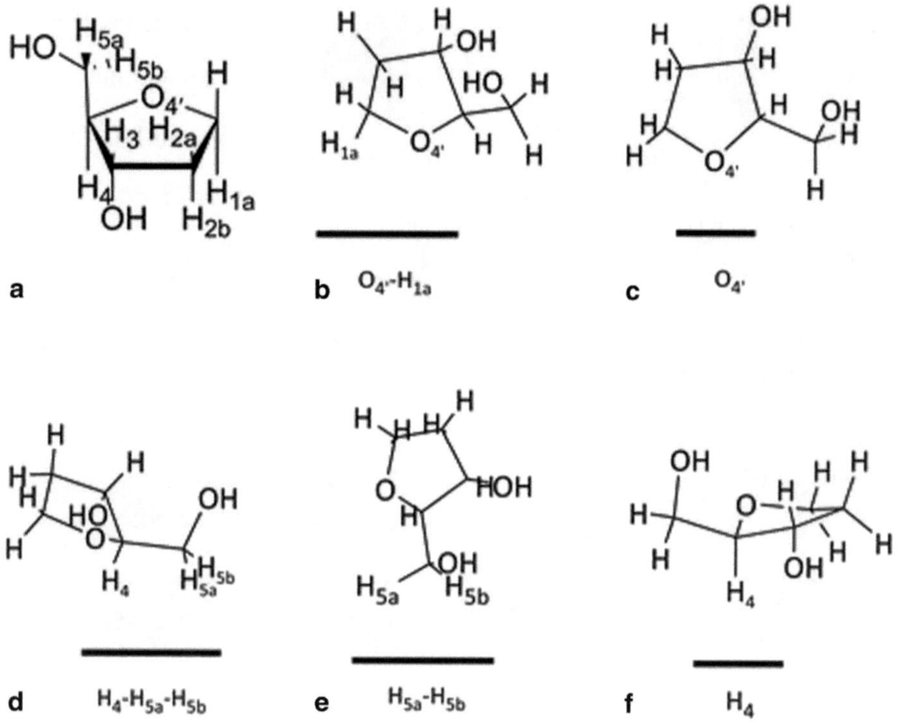

Fig. 17.11 **a** Numbering of the deoxyribose moiety and examples of different types of sugar–π interactions, namely **b** lone pair–proton, **c** lone pair, **d** face, **e** bridge and **f** single proton contacts

17.3.3 Structure of Sugar–π Interactions

Careful consideration of the large number of sugar–π interactions identified in nature reveals that these contacts can be classified according to the sugar atoms interacting with the amino acid π–system. Specifically, these interactions can be categorized as a single proton, two protons (bridged), three protons (face), a lone pair or a lone pair and a proton (lone pair–proton) contact (Fig. 17.11). Among these classifications, the bridge and face orientations are the most common (30 % each), the single–proton interactions are nearly as common (25 %), and the lone pair and the lone pair–proton interactions collectively account for the remaining 15 % of interactions (Fig. 17.12). Overall, each interaction adopts a range of relative monomer orientations (Fig. 17.13). However, the most frequent interactions occur with the C5′ edge of deoxyribose, with the most common sugar edge in each category including the H_4–H_{5a}–H_{5b} face (20 %), the H_{5a}–H_{5b} bridge (13 %) and the H_{5a} single–proton (9 %) interactions.

The amino acid interacting with the deoxyribose affects the preferred conformation of the sugar–π interaction. Due to the larger π–system, more face interactions

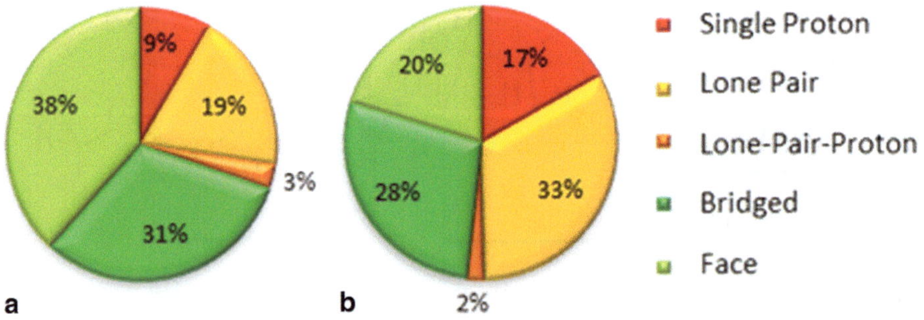

Fig. 17.12 Proportions of different categories of DNA–protein sugar–π interactions with the **a** (cyclic) aromatic and **b** acyclic amino acids

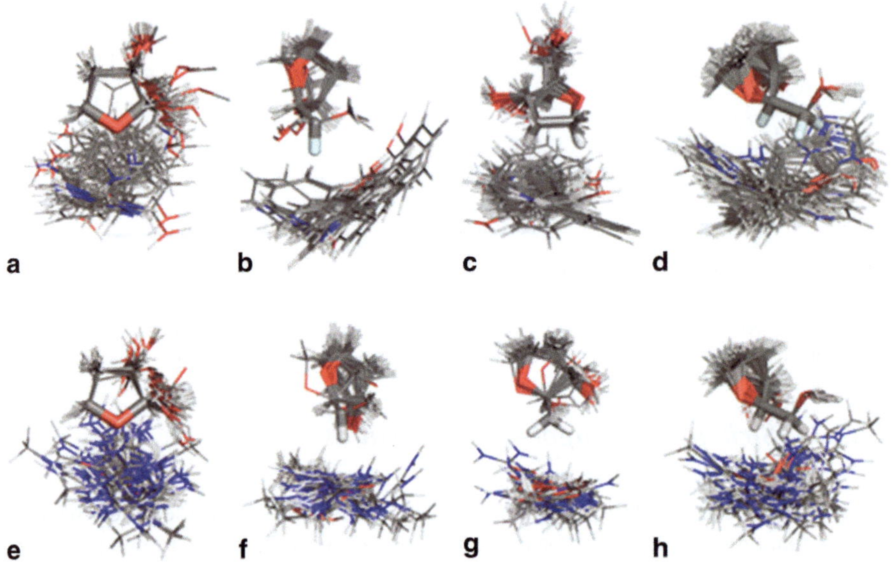

Fig. 17.13 Overlays of the most common sugar–π interactions involving the cyclic amino acids and a **a** lone–pair, **b** H_{5a}, **c** $H_{1a}H_{2b}$ or **d** $H_4H_{5a}H_{5b}$ sugar edge, and involving the acyclic amino acids and a **e** lone pair, **f** H_{5a}, **g** $H_{5a}H_{5b}$ or **h** $H_4H_{5a}H_{5b}$ sugar edge

occur with the cyclic (38 %) rather than acyclic (20 %) amino acids. Conversely, more lone pair–π interactions occur with the acyclic (33 %) than cyclic (19 %) amino acids, possibly due to stronger interactions with charged sites (Fig. 17.12). When considering the individual amino acids, 59 % of Trp interactions occur with the H_4–H_{5a}–H_{5b} face (Fig. 17.14a) and Asp commonly interacts with the H_{5a}–H_{5b} bridge (62 %, Fig. 17.14b), which is distinct from the wide variety of interactions formed with the other amino acids (Phe, Tyr, His, Arg and Glu).

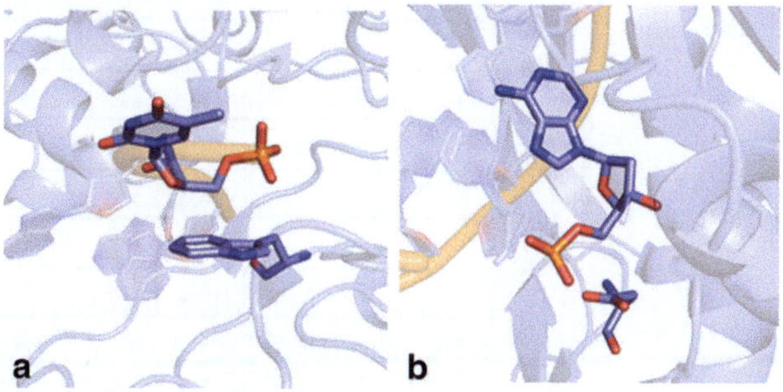

Fig. 17.14 Examples of the most common sugar–π interactions found in X-ray crystal structures of DNA–protein complexes, namely the **a** $H_4H_{5a}H_{5b}$ (4FPV; W306:T1) and **b** $H_{5a}H_{5b}$ (4ECS; Asp155:A9) interactions

17.3.4 Strength of Sugar–π Interactions

Due to the range of deoxyribose–amino acid orientations adopted in nature, the (M06-2X/6-31G(d,p)//MP2/6-31G(d)) calculated binding energies vary significantly (Fig. 17.15). The interactions with the neutral cyclic amino acids are on average stronger (up to -37.4 kJ mol^{-1}) than interactions with the neutral acyclic amino acids (up to -23.1 kJ mol^{-1}). Additionally, the cyclic amino acids adopt a greater range of energies, with 36 % of interactions involving the cyclic amino acids falling between -10 and -15 kJ mol^{-1} and 60 % of interactions involving the acyclic amino acids falling between -5 and -10 kJ mol^{-1}. The charged (cationic or anionic) interactions are significantly stronger (up to -75 kJ mol^{-1}) than the neutral sugar–π contacts. The anionic interactions can be even stronger (up to -90 kJ mol^{-1}), with most interactions falling between -50 and -55 kJ mol^{-1} (19 %). Although no one particular sugar edge leads to the strongest interactions for the neutral contacts, certain edges yield the strongest interactions with charged amino acids. Specifically, the most stable contacts involving the cationic amino acids occur with lone pair–π binding arrangements, while the strongest interactions involving the anionic amino acids occur for the (H_4–H_{5a}–H_{5b}) face and (H_{5a}–H_{5b}) bridge orientations. When compared to the nucleobase–amino acid π–π interactions, the neutral deoxyribose–amino acid sugar–π contacts are slightly weaker than the neutral nucleobase–amino acid π–π, while the charged deoxyribose–amino acid sugar–π contacts are stronger than the charged nucleobase–amino acid π–π interactions. Therefore, sugar–π interactions may contribute to DNA–protein binding to a similar extent as nucleobase–amino acid π–π interactions.

17.4 Conclusions and Future Perspectives

In summary, the computational studies completed on DNA–protein π–interactions between the amino acids and nucleobase or deoxyribose moieties to date have provided important insights into the occurrence, composition, structure, and strength of these interactions. Data mining and computational studies based on experimental X-ray crystal structures and purely theoretical scans of important geometrical

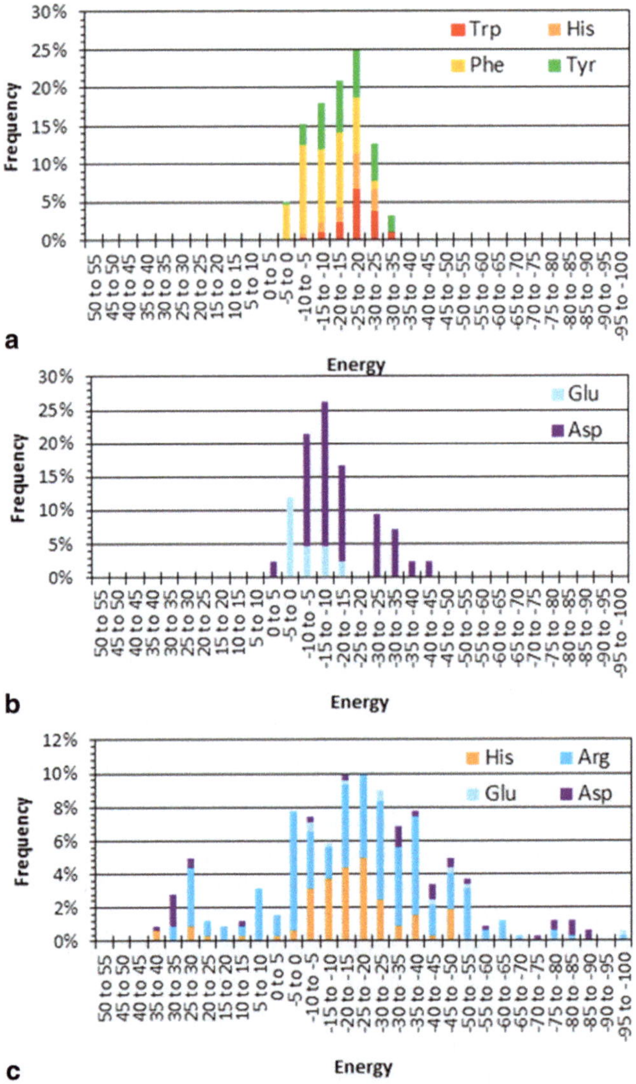

Fig. 17.15 Binding strengths for sugar–π interactions with the **a** neutral aromatic, **b** neutral acyclic, and **c** charged amino acids in experimental X-ray crystal structure geometries

parameters that dictate π–interactions between DNA and protein components have revealed both the interaction energies and their dependence on the relative monomer orientations. In fact, the generated database of binding energies obtained from computational scans allow the strength of interactions found in new X-ray structures to be estimated [75]. Most importantly, these collective studies have found that both nucleobase–amino acid π–π and deoxyribose–amino acid sugar–π interactions are common in nature and have significant strength. Therefore, both categories of DNA–protein interactions will significantly contribute to DNA–protein binding and function.

Unfortunately, the effects of various geometrical parameters on the stability of π–π interactions have yet to be studied for all nucleobase–amino acid pairings. Specifically, the acyclic amino acids have currently only been studied with C, and therefore future work must consider pairings with G, T and A. Furthermore, although the effects of some variables (R1, α and R2) on the interaction energies have been studied, the effects of systematic variations in ω have yet to be investigated. Additionally, the interaction potential energy surfaces of the DNA–protein sugar–π interactions should be mapped. Indeed, previous literature has investigated the effects of R1 and R2 on the strength of the analogous carbohydrate–π contacts [83, 97–100], and thereby revealed important information about the effects of each geometrical variable on the binding strength. In addition to expanding our investigations of known DNA–protein π–interactions, experimental structures should be further searched to identify other classes of non-covalent π–interactions. For example, analogous to DNA–protein π–π and sugar–π interactions, contacts between the RNA nucleobases or ribose and the π–containing amino acids should be investigated. Alternatively, interactions between the DNA or RNA nucleobases and other amino acid side chains (for example, Ser and Cys) should be considered to give additional information about the importance of C–H···π, O–H···π, N–H···π and/or lone pair–π (X···π, X = O, N or S) interactions. Regardless of the abundance of future work required to better understand DNA–protein π–interactions, the information gained from computational studies thus far provides substantial insight into these interactions.

References

1. Murthy PS (2006) Molecular handshake: recognition through weak noncovalent interactions. J Chem Educ 83:1010
2. Autumn K, Liang YA, Hsieh ST, Zesch W, Chan WP, Kenny TW, Fearing R, Full RJ (2000) Adhesive force of a single gecko foot-hair. Nature 405:681–685
3. Autumn K, Sitti M, Liang YA, Peattie AM, Hansen WR, Sponberg S, Kenny TW, Fearing R, Israelachvili JN, Full RJ (2002) Evidence for van der Waals adhesion in gecko setae. Proc Natl Acad Sci U S A 99:12252–12256
4. Lee H, Lee BP, Messersmith PB. (2007) A reversible wet/dry adhesive inspired by mussels and geckos. Nature 448:338–341
5. Yu J, Chary S, Das S, Tamelier J, Pesika NS, Turner KL, Israelachvili JN (2011) Gecko-inspired dry adhesive for robotic applications. Adv Funct Mater 21:3010–3018
6. Schermerhorn KM, Delaney S (2014) A chemical and kinetic perspective on base excision repair of DNA. Acc Chem Res 47:1238–1246
7. Lau AY, Wyatt MD, Glassner BJ, Samson LD, Ellenberger T (2000) Molecular basis for discriminating between normal and damaged bases by the human alkyladenine glycosylase, AAG. Proc Natl Acad Sci U S A 97:13573–13578
8. Rutledge LR, Wetmore SD (2011) Modeling the chemical step utilized by human alkyladenine DNA glycosylase: a concerted mechanism AIDS in selectively excising damaged purines. J Am Chem Soc 133:16258–16269

9. Przybylski JL, Wetmore SD (2011) A QM/QM Investigation of the hUNG2 reaction surface: the untold tale of a catalytic residue. Biochemistry 50:4218–4227
10. Sedgwick B, Bates PA., Paik J, Jacobs SC, Lindahl T (2007) Repair of alkylated DNA: recent advances. DNA Repair 6:429–442
11. Ahvazi B, Coulombe R, Delarge M, Vedadi M, Zhang L, Meighen E, Vrielink A (2000) Crystal structure of the NADP+-dependent aldehyde dehydrogenase from Vibrio harveyi: structural implications for cofactor specificity and affinity. Biochem J 349:853–861
12. An Y, Raju RK., Lu T, Wheeler SE (2014) Aromatic interactions modulate the 5′-base selectivity of the DNA-binding autoantibody ED-10. J Phys Chem B 118:5653–5659
13. Nakajima K, Yamashita A, Akama H, Nakatsu T, Kato H, Hashimoto T, Oda Ji, Yamada Y (1998) Crystal structures of two tropinone reductases: different reaction stereospecificities in the same protein fold. Proc Natl Acad Sci U S A 95:4876–4881
14. Schwans JP, Sunden F, Lassila JK, Gonzalez A, Tsai Y, Herschlag D (2013) Use of anion-aromatic interactions to position the general base in the ketosteroid isomerase active site. Proc Natl Acad Sci U S A 110:11308–11313
15. Joyce CM (1997) Choosing the right sugar: how polymerases select a nucleotide substrate. Proc Natl Acad Sci U S A 94:1619–1622
16. DeLucia AM, Grindley NDF, Joyce CM (2003) An error-prone family Y DNA polymerase (DinB homolog from Sulfolobus solfataricus) uses a 'steric gate' residue for discrimination against ribonucleotides. Nucleic Acids Res 31:4129–4137
17. Astatke M, Ng K, Grindley ND, Joyce CM (1998) A single side chain prevents *Escherichia coli* DNA polymerase I (Klenow fragment) from incorporating ribonucleotides. Proc Natl Acad Sci U S A 95:3402–3407
18. Wang W, Wu EY., Hellinga HW, Beese LS (2012) Structural factors that determine selectivity of a high fidelity DNA polymerase for Deoxy-, Dideoxy-, and ribonucleotides. J Biol Chem 287:28215–28226
19. Brown JA, Suo Z (2011) Unlocking the sugar "Steric Gate" of DNA polymerases. Biochemistry 50:1135–1142
20. Wilson KA, Kellie JL, Wetmore SD (2014) DNA-protein π-interactions in nature: abundance, structure, composition and strength of contacts between aromatic amino acids and DNA nucleobases or deoxyribose sugar. Nucleic Acids Res 42:6726–6741
21. Wilson KA, Wells RA, Abendong MN, Anderson CB, Kung RW, Wetmore SD (2015) Landscape of π–π and sugar–π Contacts in DNA-protein Interactions. J. Biomol Struct Dyn. (DOI: 10.1080/07391102.2015.1013157)
22. Cerny J, Hobza P (2007) Non-covalent interactions in biomacromolecules. Phys Chem Chem Phys 9:5291–5303
23. Sponer J, Jurecka P, Marchan I, Luque FJ, Orozco M, Hobza P (2006) Nature of base stacking: reference quantum-chemical stacking energies in ten unique B-DNA base-pair steps. Chem-Eur J 12:2854–2865
24. Kryachko ES, Ludea EV (2014) Density functional theory: foundations reviewed. Phys Reports 544:123–239
25. Raju RK, Ramraj A, Hillier IH, Vincent MA, Burton NA (2009) Carbohydrate-aromatic [small pi] interactions: a test of density functionals and the DFT-D method. Phys Chem Chem Phys 11:3411–3416
26. Hujo W, Grimme S (2012) Performance of non-local and atom-pairwise dispersion corrections to DFT for structural parameters of molecules with noncovalent interactions. J Chem Theory Comput 9:308–315
27. Burns LA, Vazquez-Mayagoitia A, Sumpter BG, Sherrill CD (2011) Density-functional approaches to noncovalent interactions: a comparison of dispersion corrections (DFT-D), exchange-hole dipole moment (XDM) theory, and specialized functionals. J Chem Phys 134
28. Wells RA, Kellie JL, Wetmore SD (2013) Significant strength of charged DNA-Protein π-π interactions: a preliminary study of cytosine. J Phys Chem B 117:10462–10474
29. Rutledge LR, Wetmore SD (2010) The assessment of density functionals for DNA-protein stacked and T-shaped complexes. Can J Chem-Rev Can De Chim 88:815–830

30. Bader RFW (1991) A quantum theory of molecular structure and its applications. Chem Rev 91:893–928
31. Matthews BW (1988) No code for recognition. Nature 335:294–295
32. Pabo CO, Sauer RT (1992) Transcription factors: structural families and principles of DNA recognition. Ann Rev Biochem 61:1053–1095
33. Suzuki M (1994) A framework for the DNA-protein recognition code of the probe helix in transcription factors: the chemical and stereochemical rules. Structure 2:317–326
34. Mandel-Gutfreund Y, Schueler O, Margalit H (1995) Comprehensive analysis of hydrogen bonds in regulatory protein DNA-complexes: in search of common principles. J Mol Biol 253:370–382
35. Luscombe NM, Laskowski RA, Thornton JM (2001) Amino acid-base interactions: a three-dimensional analysis of protein-DNA interactions at an atomic level. Nucleic Acids Res 29:2860–2874
36. Luscombe NM, Thornton JM (2002) Protein-DNA interactions: amino acid conservation and the effects of mutations on binding specificity. J Mol Biol 320:991–1009
37. Lejeune D, Delsaux N, Charloteaux B, Thomas A, Brasseur R (2005) Protein-nucleic acid recognition: statistical analysis of atomic interactions and influence of DNA structure. Proteins: Struct, Funct, Bioinform 61:258–271
38. Sathyapriya R, Vijayabaskar M, Vishveshwara S (2008) Insights into protein-DNA interactions through structure network analysis. PLoS Comput Biol 4:e1000170
39. Mao L, Wang Y, Liu Y, Hu X (2004) Molecular determinants for ATP-binding in proteins: a data mining and quantum chemical analysis. J Mol Biol 336:787–807
40. Baker CM, Grant GH (2007) Role of aromatic amino acids in protein-nucleic acid recognition. Biopolymers 85:456–470
41. Kirsanov DD, Zanegina ON, Aksianov EA, Spirin SA, Karyagina AS, Alexeevski AV (2013) NPIDB: nucleic acid-protein interaction database. Nucleic Acids Res 41:D517–D523
42. Norambuena T, Melo F (2010) The protein-DNA interface database. BMC Bioinform 11:262
43. Kumar MDS, Bava KA, Gromiha MM, Prabakaran P, Kitajima K, Uedaira H, Sarai A (2006) ProTherm and ProNIT: thermodynamic databases for proteins and protein-nucleic acid interactions. Nucleic Acids Res 34:D204–D206
44. Contreras-Moreira B (2010) 3D-footprint: a database for the structural analysis of protein-DNA complexes. Nucleic Acids Res 38:D91–D97
45. Lewis BA, Walia RR, Terribilini M, Ferguson J, Zheng C, Honavar V, Dobbs D (2011) PRIDB: a protein-RNA interface database. Nucleic Acids Res 39:D277–D282
46. Kim R, Guo J-t. (2009) PDA: an automatic and comprehensive analysis program for protein-DNA complex structures. BMC Genomics 10:S13
47. Lee S, Blundell T (2009) BIPA: a database for protein-nucleic acid interaction in 3D structures. Bioinformatics 25:1559–1560
48. Cauet E, Rooman M, Wintjens R, Lievin J, Biot C (2005) Histidine-aromatic interactions in proteins and protein-ligand complexes: quantum chemical study of X-ray and model structures. J Chem Theory Comput 1:472–483
49. Copeland KL, Anderson JA, Farley AR, Cox JR, Tschumper GS (2008) Probing phenylalanine/adenine π-stacking interactions in protein complexes with explicitly correlated and CCSD(T) computations. J Phys Chem B 112:14291–14295
50. Copeland KL, Pellock SJ, Cox JR, Cafiero ML, Tschumper GS (2013) Examination of tyrosine/adenine stacking interactions in protein complexes. J Phys Chem B 117:14001–14008
51. Gromiha MM, Santhosh C, Suwa M (2004) Influence of cation-π interactions in protein-DNA complexes. Polymer 45:633–639
52. Gromiha MM, Santhosh C, Ahmad S (2004) Structural analysis of cation-π interactions in DNA binding proteins. Int J Biol Macromol 34:203–211
53. Biot C, Buisine E, Kwasigroch JM, Wintjens R, Rooman M (2002) Probing the energetic and structural role of amino acid/nucleobase cation-π interactions in protein-ligand complexes. J Biol Chem 277:40816–40822

54. Wintjens R, Lievin J, Rooman M, Buisine E (2000) Contribution of cation-π interactions to the stability of protein-DNA complexes. J Mol Biol 302:395–410
55. Rooman M, Lievin J, Buisine E, Wintjens R (2002) Cation-π/H-bond stair motifs at protein-DNA interfaces. J Mol Biol 319:67–76
56. Biot C, Buisine E, Rooman M (2003) Free-energy calculations of protein-ligand cation-π and amino-π interactions: from vacuum to proteinlike environments. J Am Chem Soc 125:13988–13994
57. Biot C, Wintjens R, Rooman M (2004) Stair motifs at protein-DNA interfaces: Nonadditivity of H-bond, stacking, and cation-π interactions. J Am Chem Soc 126:6220–6221
58. Wintjens R, Biot C, Rooman M, Lievin J (2003) Basis set and electron correlation effects on ab initio calculations of cation-π/H-bond stair motifs. J Phys Chem A 107:6249–6258
59. Auffinger P, Bielecki L, Westhof E (2004) Anion binding to nucleic acids. Structure 12:379–388
60. Singh NJ, Min SK, Kim DY, Kim KS (2009) Comprehensive energy analysis for various types of π-interaction. J Chem Theory Comput 5:515–529
61. Zacharias N, Dougherty DA (2002) Cation-π interactions in ligand recognition and catalysis. Trends Pharmacol Sci 23:281–287
62. Dougherty DA (2012) The cation−π interaction. Acc Chem Res 46:885–893
63. Churchill CDM, Wetmore SD (2009) Noncovalent interactions involving histidine: the effect of charge on π-π stacking and T-shaped interactions with the DNA nucleobases. J Phys Chem B 113:16046–16058
64. Frontera A, Quinonero D, Costa A, Ballester P, Deya PM (2007) MP2 study of cooperative effects between cation-[small pi], anion-[small pi] and [small pi]-[small pi] interactions. New J Chem 31:556–560
65. Mahadevi AS, Sastry GN (2012) Cation−π interaction: its role and relevance in chemistry, biology, and material science. Chem Rev 113:2100–2138
66. Schottel BL, Chifotides HT, Dunbar KR (2008) Anion-[small pi] interactions. Chem Soc Rev 37:68–83
67. Jurecka P, Sponer J, Cerny J, Hobza P (2006) Benchmark database of accurate (MP2 and CCSD(T) complete basis set limit) interaction energies of small model complexes, DNA base pairs, and amino acid pairs. Phys Chem Chem Phys 8:1985–1993
68. Cysewski P (2008) A post-SCF complete basis set study on the recognition patterns of uracil and cytosine by aromatic and π-aromatic stacking interactions with amino acid residues. Phys Chem Chem Phys 10:2636–2645
69. Ebrahimi A, Habibi-Khorassani M, Gholipour AR, Masoodi HR (2009) Interaction between uracil nucleobase and phenylalanine amino acid: the role of sodium cation in stacking. Theor Chem Acc 124:115–122
70. Santos LA, da Cunha EFF, Freitas MP, Ramalho TC (2014) Hydrophobic noncovalent interactions of inosine-phenylalanine: a theoretical model for investigating the molecular recognition of nucleobases. J Phys Chem A 118:5808–5817
71. Rutledge LR, Campbell-Verduyn LS, Hunter KC, Wetmore SD (2006) Characterization of nucleobase-amino acid stacking interactions utilized by a DNA repair enzyme. J Phys Chem B 110:19652–19663
72. Rutledge LR, Wetmore SD (2008) Remarkably strong T-shaped interactions between aromatic amino acids and adenine: their increase upon nucleobase methylation and a comparison to stacking. J Chem Theory Comput 4:1768–1780
73. Rutledge LR, Durst HF, Wetmore SD (2008) Computational comparison of the stacking interactions between the aromatic amino acids and the natural or (cationic) methylated nucleobases. Phys Chem Chem Phys 10:2801–2812
74. Rutledge LR, Campbell-Verduyn LS, Wetmore SD (2007) Characterization of the stacking interactions between DNA or RNA nucleobases and the aromatic amino acids. Chem Phys Lett 444:167–175
75. Rutledge LR, Durst HF, Wetmore SD (2009) Evidence for stabilization of DNA/RNA-protein complexes arising from nucleobase-amino acid stacking and T-shaped interactions. J Chem Theory Comput 5:1400–1410

76. Churchill CDM, Navarro-Whyte L, Rutledge LR, Wetmore SD (2009) Effects of the biological backbone on DNA-protein stacking interactions. Phys Chem Chem Phys 11:10657–10670
77. Churchill CDM, Rutledge LR, Wetmore SD (2010) Effects of the biological backbone on stacking interactions at DNA-protein interfaces: the interplay between the backbone-pi and pi-pi components. Phys Chem Chem Phys 12:14515–14526
78. Rutledge LR, Navarro-Whyte L, Peterson TL, Wetmore SD (2011) Effects of extending the computational model on DNA-protein T-shaped interactions: the case of adenine-histidine dimers. J Phys Chem A 115:12646–12658
79. Biot C, Buisine E, Rooman M (2003) Free-energy calculations of protein−ligand cation−π and amino−π interactions: from vacuum to proteinlike environments. J Am Chem Soc 125:13988–13994
80. Leavens FMV, Churchill CDM, Wang S, Wetmore SD (2011) Evaluating how discrete water molecules affect protein-DNA π-π and π+-π stacking and T-shaped interactions: the case of histidine-adenine dimers. J Phys Chem B 115:10990–11003
81. Rutledge LR, Churchill CDM, Wetmore SD (2010) A preliminary investigation of the additivity of $\pi - \pi$ or $\pi + - \pi$ stacking and T-shaped interactions between natural or damaged DNA nucleobases and histidine. J Phys Chem B 114:3355–3367
82. Spiwok V, Lipovová P, Skálová T, Buchtelová E, Hašek J, Králová B (2004) Role of CH/π interactions in substrate binding by Escherichia coli β-galactosidase. Carbohydr Res 339:2275–2280
83. Spiwok V, Lipovová P, Skálová T, Vondráčková E, Dohnálek J, Hašek J, Králová B (2006) Modelling of carbohydrate-aromatic interactions: ab initio energetics and force field performance. J Comput-Aided Mol Des 19:887–901
84. Sujatha MS, Sasidhar YU, Balaji PV (2004) Energetics of galactose- and glucose-aromatic amino acid interactions: Implications for binding in galactose-specific proteins. Protein Sci 13:2502–2514
85. Wimmerová M, Kozmon S, Nečasová I, Mishra SK, Komárek J, Koča J (2012) Stacking interactions between carbohydrate and protein quantified by combination of theoretical and experimental methods. PLoS ONE 7:e46032
86. Laughrey ZR, Kiehna SE, Riemen AJ, Waters ML (2008) Carbohydrate-pi interactions: what are they worth? J Am Chem Soc 130:14625–14633
87. Bertozzi CR, Kiessling LL (2001) Chemical glycobiology. Science 291:2357–2364
88. Liang F-C, Chen RPY, Lin C-C, Huang K-T, Chan SI (2006) Tuning the conformation properties of a peptide by glycosylation and phosphorylation. Biochem Biophys Res Commun 342:482–488
89. Gabius H-J, Siebert H-C, André S, Jiménez-Barbero J, Rüdiger H (2004) Chemical biology of the sugar code. ChemBioChem 5:740–764
90. Raju RK, Ramraj A, Vincent MA, Hillier IH, Burton NA (2008) Carbohydrate-protein recognition probed by density functional theory and ab initio calculations including dispersive interactions. Phys Chem Chem Phys 10:6500–6508
91. Su Z, Cocinero EJ, Stanca-Kaposta EC, Davis BG, Simons JP (2009) Carbohydrate-aromatic interactions: a computational and IR spectroscopic investigation of the complex, methyl alpha-L-fucopyranoside · toluene, isolated in the gas phase. Chem Phys Lett 471:17–21
92. Stanca-Kaposta EC, Gamblin DP, Screen J, Liu B, Snoek LC, Davis BG, Simons JP (2007) Carbohydrate molecular recognition: a spectroscopic investigation of carbohydrate-aromatic interactions. Phys Chem Chem Phys 9:4444–4451
93. Tsuzuki S, Uchimaru T, Mikami M (2011) Magnitude and nature of carbohydrate-aromatic interactions in fucose-phenol and fucose-indole complexes: CCSD(T) level interaction energy calculations. J Phys Chem A 115:11256–11262
94. Mohamed MNA, Watts HD, Guo J, Catchmark JM, Kubicki JD (2010) MP2, density functional theory, and molecular mechanical calculations of C-H · · · π and hydrogen bond interactions in a cellulose-binding module-cellulose model system. Carbohydr Res 345:1741–1751
95. Kozmon S, Matuška R, Spiwok V, Koča J (2011) Dispersion interactions of carbohydrates with condensate aromatic moieties: theoretical study on the CH-π interaction additive properties. Phys Chem Chem Phys 13:14215–14222

96. Sujatha MS, Sasidhar YU, Balaji PV (2005) Insights into the role of the aromatic residue in galactose-binding sites: MP2/6-311G++** study on galactose− and glucose−aromatic residue analogue complexes. Biochemistry 44:8554–8562
97. Kumari M, Sunoj RB, Balaji PV (2012) Exploration of CH $\cdots\pi$ mediated stacking interactions in saccharide: aromatic residue complexes through conformational sampling. Carbohydrate Res 361:133–140
98. Kozmon S, Matuška R, Spiwok V, Koča J (2011) Three-dimensional potential energy surface of selected carbohydrates' CH/π dispersion interactions calculated by high-level quantum mechanical methods. Chem—A Eur J 17:5680–5690
99. Kumari M, Sunoj RB, Balaji PV (2012) Conformational mapping and energetics of saccharide-aromatic residue interactions: implications for the discrimination of anomers and epimers and in protein engineering. Org Biomol Chem 10:4186–4200
100. Tavagnacco L, Engström O, Schnupf U, Saboungi M-L, Himmel M, Widmalm G, Cesàro A, Brady JW (2012) Caffeine and sugars interact in aqueous solutions: a simulation and NMR study. J Phys Chem B 116:11701–11711